BIOTECHNOLOGY AND PLANT DISEASE MANAGEMENT

BIOTECHNOLOGY AND PLANT DISEASE MANAGEMENT

Edited by

Z.K. Punja

*Simon Fraser University
Department of Biological Sciences
8888 University Drive
Burnaby, BC, V5A 1S6, Canada*

S.H. De Boer

*Charlottetown Laboratory
Canadian Food Inspection Agency
93 Mount Edward Road
Charlottetown, PEI, C1A 5T1, Canada*

and

H. Sanfaçon

*Pacific Agri-Food Research Centre
Agriculture and Agri-Food Canada
4200 Highway 97, Summerland,
BC, V0H 1Z0, Canada*

CABI is a trading name of CAB International

CABI Head Office	CABI North American Office
Nosworthy Way	875 Massachusetts Avenue
Wallingford	7th Floor
Oxfordshire OX10 8DE	Cambridge, MA 02139
UK	USA
Tel: +44 (0)1491 832111	Tel: +1 617 395 4056
Fax: +44 (0)1491 833508	Fax: +1 617 354 6875
E-mail: cabi@cabi.org	E-mail: cabi-nao@cabi.org
Website: www.cabi.org	

©CAB International 2007 (except Chapters 4 and 23: ©Minister of Public Works and Government Services Canada 2007; Chapter 7: ©Her Majesty the Queen in Right of Canada, as represented by the Minister of Agriculture and Agri-Food Canada 2007; Chapter 8: ©Her Majesty the Queen in Right of Canada [Canadian Food Inspection Agency] 2007). All rights reserved. No part of this publication may be reproduced in any form or by any means, electronically, mechanically, by photocopying, recording or otherwise, without the prior permission of the copyright owners.

A catalogue record for this book is available from the British Library, London, UK.

Library of Congress Cataloging-in-Publication Data

Biotechnology and plant disease management/edited by Zamir K. Punja, Solke De Boer, and Hélène Sanfaçon.
    p. cm.
Includes bibliographical references and index.
ISBN 978-1-84593-288-6 (alk. paper) – ISBN 978-1-84593-310-4 (ebook) 1. Plant biotechnology. 2. Plant diseases. I. Punja, Zamir K. II. De Boer, S.(S. H.) III. Sanfaçon, Hélène. IV. Title.

SB106.B56B553 2008
632'.3--dc22   2007017180

ISBN-13: 978 1 84593 288 6

Typeset by SPi India Pvt Ltd, Pondicherry, India.
Printed and bound in the UK by Biddles Ltd, King's Lynn.

Contents

Contributors   ix

SECTION A: UNRAVELING MICROBE–PLANT INTERACTIONS FOR APPLICATIONS TO DISEASE MANAGEMENT

1. Signal Transduction Pathways and Disease Resistant Genes and Their Applications to Fungal Disease Control   1
   T. Xing

2. Modulating Quorum Sensing and Type III Secretion Systems in Bacterial Plant Pathogens for Disease Management   16
   C.-H. Yang and S. Yang

3. Application of Biotechnology to Understand Pathogenesis in Nematode Plant Pathogens   58
   M.G. Mitchum, R.S. Hussey, E.L. Davis and T.J. Baum

4. Interactions Between Plant and Virus Proteomes in Susceptible Hosts: Identification of New Targets for Antiviral Strategies   87
   H. Sanfaçon and J. Jovel

5. Mechanisms of Plant Virus Evolution and Identification of Genetic Bottlenecks: Impact on Disease Management   109
   M.J. Roossinck and A. Ali

6. Molecular Understanding of Viroid Replication Cycles and Identification of Targets for Disease Management   125
   R.A. Owens

SECTION B: MOLECULAR DIAGNOSTICS OF PLANT PATHOGENS FOR
DISEASE MANAGEMENT

7 Molecular Diagnostics of Soilborne Fungal Pathogens   146
  C.A. Lévesque

8 Molecular Detection Strategies for Phytopathogenic Bacteria   165
  S.H. De Boer, J.G. Elphinstone and G.S. Saddler

9 Molecular Diagnostics of Plant-parasitic Nematodes   195
  R.N. Perry, S.A. Subbotin and M. Moens

10 Molecular Diagnostic Methods for Plant Viruses   227
   A. Olmos, N. Capote, E. Bertolini and M. Cambra

11 Molecular Identification and Diversity of Phytoplasmas   250
   G. Firrao, L. Conci and R. Locci

12 Molecular Detection of Plant Viroids   277
   R.P. Singh

SECTION C: ENHANCING RESISTANCE OF PLANTS TO PATHOGENS FOR
DISEASE MANAGEMENT

13 Application of Cationic Antimicrobial Peptides   301
   for Management of Plant Diseases
   S. Misra and A. Bhargava

14 Molecular Breeding Approaches for Enhanced Resistance   321
   Against Fungal Pathogens
   R.E. Knox and F.R. Clarke

15 Protein-mediated Resistance to Plant Viruses   358
   J.F. Uhrig

16 Transgenic Virus Resistance Using Homology-dependent   374
   RNA Silencing and the Impact of Mixed Virus Infections
   M. Ravelonandro

17 Molecular Characterization of Endogenous Plant   395
   Virus Resistance Genes
   F.C. Lanfermeijer and J. Hille

18 Potential for Recombination and Creation of New Viruses   416
   in Transgenic Plants Expressing Viral Genes: Real or
   Perceived Risk?
   M. Fuchs

19 Virus-resistant Transgenic Papaya: Commercial Development   436
   and Regulatory and Environmental Issues
   J.Y. Suzuki, S. Tripathi and D. Gonsalves

SECTION D: UNDERSTANDING MICROBIAL INTERACTIONS TO ENHANCE
DISEASE MANAGEMENT

20 **Potential Disease Control Strategies Revealed by Genome Sequencing and Functional Genetics of Plant Pathogenic Bacteria** 462
    A.O. Charkowski

21 **Molecular Assessment of Soil Microbial Communities with Potential for Plant Disease Suppression** 498
    J.D. van Elsas and R. Costa

22 **Enhancing Biological Control Efficacy of Yeasts to Control Fungal Diseases Through Biotechnology** 518
    G. Marchand, G. Clément-Mathieu, B. Neveu and R.R. Bélanger

23 **Molecular Insights into Plant Virus–Vector Interactions** 532
    D. Rochon

Index 569

Colour plates for Figs 3.1 and 3.2 may be found after page 84

Colour plates for Fig. 22.3 may be found after page 532

Contributors

A. Ali, *The Samuel Roberts Noble Foundation, P.O. Box 2180, Ardmore, OK 73402, USA; Current address: Department of Biological Sciences, 600 South College Avenue Tulsa, OK 74104–3189, USA*

T.J. Baum, *Department of Plant Pathology, Iowa State University, Ames, Iowa, USA*

R.R. Bélanger, *Département de phytologie, Centre de recherche en horticulture, Faculté des sciences de l'agriculture et de l'alimentation, Université Laval, Québec, QC, G1K 7P4, Canada*

E. Bertolini, *Plant Protection and Biotechnology Centre, Instituto Valenciano de Investigaciones Agrarias (IVIA), Carretera Moncada a Náquera km 5, 46113 Moncada, Valencia, Spain*

A. Bhargava, *Department of Biochemistry and Microbiology, University of Victoria, Victoria, BC, V8N 3P6, Canada*

M. Cambra, *Plant Protection and Biotechnology Centre, Instituto Valenciano de Investigaciones Agrarias (IVIA), Carretera Moncada a Náquera km 5, 46113 Moncada, Valencia, Spain*

N. Capote, *Plant Protection and Biotechnology Centre, Instituto Valenciano de Investigaciones Agrarias (IVIA), Carretera Moncada a Náquera km 5, 46113 Moncada, Valencia, Spain*

A.O. Charkowski, *Department of Plant Pathology, University of Wisconsin-Madison, 1630 Linden Drive, Madison, WI 53706, USA*

F.R. Clarke, *Agriculture and Agri-Food Canada, Semiarid Prairie Agricultural Research Centre, Swift Current, SK, S9H 3X2, Canada*

G. Clément-Mathieu, *Département de phytologie, Centre de recherche en horticulture, Faculté des sciences de l'agriculture et de l'alimentation, Université Laval, Québec, QC, G1K 7P4, Canada*

L. Conci, *Instituto de Fitopatología y Fisiología Vegetal-INTA, Camino 60 cuadras km 5 1/2 (X5020ICA), Córdoba, Argentina*

R. Costa, *Department of Microbial Ecology, University of Groningen, Kerklaan 30, 9750RA Haren, The Netherlands*

E.L. Davis, Department of Plant Pathology, North Carolina State University, Raleigh, North Carolina, USA

S.H. De Boer, Charlottetown Laboratory, Canadian Food Inspection Agency, 93 Mount Edward Road, Charlottetown, PEI, C1A 5T1, Canada

J.G. Elphinstone, Central Science Laboratory, Sand Hutton, York, YO41 1LZ, UK

G. Firrao, Dipartimento di Biologia Applicata alla Difesa delle Piante, Università di Udine, via delle Scienze 208, 33100 Udine, Italy

M. Fuchs, Department of Plant Pathology, Cornell University, New York State Agricultural Experiment Station, Geneva, NY 14456, USA

D. Gonsalves, USDA-ARS-PWA, Pacific Basin Agricultural Research Center, 64 Nowelo Street, Hilo, Hawaii 96720, USA

J. Hille, Department of Molecular Biology of Plants, Groningen Biomolecular Sciences and Biotechnology Institute, University of Groningen, P.O. Box 14, 9750 AA, Haren, The Netherlands

R.S. Hussey, Department of Plant Pathology, University of Georgia, Athens, Georgia, USA

J. Jovel, Pacific Agri-Food Research Centre, Agriculture and Agri-Food Canada, P.O. Box 5000, 4200 Highway 97, Summerland, BC, V0H 1Z0, Canada

R. Knox, Agriculture and Agri-Food Canada, Semiarid Prairie Agricultural Research Centre, Swift Current, SK, S9H 3X2, Canada

F.C. Lanfermeijer, Laboratory of Plant Physiology, Centre for Ecological and Evolutionary Studies, University of Groningen, P.O. Box 14, 9750 AA, Haren, The Netherlands

C.A. Lévesque, Agriculture and Agri-Food Canada, Central Experimental Farm, Biodiversity, 960 Carling Ave., Ottawa, ON, K1A 0C6, Canada

R. Locci, Dipartimento di Biologia Applicata alla Difesa delle Piante, Università di Udine, via delle Scienze 208, 33100 Udine, Italy

G. Marchand, Département de phytologie, Centre de recherche en horticulture, Faculté des sciences de l'agriculture et de l'alimentation, Université Laval, Québec, QC, G1K 7P4, Canada

S. Misra, Department of Biochemistry and Microbiology, University of Victoria, Victoria, BC, V8N 3P6, Canada

M.G. Mitchum, Division of Plant Sciences, University of Missouri, Columbia, Missouri, USA

M. Moens, Institute for Agricultural and Fisheries Research, Burg. Van Gansberghelaan 96, 9280 Merelbeke, Belgium and Department of Crop Protection, Ghent University, Coupure links 653, 9000 Ghent, Belgium

B. Neveu, Département de phytologie, Centre de recherche en horticulture, Faculté des sciences de l'agriculture et de l'alimentation, Université Laval, Québec, QC, G1K 7P4, Canada

A. Olmos, Plant Protection and Biotechnology Centre, Instituto Valenciano de Investigaciones Agrarias (IVIA), Carretera Moncada a Náquera km 5, 46113 Moncada, Valencia, Spain

R.A. Owens, Molecular Plant Pathology Laboratory – USDA/ARS, Beltsville Agricultural Research Center, 10300 Baltimore Avenue, Beltsville, MD 20705, USA

R.N. Perry, *Plant Pathogen Interaction Division, Rothamsted Research, Harpenden, Hertfordshire AL5 2JQ, UK and Biology Department, Ghent University, K.L. Ledeganckstraat 35, 9000 Ghent, Belgium*

M. Ravelonandro, *INRA-Bordeaux, UMR GDPP-1090, BP 81, F-33881 Villenave d'ornon, France*

D. Rochon, *Agriculture and Agri-Food Canada, Pacific Agri-Food Research Centre, 4200 Highway 97, Summerland, BC, V0H 1Z0, Canada*

M.J. Roossinck, *The Samuel Roberts Noble Foundation, P.O. Box 2180, Ardmore, OK 73402, USA*

G.S. Saddler, *Scottish Agricultural Science Agency, Edinburgh, EH12 9FJ, UK*

H. Sanfaçon, *Pacific Agri-Food Research Centre, Agriculture and Agri-Food Canada, 4200 Highway 97, Summerland, BC, V0H 1Z0, Canada*

R.P. Singh, *Agriculture and Agri-Food Canada, Potato Research Centre, 850 Lincoln Road, P.O. Box 28280, Fredericton, NB, E3B 4Z7, Canada*

S.A. Subbotin, *University of California, Riverside, CA 92521, USA and Biology Department, Ghent University, K.L. Ledeganckstraat 35, 9000 Ghent, Belgium*

J.Y. Suzuki, *USDA-ARS-PWA, Pacific Basin Agricultural Research Center, 64 Nowelo Street, Hilo, HI 96720, USA*

S. Tripathi, *USDA-ARS-PWA, Pacific Basin Agricultural Research Center, 64 Nowelo Street, Hilo, HI 96720, USA*

J.F. Uhrig, *University of Cologne, Botanical Institute III, Gyrhofstr. 15, D-50931 Cologne, Germany*

J.D. van Elsas, *Department of Microbial Ecology, University of Groningen, Kerklaan 30, 9750RA Haren, The Netherlands*

T. Xing, *Department of Biology and Institute of Biochemistry, Carleton University, Ottawa ON, K1S 5B6, Canada*

C.-H. Yang, *Department of Biological Sciences, University of Wisconsin, Milwaukee, WI 53211, USA*

S. Yang, *Department of Biological Sciences, University of Wisconsin, Milwaukee, WI 53211, USA*

1 Signal Transduction Pathways and Disease Resistant Genes and Their Applications to Fungal Disease Control

T. XING

Abstract

Plants must continuously defend themselves against attack from fungi, bacteria, viruses, invertebrates and even other plants. The regulation mechanisms of any plant–pathogen interaction are complex and dynamic. The application of biochemical and molecular genetic techniques has resulted in major advances in elucidating the mechanisms that regulate gene expression and in identifying components of many signal transduction pathways in diverse physiological systems. Advances in genomics and proteomics have profoundly altered the ways in which we select and approach research questions and have offered opportunities to view signal transduction events in a more systemic way. Although many disease resistant genes and signalling mechanisms are now characterized, it is still ambiguous whether and how they can be engineered to enhance disease resistance. Caution is needed when assessing manipulation strategies so that the manipulations will achieve the desired results without having detrimental effects on plant growth and development. This chapter discusses some other effective approaches for identification of signal transduction components, such as RNA interference (RNAi), yeast two-hybrid system and proteomics approaches.

Introduction

Plant diseases have been present from the very beginning of organized agriculture. In nature, plants encounter pathogen challenges and have to defend themselves. Because their immobility precludes escape, plants possess both a preformed and an inducible defence capacity. This is in striking contrast to the vertebrate immune system, in which specialized cells devoted to defence are rapidly mobilized to the infection site to kill pathogens or limit pathogen growth. The lack of such a circulatory system requires a strategy by the plant to minimize infections. It is often observed that in wild populations, most plants are healthy most of the time; if disease occurs, it is usually restricted only to a small amount of tissue.

Plant defence involves signal perception, signal transduction, signal response and termination of signalling events. Many components of the perception systems and transduction pathways are now characterized and the underlying genes are known. As our knowledge of the cellular and genetic mechanisms of plant disease resistance increases, so does the potential for modifying these processes to achieve broad-spectrum and durable disease resistance. In this chapter, the current understanding of the defence systems, including perception of pathogen signals, transduction of the signals, transcriptional, translational and post-translational regulations in host plants will be reviewed. Examples to indicate the applications of some approaches to disease control will be provided. A discussion of how new technologies can be applied in helping to further understand the mechanisms which may lead to new strategies in the development of plant disease management approaches will also be reviewed.

Disease Resistant Genes and Signal Perception

Disease resistance is usually mediated by dominant genes, but some recessive resistant genes also exist. Harold H. Flor developed the 'gene-for-gene' concept in the 1940s from the studies on flax and the flax rust pathogen interactions. In this model, for resistance to occur (incompatibility), complementary pairs of dominant genes, one in the host and one in the pathogen, are required. An alteration or loss of the plant resistance gene (R changing to r) or of the pathogen avirulence gene (*Avr* changing to *avr*) leads to disease (compatibility). The model holds true for most biotrophic plant–pathogen interactions. According to the structural characteristics of their proteins, R genes are grouped into three classes. Data from the genetic and molecular analysis support the model.

NBS–LRR genes

This class contains a large number of proteins having C-terminal leucine-rich repeats (LRRs), a putative nucleotide-binding site (NBS) and an N-terminal Toll/interleukin-1 receptor (TIR) homology region or the coiled-coil (CC) sequence. Although extremely divergent in DNA sequence, at the amino acid level, they are readily identified by these motifs. Genome search indicates that the NBS–LRR class of R genes represents as much as 1% of the *Arabidopsis* genome (Meyers *et al.*, 1999). The proteins RPS2, RPP5 and RPM1 from *Arabidopsis*, N from tobacco, and L6 and M from flax are some members of this class. Although these R proteins do not appear to have intrinsic kinase activity, they can bind ATP or GTP and then activate the defence response. Mutations in NBS destroy R protein function.

Extracellular LRR genes

The extracellular LRR class includes the rice *Xa21* gene and the tomato *Cf* genes. *Xa21* encodes an active serine/threonine receptor-like kinase (RLK) with a putative extracellular domain composed of 23 LRRs, and an intracellular domain (*Xa21K*) comprised mainly of invariant amino acid residues characteristic of serine/threonine protein kinases (Liu *et al.*, 2002). The *Xa21K* intracellular domain is believed to become autophosphorylated through homodimerization or heterodimerization of *Xa21K* with a second receptor kinase that transphosphorylates the *Xa21K* serine and threonine residues following the extracellular pathogen reception (Liu *et al.*, 2002).

The *Cf* gene products contain extracelluar LRRs and a transmembrane domain, but lack a significant intracellular region that could relay the signal (e.g. a protein kinase domain). Studies have suggested some possibilities on how the Cf receptors transduce signals across the plasma membrane. In one study, *Avr9* binds to Cf-9 indirectly through a high-affinity Avr9-binding site and a third protein subunit of a membrane-associated protein complex (Rivas *et al.*, 2004). In its activated form, this additional transmembrane protein containing an extracellular interacting domain (ID) and an intracellular signalling domain (SD) is suspected to interact with the complex as Cf-9 lacks any suitable domains for signal transduction (Rivas *et al.*, 2004). Yeast two-hybrid screens using the cytoplasmic domain of Cf-9 revealed a thioredoxin homologue known as CITRX that binds to the C-terminal domain of Cf-9 (Rivas *et al.*, 2004). Further studies on CITRX suggest a potential role in negative regulation of Cf-9/Avr9 pathogen defence responses in early signal transduction through its interaction with the SD region of the signalling protein.

Pto gene

As in the case of *Xa21*, phosphorylation of a protein kinase by an upstream signal is a representative approach for signal amplification. *Pto* was identified in tomato plants as a unique *R* gene due to its cytoplasmic location and lack of an LRR motif. Transduction of the *Pto–avePto* interaction requires *Prf*, a gene that encodes a protein with leucine-zipper, NBS and LRR motifs. The binding of *avrPto* to *Pto* induces a structural change through overlapping surface areas, which allows for the interaction of Prf as an initial stage in the activation of subsequent phosphorylation cascades (Xiao *et al.*, 2003a; Mucyn *et al.*, 2006).

Two different classes of Pto-interactive proteins, i.e. Pti1 and Pti4/5/6, were identified. Pti1 is an Ser/Thr kinase. The major Pti1 site that is phosphorylated by Pto was Thr233. The phosphorylation of this site is required for Pto–Pti1 physical interaction in the yeast two-hybrid analysis. This interaction leads to the hypersensitive response (HR). Pti4, Pti5 and Pti6 are transcription factors and they are activated by phosphorylation.

Interaction of Pto and Pti4/5/6 activates pathogenesis-related (PR) genes (Martin, 1999; Martin et al., 2003).

Another significant component in the Pto defence system is the Prf protein. This is an NBS–LRR protein which detects and potentially 'guards' the Pto–AvrPto physical interaction (Dangl and Jones, 2001; McDowell and Woffenden, 2003). This model predicts that R proteins activate resistance when they interact with another plant protein (a guardee) that is targeted and modified by the pathogen in its quest to create a favourable environment. Resistance is triggered when the R protein detects an attempt to attack its guardee, which might not necessarily involve direct interaction between the R and Avr proteins. Prf acts to guard Pto and activates plant defences when it detects avrPto–Pto complexes. In terms of signal detection, Prf can be taken as the true *R* gene. Compelling evidence for this model was also reported for an *Arabidopsis* R protein. Here, RIN4 interacts with both RPM1 and its cognate avirulence proteins, AvrRPM1 and AvrB, to activate disease resistance (Mackey et al., 2002).

Signal Transduction

Parallel pathways and signal convergence

Multiple types of defence reactions are activated by pathogen attacks. Defence responses triggered by different R proteins are common to a large array of plant–pathogen interactions. Parallel or converging signalling pathways exist and a network of multiple interconnected signalling pathways acting together may amplify R gene-mediated signals (Xing and Jordan, 2000; Xing et al., 2002). For example, the genes *RPS2*, *RPP4* and *RPP5* share a similar requirement as *EDS1*, *PAD4* and *RAR1*. However, *RPP4* and *RPP5* differ from *RPS2* in their requirement for *SGT1*, suggesting the existence of interconnected rather than linear pathways. The *NDR1* gene represents a convergence point for cascades specified by *R* genes of the CC–NBS–LRR class (Century et al., 1997; Aarts et al., 1998), whereas EDS1 and PAD4 are convergence points for pathways originating from *R* genes of the TIR–NBS–LRR class (Aarts et al., 1998; Feys et al., 2001). EDS1 and PAD4 function in close proximity in the signalling pathway, as the proteins they encode physically interact *in vitro* and co-immunoprecipitate from plant extracts (Feys et al., 2001). However, they fulfil distinct roles in resistance: EDS1 is essential for the oxidative burst and HR elicitation, while PAD4 is required for phytoalexin, PR1 and salicylic acid accumulation (Rogers and Ausubel, 1997; Zhou et al., 1998; Rusterucci et al., 2001). The study of the requirement for NDR1 and EDS1 by the CC–NBS–LRR *R* genes RPP7 and RPP8 revealed that the utilization by a specific *R* gene of either an EDS1/PAD4 or NDR1 pathway is not mutually exclusive (McDowell et al., 2000). Certain pathways can operate additively through EDS1, NDR1 and additional unknown signalling components. Similar convergence exists further downstream in signalling pathways, e.g. at mitogen-activated protein kinase (MAPK) cascade levels (Romeis et al., 1999; Xing et al., 2001a, 2003).

Protein phosphorylation

Within the *Arabidopsis* genome, there are approximately 1000 protein kinase genes and 200 phosphatase genes (Xing *et al.*, 2002). The large pool of kinases and phosphatases indicates the importance of phosphorylation and dephosphorylation mechanisms in the growth and development of *Arabidopsis*. Some of the R proteins (Pto, Xa21 and Rpg1) are virtually protein kinases or have kinase catalytic domains as already discussed, and several R gene-mediated signalling components encode protein kinases. Members of the calmodulin domain-like protein kinase (CDPK) family also participate in *R* gene-mediated disease resistance. Two tobacco CDPKs, NtCDPK2 and NtCDPK3, are rapidly phosphorylated and activated in cell cultures in a *Cf-9/Avr9*-dependent manner (Romeis *et al.*, 2000, 2001). CDPK also regulates *R* gene-mediated production of reactive oxygen species (ROS) (Xing *et al.*, 1997, 2001b). Silencing CDPK caused a reduced elicitation of the HR mediated by the *Cf-9 R* genes (Romeis *et al.*, 2001).

Multiple levels of regulation

At each level of signalling events, many of the signalling components can be regulated at transcriptional, translational and post-translational levels. For example, many protein kinases involved in plant signalling are regulated at the post-translational level. However, kinases are also regulated at the transcriptional level, such as the rapid activation of maize CPK kinase ZmCPK10 (Murillo *et al.*, 2001). In fact, each regulation at transcriptional, translational and post-translational levels is very important, and the relative contribution of each level to the overall response may vary. The tobacco WIPK (an MAPK) gene is activated at multiple levels during the induction of cell death by fungal elicitins (Zhang *et al.*, 2000). De novo transcription and translation were shown to be necessary for the activation of the kinase activity and the onset of HR-like cell death. In the same study, a fungal cell wall elicitor that did not cause cell death induced WIPK mRNA and protein to similar levels as those observed with the elicitins. However, no corresponding increase in WIPK activity was detected. This demonstrated that post-translational control is also critical in elicitin-induced cell death. Plant WIPK is a perfect example demonstrating that the multiple levels of regulation of kinases contribute to the final effectiveness of signalling pathways.

Protein degradation in defence signalling

Roles of protein degradation in *R* gene-mediated signalling are emerging from the characterization of the RAR1 and SGT1 proteins and their interaction with components of the SCF (Skp1, Cullin and F-box) E3 ubiquitin ligase complex and with subunits of the COP9 signalosome (see Martin *et al.*, 2003). SCF complex mediates degradation of proteins involved in diverse signalling pathways through a ubiquitin proteasome pathway. Plant SGT1, which is essential for several *R* gene-mediated pathways, physically interacts with RAR1 in yeast two-hybrid screens and in plant extracts (see Martin *et al.*, 2003). Tight control of the levels of resistance proteins is critical for the homeostasis of plants. Overexpression of resistance genes can lead to deleterious effects on plant growth and development and constitutive activation of plant defence (Tang *et al.*, 1999; Xiao *et al.*, 2003b). Increasing evidence has suggested that regulation of protein stability is an important mechanism to control the steady-state levels of plant resistance proteins. Accumulation of the *Arabidopsis* resistance protein RPM1 requires three other proteins (RIN4, AtRAR1 and HSP90) that interact with RPM1 (Mackey *et al.*, 2002; Tornero *et al.*, 2002; Hubert *et al.*, 2003). The steady-state levels of the barley resistance proteins MLA1 and MLA6 were reduced when the *RAR1* gene was mutated (Bieri *et al.*, 2004). These findings suggest that direct or indirect protein–protein interactions play an important role in the stabilization of resistance proteins. Thus, the proteolytic activity may represent a security system to prevent XA21 from overaccumulating. A recent study has suggested that the proteolytic activity could be developmentally regulated, and autophosphorylation of Ser686, Thr688 and Ser689 residues in the intracellular juxtamembrane domain of XA21 may stabilize XA21 against such developmentally controlled proteolytic activity (Xu *et al.*, 2006).

Signal Responses

Massive changes in gene expression

Plant response to pathogen infection is associated with massive changes in gene expression. An array representing about 8000 genes (Zhu and Wang, 2000), nearly one-third of the total number of protein-encoding genes in *Arabidopsis*, was used to study the gene-for-gene resistance response to the bacterial pathogen *Pseudomonas syringae* (Glazebrook *et al.*, 2003; Tao *et al.*, 2003). More than 2000 genes changed expression levels within 9 h of inoculation with the pathogen (Tao *et al.*, 2003). Although it is not known how this information on 8000 genes will extrapolate to all of the protein-encoding genes in *Arabidopsis*, more than 2000 genes is still a massive change even at the whole-genome level. Overall qualitative similarities in defence responses between compatible and incompatible interactions were demonstrated by global expression profiling. Another question is whether all of the genes whose expression changes in response to a given pathogen are involved in resistance against that pathogen? When a plant detects a pathogen, it may not tailor its response to the pathogen at hand. Instead, it turns on many of the defence mechanisms it has, among which some may be effective against a particular pathogen. Many of the pathogen-responsive genes are not part of the defence response, but undergo expression changes in response to alterations in cellular state that result from the actions of the pathogen. For example, turning on defence mechanisms is energy-intensive, and some genes might be induced or repressed to promote efficient energy utilization during defence. This change could occur in response to a decrease in the energy reserve, which is an altered cell state. Thus, in global analysis, low false-negative rates are also important. A low false-positive rate is associated with a high false-negative rate. When the false-negative rate is high, a large number of genes that are associated with the global response are excluded from the analysis, so the results of such an analysis could be highly biased. The statistical criteria chosen for defining genes with significant changes in expression level should provide a balance between false-positive and false-negative rates.

Qualitative similarity in expression profiles from different pathogen interactions

For quite a long time, some scientists anticipated that resistance was associated with resistance-specific responses. Gene profiling studies have clearly indicated that although resistance-specific responses certainly exist, large sections of the global changes are qualitatively similar in resistant and susceptible responses. In *P. syringae*-induced responses, quantitative or kinetic differences in defence responses appeared to be important for determining resistance or susceptibility to the pathogen (Tao *et al.*, 2003). This observation is consistent with the fact that most of the known mutants that affect gene-for-gene resistance, except for those that affect pathogen recognition directly, also affect basal resistance (Glazebrook, 2001).

The resistance of *Arabidopsis* to *P. syringae* is mainly controlled by salicylic acid-mediated signalling mechanisms and the resistance to the necrotrophic fungal pathogen *Alternaria brassicicola* is mainly controlled by signalling mechanisms that are dependent on jasmonic acid (JA) (Thomma *et al.*, 1998; Glazebrook, 2001). However, the *Arabidopsis* genes that are induced by these two pathogens overlap substantially (about 50% of the responding genes are common for both pathogens) (van Wees *et al.*, 2003). Here, although the responses that are crucial for resistance against these pathogens are quite different, the overall signalling mechanisms that control changes in gene expression after infection have much in common and the level of specialization is low.

Signal Termination

The signal should be terminated when it has been transduced and responded to. This is particularly important to host plants when the response involves changes to a critical component of plant growth and development. For example, fungal elicitors have been shown to induce changes in the phosphorylation status of proteins in tomato cells. The dephosphorylation of the host plasma membrane H^+-ATPase occurred soon after treatment with elicitors from incompatible races of the fungal pathogen *Cladosporium fulvum* (Xing et al., 1996). The rephosphorylation followed soon after the dephosphorylation and at least two different protein kinases, a protein kinase C (PKC) and a Ca^{2+}/CaM-dependent protein kinase, were involved successively (Xing et al., 1996, 2001b). The protein kinases might act as negative elements and be responsible for ensuring an elicitor-induced response that would be quantitatively appropriate, correctly timed, highly coordinated with other activities of the host cells and probably more specifically terminated when the elicitor-induced signal transduction is completed; otherwise, the prolonged membrane potential change would harm host cells.

Applications to Fungal Disease Control

Resistance genes can be bred into crop plants to control diseases, but this approach has only limited success. Recent studies have also indicated that pathogens have evolved mechanisms to counteract plant defence responses, including: (i) modification of the elicitor proteins by mutations, or by deletion of the *Avr* genes, or by downregulation of *Avr* gene expression; (ii) secretion of enzymes that detoxify defence compounds (e.g. phytoalexins); (iii) use of ATP-binding cassette (ABC) transporters to mediate the efflux of toxic compounds; and (iv) secretion of glucanase-inhibitor proteins, which inhibit the endoglucanase activity of host plants.

Transforming susceptible plants with cloned *R* genes may provide pathogen resistance. When a susceptible tomato cultivar was transformed with the *Pto* gene, the plant became resistant to the bacterial pathogen *P. syringae* (Tang et al., 1999). Once it was thought that a major drawback of most *R* genes is their extreme specificity of action towards a single *avr* gene of one specific microbial species. However, *Pto* overexpression in plants constitutively activates defence responses and results in general resistance in the absence of the *avrPto* gene as it also gained resistance against the fungal pathogen *C. fulvum* (Tang et al., 1999).

Another strategy is to manipulate key signal transduction components. It has been argued that key component manipulation is promising for the following reasons: (i) interspecies transferability; (ii) high potential for broad-spectrum resistance; (iii) new alternatives in systems, such as wheat-*Fusarium* head blight, where information about resistance genes is limited; (iv) pathway sharing or interaction between abiotic and biotic stresses; (v) multiple barriers; and (vi) reduction in the possibility that pathogens will evolve new strategies to overcome resistance in transgenic plants generated by conventional approaches (Xing et al., 2002). A constitutively activated tomato MAPK kinase gene, $tMEK2^{MUT}$, was created to ensure the production of transcripts of MAPKs and the status of phosphorylation (Xing et al., 2001a). When overexpressed in tomato and wheat, $tMEK2^{MUT}$ increased resistance to the bacterial pathogen *P. syringae* pv. *tomato* and to the fungal pathogen *Puccinia triticina* (Xing et al., 2003; Jordan et al., 2006).

Fig. 1.1. Overexpression of *tMEK2* and the corresponding resistance to biotic stresses. (A) Reduced disease symptoms on leaves of transgenic tomato 5 days after inoculation with *Pseudomonas syringae* pv. *tomato*. Shown are a non-transgenic line (control) and a representative *tMEK2^{MUT}* transgenic line. (B) Leaf rust reaction of: (i) wild-type wheat cv. 'Fielder' (susceptible); (ii) transgenic 'Fielder' expressing *tMEK2^{MUT}*; and (iii) wild-type wheat cv. 'Superb' (resistant).

Our data also suggest that MAPK pathways mediate defence-related signal transduction in both the dicotyledonous (tomato) and the monocotyledonous (wheat) plants. The above results are shown in Fig. 1.1.

New Technologies

Short interfering RNA (siRNA) is responsible for the phenomenon of RNA interference (RNAi). The phenomenon of RNAi was first observed in

Petunia plants, although the mechanism was not understood at the time. In an attempt to produce *Petunia* flowers with a deep purple colour, the plants were transformed with extra copies of the gene for chalcone synthase, a key enzyme in the synthesis of anthocyanin pigments. But instead of dark purple flowers, the transformants produced only white flowers. The tendency of extra copies of a gene to induce the suppression of the native gene was termed cosuppression (Baulcombe, 2004). A related phenomenon was discovered by plant virologists studying viral resistance mechanisms. The genomes of most plant viruses consist of single-stranded RNA (ssRNA). Plants expressing viral proteins exhibited increased resistance to viruses, but it was subsequently found that even plants expressing short, non-coding regions of viral RNA sequences became resistant to the virus. The short viral sequences were somehow able to attack the incoming viruses (Baulcombe, 2004). RNAi has been used to understand defence-related mechanisms (e.g. Shen *et al.*, 2003; Seo *et al.*, 2007).

Yeast two-hybrid systems can generate information on protein–protein interactions. The system has been used to identify proteins that interact with the Pto kinase (Zhou *et al.*, 1997) such as Pti4, Pti5 and Pti6. In recent genomics efforts, a high-throughput yeast two-hybrid system has been developed (Uetz *et al.*, 2000) that offers extra advantages as follows: (i) we can identify interactions that place functionally unclassified proteins into a biological context; (ii) it offers insight into novel interactions between proteins involved in the same biological function; and (iii) novel interactions that connect biological functions to larger cellular processes might be discovered. Since $tMEK2^{MUT}$-transgenic wheat gained partial resistance to wheat leaf rust (Fig. 1), the mechanisms of tMEK2 function were studied. Heterologous screening for tomato tMEK2 interactive proteins in a wheat yeast two-hybrid library identified 46 positive colonies. Interaction of tMEK2 with three proteins has been confirmed in yeast. Heterologous yeast two-hybrid screening indicated the interaction of tMEK2 with a cytosolic glutamine synthetase (GS), a high mobility group (HMG)-like protein and a novel protein. Cytosolic and chloroplast GS are key enzymes in ammonium assimilation and their genetic engineering was shown to change plant development and response to various abiotic stresses (Vincent *et al.*, 1997; Harrison *et al.*, 2000). HMG proteins facilitate gene regulation through interactions with chromatin and other protein factors (Bustin and Reeves, 1996). Klosterman and Hadwiger (2002) reviewed the role of plant HMG-I/Y, one of the three groups of HMG proteins under the classification of mammalian HMGs, in the regulation of developmental and defence genes. The interaction with GS and HMG-like protein may suggest that tMEK2 is involved in response to abiotic and biotic stresses.

Proteomics has mostly been used to seek out underexpression and overexpression of proteins separated by two-dimensional electrophoresis (2DE), in experiments that are comparable to nucleic acid microarray experiments in genomics (e.g. Xing *et al.*, 2003). A proteomic approach is valuable in understanding regulatory networks because it deals with identifying new proteins in relation to their function and ultimately aims to unravel how their expression and modification is controlled. The 2D gel is in fact a protein array with molecular weight and isoelectric point dimensions, and proteins from it can usually be identified successfully by peptide mass fingerprinting or de novo sequencing (Standing, 2003), in either case using a matrix-assisted laser desorption ionization (MALDI) time-of-flight (TOF) mass spectrometer (MS). There are many examples in the literature of its successful application to plant pathology (e.g. Ventelon-Debout *et al.*, 2004).

One of the major control mechanisms for protein activity in plant–pathogen interactions is protein phosphorylation. However, studying protein phosphorylation cascades in plants presents two major technical challenges: (i) many of the signalling components are present at very low copy numbers, which makes them difficult to detect; and (ii) they are difficult to identify because there are currently only three plants with a complete genome sequence, i.e. *Arabidopsis thaliana*, *Populus* (poplar) and *Oryza sativa* (rice). Approaches to phosphoprotein discovery in plants have recently been reviewed (Rampitsch *et al.*, 2005; Thurston *et al.*, 2005). These include the use of anti-phosphotyrosine antibodies, ^{32}P labelling, and a phospho amino acid-sensitive fluorescent stain to label spots of interest on a 2D gel.

Perspective

Many exciting insights have emerged from recent research on plant defence signalling. The advantages of successfully engineering plants for disease resistance response are evident: increased yields and improved quality, avoidance of grain contamination by toxic secondary metabolites associated with certain fungal diseases and reduction of fungicide use and chemical release into the environment (Punja, 2004; Gilbert *et al.*, 2006). However, along with the recent research, we have realized that our understanding of the plant disease resistance response is still very fragmentary. We know very little about the structural basis of pathogen recognition. We are less sure than before about what R proteins actually recognize (Avr proteins, modified guardees or complexes that include both?). Furthermore, many gaps remain in our models of the defence signal transduction network. With the progress made so far, we expect that additional useful *R* genes and R protein-interactive proteins will be cloned or identified, and that models of resistance signalling developed in *Arabidopsis*, tomato, tobacco, rice and wheat will continue to be evaluated for applicability in other crops.

It is the time for systems biology approaches, i.e. the exercise of integrating the existing knowledge about biological components, building a model of the system as a whole and extracting the unifying organizational principles that explain the form and function of living organisms. With this approach, we will greatly increase our understanding of plant–pathogen interactions as well as obtain a holistic view of the form and function of biological systems. Such a strategy will greatly accelerate the pace of discovery and provide new insights into interactions between defence signalling and other plant processes, which is critical when new rational approaches are adopted in the manipulation of disease resistance.

Acknowledgements

This work was supported by research grants from the Natural Sciences and Engineering Research Council of Canada.

References

Aarts, N., Metz, M., Holub, E., Staskawicz, B.J., Daniels, M.J. and Parker, J.E. (1998) Different requirements for EDS1 and NDR1 by disease resistance genes define at least two R gene-mediated signaling pathways in *Arabidopsis*. *Proceedings of the National Academy of Sciences of the United States of America* 95, 10306–10311.

Baulcombe, D. (2004) RNA silencing in plants. *Nature* 431, 356–363.

Bieri, S., Mauch, S., Shen, Q.-H., Peart, J., Devoto, A., Casais, C., Ceron, F., Schulze, S., Steinbiß, H.-H., Shirasu, K. and Schulze-Lefert, P. (2004) RAR1 positively controls steady state levels of barley MLA resistance proteins and enables sufficient MLA6 accumulation for effective resistance. *The Plant Cell* 16, 3480–3495.

Bustin, M. and Reeves, R. (1996) High mobility group chromosomal proteins: architectural components that facilitate chromatin function. *Progress in Nucleic Acids Research and Molecular Biology* 54, 35–100.

Century, K.S., Shapiro, A.D., Repetti, P.P., Dahlbeck, D., Holub, E. and Staskawicz, B.J. (1997) NDR1, a pathogen-induced component required for *Arabidopsis* disease resistance. *Science* 278, 1963–1965.

Dangl, J.L. and Jones, J.D.G. (2001) Plant pathogens and integrated defense responses to infection. *Nature* 411, 826–833.

Feys, B.J., Moisan, L.J., Newman, M.A. and Parker, J.E. (2001) Direct interaction between the *Arabidopsis* disease resistance signaling proteins, EDS1 and PAD4. *The EMBO Journal* 20, 5400–5411.

Gilbert, J., Jordan, M., Somers, D., Xing, T. and Punja, Z. (2006) Engineering plants for durable disease resistance. In: Tuzun, S. and Bent, E. (eds) *Multigenic and Induced Systemic Resistance in Plants*. Kluwer Academic/Plenum Publishers, New York, pp. 415–455.

Glazebrook, J. (2001) Genes controlling expression of defense responses in *Arabidopsis* – 2001 status. *Current Opinion in Plant Biology* 4, 301–308.

Glazebrook, J., Chen, W., Estes, B., Chang, H.S., Nawrath, C., Metraux, J.P., Zhu, T. and Katagiri, F. (2003) Topology of the network integrating salicylate and jasmonate signal transduction derived from global expression phenotyping. *The Plant Journal* 34, 217–228.

Harrison, J., Brugière, N., Phillipson, B., Ferrario-Mery, S., Becker, T., Limami, A. and Hirel, B. (2000) Manipulating the pathway of ammonia assimilation through genetic engineering and breeding: consequences to plant physiology and plant development. *Plant and Soil* 221, 81–93.

Hubert, D.A., Tornero, P., Belkhadir, Y., Krishna, P., Takahashi, A., Shirasu, K. and Dangl, J.L. (2003) Cytosolic HSP90 associates with and modulates the *Arabidopsis* RPM1 disease resistance protein. *The EMBO Journal* 22, 5679–5689.

Jordan, M., Cloutier, S., Somers, D., Procunier, D., Rampitsch, C. and Xing, T. (2006) Beyond R genes: dissecting disease-resistance pathways using genomics and proteomics. *Canadian Journal of Plant Pathology* 28, S228–232.

Klosterman, S. and Hadwiger, L.A. (2002) Plant HMG proteins bearing the AT-hook motif. *Plant Science* 162, 855–866.

Liu, G., Pi, L., Walker, J., Ronald, P. and Song, W. (2002) Biochemical characterization of the kinase domain of the rice disease resistance receptor-like kinase XA21. *Journal of Biological Chemistry* 277, 20264–20269.

Mackey, D., Holt, B.F., Wiig, A. and Dangl, J.L. (2002) RIN4 interacts with *Pseudomonas syringae* type III effector molecules and is required for RPM1-mediated disease resistance in *Arabidopsis*. *Cell* 108, 743–754.

Martin, G.B. (1999) Functional analysis of plant disease resistance genes and their downstream effectors. *Current Opinion in Plant Biology* 2, 273–279.

Martin, G.B., Bogdanove, A.J. and Sessa, G. (2003) Understanding the functions of plant disease resistance proteins. *Annual Review of Plant Biology* 54, 23–61.

McDowell, J.M. and Woffenden, B.J. (2003) Plant disease resistance genes: recent insights and potential applications. *Trends in Plant Science* 21, 178–183.

McDowell, J.M., Cuzick, A., Can, C., Beynon, J., Dangl, J.L. and Holub, E.B. (2000) Downy mildew (*Peronospora parasitica*) resistance genes in *Arabidopsis* vary in functional requirements for NDR1, EDS1, NPR1 and salicylic acid accumulation. *The Plant Journal* 22, 523–529.

Meyers, B.C., Dickerman, A.W., Michelmore, R.W., Sivaramakrishnan, S., Sobral, B.W. and Young, N.D. (1999) Plant disease resistance genes encode members of an ancient and diverse protein family within the nucleotide-binding superfamily. *The Plant Journal* 20, 317–332.

Mucyn, T.S., Clemente, A., Andriotis, V., Balmuth, A., Oldroyd, G., Staskawicz, B. and Rathjen, J. (2006) The tomato NBARC-LRR protein Prf interacts with Pto kinase in vivo to regulate specific plant immunity. *The Plant Cell* 18, 2792–2806.

Murillo, I., Jaeck, E., Cordero, M.J. and Segundo, B.S. (2001) Transcriptional activation of a maize calcium-dependent protein kinase gene in response to fungal elicitors and infection. *Plant Molecular Biology* 45, 145–158.

Punja, Z.K. (2004) Genetic engineering of plants to enhance resistance to fungal pathogens. In: Punja, Z.K. (ed.) *Fungal Disease Resistance in Plants: Biochemistry, Molecular Biology and Genetic Engineering*. Haworth Press, New York, pp. 207–258

Rampitsch, C., Bykova, N., Xing, T. and Ens, W. (2005) Phosphoproteomics: approaches to studying MAP kinase signalling in the tomato leaf. *Recent Research and Development in Biochemistry* 5, 291–306.

Rivas, S., Rougon-Cardoso, A., Smoker, M., Schauser, L., Yoshioka, H. and Jones, J. (2004) CITRX thioredoxin interacts with the tomato Cf-9 resistance protein and negatively regulates defence. *The EMBO Journal* 23, 2156–2165.

Rogers, E.E. and Ausubel, F.M. (1997) *Arabidopsis* enhanced disease susceptibility mutants exhibit enhanced susceptibility to several bacterial pathogens and alterations in *PR-1* gene expression. *The Plant Cell* 9, 305–316.

Romeis, T., Piedras, P., Zhang, S., Klessig, D.F., Hirt, H. and Jones, J.D. (1999) Rapid Avr9- and Cf-9-dependent activation of MAP kinases in tobacco cell cultures and leaves: convergence of resistance gene, elicitor, wound, and salicylate responses. *The Plant Cell* 11, 273–287.

Romeis, T., Piedras, P. and Jones, J.D. (2000) Resistance gene-dependent activation of a calcium-dependent protein kinase in the plant defense response. *The Plant Cell* 12, 803–816.

Romeis, T., Ludwig, A.A., Martin, R. and Jones, J.D. (2001) Calcium-dependent protein kinases play an essential role in a plant defence response. *The EMBO Journal* 20, 5556–5567.

Rusterucci, C., Aviv, D.H., Holt, B.F., Dangl, J.L. and Parker, J.E. (2001) The disease resistance signaling components EDS1 and PAD4 are essential regulators of the cell death pathway controlled by LSD1 in *Arabidopsis*. *The Plant Cell* 13, 2211–2224.

Seo, S., Katou, S., Seto, H., Gomi, K. and Ohashi, Y. (2007) The mitogen-activated protein kinases WIPK and SIPK regulate the levels of jasmonic and salicylic acids in wounded tobacco plants. *The Plant Journal* 49, 899–909.

Shen, Q.H., Zhou, F.S., Bieri, S., Haizel, T., Shirasu, K. and Schulze-Lefert, P. (2003) Recognition specificity and RAR1/SGT1 dependence in barley *Mla* disease resistance genes to the powdery mildew fungus. *The Plant Cell* 15, 732–744.

Standing, K. (2003) Peptide and protein sequencing by mass spectrometry. *Current Opinion in Structural Biology* 13, 595–601.

Tang, X., Xie, M., Kim, Y.J., Zhou, J., Klessig, D.F. and Martin, G.B. (1999) Overexpression of Pto activates defense responses and confers broad resistance. *The Plant Cell* 11, 15–29.

Tao, Y., Xie, Z., Chen, W., Glazebrook, J., Chang, H.S., Han, B., Zhu, T., Zou, G. and Katagiri, F. (2003) Quantitative nature of *Arabidopsis* responses during compatible and incompatible interactions with the bacterial pathogen *Pseudomonas syringae*. *The Plant Cell* 15, 317–330.

Thomma, B., Eggermont, K., Penninckx, I., Mauch-Mani, B., Vogelsang, R., Cammue, B.P.A. and Broekaert, W.F. (1998) Separate jasmonate-dependent and salicylate-dependent defense-response pathways in *Arabidopsis* are essential for resistance to distinct microbial pathogens. *Proceedings of the National Academy of Sciences of the United States of America* 95, 15107–15111.

Thurston, G., Regan, S., Rampitsch, C. and Xing, T. (2005) Proteomic and phosphoproteomic approaches to understand plant–pathogen interactions. *Physiological and Molecular Plant Pathology* 66, 3–11.

Tornero, P., Merritt, P., Sadanandom, A., Shirasu, K., Innes, R.W. and Dangl, J.L. (2002) RAR1 and NDR1 contribute quantitatively to disease resistance in *Arabidopsis*, and their relative contributions are dependent on the *R* gene assayed. *The Plant Cell* 14, 1005–1015.

Uetz, P., Giot, L., Cagney, G., Mansfield, T.A., Judson, R.S., Knight, J.R., Lockshon, D., Narayan, V., Srinivasan, M., Pochart, P., Qureshi-Emili, A., Li, Y., Godwin, B., Conover, D., Kalbfleisch, T., Vijayadamodar, G., Yang, M., Johnston, M., Fields, S. and Rothberg, J.M. (2000) A comprehensive analysis of protein–protein interactions in *Saccharomyces cerevisiae*. *Nature* 403, 623–627.

van Wees, S.C., Chang, H.S., Zhu, T. and Glazebrook, J. (2003) Characterization of the early response of *Arabidopsis* to *Alternaria brassicicola* infection using expression profiling. *Plant Physiology* 132, 606–617.

Ventelon-Debout, M., Delalande, F., Brizard, J.P., Diemer, H., Dorsselaer, A.V. and Brugidou, C. (2004) Proteome analysis of cultivar-specific deregulations of *Oryza sativa* indica and *O. sativa* japonica cellular suspensions undergoing rice yellow mottle virus infection. *Proteomics* 4, 216–225.

Vincent, R., Fraisier, V., Chaillou, S., Limami, M.A., Deleens, E., Phillipson, B., Douat, C., Boutin, J.P., and Hirel, B. (1997) Overexpression of a soybean gene encoding cytosolic glutamine synthetase in shoots of transgenic *Lotus corniculatus* L. plants triggers changes in ammonium assimilation and plant development. *Planta* 201, 424–433.

Xiao, F., Lu, M., Li, J., Zhao, T., Yi, S., Thara, V.K., Tang, X. and Zhou, J. (2003a) *Pto* mutants differentially activate *Prf*-dependent, *avrPto*-independent resistance and gene-for-gene resistance. *Plant Physiology* 131, 1239–1249.

Xiao, S., Brown, S., Patrick, E., Brearley, C. and Turner, J.G. (2003b) Enhanced transcription of the *Arabidopsis* disease resistance genes RPW8.1 and RPW8.2 via a salicylic acid-dependent amplification circuit is required for hypersensitive cell death. *The Plant Cell* 15, 33–45.

Xing, T. and Jordan, M. (2000) Genetic engineering of signal transduction mechanisms. *Plant Molecular Biology Reporter* 18, 309–318.

Xing, T., Higgins, V.J. and Blumwald, E. (1996) Regulation of plant defense responses to fungal pathogens: two types of protein kinases in the reversible phosphorylation of the host-plasma membrane H^+-ATPase. *The Plant Cell* 8, 555–564.

Xing, T., Higgins, V.J. and Blumwald, E. (1997) Race-specific elicitors of *Cladosporium fulvum* promote translocation of cytosolic components of NADPH oxidase to plasma membrane of tomato cells. *The Plant Cell* 9, 249–259.

Xing, T., Malik, K., Martin, T. and Miki, B.L. (2001a) Activation of tomato PR and wound-related genes by a mutagenized tomato MAP kinase kinase through divergent pathways. *Plant Molecular Biology* 46, 109–120.

Xing, T., Wang, X.J., Malik, K. and Miki, B.L. (2001b) Ectopic expression of a CDPK enhanced NADPH oxidase activity and oxidative burst. *Molecular Plant–Microbe Interactions* 14, 1261–1264.

Xing, T., Ouellet, T. and Miki, B. (2002) Towards genomics and proteomics studies of protein phosphorylation in plant–pathogen interactions. *Trends in Plant Science* 7, 224–230.

Xing, T., Rampitsch, C., Miki, B.L., Mauthe, W., Stebbing, J., Malik, K. and Jordan, M. (2003) MALDI-TOF-MS and transient gene expression analysis indicated co-enhancement of β-1,3-glucanase and endochitinase by tMEK2 through divergent pathways in transgenic tomato. *Physiological and Molecular Plant Pathology* 62, 209–217.

Xu, W.H., Wang, Y.S., Liu, G.Z., Chen, X.H., Tinjuangjun, P., Pi, L.Y. and Song, W.Y. (2006) The autophosphorylated Ser686, Thr688, and Ser689 residues in the intracellular juxtamembrane domain of XA21 are implicated in stability control of rice receptor-like kinase. *The Plant Journal* 45, 740–751.

Zhang, S., Liu, Y. and Klessig, D.F. (2000) Multiple levels of tobacco WIPK activation during the induction of cell death by fungal elicitins. *The Plant Journal* 23, 339–347.

Zhou, J., Tang, X. and Martin, G.B. (1997) The Pto kinase conferring resistance to tomato bacterial speck disease interacts with proteins that bind a cis-element of pathogenesis-related genes. *The EMBO Journal* 16, 3207–3218.

Zhou, N., Tootle, T.L., Tsui, F., Klessig, D.F. and Glazebrook, J. (1998) PAD4 functions upstream from salicylic acid to control defense responses in *Arabidopsis*. *The Plant Cell* 10, 1021–1030.

Zhu, T. and Wang, X. (2000) Large-scale profiling of the *Arabidopsis* transcriptome. *Plant Physiology* 124, 1472–1476.

2 Modulating Quorum Sensing and Type III Secretion Systems in Bacterial Plant Pathogens for Disease Management

C.-H. YANG AND S. YANG

Abstract
In this chapter, we briefly describe three major classes of quorum sensing (QS) systems as well as the structural components, substrates, chaperones, signals and regulation of the type III secretion system (T3SS). In addition, we discuss current knowledge about the regulatory network between QS and T3SS, including examples of QS controlling T3SS, and the connection between T3SS and QS through the regulator of secondary metabolism (Rsm) system, GacS/GacA two-component signal transduction system (TCSTS) and other regulators, as well as the interrelationships among these systems. In addition to the QS modulation mechanisms, we discuss disease management strategies by targeting the QS and T3SS. Finally, the current application of TCSTS histidine kinase inhibitor and QS interference (QSI) for disease management is further discussed. Future directions to enhance our understanding of the QS and T3SS systems themselves, as well as managing bacterial plant diseases by modulating the QS and T3SS systems, are suggested and the potential problems associated with the application of QS and T3SS in plant disease management are also briefly discussed.

Introduction

Bacteria live unicellularly and were previously thought of as solitary cells without communication with others. It is now becoming clear that bacteria act as multicellular organisms in communication with their extracellular environment and intracellular physiological conditions. Bacteria respond rapidly to changes by integrating the signals of small-molecule mixtures into the regulatory network and synchronizing the activities of large groups of cells to benefit the whole community. One well-studied example of cell-to-cell communication is quorum sensing (QS). QS is a cell density-dependent process in which bacteria communicate through the secreted signal molecules named autoinducers (AIs) to regulate gene expression collectively; this involves production, release and perception of the signalling molecules. QS was first described in the late 1970s in two bioluminescent marine bacteria, *Vibrio harveyi* and *Vibrio fisheri*. The QS characterized in the *Vibrio* spp. has become the paradigm of QS. Since then, it has been found to be a widespread mechanism with many different QS in different bacterial species controlling multiple cellular functions (Miller and Bassler, 2001; Taga and Bassler, 2003; Henke and Bassler, 2004). Numerous animal and plant pathogens regulate virulence factor expression by using QS, which allows microorganisms to elicit an overwhelming attack before host cells can mount an effective defence.

The term type III secretion system (T3SS) was first used to describe one of the mechanisms by which Gram-negative bacteria export proteins from the cell through a Sec-independent secretion system. The study of T3SS has expanded rapidly in recent years. More than 2350 publications on this subject were listed in the ISI Web of Knowledge database, with half of them published in the last 5 years. T3SS is a secretion system to translocate effectors directly into the cytosol of eukaryotic host cells, where the effectors facilitate bacterial pathogenesis or symbiosis by specifically interfering with host cell signal transduction and other cellular processes. T3SS allows a fast and efficient translocation of effector proteins across the barriers of the bacterial inner membrane, periplasm, outer membrane, lipopolysacharide (LPS) layer and the eukaryotic cell membrane in a single step. The genes encoding T3SS in bacteria are clustered on certain pathogenicity islands of the chromosome and/or plasmids, which may have been acquired by horizontal genetic transfer (Galan and Collmer, 1999).

Various pathogens have been reported to utilize T3SS as a conserved basic virulence mechanism, including the animal pathogens *Chlamydia* spp., *Escherichia coli*, *Pseudomonas aeruginosa*, *Salmonella* spp., *Shigella* spp., *Vibrio parahaemolyticus*, *Yersinia* spp. and the plant pathogens such as *Pseudomonas syringae*, *Pectobacterium* spp., *Pantoea* spp., *Ralstonia solanacearum*, *Xanthomonas campestris* and *Rhizobium* spp. Some components of the T3SS apparatus (T3SA) are conserved. Most T3SS substrates share a common N-terminal secretion signal, require the presence of the specific chaperones for secretion, and T3SS secretion activity is tightly regulated. Phytopathogenic bacteria including *Pectobacterium*, *Pseudomonas*, *Pantoea*, *Xanthomonas* and *Ralstonia* cause diverse diseases in many different plants, but they all colonize intercellular spaces of susceptible plants and are capable of killing plant cells. The ability of these bacteria to multiply inside their hosts and produce necrotic symptoms is dependent on T3SS and the effector proteins secreted by this system. T3SS is required for bacterial pathogenicity on host plants by compatible pathogens and the elicitation of the hypersensitive response (HR), a programmed death of the plant cells at the site of pathogen invasion associated with plant defence, in non-host plants (Galan and Collmer, 1999; Mota *et al.*, 2005; Buttner and Bonas, 2006).

Over the last 15 years, our knowledge of bacterial virulence factors has been accumulated from a few specialized toxins and adhesions to a hidden landscape of sophisticated and diverse virulence systems in bacteria. The pathogenic factors have been characterized genetically and biochemically as to which could be used by pathogens to precisely target specific host cell activities, such as cytoskeletal reshuffle, cell cycle progression, vesicular trafficking and apoptosis (Stebbins, 2005). Over the past few years, many genes involved in the make-up of the complex QS and T3SS systems and the regulation of their expression and activity have been identified and characterized. In addition, the structural biology and biochemistry on the core T3SA and the needle, as well as the T3SS chaperones, have shed new light on the assembly process and the effector manner of translocation and regulation. The coming years in QS and T3SS research are expected not only to provide insight into the mechanisms of manipulation of host cell functions by bacterial pathogens, resulting in a better understanding of the system itself, but also to lead to the discovery of new concepts in molecular biology, microbiology and biochemistry, and to findings that may provide a unique target for the development of therapeutic agents and contribute to the design of new drugs to combat many important bacterial pathogens (Mota *et al.*, 2005). Finally, significant progress in research work on QS and T3SS of animal pathogens has been made and, along with plant pathogens, some of these related studies on animal pathogens have been included here.

Quorum Sensing Systems

QS systems consist of three components, which include the AI signal, the signal synthetase and the corresponding regulator to produce and perceive the signal. Based on the signal, regulator and the circuit, QS used by bacteria can be divided into four different classes. A summary of different QS systems in different bacterial species and their functions as well as the corresponding QS interference (QSI) are listed in Table 2.1 and are described in the reviews mentioned above.

Gram-negative LuxI/R class

The LuxI/R-type system was first investigated in the marine bacteria *Vibrio* spp. and primarily used by Gram-negative bacteria. The AIs acyl homoserine lactones (AHLs) are produced by a LuxI family AHL synthetase and function as a ligand for the cognate LuxR-type transcriptional regulator to modulate gene expression of the bacteria. Production of AHL QS signals is widespread among Gram-negative bacteria. Over 70 species of Gram-negative bacteria, including *Burkholderia cepacia, Clostridium difficile, E. coli, P. aeruginosa, Rhizobium leguminosarum, R. solanacearum, Yersinia pseudotuberculosis, Pectobacterium* spp. and *Pseudomonas* spp., have been identified as using the LuxI/R analogue as a gene regulatory mechanism to control various phenotypes (Fuqua *et al.*, 2001; Fuqua and Greenberg, 2002; Federle and Bassler, 2003).

Table 2.1. Summary of different QS in different bacterial species and their functions as well as the corresponding QS interference.

QS systems	Gram-negative QS	Gram-positive QS	LuxS/AI-2 QS
Circuit			
Examples	LuxI/R, LasI/R, RhlI/R, ExpI/R, YenI/R, EsaI/R	Agr system, ComAP system	LuxP/Q, Lsr
Species	Agrobacterium spp., Pectobacterium spp., Pantoea spp., Pseudomonas spp., Ralstonia spp., Rhizobium spp., Xanthomonas spp.	Streptococcus spp., Bacillus spp., Lactococcus spp., Staphylococcus spp.	Bacillus spp., Listeria spp., Mycobacterium spp., Staphylococcus spp., Streptococcus spp., Escherichia coli, Helicobacter pylori, Streptococcus pneumoniae, Shigella spp., Vibrio spp., Pectobacterium spp., Salmonella spp.
Signal (signal synthetase/ response regulator)	C4-HSL (RhlI/R), C6-HSL (PheI/R), C8-HSL (TraI/R), 3-OH-C4-HSL (LuxLM/LuxN), 3-O-C6-HSL (LuxI/R, AhlI/R), 3-O-C12-HSL (LasI/R), 3-O-C14:1-HSL (HdtS/?)	CSF (PhrC), oligopeptide (comX/A), AIP (AgrD/A) CSP, A-factor	S-THMF-borate AI-2 (LuxS/LuxR); R-THMF AI-2 (LuxS/LsrR)

Continued

Table 2.1. Continued

QS systems	Gram-negative QS	Gram-positive QS	LuxS/AI-2 QS
Signal transporter and sensor	NA/LuxN	ABC transporter (Opp)/ComP, AgrB permease/AgrC	NA/LuxP, LsrAC/LsrB
Functions	Bioluminescence, exoenzyme production, pigment and antibiotic biosynthesis, biofilm, conjugation, virulence	Competence, sporulation, antibiotic biosynthesis, virulence, biofilm	Bioluminescence, protease production, toxin and antibiotic biosynthesis, biofilm, T3SS, motility, iron acquisition, virulence
QS interference	Signal degradation: AHL-lactonasae (*aiiA, attM, ahlD, aiiB, ahlK, PONs*) AHL-acylase (*aiiD, pvdQ, ACY1*) Signal generation inhibition: fatty acid biosynthesis inhibitor (triclosan) AHL analogues: 3-O-C12-(2-aminocyclohexanone) R protein degradation: halogenated furanones	Sensor kinase inhibition: kinase inhibitor (closantel, RWJ-49815) Receptor competition: AIP analogues (truncated AIP-II, furanone C-30)	Histidine kinase inhibition: closantel AI-2 mimics: epinephrine, norepinephrine

The AHLs are highly conserved. The AI AHLs of this class are characterized by a common homoserine lactone (HSL) moiety ligated to a variable acyl side chain and substitution (carbonyl or hydroxyl) at the C-3 carbon (Fuqua and Greenberg, 2002). Different LuxI homologue proteins catalyse the synthesis of a range of specific AHLs by connecting the homocysteine moiety of S-adenosylmethionine (SAM) to the acyl side chain from the appropriately charged acyl–acyl carrier protein (acyl-ACP) or acyl-coenzyme A (acyl-CoA). More than one AHL can be produced by utilizing alternative acyl-ACP or acyl-CoA side chain precursors. Although the LuxR-type regulators from different species can bind to *lux* boxes, a similar DNA sequence in the promoter region of the LuxR-type regulator targeted gene (Taga and Bassler, 2003), the interaction between the AI AHL to its cognate LuxR-type regulator is specific. The AHL produced by one species of bacteria can rarely interact with the LuxR-type regulator of another species (Fuqua *et al.*, 2001). Signal specificity is conferred by the length of the acyl side chain which ranges from 4 to 18 carbons, the nature of the substitution at C-3 and unsaturations within the acyl chain. Meanwhile, the function of LuxR homologues as quorum sensors has been suggested to be mediated by the binding of AHL signal molecules to the N-terminal receptor site of the proteins (Koch *et al.*, 2005).

Beyond AHLs, there is evidence that other signalling molecules, including peptides and cyclic dipeptides, exist that could be involved in intraspecies communication in Gram-negative bacteria. One known molecule is the 2-heptyl-3-hydroxy-4-quinolone, the *Pseudomonas* quinolone signal (PQS), which is involved in the QS pathways of *P. aeruginosa*. Other examples of different signalling molecules are 3-hydroxypalmitic acid methyl ester (3-OH PAME) involved in virulence regulation in *R. solanacearum* and a molecule named bradyoxetin involved in symbiosis in *Bradyrhizobium japonicum* (Lyon and Muir, 2003). It is likely that a massive number of other unknown compounds and signalling cascades are just waiting to be discovered, which will then open the door for further anti-infective drug discovery efforts aimed at the inhibition of these pathways (Fast, 2003).

Gram-positive oligopeptide/TCSTS class

It is primarily Gram-positive bacteria that use the modified oligopeptide AIs, which are detected by a two-component signal transduction system (TCSTS). The Gram-positive QS system generally consists of three components: (i) a modified oligopeptide as AI signalling molecules; (ii) TCSTS for signal detection; and (iii) the response regulator for target gene regulation. The AI signalling molecule oligopeptide, which is also called pheromone, is derived by post-translational processing of a larger precursor peptide in the cytoplasm. The precursor autoinducing polypeptides (AIPs) are synthesized, and subsequently processed and modified to make the mature oligopeptide AI molecule, which is then exported extracellularly via an ATP-binding cassette (ABC) transporter complex. The AIs are detected via TCSTS in which the external portion of a membrane-bound histidine kinase sensor detects the AIs and then transduces the signal to the corresponding intracellular response regulator via a conserved phosphorylation–dephosphorylation mechanism. The regulator then binds to DNA and regulates the target gene transcription (Sturme et al., 2002).

QS in Gram-positive bacteria has been found to regulate a number of physiological activities, including competence development in *Streptococcus pneumoniae* and *Streptococcus mutans*, sporulation in *Bacillus subtilis*, antibiotic biosynthesis in *Lactococcus lactis* and virulence factor induction in *Staphylococcus aureus* and biofilm formation in *S. mutans* and *S. intermedius*. The prototype of Gram-positive oligopeptide/TCSTS class is the Agr (accessory gene regulator) QS system in *S. aureus*, which regulates virulence gene expression and biofilm formation. The oligopeptide signal is produced by AgrD and modified by AgrB. The resulting AI, which is eight or nine amino acids long and contains thiolactone rings, is then detected by the sensor AgrC and activates the regulator AgrA for the subsequent gene regulation (Zhang et al., 2004; Abraham, 2006).

Interspecies LuxS/AI-2 class

LuxS/AI-2 signalling has been proposed to be a universal signal system found in both Gram-negative and Gram-positive bacteria with the autoinducer AI-2 produced by a LuxS family synthetase for interspecies communication. This QS system was initially characterized in *V. harveyi*, and has been detected in more than 55 species by sequence analysis or functional assays. The biosynthetic pathways and the biochemical intermediates in AI-2 biosynthesis are identical in several Gram-negative bacteria. Considering the occurrence both in Gram-positive and Gram-negative species and the broad representation of *luxS* among bacteria, AI-2 has been proposed to be a universal interspecies signal for communication between and/or among species (Sperandio et al., 2003; Henke and Bassler, 2004; Kaper and Sperandio, 2005; Xavier and Bassler, 2005).

The LuxS acts as a key AI-2 synthetase in the activated methyl cycle and converts S-ribosylhomocysteine to homocysteine and AI-2. However, instead of a single chemical entity, AI-2 is a collective term used for a group of furanone derivatives that form spontaneously from the same precursor 4,5-dihydroxy-2,3-pentanedione (DPD). AI-2 has been reported in the literature to control an assortment of apparently 'niche-specific' genes as a ligand with other regulators (Winzer et al., 2003). Genes potentially regulated by AI-2 in other species have been identified by constructing a *luxS* mutant of a test species and comparing gene expression in the wildtype and the *luxS* mutant. Among the phenotypes and functions affected by *luxS* mutations are T3SS and flagellum expression in enterohaemorrhagic *E. coli* (EHEC) O157:H7, T3SS in *V. harveyi* and *V. parahaemolyticus*, toxin production in *C. perfringens*, cell attachment during biofilm formation in *Listeria monocytogenes*, mixed-species biofilm formation of *Porphyromonas gingivalis* and *S. gordonii*, SpeB cysteine protease secretion in *S. pyogenes* and the regulation of acid and oxidative stress tolerance and biofilm formation in *S. mutans* (McNab *et al.*, 2003; Sperandio *et al.*, 2003; Henke and Bassler, 2004).

AI-2, AI-3 and prokaryotic–eukaryotic communication

It has been clear that AI-2 production is widespread in the bacterial kingdom. However, LuxP homologues, the periplasmic receptor which binds to AI-2, as well as homologues from this signalling cascade, have been found only in *Vibrio* spp. In non-*Vibrio* species, the only genes shown to be directly regulated by AI-2 encode an ABC transporter named Lsr (LuxS regulated) in *S. typhimurium* and *E. coli*, which is responsible for the AI-2 uptake by these species. AI-2 binds to LsrB and is transported inside the cell, where it is phosphorylated by LsrK and proposed to interact with LsrR, a SorC-like transcription factor involved in repressing expression of the *lsr* operon (Taga and Bassler, 2003; Kaper and Sperandio, 2005). Meanwhile, the role of AI-2 as a universal signal in bacteria other than *V. harveyi* has not been readily established. Several groups have been unable to detect the AI-2 furanosyl-borate diester in purified fractions containing AI-2 activity from *Salmonella* and *E. coli*. The furanosyl compounds identified from these fractions did not contain boron. Actually, instead of a furanosyl-borate diester, the ligand of the receptor LsrB was a furanone (($2R,4S$)-2-methyl-2,3,3,4-tetrahydroxytetrahydrofuran), which has been co-crystallized with LsrB in *Salmonella*. The results are consistent with what has been observed in AI-2 fractions of *Salmonella* and *E. coli* (Sperandio *et al.*, 2003; Winzer *et al.*, 2003; Miller *et al.*, 2004). These differences from AI-2 detection in *V. harveyi* raise the question of whether all bacteria may actually use AI-2 as a signalling compound or whether it is released as a waste product or used as a metabolite by some bacteria. Furthermore, the effects of *luxS* inactivation are species dependent and sometimes even strain dependent, and it is often not clear whether the mutant phenotypes observed are the result of a signalling defect, such as the loss of AI-2 or the metabolic perturbations caused by the disruption of the activated methyl cycle, since a *luxS* mutation could interrupt the methionine metabolic pathway, and thereby change the whole metabolism of the bacteria (Winzer *et al.*, 2003).

A recent breakthrough in distinguishing the potential cell signalling functions from general metabolic functions was the discovery of a new signalling molecule called AI-3, whose synthesis is dependent on LuxS. Building on their previous studies showing that a *luxS* mutant of EHEC was deficient in T3SS and flagellum production, Sperandio *et al.* (2003) showed that purified AI-2, which was synthesized *in vitro*, was unable to restore these phenotypes when it was added to the mutant. The AI responsible for this signalling is dependent on the presence of the *luxS* gene for its synthesis, but it is different from AI-2, and a novel compound named AI-3 was identified by the Sperandio team. Further, they raised questions about the validity of some of the phenotypes attributed to AI-2 signalling since LuxS is not devoted to AI-2 production, which, in fact, is an important enzyme affecting the metabolism of SAM and various amino acid pathways. Consequently, altered gene expression in a *luxS* mutation may include both genes affected by QS itself and the interruption of metabolic pathway.

Furthermore, one also has to take into consideration that a knockout of *luxS* seems to affect the synthesis of at least two AIs, AI-2 and AI-3. The activities of the two signals can be uncoupled by utilizing biological tests specific to each signal. For example, AI-3 shows no activity in the *V. harveyi* bioluminescence assay for AI-2 production. On the other hand, AI-3 activates the transcription of the EHEC T3SS genes, while AI-2 has no effect in this assay. The only two phenotypes shown to be AI-2 dependent, using either purified or *in vitro*-synthesized AI-2, are bioluminescence in *V. harveyi* and expression of the *lsr* operon in *S. typhimurium* (Sperandio et al., 2003; Taga and Bassler, 2003).

There are several examples of prokaryotic–eukaryotic communication in which bacterial signals can modulate expression of eukaryotic genes or vice versa in cross-kingdom communication. One of the AIs of *P. aeruginosa*, 3-oxo-C12-HSL, has been reported to have an immunomodulatory activity, and it can downregulate tumour necrosis factor alpha and interleukin-12 production in leukocytes as well as upregulate expression of the proinflammatory cytokine gamma interferon (Smith et al., 2002). Meanwhile, eukaryotic factors also can affect prokaryotic gene transcription, which has been demonstrated by the effect of epinephrine and norepinephrine on transcription of genes encoding the T3SS and flagella in EHEC and enteropathogenic *E. coli* (EPEC). Another example is the halogenated furanones produced by the red alga *Delisea pulchra* which can inhibit QS mechanisms of the plant pathogen *Pectobacterium carotovorum* (Manefield et al., 2002).

Type III Secretory System

Owing to space limitation, please see Table 2.2 and reviews for further detail (Galan and Collmer, 1999; Francis et al., 2002; Page and Parsot, 2002; Feldman and Cornelis, 2003; Mota et al., 2005; Buttner and Bonas, 2006; Tang et al., 2006; Yip and Strynadka, 2006).

T3SA component and structure

Over the past few years, our view of the structure and function of the T3SA has changed with the data of the cryo-electron microscopy maps of the core T3SA and extracellular structures as well as the detailed biochemical and high-resolution crystal structural characterizations of the T3SA components, which have shed light on the molecular organization, assembly process, effectors translocation manner and overall structure and

Table 2.2. Summary of the proteins of T3SS components, substrates, chaperones, signals and regulators in different bacterial species and their host target and/or functions.

T3SS systems	Group/subgroup	Protein or signal	Species	Host target/functions
T3SA	Translocon	YopB, YopD, LcrV; HrpF	Yersinia, Xanthomonas	
	Needle extension/filament	LcrV, SseB	Yersinia, Escherichia coli, Salmonella	
	Needle/pilus	YscF, MxiH, PrgI, EspA/EscF, HrpA	Yersinia, Shigella, E. coli, Salmonella, Pseudomonas syringae	
	Outer membrane rings	YscC, MxiD, EscC, InvG, HrcC	Yersinia, Shigella, E. coli, Salmonella, P. syringae	
	Outer membrane	M9-xiM	Shigella	
	Inner membrane ring	EscJ, PrgH	E. coli, Salmonella, Shigella, Yersinia	
		PrgK, MxiG, YscJ, MxiJ		
		YscV/InvA/EscV, YscU/SpaS/EscU,		
	Inner membrane	YscR/SpaP/EscR, YscS/SpaQ/EscS, YscT/SpaR/EscT	Yersinia, Salmonella, E. coli	
	Inner membrane rod	PrgJ	Salmonella, Shigella	
	Cytoplasmic ring	Spa33, YscQ	Shigella, Yersinia	
	ATPase	YscN, Spa47, EscN, InvC	Yersinia, Shigella, E. coli, Salmonella, P. syringae	
T3S substrate	T3SA component	YopB, YopD; LcrV, SseB; YscF, MxiH, PrgI, EspA/EscF, HrpA	Yersinia, Shigella, E. coli, Salmonella, P. syringae	

Continued

Table 2.2. *Continued*

T3SS systems	Group/subgroup	Protein or signal	Effector	Species	Host target/functions
	Effector	ADP-ribosyltransferase	ExoS, ExoT	*Pseudomonas aeruginosa*	Inhibition of phagocytosis; cytotoxicity
		Adenylate cyclase	ExoY	*P. aeruginosa*	Cytotoxicity
		Kinase	YpkA (or YpoO)	*Yersinia*	Actin cytoskeleton reorganization, stimulation of Cl- secretion; inhibition of phagocytosis
		Phosphotase	SopB (SigA), SptP, IpgD, YopH	*Salmonella, Shigella, Yersinia*	
		Exchange factor for Rho GTPases	SopE	*Salmonella*	CDC42, Rac/actin cytoskeleton reorganization; activation of MAP kinase pathways
		Others	Tir	*E. coli*	Receptor for intimin/ effacement of the microvilli of the intestinal brush border
			SipA	*Salmonella*	Actin cytoskeleton reorganization
			IpaB, IpaA	*Shigella* spp.	Activation of caspase-1, binds integrins and CD44; apoptosis; stimulation of bacterial entry
			YopE, YopJ	*Yersinia* spp.	Disruption of the actin cytoskeleton; inhibition of phagocytosis; apoptosis
			AvrPto	*P. syringae* pv. *tomato*	Pto/activates Pto serine/threonine kinase signalling pathway, HR

Quorum Sensing and Type III Secretion Systems

Category	Subtype	Components	Organisms	Function
		AvrBs2	Xanthomonas campestris	Bs2 R/synthesize or hydrolyse phosphodiester linkages, HR
		AvrBs3 family	X. campestris	Localized to plant nuclei/transcription factors, HR
T3SS chaperone	IA	SycE, SycH, SycT, SycN and YscB; IpgE; SicP; SigE; CesT, CesF; SpcU, Orf1; DspB/F; ShcA, ShcB1, AvrF; ShcS1, ShcS2, ShcO1, ShcF, ShcM, ShcV	Yersinia, Shigella, E. coli, Salmonella, P. aeruginosa, P. Amylovorum; P. syringae	Antifolding factors; T3SS substrates targeting signals; anti-aggregation and stabilizing factors; T3SS component expression regulators
	IB	YsaK, Spa15, InvB, HpaB	Yersinia, Shigella, Salmonella, X. campestris	
	II	SycD, IpgC, CesD, SicA	Yersinia, Shigella, E. coli, Salmonella	
T3SS signal		Ion (Mg^{2+}, Ca^{2+}, Pi), O_2, osmolarity pH, temperature, growth phase, cell contact, plant factors		
	TCSTS	PhoP/Q, PhoR/B, Bar/SirA, OmpR/EnvZ, SsrA/B, CpxA/R, HrpX/Y, GacS/A, HrpG	Yersinia, Shigella, E. coli, Salmonella, P. aeruginosa, Pectobacterium spp., P. syringae, X. campestris, Ralstonia solanacearum	
T3SS regulator	AraC family	HilC, HilD, InvF, VirF, LcrF, ExsA, Per, GadX, HrpX, HrpB		
	Other transcriptional regulators	H-NS, IHF, FIS, CsrA (RsmA), LuxS Ler; HrpL, HrpS, HrpR; HrpV; PrhA, PrhI, PrhJ, PrhR; LcrQ, LcrH, SycH, LcrV		

function of the T3SA. In contrast with the secreted virulence factors, most of the 15–20 membrane-associated proteins that constitute the T3SA are conserved among different pathogens. They are involved in constructing a macromolecular complex that spans the bacterial inner membrane, the periplasmic space, the peptidoglycan layer, the bacterial outer membrane, the extracellular space and the host cellular membrane, which serves as a hollow conduit to provide a continuous and direct path for effectors to rapidly translocate from the bacterial cytoplasm into the host cells in one step (Mota et al., 2005; Yip and Strynadka, 2006).

T3SA consists of three distinct structural components, a secretion nanomachine, a needle extension and a translocon. The secretion nanomachine is a complex organelle composed of a base and a needle, which is made of approximately 25 components and called as injectisome, or needle complex in animal pathogens, or hrp (HR and pathogenicity) in phytopathogens (Mota et al., 2005). To serve as a long-distance transport device for T3SS substrates across the plant cell wall, the phytopathogenic hrp pilus is longer than the needle from animal pathogens. The base of the secretion nanomachine contains two pairs of rings spanning the inner and outer bacterial membranes (inner membrane ring and outer membrane ring) joined together by a rod. The needle is a hollow, elongated and rigid helical polymer structure made from a few hundred copies of a single protein of the YscF family with a diameter of approximately 25 Å and the length between 45 and 80 nm according to the bacterial species. Molecular ruler proteins control the needle length called YscP in *Yersinia*, Spa32 in *Shigella* and InvJ in *Salmonella*. The precise needle length seems to have been attuned with regard to the dimensions of other structures such as adhesion or lipopolysaccharide at the bacterial surface, which reflects an adaptation to sense host cells and to fit the physical and chemical environments of the bacteria–host interface (Mota et al., 2005; Yip and Strynadka, 2006). Depending on the species, the needle extension consists of one sequence-divergent protein that forms either a bell-shaped tip complex or a filamentous structure. The needle extensions are attached to the tip of the needles and are involved in connecting to and mediating formation of the translocation pore. The translocon is a hetero-oligomeric protein translocation channel inserted into the host cell membrane formed by the T3S substrates called the 'translocators' from the YopB and YopD family. The translocon is continuous with the secretion nanomachine and the translocation of effector proteins from the cytoplasm of bacteria into the eukaryotic host cell occurs in one step (Galan and Collmer, 1999; Mota et al., 2005; Buttner and Bonas, 2006; Yip and Strynadka, 2006).

T3SS substrates

T3SS substrates include the effectors and T3SS structural components such as the units of the needle, the needle extension and the translocon (Yip and Strynadka, 2006). Although T3SA is structurally and functionally conserved, the types of effector molecules they deliver vary greatly among different species, which is consistent with the different phenotypes associated with different T3SS. Structural studies have revealed the effector functions at the molecular level, which suggest horizontal and convergent evolution of host mimicry as well as completely novel methods to manipulate the host. T3SS effectors possess various biochemical activities including kinase, protease, protein and lipid phosphatase, nucleotide exchange factor and Rho family GTPase, actin-polymerizing and actin-bundling factor and tubulin-binding protein. Effectors stimulate or interfere with host cellular processes enabling the bacteria to modulate directly the host environments (Mota et al., 2005). For example, inside the eukaryotic cells, the effectors can disrupt the cytoskeleton, cause apoptosis or modify the intercellular signalling cascades.

Although heterogonous secretion of proteins by different T3SS indicates that the effector secretion signal should be conserved among different T3SS, the exact molecular determinants that allow the T3SS to recognize the T3SS substrates among all the other bacterial proteins are still to be elucidated. T3SS substrates lack a single, defined secretion signal, and at least three independent secretion signals that direct substrates for secretion through the T3SS have been proposed: the 5' region of the mRNA, the N-terminus of the substrate and the ability of a secretion chaperone to bind the substrate before secretion. The use of different or multiple targeting mechanisms may determine the timing of the secretion of some effectors and contribute to the robust secretion of others (Galan and Collmer, 1999). In addition to the secretion signal, T3SS effectors also contain a translocation signal within their 50–100 N-terminal amino acids to target the effectors across the plant plasma membrane; and the dual activity of HpaB during type III-dependent protein translocation suggests that secretion and/or translocation of translocon units and effector proteins is differentially regulated (Buttner and Bonas, 2006).

Animal pathogens utilize T3SS to deliver a few anti-host effectors into mammalian cells, resulting in a cascade of regulated events to overcome animal defence mechanisms for establishment of bacterial infection. For example, the *P. aeruginosa* T3SS translocates four effectors, ExoS, ExoT, ExoU and ExoY, to host cells to promote pathogenesis by disabling the local immune response, destroying epithelia and inhibiting wound healing processes. In contrast to animal pathogenic bacteria, phytopathogenic bacteria possess a large arsenal of effectors. For example, there are 29 effectors in *P. syringae* pv. *tomato* DC3000 and only six known effectors in *Yersinia*. Plant pathogens utilize T3SSs to cause disease in susceptible plants and trigger the HR in resistant plants. Although the primary function of effector proteins is to promote pathogenicity, some phytopathogen effector proteins are also named Avr (avirulence) proteins, because they can betray the pathogens to the plant R (resistance) protein surveillance system eliciting specific defence responses in plants expressing a cognate resistance (R) gene and render the pathogen avirulent through the HR. The HR is a rapid, programmed host cell death at the infection site to restrict pathogen growth. In addition to the effectors, phytopathogenic T3SS is also required for the secretion of T3SS structural components and helper proteins such as harpins, the glycine-rich, cysteine-lacking, heat-stable proteins, which do not need to be translocated inside the host cell to exert their function and can elicit the HR when delivered to the host surface. Secretion and translocation of the effector proteins are coordinately regulated and require contact with a plant cell. In contrast, harpins and proteins that function as components of the extracellular T3SA seem to be readily secreted in culture media and can be detected in the culture supernatants in large amounts (Galan and Collmer, 1999; Brencic and Winans, 2005; Buttner and Bonas, 2006; Tang et al., 2006).

T3SS chaperones

Generally, chaperones are involved in folding proteins or redirecting misfolded proteins to degradation pathways transiently, but do not take part in the final function of their substrates. The efficient secretion and translocation of T3SS substrates not only depend on protein signals, but also require binding to their cognate chaperones (Buttner and Bonas, 2006). The majority of T3SS substrates bind to chaperones in the bacterium before delivery into the host. Typical T3SS chaperones are low molecular mass (<15 kDa), acidic (pI < 5), and leucine-rich dimer-forming proteins that specifically bind to N-terminal domains of the T3SS substrates and are usually encoded adjacently to their cognate substrate. In general, the chaperones do not have ATP-binding domains and share little detectable sequence similarity although they share a very similar fold and general T3SS substrate-binding mode as well as a predicted C-terminal amphipathic helix. The absence of the chaperone often results in premature degradation or inappropriate interaction of its substrate with a dramatic decrease in translocation into the host (Page and Parsot, 2002).

Three classes of T3SS chaperones have been distinguished based on their substrate specificities: whether they associate with one effector, several effectors or the translocators. Class IA chaperones interact with one or several homologous effectors and are often encoded next to their interaction partners, whereas class IB chaperones bind to a wide range of effectors. Chaperones of translocon proteins are termed class II. Although the roles of chaperones on effector proteins are varied and not yet fully understood, their functions can be summarized by the following four aspects. First, they act as the antifolding factors to keep the T3SS substrates in an unfolded secretion-competent state for transit through the narrow channel of the T3SS, which is between 2 and 3 nm. Second, they serve as signals to target their substrates to the T3SS, preventing them from secretion through the flagellar T3SS. It has also been proposed that the conserved three-dimensional path of the polypeptide in the chaperone–effector complexes may be a universal conformational signal recognized by the T3SS, and the binding of the chaperones may confer to the effector a competitive advantage in a hierarchy of secretion of the multiple effector proteins. Third, they act as anti-aggregation and stabilizing factors to prevent inappropriate premature interactions with other proteins. Fourth, they act as the regulators for the expression of some components of T3SS (Page and Parsot, 2002; Feldman and Cornelis, 2003; Buttner and Bonas, 2006).

T3SS signals and regulation

Pathogenic bacteria inhabit different environments, and face unique environmental pressures. The expression of the T3SS is affected by physiological signals, such as the availability of nutrients and growth phase as well as a vast array of environmental signals, including temperature, osmolarity, pH, divalent cations and the physical contact with the host cells (Table 2.2; Galan and Collmer, 1999; Francis et al., 2002; Tang et al., 2006). Recent reports suggest that the bacteria could sense cholesterol in eukaryotic cell lipid membranes to trigger T3SS through the cholesterol-binding proteins IpaB and SipB. Since the induction of T3SS by cell contact is due to the presence of the cholesterol, they cannot translocate effectors into cholesterol-depleted host cells (Mota et al., 2005).

The expression patterns of T3SS genes in five phytopathogenic genera (*Pectobacterium*, *Pseudomonas*, *Pantoea*, *Ralstonia* and *Xanthomonas*) are very similar. The T3SS genes are repressed when bacteria are cultured in complex media, but are induced when bacteria are grown under conditions mimicking plant apoplast. Factors affecting T3SS gene expression are osmolarity, pH, temperature, nitrogen and carbon sources. Since the expression of T3SS is a highly energy-consuming process, it would benefit the bacteria to fully induce the T3SS genes only when they are in close contact with host cells. T3SSs are indeed induced at a high level in close contact with plant cells. For example, the HrpB transcriptional activator controlling T3SS gene expression in *R. solanacearum* is activated through a three-component signalling module by an outer membrane protein (PhrA) that acts as a contact-dependent sensor for a plant cell wall signal and an inner membrane protein (PrhR) as well as the extracytoplasmic factor (ECF) family sigma factor PrhI. Although the plant factors are required for the induction of the phytopathogenic T3SS, no specific plant inducers of T3SS gene expression have thus far been characterized, and the nature of the signals inducing T3SS gene expression *in planta* is not clearly defined (Francis et al., 2002; Brencic and Winans, 2005; Mota et al., 2005; Buttner and Bonas, 2006; Tang et al., 2006).

Expression of T3SS is coordinately regulated by networks of transcription factors in response to environmental stimuli. With the complexity of T3SS, it is not surprising that the integration of multifaceted regulatory pathways is required to impart spatial and temporal control of T3SS gene expression for controlling the T3SA synthesis and assembly as well as the substrate synthesis and secretion. It is apparent that regulation occurs in at least two distinct steps: expression of genes required for assembly of the T3SA followed by expression of genes whose products are substrates for T3SS (Francis et al., 2002). Regulation takes place at both the transcriptional and the post-translational levels. Transcriptional regulation is accomplished by one or several specific transcription factors as well as by components of global regulatory networks that control the expression of T3SS in response to a variety of environmental cues mentioned earlier. In addition, as described in *Yersinia* spp., T3SS gene expression is also controlled by sensing the secretion process itself, which relies on the secretion of a negative regulator through the T3SS, thereby coupling the transcription and secretion processes. The post-translational regulation of the secretion process is less well understood. It appears that, at least in some systems, the physiological signal that stimulates this regulatory pathway involves contact with host factors. This regulation is post-translational because inhibition of novel bacterial protein synthesis does not prevent the host cell responses stimulated by T3SS and translocation. Nevertheless, the coupling of secretion to the transcriptional control mechanisms indicates that the post-translational stimulation of secretion will eventually result in the activation of transcription of T3SS genes (Galan and Collmer, 1999).

The regulation of T3SS through the regulatory proteins such as AraC-like transcriptional activators, histone-like proteins and TCSTS is thought to ensure expression and assembly of the T3SA and a certain level of expression of T3SS substrates. Most bacteria employ phosphorelay mechanisms to sense the surrounding environment. In animal pathogens, these signals are most often routed to AraC-like transcriptional activators that establish a cascade of type III gene activation. The examples are the VirF responding to the temperature shift by activating a regulatory cascade mediated by VirB in *Shigella* spp. and the HrpX, HrpB of *X. campestris* and *R. solanacearum* (Francis et al., 2002; Tang et al., 2006). Another class of regulator is the nucleoid-associated proteins like H-NS, which bind to DNA and change its topological structure and thereby regulate the T3SS gene expression exemplified by the negative regulation of temperature-dependent expression of T3SS genes through the repression of the promoter of the primary regulator such as VirF in *Shigella* spp. (Falconi et al., 2001; Francis et al., 2002). In addition to H-NS, other nucleoid-associated proteins are emerging as potential modulators of T3SS gene expression. For instance, Falconi et al. (2001) indicate that factor inversion stimulation (FIS) antagonizes the repressive function of H-NS bound to the VirF regulator in a temperature-dependent manner. They further propose that modulation of the DNA architecture has been universally evolved by pathogens as a strategy to regulate virulence gene expression in response to environmental stimuli. It follows that several signals encountered by a pathogen during infection, such as temperature, osmolarity, pH and oxygen tension, have all been implicated to cause topological changes in local DNA structure. TCSTSs such as GacS/GacA system also are involved in the regulation of T3SS and are discussed elsewhere in this chapter. Interestingly, the cyclic AMP (cAMP) signalling cascade plays a regulatory role on the regulation of T3SS in *P. aeruginosa* by the induction of the low Ca^{2+}, which is mediated by at least two distinct signalling pathways. Under low Ca^{2+} conditions, a membrane-associated adenylate cyclase (CyaB) catalyses the formation of cAMP. The rise in intracellular cAMP activates the cAMP-dependent transcriptional factor Vfr resulting in transcription of genes encoding the T3SS (Wolfgang *et al.*, 2003).

In addition, *Pseudomas* spp., *Pectobacterium* spp. and *Ralstonia* spp. employ ECF family alternative sigma factors for the regulation of T3SS. Expression of the alternative sigma factor requires RpoN and enhancer-binding proteins that function as response regulators (Francis *et al.*, 2002; Brencic and Winans, 2005; Tang *et al.*, 2006). Meanwhile, a novel activator/anti-activator/anti-anti-activator system, ExsA/ExsD/ExsC, controlling transcription of a T3SS has been identified recently in *P. aeruginosa*. In *P. aeruginosa*, two proteins, ExsA and ExsD, are shown to play a role in coupling transcription to secretion. ExsA is an activator of T3SS gene transcription, and ExsD is an anti-activator of ExsA. In the absence of environmental secretion cues, ExsD binds ExsA and inhibits transcription. Dasgupta *et al.* (2004) further characterized the ExsC as an anti-anti-activator of T3SS expression and proposed a model that the anti-anti-activator (ExsC) binds to and sequesters the anti-activator (ExsD) under low Ca^{2+} conditions, freeing ExsA and allowing for transcription of the T3SS. Homologues of both *exsC* and *exsD* have been identified in the *V. parahaemolyticus* and *Photorhabdus luminescens* genomes, which suggests that other organisms may use a similar mechanism to regulate the transcription of T3SS genes.

The secretion activity in T3SS can also modulate the expression of some T3SS genes encoding secreted proteins. As discussed above, in addition to the roles of chaperones in stability and secretion of their substrates, a role in regulation also has been demonstrated, including the effector chaperone SycH and the translocator chaperones SycD/LcrH of *Yersinia* spp., SicA and IpgC of *Shigella* spp. Regulation by chaperones SicA and IpgC occurs at the transcriptional level. It seems that T3SS chaperones act as sensors of the intracellular levels of their cognate proteins. The expression of most of the genes of the T3SS is known to be activated upon contact with a eukaryotic cell or in conditions mimicking such contact *in vitro*. Upon contact with the eukaryotic cells, secretion of the effectors and translocators starts and the intracellular amount of free chaperone increases, allowing them to activate or inhibit the transcriptional regulators. In *Yersinia*, the feedback control mechanism that keeps T3SS expression at low levels is relieved by the export of the LcrQ/YscM protein (Francis *et al.*, 2002; Feldman and Cornelis, 2003; Brencic and Winans, 2005; Mota *et al.*, 2005).

In phytopathogens, induction of T3SS genes in bacteria occurs early after contact with a plant. In all cases, T3SS gene expression is induced *in planta* or in certain minimal media. Transcription of T3SS genes is controlled by multicomponent regulatory networks that integrate diverse sets of environmental cues. The external activating stimulus is transmitted to a cascade of regulatory proteins, such as TCSTS regulators, alternative sigma factors and AraC-type transcriptional activators. These regulatory networks, however, differ significantly among different species. Based on differences in regulation, T3SS genes can be divided into two groups. Group I T3SS is found in *P. syringae*, *Pectobacterium* spp. and *Pantoea* spp., where T3SS genes are activated by a member of the ECF family sigma factor called HrpL. Group II T3SS is found in *X. campestris* and *R. solanacearum*, where transcription of T3SS-associated genes is regulated by members of the AraC family of proteins (Francis *et al.*, 2002; Brencic and Winans, 2005; Mota *et al.*, 2005; Tang *et al.*, 2006).

Regulatory Network Between QS and T3SS

QS controlling T3SS gene expression

To our knowledge, there is no report yet that T3SS controls QS: it appears that QS works upstream of the T3SS regulatory network and plays a role in the regulation of T3SS. Sperandio *et al.* (2003) first reported the connection between QS and T3SS in EHEC and EPEC, showing that a different QS system from the HSL system activates T3SS at high cell density via a LuxS protein and its cognate autoinducer AI-3. In contrast to the positive regulatory role of QS on T3SS in the EHEC and EPEC, Henke and Bassler (2004) provided evidence that the expression of the genes encoding the T3SS requires an intact QS signal transduction cascade and QS represses T3SS in *V. harveyi* and *V. parahaemolyticus*. Similarly, following the report of QS-dependent control of T3SS effector ExoS (Hogardt *et al.*, 2004), Bleves *et al.* (2005) further examined the effect of QS on the expression of T3SS regulon and identified that, except for the regulatory operon *exsCBA*, T3SS is negatively regulated by RhlI/R-C_4-HSL in *P. aeruginosa*. The QS repression of the T3SS regulon suggests that the associated virulence functions of T3SS are likely to be required at early stages of bacterial infection, prior to the establishment of a high-cell-density bacterial population and the further development of a chronic infection.

In plant pathogens, the QS regulon also has been reported to act on the T3SS. The QS signal, *N*-[3-oxohexanoyl]-L-homoserine lactone (OHL), is required for the expression of the T3SS regulon and the pathogenicity of *P. carotovorum* subsp. *carotovorum* (Koiv and Mae, 2001). However, QS did not affect the expression pattern of T3SS and plant cell wall-degradation enzymes in *Dickeya dadantii*. Mutation of *luxI* homologue of *expI* had no apparent effect on the growth phase-dependent expression of *hrpN* and *pelE*, or on the virulence of *D. dadantii* in chicory leaves (Ham *et al.*, 2004).

Connection between T3SS and QS through Rsm, GacS/GacA and other regulators

Rsm system

Regulator of secondary metabolism (Rsm) is a novel type of post-transcriptional regulatory system mediated by the RsmA–rsmB pair (CsrA and *csrB* in *E. coli*). The Rsm system plays a critical role in gene expression and has a profound effect on bacterial metabolism and behaviour in many prokaryotic species. RsmA, *rsmB* and RsmC are the major components of this global regulatory system (Fig. 2.1). RsmA is a small RNA (sRNA)-binding protein that acts by repressing translation and by lowering the half-life of the mRNA species. *rsmB* is an untranslated regulatory RNA that binds RsmA and neutralizes its negative regulatory effect by forming an inactive ribonucleoprotein complex. RsmC controls the production of RsmA and *rsmB* RNA by positively regulating *rsmA* and negatively controlling *rsmB* (Liu *et al.*, 1998; Cui *et al.*, 1999).

The Rsm regulatory system is conserved in many prokaryotes. There are substantial data indicating the existence of the Rsm system in various pathogens such as *Enterobacter aerogenes*, *E. coli*, *S. typhimurium*, *S. flexneri*, *Serratia marcescens*, *Y. pseudotuberculosis*; and plant pathogen *Pectobacterium* spp. *rsmA* homologues have been cloned from *P. aeruginosa* and *P. fluorescens*. Moreover, *rsmA* homologues are present in *B. subtilis*, *C. jejuni*, *C. acetabutylicum*, *C. difficile*, *Haemophilus influenzae*, *Helicobacter pylori*, *Legionella pneumophila*, *Proteus mirabilis*, *Pasturella multocida*, *S. marcescens*, *Shewanella putrefaciens*, *Thermotoga maritime*, *Treponema pallidum* and *V. cholerae*. The Rsm system has been reported

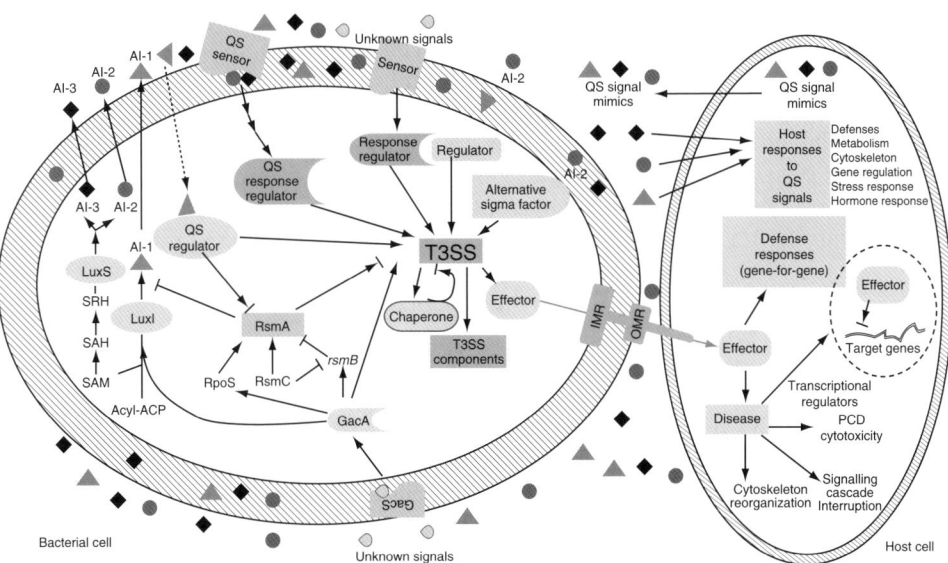

Fig. 2.1. Schematic representation of the interaction between quorum sensing and type III secretion system and the host–bacteria interaction.

to control different gene products contributing to the pathogenicity and the host–pathogen interactions. It affects the production of different extracellular enzymes like pectinases, proteases, cellulases and secondary metabolites such as phytohormones, antibiotics, pigments and polysaccharides. It is also involved in flagella biosynthesis, motility and biofilm formation (Cui et al., 1999; Chatterjee et al., 2002; Heeb et al., 2002; Valverde et al., 2003; Kay et al., 2005; Burrowes et al., 2006; Mulcahy et al., 2006). The expression of T3SS elicitor HrpN and other T3SS genes in P. carotovorum is repressed by the RsmA and RsmC. Meanwhile, the Rsm system affects the expression of the HrpL by RsmA-promoted decay of $hrpL_{(Ecc)}$ RNA (Fig. 2.1; Cui et al., 1999; Cui et al., 2001; Chatterjee et al., 2002). Interestingly, recent work by Mulcahy et al. (2006) provided a positive role of RsmA on T3SS in P. aeruginosa. The rsmA mutant of P. aeruginosa is defective in the production of key effector and translocation proteins and shows decreased expression of T3SS regulators. RsmA promoted actin depolymerization, cytotoxicity and anti-internalization of P. aeruginosa by positively regulating the virulence-associated T3SS. A fuller understanding of the broader impact of RsmA on cellular activities has been further addressed by comparing the transcriptome profiles of P. aeruginosa PAO1 and an rsmA mutant. Loss of RsmA altered the expression of genes involved in a variety of pathways and systems important for virulence, including iron acquisition, PQS biosynthesis, multidrug efflux pump formation and motility, although not all of these effects can be explained through the established regulatory roles of RsmA (Burrowes et al., 2006).

Moreover, the Rsm system influences the levels of the QS signal. For example, RsmA negatively regulates lasI, rhlI in P. aeruginosa, and the biosynthesis of the PQS is also affected by RsmA (Burrowes et al., 2005, 2006). Rsm also controls the levels of the QS signal OHL in various Pectobacterium spp. In P. carotovorum, RsmA reduces the levels of transcripts of hslI, a luxI homologue required for HSL biosynthesis. The finding that HSL is required for extracellular enzyme production and pathogenicity in soft-rotting Pectobacterium spp. supports the hypothesis that RsmA controls these traits by modulating the QS signal level of the bacteria. In the plant-beneficial rhizosphere bacterium P. fluorescens CHA0, the sRNA rsmX together with rsmY and rsmZ, forms a triad of GacA-dependent sRNAs, which sequester the RNA-binding proteins RsmA and RsmE and thereby antagonize translational repression exerted by these proteins. This sRNA triad was found to positively regulate the synthesis of a QS signal and autoinduce the Gac–Rsm cascade of P. fluorescens (Kay et al., 2005).

On the other hand, QS also has been reported to regulate the Rsm system. For example, Koiv and Mae (2001) reported that AHL represses the expression of rsmA, which in turn leads to the activation of plant cell wall-degradation enzyme production in P. carotovorum, and the deficiency of AHL led to increased production of both rsmA and rsmB RNAs transcripts. Chatterjee et al. (2005) further showed that the ExpR of P. carotovorum binds rsmA DNA and activates rsmA transcription. The

ExpR-mediated activation of *rsmA* expression and ExpR binding with *rsmA* DNA are inhibited by AHL.

GacS/GacA system
GacS/GacA is a TCSTS, which is widely distributed in many bacteria to respond to environmental stimuli and adapt to different environmental conditions. GacS is the putative histidine kinase sensor and GacA is the response regulator. Although the signal molecule for GacS autophosphorylation is still unknown, the GacS probably activates GacA and the activated GacA works as a transcriptional activator and further activates targeted genes in bacteria. The homologues of the TCSTS of GacS/GacA have been reported in a variety of Gram-negative bacteria, including *E. coli* (BarA/UvrY), *Pectobacterium* spp., *S. typhimurium* (BarA/SirA), *Pseudomonas* spp. (GacS/GacA), and *L. pneumophila* (LetS/LetA), *Vibrio* spp. (Cui *et al.*, 2001; Heeb *et al.*, 2002; Reimmann *et al.*, 2005). GacS/GacA plays important roles in many biological functions. For example, in plant-beneficial bacteria, the GacS/GacA homologue is essential for expressing disease biocontrol factors (Chancey *et al.*, 2002). In plant pathogens like *P. carotovorum* and *P. syringae*, this TCSTS has been intensively studied, and many virulence factors including regulatory RNA, QS signals, T3SS genes, pectate lyases, proteases, toxins-like syringomycin, etc. are found to be regulated by GacS/GacA (Cui *et al.*, 2001; Chatterjee *et al.*, 2003).

GacS/GacA systems are located at the top of a regulatory cascade and function as a central regulator by controlling an assortment of transcriptional and post-transcriptional factors. GacS/GacA affects the T3SS gene expression in different bacterial species. For example, GacA positively regulates the expression of the T3SS alternative sigma factor HrpL as well as several T3SS genes of *avr, hrp* and *hop* controlled by HrpL in *P. syringae*. The *gacA* mutant of *P. syringae* pv. *tomato* DC3000 produces lower levels of transcripts of the T3SS regulator gene *hrpL, hrpR* and *hrpS* as well as the a*vr, hrp* and *hop* genes, causing drastic changes in bacterial virulence towards *Arabidopsis thaliana* and tomato, as well as the multiplication *in planta* and the efficiency of HR induction (Chatterjee *et al.*, 2003). Recently, we characterized the GacA effect on the T3SS gene expression and identified GacA as a master regulator controlling *hrpL* and several T3SS gene in *D. dadantii* (Yang *et al.*, 2006).

The GacS/GacA system also has been reported to influence the expression of QS signal synthetase and therefore works upstream of the QS affecting the entire QS regulon. For instance, GacA was required for the transcription of *ahlI* and acted as activators of the AhlI/AhlR QS system in *P. syringae* (Chatterjee *et al.*, 2003; Quinones *et al.*, 2005; Reimmann *et al.*, 2005). In *P. aureofaciens*, GacA is also required for the transcription of the AHL synthetase gene *phzI*. Finally, the GacS/GacA homologue VarS/VarA of *V. cholerae* has also been reported to affect the expression of the entire QS regulon (Chancey *et al.*, 1999; Lenz *et al.*, 2005).

The influence of the GacS/GacA system on QS and T3SS is channelled, at least in part, through the Rsm system by activating the *rsmB* or *rsmB* homologue, which binds to and inhibits the mRNA decay effect of RsmA. The examples are widespread in different species. For instance, GacS/GacA controls *rsmB* transcription in *Pectobacterium* spp., the *rsmB* homologue *rsmZ* transcription in *P. aeruginosa*, the expression of a *rsmB* homologue *prrB* RNA and consequent secondary metabolite production in *P. fluorescens*. In the *Pseudomonas* spp., the GacS/GacA activates the transcription of *rsm* genes encoding sRNAs like *rsmB* and *rsmZ* RNA production that inhibits RsmA (Cui *et al.*, 2001; Heeb *et al.*, 2002; Chatterjee *et al.*, 2003; Valverde *et al.*, 2003; Goodman *et al.*, 2004; Kay *et al.*, 2005; Reimmann *et al.*, 2005; Burrowes *et al.*, 2005). Recently, GacS/GacA homologue VarS/VarA in *V. cholerae* has been reported to work in parallel with CAI-1-CqsS and AI-2-LuxPQ to control the expression of redundant regulatory sRNAs called quorum regulatory RNAs (Qrr) and thus the expression of the entire QS regulon. The function of VarS/VarA is through the activation of the three redundant sRNAs CsrB, CsrC and CsrD to control the RsmA homologue CsrA, which in turn controls the expression of Qrrs and the entire QS regulon through the effect of CsrA on LuxO (Lenz *et al.*, 2005).

Interaction Between Gac–Rsm and Other Regulators Controlling T3SS and QS

Other regulatory proteins also play important roles in controlling the T3SS and QS systems by distinct mechanisms. Since the Gac–Rsm system is so critical, it was predicted that *rsmA* and *rsmB* expression would be rigorously controlled. Indeed, studies of *P. carotovorum* have disclosed that several transcriptional factors control the expression of the *rsmA* and *rsmB* genes, including a LysR-type regulator HexA, and an IclR-type repressor KdgR, which can positively control RsmA production and negatively control the levels of *rsmB* RNA (Mukherjee *et al.*, 2000; Cui *et al.*, 2001). Other examples include the identification of AefR as an important and novel regulator of QS in *P. syringae* from work in Lindow's lab, proposing that both AefR and GacA act as activators of the AhlI/AhlR QS system via independent pathways (Quinones *et al.*, 2005).

The stress sigma factor RpoS is an alternate sigma factor responsible for the activation of many genes expressed mainly during the stationary phase. RpoS participates in Gac–Rsm-mediated resistance to oxidative stress in *P. fluorescens*; the expression of RpoS was controlled positively by GacA and negatively by RsmA (Heeb *et al.*, 2005). In *P. aeruginosa*, Hfq has also been reported to exert a moderate stimulatory effect on translation of the *rhlR* and *qscR* genes as well as a stimulatory effect on *rhlI* expression, which might be through the Rsm system since Hfq can bind to and stabilize the RsmY, which is further shown to bind to RsmA (Sonnleitner *et al.*, 2006).

Another regulator influencing Rsm is MvfR in *P. aeruginosa*, which regulates RsmA post-transcriptionally. MvfR is required for the function of QS-regulated virulence and PQS. Recently, a hybrid sensor kinase/response regulator RetS (RtsM) influencing the RsmA levels has been characterized in *P. aeruginosa*. Transcriptome profiling of *retS* mutant revealed a decrease in the expression of genes involved in T3SS and virulence (Goodman *et al.*, 2004). Ventre *et al.* (2006) identified a signal sensor LadS, a hybrid sensor kinase that controls the reciprocal expression of genes for T3SS and biofilm-promoting polysaccharides that repress T3SS gene expression. They provided evidence that LadS counteracts the activities of RetS. LadS and RetS exert opposite effects on the RsmZ. They (Ventre *et al.*, 2006) further proposed a signal transduction network in which the activities of signal-receiving sensor kinases LadS, RetS and GacS regulate expression of virulence genes associated with acute or chronic infection by transcriptional and post-transcriptional mechanisms.

Disease Management by Targeting QS and T3SS

Although industries are losing interest in antibacterial drug discovery due to the challenge of developing new, effective antibiotics against the antibiotic-resistant organisms, an intriguing approach is to investigate bacterial pathogenesis along with the development of reagents and strategies for disease control. The conserved virulence mechanisms utilized by a range of pathogens to cause infection, such as TCSTS, QS, T3SS, Gac–Rsm system and biofilm formation, are becoming targets for disease management. The virulence mechanisms have been elucidated and various novel strategies have been developed. Virulence inhibitors that target particular virulence determinants, such as adhesion or T3SS (Lyon and Muir, 2003), as well as regulator and QS-based pathways without detrimental effects on bacterial growth, have become useful chemical probes for studies on bacterial virulence and drug development. Although the virulence-target-based therapies may not be powerful enough to clear an existing infection alone, they may become an efficient strategy if used as an accessory therapy to existing antibiotics or as potentiators of the host immune response.

QS modulation mechanisms

AHL inhibition
There are different strategies aimed at the interruption of bacterial QS circuits based on QS signal generation, signal dissemination and signal detection. First, knowledge about signal generation can be exploited to develop QS inhibitor molecules that target QS signal generation. For example, LuxI family synthetase uses SAM to produce the AHL signal. Since the reaction chemistry of AHL synthetase with SAM appears to be unique, SAM analogues could be used as specific inhibitors of QS signal generation without affecting eukaryotic enzymes that use SAM as a substrate. Various analogues of SAM, such as *S*-adenosylhomocysteine,

S-adenosylcysteine and sinefungin, have been demonstrated to be potent inhibitors of AHL synthesis catalysed by the *P. aeruginosa* RhlI protein (Hentzer and Givskov, 2003).

Second, the QS inhibition could be fulfilled by the inhibition of QS signal dissemination through the decay of the active QS signal concentration in the environment, which is also referred to as QS quenching. QS signal decay might be a consequence of a non-enzymatic reaction such as the alkaline hydrolysis of AHL signals at high pH values (Yates *et al.*, 2002), or the specific degradation of QS signal by QS degradation enzymes secreted by other bacterial species or the host. There are three possible routes to inactivate the AHL, including the lactone ring hydrolysis, the amide bond hydrolysis and the racemization to give *N*-3-oxohexanoyl-D-homoserine lactone. The AHL-degrading enzymes identified so far fall into two groups according to the cleavage site of AHL. AHLases degrade AHLs by hydrolysing the lactone ring of AHLs and produce corresponding acyl homoserine molecules, which include the AiiA from the *Bacillus* spp., and the lactonase in *Variovorax paradoxus* and *Rhodococcus* spp. (Leadbetter and Greenberg, 2000; Dong *et al.*, 2001). In addition, the AHL lactonase AhlD from *Arthrobacter* IBN110 has recently been found to be involved in the utilization of AHLs as a nutrient source. In another study, a *V. paradoxus* strain able to grow using 3-oxo-C6-*N*-homoserine lactone as the sole energy and nitrogen source was isolated from a soil sample (Abraham, 2006). Another group is AHL-acylases, which hydrolyse the amide bond of AHLs to release HSL and the acyl chain, which could be further metabolized by *V. paradoxus* and other species. Meanwhile, *R. erythropolis* W2 also was reported to degrade AHLs by both oxido-reductase and AHL-acylase, and bromoperoxidase in *Laminaria digitata* forms hypobromous acid, which deactivates the signalling of 3-oxohexanoyl-homoserine lactone by oxidation. The QS quenching enzyme exists in different environments, such as rhizosphere and phyllosphere, including Proteobacteria, low G + C Gram-positive bacteria, and high G + C Gram-positive bacteria. For example, bacteria degrading the QS signal AHL were isolated from a tobacco rhizosphere, and included members of the genera *Pseudomonas, Comamonas, Variovorax* and *Rhodococcus* (Leadbetter and Greenberg, 2000; Dong *et al.*, 2001; Uroz *et al.*, 2003). AHL-degrading enzymes are of potential agricultural and clinical interest for use in the prevention of diseases caused by QS-proficient bacterial populations.

Third, the QS inhibition could result from inhibition of QS signal reception, which is also called QS mimicry. These compounds consist of two main groups: competitive inhibitors and non-competitive inhibitors. Competitive inhibitors are structural analogues of the native QS signal, which could bind to and occupy the QS signal-binding site, but fail to activate the QS receptor, and non-competitive inhibitors bind to different sites on the receptor protein lacking structural similarity to QS signals.

The structure–activity relationship analyses of the AHLs demonstrated that the length of the side chain, the C-3 carbonyl group, as well as the ring structure, influence binding of the signal molecule to the receptor protein (Lyon and Muir, 2003; Martinelli *et al.*, 2004). Many studies have been carried out on the AHL QS signal analogue synthesis and their mechanism characterization, which provided the QS antagonist candidates for disease control as well as the future development of novel antagonists. One such study showed that the AI of TraR in *A. tumefaciens*, *N*-(3-oxooctanoyl)-L-homoserine lactone, can be converted into an antagonist of similar potency by simply replacing the carbonyl at position 3 with a methylene to form *N*-(octanoyl)-L-homoserine lactone, which indicates that the 3-oxo group plays an important role in TraR activation, but is unnecessary for TraR binding (Lyon and Muir, 2003). Several other antagonists of AHL have been synthesized and the molecular mechanisms of these antagonists have been proposed (Reverchon *et al.*, 2002; Castang *et al.*, 2004; Frezza *et al.*, 2006). Reverchon *et al.* (2002) evaluated a series of 22 synthetic AHL analogues with the modification of the acyl chain for both their inducing activity and ability to competitively inhibit the action of 3-oxohexanoyl-L-homoserine lactone. They found that most of the analogues bearing either acyclic or cyclic alkyl substituents showed inducing activity, but the phenyl-substituted analogues displayed significant antagonist activity. Castang *et al.* (2004) synthesized a series of 11 AHL analogues with the carboxamide bond replaced by a sulfonamide one, and found that several compounds had antagonist activity. Recently, Frezza *et al.* (2006) synthesized a series of 15 racemic alkyl- and aryl-*N*-substituted ureas, derived from AHL. *N*-alkyl ureas with an alkyl chain of at least four carbon atoms, as well as certain ureas bearing a phenyl group at the extremity of the alkyl chain, were found to be significant antagonists. They further proposed that the antagonist activity of these AHL analogues was related to the inhibition of the dimerization of the N-terminal domain of LuxR homologue ExpR resulting from the formation of an additional hydrogen bond in the protein AHL-binding cavity (Reverchon *et al.*, 2002; Castang *et al.*, 2004; Frezza *et al.*, 2006). Antagonists generated by changing the macrocyclic part of AHLs have been reported and some of these analogues inhibit the QS. For example, the lactam analogue of the *P. aeruginosa* AI *N*-(3-oxododecanoyl)-L-homoserine lactone had markedly reduced activity. Synthetic molecules with carbamate substituents of the HSL ring of 3-oxo-C8-HSL also had vastly decreased activity. In addition, a library of 96 AHL analogues in which the macrocycle was systematically altered has been synthesized, and some of the tested analogues such as 3-oxo-C12-(2-aminocyclohexanone) are antagonists of QS *in vitro*, and inhibit biofilm formation (Lyon and Muir, 2003).

AIP mediated
In addition to the synthesis of Gram-negative bacterial AHL QS signal analogue, the antagonists of the Gram-positive bacterial QS signal have been reported. For example, Otto *et al.* (1999) demonstrated that a synthetic *S. epidermidis* pheromone is a competent inhibitor of *agr* QS system of *S. aureus*. Structure–activity studies on the AIPs provided important insights to allow the rational design of AIP analogues that are global

inhibitors of *S. aureus* virulence. Lyon *et al.* (2000) demonstrated that the truncated version of AIP-II (trAIP-II) without the tail of the signal AIP-II was an inhibitor of all four *S. aureus* groups as well as some other *Staphylococci* spp. and the cyclic peptides such as trAIP-II are excellent starting points for peptidomimetic-type strategies designed to improve the bioavailability or potency of the initial compounds. By exploiting the unique chemical architecture of the naturally occurring AIP-1, several potent inhibitors of staphylococcal QS were designed (Scott *et al.*, 2003). Vieira-da-Motta *et al.* (2001) reported that the synthetic analogues of the RNAIII-inhibiting peptide (RIP) to its target molecule TRAP function *in vitro* as efficient suppressors of QS *agr*-regulated exotoxin production by *S. aureus*. Yang *et al.* (2003) selected two RAP-binding peptides (RBPs) from a random 12-mer phage-displayed peptide library capable of inhibiting RNAIII production *in vitro* and protecting mice from a *S. aureus* infection *in vivo*. Interestingly, Qazi *et al.* (2006) recently reported that long-chain 3-oxo-substituted AHLs, such as 3-oxo-C12-HSL, are capable of interacting with the *S. aureus* cytoplasmic membrane in a saturable, specific manner and at sub-growth-inhibitory concentrations, antagonizing the QS in *S. aureus* through both *sarA* and *agr* QS systems to downregulate the exotoxin production.

AI-2- and AI-3-mediated QSI
Most QS AIs promote intraspecies communication, but AI-2 is produced and detected by a wide variety of bacteria and may allow interspecies communication. Some species of bacteria can manipulate AI-2 signalling and interfere with other species' ability to assess and respond correctly to changes in cell population density. AI-2 signalling and AI-2 QSI could have important consequences for eukaryotes to protect them from pathogenic bacteria (Xavier and Bassler, 2005). Two AI-2 QS substrate analogues, *S*-anhydroribosyl-L-homocysteine and *S*-homoribosyl-L-cysteine, have been synthesized that can prevent the initial and final step of the AI-2 synthesis, respectively (Alfaro *et al.*, 2004).

EHEC uses a QS regulatory system to sense whether it is within the intestines and to activate genes essential for intestinal colonization. The LuxS/AI-2 QS system used by EHEC is extensively involved in interspecies communication. Sperandio *et al.* (2003) show that an EHEC *luxS* mutant, unable to produce the bacterial AI, still responds to a eukaryotic cell signal to activate expression of its virulence genes, and identified this signal as the hormone epinephrine and showed that beta- and alpha-adrenergic antagonists can block the bacterial response to this hormone. Furthermore, using purified and *in vitro* synthesized AI-2, it shows that AI-2 is not the AI involved in bacterial signalling. EHEC produces another, previously undescribed AI (AI-3), whose synthesis depends on the presence of LuxS. These results imply a potential opportunity to develop QSIs to block the cross-communication between the luxS/AI-3 bacterial QS system and the epinephrine host signalling system. Given that eukaryotic cell-to-cell signalling typically occurs through hormones, and that bacterial cell-to-cell signalling occurs through QS, QS might be a language by which bacteria and host cells communicate.

QS inhibitors expressed by higher organisms
A number of reports describe the ability of higher organisms to interfere with QS through the release of compounds that mimic the activity of AHL signals. The best-characterized example is the Australian macroalga *D. pulchra*. *D. pulchra* furanone compounds consist of a furan ring structure with a substituted acyl chain at the C-3 position and a bromine substitution at the C-4 position. The substitution at the C-5 position may vary in terms of side chain structure. The natural furanone is halogenated at various positions by bromine, iodide or chloride. *D. pulchra* produces at least 30 different species of halogenated furanone compounds, which are stored in specialized vesicles and are released at the surface of the thallus. Furanones of *D. pulchra* constitute a specific means of eukaryotic interference with bacterial signalling processes. The discovery of the furanone-mediated displacement of radiolabelled AHL molecules from LuxR suggests that furanone compounds compete with the cognate AHL signal for the LuxR receptor site (Hentzer and Givskov, 2003). Other work suggests that AHLs function by stabilizing unstable LuxR-type proteins. Rather than displacing the AHL signal from LuxR, the interaction of furanones with LuxR produces conformational changes that result in rapid proteolytic turnover of the complex (Fast, 2003; Lyon and Muir, 2003). Furanones and their synthetic analogues are able to antagonize QS-controlled gene expression, including swarming motility of *S. liquefaciens*; biofilm formation and virulence factor production, and pathogenesis in *P. aeruginosa*; bioluminescence and toxin production of *Vibrio* spp.; and carbapenem and exoenzyme production of *P. carotovorum*. Finally, furanones are also produced by marine green, red or brown algae, by sponges, fungi and ascidians (Hentzer and Givskov, 2003; Hjelmgaard *et al.*, 2003; Martinelli *et al.*, 2004).

It is now apparent that QS signals are used for regulating diverse behaviours in epiphytic, rhizosphere-inhabiting and plant pathogenic bacteria. Plants may produce their own metabolites that interfere with this signalling. Teplitski *et al.* (2000) showed that several plants secrete substances that mimic bacterial AHL signal activities and affect QS-regulated behaviours in associated bacteria. They found that a large number of plant extracts contained QS-inhibitory activities and AHL-producing bacteria associated with these plants and their roots. The interplay of signals and signal inhibitors enables a stable coexistence of the eukaryotic host and the bacteria as long as the plant or root produces a sufficient inhibitor to block the QS systems of the colonizing organisms (Hentzer and Givskov, 2003).

Modulation of QS for disease management

QS is a community genetic regulation mechanism employed by many bacteria to control various microbiological functions. The discovery of microbial QS signals and the signalling mechanisms led to the identification of numerous enzymatic and non-enzymatic signal interference mechanisms as discussed above, which can be developed as promising approaches to control pathogenic infections. In addition, these mechanisms exist not only in microorganisms, but also in the hosts of bacterial pathogens, highlighting their potential implications in microbial ecology and in host–pathogen interactions. A large variety of synthetic AHL analogues and natural products libraries have been screened and a number of QS inhibitors (QSI) have been identified for the control of bacterial infections through the inhibition of bacterial cell-to-cell communication systems that are involved in the regulation of virulence factor production, host colonization and biofilm formation. Promising QSI compounds have been shown to make biofilms more susceptible to antimicrobial treatments and are capable of reducing mortality and virulence as well as promoting clearance of bacteria in experimental animal models of infection. Meanwhile, virulence-targeted disease control through the interruption of the QS for chemical attenuation of bacterial activities rather than bactericidal or bacteriostatic strategies is a less powerful selection for the evolution of resistance. Therefore, it is predictable that further work on QS and signal interference mechanisms will significantly broaden the scope of research in microbial pathogenicity mechanisms and disease management strategies (Hentzer and Givskov, 2003; Zhang and Dong, 2004).

Modulation of T3SS and programmed cell death for plant disease management

Programmed cell death (PCD), also called apoptosis, is present in both animals and plants. Plants endure PCD for growth, development and in response to environmental insults. The plant disease resistance response in most cases is accompanied by the HR-linked PCD at the infection site. Sphingolipid and its phosphorylated derivatives, synthesized in plants through the ceramide biosynthetic pathway, are the signal molecules perceived by the pathogens to trigger the endogenous PCD during susceptible disease response. During infection, the Avr effectors secreted by T3SS may be recognized by the plant R proteins and elicit a plant resistance response in recognition of a number of intrinsic signals, such as caspases, reactive oxygen/nitrogen species, salicylic acid, mitogen-activated protein kinase (MAPK) and membrane ion channels, cascading either alone or in combinations through a coordinated signal transduction pathway (Khurana et al., 2005).

The characterization of the functions and the mechanisms of the T3SS effectors like tyrosine phophatase YopH had led to the identification of the complexes with phosphopeptide and small molecule inhibitor activity for the rational design of inhibitors of virulence. Such compounds have the potential to be used as therapeutic agents. Transgenic expression of animal anti-apoptotic genes in plants provides broad-spectrum disease resistance against compatible obligate pathogens. Similarly, plant- and animal-derived proapoptotic genes can be engineered to be expressed in plants for activating HR-linked resistant disease response against pathogens. The modulation of PCD provides a promising, yet challenging, strategy for plant disease management (Khurana *et al.*, 2005). Although many studies have focused on screening for compounds that target traditional pathways or particular virulence determinants, such as adhesion or T3SS, and selective inhibition or induction of PCD that has been successfully employed to control plant diseases caused by necrotrophic or biotrophic pathogens, respectively (Lyon and Muir, 2003; Khurana *et al.*, 2005), the disease management strategy targeting the T3SS has been rarely reported and needs more research.

Development of New Technologies for Disease Control

TCSTS histidine kinase inhibitor and the whole signalling system

TCSTS is widespread in bacterial pathogens. For example, there are 9, 13, 17 and 29 putative histidine kinase–response regulator pairs in *E. faecalis, S. pneumoniae, S. aureus* and *B. subtilis*, respectively (Fast, 2003). Some of these are considered essential for growth and pathogenesis. For example, eight out of the 13 TCSTSs found in the genome of *S. pneumoniae* are required for virulence in a murine respiratory tract model (Throup *et al.*, 2000). TCSTS is also involved in QS in Gram-positive bacteria, and the GacS/GacA TCSTS system regulates the T3SS and QS, playing a central role in virulence as discussed earlier. The central role of these systems in bacterial pathogenesis promotes substantial interest in the development of broad-spectrum two-component inhibitors for disease control. Lyon *et al.* (2002) demonstrated that activators and inhibitors interact at a common site on the receptor, and suggested that molecules designed to compete with natural agonists for binding at receptor–histidine kinase sensor domains could represent a general approach to the inhibition of receptor–histidine kinase signalling. The expression of many staphylococcal virulence factors is regulated by the *agr* locus via a TCSTS, which is activated in response to a secreted AIP. By exploiting the unique chemical architecture of the naturally occurring AIP-1, several potent inhibitors of staphylococcal TCSTS have been designed and synthesized using either a linear or branched solid-phase approach. These inhibitors are competitive binders and contain the crucial 16-membered side chain to tail thiolactone peptide pharmacophore (Scott *et al.*, 2003). A number of potent competitive AgrC antagonists, i.e. *agr* TCSTS inhibitors, have been reported and reviewed (Matsushita and Janda, 2002; Chan *et al.*, 2004).

Nevertheless, some studies have demonstrated that many histidine kinase inhibitors identified using high-throughput *in vitro* screens act non-specifically in cells either by disrupting membrane integrity or by causing protein aggregation. The recent crystal structure of the nucleotide-binding domain of a sensor kinase (CheA) from *T. maritima* in complex with ADP and various analogues of ATP bodes well for future structure-based design of inhibitors specifically targeting the autokinase domain of sensor kinases and avoiding the homologous domains of host ATPases (Bilwes et al., 2001). For comprehensive reviews of the growing field of two-component inhibition, see Matsushita and Janda (2002) and Stephenson and Hoch (2004). The development of both selective and global inhibitors of TCSTSs will make progress with future studies using appropriate experimental plant and animal infection models (Chan et al., 2004).

Application of QS for disease control

QS quenching application

An attractive strategy for QS control is to use non-native AHL derivatives that block natural AHL signals. For the animal or human pathogen, the QS signal analogues have been widely tested for the purpose of therapy and much progress has been made. For example, some derivatives of the D. pulchra furanone compounds were shown to repress QS in P. aeruginosa and to reduce virulence factor expression, and furanone-treated biofilms were more susceptible to killing by antibiotic tobramycin than their untreated counterparts. QS-inhibitory compounds might constitute a new generation of antimicrobial agents with applications in many fields, including agriculture, medicine, and the food industry. In recent years, a number of biotechnology companies have emerged that specifically aim at developing anti-QS and anti-biofilm drugs (Hentzer and Givskov, 2003).

As discussed earlier, QS quenching enzymes existing in different environments are produced by different species, including the bacteria in rhizosphere, phyllosphere, etc. The application of QS quenching enzymes for disease management has been reported using the engineered QS quenching enzyme and the bacterial strain itself. For example, expression of the aiiA gene of Bacillus spp. in the plant pathogen P. carotovorum resulted in reduced release of AHL signals, decreased extracellular pectolytic enzyme activity and attenuated soft rot disease symptoms in all plants tested. Moreover, transgenic tobacco and potato plants expressing AiiA lactonase from B. cereus increased host resistance and were less susceptible to infection by P. carotovorum, which highlights a promising potential to use QS signals as molecular targets for disease control, thereby broadening current approaches for prevention of bacterial infections (Hentzer and Givskov, 2003). Another example is the application of R. erythropolis strain W2 to quench QS-regulated functions of other microbes. In vitro, R. erythropolis strongly interfered with violacein production by Chromobacterium violaceum, and transfer of pathogenicity in A. tumefaciens. In planta, R. erythropolis W2 markedly reduced the pathogenicity of P. carotovorum subsp. carotovorum in potato tubers (Uroz et al., 2003).

However, some questions are raised for the application of the QS quenching mechanism in disease management. Smadja et al. (2004) indicated that QS regulation is involved in the second maceration process leading to the development of disease instead of the initial plant colonization by P. atrosepticum. They pointed out that the use of QS quenching strategies for biological control of *P. atrosepticum* cannot prevent initial infection and multiplication.

Biocontrol based on QS
Biological control is an accepted and important component of current plant disease management strategies. The introduction of bacterized seeds carrying bacterial isolates with proven growth promotion capabilities and antagonistic characteristics offers attractive alternatives or supplements to the use of conventional methods such as chemical protectants for plant disease management. Pathogens are typically affected by certain modes of actions and not by others according to their nature (i.e. biotrophs vs necrotrophs). Resistance in the host plant may be induced locally or systemically by either live or dead cells of the biocontrol agent and may affect pathogens of various groups (Elad, 2003).

Valdez *et al.* (2005) evaluated the ability of the probiotic organism *Lactobacillus plantarum* to inhibit the pathogenic activity of *P. aeruginosa* and found that *L. plantarum* and/or its by-products are potential therapeutic agents for the local treatment of *P. aeruginosa* burn infections. The cultivation of beneficial plant-growth-promoting or biocontrol bacteria, or the selective expression of useful genes in the plant environment, is of significant biotechnological interest. AHLs produced by indigenous bacteria of the tomato rhizosphere also can diffuse within the rhizosphere and are capable of modulating the ecology of rhizosphere bacterial populations. AHL signal molecules produced by bioengineered plants represent an approach to improve the ability of plants to communicate with and to regulate beneficial genes in introduced bacteria (Scott *et al.*, 2006). An effect of AHL degradation products on plant physiology was reported by Joseph and Phillips (2003), who demonstrated that treatment of bean roots with physiologically relevant concentrations of HSL and homoserine significantly increased stomatal conductance and transpiration in the plant. Transpiration also enhances the availability of mineral nutrients for the growth of plants and root-associated bacteria. It is conceivable that the bacterial production of AHLs and their degradation to HSL and homoserine may represent a plant–microbe interaction in which both the plant and root-associated bacteria benefit from the production of QS signals in the rhizosphere.

The concept of bioengineering a plant host to produce AHLs was demonstrated previously by the expression of the *yenI* gene from *Yersinia enterocolitica* in tobacco and potato as well as *expI* from *P. carotovorum* in tobacco (Toth *et al.*, 2004). Mae *et al.* (2001) showed that the OHL-producing transgenic tobacco lines as well as the wild-type plant with the exogenous addition of OHL exhibited enhanced resistance to infection by wild-type *P. carotovorum*. The constitutive expression of *yenI* in tobacco and potato leads to the endogenous accumulation of the major AHLs for YenI synthetase, namely *N*-(3-oxohexanoyl)-L-homoserine lactone (3-oxo-C6-HSL) and *N*-hexanoyl-L-homoserine lactone (C6-HSL) (Toth *et al.*, 2004). These plants have dramatically altered susceptibilities to infection by pathogenic *Pectobacterium* spp.. Scott *et al.* (2006)

further reported that transgenic plants expressing the LasI and YenI QS signal synthetases individually or in combination within plant cell plastids produce long- and short-chain AHLs that are readily detectable in the rhizosphere and phyllosphere. Moreover, they also show that plant-produced AHLs become an active component of rhizosphere and non-rhizosphere soil. Accordingly, their work demonstrates the feasibility of designing AHL-specific plant–bacteria biosystems that can enhance beneficial plant–microbe interactions or inhibit harmful interactions and it is practical to utilize bioengineered plants to supplement soils with specific AHLs to modulate bacterial phenotypes (Scott *et al.*, 2006).

Combination of biocontrol and disease management
Large-scale use of biocontrol is still limited because of the variability and inconsistency of biocontrol activity. In some cases, this may be caused by sensitivity of the biocontrol agents to environmental influences. Methods to overcome biocontrol limitations and to improve its efficacy are: (i) integration of biocontrol with chemical protectants; and (ii) introduction of two or more biocontrol agents in a mixture, assuming that each of them has different ecological requirements and/or different modes of action. Implementation of one (or more) of these approaches has lowered the variability and increased the consistency of disease suppression.

Zhu *et al.* (2006) provided a new strategy for developing a genetically engineered multifunctional *Bacillus thuringiensis* strain that possesses insecticidal activity together with restraint of bacterial pathogenicity for biocontrol and disease management. The genetically modified *B. thuringiensis* strain BMB821A expressing an AHL lactonase gene *aiiA* produced 2.4-fold more AHL lactonase and could degrade more AHLs than the original strain BMB-005. The BMB821A strain strongly restrained *P. carotovorum* infection on potato slices and cactus stems, and retained the insecticidal activity to the lepidopteran *Spodoptera exigua*, although the toxicity was a little reduced.

Future Directions and Potential Problems

Although we are beginning to understand some aspects of the function of QS and T3SS machineries, some future research directions are suggested in this chapter to enhance our understanding of managing bacterial plant disease by modulating the QS and T3SS systems.

Signals for T3SS, TCSTS and the integration of these signals for gene regulation at individual cell and community levels

There are many other signalling systems being discovered in various bacteria, including T3SS, QS, contact-dependent signalling and two-component signalling, with many more to be discovered. Furthermore, many of these new molecules will signal through novel uncharacterized proteins, which has been suggested in the case of cytolysin induction in *E. faecalis* (Haas *et al.*, 2002; Fast, 2003). Questions remaining to be solved in the future are: what are the signals for the T3SS, GacS/GacA TCSTS? Is there any connection among the signals of QS, T3SS and GacS/GacA TCSTS, and how do the bacteria detect, distinguish and respond to these signals and integrate all the signals and the signal regulatory pathways in harmony individually and at a community level to benefit the whole population? Do all cells have the same response to these signals or do different subgroups of bacteria respond to different signals and cooperate to deal with the complex signals the whole community encounters? What is the response of an individual cell to signal stimuli and are there any differences in gene expression profiles between the individual cells and the whole population? In particular, the search for T3SS, GacS/GacA and other TCSTS signals as well as investigation of the connection among these signals will shed light on our understanding of the virulence mechanisms and the development of novel strategies for disease management. For example, during the competence and sporulation, a few cells of *Bacillus* retain competence ability while most cells are undergoing the sporulation.

Structure investigation, mathematic modelling and computer-based prediction for novel compound design

The investigation of the structure, activity and turnover of the QS signal has helped in the elucidation of the QS mechanism and the development of the QS inhibitor as a therapeutic strategy. For example, a series of staphylococcal AIP signal molecule analogues (including the L-alanine- and D-amino acid-scanned peptides) have been synthesized to determine the functionally critical residues within the *S. aureus* group I AIP and the addition of exogenous synthetic AIPs to *S. aureus* inhibited the production of toxic shock syndrome toxin and enterotoxin C3, confirming the potential of a QS blockade as a therapeutic strategy (MDowell *et al.*, 2001). Hjelmgaard *et al.* (2003) reported the interesting structure–activity relationships of the furanone-based natural product analogues towards the QS systems. During the course of structure–activity studies on AIP1 and AIP2, Chan *et al.* (2004) found that a number of unexpected observations could be useful pointers for the design of new AgrC antagonists.

Fagerlind *et al.* (2005) developed a mathematical model of the QS system in *P. aeruginosa* to virtually add 3-O-C12-HSL antagonists that differed in their affinity for the receptor protein and for their ability to mediate degradation of the receptor. Their model suggests that very small differences in these parameters for different 3-O-C12-HSL antagonists can greatly affect the success of QS blocker (QSB)-based inhibition of the QS system in *P. aeruginosa*. Most importantly, the ability of the 3-O-C12-HSL antagonist to mediate degradation of LasR is the core parameter for successful QSB-based inhibition of the QS system in *P. aeruginosa*, and QSBs can shift the system to an uninduced state and the use of 3-O-C12-HSL antagonists may constitute a promising therapeutic approach against *P. aeruginosa* infections. Based on the structural information and with the aid of computer-based prediction, Riedel *et al.* (2006) designed novel compounds that specifically inhibit the AHL-dependent QS system of the genus *Burkholderia*, which efficiently inhibits the expression of virulence factors and attenuates the pathogenicity of the organism.

With an integration of different areas like mathematics, statistics, chemistry, biology, bioinformatics and computer science, future work on signal structure investigation and QS modelling will help unravel the virulence mechanisms of the QS, T3SS and TCSTS, and develop new strategies for disease management.

Integration of genomics, functional genomics, proteomics and bioinformatics with the traditional and novel methods to reveal disease mechanisms

The well-established Tn*phoA* transposon was continuouslyused to recognize QS-dependent genes and to unravel the QS mechanisms. For example, a phosphatase (*phoA*)-deficient *P. carotovorum* was mutagenized by the Tn*phoA* transposon to identify OHHL-regulated genes that encode proteins that are important in the soft rot *Pectobacterium*–plant interaction. The expression of the reporter gene fusion was then assessed in the presence and absence of OHHL, and OHHL-responsive fusions were isolated and seven novel QS-dependent genes were identified (Pemberton *et al.*, 2005).

Furanone penetration and half-life were estimated by using the green fluorescent protein (GFP)-based single-cell technology in combination with scanning confocal laser microscopy, enabling scientists to identify synthetic compounds that not only inhibited the quorum sensors in the majority of the cells, but also led to the formation of flat, undifferentiated biofilms that eventually detached. By means of AHL monitors built on the *P. aeruginosa* quorum sensors and the *lasB-gfp* target gene, the efficacy of these compounds was measured via GFP expression (Hentzer and Givskov, 2003).

Studying bacterial communication will further our understanding of the extraordinary diversity in the surrounding environment and will provide novel strategies against bacterial infections. Efforts to mine the genomes of the bacterial world for unusual and interesting natural products have already yielded and will continue to yield new avenues for disease control (Fast, 2003).

Genome-wide screens for QS-controlled genes within some organisms have been performed using genome microarray and proteomic studies. Microarray technology has been widely applied to discover the basic virulence mechanisms and drug development in research and pharmaceutical labs on different pathogens. For example, the cDNA microarray of *P. aeruginosa* (Affymetrix Inc., California) has been used to demonstrate that furanone compounds specifically repress expression of QS-controlled genes in *P. aeruginosa* and the target specificity of certain first-generation antipathogenic drugs (Hentzer and Givskov, 2003).

Potential problems

There are limited reports of plant disease management by targeting T3SS. Although some progress of the application of QS on disease control has been accomplished, there are still problems to be solved for the development of efficient disease management strategies. For example, although a robust solid-phase synthetic route to both natural and non-natural AHLs in high purity has been developed and a set of non-native AHLs that are among the most potent inhibitors of bacterial QS has been identified, the toxicity of the QS inhibitors may prevent them from use for the treatment of bacterial infections (Geske *et al.*, 2005). Another disadvantage associated with QS antagonists is the narrow spectrum of antagonists, especially the AHL antagonists. Specific antagonists have to be developed for each targeted organism. In addition, new synthetic approaches for the generation of QS analogues and the systematic evaluation of the effects of QS ligand structure on QS are still required.

Meanwhile, the presence of complex microbial species and the variety of QS signalling molecules and/or the signal inhibitors that might be produced by the microorganisms or even the host greatly complicate the application of anti-QS therapy. It will require a specific QS inhibitor only to attenuate a single, pathogenic organism living in a mixed population of normal bacterial flora while leaving the rest of the bacterial population unaffected. Other factors influencing the QS-based disease control include the impacts of technological, environmental, socio-economic and climatic changes on plant hosts, which could alter stages and rates of development of the pathogen, modify host resistance and result in changes in the physiology of host–pathogen interactions, all of which add complexity and uncertainty onto a system that is already exceedingly difficult to manage on a sustainable basis. Unfortunately, most recent work has been concentrated on the effects of a single variable on the host, pathogen or the interaction of the two under controlled conditions. Intensified research on multiple issues could result in an improved understanding and management of plant diseases in the future (Nalca *et al.*, 2006).

Acknowledgements

We acknowledge the critical comments from our colleague Mary Lynne P. Collins and financial help from the Research Growth Initiative of the University of Wisconsin-Milwaukee and the National Science Foundation, USA.

References

Abraham, W.R. (2006) Controlling biofilms of gram-positive pathogenic bacteria. *Current Medicinal Chemistry* 13, 1509–1524.

Alfaro, J.F., Zhang, T., Wynn, D.P., Karschner, E.L. and Zhou, Z.S. (2004) Synthesis of LuxS inhibitors targeting bacterial cell–cell communication. *Organic Letters* 6, 3043–3046.

Bilwes, A.M., Quezada, C.M., Croal, L.R., Crane, B.R. and Simon, M.I. (2001) Nucleotide binding by the histidine kinase CheA. *Nature Structural Biology* 8, 353–360.

Bleves, S., Soscia, C., Nogueira-Orlandi, P., Lazdunski, A. and Filloux, A. (2005) Quorum sensing negatively controls type III secretion regulon expression in *Pseudomonas aeruginosa* PAO1. *Journal of Bacteriology* 187, 3898–3902.

Brencic, A. and Winans, S.C. (2005) Detection of and response to signals involved in host–microbe interactions by plant-associated bacteria. *Microbiology and Molecular Biology Reviews* 69, 155–174.

Burrowes, E., Abbas, A., O'Neill, A., Adams, C. and O'Gara, F. (2005) Characterisation of the regulatory RNA *rsmB* from *Pseudomonas aeruginosa* PAO1. *Research in Microbiology* 156, 7–16.

Burrowes, E., Baysse, C., Adams, C. and O'Gara, F. (2006) Influence of the regulatory protein RsmA on cellular functions in *Pseudomonas aeruginosa* PAO1, as revealed by transcriptome analysis. *Microbiology* 152, 405–418.

Buttner, D. and Bonas, U. (2006) Who comes first? How plant pathogenic bacteria orchestrate type III secretion. *Current Opinion in Microbiology* 9, 193–200.

Castang, S., Chantegrel, B., Deshayes, C., Dolmazon, R., Gouet, P., Haser, R., Reverchon, S., Nasser, W., Hugouvieux-Cotte-Pattat, N. and Doutheau, A. (2004) N-Sulfonyl homoserine lactones as antagonists of bacterial quorum sensing. *Bioorganic and Medicinal Chemistry Letters* 14, 5145–5149.

Chan, W.C., Coyle, B.J. and Williams, P. (2004) Virulence regulation and quorum sensing in staphylococcal infections: competitive AgrC antagonists as quorum sensing inhibitors. *Journal of Medicinal Chemistry* 47, 4633–4641.

Chancey, S.T., Wood, D.W. and Pierson, L.S. III (1999) Two-component transcriptional regulation of N-acyl-homoserine lactone production in *Pseudomonas aureofaciens*. *Applied and Environmental Microbiology* 65, 2294–2299.

Chancey, S.T., Wood, D.W., Pierson, E.A. and Pierson, L.S. III (2002) Survival of GacS/GacA mutants of the biological control bacterium *Pseudomonas aureofaciens* 30–84 in the wheat rhizosphere. *Applied and Environmental Microbiology* 68, 3308–3314.

Chatterjee, A., Cui, Y. and Chatterjee, A.K. (2002) Regulation of *Erwinia carotovora hrpL*(Ecc) (sigma-L(Ecc)), which encodes an extracytoplasmic function subfamily of sigma factor required for expression of the HRP regulon. *Molecular Plant-Microbe Interactions* 15, 971–980.

Chatterjee, A., Cui, Y., Yang, H., Collmer, A., Alfano, J.R. and Chatterjee, A.K. (2003) GacA, the response regulator of a two-component system, acts as a master regulator in *Pseudomonas syringae* pv. *tomato* DC3000 by controlling regulatory RNA, transcriptional activators, and alternate sigma factors. *Molecular Plant-Microbe Interactions* 16, 1106–1117.

Chatterjee, A., Cui, Y., Hasegawa, H., Leigh, N., Dixit, V. and Chatterjee, A.K. (2005) Comparative analysis of two classes of quorum-sensing signaling systems that control production of extracellular proteins and secondary metabolites in *Erwinia carotovora* subspecies. *Journal of Bacteriology* 187, 8026–8038.

Cui, Y., Mukherjee, A., Dumenyo, C.K., Liu, Y. and Chatterjee, A.K. (1999) *rsmC* of the soft-rotting bacterium *Erwinia carotovora* subsp. *carotovora* negatively controls extracellular enzyme and *harpin*(Ecc) production and virulence by modulating levels of regulatory RNA (*rsmB*) and RNA-binding protein (RsmA). *Journal of Bacteriology* 181, 6042–6052.

Cui, Y., Chatterjee, A. and Chatterjee, A.K. (2001) Effects of the two-component system comprising GacA and GacS of *Erwinia carotovora* subsp. *carotovora* on the production of global regulatory *rsmB* RNA, extracellular enzymes, and *harpin*Ecc. *Molecular Plant-Microbe Interactions* 14, 516–526.

Dasgupta, N., Lykken, G.L., Wolfgang, M.C. and Yahr, T.L. (2004) A novel anti-anti-activator mechanism regulates expression of the *Pseudomonas aeruginosa* type III secretion system. *Molecular Microbiology* 53, 297–308.

Dong, Y.H., Wang, L.H., Xu, J.L., Zhang, H.B., Zhang, X.F. and Zhang, L.H. (2001) Quenching quorum-sensing-dependent bacterial infection by an N-acyl homoserine lactonase. *Nature* 411, 813–817.

Elad, Y. (2003) Biocontrol of foliar pathogens: mechanisms and application. *Communications in Agricultural and Applied Biological Sciences* 68, 17–24.

Fagerlind, M.G., Nilsson, P., Harlen, M., Karlsson, S., Rice, S.A. and Kjelleberg, S. (2005) Modeling the effect of acylated homoserine lactone antagonists in *Pseudomonas aeruginosa*. *Biosystems* 80, 201–213.

Falconi, M., Prosseda, G., Giangrossi, M., Beghetto, E. and Colonna, B. (2001) Involvement of FIS in the H-NS-mediated regulation of *virF* gene of *Shigella* and enteroinvasive *Escherichia coli*. *Molecular Microbiology* 42, 439–452.

Fast, W. (2003) Molecular radio jamming: autoinducer analogs. *Chemistry and Biology* 10, 1–2.

Federle, M.J. and Bassler, B.L. (2003) Interspecies communication in bacteria. *The Journal of Clinical Investigation* 112, 1291–1299.

Feldman, M.F. and Cornelis, G.R. (2003) The multitalented type III chaperones: all you can do with 15 kDa. *FEMS Microbiology Letters* 219, 151–158.

Francis, M.S., Wolf-Watz, H. and Forsberg, A. (2002) Regulation of type III secretion systems. *Current Opinion in Microbiology* 5, 166–172.

Frezza, M., Castang, S., Estephane, J., Soulere, L., Deshayes, C., Chantegrel, B., Nasser, W., Queneau, Y., Reverchon, S. and Doutheau, A. (2006) Synthesis and biological evaluation of homoserine lactone derived ureas as antagonists of bacterial quorum sensing. *Bioorganic and Medicinal Chemistry* 14, 4781–4791.

Fuqua, C. and Greenberg, E.P. (2002) Listening in on bacteria: acyl-homoserine lactone signalling. *Nature Reviews Molecular Cell Biology* 3, 685–695.

Fuqua, C., Parsek, M.R. and Greenberg, E.P. (2001) Regulation of gene expression by cell-to-cell communication: acyl-homoserine lactone quorum sensing. *Annual Review of Genetics* 35, 439–468.

Galan, J.E. and Collmer, A. (1999) Type III secretion machines: bacterial devices for protein delivery into host cells. *Science* 284, 1322–1328.

Geske, G.D., Wezeman, R.J., Siegel, A.P. and Blackwell, H.E. (2005) Small molecule inhibitors of bacterial quorum sensing and biofilm formation. *Journal of the American Chemical Society* 127, 12762–12763.

Goodman, A.L., Kulasekara, B., Rietsch, A., Boyd, D., Smith, R.S. and Lory, S. (2004) A signaling network reciprocally regulates genes associated with acute infection and chronic persistence in *Pseudomonas aeruginosa*. *Developmental Cell* 7, 745–754.

Haas, W., Shepard, B.D. and Gilmore, M.S. (2002) Two-component regulator of *Enterococcus faecalis* cytolysin responds to quorum-sensing autoinduction. *Nature* 415, 84–87.

Ham, J.H., Cui, Y., Alfano, J.R., Rodriguez-Palenzuela, P., Rojas, C.M., Chatterjee, A.K. and Collmer, A. (2004) Analysis of *Erwinia chrysanthemi* EC16 *pelE::uidA, pelL::uidA*, and *hrpN::uidA* mutants reveals strain-specific atypical regulation of the Hrp type III secretion system. *Molecular Plant-Microbe Interactions* 17, 184–194.

Heeb, S., Blumer, C. and Haas, D. (2002) Regulatory RNA as mediator in GacA/RsmA-dependent global control of exoproduct formation in *Pseudomonas fluorescens* CHA0. *Journal of Bacteriology* 184, 1046–1056.

Heeb, S., Valverde, C., Gigot-Bonnefoy, C. and Haas, D. (2005) Role of the stress sigma factor RpoS in GacA/RsmA-controlled secondary metabolism and resistance to oxidative stress in *Pseudomonas fluorescens* CHA0. *FEMS Microbiology Letters* 243, 251–258.

Henke, J.M. and Bassler, B.L. (2004) Bacterial social engagements. *Trends in Cell Biology* 14, 648–656.

Hentzer, M. and Givskov, M. (2003) Pharmacological inhibition of quorum sensing for the treatment of chronic bacterial infections. *The Journal of Clinical Investigation* 112, 1300–1307.

Hjelmgaard, T., Persson, T., Rasmussen, T.B., Givskov, M. and Nielsen, J. (2003) Synthesis of furanone-based natural product analogues with quorum sensing antagonist activity. *Bioorganic and Medicinal Chemistry* 11, 3261–3271.

Hogardt, M., Roeder, M., Schreff, A.M., Eberl, L. and Heesemann, J. (2004) Expression of *Pseudomonas aeruginosa exoS* is controlled by quorum sensing and RpoS. *Microbiology* 150, 843–851.

Joseph, C.M. and Phillips, D.A. (2003) Metabolites from soil bacteria affect plant water relations. *Plant Physiology and Biochemistry* 41, 189–192.

Kaper, J.B. and Sperandio, V. (2005) Bacterial cell-to-cell signaling in the gastrointestinal tract. *Infection and Immunity* 73, 3197–3209.

Kay, E., Dubuis, C. and Haas, D. (2005) Three small RNAs jointly ensure secondary metabolism and biocontrol in *Pseudomonas fluorescens* CHA0. *Proceedings of the National Academy of Sciences of the United States of America* 102, 17136–17141.

Khurana, S.M.P., Pandey, S.K., Sarkar, D. and Chanemougasoundharam, A. (2005) Apoptosis in plant disease response: a close encounter of the pathogen kind. *Current Science* 88, 740–752.

Koch, B., Liljefors, T., Persson, T., Nielsen, J., Kjelleberg, S. and Givskov, M. (2005) The LuxR receptor: the sites of interaction with quorum-sensing signals and inhibitors. *Microbiology* 51, 3589–3602.

Koiv, V. and Mae, A. (2001) Quorum sensing controls the synthesis of virulence factors by modulating *rsmA* gene expression in *Erwinia carotovora* subsp. *carotovora*. *Molecular Genetics and Genomics* 265, 287–292.

Leadbetter, J.R. and Greenberg, E.P. (2000) Metabolism of acyl-homoserine lactone quorum-sensing signals by *Variovorax paradoxus*. *Journal of Bacteriology* 182, 6921–6926.

Lenz, D.H., Miller, M.B., Zhu, J., Kulkarni, R.V. and Bassler, B.L. (2005) CsrA and three redundant small RNAs regulate quorum sensing in *Vibrio cholerae*. *Molecular Microbiology* 58, 1186–1202.

Liu, Y., Cui, Y., Mukherjee, A. and Chatterjee, A.K. (1998) Characterization of a novel RNA regulator of *Erwinia carotovora* ssp. *carotovora* that controls production of extracellular enzymes and secondary metabolites. *Molecular Microbiology* 29, 219–234.

Lyon, G.J. and Muir, T.W. (2003) Chemical signaling among bacteria and its inhibition. *Chemistry and Biology* 10, 1007–1021.

Lyon, G.J., Mayville, P., Muir, T.W. and Novick, R.P. (2000) Rational design of a global inhibitor of the virulence response in *Staphylococcus aureus*, based in part on localization of the site of inhibition to the receptor-histidine kinase, AgrC. *Proceedings of the National Academy of Sciences of the United States of America* 97, 13330–13335.

Lyon, G.J., Wright, J.S., Christopoulos, A., Novick, R.P. and Muir, T.W. (2002) Reversible and specific extracellular antagonism of receptor-histidine kinase signaling. *Journal of Biological Chemistry* 277, 6247–6253.

Mae, A., Montesano, M., Koiv, V. and Palva, E.T. (2001) Transgenic plants producing the bacterial pheromone N-acyl-homoserine lactone exhibit enhanced resistance to the bacterial phytopathogen *Erwinia carotovora*. *Molecular Plant-Microbe Interactions* 14, 1035–1042.

Manefield, M., Rasmussen, T.B., Henzter, M., Andersen, J.B., Steinberg, P., Kjelleberg, S. and Givskov, M. (2002) Halogenated furanones inhibit quorum sensing through accelerated LuxR turnover. *Microbiology* 148, 1119–1127.

Martinelli, D., Grossmann, G., Sequin, U., Brandl, H. and Bachofen, R. (2004) Effects of natural and chemically synthesized furanones on quorum sensing in *Chromobacterium violaceum*. *BMC Microbiology* 4, 25.

Matsushita, M. and Janda, K.D. (2002) Histidine kinases as targets for new antimicrobial agents. *Bioorganic and Medicinal Chemistry* 10, 855–867.

McNab, R., Ford, S.K., El-Sabaeny, A., Barbieri, B., Cook, G.S. and Lamont, R.J. (2003) LuxS-based signaling in *Streptococcus gordonii*: autoinducer 2 controls carbohydrate metabolism and biofilm formation with *Porphyromonas gingivalis*. *Journal of Bacteriology* 185, 274–284.

MDowell, P., Affas, Z., Reynolds, C., Holden, M.T., Wood, S.J., Saint, S., Cockayne, A., Hill, P.J., Dodd, C.E., Bycroft, B.W., Chan, W.C. and Williams, P. (2001) Structure, activity and evolution of the group I thiolactone peptide quorum-sensing system of *Staphylococcus aureus*. *Molecular Microbiology* 41, 503–512.

Miller, M.B. and Bassler, B.L. (2001) Quorum sensing in bacteria. *Annual Review of Microbiology* 55, 165–199.

Miller, S.T., Xavier, K.B., Campagna, S.R., Taga, M.E., Semmelhack, M.F., Bassler, B.L. and Hughson, F.M. (2004) *Salmonella typhimurium* recognizes a chemically distinct form of the bacterial quorum-sensing signal AI-2. *Molecular Cell* 15, 677–687.

Mota, L.J., Sorg, I. and Cornelis, G.R. (2005) Type III secretion: the bacteria-eukaryotic cell express. *FEMS Microbiology Letters* 252, 1–10.

Mukherjee, A., Cui, Y., Ma, W.L., Liu, Y. and Chatterjee, A.K. (2000) *hexA* of *Erwinia carotovora* ssp *carotovora* strain Ecc71 negatively regulates production of RpoS and *rsmB* RNA, a global regulator of extracellular proteins, plant virulence and the quorum-sensing signal, N-(3-oxo-hexanoyl)-L-homoserine lactone. *Environmental Microbiology* 2, 203–215.

Mulcahy, H., O'Callaghan, J., O'Grady, E.P., Adams, C. and O'Gara, F. (2006) The posttranscriptional regulator RsmA plays a role in the interaction between *Pseudomonas aeruginosa* and human airway epithelial cells by positively regulating the type III secretion system. *Infection and Immunity* 74, 3012–3015.

Nalca, Y., Jansch, L., Bredenbruch, F., Geffers, R., Buer, J. and Haussler, S. (2006) Quorum-sensing antagonistic activities of azithromycin in *Pseudomonas aeruginosa* PAO1: a global approach. *Antimicrobial Agents and Chemotherapy* 50, 1680–1688.

Otto, M., Sussmuth, R., Vuong, C., Jung, G. and Gotz, F. (1999) Inhibition of virulence factor expression in *Staphylococcus aureus* by the *Staphylococcus epidermidis agr* pheromone and derivatives. *FEBS Letters* 450, 257–262.

Page, A.L. and Parsot, C. (2002) Chaperones of the type III secretion pathway: jacks of all trades. *Molecular Microbiology* 46, 1–11.

Pemberton, C.L., Whitehead, N.A., Sebaihia, M., Bell, K.S., Hyman, L.J., Harris, S.J., Matlin, A.J., Robson, N.D., Birch, P.R., Carr, J.P., Toth, I.K. and Salmond, G.P. (2005) Novel quorum-sensing-controlled genes in *Erwinia carotovora* subsp. *carotovora*: identification of a fungal elicitor homologue in a soft-rotting bacterium. *Molecular Plant-Microbe Interactions* 18, 343–353.

Qazi, S., Middleton, B., Muharram, S.H., Cockayne, A., Hill, P., O'Shea, P., Chhabra, S.R., Camara, M. and Williams, P. (2006) N-acylhomoserine lactones antagonize virulence gene expression and quorum sensing in *Staphylococcus aureus*. *Infection and Immunity* 74, 910–919.

Quinones, B., Dulla, G. and Lindow, S.E. (2005) Quorum sensing regulates exopolysaccharide production, motility, and virulence in *Pseudomonas syringae*. *Molecular Plant-Microbe Interactions* 18, 682–693.

Reimmann, C., Valverde, C., Kay, E. and Haas, D. (2005) Posttranscriptional repression of GacS/GacA-controlled genes by the RNA-binding protein RsmE acting together with RsmA in the biocontrol strain *Pseudomonas fluorescens* CHA0. *Journal of Bacteriology* 187, 276–285.

Reverchon, S., Chantegrel, B., Deshayes, C., Doutheau, A. and Cotte-Pattat, N. (2002) New synthetic analogues of N-acyl homoserine lactones as agonists or antagonists of transcriptional regulators involved in bacterial quorum sensing. *Bioorganic and Medicinal Chemistry Letters* 12, 1153–1157.

Riedel, K., Kothe, M., Kramer, B., Saeb, W., Gotschlich, A., Ammendola, A. and Eberl, L. (2006) Computer-aided design of agents that inhibit the *cep* quorum-sensing system of *Burkholderia cenocepacia*. *Antimicrobial Agents and Chemotherapy* 50, 318–323.

Scott, R.J., Lian, L.Y., Muharram, S.H., Cockayne, A., Wood, S.J., Bycroft, B.W., Williams, P. and Chan, W.C. (2003) Side-chain-to-tail thiolactone peptide inhibitors of the staphylococcal quorum-sensing system. *Bioorganic and Medicinal Chemistry Letters* 13, 2449–2453.

Scott, R.A., Weil, J., Le, P.T., Williams, P., Fray, R.G., von Bodman, S.B. and Savka, M.A. (2006) Long- and short-chain plant-produced bacterial *N*-acyl-homoserine lactones become components of phyllosphere, rhizosphere, and soil. *Molecular Plant-Microbe Interactions* 19, 227–239.

Smadja, B., Latour, X., Faure, D., Chevalier, S., Dessaux, Y. and Orange, N. (2004) Involvement of *N*-acylhomoserine lactones throughout plant infection by *Erwinia carotovora* subsp. *atroseptica* (*Pectobacterium atrosepticum*). *Molecular Plant-Microbe Interactions* 17, 1269–1278.

Smith, R.S., Harris, S.G., Phipps, R. and Iglewski, B. (2002) The *Pseudomonas aeruginosa* quorum-sensing molecule *N*-(3-oxododecanoyl) homoserine lactone contributes to virulence and induces inflammation *in vivo*. *Journal of Bacteriology* 184, 1132–1139.

Sonnleitner, E., Schuster, M., Sorger-Domenigg, T., Greenberg, E.P. and Blasi, U. (2006) Hfq-dependent alterations of the transcriptome profile and effects on quorum sensing in *Pseudomonas aeruginosa*. *Molecular Microbiology* 59, 1542–1558.

Sperandio, V., Torres, A.G., Jarvis, B., Nataro, J.P. and Kaper, J.B. (2003) Bacteria-host communication: the language of hormones. *Proceedings of the National Academy of Sciences of the United States of America* 100, 8951–8956.

Stebbins, C.E. (2005) Structural microbiology at the pathogen–host interface. *Cellular Microbiology* 7, 1227–1236.

Stephenson, K. and Hoch, J.A. (2004) Developing inhibitors to selectively target two-component and phosphorelay signal transduction systems of pathogenic microorganisms. *Current Medicinal Chemistry* 11, 765–773.

Sturme, M.H., Kleerebezem, M., Nakayama, J., Akkermans, A.D., Vaugha, E.E. and de Vos, W.M. (2002) Cell to cell communication by autoinducing peptides in gram-positive bacteria. *Antonie Van Leeuwenhoek* 81, 233–243.

Taga, M.E. and Bassler, B.L. (2003) Chemical communication among bacteria. *Proceedings of the National Academy of Sciences of the United States of America* 100, 14549–14554.

Tang, X., Xiao, Y. and Zhou, J.M. (2006) Regulation of the type III secretion system in phytopathogenic bacteria. *Molecular Plant-Microbe Interactions* 19, 1159–1166.

Teplitski, M., Robinson, J.B. and Bauer, W.D. (2000) Plants secrete substances that mimic bacterial *N*-acyl homoserine lactone signal activities and affect population density-dependent behaviors in associated bacteria. *Molecular Plant-Microbe Interactions* 13, 637–648.

Throup, J.P., Koretke, K.K., Bryant, A.P., Ingraham, K.A., Chalker, A.F., Ge, Y., Marra, A., Wallis, N.G., Brown, J.R., Holmes, D.J., Rosenberg, M. and Burnham, M.K. (2000) A genomic analysis of two-component signal transduction in *Streptococcus pneumoniae*. *Molecular Microbiology* 35, 566–576.

Toth, I.K., Newton, J.A., Hyman, L.J., Lees, A.K., Daykin, M., Ortori, C., Williams, P. and Fray, R.G. (2004) Potato plants genetically modified to produce N-acylhomoserine lactones increase susceptibility to soft rot *Erwiniae*. *Molecular Plant-Microbe Interactions* 17, 880–887.

Uroz, S., ngelo-Picard, C., Carlier, A., Elasri, M., Sicot, C., Petit, A., Oger, P., Faure, D. and Dessaux, Y. (2003) Novel bacteria degrading *N*-acylhomoserine lactones and their use as quenchers of quorum-sensing-regulated functions of plant-pathogenic bacteria. *Microbiology* 149, 1981–1989.

Valdez, J.C., Peral, M.C., Rachid, M., Santana, M. and Perdigon, G. (2005) Interference of *Lactobacillus plantarum* with *Pseudomonas aeruginosa in vitro* and in infected burns: the potential use of probiotics in wound treatment. *Clinical Microbiology and Infection* 11, 472–479.

Valverde, C., Heeb, S., Keel, C. and Haas, D. (2003) RsmY, a small regulatory RNA, is required in concert with RsmZ for GacA-dependent expression of biocontrol traits in *Pseudomonas fluorescens* CHA0. *Molecular Microbiology* 50, 1361–1379.

Ventre, I., Goodman, A.L., Vallet-Gely, I., Vasseur, P., Soscia, C., Molin, S., Bleves, S., Lazdunski, A., Lory, S. and Filloux, A. (2006) Multiple sensors control reciprocal expression of *Pseudomonas aeruginosa* regulatory RNA and virulence genes. *Proceedings of the National Academy of Sciences of the United States of America* 103, 171–176.

Vieira-da-Motta, O., Ribeiro, P.D., as da, S.W. and Medina-Acosta, E. (2001) RNAIII inhibiting peptide (RIP) inhibits *agr*-regulated toxin production. *Peptides* 22, 1621–1627.

Winzer, K., Hardie, K.R. and Williams, P. (2003) LuxS and autoinducer-2: their contribution to quorum sensing and metabolism in bacteria. *Advances in Applied Microbiology* 53, 291–396.

Wolfgang, M.C., Lee, V.T., Gilmore, M.E. and Lory, S. (2003) Coordinate regulation of bacterial virulence genes by a novel adenylate cyclase-dependent signaling pathway. *Developmental Cell* 4, 253–263.

Xavier, K.B. and Bassler, B.L. (2005) Interference with AI-2-mediated bacterial cell–cell communication. *Nature* 437, 750–753.

Yang, G., Cheng, H., Liu, C., Xue, Y., Gao, Y., Liu, N., Gao, B., Wang, D., Li, S., Shen, B. and Shao, N. (2003) Inhibition of *Staphylococcus aureus* pathogenesis *in vitro* and *in vivo* by RAP-binding peptides. *Peptides* 24, 1823–1828.

Yates, E.A., Philipp, B., Buckley, C., Atkinson, S., Chhabra, S.R., Sockett, R.E., Goldner, M., Dessaux, Y., Camara, M., Smith, H. and Williams, P. (2002) N-acylhomoserine lactones undergo lactonolysis in a pH-, temperature-, and acyl chain length-dependent manner during growth of *Yersinia pseudotuberculosis* and *Pseudomonas aeruginosa*. *Infection and Immunity* 70, 5635–5646.

Yip, C.K. and Strynadka, N.C. (2006) New structural insights into the bacterial type III secretion system. *Trends in Biochemical Sciences* 31, 223–230.

Zhang, L.H. and Dong, Y.H. (2004) Quorum sensing and signal interference: diverse implications. *Molecular Microbiology* 53, 1563–1571.

Zhang, L., Lin, J. and Ji, G. (2004) Membrane anchoring of the AgrD N-terminal amphipathic region is required for its processing to produce a quorum-sensing pheromone in *Staphylococcus aureus*. *Journal of Biological Chemistry* 279, 19448–19456.

Zhu, C., Yu, Z. and Sun, M. (2006) Restraining *Erwinia* virulence by expression of N-acyl homoserine lactonase gene *pro3A-aiiA* in *Bacillus thuringiensis* subsp *leesis*. *Biotechnology and Bioengineering* 95, 526–532.

3 Application of Biotechnology to Understand Pathogenesis in Nematode Plant Pathogens

M.G. Mitchum, R.S. Hussey, E.L. Davis and T.J. Baum

Abstract

Substantial progress has been made over the past decade in our understanding of nematode parasitism of plants that is now being applied to devise novel management strategies for nematodes through the use of biotechnology. The main research focus has been on the discovery and functional analysis of genes that enable nematodes to parasitize plants. Plant-parasitic nematodes are equipped with an arsenal of parasitism-associated genes which encode for secreted proteins that are expressed in their oesophageal glands. The expressed proteins are released into the apoplast or cytoplasm of host cells through the nematode stylet for the establishment of successful parasitic associations. Understanding the nature of the secreted products of nematode parasitism genes is expanding our knowledge of what makes a nematode a plant parasite. Moreover, this knowledge has unveiled key targets for the genetic manipulation of plants for developing novel resistance strategies against nematodes.

Introduction

Plant-parasitic nematodes are microscopic roundworms that cause extensive damage to most plant species and continue to pose major challenges to agriculture. Effective management has relied on the traditional approaches of natural host plant resistance, crop rotation and the use of nematicides, but each of these methods has its limitations. Natural plant resistance is available only in a few crops for a limited number of nematode species and genetic variability within nematode field populations presents a continued threat to the durability of plant resistance genes. Crop rotation is only effective against nematodes with narrow host ranges and toxicity issues associated with nematicides have resulted in their constrained use. Thus, there is a strong demand for the development of alternative nematode control strategies including ones developed using biotechnology.

Understanding the molecular mechanisms of nematode pathogenesis can aid in devising novel approaches for nematode control. Dissecting the genetic basis of nematode parasitism has been inherently difficult due to the obligate nature of these parasites, complicated genetics, limited genomic resources and associated technical difficulties in working with plant-parasitic nematodes. However, in recent years, substantial progress in our understanding of nematode pathogenesis has been made through the use of novel biotechnological approaches for gene discovery, the increasing availability of model biological species and genomic resources and advances in analyses of gene function. Because the agriculturally significant sedentary endoparasitic cyst nematodes (*Heterodera* spp. and *Globodera* spp.) and root-knot nematodes (*Meloidogyne* spp.) have been the focus of molecular studies to understand the mechanisms of nematode pathogenesis, the data given in this chapter are largely derived from these pathosystems.

The application of contemporary biotechnology techniques has contributed to significant advancements that have been made in identifying genes encoding products involved in parasitism by the cyst and root-knot nematodes over the last decade. Although the function of the majority of identified parasitism genes remains speculative at the present time, the adoption of novel reverse genetic techniques such as RNA interference (RNAi) (Fire *et al.*, 1998; Wesley *et al.*, 2001) is opening the way for the direct assessment of nematode gene function and will greatly enhance our understanding of the molecular basis of nematode pathogenesis. This chapter focuses on recent discoveries that have contributed to our understanding of nematode parasitism of plants and which have simultaneously revealed potential targets for engineering resistance. Since considerable data on the molecular response of host plants to infection by these nematodes have been generated and summarized (Gheysen and Fenoll, 2002), the emerging identification of nematode secretions that promote plant parasitism will be the emphasis in this chapter.

Nematode Pathogenesis

Plant-parasitic nematodes are obligate parasites of plants that have evolved diverse feeding strategies to obtain nutrients from host tissues to support their development and reproduction. All plant-parasitic nematodes use a hollow protrusible mouth spear, called a stylet, to pierce the plant cell wall, release oesophageal gland secretions into plant tissues and withdraw nutrients from the cytoplasm of host cells. Some parasitic nematodes feed while remaining outside the root (ectoparasites), while others penetrate completely into the root tissues (endoparasites). Nematodes either remain as migratory feeders or feed for prolonged periods from a single plant cell or group of cells. The simplest feeding strategy is that of the *migratory ectoparasites*. These nematodes migrate along the exterior of the root surface and move from cell to cell using their stylet to withdraw cellular contents. In contrast, *sedentary ectoparasites* remain exterior of the root, but feed for extended periods from the same site, usually in the cortex. *Migratory endoparasites* also feed transiently; however, these nematodes completely penetrate and feed from within the root, typically causing extensive root damage. The most complex and sophisticated feeding strategy is that of the *sedentary endoparasites*. These parasites modify specific cells of the root vasculature into highly metabolically active feeding cells to sustain their growth and development. Root-knot and cyst nematodes are sedentary endoparasites that establish feeding cells, called giant cells and syncytia, respectively (Hussey and Grundler, 1998).

Cyst and root-knot nematodes follow the same general nematode life cycle (Abad *et al.*, 2003; Lilley *et al.*, 2005). Infective second-stage juveniles hatch from eggs in the soil and find their way to host plant roots by attraction to diffusates. Using their stylets, juveniles mechanically penetrate the cell wall while secreting cell wall-hydrolysing enzymes that aid the nematode as it migrates towards the root vasculature. Once the juvenile reaches a specific root locale, usually in the vicinity of the vasculature, it selects a specific plant cell type to transform into a unique feeding structure. Concomitant with feeding, the juvenile swells and becomes sedentary as its somatic musculature degenerates and, subsequently, is completely dependent on the formation of feeding cells for nutrient acquisition and completion of its life cycle. As the nematode begins feeding, it proceeds through three more molts to the adult life stage. Cyst nematodes reproduce sexually and the males migrate out of the root to fertilize the protruding females. The adult cyst female deposits some eggs in a gelatinous matrix outside her body; however, the majority of eggs are retained inside the uterus. When the female dies, her body serves as a cyst to protect the eggs in the soil. Root-knot nematode species reproduce primarily by parthenogenesis, and adult females deposit all their eggs in an egg mass, a gelatinous matrix found on the surface of nematode-induced root galls.

The ability of the nematodes to parasitize plants is reflected by two major evolutionary adaptations: (i) the stylet, without which a nematode cannot penetrate into a plant and (ii) the products of parasitism genes expressed in the oesophageal glands and secreted through the stylet into plant tissues to facilitate migration through roots, establishment of feeding sites and feeding (Davis *et al.*, 2000). These glands are present in the oesophagus of plant-parasitic nematodes and consist of three large and complex secretory cells. Two subventral (SvG) and one dorsal (DG) oesophageal gland cell synthesize secretory proteins that are packaged into spherical membrane-bound Golgi-derived granules. Our understanding of the function of the SvG and DG in root-knot and cyst nematodes during the parasitic process has evolved and significant progress has been made in studying parasitism genes and the mechanisms of pathogenesis. In the early studies of gland morphology, because the SvG released their secretions in the oesophageal lumen at the base of the metacorpus pump chamber, which actively pumps only during food ingestion, it was generally assumed that secretions from these glands moved only posteriorly towards the intestine and, therefore, would not be secreted through the stylet into host tissues. In contrast, secretions from the DG, which connects to the oesophageal lumen near the base of the stylet, were considered to be secreted through the stylet into plant tissue and, therefore, play a significant role in the interaction of the nematode with its hosts. This point of view has changed with the identification of the first parasitism genes, which were expressed in the SvG glands and whose corresponding parasitism proteins were secreted through the stylet during migration inside the plant tissue (Smant et al., 1998; Wang et al., 1999). Other studies revealed that changes in the ultrastructure and morphology of the two types of oesophageal gland cells were correlated with developmental phases in the life cycle of the root-knot and cyst nematodes. The SvG cells are the most active oesophageal glands in infective second-stage juveniles (i.e. packed with secretory granules) and become smaller and contain fewer secretory granules in later parasitic stages. The DG, on the other hand, is stimulated to increase synthesis of secretory proteins after the onset of parasitism (penetration into host plant tissues), to become the most predominant gland in the parasitic stages. However, in root-knot nematodes, the SvG are actively expressing parasitism genes throughout the second-stage juvenile stage (10–12 days after root penetration) and in the cyst nematodes through the juvenile stages. Therefore, the roles of the two gland types clearly differ during the parasitic cycle, which involves root penetration, migration, feeding-cell induction and maintenance, and feeding-tube formation.

Proteins secreted through the stylet of root-knot and cyst nematodes are used to metabolically and developmentally reprogramme normal root cells for the formation of specialized feeding cells (Fig. 3.1). Cyst nematodes typically transform cells near the vasculature into a syncytium. The syncytium forms, by coordinated dissolution of plant cell walls, a process requiring extensive modifications to the cell wall architecture. Ultimately, protoplasts of adjacent cells coalesce to form a multinucleate syncytium made up of hundreds of cells. The nuclei within the syncytium enlarge, develop an amoeboid appearance, have a prominent nucleolus and are polyploid. The syncytium is metabolically highly active and there is an associated increase in cytoplasmic density, the large central vacuole is reduced to several smaller vacuoles, organelles proliferate and cell walls thicken (Hussey and Grundler, 1998). Root-knot nematodes, on the other hand, induce the formation of several, so-called giant-cells, around their heads. Selected cells enlarge up to 100× their size and undergo repeated nuclear divisions without cell divisions to generate a unique multinucleate cell type. Similar to syncytia, giant-cells become highly metabolically active with a dense granular cytoplasm, small vacuoles, increased numbers of organelles and thickened walls. The nuclei are polyploid from repeated rounds of endoreduplication reflecting alterations to the cell cycle. The cells surrounding the giant cells undergo hyperplasia to form a characteristic root-knot (gall).

Syncytia and giant-cells serve as major sinks for metabolites that are withdrawn by the feeding nematodes. Finger-like protuberances (ingrowths) are formed along the walls adjacent to the vasculature to increase the surface area of the plasmamembrane for solute uptake, typical of transfer cells (Jones and Northcote, 1972). The parasitized root cells are metabolically reprogrammed to support increased energy demands which are reflected in increased rates of metabolism by the glycolytic and pentose phosphate pathways (Favery et al., 1998; Mazarei et al., 2003). Feeding-cell formation is the result of these processes and is accompanied by drastic changes in plant gene expression. Alterations in plant gene expression within developing feeding cells have been studied extensively to gain insight into the molecular mechanisms underlying syncytia and giant-cell formation. The changes in plant gene expression within feeding cells have been described and summarized by Gheysen and Fenoll (2002). Despite these advances, a thorough understanding of nematode pathogenesis will require the identification of the nematode signal or signals that trigger the initiation of feeding cells. As of yet, such signals remain elusive. The following sections highlight the approaches taken to identify nematode parasitism proteins and the insights being gained with regard to their potential function in stimulating changes to basic plant cellular processes for nematode pathogenesis.

Unlocking the Nematode Toolbox through Biotechnology

Over the course of the last decade, advances in biotechnology have facilitated the identification of nematode 'parasitism' genes. Parasitism genes are defined as genes that encode secretions from a nematode that have a direct role in parasitism (Davis et al., 2004). Nematode stylet secretions have long been considered to have a direct role in parasitism by facilitating infection and provoking the observed physiological and molecular changes in host root cells for feeding-cell formation (Hussey, 1989; Williamson and Hussey, 1996). Consequently, the three oesophageal gland cells have been a focal point for research because these cells are the source of secretory products released directly into host plant tissues through the stylet. Indeed, identifying the secretory proteins produced in the gland cells is revealing key molecules involved in parasitism. The tremendous advancements made in identifying nematode parasitism gene candidates (PGCs) are the result of a combination of biotechnological approaches. Initially, the difficulties in isolating sufficient quantities of nematode secretions for direct analysis led to the development of indirect methods. Consequently, early efforts focused on the chemical stimulation of stylet secretions *in vitro* for the production of panels of monoclonal antibodies that bound specifically to secretory granules in oesophageal gland cells in immunolocalization studies (Hussey et al., 1990; Davis et al., 1994; Goverse et al., 1994). The monoclonal antibodies were then used to isolate corresponding antigens or screen cDNA expression libraries to identify the corresponding genes. This approach proved successful for cloning β-1,4-endoglucanases genes (*Eng* genes; reviewed in Davis et al., 2000), the first nematode parasitism genes identified in the SvG cells of cyst nematodes (Smant et al., 1998).

Of particular interest was that the cyst nematode endoglucanases shared greatest similarity with those of soil bacteria (Smant *et al.*, 1998), presenting some of the earliest evidence for potential horizontal gene transfer between prokaryotes and eukaryotes. Despite the successes, this MAb-based method was slow and expensive. Therefore, researchers instituted refined gene expression analysis approaches that would promote efficient parasitism gene identification. RNA fingerprinting, cDNA-amplified fragment length polymorphism (AFLP) and suppressive subtraction analyses are among the methods that have been used successfully to identify PGCs (Ding *et al.*, 1998, 2000; Lambert *et al.*, 1999; Qin *et al.*, 2000; Grenier *et al.*, 2002; Tytgat *et al.*, 2004; Blanchard *et al.*, 2005).

More recently, genomic technologies, including high-throughput sequencing and global analysis of gene expression in nematodes through cDNA library construction and generation of life stage-specific collections of expressed sequence tags (ESTs), have also been successfully exploited for the identification of PGCs (Popeijus *et al.*, 2000a; Dautova *et al.*, 2001; Vanholme *et al.*, 2005). With technological advancements, proteomic approaches for the direct analysis of stylet secretions have been employed to determine the identity of nematode-secreted proteins (de Meutter *et al.*, 2001; Jaubert *et al.*, 2002a,b). However, these approaches still suffer from several shortcomings including a lack of genome information and the fact that secretions can only be isolated *in vitro* from pre-parasitic juvenile stages. Despite the drawbacks, *in vitro* production of stylet secretions coupled with two-dimensional (2D) sodium dodecyl sulphate polyacrylamide gel electrophoresis (SDS-PAGE) and either microsequencing or mass spectrometry has successfully identified several secreted proteins. Using this approach, de Meutter *et al.* (2001) identified β-1,4-endoglucanases and an unknown protein from *Heterodera schachtii*. In addition, a gland-expressed calreticulin was identified from *Meloidogyne incognita* (Jaubert *et al.*, 2002b) that was later shown to be secreted into plant tissues (Jaubert *et al.*, 2005). However, the most successful strategy employed to date directly targeted the oesophageal gland cells for the identification of PGCs (Fig. 3.2). In this approach, which directly analyses the transcriptome of the oesophageal gland cells, microaspiration of the cytoplasm of the gland cells was coupled with cDNA library construction. The gland-enriched cDNA libraries were sequenced and data mining tools were used to identify predicted secretion signal peptides, the presence of which suggests that the gene products may be secreted. Gland-specific expression was confirmed by *in situ* mRNA hybridization which led to identification of more than 60 PGCs for the cyst nematode, *Heterodera glycines* (Gao *et al.*, 2001a, 2003; Wang *et al.*, 2001) and at least 48 candidates for the root-knot nematode, *M. incognita* (Huang *et al.*, 2003, 2004).

A current list of known and candidate nematode parasitism genes with predicted functions that are expressed specifically within the oesophageal gland cells of *Meloidogyne* spp., *Heterodera* spp. and *Globodera* spp. is presented in Table 3.1. Interestingly, more than 70% of the PGCs identified to date do not have any homology to sequences in existing databases and have been termed 'pioneers'. A list of 'pioneer' PGCs can be found in

Table 3.1. Cyst and root-knot nematode parasitism genes with predicted functions[a].

Parasitism gene	Species	Gland cell	References
Cell wall-modifying proteins			
Endo-1,4-β-glucanase			
Gr-Eng-1; Gr-Eng-2	Globodera rostochiensis	SvG	Smant et al. (1998)
Gt-Eng-1; Gt-Eng-2	Globodera tabacum	SvG	Goellner et al. (2000)
Hg-Eng-1; Hg-Eng-2; Hg-Eng-3; Hg-Eng-4; Hg-Eng-5; Hg-Eng-6	Heterodera glycines	SvG	Smant et al. (1998); Yan et al. (2001); Gao et al. (2004a)
Hs-Eng-1; Hs-Eng-2	Heterodera schachtii	nd	de Meutter et al. (2001); Vanholme et al. (2005)
Mi-Eng-1; Mi-Eng-2 (5A12B); Mi-Eng-3 (8E08B)	Meloidogyne incognita	SvG	Rosso et al. (1999); Huang et al. (2003)
Cellulose-binding domain protein			
Mi-Cbp-1	M. incognita	SvG	Ding et al. (1998)
Hg-Cbp-1	H. glycines	SvG	Gao et al. (2004b)
Hs-Cbp	H. schachtii	nd	Vanholme et al. (2005a)
Pectate lyase			
Gr-Pel-1	G. rostochiensis	SvG	Popeijus et al. (2000b)
Hg-Pel-1	H. glycines	SvG	de Boer et al. (2002a)
Hs-Pel	H. schachtii	nd	Vanholme et al. (2005)
Mj-Pel-1	Meloidogyne javanica	SvG	Doyle and Lambert (2002)
Mi-Pel-1; Mi-Pel-2 (2B02B)	M. incognita	SvG	Huang et al. (2003); Huang et al. (2005a)
Polygalacturonase			
Mi-Pg-1	M. incognita	SvG	Jaubert et al. (2002a)
Expansin			
Gr-Expβ-1	G. rostochiensis	SvG	Qin et al. (2004); Kudla et al. (2005)
Endo-1,4-β-xylanase			
Mi-Xyl-1	M. incognita	SvG	Mitreva-Dautova et al. (2006)
Arabinogalactan endo-1,4-β-galactosidase			
Hs-Gal-1	H. schachtii	nd	Vanholme et al. (2005)
Annexin			
Hg-Ann-1	H. glycines	DG	Gao et al. (2003)
Chitinase			
Hg-Chi-1	H. glycines	SvG	Gao et al. (2002b)
Hs-Chi-1	H. schachtii	nd	Vanholme et al. (2005)
Chorismate mutase			
Mj-Cm-1	M. javanica	SvG	Lambert et al. (1999)
Hg-Cm-1; Hg-Cm-2	H. glycines	DG	Gao et al. (2003); Bekal et al. (2003)
Gp-Cm-1	Globodera pallida	SvG	Jones et al. (2003)
Mi-Cm-1; Mi-Cm-2	M. incognita	SvG	Huang et al. (2005b)

Continued

Table 3.1. *Continued*

Parasitism gene	Species	Gland cell	References
Signalling peptide			
CLAVATA3/ESR-like (CLE)			
Hg-Cle-1 (4G12); Hg-Cle-2 (SYV46;2B10)	H. glycines	DG	Wang et al. (2001); Gao et al. (2003)
Hs-Cle-1; Hs-Cle-2	H. schachtii	DG	Wang et al. (2006)
Gr-Cle-1; Gr-Cle-4	G. rostochiensis	DG	Lu and Wang (2006)
Other			
Mi-16D10	M. incognita	SvG	Huang et al. (2006a)
Cell cycle regulation			
Ran-binding protein			
Gr-Rbp-1	G. rostochiensis	DG	Qin et al. (2000); Rehman et al. (2006)
Gr-Rbp-2	G. rostochiensis	DG	Qin et al. (2000); Rehman et al. (2006)
Gr-Rbp-3	G. rostochiensis	DG	Qin et al. (2000); Rehman et al. (2006)
Gp-Rbp-1	G. pallida	DG	Blanchard et al. (2005)
Gm-Rbp-1	Globodera mexicana	DG	Blanchard et al. (2005)
Ubiquitination pathway components			
Ubiquitin extension protein			
Hg-Ubi-1; Hg-Ubi-2	H. glycines	DG	Gao et al. (2003)
Hs-Ubi-1; Hs-Ubi-2	H. schachtii	DG	Tytgat et al. (2004)
Other			
Hg-Skp-1	H. glycines	DG	Gao et al. (2003)
Hg-Ring-H2	H. glycines	DG	Gao et al. (2003)
Elcitor-like protein			
Small cysteine-rich proteins			
Hg-4E02	H. glycines	SvG	Gao et al. (2003)
Avirulence protein			
Gp-Rbp-1	G. pallida	DG	Moffett and Sacco (2006)
Venom allergen-like protein			
Hg-Vap-1; Hg-Vap-2	H. glycines	SvG	Gao et al. (2001b)
Hs-Vap-1; Hs-Vap-2	H. schachtii	nd	Vanholme et al. (2005)
Mi-Vap-1 (Mi-Msp-1)	M. incognita	SvG	Ding et al. (2000)
Calreticulin			
Mi-Crt	M. incognita	SvG; DG	Jaubert et al. (2005)
Phosphatase			
Acid phosphatase			
Mi-Ap	M. incognita	SvG	Huang et al. (2003)

ªAll genes encoded a predicted secretion signal peptide.
nd = not determined.

several primary articles (Gao *et al.*, 2001a, 2003; Wang *et al.*, 2001; Huang *et al.*, 2003, 2004; Vanholme *et al.*, 2005). These data revealed that a large number of cell wall-modifying enzymes are produced and secreted from the SvG cells (Smant *et al.*, 1998; Rosso *et al.*, 1999; de Boer *et al.*, 2002;

Gao et al., 2004; Qin et al., 2004; Mitreva-Dautova et al., 2006). This is consistent with the synthesis and accumulation of secretory granules in the SvG during the penetration and migration phase of the nematode life cycle. Both SvG and DG are active during the initiation of feeding cells, and subsequently, the SvG begin to decline in activity as feeding cells mature while the single DG cell becomes highly active. Thus, proteins secreted by the SvG may also play important roles in early feeding-cell induction. Using several of the aforementioned approaches, a large number of PGCs expressed in the DG cell have now been identified (Qin et al., 2000; Wang et al., 2001; Gao et al., 2003; Huang et al., 2003, 2004) and likely play important roles in feeding-cell induction, function, maintenance and the formation of feeding tubes. Interestingly, cyst and root-knot nematodes have many unique components in their toolboxes which may be reflective of the ontological differences between syncytia and giant-cells. Similarly, very few genes are expressed in both types of gland cells, supporting earlier speculations that secretory proteins produced in each cell type likely have distinct roles during parasitism. The following sections summarize the insights we are gaining in our understanding of how nematodes use the secreted products of parasitism genes to manipulate various aspects of plant cell biology. Recent advances in the identification of small molecules and virulence genes in plant-parasitic nematodes are also discussed.

Parasitism Genes

Cell wall modification

Sedentary endoparasitic cyst and root-knot nematodes must penetrate through root epidermal cells and migrate in the cortex towards the vasculature where they establish permanent feeding sites. One of the first obstacles that these parasites must overcome is the structural barrier of the plant cell wall. Cyst and root-knot nematodes have evolved sophisticated mechanisms to breach this structural barrier, including a stylet for mechanical penetration and, as mentioned above, the ability to secrete a battery of cell wall-modifying proteins (CWMPs; Table 3.1) to break down cell wall components and facilitate penetration and migration through host root tissues. Genes encoding β-1,4-endoglucanases (ENGs; EGases) and pectate lyases have been cloned from both groups of nematodes (Smant et al., 1998; Yan et al., 1998; Rosso et al., 1999; Bera-Maillet et al., 2000; Goellner et al., 2000; Popeijus et al., 2000b; Yan et al., 2001; de Boer et al., 2002; Doyle and Lambert, 2002; Gao et al., 2002a,b; Gao et al., 2004; Huang et al., 2005a,b; or see Baum et al., 2007 for review) and their secretion has been detected along the migratory path of juveniles invading host roots (Wang et al., 1999; Goellner et al., 2001; Doyle and Lambert, 2002). Expression of cyst nematode *Eng* genes declines in parasitic juvenile stages as the nematode establishes a feeding site suggesting that any potential role in feeding-cell formation may be minimal, if any (de Boer *et al.*, 1999; Goellner *et al.*, 2000, 2001; Gao *et al.*, 2004). Consistent with a role in facilitating nematode migration through root tissue, the *Eng* genes are expressed in adult male cyst nematodes that regain mobility for migration out of the root to fertilize the females (de Boer *et al.*, 1999; Goellner *et al.*, 2000). *Eng-1* is also expressed in the rectal glands of adult female root-knot nematodes which may reflect a requirement of CWMPs to assist in cell wall loosening in cells surrounding the expanding female as it develops (Rosso *et al.*, 1999). The production of cellulolytic enzymes appears to be a common mechanism for parasitism by plant-parasitic nematodes, and suggests that their potential acquisition from soil microbes (Davis *et al.*, 2000) was a pivotal event in adaptations of nematodes for plant parasitism. Two β-1,4-endoglucanase gene sequences have also been isolated from the migratory endoparasitic root lesion nematode, *Pratylenchus penetrans* (Uehara *et al.*, 2001). Similarly, a family of glycosyl hydrolase 45 cellulase gene sequences was cloned from the pine wilt nematode, *Bursaphelenchus xylophilus* (Kikuchi *et al.*, 2004). Gene sequences coding for secreted cellulose-binding domain proteins have also been identified in both cyst and root-knot nematodes (Ding *et al.*, 1998; Gao *et al.*, 2004b; Vanholme *et al.*, 2005). The role of cellulose-binding protein in plants is unknown; however, recombinant cellulose-binding protein has been shown to modulate plant cell elongation *in vitro* and overexpression of a bacterial carbohydrate-binding module family III *in planta* can stimulate plant growth (Shpigel *et al.*, 1998; Safra-Dassa *et al.*, 2006). Root-knot nematodes have also been shown to secrete polygalacturonases from the SvG cells and, thus, appear to be equipped with a mixture of pectin-degrading enzymes to hydrolyse the middle lamella during intercellular migration (Jaubert *et al.*, 2002a). Recently, two additional classes of CWMPs were identified in plant-parasitic nematodes. An expansin gene was cloned from the potato cyst nematode and an endo-1,4-β-xylanase was cloned from root-knot nematode, both were shown to be expressed within the SvG (Qin *et al.*, 2004; Kudla *et al.*, 2005; Mitreva-Dautova *et al.*, 2006). DNA blot analysis and searches of EST databases suggest that potato cyst nematode expansins belong to a multigene family and are present in other cyst nematodes. Similarly, database searches revealed xylanase homologues in *Meloidogyne javanica*, *Meloidogyne arenaria* and *Meloidogyne chitwoodi* but not *Meloidogyne hapla* or cyst nematodes (Mitreva-Dautova *et al.*, 2006). Endo-1,4-β-xylanases function as bacterial and fungal virulence factors on monocots, and the same has been postulated for root-knot nematode xylanases (Mitreva-Dautova *et al.*, 2006). An arabinogalactan endo-1,4-β-galactosidase was also identified recently from *H. schachtii* and *H. glycines* ESTs (Vanholme *et al.*, 2005). Although nematodes clearly secrete a complex mixture of CWMPs to facilitate penetration and migration through host root tissues, there is very little evidence to support a role for these proteins in the induction and formation of syncytia and giant cells. On the contrary, mounting molecular evidence suggests that the CWMPs involved in the extensive wall modifications within feeding cells are of plant origin (Goellner *et al.*, 2001).

Metabolic reprogramming

Plant-parasitic nematodes appear to have direct control over redirecting the metabolic activity of feeding cells. Cyst and root-knot nematodes secrete chorismate mutase (Popeijus et al., 2000a; Bekal et al., 2003; Doyle and Lambert, 2003; Gao et al., 2003; Jones et al., 2003; Huang et al., 2005b; Lu and Wang, 2006), an enzyme of the shikimate pathway, directly into the cytoplasm of plant cells, thereby potentially altering the regulation of this pathway for the benefit of the parasite. The shikimate pathway produces essential amino acids required by the nematode that can only be obtained from their diet. In the shikimate pathway, the products of glycolysis and pentose phosphate pathway are converted to chorismate, a branch-point metabolite for the production of aromatic amino acids. Chorismate is produced in the plastid where it is then converted by chorismate mutase to prephenate to provide precursors for the synthesis of the aromatic amino acids such as phenylalanine, tryrosine and tryptophan. Tryptophan serves as the precursor of indole-3-acetic acid (IAA) and phenylalanine is a precursor for the production of flavonoids, salicyclic acid and phytoalexins, each of which has established roles in plant–microbe interactions. For example, salicyclic acid has been shown to play a role in mediating resistance responses to root-knot nematodes (Branch et al., 2004). Chorismate is also utilized in the cytosol and it is hypothesized that increases in cytosolic chorismate mutase produced by the nematode may increase the flow through the cytosolic branch of the shikimate pathway, thereby decreasing the biosynthesis of plastid-derived phenolics. Overexpression of a nematode chorismate mutase gene in roots caused aborted lateral roots and impaired vasculature development that could be rescued by exogenous IAA, supporting decreased auxin levels in the roots (Doyle and Lambert, 2003). An early increase in auxin concentrations within feeding cells that declines by 96 h post-infection has been suggested using auxin-responsive promoter elements (Hutangura et al., 1999; Karczmarek et al., 2004). Similarly, the induction and morphogenesis of syncytia is impaired on polar auxin transport mutants, suggesting that an auxin balance is important for feeding-cell formation. Increased flux through the cytosolic branch of the shikimate pathway may be one strategy the nematode uses to decrease accumulation of plastid-derived phenolic compounds known to mediate plant defence responses, thereby suppressing plant defence. Additional studies measuring the metabolite concentrations in roots overexpressing nematode chorismate mutase will be needed to determine exactly how nematodes may be contributing to the metabolic reprogramming of host plant cells leading to a compatible interaction.

Secreted signalling peptides

Nematodes also have evolved mimics of plant signals to reprogramme host cell development for the successful establishment of parasitic associations with their hosts. The complex process of dedifferentiating plant cells into feeding sites likely requires an exchange of signals between the nematode and recipient host cells. Insight into the intriguing question of how plant-parasitic nematodes induce this host response and the 'putative' signal or signals that are exchanged is beginning to emerge. Secreted peptide signals have been identified as potential candidates for mediating the signal exchange between nematodes and plants. Goverse *et al.* (1999) demonstrated that naturally induced secretions of potato cyst nematode can co-stimulate the proliferation of tobacco leaf protoplasts and human peripheral blood mononuclear cells. The unidentified active component of the secretions was shown to be <3 kD in size and susceptible to pronase degradation, thus implicating small, secreted, mitogenic oligopeptides as prime candidates (Goverse *et al.*, 1999). Since then, several different classes of signalling peptides have been identified from plant-parasitic nematodes (Wang *et al.*, 2001, 2005; Huang *et al.*, 2006a; van Bers *et al.*, 2006).

The first nematode gene encoding a signalling peptide was identified from *H. glycines* (Wang *et al.*, 2001) and contained a C-terminal domain with similarity to plant CLAVATA3/ESR-related (CLE) peptides (Olsen and Skriver, 2003). Significantly, this was the first identification of a CLE domain outside the plant kingdom, and the uniqueness of the full-length protein suggested that cyst nematodes may have independently evolved a domain that could developmentally reprogramme host plant cells through functional ligand mimicry. *CLE* genes have been identified from a number of different monocotyledenous and dicotyledenous plant species, including rice, wheat, soybean, maize and Arabidopsis (Cock and McCormick, 2001). In recent years, their role in peptide signalling in plant biology has been shown to be increasingly important in regulating various aspects of plant development (Lindsey *et al.*, 2002). All of the plant *CLE* genes encode small peptides with a secretion signal peptide, indicating an extracellular function and maintain a conserved C-terminal CLE motif. One of the best studied plant CLEs is CLV3 which has been shown to play an integral role in the maintenance of stem cells in the shoot and floral meristems of Arabidopsis. CLV3 belongs to a 31-member (CLE) gene family in Arabidopsis (Sharma *et al.*, 2003; Strabala *et al.*, 2006) and functions as a ligand for the CLV1/CLV2 receptor-like kinase complex. It is likely that nematode CLE peptides function as secreted ligand mimics of natural plant signalling peptides to co-opt developmental programmes for the dedifferentiation of host plant cells into syncytia (Wang *et al.*, 2005). Nematode CLEs are expressed in the DG cell of feeding life stages of *H. glycines* (Wang *et al.*, 2001) when the nematode is actively initiating or feeding from syncytia and they are absent from non-feeding adult males (M.G. Mitchum, 2006), which supports a role for CLEs in syncytium induction, function or maintenance. Moreover, constitutive overexpression of the *H. glycines* CLE gene in Arabidopsis mimics CLV3 overexpression (Wang *et al.*, 2005) causing a characteristic *wuschel* (*wus*) phenotype (Brand *et al.*, 2000; Hobe *et al.*, 2003).

Consistently, the *H. glycines* CLE could complement the *clv3* mutant (Wang et al., 2005). These preliminary studies suggest that the nematode CLE peptide may have functional similarity to the plant CLV3 peptide. Several Arabidopsis *CLE* gene family members have been shown to complement CLV3, which indicates a certain level of functional redundancy among CLE family members. This, along with the fact that the function and receptor partners of other plant *CLE* gene family members have not yet been determined, makes for an exciting future challenge to discover nematode CLE receptors in plants to determine host regulatory pathways co-opted by nematodes to facilitate parasitism. CLE genes have also been cloned from other cyst nematodes, including *H. schachtii* and *Globodera rostochiensis* (Lu and Wang, 2006; Wang et al., 2006) suggesting that ligand mimicry may be a conserved molecular tool for cyst nematode parasitism.

A signalling peptide encoded by an SvG parasitism gene in *M. incognita* was shown to directly interact with an intracellular plant regulatory protein (Huang et al., 2006a). The gene referred to as *16D10* encodes a small novel secreted peptide of 13 amino acids, including a 30 amino acid secretion signal peptide and is conserved in *Meloidogyne* spp. (Huang et al., 2003, 2006a). Despite the fact that the16D10 peptide has 6 of the 14 amino acids conserved with the plant CLV3 motif, it could not complement the *clv3* mutant (Huang et al., 2006a). Overexpressing *16D10* in Arabidopsis significantly stimulated root growth with normal differentiation and without any above-ground phenotypes (Huang et al., 2006a). Interestingly, in a yeast two-hybrid screen for 16D10-interacting proteins it was shown to bind to the SAW domain of two Arabidopsis SCARECROW-like (SCL) transcription factors, AtSCL6 and AtSCL21 (Huang et al., 2006a). SCL transcription factors belong to the GRAS protein family, members of which play important roles in plant development and signalling (Bolle, 2004). Although the function of AtSCL6 and AtSCL21 is unknown, homologues of AtSCL6 function in mitotic cell cycle and cell cycle control, and AtSCL21 homologues are involved in phytochrome signalling. Because the conserved root-knot nematode-secreted signalling peptide is strongly expressed in the SvG cells of J2 at the time when the giant-cells are being developed and the peptide binds to specific plant transcription factor proteins, *16D10* has been postulated to have a role in the reprogramming of gene expression required for giant-cell formation. Importantly, an essential role for *16D10* in root-knot nematode parasitism was recently confirmed by *in planta* delivery of dsRNA for RNAi of *16D10* (Huang et al., 2006b).

A third class of genes, the *dgl1* (dorsal gland-specific 1) gene family, encoding secreted signalling peptides was recently identified from the potato cyst nematode, *G. rostochiensis* (van Bers et al., 2006). At least six members encode for small transcripts (353–573 bp) that share conserved 3'UTR and secretion signal peptide sequences, but exhibit a high level of sequence divergence in the mature peptides. Overexpression of *dgl1* members in potato hairy roots has identified one member that dramatically increases the formation of lateral roots. The fact that DGL1 family members maintain a positive charge and high predicted protein-binding potential suggests that DGL1s may also function as ligands for plant receptors to redirect plant signalling pathways for parasitism (van Bers *et al.*, 2006).

Cell cycle manipulation

One of the earliest responses in feeding-cell formation is the observed reactivation of the cell cycle (Goverse *et al.*, 2000). Nuclei of cells incorporated into syncytia take on an enlarged and amoeboid appearance and molecular studies suggest that cell cycle activation is important for feeding-cell formation (Engler *et al.*, 1999). These nuclei undergo repeated rounds of endoreduplication, leading to polyploidy without mitosis. In contrast, giant-cell nuclei are polyploid and repeatedly divide in the absence of cell division to give rise to multinucleate feeding cells. The trigger that stimulates feeding-cell nuclei to re-enter into aberrant cell cycles has not been identified; however, nematode-secreted products are likely candidates. As mentioned above, low-molecular weight mitogenic peptides have been putatively identified in nematode secretions (Goverse *et al.*, 1999) and may function in this capacity. Additional candidates include secreted homologues of ran-binding proteins in the microtubule-organizing centre (RanBPMs) which are expressed in the DG cell of cyst nematodes (Qin *et al.*, 2000; Blanchard *et al.*, 2005). RanBPMs have been shown to cause microtubule nucleation in overexpression studies (Nakamura *et al.*, 1998). Thus, one possibility is that nematode-secreted RanBPMs function in feeding-cell formation by stabilizing the microtubule network, resulting in the observed shunting of the M-phase in syncytia (reviewed in Davis *et al.*, 2004). Further functional investigations will be required to correlate a role for RanBPMs with cell cycle modulation in feeding cells. Interestingly, there is also recent evidence to suggest that nematode RanBPMs may function as avirulence determinants (see below).

Targeted protein degradation

Targeted protein degradation by the ubiquitination pathway is an important mechanism for post-translational gene regulation in eukaryotes. Ubiquitination-mediated protein degradation has been shown to play a role in regulating defence responses in plants and more recent evidence indicates that pathogens can deliver effectors that can mimic the function of components of host E3 ubiquitin ligase complexes to interfere with plant defence responses (Zeng *et al.*, 2006). Studies on nematode parasitism genes suggest that this may also be a mechanism for nematode parasitism. Candidate parasitism genes encoding secreted proteins with homology to components of the ubiquitination pathway including ubiquitin extension,

S-phase kinase-associated protein 1 (SKP-1)-like, and RING-H2 proteins have been identified from cyst nematodes (Gao et al., 2003). The presence of a signal peptide on these proteins is unfounded in plants. Ubiquitin extension proteins contain an ubiquitin monomer normally fused to a C-terminal extension peptide that is cleaved by ubiquitin C-terminal hydrolases and incorporated into ribosomes. The function of the monoubiquitin domain is less clear although it may play a role in targeting proteins for degradation. In nematode ubiquitin extension proteins, the signal peptide is followed by a monoubiquitin domain and a novel C-terminal extension peptide (22 amino acids for *H. schachtii*; 19 amino acids for *H. glycines*) (Gao et al., 2003; Tytgat et al., 2004). A nuclear localization signal also targets the nematode extension peptide to the nucleolus in tobacco BY-2 cells (Tytgat et al., 2004). Nematode ubiquitin extension proteins are only expressed in the DG cell of feeding life stages. Detection of these proteins along the glandular extensions provides strong evidence for secretion and is suggestive of a potential role in feeding-cell formation (Tytgat et al., 2004). SKP-1 and RING-H2 proteins are components of E3 ubiquitin ligase complexes involved in targeting proteins for ubiquitination and subsequent degradation by the proteosome to regulate a wide range of signalling pathways in plants. To date, nematode-secreted components of the ubiquitination pathway have not been identified from root-knot nematodes. It is tempting to speculate that plant-parasitic cyst nematodes have evolved the unique ability to manipulate the ubiquitination pathway of host cells for parasitism. However, another possibility to explore is that the novel C-terminal extension peptide of the cyst nematode ubiquitins might function as a signalling peptide when cleaved from the ubiquitin domain in the parasitized host cell.

Unclassified parasitism gene candidates

A subset of PGCs was identified, which encode proteins with homology to proteins of other nematode species, yet their function in plant–nematode interactions remains obscure. For example, several PGCs encoding proteins similar to the venom allergen antigen 5 family found in hymenoptera insects and other nematode species, including *Caenorhabditis elegans*, *Ancylostoma caninum*, *Dirofilaria immitis*, *Onchocerca volvulus* and *Brugia malayi*, have been identified from *Meloidogyne* spp. and *H. glycines* (Ding et al., 2000; Gao et al., 2001b). These genes are expressed during the transition to parasitism in the animal parasitic nematode *Ancylostoma*, although their function is unknown (Hawdon et al., 1999). The expression of these genes in the SvG cells of the infective stage of plant-parasitic nematodes suggests a role for their secretory products in the early stages of plant parasitism. The venom allergen proteins have some similarity (25%) to pathogenesis-related proteins and cysteine-rich secretory proteins (CRISP), but the biological relevance of this finding is unclear at this time.

Roles of Subventral Oesophageal Glands and Dorsal Gland in Plant Parasitism

The cloning and expression analyses of parasitism genes expressed in the SvG and DG cells are now enabling us to assign different roles for the two types of oesophageal glands in parasitism and clearly show that secretions of both types of glands are essential for nematode parasitism of plants. Only the SvG cells express parasitism genes encoding cell wall-digesting enzymes, which are used by nematodes during penetration and migration within roots (Table 3.1). However, the role of the SvG cells in plant parasitism is not limited to facilitating migration by the nematodes as a large number of secretory proteins synthesized in these glands have no clear role in second-stage juvenile migration in roots (Gao et al., 2003; Huang et al., 2003). Furthermore, chorismate mutase, an enzyme involved in aromatic amino acid synthesis, is also produced in these glands (Table 3.1). In addition, a novel root-knot nematode-secreted parasitism peptide synthesized in the SvG cells has been shown to function as a signalling molecule by specifically targeting a host plant regulatory protein (Huang et al., 2006a). These discoveries provide convincing evidence that secretions from the SvG have functions other than nematode migration and, in fact, may mediate early signalling events in plant–nematode interactions including a possible role in feeding-cell induction. The DG cell is the principal functional gland in the adult females of root-knot and cyst nematodes and the majority of the parasitism genes that have been discovered are expressed in this gland (Gao et al., 2003; Huang et al., 2003). In the potato cyst nematode, however, the DG becomes active in second-stage juveniles in the egg after stimulation by potato root diffusate and is packed with secretory granules in the hatched infective second-stage juveniles (Blair et al., 1999 and references therein). This may indicate that the DG in the potato cyst nematode has a more prominent role in feeding-cell initiation. The function of the products of the DG parasitism genes in feeding-cell development, maintenance and feeding-tube formation has not been resolved. However, the large number of parasitism genes expressed in the DG suggests that feeding-cell maintenance is a complex process.

Symbiont Mimics

Gross phenotypic similarities between galls induced by *Meloidogyne* spp. and *Rhizobium*-induced nodules have drawn considerable interest that has led to comparative studies between these two processes (reviewed in Davis and Mitchum, 2005). Nod factors (NF), which comprise a family of lipo-chito-oligosaccharide molecules specific to *Rhizobium* spp., have been identified as the bacteria-derived signalling molecules that are required to induce nodule formation (Riely et al., 2004). Analogous molecules have not been identified from plant-parasitic nematodes, although recent evidence suggests that molecular mimics of these may exist. In

Lotus japonicus, it was recently shown that root-knot nematodes and bacterial NFs can elicit common signal transduction events (Weerasinghe *et al.*, 2005). Cytoskeletal responses induced in root hairs in response to root-knot nematodes were identical to those observed in response to NF. Moreover, the early host responses of root hair waviness and branching that precede rhizobial infection were also observed in response to root-knot nematodes. The use of perfusion chambers prevented any physical contact between the nematodes and root hairs suggesting that the signal, referred to as NemF, can function at a distance. Consistently, nematodes treated with sodium azide did not induce any host responses. The inability of root-knot nematodes to reproduce on *Lotus* plants with mutations in the NF receptor genes *nfr1*, *nfr5* and *symRK* suggests that the nematode-derived signal may be functionally equivalent to NF (Weerasinghe *et al.*, 2005). Interestingly, one of the several horizontal gene transfer candidates identified from *Meloidogyne* ESTs has similarity to the *NodL* gene that encodes an *N*-acetyltransferase in the biosynthetic pathway of NF in *Rhizobium* spp. (Scholl *et al.*, 2003). The isolation and characterization of NemF presents a future challenge to determine if components of root-knot nematode pathogenesis may have evolved by conscription of symbiotic pathways (Weerasinghe *et al.*, 2005).

Virulence Genes

The molecular nature of nematode virulence continues to elude researchers yet the ability of nematodes to either evade or overcome host plant resistance is a widespread problem for management practices that rely on natural host plant resistance. In certain cases, the interactions between nematodes and plants on a genetic level have been shown to follow the gene-for-gene hypothesis. This has been demonstrated for the potato cyst nematode (*G. rostochiensis*) interaction with potato carrying the *H1* resistance gene (Janssen *et al.*, 1991). Consistently, several canonical plant resistance genes that confer resistance to plant-parasitic cyst and root-knot nematodes have been cloned and characterized to date (Milligan *et al.*, 1998; van der Vossen *et al.*, 2000; Ernst *et al.*, 2002; Paal *et al.*, 2004). In contrast, the identification of the corresponding avirulence genes from nematodes has not been trivial. The difficulty in using genetic approaches to map genes for (a)virulence and the inability to transform plant-parasitic nematodes has slowed progress in this area compared with other plant pathogens. Using classical genetic analysis, several dominant and recessive *H. glycines ror* (for reproduction on a resistant host) genes were identified in pure lines that carried fixed genes for parasitic ability on soybean (Dong and Opperman, 1997; Dong *et al.*, 2005). However, the genes controlling *H. glycines* parasitism on resistant soybean at these loci were not determined. *M. incognita* reproduces exclusively by mitotic parthenogenesis, making classical genetic analysis impossible. Alternative strategies to identify potential (a)virulence candidates in *M. incognita* have included comparative analyses between near-isogenic lines differing only in their ability to parasitize resistant plants. Using an AFLP approach, Semblat et al. (2001) identified the *Map-1* gene by comparing near-isogenic *M. incognita* lines that were avirulent or virulent on tomato containing the *Mi* resistance gene. *Map-1* encodes an unknown protein that is found in nematode amphidial secretions. However, its function in avirulence remains to be shown. cDNA-AFLP approaches have also identified several sequences that are differentially expressed between avirulent and virulent near-isogenic lines of *M. incognita* (Semblat et al., 2001; Neveu et al., 2003). The majority of the differentially expressed sequences correspond to pioneer proteins, which provides no indication of function. Thus, the adoption of functional analyses such as RNAi (see below) will be required to demonstrate a role for these genes in (a)virulence. As genetic mapping strategies are developed (see below), this too should help lead to the identification of (a)virulence candidates in nematodes.

Recent hypotheses suggest that variants of plant-pathogen effector molecules with a central role in virulence may function as avirulence factors depending on the genetic context of the host plant (Birch et al., 2006). Thus, effector molecules encoded by nematode parasitism genes are strong avirulence gene candidates. Small cysteine-rich proteins (<150 amino acids) are among the gland-expressed PGCs (Gao et al., 2003). Effector molecules with an even number of cysteine residues have been identified in bacterial, fungal and oomycete pathogens of plants and shown to function in defence induction and as avirulence factors (Birch et al., 2006). The *H. glycines* chorismate mutase gene (*Hg-cm-1*) is also a strong candidate virulence gene. Polymorphisms in *Hg-cm-1* were identified that correlate with virulence on a set of soybean cyst nematode-resistant soybean lines (Bekal et al., 2003) and the *Hg-cm-1A* allele was preferentially selected on the germplasm PI88788, the most common source of soybean cyst nematode resistance (Lambert et al., 2005). Similarly, polymorphisms have been identified in the *HgCLE* genes among soybean cyst nematode inbred populations that differ in virulence on resistant soybean implicating their role in virulence (J. Wang and M.G. Mitchum, 2005). Moreover, in a recent study to identify interacting proteins of the potato *Rx* resistance gene (which confers resistance to potato virus X), an RanGAP protein was identified that could also interact with the GPA2 resistance gene product. Rx is very similar to GPA2, an NBS–LRR protein that confers resistance to the potato cyst nematode, *G. pallida*. As mentioned above, among the encoded products of nematode parasitism genes are RanBPMs. Transient expression of GPA2 and a *G. pallida* RanBPM in tobacco leaf tissue resulted in a strong hypersensitive response, suggesting that the nematode RanBPM may be the corresponding avirulence factor for GPA2 (Moffett and Sacco, 2006). Future work to elucidate the molecular mechanisms of nematode virulence promises to significantly advance our understanding of nematode–plant interactions and provide essential knowledge for both the development and deployment of nematode-resistant plants.

Nematode Genomics

Combined with the extensive genetic and genomic analyses of *C. elegans* (Lee *et al.*, 2004), the increase in genomic analyses of parasitic nematodes in recent years promises to provide an unprecedented understanding of nematode biology and pathogenesis. Single pass sequencing of random clones from cDNA libraries to generate ESTs from different nematode life stages has become a powerful tool for identifying genes important in nematode–host interactions. This approach has contributed more than 400,000 publicly available parasitic nematode-expressed sequences to databases and is enabling the identification of nematode-specific gene families (Popeijus *et al.*, 2000a,b; Dautova *et al.*, 2001; Parkinson *et al.*, 2003, 2004). The large amount of parasitic nematode EST data has been coupled with bioinformatic pipelines for secretion signal peptide detection to identify additional nematode PGCs and distinguish those that may have been acquired via horizontal gene transfer (McCarter *et al.*, 2003; Scholl *et al.*, 2003; Vanholme *et al.*, 2005). Large-scale EST data sets have also been applied to develop microarray platforms for comprehensive profiling of nematode gene expression changes during parasitism (Ithal *et al.*, 2006). This approach has identified coordinated regulation of *H. glycines* parasitism genes and additional novel PGCs (Ithal *et al.*, 2006). To elucidate nematode genome organization, plant-parasitic nematode genome sequencing projects for both root-knot (*M. hapla* and *M. incognita*) and cyst (*H. glycines*) nematodes are currently underway (D. Bird, P. Abad and K. Lambert, North Carolina, 2006, personal communication) and upon completion will provide genome-wide catalogues of nematode PGCs for both functional and comparative analysis that should reveal additional novel insight into mechanisms of nematode pathogenesis. As genome sequences of plant-parasitic nematodes become available, gene structure, organization and function can be compared across different genomes to facilitate evolutionary analyses within the phylum Nematoda (Mitreva *et al.*, 2005).

Genetic Analysis of Parasitism

The use of forward genetic strategies to add to our understanding of nematode parasitism and virulence has been complicated by the obligate nature of plant-parasitic nematodes, parthenogenic mode of reproduction of some species and a lack of tools for genetic mapping. Consequently, this approach has only been used in a few studies as mentioned earlier. More recently, however, several groups have initiated the development of genetically homogeneous nematode populations, molecular markers and genetic maps to facilitate map-based cloning efforts in both cyst and root-knot nematodes (Liu and Williamson, 2004; Atibalentja *et al.*, 2005). A genetic linkage map for *H. glycines* has been constructed and its utility was demonstrated by mapping the *Hg-Cm-1* gene (Atibalentja *et al.*, 2005). Of the four most damaging *Meloidogyne* spp., *M. hapla* has been found to reproduce by facultative meiotic parthenogenesis (Triantaphyllou, 1985; Van der Beek *et al.*, 1998) such that both selfed and outcrossed progeny can be generated for classical genetic studies and, therefore, *M. hapla* has been adopted as a model system. In addition to the *M. hapla* genome sequencing project, a genetic map of *M. hapla* is under construction and F2 mapping populations have been generated (Liu and Williamson, 2004). With the development of these new tools, researchers will soon be equipped with the ability to identify the genes controlling various traits such as nematode virulence and host range to add to our current understanding of nematode pathogenesis.

Reverse Genetic Strategies

The application of large-scale genome-wide data on plant-parasitic nematodes to an understanding of nematode pathogenesis will require the functional analysis of gene products. In the past years, the lack of reverse genetic approaches in plant-parasitic nematodes has created a major bottleneck in interpreting large-scale functional genomic data. RNAi through post-transcriptional gene silencing can be achieved by introducing double-stranded RNA (dsRNA) complementary to the MRNA of a gene of interest (Fire *et al.*, 1998). This mechanism was first demonstrated in *C. elegans* (Fire *et al.*, 1998) and is now used in a wide variety of eukaryotes to assess gene function. RNAi 'soaking' methodologies to knock-down genes in *C. elegans* and other nematodes have recently been adapted for the plant-parasitic cyst and root-knot nematodes (Bakhetia *et al.*, 2005). Using RNAi approaches, the relative importance of PGCs can be assessed. These methods promise to provide essential functional information to identify key components of the nematode toolbox. Gene knock-down has been achieved in *H. glycines*, *M. incognita*, *G. pallida* and *G. rostochiensis* using a modified soaking method that requires the use of either octopamine (cyst) or resorcinol (root-knot nematode) to stimulate dsRNA ingestion *in vitro* (Urwin *et al.*, 2002; Rosso *et al.*, 2005). This approach has been used to demonstrate the role of several gland-expressed genes in nematode pathogenicity. Chen *et al.* (2005) showed that knocking out the *G. rostochiensis Eng* genes reduced the ability of the juveniles to invade host plant roots. Similarly, the application of RNAi to two genes expressed in *M. incognita* SvG cells, calreticulin (*Mi-Crt*) and polygalacturonase (*Mi-Pg-1*) led to their effective silencing (Rosso *et al.*, 2005) and a reduction in gall number and size in subsequent infection assays (Rosso *et al.*, 2005).

More recently, the potential for using transgenic *in planta* RNAi technology (Wesley *et al.*, 2001), i.e. the delivery of dsRNA or siRNA from plant cells to nematodes during feeding, has shown promise as an effective approach for bioengineering plant resistance to parasitic nematodes (Huang *et al.*, 2006b; Yadav *et al.*, 2006). Target genes may interfere with fundamental aspects of nematode biology such as those encoding integrase (Yadav *et al.*, 2006), or they may target essential nematode parasitism genes to disrupt the infection process (Huang *et al.*, 2006b). The siRNA to the 16D10 parasitism gene of root-knot nematodes produced in transgenic plants (Huang *et al.*, 2006a,b) did not exceed the size exclusion limit for ingestion through the nematode feeding tube (Hussey and Grundler, 1998) and produced dramatic inhibition of successful nematode infection. Since 16D10 is conserved in *Meloidogyne* spp., silencing the gene resulted in transgenic plants that were resistant to multiple root-knot nematode species. Because no known natural resistance gene has this wide effective range of root-knot resistance, RNAi silencing of parasitism genes could lead to the development of transgenic crops with effective broad host resistance to this agriculturally important pathogen and provide a strategy for developing root-knot nematode-resistant crops for which natural resistance genes do not exist. The emerging data demonstrate the potential of using RNAi not only for the elucidation of gene function, but also for engineering novel and durable nematode resistance in transgenic crop plants.

Conclusions

Over the last decade, plant nematologists have made significant progress towards the identification of the complete profile of parasitism genes expressed in the nematode oesophageal gland cells during parasitism of plants. Consequently, our understanding of the molecular basis of nematode parasitism on plants and what makes a nematode a 'plant parasite' has advanced considerably. The current exciting challenge is to elucidate the function of the secreted products of nematode parasitism genes. Indeed, the ultimate question of how they function in consort to elicit host cell responses for successful parasitism is beginning to be revealed. Expression of parasitism genes in host plant cells, RNAi and protein–protein interaction studies has been successfully applied in studying the function of parasitism genes in nematode–plant interactions. The completion of nematode genome sequences on the horizon promises to contribute even more to our current understanding of nematode pathogenesis.

Acknowledgements

The authors gratefully acknowledge the members of their laboratories for their contributions to this work. Research support for the authors has been awarded by the National Research Initiative of the United States Department of Agriculture Cooperative State Research, Education, and Extension Service, United Soybean Board, Missouri Soybean Merchandising Council, Iowa Soybean Promotion Board and Experiment Stations of University of Missouri, University of Georgia, North Carolina State University and Iowa State University.

References

Abad, P., Favery, B., Rosso, M.N. and Castagnone-Sereno, P. (2003) Root-knot nematode parasitism and host response: molecular basis of a sophisticated interaction. *Molecular Plant Pathology* 4, 217–224.

Atibalentja, N., Bekal, S., Domier, L.L., Niblack, T.L., Noel, G.R. and Lambert, K.N. (2005) A genetic linkage map of the soybean cyst nematode *Heterodera glycines*. *Molecular Genetics and Genomics* 273, 273–281.

Bakhetia, M., Charlton, W.L., Urwin, P.E., McPherson, M.J. and Atkinson, H.J. (2005) RNA interference and plant-parasitic nematodes. *Trends in Plant Science* 10, 362–367.

Baum, T.J., Hussey, R.S. and Davis, E.L. (2007) Root-knot and cyst nematode parasitism genes: the molecular basis of plant parasitism. In: Setlow, J.K. (ed.) *Genetic Engineering,* Volume 28. Springer, Heidelberg, pp. 17–42.

Bekal, S., Niblack, T.L. and Lambert, K.N. (2003) A chorismate mutase from the soybean cyst nematode *Heterodera glycines* shows polymorphisms that correlate with virulence. *Molecular Plant–Microbe Interactions* 16, 439–446.

Bera-Maillet, C., Arthaud, L., Abad, P. and Rosso, M.N. (2000) Biochemical characterization of MI-ENG1, a family 5 endoglucanase secreted by the root-knot nematode *Meloidogyne incognita*. *European Journal of Biochemistry* 267, 3255–3263.

Birch, P.R.J., Rehmany, A.P., Pritchard, L., Kamoun, S. and Beynon, J.L. (2006) Trafficking arms: oomycete effectors enter host plant cells. *Trends in Microbiology* 14, 8–11.

Blair, L., Perry, R.N., Oparka, K. and Jones, J.T. (1999) Activation of transcription during the hatching process of the potato cyst nematode *Globodera rostochiensis*. *Nematology* 1, 103–111.

Blanchard, A., Esquibet, M., Fouville, D. and Grenier, E. (2005) RanBPM homologue genes characterised in the cyst nematodes *Globodera pallida* and *Globodera 'mexicana'*. *Physiological and Molecular Plant Pathology* 67, 15–22.

Bolle, C. (2004) The role of GRAS proteins in plant signal transduction and development. *Planta* 218, 683–692.

Branch, C., Hwang, C.F., Navarre, D.A. and Williamson, V.M. (2004) Salicylic acid is part of the Mi-1-mediated defense response to root-knot nematode in tomato. *Molecular Plant–Microbe Interactions* 17, 351–356.

Brand, U., Fletcher, J.C., Hobe, M., Meyerowitz, E.M. and Simon, R. (2000) Dependence of stem cell fate in *Arabidopsis* on a feedback loop regulated by *CLV3* activity. *Science* 289, 617–619.

Chen, Q., Rehman, S., Smant, G. and Jones, J.T. (2005) Functional analysis of pathogenicity proteins of the potato cyst nematode *Globodera rostochiensis* using RNAi. *Molecular Plant–Microbe Interactions* 18, 621–625.

Cock, J.M. and McCormick, S. (2001) A large family of genes that share homology with *CLAVATA3*. *Plant Physiology* 126, 939–942.

Dautova, M., Rosso, M.N., Abad, P., Gommers, F.J., Bakker, J. and Smant, G. (2001) Single pass cDNA sequencing – a powerful tool to analyse gene expression in preparasitic juveniles of the southern root-knot nematode *Meloidogyne incognita*. *Nematology* 3, 129–139.

Davis, E.L. and Mitchum, M.G. (2005) Nematodes: sophisticated parasites of legumes. *Plant Physiology* 137, 1182–1188.

Davis, E.L., Allen, R. and Hussey, R.S. (1994) Developmental expression of esophageal gland antigens and their detection in stylet secretions of *Meloidogyne incognita*. *Fundamental and Applied Nematology* 17, 255–262.

Davis, E.L., Hussey, R.S., Baum, T.J., Bakker, J. and Schots, A. (2000) Nematode parasitism genes. *Annual Review of Phytopathology* 38, 365–396.

Davis, E.L., Hussey, R.S. and Baum, T.J. (2004) Getting to the roots of parasitism by nematodes. *Trends in Parasitology* 20, 134–141.

de Boer, J.M., Yan, Y.T., Wang, X.H., Smant, G., Hussey, R.S., Davis, E.L. and Baum, T.J. (1999) Developmental expression of secretory beta-1,4-endoglucanases in the subventral esophageal glands of *Heterodera glycines*. *Molecular Plant–Microbe Interactions* 12, 663–669.

de Boer, J.M., McDermott, J.P., Davis, E.L., Hussey, R.S., Popeijus, H., Smant, G. and Baum, T.J. (2002) Cloning of a putative pectate lysase gene expressed in the subventral esophageal glands of *Heterodera glycines*. *Journal of Nematology* 34, 9–11.

de Meutter, J., Vanholme, B., Bauw, G., Tygat, T. and Gheysen, G. (2001) Preparation and sequencing of secreted proteins from the pharyngeal glands of the plant-parasitic nematode *Heterodera schachtii*. *Molecular Plant Pathology* 2, 297–301.

Ding, X., Shields, J., Allen, R. and Hussey, R.S. (1998) A secretory cellulose-binding protein cDNA cloned from the root-knot nematode (*Meloidogyne incognita*). *Molecular Plant–Microbe Interactions* 11, 952–959.

Ding, X., Shields, J., Allen, R. and Hussey, R.S. (2000) Molecular cloning and characterisation of a venom allergen AG5-like cDNA from *Meloidogyne incognita*. *International Journal for Parasitology* 30, 77–81.

Dong, K. and Opperman, C.H. (1997) Genetic analysis of parasitism in the soybean cyst nematode *Heterodera glycines*. *Genetics* 146, 1311–1318.

Dong, K., Barker, K.R. and Opperman, C.H. (2005) Virulence genes in *Heterodera glycines*: allele frequencies and ror gene groups among field isolates and inbred lines. *Phytopathology* 95, 186–191.

Doyle, E.A. and Lambert, K.N. (2002) Cloning and characterization of an esophageal-gland-specific pectate lyase from the root-knot nematode *Meloidogyne javanica*. *Molecular Plant–Microbe Interactions* 15, 549–556.

Doyle, E.A. and Lambert, K.N. (2003) *Meloidogyne javanica* chorismate mutase 1 alters plant cell development. *Molecular Plant–Microbe Interactions* 16, 123–131.

Engler, J.D., De Vleesschauwer, V., Burssens, S., Celenza, J.L., Inze, D., Van Montagu, M., Engler, G. and Gheysen, G. (1999) Molecular markers and cell-cycle inhibitors show the importance of cell cycle progression in nematode-induced galls and syncytia. *The Plant Cell* 11, 793–807.

Ernst, K., Kumar, A., Kriseleit, D., Kloos, D.U., Phillips, M.S. and Ganal, M.W. (2002) The broad-spectrum potato cyst nematode resistance gene (*Hero*) from tomato is the only member of a large gene family of NBS-LRR genes with an unusual amino acid repeat in the LRR region. *The Plant Journal* 31, 127–136.

Favery, B., Lecomte, P., Gil, N., Bechtold, N., Bouchez, D., Dalmasso, A. and Abad, P. (1998) *RPE*, a plant gene involved in early developmental steps of nematode feeding cells. *The EMBO Journal* 17, 6799–6811.

Fire, A., Xu, S.Q., Montgomery, M.K., Kostas, S.A., Driver, S.E. and Mello, C.C. (1998) Potent and specific genetic interference by double-stranded RNA in *Caenorhabditis elegans*. *Nature* 391, 806–811.

Gao, B., Allen, R., Maier, T., Davis, E.L., Baum, T.J. and Hussey, R.S. (2001a) Identification of putative parasitism genes expressed in the esophageal gland cells of the soybean cyst nematode *Heterodera glycines*. *Molecular Plant–Microbe Interactions* 14, 1247–1254.

Gao, B., Allen, R., Maier, T., Davis, E.L., Baum, T.J. and Hussey, R.S. (2001b) Molecular characterisation and expression of two venom allergen-like protein genes in *Heterodera glycines*. *International Journal of Parasitology* 31, 1617–1625.

Gao, B., Allen, R., Maier, T., Davis, E.L., Baum, T.J. and Hussey, R.S. (2002a) Identification of a new beta-1,4-endoglucanase gene expressed in the esophageal subventral gland cells of *Heterodera glycines*. *Journal of Nematology* 34, 12–15.

Gao, B., Allen, R., Maier, T., McDermott, J., Davis, E.L., Baum, T.J. and Hussey, R.S. (2002b) Characterisation and developmental expression of a chitinase gene in *Heterodera glycines*. *International Journal of Parasitology* 32, 1293–1300.

Gao, B., Allen, R., Maier, T., Davis, E.L., Baum, T.J. and Hussey, R. (2003) The parasitome of the phytonematode *Heterodera glycines*. *Molecular Plant–Microbe Interactions* 16, 720–726.

Gao, B., Allen, R., Davis, E.L., Baum, T.J. and Hussey, R.S. (2004a) Developmental expression and biochemical properties of a beta-1,4-endoglucanase family in the soybean cyst nematode, *Heterodera glycines*. *Molecular Plant Pathology* 5, 93–104.

Gao, B., Allen, R., Davis, E.L., Baum, T.J. and Hussey, R.S. (2004b) Molecular characterisation and developmental expression of a cellulose-binding protein gene in the soybean cyst nematode *Heterodera glycines*. *International Journal of Parasitology* 34, 1377–1383.

Gheysen, G. and Fenoll, C. (2002) Gene expression in nematode feeding sites. *Annual Review of Phytopathology* 40, 124–168.

Goellner, M., Smant, G., de Boer, J.M., Baum, T.J. and Davis, E.L. (2000) Isolation of beta-1,4-endoglucanase genes from *Globodera tabacum* and their expression during parasitism. *Journal of Nematology* 32, 154–165.

Goellner, M., Wang, X.H. and Davis, E.L. (2001) Endo-beta-1,4-glucanase expression in compatible plant–nematode interactions. *The Plant Cell* 13, 2241–2255.

Goverse, A., Davis, E.L. and Hussey, R.S. (1994) Monoclonal antibodies to the esophageal glands and stylet secretions of *Heterodera glycines*. *Journal of Nematology* 26, 251–259.

Goverse, A., van der Voort, J.R., van der Voort, C.R., Kavelaars, A., Smant, G., Schots, A., Bakker, J. and Helder, J. (1999) Naturally induced secretions of the potato cyst nematode co-stimulate the proliferation of both tobacco leaf protoplasts and human peripheral blood mononuclear cells. *Molecular Plant–Microbe Interactions* 12, 872–881.

Goverse, A., Engler, J.D., Verhees, J., van der Krol, S., Helder, J. and Gheysen, G. (2000) Cell cycle activation by plant-parasitic nematodes. *Plant Molecular Biology* 43, 747–761.

Grenier, E., Blok, V.C., Jones, J.T., Fouville, D. and Mugniery, D. (2002) Identification of gene expression differences between *Globodera pallida* and *G-'mexicana'* by suppression subtractive hybridization. *Molecular Plant Pathology* 3, 217–226.

Hawdon, J.M., Narasimhan, S. and Hotez, P.J. (1999) Anclyostoma secreted protein 2: cloning and characterization of a second member of a family of nematode secreted proteins from *Ancylostoma caninum*. *Molecular and Biochemical Parasitology* 99, 149–165.

Hobe, M., Muller, R., Grunewald, M., Brand, U. and Simon, R. (2003) Loss of CLE40, a protein functionally equivalent to the stem cell restricting signal CLV3, enhances root waving in *Arabidopsis*. *Development Genes and Evolution* 213, 371–381.

Huang, G.Z., Gao, B.L., Maier, T., Allen, R., Davis, E.L., Baum, T.J. and Hussey, R.S. (2003) A profile of putative parasitism genes expressed in the esophageal gland cells of the root-knot nematode *Meloidogyne incognita*. *Molecular Plant–Microbe Interactions* 16, 376–381.

Huang, G.Z., Dong, R.H., Maier, T., Allen, R., Davis, E.L., Baum, T.J. and Hussey, R.S. (2004) Use of solid-phase subtractive hybridization for the identification of parasitism gene candidates from the root-knot nematode *Meloidogyne incognita*. *Molecular Plant Pathology* 5, 217–222.

Huang, G.Z., Dong, R.H., Allen, R., Davis, E.L., Baum, T.J. and Hussey, R.S. (2005a) Developmental expression and molecular analysis of two *Meloidogyne incognita* pectate lyase genes. *International Journal of Parasitology* 35, 685–692.

Huang, G.Z., Dong, R.H., Allen, R., Davis, E.L., Baum, T.J. and Hussey, R.S. (2005b) Two chorismate mutase genes from the root-knot nematode *Meloidogyne incognita*. *Molecular Plant Pathology* 6, 23–30.

Huang, G.Z., Dong, R.H., Allen, R., Davis, E.L., Baum, T.J. and Hussey, R.S. (2006a) A root-knot nematode secretory peptide functions as a ligand for a plant transcription factor. *Molecular Plant–Microbe Interactions* 19, 463–470.

Huang, G., Dong, R., Allen, R., Davis, E.L., Baum, T.J. and Hussey, R.S. (2006b) Engineering broad root-knot resistance in transgenic plants by RNAi silencing of a conserved and essential root-knot nematode parasitism gene. *Proceedings of the National Academy of Sciences of the United States of America* 103, 14302–14306.

Hussey, R.S. (1989) Disease-inducing secretions of plant-parasitic nematodes. *Annual Review of Phytopathology* 27, 123–141.

Hussey, R.S. and Grundler, F.M. (1998) Nematode parasitism of plants. In: Perry, R.N. and Wright, J. (eds) *Physiology and Biochemistry of Free-living and Plant-parasitic Nematodes.* CAB International, Wallingford, UK, pp. 213–243.

Hussey, R.S., Pagio, O.R. and Seabury, F. (1990) Localization and purification of a secretory protein from the esophageal glands of *Meloidogyne incognita* with a monoclonal antibody. *Phytopathology* 80, 709–714.

Hutangura, P., Mathesius, U., Jones, M.G.K. and Rolfe, B.G. (1999) Auxin induction is a trigger for root gall formation caused by root-knot nematodes in white clover and is associated with the activation of the flavonoid pathway. *Australian Journal of Plant Physiology* 26, 221–231.

Ithal, N., Recknor, J., Nettleton, D., Hearne, L., Maier, T., Baum, T.J. and Mitchum, M.G. (2006) Parallel genome-wide expression profiling of host and pathogen during soybean cyst nematode infection of soybean. *Molecular Plant–Microbe Interactions* 20, 293–305.

Janssen, J., Bakker, J. and Gommers, F.J.G. (1991) Mendelian proof for a gene-for-gene relationship between virulence of *Globodera rostochiensis* and the *H1* resistance gene in *Solanum tuberosum* ssp. *andigena* CPC 1673. *Review of Nematology* 14, 207–211.

Jaubert, S., Laffaire, J.B., Abad, P. and Rosso, M.N. (2002a) A polygalacturonase of animal origin isolated from the root-knot nematode *Meloidogyne incognita*. *FEBS Letters* 522, 109–112.

Jaubert, S., Ledger, T.N., Laffaire, J.B., Piotte, C., Abad, P. and Rosso, M.N. (2002b) Direct identification of stylet secreted proteins from root-knot nematodes by a proteomic approach. *Molecular and Biochemical Parasitology* 121, 205–211.

Jaubert, S., Milac, A.L., Petrescu, A.J., de Almelda-Engler, J., Abad, P. and Rosso, M.N. (2005) In planta secretion of a calreticulin by migratory and sedentary stages of root-knot nematode. *Molecular Plant–Microbe Interactions* 18, 1277–1284.

Jones, J.T., Furlanetto, C., Bakker, E., Banks, B., Blok, V., Chen, Q., Phillips, M. and Prior, A. (2003) Characterization of a chorismate mutase from the potato cyst nematode *Globodera pallida*. *Molecular Plant Pathology* 4, 43–50.

Jones, M.G.K. and Northcote, D.H. (1972) Nematode induced syncytium – multinucleate transfer cell. *Journal of Cell Science* 10, 789–809.

Karczmarek, A., Overmars, H., Helder, J. and Goverse, A. (2004) Feeding cell development by cyst and root-knot nematodes involves a similar early, local and transient activation of a specific auxin-inducible promoter element. *Molecular Plant Pathology* 5, 343–346.

Kikuchi, T., Jones, J.T., Aikawa, T., Kosaka, H. and Ogura, N. (2004) A family of glycosyl hydrolase family 45 cellulases from the pine wood nematode *Bursaphelenchus xylophilus*. *FEBS Letters* 572, 201–205.

Kudla, U., Qin, L., Milac, A., Kielak, A., Maissen, C., Overmars, H., Popeijus, H., Roze, E., Petrescu, A., Smant, G., Bakker, J. and Helder, J. (2005) Origin, distribution and 3D-modeling of Gr-EXPB1, an expansin from the potato cyst nematode *Globodera rostochiensis*. *FEBS Letters* 579, 2451–2457.

Lambert, K.N., Allen, K.D. and Sussex, I.M. (1999) Cloning and characterization of an esophageal-gland-specific chorismate mutase from the phytoparasitic nematode *Meloidogyne javanica*. *Molecular Plant–Microbe Interactions* 12, 328–336.

Lambert, K.N., Bekal, S., Domier, L.L., Niblack, T.L., Noel, G.R. and Smyth, C.A. (2005) Selection of *Heterodera glycines* chorismate mutase-1 alleles on nematode-resistant soybean. *Molecular Plant–Microbe Interactions* 18, 593–601.

Lee, J., Nam, S., Hwang, S.B., Hong, M.G., Kwon, J.Y., Joeng, K.S., Im, S.H., Shim, J.W. and Park, M.C. (2004) Functional genomic approaches using the nematode *Caenorhabditis elegans* as a model system. *Journal of Biochemistry and Molecular Biology* 37, 107–113.

Lilley, C.J., Atkinson, H.J. and Urwin, P.E. (2005) Molecular aspects of cyst nematodes. *Molecular Plant Pathology* 6, 577–588.

Lindsey, K., Casson, S. and Chilley, P. (2002) Peptides: new signalling molecules in plants. *Trends in Plant Science* 7, 78–83.

Liu, Q. and Williamson, V.M. (2004) Genetics of virulence in root-knot nematode, *Meloidogyne hapla*. *Phytopathology* 94, S62.

Lu, S. and Wang, X. (2006) A new class of CLAVATA3/ESR (CLE)-like peptides expressed in the esophageal gland cells of the potato cyst nematode *Globodera rostochiensis*. *Phytopathology* 96, S70.

Mazarei, M., Lennon, K.A., Puthoff, D.P., Rodermel, S.R. and Baum, T.J. (2003) Expression of an *Arabidopsis* phosphoglycerate mutase homologue is localized to apical meristems, regulated by hormones, and induced by sedentary plant-parasitic nematodes. *Plant Molecular Biology* 53, 513–530.

McCarter, J.P., Mitreva, M.D., Martin, J., Dante, M., Wylie, T., Rao, U., Pape, D., Bowers, Y., Theising, B., Murphy, C.V., Kloek, A.P., Chiapelli, B.J., Clifton, S.W., Bird, D.M. and Waterston, R.H. (2003) Analysis and functional classification of transcripts from the nematode *Meloidogyne incognita*. *Genome Biology* 4, R26.

Milligan, S.B., Bodeau, J., Yaghoobi, J., Kaloshian, I., Zabel, P. and Williamson, V.M. (1998) The root-knot nematode resistance gene *Mi* from tomato is a member of the leucine zipper, nucleotide binding, leucine-rich repeat family of plant genes. *The Plant Cell* 10, 1307–1319.

Mitreva, M., Blaxter, M.L., Bird, D.M. and McCarter, J.P. (2005) Comparative genomics of nematodes. *Trends in Genetics* 21, 573–581.

Mitreva-Dautova, M., Roze, E., Overmars, H., de Graaff, L., Schots, A., Helder, J., Goverse, A., Bakker, J. and Smant, G. (2006) A symbiont-independent endo-1,4-beta-xylanase from the plant-parasitic nematode *Meloidogyne incognita*. *Molecular Plant–Microbe Interactions* 19, 521–529.

Moffett, P. and Sacco, M. (2006) A nematode avirulence determinant and a co-factor for a disease resistance protein converge on the Ran signaling pathway. *Phytopathology* 96, S80.

Nakamura, M., Masuda, H, Horii, J., Kuma, K., Yokohama, N., Ohba, T., Nishitani, H., Miyata, T., Tanaka, M. and Nishimoto, T. (1998) When overexpressed, a novel centrosomal protein, RanBPM, causes ectopic microtubule nucleation similar to γ-tubulin. *The Journal of Cell Biology* 143, 1041–1052.

Neveu, C., Jaubert, S., Abad, P. and Castagnone-Sereno, P. (2003) A set of genes differentially expressed between avirulent and virulent *Meloidogyne incognita* near-isogenic lines encode secreted proteins. *Molecular Plant–Microbe Interactions* 16, 1077–1084.

Olsen, A.N. and Skriver, K. (2003) Ligand mimicry? Plant-parasitic nematode polypeptide with similarity to CLAVATA3. *Trends in Plant Science* 8, 55–57.

Paal, J., Henselewski, H., Muth, J., Meksem, K., Menendez, C.M., Salamini, F., Ballvora, A. and Gebhardt, C. (2004) Molecular cloning of the potato *Gro1-4* gene conferring resistance to pathotype Ro1 of the root cyst nematode *Globodera rostochiensis*, based on a candidate gene approach. *The Plant Journal* 38, 285–297.

Parkinson, J., Mitreva, M., Hall, N., Blaxter, M. and McCarter, J.P. (2003) 400,000 Nematode ESTs on the Net. *Trends in Parasitology* 19, 283–286.

Parkinson, J., Mitreva, M., Whitton, C., Thomson, M., Daub, J., Martin, J., Schmid, R., Hall, N., Barrell, B., Waterston, R.H., McCarter, J.P. and Blaxter, M.L. (2004) A transcriptomic analysis of the phylum Nematoda. *Nature Genetics* 36, 1259–1267.

Popeijus, M., Blok, V.C., Cardle, L., Bakker, E., Phillips, M.S., Helder, J., Smant, G. and Jones, J.T. (2000a) Analysis of genes expressed in second stage juveniles of the potato cyst nematodes *Globodera rostochiensis* and *Globodera pallida* using the expressed sequence tag approach. *Nematology* 2, 567–574.

Popeijus, H., Overmars, H., Jones, J., Blok, V., Goverse, A., Helder, J., Schots, A., Bakker, J. and Smant, G. (2000b) Enzymology – degradation of plant cell walls by a nematode. *Nature* 406, 36–37.

Qin, L., Overmars, B., Helder, J., Popeijus, H., van der Voort, J.R., Groenink, W., van Koert, P., Schots, A., Bakker, J. and Smant, G. (2000) An efficient cDNA-AFLP-based strategy for the identification of putative pathogenicity factors from the potato cyst nematode *Globodera rostochiensis*. *Molecular Plant–Microbe Interactions* 13, 830–836.

Qin, L., Kudla, U., Roze, E.H.A., Goverse, A., Popeijus, H., Nieuwland, J., Overmars, H., Jones, J.T., Schots, A., Smant, G., Bakker, J. and Helder, J. (2004) Plant degradation: a nematode expansin acting on plants. *Nature* 427, 30.

Riely, B.K., Ane, J.M., Penmetsa, R.V. and Cook, D.R. (2004) Genetic and genomic analysis in model legumes bring Nod-factor signaling to center stage. *Current Opinion in Plant Biology* 7, 408–413.

Rosso, M.N., Favery, B., Piotte, C., Arthaud, L., de Boer, J.M., Hussey, R.S., Bakker, J., Baum, T.J. and Abad, P. (1999) Isolation of a cDNA encoding a beta-1,4-endoglucanase in the root-knot nematode *Meloidogyne incognita* and expression analysis during plant parasitism. *Molecular Plant–Microbe Interactions* 12, 585–591.

Rosso, M.N., Dubrana, M.P., Cimbolini, N., Jaubert, S. and Abad, P. (2005) Application of RNA interference to root-knot nematode genes encoding esophageal gland proteins. *Molecular Plant–Microbe Interactions* 18, 615–620.

Safra-Dassa, L., Shani, Z., Danin, A., Roiz, L., Shoseyov, O. and Wolf, S. (2006) Growth modulation of transgenic potato plants by heterologous expression of bacterial carbohydrate-binding module. *Molecular Breeding* 17, 355–364.

Scholl, E.H., Thorne, J.L., McCarter, J.P. and Bird, D.M. (2003) Horizontally transferred genes in plant-parasitic nematodes: a high-throughput genomic approach. *Genome Biology* 4(6), R39.

Semblat, J.P., Rosso, M.N., Hussey, R.S., Abad, P. and Castagnone-Sereno, P. (2001) Molecular cloning of a cDNA encoding an amphid-secreted putative avirulence protein from the root-knot nematode *Meloidogyne incognita*. *Molecular Plant–Microbe Interactions* 14, 72–79.

Sharma, V.K., Ramirez, J. and Fletcher, J.C. (2003) The *Arabidopsis CLV3*-like (*CLE*) genes are expressed in diverse tissues and encode secreted proteins. *Plant Molecular Biology* 51, 415–425.

Shpigel, E., Roiz, L., Goren, R. and Shoseyov, O. (1998) Bacterial cellulose-binding domain modulates *in vitro* elongation of different plant cells. *Plant Physiology* 117, 1185–1194.

Smant, G., Stokkermans, J.P.W.G., Yan, Y.T., de Boer, J.M., Baum, T.J., Wang, X.H., Hussey, R.S., Gommers, F.J., Henrissat, B., Davis, E.L., Helder, J., Schots, A. and Bakker, J. (1998) Endogenous cellulases in animals: isolation of beta-1,4-endoglucanase genes from two species of plant-parasitic cyst nematodes. *Proceedings of the National Academy of Sciences of the United States of America* 95, 4906–4911.

Strabala, T.J., O'Donnell, P.J., Smit, A.M., Ampomah-Dwamena, C., Martin, E.J., Netzler, N., Nieuwenhuizen, N.J., Quinn, B.D., Foote, H.C.C. and Hudson, K.R. (2006) Gain-of-function phenotypes of many CLAVATA3/ESR genes, including four new family members, correlate

Fig 3.1

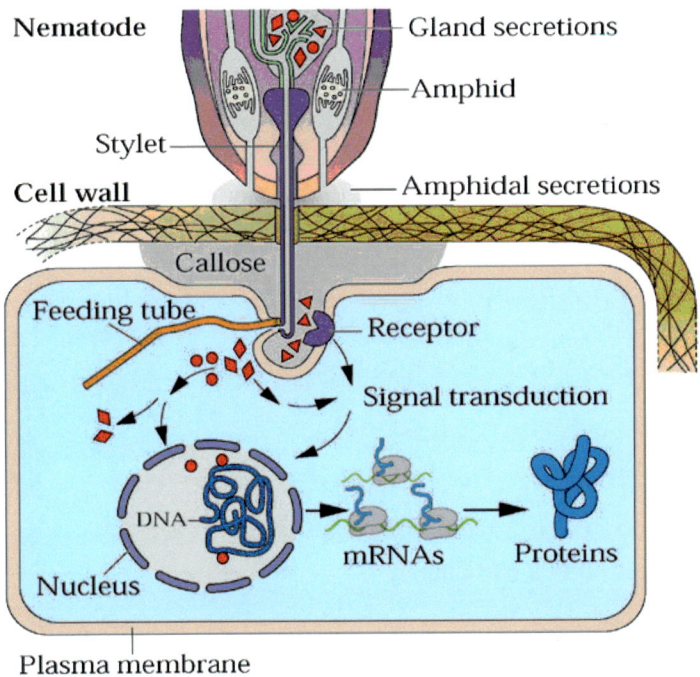

Fig. 3.1
A model of the interactions of the secreted products of nematode parasitism genes with recipient host plant cells. The nematode stylet penetrates through the cell wall, but does not pierce the plasma membrane which invaginates around the tip of the stylet. Gland secretions originating in the oesophageal gland cells are released through the stylet into the apoplast or the cytoplasm (through a small pore in the plasma membrane at the stylet orifice) of selected plant cells where they interact with receptors or other host proteins for the induction and maintenance of feeding cells and the formation of feeding tubes. Secreted products containing nuclear localization signals may be targeted to the host cell nucleus to directly regulate changes in host gene expression. (Reproduced from Baum et al., 2007, with permission from the publisher.)

Fig. 3.2

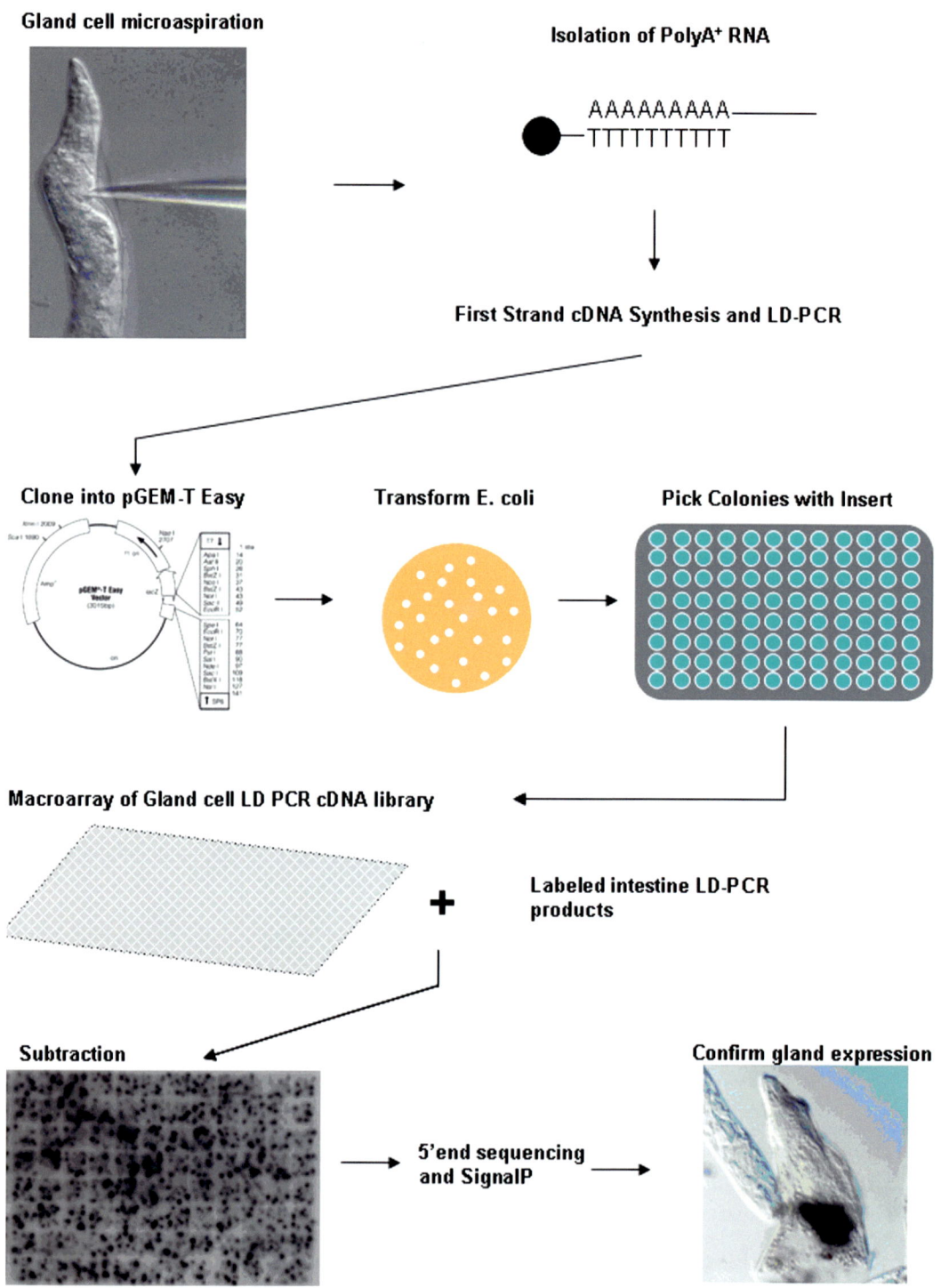

Fig. 3.2
Schematic presentation of the method used for the construction of gland-enriched cDNA libraries. LD-PCR, long distance polymerase chain reaction. pGEM-T easy is a product of Promega Corporation (Madison, Wisconsin).

with tandem variations in the conserved CLAVATA3/ESR domain. *Plant Physiology* 140, 1331–1344.
Triantaphyllou, A.C. (1985) Cytogenetics, cytotaxonomy and phylogeny of root-knot nematodes. In: Sasser, J. and Carter, C. (eds) *An Advanced Treatise on Meloidogyne*, Volume I. North Carolina University Press, Raleigh, NC, pp. 113–126.
Tytgat, T., Vanholme, B., de Meutter, J., Claeys, M., Couvreur, M., Vanhoutte, I., Gheysen, G., Van Criekinge, W., Borgonie, G., Coomans, A. and Gheysen, G. (2004) A new class of ubiquitin extension proteins secreted by the dorsal pharyngeal gland in plant-parasitic cyst nematodes. *Molecular Plant–Microbe Interactions* 17, 846–852.
Uehara, T., Kushida, A., and Momota, Y. (2001) PCR-based cloning of two β-1,4-endoglucanases from the root-lesion nematode *Pratylenchus penetrans*. *Nematology* 3, 335–341.
Urwin, P.E., Lilley, C.J. and Atkinson, H.J. (2002) Ingestion of double-stranded RNA by preparasitic juvenile cyst nematodes leads to RNA interference. *Molecular Plant–Microbe Interactions* 15, 747–752.
van Bers, N., Nouws, J., Overmars, H., Qin, L., Goverse, A., Bakker, J. and Smant, G. (2006) Overexpression of small polypeptide genes from the potato cyst nematode affects lateral root formation in *Solanum tuberosum*. XXVIII European Society of Nematology (Meeting Abstract).
Vanholme, B., Mitreva, M.D., Van Criekinge, W., Logghe, M., Bird, D., McCarter, J.P. and Gheysen, G. (2005) Detection of putative secreted proteins in the plant-parasitic nematode *Heterodera schachtii*. *Parasitology Research* 98, 414–424.
Van der Beek, J., Los, J. and Pijanacker, L. (1998) Cytology of parthogenesis of five *Meloidogyne* species. *Fundamental and Applied Nematology* 21, 393–399.
van der Vossen, E.A.G., van der Voort, J.N.A.M.R., Kanyuka, K., Bendahmane, A., Sandbrink, H., Baulcombe, D.C., Bakker, J., Stiekema, W.J. and Klein-Lankhorst, R.M. (2000) Homologues of a single resistance-gene cluster in potato confer resistance to distinct pathogens: a virus and a nematode. *The Plant Journal* 23, 567–576.
Wang, J., Replogle, A., Wang, X., Davis, E.L. and Mitchum, M.G. (2006) Functional analysis of nematode secreted CLAVATA3/ESR (CLE)-like peptides of the genus *Heterodera*. *Phytopathology* 96, S120.
Wang, X.H., Meyers, D., Yan, Y.T., Baum, T., Smant, G., Hussey, R. and Davis, E. (1999) In planta localization of a beta-1,4-endoglucanase secreted by *Heterodera glycines*. *Molecular Plant–Microbe Interactions* 12, 64–67.
Wang, X.H., Allen, R., Ding, X.F., Goellner, M., Maier, T., de Boer, J.M., Baum, T.J., Hussey, R.S. and Davis, E.L. (2001) Signal peptide-selection of cDNA cloned directly from the esophageal gland cells of the soybean cyst nematode *Heterodera glycines*. *Molecular Plant–Microbe Interactions* 14, 536–544.
Wang, X.H., Mitchum, M.G., Gao, B.L., Li, C.Y., Diab, H., Baum, T.J., Hussey, R.S. and Davis, E.L. (2005) A parasitism gene from a plant-parasitic nematode with function similar to CLAVATA3/ESR (CLE) of *Arabidopsis thaliana*. *Molecular Plant Pathology* 6, 187–191.
Weerasinghe, R.R., Bird, D.M. and Allen, N.S. (2005) Root-knot nematodes and bacterial Nod factors elicit common signal transduction events in *Lotus japonicus*. *Proceedings of the National Academy of Sciences of the United States of America* 102, 3147–3152.
Wesley, S.V., Helliwell, C.A., Smith, N.A., Wang, M., Rouse, D.T., Liu, Q., Gooding, P.S., Singh, S.P., Abbott, D., Stoutjesdijk, P.A., Robinson, S.P., Gleave, A.P., Green, A.G. and Waterhouse, P.M. (2001) Construct design for efficient, effective and high-throughput gene silencing in plants. *The Plant Journal* 27, 581–590.
Williamson, V.M. and Hussey, R.S. (1996) Nematode pathogenesis and resistance in plants. *The Plant Cell* 8, 1735–1745.
Yadav, B.C., Veluthambi, K. and Subramaniam, K. (2006) Host-generated double-stranded RNA induces RNAi in plant-parasitic nematodes and protects the host from infection. *Molecular and Biochemical Parasitology* 148, 219–222.

Yan, Y.T., Smant, G., Stokkermans, J., Qin, L., Helder, J., Baum, T., Schots, A. and Davis, E. (1998) Genomic organization of four beta-1,4-endoglucanase genes in plant-parasitic cyst nematodes and its evolutionary implications. *Gene* 220, 61–70.

Yan, Y.T., Smant, G. and Davis, E. (2001) Functional screening yields a new beta-1,4-endoglucanase gene from *Heterodera glycines* that may be the product of recent gene duplication. *Molecular Plant–Microbe Interactions* 14, 63–71.

Zeng, L.-R., Vega-Sanchez, M.E., Zhu, T. and Wang, G.-L. (2006) Ubiquitination-mediated protein degradation and modification: an emerging theme in plant–microbe interactions. *Cell Research* 16, 413–426.

4 Interactions Between Plant and Virus Proteomes in Susceptible Hosts: Identification of New Targets for Antiviral Strategies

H. Sanfaçon and J. Jovel

Abstract

Plant viruses are obligate parasites that depend on their hosts for each step of their replication cycle. Host factors that facilitate the replication of the viral genome and its subsequent spread throughout the plant have been characterized and include, but are not limited to: translation factors, transcription factors, cell cycle regulators, integral membrane proteins and modifying enzymes, such as kinases, chaperones and components of the proteasome. The design of resistance strategies based on the mutation or silencing of such host factors requires an in-depth understanding of plant–virus interactions in compatible hosts. In several cases, mutation of a specific host factor has been shown to confer increased resistance to virus infection without significantly affecting the physiology of the plant. Positive-sense single-stranded RNA viruses and single-stranded DNA viruses represent the vast majority of economically important plant viruses that infect cultivated plants worldwide. Using these two large groups of plant viruses as examples, various levels of plant–virus interactions that allow successful virus infection in compatible hosts are discussed. The potential and limitations of emerging technologies for engineering resistance to plant viruses that are based on the inactivation of the characterized host factors are also described.

Introduction

Although the origin of plant viruses remains uncertain, it is well accepted that they have co-evolved with their hosts over a long period of time, a feature they share with other obligate parasites. Plant viruses have a relatively small genome (~3–19 kb), encompassing a limited number of genes that code for proteins that, although evolutionarily optimized to achieve one or several functions, are clearly insufficient to accomplish the viral replication cycle. To compensate for this limitation, viruses sequester host factors to replicate and spread through the plant (Whitham and Wang, 2004; Boguszewska-Chachulska and Haenni, 2005; Nelson and Citovsky, 2005).

For example, they usurp components of the host replication, transcription or translation machinery to assist them in the translation and/or replication of their genome (Ahlquist *et al.*, 2003; Schneider and Mohr, 2003; Hanley-Bowdoin *et al.*, 2004). Plant viruses must also counteract the various layers of defence deployed by their hosts. These include the ubiquitous post-transcriptional gene silencing pathway that degrades their genome or the more specific defence responses directed by dominant resistance genes (see Chapters 16 and 17, this volume). Specific interactions between viral components and host factors (proteins, intracellular membranes and small RNAs) play a key role in the successful establishment of viral infection. Thus, the identification of contact points between the virus and its host will undoubtedly reveal new targets for virus control.

This chapter discusses recent scientific developments aimed at unraveling the complex network of protein–protein interactions required for virus replication and the subsequent invasion of the plant in compatible interactions. Selected success stories, in which plants were manipulated to prevent specific virus–host interactions to result in increased resistance to plant viruses, are also discussed. We have chosen to focus our attention on two large groups of plant viruses of economic importance: (i) viruses that have a positive-sense single-stranded RNA [(+)ssRNA] genome; and (ii) viruses that have a single-stranded DNA (ssDNA) genome. (+)ssRNA viruses represent the vast majority of plant viruses. This group includes well-characterized models in several genera, such as tobacco mosaic virus (TMV, genus *Tobamovirus*), brome mosaic virus (BMV, genus *Bromovirus*), tobacco etch virus (TEV, genus *Potyvirus*) and tomato bushy stunt virus (TBSV, genus *Tombusvirus*). ssDNA viruses include members of the families *Geminiviridae* and *Nanoviridae*, many of which cause devastating diseases in crops cultivated in tropical and subtropical regions of the world. Although we will not discuss other groups of plant viruses in detail, we acknowledge that they include many well-characterized model systems and economically important viruses. The methodologies described in this chapter to identify plant–virus interactions and to develop disease resistance strategies are applicable to these groups of plant viruses.

Background

Requirements for a successful resistance strategy

In nature, recessive resistance to plant viruses has been linked to the loss or mutation of a host factor required for the virus replication cycle (Kang *et al.*, 2005; Soosaar *et al.*, 2005; Robaglia and Caranta, 2006). For example, analysis of recessive resistance has led to the identification of translation factors that enhance the replication cycle of (+)ssRNA viruses. However, many other plant factors interact with viral proteins or the viral RNA and positively influence the outcome of viral infection in susceptible hosts. The identification of these factors is likely to reveal new targets for disease control strategies. For the successful development of new virus resistance strategies, two important points must be considered. First, the mutation or inhibition of a plant housekeeping gene is likely to have a negative impact on the physiology of the plant. Thus, only plant genes that are essential for the virus replication cycle, but dispensable for the plant, should be targeted. Second, the durability of the resistance is likely to be affected by the ability of plant viruses to evolve in response to selective pressure, such as engineered resistance (see Chapter 5, this volume). This can be minimized by using strategies that target early steps in the virus replication cycle and prevent the initial accumulation of the virus. Also, the use of knock-out mutations that eliminate a required factor is likely more difficult for viruses to overcome than point mutations. Finally, it should be noted that although silencing of a host gene using interfering RNA (RNAi) is an *a priori* attractive approach, many plant viruses encode potent suppressors of silencing (see Chapter 16, this volume) that may affect the durability of the resistance.

The hunt for host factors in the proteomic era

Many viral proteins are active within large protein complexes that contain host and viral components. Viral replication complexes were the first viral–host protein complexes isolated (reviewed in Boguszewska-Chachulska and Haenni, 2005). Purification of these complexes from plant extracts was initially facilitated by the fact that virus-specific replication activity can be assayed *in vitro*. The association of host factors with replication complexes was reported for the bacteriophage Qβ as early as 1979. The presence of host proteins associated with a plant virus replication complex was first suggested in 1984 (Mouches *et al.*, 1984). However, until 1993, the identity of a plant protein associated with a viral replication complex was not determined unequivocally (Quadt *et al.*, 1993). Through the use of small epitope tags that can be fused to individual viral proteins, the purification of viral–host protein complexes is no longer limited to proteins for which measurable enzymatic activity assays or specific antibodies are available (Figeys *et al.*, 2001). Tagged viral proteins can be expressed independently using agroinfiltration or inserted into viral infectious clones and analysed in the context of viral infection. Protein complexes can be isolated from plant extracts by immunoprecipitation with tag-specific antibodies or by affinity chromatography. Recent progress in proteomic research, e.g. the improvement of protein separation and microsequencing methods combined with the availability of large databases of plant genomic sequences, has facilitated the identification of plant proteins that interact with tagged viral proteins (Serva and Nagy, 2006).

The yeast two-hybrid system has been used to identify interacting protein partners. Although several variations of this technique exist, they all rely on the same principle of splitting a biologically active protein into two non-functional halves which are fused to interacting partners (Fields,

2005). Interaction of the two partners reconstitutes the activity of the split protein. Although the assay was initially limited to soluble proteins that could be fused to distinct domains of a transcription factor, the technique has recently been expanded to the study of membrane proteins (Fields, 2005). Specific viral proteins are used as bait to screen for interacting partners using large libraries of expressed plant genes. However, caution is required when interpreting the results of such screens, since false positives are notorious and many genuine interactions can remain undetected due to difficulties encountered in expressing proteins that are toxic to yeast cells. Detection of an interaction also requires that both partners are functional when expressed as fusion proteins. In spite of the limitations of the method, many important interactions were first discovered using the yeast two-hybrid system. The principle of reconstituting a split protein was recently extended to allow *in vivo* examination of protein–protein interactions by using the bimolecular fluorescence complementation technique. Here, the fluorescence of a reconstituted split protein can be visualized by confocal microscopy, thus enabling the identification of the subcellular site of the interaction (Bhat et al., 2006).

The analysis of plant mutants present in natural populations or induced by ethylene methane-sulfonate (ems), transposon tagging or TDNA knock-out insertions can confirm the biological importance of known host factors or assist in the identification of new factors. Moreover, when visible symptoms are induced by viruses in susceptible plants, mutations that result in resistance to viral infection can be rapidly screened from large mutant populations (Lellis et al., 2002). The main advantage of such screening techniques is that mutations that do not affect the plant morphology but hinder the virus infection process can be identified, directly providing potential targets for antiviral strategies. *Arabidopsis thaliana*, a model plant with a rapid life cycle and for which the entire genomic sequence has been established, has proven to be a useful host for these experiments.

Several plant viruses can also replicate in yeast, an organism for which powerful genetic tools are available (Alves-Rodrigues et al., 2006; Nagy and Pogany, 2006). The method has been developed for the study of the replication of (+)ssRNA viruses that have a relatively simple genome (e.g. BMV, TBSV). Viral replication proteins are synthesized from yeast expression vectors and are able to replicate *in trans* a defective viral RNA. Incorporation of a reporter gene into the replicating RNA allows for easy monitoring of virus replication. Recently, a geminivirus was also reported to replicate in yeast (Raghavan et al., 2004). Analysis of large collections of yeast mutants has resulted in the discovery of many host factors that are involved in the replication of BMV and TBSV, but only a fraction of them have been characterized. Many factors identified in yeast have plant homologues, suggesting that their functions are conserved. However, functional studies demonstrating the importance of the identified factors in the natural hosts are necessary. Furthermore, other steps of the replication cycle, e.g. cell-to-cell and systemic movement, seed transmission, symptom induction or host specificity, must be studied in plants. It also remains to be established whether efficient yeast replication systems can be optimized for viruses that have more complex genomes (Alves-Rodrigues *et al.*, 2006).

The importance of specific host factors in the establishment of the virus disease can also be confirmed using an RNAi approach. In this method, sequence-specific silencing of the target gene is induced in transgenic plants that express a small portion of the gene in a hairpin conformation (Watson *et al.*, 2005). For fast throughput, the technique of virus-induced gene silencing (VIGS) has been developed (Lu *et al.*, 2003). In this technique, plant viruses (e.g. potato virus X or tobacco rattle virus) are used as carriers of the silencing sequence. The plant is subsequently infected with the test virus and the biological importance of the plant factors can be rapidly confirmed. A possible limitation of this technique is that initial infection by the carrier virus may alter the physiology of the plant.

Host Factors Involved in the Replication Cycle of (+)ssRNA Viruses

Replication cycle of (+)ssRNA viruses

(+)ssRNA viruses encode the RNA-dependent RNA polymerase (RdRp) that replicates their genome but do not package the polymerase into their virions. As a result, the first step in their replication cycle is the uncoating of the viral RNA followed by its translation by the host machinery. After viral replication proteins are produced, the genomic RNA becomes a template for the synthesis of progeny viral RNA molecules. As mentioned above, this process occurs in large complexes that contain viral and host proteins and the viral RNA. The complexes are associated with various host intracellular membranes (Salonen *et al.*, 2005; Sanfaçon, 2005). Viruses induce structural alterations in these membranes such as invaginations or vesicles that are protected environments wherein virus replication takes place. Replication of the RNA genome occurs in two steps: (i) an intermediate negative-sense single-strand RNA [(−)ssRNA] is produced using the genomic RNA as a template; and (ii) the (−)ssRNA serves as a template for the synthesis of progeny (+)ssRNA. The newly synthesized viral RNA must migrate to the periphery of the cell, move to an adjacent cell (cell-to-cell movement) and eventually spread throughout the entire plant (systemic movement).

A plant may be seen as a symplastic entity through which plasmodesmata provide continuity between adjacent cells, thus enabling the traffic of molecules and ions (Lough and Lucas, 2006). Plant viruses encode one or more movement proteins that facilitate their symplastic movement, thereby avoiding crossing additional membranes. Movement proteins also have the ability to modify the size exclusion limit of the plasmodesmata, thus facilitating virus intercellular transport. Two general mechanisms of virus movement have been described. In the first mechanism, movement proteins bind to the viral RNA and direct the ribonucleoprotein complex, termed movement complex, to the plasmodesmata (Boevink and Oparka, 2005). For some viruses, the movement protein is the only requirement for successful intercellular spread. For others, the coat protein (CP) is also required. In the second mechanism, viruses apparently move as intact virus particles through the formation of tubular structures that are embedded in highly modified plasmodesmata and traverse the cell wall (Scholthof, 2005). The viral movement protein is a structural component of these tubules.

The type of cells a virus can colonize in a host defines its tissue tropism and is a fundamental component of the fitness of a virus. Many phloem-limited viruses (e.g. poleroviruses and several geminiviruses) are unable to cross the phloem parenchyma-bundle sheath boundary. The presence of a checkpoint at the interface between companion cells and sieve elements has been suggested (Oparka and Turgeon, 1999). Thus, specific protein–protein interactions may be required to cross such a boundary (Lough and Lucas, 2006).

In the following sections, the role of selected plant factors in the replication cycle of (+)ssRNA viruses is discussed. Several of these factors play multiple roles in more than one step of the virus replication cycle. Many other host proteins, as yet uncharacterized, are required to accomplish the viral replication cycle. In fact, large-scale analyses of yeast mutants have revealed the involvement of an extensive variety of genes required for viral RNA replication (Kushner et al., 2003; Panavas et al., 2005). The host genes required for replication of TBSV and BMV, two unrelated (+)ssRNA viruses, are largely different. Thus, rigorous examination of plant–virus interactions is required for each virus family before strategies for virus control can be designed.

Translation factors

Viral genomic RNA acts as a messenger RNA (mRNA) but often lacks the 5′ end cap, the 3′ end poly (A) tail or both. Since these structures play a key role in recruiting translation factors to cellular mRNAs, viruses have developed alternate strategies to reroute host translation factors to their own RNAs, using complex networks of protein–protein, protein–RNA and/or RNA–RNA interactions (Thivierge et al., 2005; Dreher and Miller, 2006). Because the genomic RNA serves both as an mRNA for translation and as a template for replication, the two processes are tightly linked. Perhaps, not surprisingly, several host translation factors were found to co-purify with viral replication complexes, interact with viral replication proteins and regulate the efficiency of viral RNA replication or the switch from translation to replication (for reviews see Noueiry and Ahlquist, 2003; Sanfaçon, 2005).

Of particular interest for the development of resistance strategies are cases in which viruses sequester translation factors that are dispensable for the translation of cellular mRNAs. One example is the interaction of potyviruses with translation initiation factor 4E (eIF4E), the cap-binding protein (Thivierge et al., 2005; Robaglia and Caranta, 2006). eIF4E directs the assembly of the translation complex on cellular mRNAs. In plants, the protein exists in two isoforms (eIF4E and eIF(iso)4E) which, although divergent in their affinity for different mRNA substrates, are functionally interchangeable (Robaglia and Caranta, 2006). Thus, mutation of one isoform does not significantly affect the physiology of the plant (Duprat et al., 2002; Yoshii et al., 2004). Initially, an interaction between A. thaliana eIF(iso)4E and the turnip mosaic virus (TuMV) VPg protein was discovered (Wittmann et al., 1997). Similar interactions between one or both isoforms of eIF4E and the VPg of several potyviruses were subsequently described (reviewed in Thivierge et al., 2005). Mutations in VPg that hinder its interaction with eIF(iso)4E render TuMV non-infectious, suggesting that this interaction is essential for the virus (Duprat et al., 2002). This idea is further supported by the observation that mutation of eIF4E isoforms results in resistance to potyviruses (Robaglia and Caranta, 2006). The biological function of the interaction between eIF4E isoforms and VPg is not completely understood, but it is likely that it plays a role in promoting viral translation and/or shutting off host translation (Plante et al., 2004; Michon et al., 2006). A function in viral replication has also been suggested (Lellis et al., 2002; Beauchemin et al., 2007). Bimolecular fluorescence complementation analysis revealed that interaction of eIF(iso)4E with different polyprotein precursors of the TuMV VPg occurred at distinct subcellular locations, reinforcing the idea that the interaction plays multiple roles in the viral replication cycle (Beauchemin et al., 2007). A function in cell-to-cell movement of pea seed-borne mosaic virus (PSbMV) has also been attributed to eIF4E (Gao et al., 2004). In complementation experiments using PSbMV-resistant and PSbMV-susceptible pea lines, impaired movement in resistant plants was partially alleviated by co-expression of eIF4E from a susceptible line. It has been proposed that eIF4E acts as a guide molecule for movement through the plasmodesmata, probably with the cooperation of the ancillary eIF4G, which has a high affinity for microtubules (Gao et al., 2004).

Another notable example of interaction of a plant virus with a non-essential translation factor is the dependence of BMV on DED1 for the translation of one of its RNAs in yeast. DED1 is an RNA helicase related to eIF4A, a member of the translation initiation complex in eukaryotes. Mutation of DED1 results in the inhibition of viral replication, but does not affect yeast growth (Noueiry et al., 2000). BMV also recruits LSM1/LSM7, a series of deadenylation-dependent decapping factors for the translation of its genomic RNAs in yeast (Noueiry et al., 2003). LSM1 has a positive effect on viral replication and may be involved in the recruitment of this RNA directly from active translation complexes to the replication complex (Noueiry et al., 2003). LSM1/LSM7 are involved in cellular mRNA turnover, but are not essential for their translation. It is not known whether mutation of plant genes homologous to DED1 or LSM1 would also result in resistance to BMV without affecting the host.

Finally, a cucumber recessive resistance gene to cucumber mosaic virus (genus *Cucumovirus*) was mapped to a mutation in the gene coding for eIF4G, a multi-adaptor protein of the translation initiation complex that simultaneously binds eIF4E, eIF4A and other translation factors (eIF3 and the polyA-binding protein) (Yoshii et al., 2004). The mutation also conferred partial resistance to a carmovirus without affecting the physiology of the plant (Yoshii et al., 2004).

Membrane proteins and enzymes involved in membrane synthesis

Plant viruses have developed various means to anchor their replication complexes to specific intracellular membranes (Sanfaçon, 2005). In many cases, viral replication proteins interact directly with the lipid bilayer of the membranes and bring other viral and plant proteins to the replication complex through protein–protein interactions. However, in the case of tobamoviruses, host integral membrane proteins have been suggested to play a critical role in the assembly or maintenance of the replication complex in association with endoplasmic reticulum (ER) membranes. The analysis of mutants of *A. thaliana* impaired in their ability to support tomato mosaic virus (ToMV) replication revealed that they have defects in genes coding for host transmembrane proteins (Yamanaka et al., 2000; Tsujimoto et al., 2003). One of these proteins is Tom1, a seven-pass transmembrane protein that co-fractionates with the replication complex and interacts with ToMV replication proteins (Yamanaka et al., 2000; Hagiwara et al., 2003; Nishikiori et al., 2006). Another protein is Tom2, a four-pass transmembrane protein that interacts with Tom1 but not with ToMV replication proteins (Tsujimoto et al., 2003). Although ToMV replication proteins do not contain obvious hydrophobic domains to mediate their interaction with the lipid bilayer of the membranes, they are targeted to the ER when they are expressed independently of other viral proteins (dos Reis Figueira et al., 2002). Taken together, these results suggest that Tom1 is a membrane anchor for the ToMV replication complex. Tom2 may play an accessory role in viral replication by promoting proper folding of Tom1 (Tsujimoto et al., 2003).

In addition to protecting the replication complex, host membranes probably play an active role in viral replication. Indeed, the level of BMV (−)ssRNA synthesis is drastically reduced in a yeast mutant that carries a defect in δ9 fatty acid desaturase which results in alteration of the lipid composition of the membrane (Lee et al., 2001). It was also shown that although only a small population of ToMV RdRp is membrane-associated in infected cells, only this subpopulation is active in interacting with host factors and in replicating the viral RNA (Nishikiori et al., 2006). Large-scale analysis of yeast mutants deficient in BMV replication revealed the identity of several proteins involved in membrane modification pathways, although their functions in the viral replication cycle have not been characterized (Kushner et al., 2003).

Intracellular membranes are also probably involved in facilitating viral cell-to-cell movement. A critical early step for virus movement is the localization of plasmodesmal gates. This explains why many viruses move in close association with the endomembrane system and/or the cytoskeleton. The ER network, which terminates in desmotubules that guide the traffic of macromolecules through the plasmodesmata, has been implicated in the cell-to-cell movement of plant viruses and many viral movement proteins interact directly or indirectly with the ER (Boevink and Oparka, 2005). In tubule-forming viruses like grapevine fanleaf virus (GFLV, genus *Nepovirus*) and cowpea mosaic virus (CPMV, genus *Comovirus*), viral tubule formation leads to the disintegration of desmotubules and rupture of ER–plasmodesmata junctions (Pouwels et al., 2002; Laporte et al., 2003). Thus, protein–protein interactions at or near the plasmodesmata may be a converging point for viruses deploying different movement strategies.

The first report of an interaction between a host factor and a viral movement protein refers to a pectin methylesterase (PME) and the TMV movement protein (Dorokhov et al., 1999). TMV mutants harbouring a mutation in the movement protein that abolishes interaction with PME show limited movement in infected tissues (Chen et al., 2000). The biological significance of this interaction is still uncertain. However, it has been suggested that PME anchors the movement protein to the ER and directs the transport of the movement complex towards the cell wall and the plasmodesmata (Waigmann et al., 2004). A role for PME in TMV systemic movement has also been suggested. Although most attempts to suppress PME expression were unviable, analysis of transgenic tobacco plants expressing reduced levels of PME in the vasculature revealed that TMV is able to enter and traffic into the vascular system, but its unloading into mesophyll and epidermal cells is limited, suggesting that systemic transport of viruses is a polar process tightly regulated by protein–protein interactions (Chen and Citovsky, 2003). The movement proteins of tobamovirus and pararetrovirus also bind PME suggesting that recruiting of PME is a widespread strategy deployed by diverse viruses to accomplish their intercellular transport (Chen et al., 2000).

PME is not the only protein that has been suggested to assist targeting of movement proteins to the plasmodesmata. Calreticulin, a Ca^+-sequestering protein that accumulates in the plasmodesmata and harbours a peptide signal for ER localization, has been shown to interact with the TMV movement protein (Chen et al., 2005). Interestingly, overexpression of calreticulin in transgenic plants provokes a delay in virus cell-to-cell movement, possibly as a result of competition between unloaded calreticulin molecules and those loaded with the TMV movement protein for ER and/or plasmodesmata targeting (Chen et al., 2005). Overexpression of calreticulin also redirects the TMV movement protein to the microtubules network (Chen et al., 2005). The interaction with the microtubule network is possibly a movement-unrelated phenomenon leading to degradation of the movement protein (Chen et al., 2005 and references therein). In support of this suggestion, a microtubule-associated protein (MPB2C) was found to interact with the TMV movement protein and to interfere with its cell-to-cell movement (Kragler et al., 2003).

GFLV movement protein-induced tubular structures are formed at discrete intercellular sites that colocalize with calreticulin. Thus, calreticulin may act as a receptor for the movement protein (Laporte et al., 2003). In addition, the GFLV movement protein interacts directly or indirectly with KNOLLE, an integral membrane protein predominantly expressed during cytokinesis (Laporte et al., 2003). KNOLLE may help mobilize the GFLV movement protein to the cell plate during cell division. Mutation or silencing of KNOLLE and calreticulin will be necessary to clarify their roles in GFLV subcellular localization and transport.

Enzymes involved in callose synthesis

Regulation of the size exclusion limit of the plasmodesmata occurs at least in part by accumulation of callose (1,3-β-glucan), a structural component of plasmodesmata channels that is regulated by the action of β-1,3-glucanases. A role of callose in antiviral defence was initially suggested by the observation that antisense tobacco plants deficient in β-1,3-glucanase synthesis are less susceptible to TMV and tobacco necrosis virus infection than their wild-type counterpart (Beffa et al., 1996). Cell-to-cell movement of TMV, PVX and the movement protein of a cucumovirus is delayed in these transgenic plants (Iglesias and Meins, 2000).

An interaction was identified between a potexvirus movement protein and three tobacco proteins of the ankyrin repeat-containing proteins HBP1 family (Fridborg et al., 2003). These three proteins also interact with β-1,3-glucanase and may mediate the co-localization of the viral movement protein and β-1,3-glucanases in the vicinity of the plasmodesmata. Thus, traversing of the PVX movement complex through the plasmodesmata would be facilitated by this interaction.

Modifying enzymes

Recent evidence suggests that the activity of replication proteins, their ability to interact with each other or their stability is controlled by their state of phosphorylation, although the specific cellular kinases and phosphatases involved in these regulations have not yet been identified (see Sanfaçon, 2005). Similarly, many viral movement proteins are susceptible to phosphorylation and their function may be modulated through phosphorylation and dephosphorylation events mediated by kinases and phosphatases. For a detailed discussion about phosphorylation of movement proteins, the reader is advised to see Waigmann et al. (2004).

Host chaperones have also been implicated in regulating assembly of the replication complex. Analysis of yeast mutants has demonstrated that the

YDJ1 chaperone and an HSP70 homologue enhance the replication of BMV and CNV, respectively (Tomita et al., 2003; Serva and Nagy, 2006). In addition, the HSP70 protein was found in association with ToMV replication proteins within the active replication complex (Nishikiori et al., 2006).

Host Factors Involved in the Replication Cycle of ssDNA Viruses

Replication cycle of ssDNA viruses

Geminiviruses do not possess a *bona fide* polymerase. Instead, the replication-associated protein Rep provides the minimal requirements for initiation and termination of virus replication by the rolling circle mechanism (Hanley-Bowdoin et al., 2004). These viruses consequently rely on the cellular machinery to replicate their DNA. Furthermore, since geminiviruses are excluded from meristematic tissues (Hanley-Bowdoin et al., 2000) where DNA polymerases are active, they must induce quiescent cells to re-enter into the synthetic (S) phase of the cell cycle, thereby reinstating DNA replication. In the rolling circle mechanism, double-stranded DNA is generated by the action of nuclear host polymerases. Rep then nicks the (+)-strand of the duplex DNA at a highly conserved nonanucleotide and remains covalently attached to its 5' end. Unwinding of the parental (+)-strand DNA is mediated by a helicase while a DNA polymerase III adds nucleotides to the free 3'-OH end. This process is called rolling circle because at the same time that the (+)-strand is elongated, the (−)-strand is rotated on its own axis. Once a round of replication is completed, the new origin of replication is cleaved and a monomeric molecule is released. The newly synthesized DNA molecule is circularized through the ligase activity of Rep (reviewed in Hanley-Bowdoin et al., 2000).

To accomplish their trafficking within infected plants, geminiviruses require the concerted action of two proteins: (i) the nuclear shuttle protein; and (ii) the cell-to-cell movement protein. One of the models for intracellular and intercellular transport largely substantiated by experimental evidence proposes that a subpopulation of newly synthesized ssDNA is trapped, although not encapsidated, by the CP thereby preventing re-entrance into the rolling circle replication pool. The ssDNA is then shuttled to the cytoplasm by the nuclear shuttle protein and directed to the cell wall and plasmodesmata by the movement protein. In neighbouring cells, the movement protein is dissociated from the complex and the nuclear shuttle protein mobilizes the ssDNA into the nucleus to start a new cycle of replication (Lazarowitz and Beachy, 1999).

Given their dependence on cellular factors to accomplish their life cycle, ssDNA viruses have become ideal models for the study of host DNA replication, transcription and cell cycle regulation (Hanley-Bowdoin et al., 2000). This has spurred an intense search for host proteins that facilitate their replication and gene expression, as well as their intercellular and systemic spread through the infected plant.

Cell cycle regulators and DNA replication enzymes

E2F transcription factors play a pivotal role in regulating the G1/S transition in the cell cycle of higher eukaryotes. In quiescent cells, DNA replication is inhibited when a repressor complex is bound to the promoter of the gene coding for the proliferating cell nuclear antigen (PCNA), a processivity factor of the DNA polymerase δ. E2F binds to the PCNA promoter and to the retinoblastoma protein, a suppressor of chromatin remodelling activities, thereby creating the repressor complex. Mammalian DNA tumor-inducing viruses bind to the retinoblastoma protein, disrupt its interaction with E2F and relieve PCNA transcriptional repression.

Many geminiviruses alter the cell cycle progression by forcing an endocycle that skips the G2 and M phases and allows continuous DNA synthesis. Other geminiviruses seem to induce cell proliferation (Hanley-Bowdoin et al., 2004). Although PCNA is normally expressed in young but not mature leaves of healthy plants, it is detected in mature cells infected with tomato golden mosaic virus (TGMV) suggesting that PCNA transcriptional repression is relieved as a consequence of virus infection (Nagar et al., 1995). Two geminivirus replication-associated proteins (Rep and REn) were shown to bind pRBR, a plant homologue of the retinoblastoma protein, and disrupt its interaction with E2F. The biological significance of the Rep–pRBR interaction is evidenced by the fact that TGMV mutants impaired in their ability to bind pRBR induce only a moderate symptomatic phenotype (Nagar et al., 1995). Also, PCNA accumulates in differentiated cells of transgenic plants expressing the Rep protein (Hanley-Bowdoin et al., 2004). Similarly, the nanovirus CLINK protein, an accessory replication protein, was shown to interact with pRBR and with the core component of a particularly versatile class of E3 ubiquitin ligases referred to as MsSKP1 (Aronson et al., 2000). Although it is possible that the dual interaction of CLINK with MsSKP1 and pRBR contributes to the release of transcriptional repression by facilitating pRBR degradation, further experimental evidence is required to confirm this hypothesis.

Other protein–protein interactions may also be involved in the alteration of the plant cell cycle. The TGMV Rep protein interacts with a plant kinesin termed GRIMP (Kong and Hanley-Bowdoin, 2002). Although GRIMP is expressed constitutively in plants, it is associated with the spindle apparatus and condensed chromosomes during mitosis. It has been suggested that the Rep–GRIMP interaction prevents phosphorylation of GRIMP and alters the cell cycle (Kong and Hanley-Bowdoin, 2002).

Geminiviruses must recruit the host DNA replisome to the virus genome and interactions between viral replication-associated proteins and host processivity factors, such as PCNA, have been suggested to play a key role in this process. For example, the geminivirus Rep protein and REn protein interact with PCNA (Castillo et al., 2003). Similarly, the Rep protein of wheat dwarf virus interacts with replication factor C (RFC), another processivity factor that facilitates loading of PCNA and polymerase δ onto the replisome (Luque et al., 2002). An interaction was also identified between the TGMV Rep protein and histone H3 of *Nicotiana benthamiana* (Kong and Hanley-Bowdoin, 2002). Since the geminiviral DNA is packaged into minichromosomes, the Rep–Histone H3 association may permit displacement of nucleosomes thus allowing access of the replication and transcription machinery to the regulatory elements of the geminiviral genome.

Because the host proteins described above are central regulatory proteins involved in plant development, cell cycle regulation and DNA replication and repair, it is difficult to envisage how they could be appropriate targets for antiviral strategies. In fact, it is likely that inactivation of any of these proteins would severely affect the plant physiology. However, a basic understanding of the mechanism by which geminiviruses replicate in quiescent plant cells is the keystone for the identification of other possibly more specific host factors.

Transcription factors

In two separate studies, the Rep or REn protein of geminiviruses were shown to interact with members of the nascent polypeptide-associated complex (NAC) domain protein family (Xie *et al.*, 1999; Selth *et al.*, 2005). This large family of proteins is involved in many essential processes such as plant development, senescence and plant defence (Selth *et al.*, 2005 and references therein). This may explain why different NAC proteins shown to interact with geminivirus proteins have apparently opposite effects on virus replication. The Rep protein of wheat dwarf virus interacts with several members of the NAC domain family (Xie *et al.*, 1999). Overexpression of these proteins in cultured wheat cells exerts an inhibitory effect on viral DNA replication, suggesting that they repress the expression of viral genes or of cellular factors required for virus multiplication (Xie *et al.*, 1999).

The REn protein of tomato leaf curl virus (TLCV) was shown to interact with tomato SlNAC1 and upregulate its expression (Selth *et al.*, 2005). In contrast to the G-related α2-macroglobulin-binding (GRAB) proteins, transient expression of SlNAC1 increases virus accumulation in *N. benthamiana* (Selth *et al.*, 2005). One possible interpretation of this result is that SlNAC1 promotes the transcription of genes required for initiation and/or maintenance of the S phase of the cell cycle. Because SlNAC1 mRNA is detected in TLCV-infected cells but not in non-infected cells of *N. benthamiana*, it is a potential candidate for antiviral strategies (Selth *et al.*, 2005).

Cell wall synthesis enzymes

The enzyme SlUTPG1 from tomato is involved in the synthesis of cell wall components and interacts with the V1 protein of three different geminiviruses (Selth *et al.*, 2006). SlUTPG1 also enhances virus accumulation. TLCV V1 mutants unable to interact with SlUTPG1 exhibit reduced cell-to-cell movement suggesting that the contribution of this host factor to virus accumulation is due, at least partially, to enhancement of virus transport (Selth et al., 2006). Further experiments will be necessary to determine whether inactivation of SlUTPG1 results in increased resistance to TLCV without impairing essential plant functions.

Plant-modifying enzymes

Interaction of geminivirus proteins with plant protein kinases plays a central role in modulating the functional state of viral and host proteins. Some of these kinases have been shown to enhance virus accumulation. For example, the nuclear shuttle protein interacts with NsAK, a proline-rich extensin-like receptor protein kinase (PERK) (Florentino et al., 2006). This kinase seems to enhance geminivirus infectivity apparently through phosphorylation of the nuclear shuttle protein. Disruption of NsAK activity in T-DNA insertion mutants attenuates, although does not abolish, viral infectivity. The TGMV Rep protein interacts with a serine–threonine kinase termed GRIK (Kong and Hanley-Bowdoin, 2002). Although GRIK is normally expressed in young but not mature tissues, TGMV infection induces the expression of GRIK in mature leaves. Since Rep itself is not phosphorylated by GRIK, the Rep–GRIK interaction may facilitate the recruiting of proteins that are phosphorylated by GRIK and required for geminivirus replication (Kong and Hanley-Bowdoin, 2002).

In other cases, geminivirus proteins interact with kinases that are involved in a general plant defence response. These kinases have a negative effect on virus accumulation. For example, the geminivirus transcriptional transactivator protein (TrAP) interacts with and inhibits the activity of SNF1 and ADK. These two cellular kinases are normally activated in response to pathogen infection. Downregulation of SNF1 increases BCTV pathogenicity, while its constitutive expression confers resistance to viral infection (Hao et al., 2003; Wang et al., 2003). Similarly, the geminivirus nuclear shuttle protein interacts with several kinases, which structurally resemble receptor-like protein kinases (RLKs) involved in signal transduction during stress responses (Mariano et al., 2004). A. thaliana loss-of-function mutants are highly susceptible to virus infection, suggesting that the nuclear shuttle protein blocks a signal cascade mediated by these kinases and activated in response to virus invasion (Fontes et al., 2004).

Other types of modifying enzymes have been implicated in the replication cycle of geminiviruses. Acetyltransferases are involved in essential processes in plants ranging from DNA replication, transcription, recombination and repair to apoptosis (Carvalho et al., 2006). The nuclear shuttle protein of a geminivirus interacts with a nuclear acetyltransferase, AtNSI, and inhibits its activity (McGarry et al., 2003; Carvalho et al., 2006). However, virus accumulation is enhanced in transgenic plants overexpressing AtNSI, indicating that AtNSI is required for virus infectivity. To reconcile these apparently contradictory results, a dual role in geminivirus infection has been proposed for AtNSI (Carvalho et al., 2006). On the one hand, the nuclear shuttle protein may sequester AtNSI to the ssDNA–CP complex and promote acetylation of the CP, a process which has been suggested to facilitate the export of the ssDNA from the nucleus. On the other hand, sequestering of AtNSI by the nuclear shuttle protein may inhibit its ability to acetylate host proteins involved in cell differentiation. It is not known whether mutation or silencing of AtNSI would inhibit virus accumulation in plants.

Sumoylation, or covalent attachment of a ubiquitin-like polypeptide called SUMO, can regulate the activity or localization of target proteins. The Rep of two geminiviruses interacts with a SUMO-conjugating enzyme, NbSCE1 (Castillo et al., 2004). Sumoylation has been implicated in diverse physiological processes activated during stress responses. Perhaps not surprisingly, both upregulation and downregulation of SUMO in transgenic Nicotiana tabacum, by sense and antisense expression of the tomato ortholog LeSUMO, lead to a reduction in viral DNA accumulation. This effect was also observed in a transient leaf disk assay, suggesting that it is acting at the level of replication rather than cell-to-cell movement. Although many proteins that interact with SUMO-conjugating enzymes appear to be sumoylated, it is still unknown whether Rep itself is a substrate for sumoylation by NbSCE1 in infected plants. Three potential sumoylation sites were delineated in domains which are essential for Rep functions. The second site overlaps with the motif involved in interaction with pRBR (Castillo et al., 2004). Thus, sumoylation may interfere with the ability of Rep to bind pRBR and disrupt transcriptional repression.

Development of new technologies

We have already mentioned several examples in which mutations of genes coding for specific host factors result in reduced susceptibility or even resistance to various plant viruses. To illustrate the potential and limitations of these approaches, we discuss in detail two specific examples of engineered virus resistance obtained by inactivating well-characterized host factors.

Inactivation of eIF4E isoforms confers resistance against potyviruses

The number of genes that encode proteins related to the eIF(iso)4E or eIF4E families in plants varies from species to species. In A. thaliana, there is one eIF(iso)4E gene and three eIF4E genes, although only eIF4E1 has been conclusively demonstrated to encode a cap-binding protein. As mentioned earlier, knock-out mutations of either eIF4E1 or eIF(iso)4E are not deleterious for A. thaliana (Duprat et al., 2002; Yoshii et al., 2004). In fact, many recessive resistance genes correspond to mutations of eIF4E isoforms (reviewed in Robaglia and Caranta, 2006). The importance of eIF4E in virus infection is not limited to potyviruses. Resistance to a cucumovirus and a bymovirus is also conferred by mutation of eIF4E (Yoshii et al., 2004; Stein et al., 2005). The interaction between eIF4E and potyvirus VPgs is, in general, highly specific as evidenced by the observation that within a plant species, some potyviruses interact with eIF4E while others interact with eIF(iso)4E (Robaglia and Caranta, 2006). For instance, in *A. thaliana*, mutation of eIF(iso)4E confers resistance to TuMV, lettuce mosaic virus (LMV) and TEV, but not to clover yellow vein virus (ClYVV) while mutation of eIF4E1 results in resistance to ClYVV, but not to TuMV (Duprat et al., 2002; Lellis et al., 2002; Sato et al., 2005). In some cases, the interaction between the potyvirus VPg and eIF4E is even strain specific (Schaad et al., 2000). A given potyvirus may also depend on distinct isoforms of eIF4E in different hosts. For example, LMV interacts with eIF(iso)4E in *A. thaliana* but with eIF4E in lettuce (Duprat et al., 2002; Nicaise et al., 2003). One exception to this rule is the recent observation that pepper veinal mottle virus can use either eIF4E or eIF(iso)4E in pepper and that mutation of both isoforms is necessary to obtain resistance to the virus (Ruffel et al., 2006). Thus, although mutation or inactivation of specific eIF4E genes is an attractive strategy to engineer resistance to potyviruses, careful examination of the specific interaction must be conducted for each potyvirus–host combination before effective resistance can be engineered. Finally, it should be noted that hypervirulent potyvirus isolates that are able to overcome resistant cultivars harbouring point mutations of eIF4E isoforms have been reported. In most cases, the increased virulence was linked to mutations in the VPg protein (reviewed in Robaglia and Caranta, 2006). Although it is likely that knocking out the expression of an eIF4E isoform will be more difficult for the virus to overcome, the possible emergence of virulent isolates in response to knock-out mutations has not been studied.

Inactivation of host membrane proteins confers resistance against tobamoviruses

As mentioned earlier, the Tom1 protein is probably the membrane anchor for the ToMV replication complex. Two homologues of Tom1, i.e. Tom3 and THH1, have been identified in *A. thaliana*. These proteins can functionally replace Tom1 although they cannot support ToMV replication with equal efficiency (Yamanaka et al., 2002; Fujisaki et al., 2006). Analysis of single, double and triple mutants of Tom1, Tom3 and THH1 revealed that Tom1 is the primary determinant of ToMV replication. When Tom1 is mutated, Tom3 is more efficient than THH1 at supporting viral replication. In fact, viral replication is completely abolished in a double Tom1–Tom3 mutant (Yamanaka et al., 2002). Of practical importance, the dependence of tobamoviruses on Tom1/Tom3 is not limited to the ToMV–*A. thaliana* interaction, but is applicable to other host–tobamovirus combinations. Recently, *N. tabacum* orthologs of Tom1 and Tom3 have been identified. Transgenic expression of either one of the *N. tabacum* genes in the *A. thaliana tom1/tom3* double mutant restores its ability to support

ToMV replication (Asano *et al.*, 2005). Tom1/Tom3 homologues also exist in tomato, melon and rice, and the tomato and melon genes can complement the *A. thaliana* mutants (Asano *et al.*, 2005). To confirm the activity of Tom1/Tom3 in *Nicotiana* spp., an RNAi approach was used to selectively silence the expression of one or both genes in the *N. tabacum* var. Samsun, a host highly susceptible to many tobamoviruses (Asano *et al.*, 2005). While silencing of only one gene results in moderate resistance to ToMV, simultaneous silencing of both genes confers to the plant high resistance not only to ToMV, but also to three other tobamoviruses. The doubly silenced plant remained highly susceptible to a cucumovirus, confirming that the Tom1/Tom3 proteins are specifically required for tobamovirus replication. However, resistance to the tobamoviruses is not complete, as low levels of virus accumulation are observed in systemic leaves after longer incubations (40 days post-inoculation) and this is accompanied by mild mosaic. This is not due to an emerging mutated population of the virus, since virus recovered from the mildly infected leaves does not show increased multiplication when reinoculated on the doubly silenced plants. In fact, virulent mutants of the virus are not observed even after several successive passages through the resistant plants (Asano *et al.*, 2005). It is not known whether viral suppressors of silencing are responsible at least in part for the partial breaking of the resistance conferred by the RNAi approach. The plant morphology is not significantly affected by the inactivation of the two genes in *N. tabaccum* or *A. thaliana*, at least in a greenhouse experimental set-up, although the authors noted an occasional slight reduction in the plant growth (Yamanaka *et al.*, 2002; Asano *et al.*, 2005). These results are promising as they suggest that broad-spectrum resistance to tobamoviruses can be engineered in a variety of crops through the mutation or inactivation of the Tom1/Tom3 host genes.

Conclusions and Future Directions

The last 10 years of research in molecular plant virology have unraveled an unforeseen degree of complexity in the nature of interactions between plant viruses and their hosts. With the increased speed of discovery of new interactions, it is likely that large networks of protein–protein interactions will be mapped in the not-too-distant future. The next challenge is to unravel the biological significance of these interactions not only in model hosts (e.g. yeast, *A. thaliana*), but also in economically important crops which may differ in their ability to provide a specific host factor. Unfortunately, many important crops have been less well characterized at the genomic level and are also less amenable to genetic manipulation, making reverse genetic approaches more challenging.

The potential of antiviral strategies based on the modification of the expression pattern of host proteins involved in the plant virus replication cycle has been demonstrated. However, the durability of such approaches must be rigorously tested. The continued study of plant–virus interactions will allow the design of improved strategies based on the targeting of multiple host factors that would affect more than one step of the replication cycle, thereby reducing the ability of the virus to overcome these introduced barriers.

Acknowledgements

We wish to thank Drs J.F. Laliberte and M. Ishikawa for sharing results prior to publication. We would also like to thank Joan Chisholm for critical reading of the manuscript. Because of space limitation, we were forced to limit ourselves to a few selected examples of protein–protein interactions between plant viruses and their hosts. We apologize to our readers for this shortcoming and to the authors of many excellent studies which could not be included. Work in the Sanfaçon laboratory is supported in part by an NSERC Discovery Grant and by targeted funding from the plum pox virus national initiative.

References

Ahlquist, P., Noueiry, A.O., Lee, W.M., Kushner, D.B. and Dye, B.T. (2003) Host factors in positive-strand RNA virus genome replication. *Journal of Virology* 77, 8181–8186.

Alves-Rodrigues, I., Galao, R.P., Meyerhans, A. and Diez, J. (2006) *Saccharomyces cerevisiae*: a useful model host to study fundamental biology of viral replication. *Virus Research* 120, 49–56.

Aronson, M.N., Meyer, A.D., Gyorgyey, J., Katul, L., Vetten, H.J., Gronenborn, B. and Timchenko, T. (2000) Clink, a nanovirus-encoded protein, binds both pRB and SKP1. *Journal of Virology* 74, 2967–2972.

Asano, M., Satoh, R., Mochizuki, A., Tsuda, S., Yamanaka, T., Nishiguchi, M., Hirai, K., Meshi, T., Naito, S. and Ishikawa, M. (2005) Tobamovirus-resistant tobacco generated by RNA interference directed against host genes. *FEBS Letters* 579, 4479–4484.

Beauchemin, C., Boutet, N. and Laliberte, J.F. (2007) Visualisation of the interaction between the precursors of the viral protein linked to the genome (VPg) of *Turnip mosaic virus* and the translation eukaryotic initiation factor iso 4E in planta. *Journal of Virology* 81, 775–782.

Beffa, R.S., Hofer, R.M., Thomas, M. and Meins, F. Jr (1996) Decreased susceptibility to viral disease of [beta]-1,3-glucanase-deficient plants generated by antisense transformation. *The Plant Cell* 8, 1001–1011.

Bhat, R.A., Lahaye, T. and Panstruga, R. (2006) The visible touch: in planta visualization of protein–protein interactions by fluorophore-based methods. *Plant Methods* 2, 12.

Boevink, P. and Oparka, K.J. (2005) Virus–host interactions during movement processes. *Plant Physiology* 138, 1815–1821.

Boguszewska-Chachulska, A.M. and Haenni, A.L. (2005) RNA viruses redirect host factors to better amplify their genome. *Advances in Virus Research* 65, 29–61.

Carvalho, M.F., Turgeon, R. and Lazarowitz, S.G. (2006) The geminivirus nuclear shuttle protein NSP inhibits the activity of AtNSI, a vascular-expressed Arabidopsis acetyltransferase regulated with the sink-to-source transition. *Plant Physiology* 140, 1317–1330.

Castillo, A.G., Collinet, D., Deret, S., Kashoggi, A. and Bejarano, E.R. (2003) Dual interaction of plant PCNA with geminivirus replication accessory protein (Ren) and viral replication protein (Rep). *Virology* 312, 381–394.

Castillo, A.G., Kong, L.J., Hanley-Bowdoin, L. and Bejarano, E.R. (2004) Interaction between a geminivirus replication protein and the plant sumoylation system. *Journal of Virology* 78, 2758–2769.

Chen, M.H. and Citovsky, V. (2003) Systemic movement of a tobamovirus requires host cell pectin methylesterase. *The Plant Journal* 35, 386–392.

Chen, M.H., Sheng, J., Hind, G., Handa, A.K. and Citovsky, V. (2000) Interaction between the tobacco mosaic virus movement protein and host cell pectin methylesterases is required for viral cell-to-cell movement. *The EMBO Journal* 19, 913–920.

Chen, M.H., Tian, G.W., Gafni, Y. and Citovsky, V. (2005) Effects of calreticulin on viral cell-to-cell movement. *Plant Physiology* 138, 1866–1876.

Dorokhov, Y.L., Makinen, K., Frolova, O.Y., Merits, A., Saarinen, J., Kalkkinen, N., Atabekov, J.G. and Saarma, M. (1999) A novel function for a ubiquitous plant enzyme pectin methylesterase: the host-cell receptor for the tobacco mosaic virus movement protein. *FEBS Letters* 461, 223–228.

dos Reis Figueira, A., Golem, S., Goregaoker, S.P. and Culver, J.N. (2002) A nuclear localization signal and a membrane association domain contribute to the cellular localization of the Tobacco mosaic virus 126-kDa replicase protein. *Virology* 301, 81–89.

Dreher, T.W. and Miller, W.A. (2006) Translational control in positive strand RNA plant viruses. *Virology* 344, 185–197.

Duprat, A., Caranta, C., Revers, F., Menand, B., Browning, K.S. and Robaglia, C. (2002) The Arabidopsis eukaryotic initiation factor (iso)4E is dispensable for plant growth but required for susceptibility to potyviruses. *The Plant Journal* 32, 927–934.

Fields, S. (2005) High-throughput two-hybrid analysis. The promise and the peril. *FEBS Journal* 272, 5391–5399.

Figeys, D., McBroom, L.D. and Moran, M.F. (2001) Mass spectrometry for the study of protein–protein interactions. *Methods* 24, 230–239.

Florentino, L.H., Santos, A.A., Fontenelle, M.R., Pinheiro, G.L., Zerbini, F.M., Baracat-Pereira, M.C. and Fontes, E.P. (2006) A PERK-like receptor kinase interacts with the geminivirus nuclear shuttle protein and potentiates viral infection. *Journal of Virology* 80, 6648–6656.

Fontes, E.P., Santos, A.A., Luz, D.F., Waclawovsky, A.J. and Chory, J. (2004) The geminivirus nuclear shuttle protein is a virulence factor that suppresses transmembrane receptor kinase activity. *Genes and Development* 18, 2545–2556.

Fridborg, I., Grainger, J., Page, A., Coleman, M., Findlay, K. and Angell, S. (2003) TIP, a novel host factor linking callose degradation with the cell-to-cell movement of *Potato virus X*. *Molecular Plant-Microbe Interactions* 16, 132–140.

Fujisaki, K., Ravelo, G.B., Naito, S. and Ishikawa, M. (2006) Involvement of THH1, an *Arabidopsis thaliana* homologue of the TOM1 gene, in tobamovirus multiplication. *Journal of General Virology* 87, 2397–2401.

Gao, Z., Johansen, E., Eyers, S., Thomas, C.L., Noel Ellis, T.H. and Maule, A.J. (2004) The potyvirus recessive resistance gene, sbm1, identifies a novel role for translation initiation factor eIF4E in cell-to-cell trafficking. *The Plant Journal* 40, 376–385.

Hagiwara, Y., Komoda, K., Yamanaka, T., Tamai, A., Meshi, T., Funada, R., Tsuchiya, T., Naito, S. and Ishikawa, M. (2003) Subcellular localization of host and viral proteins associated with tobamovirus RNA replication. *The EMBO Journal* 22, 344–353.

Hanley-Bowdoin, L., Settlage, S.B., Orozco, B.M., Nagar, S. and Robertson, D. (2000) Geminiviruses: models for plant DNA replication, transcription, and cell cycle regulation. *Critical Reviews in Biochemistry and Molecular Biology* 35, 105–140.

Hanley-Bowdoin, L., Settlage, S.B. and Robertson, N. (2004) Reprogramming plant gene expression: a prerequisite to geminivirus DNA replication. *Molecular Plant Pathology* 5, 149–156.

Hao, L., Wang, H., Sunter, G. and Bisaro, D.M. (2003) Geminivirus AL2 and L2 proteins interact with and inactivate SNF1 kinase. *The Plant Cell* 15, 1034–1048.

Iglesias, V.A. and Meins, F. Jr (2000) Movement of plant viruses is delayed in a beta-1,3-glucanase-deficient mutant showing a reduced plasmodesmatal size exclusion limit and enhanced callose deposition. *The Plant Journal* 21, 157–166.

Kang, B.C., Yeam, I. and Jahn, M.M. (2005) Genetics of plant virus resistance. *Annual Review of Phytopathology* 43, 581–621.

Kong, L.J. and Hanley-Bowdoin, L. (2002) A geminivirus replication protein interacts with a protein kinase and a motor protein that display different expression patterns during plant development and infection. *The Plant Cell* 14, 1817–1832.

Kragler, F., Curin, M., Trutnyeva, K., Gansch, A. and Waigmann, E. (2003) MPB2C, a microtubule-associated plant protein binds to and interferes with cell-to-cell transport of tobacco mosaic virus movement protein. *Plant Physiology* 132, 1870–1883.

Kushner, D.B., Lindenbach, B.D., Grdzelishvili, V.Z., Noueiry, A.O., Paul, S.M. and Ahlquist, P. (2003) Systematic, genome-wide identification of host genes affecting replication of a positive-strand RNA virus. *Proceedings of the National Academy of Sciences of the United States of America* 100, 15764–15769.

Laporte, C., Vetter, G., Loudes, A.M., Robinson, D.G., Hillmer, S., Stussi-Garaud, C. and Ritzenthaler, C. (2003) Involvement of the secretory pathway and the cytoskeleton in intracellular targeting and tubule assembly of *Grapevine fanleaf virus* movement protein in tobacco BY-2 cells. *The Plant Cell* 15, 2058–2075.

Lazarowitz, S.G. and Beachy, R.N. (1999) Viral movement proteins as probes for intracellular and intercellular trafficking in plants. *The Plant Cell* 11, 535–548.

Lee, W.M., Ishikawa, M. and Ahlquist, P. (2001) Mutation of host delta9 fatty acid desaturase inhibits brome mosaic virus RNA replication between template recognition and RNA synthesis. *Journal of Virology* 75, 2097–2106.

Lellis, A.D., Kasschau, K.D., Whitham, S.A. and Carrington, J.C. (2002) Loss-of-susceptibility mutants of *Arabidopsis thaliana* reveal an essential role for eIF(iso)4E during potyvirus infection. *Current Biology* 12, 1046–1051.

Lough, T.J. and Lucas, W.J. (2006) Integrative plant biology: role of phloem long-distance macromolecular trafficking. *Annual Review of Plant Biology* 57, 203–232.

Lu, R., Martin-Hernandez, A.M., Peart, J.R., Malcuit, I. and Baulcombe, D.C. (2003) Virus-induced gene silencing in plants. *Methods* 30, 296–303.

Luque, A., Sanz-Burgos, A.P., Ramirez-Parra, E., Castellano, M.M. and Gutierrez, C. (2002) Interaction of geminivirus Rep protein with replication factor C and its potential role during geminivirus DNA replication. *Virology* 302, 83–94.

Mariano, A.C., Andrade, M.O., Santos, A.A., Carolino, S.M., Oliveira, M.L., Baracat-Pereira, M.C., Brommonshenkel, S.H. and Fontes, E.P. (2004) Identification of a novel receptor-like protein kinase that interacts with a geminivirus nuclear shuttle protein. *Virology* 318, 24–31.

McGarry, R.C., Barron, Y.D., Carvalho, M.F., Hill, J.E., Gold, D., Cheung, E., Kraus, W.L. and Lazarowitz, S.G. (2003) A novel Arabidopsis acetyltransferase interacts with the geminivirus movement protein NSP. *The Plant Cell* 15, 1605–1618.

Michon, T., Estevez, Y., Walter, J., German-Retana, S. and Le Gall, O. (2006) The potyviral virus genome-linked protein VPg forms a ternary complex with the eukaryotic initiation factors eIF4E and eIF4G and reduces eIF4E affinity for a mRNA cap analogue. *FEBS Journal* 273, 1312–1322.

Mouches, C., Candresse, T. and Bove, J.M. (1984) Turnip yellow mosaic virus RNA-replicase contains host and virus-encoded subunits. *Virology* 134, 78–89.

Nagar, S., Pedersen, T.J., Carrick, K.M., Hanley-Bowdoin, L. and Robertson, D. (1995) A geminivirus induces expression of a host DNA synthesis protein in terminally differentiated plant cells. *The Plant Cell* 7, 705–719.

Nagy, P.D. and Pogany, J. (2006) Yeast as a model host to dissect functions of viral and host factors in tombusvirus replication. *Virology* 344, 211–220.

Nelson, R.S. and Citovsky, V. (2005) Plant viruses – invaders of cells and pirates of cellular pathways. *Plant Physiology* 138, 1809–1814.

Nicaise, V., German-Retana, S., Sanjuan, R., Dubrana, M.P., Mazier, M., Maisonneuve, B., Candresse, T., Caranta, C. and LeGall, O. (2003) The eukaryotic translation initiation factor 4E controls lettuce susceptibility to the Potyvirus *Lettuce mosaic virus*. *Plant Physiology* 132, 1272–1282.

Nishikiori, M., Dohi, K., Mori, M., Meshi, T., Naito, S. and Ishikawa, M. (2006) Membrane-bound tomato mosaic virus replication proteins participate in RNA synthesis and are associated with host proteins in a pattern distinct from those that are not membrane bound. *Journal of Virology* 80, 8459–8468.

Noueiry, A.O. and Ahlquist, P. (2003) Brome mosaic virus RNA replication: revealing the role of the host in RNA virus replication. *Annual Review of Phytopathology* 41, 77–98.

Noueiry, A.O., Chen, J. and Ahlquist, P. (2000) A mutant allele of essential, general translation initiation factor DED1 selectively inhibits translation of a viral mRNA. *Proceedings of the National Academy of Sciences of the United States of America* 97, 12985–12990.

Noueiry, A.O., Diez, J., Falk, S.P., Chen, J. and Ahlquist, P. (2003) Yeast Lsm1p-7p/Pat1p deadenylation-dependent mRNA-decapping factors are required for brome mosaic virus genomic RNA translation. *Molecular and Cellular Biology* 23, 4094–4106.

Oparka, K.J. and Turgeon, R. (1999) Sieve elements and companion cells-traffic control centers of the phloem. *The Plant Cell* 11, 739–750.

Panavas, T., Serviene, E., Brasher, J. and Nagy, P.D. (2005) Yeast genome-wide screen reveals dissimilar sets of host genes affecting replication of RNA viruses. *Proceedings of the National Academy of Sciences of the United States of America* 102, 7326–7331.

Plante, D., Viel, C., Leonard, S., Tampo, H., Laliberte, J.F. and Fortin, M.G. (2004) Turnip mosaic virus VPg does not disrupt the translation initiation complex but interfere with cap binding. *Physiological and Molecular Plant Pathology* 64, 219–226.

Pouwels, J., Carette, J.E., Van Lent, J. and Wellink, J. (2002) Cowpea mosaic virus: effects on host cell processes. *Molecular Plant Pathology* 3, 411–418.

Quadt, R., Kao, C.C., Browning, K.S., Hershberger, R.P. and Ahlquist, P. (1993) Characterization of a host protein associated with brome mosaic virus RNA-dependent RNA polymerase. *Proceedings of the National Academy of Sciences of the United States of America* 90, 1498–1502.

Raghavan, V., Malik, P.S., Choudhury, N.R. and Mukherjee, S.K. (2004) The DNA-A component of a plant geminivirus (Indian mung bean yellow mosaic virus) replicates in budding yeast cells. *Journal of Virology* 78, 2405–2413.

Robaglia, C. and Caranta, C. (2006) Translation initiation factors: a weak link in plant RNA virus infection. *Trends in Plant Sciences* 11, 40–45.

Ruffel, S., Gallois, J.L., Moury, B., Robaglia, C., Palloix, A. and Caranta, C. (2006) Simultaneous mutations in translation initiation factors eIF4E and eIF(iso)4E are required to prevent pepper veinal mottle virus infection of pepper. *Journal of General Virology* 87, 2089–2098.

Salonen, A., Ahola, T. and Kaariainen, L. (2005) Viral RNA replication in association with cellular membranes. *Current Topics in Microbiology and Immunology* 285, 139–173.

Sanfaçon, H. (2005) Replication of positive-strand RNA viruses in plants: contact points between plant and virus components. *Canadian Journal of Botany* 83, 1529–1549.

Sato, M., Nakahara, K., Yoshii, M., Ishikawa, M. and Uyeda, I. (2005) Selective involvement of members of the eukaryotic initiation factor 4E family in the infection of *Arabidopsis thaliana* by potyviruses. *FEBS Letters* 579, 1167–1171.

Schaad, M.C., Anderberg, R.J. and Carrington, J.C. (2000) Strain-specific interaction of the tobacco etch virus NIa protein with the translation initiation factor eIF4E in the yeast two-hybrid system. *Virology* 273, 300–306.

Schneider, R.J. and Mohr, I. (2003) Translation initiation and viral tricks. *Trends in Biochemical Sciences* 28, 130–136.

Scholthof, H.B. (2005) Plant virus transport: motions of functional equivalence. *Trends in Plant Sciences* 10, 376–382.

Selth, L.A., Dogra, S.C., Rasheed, M.S., Healy, H., Randles, J.W. and Rezaian, M.A. (2005) ANAC domain protein interacts with tomato leaf curl virus replication accessory protein and enhances viral replication. *The Plant Cell* 17, 311–325.

Selth, L.A., Dogra, S.C., Rasheed, M.S., Randles, J.W. and Rezaian, M.A. (2006) Identification and characterization of a host reversibly glycosylated peptide that interacts with the *Tomato leaf curl virus* V1 protein. *The Plant Molecular Biology* 61, 297–310.

Serva, S. and Nagy, P.D. (2006) Proteomics analysis of the tombusvirus replicase: Hsp70 molecular chaperone is associated with the replicase and enhances viral RNA replication. *Journal of Virology* 80, 2162–2169.

Soosaar, J.L., Burch-Smith, T.M. and Dinesh-Kumar, S.P. (2005) Mechanisms of plant resistance to viruses. *Nature Reviews Microbiology* 3, 789–798.

Stein, N., Perovic, D., Kumlehn, J., Pellio, B., Stracke, S., Streng, S., Ordon, F. and Graner, A. (2005) The eukaryotic translation initiation factor 4E confers multiallelic recessive Bymovirus resistance in *Hordeum vulgare* (L.). *The Plant Journal* 42, 912–922.

Thivierge, K., Nicaise, V., Dufresne, P.J., Cotton, S., Laliberte, J.F., Le Gall, O. and Fortin, M.G. (2005) Plant virus RNAs. Coordinated recruitment of conserved host functions by (+)ssRNA viruses during early infection events. *Plant Physiology* 138, 1822–1827.

Tomita, Y., Mizuno, T., Diez, J., Naito, S., Ahlquist, P. and Ishikawa, M. (2003) Mutation of host DnaJ homolog inhibits brome mosaic virus negative-strand RNA synthesis. *Journal of Virology* 77, 2990–2997.

Tsujimoto, Y., Numaga, T., Ohshima, K., Yano, M.A., Ohsawa, R., Goto, D.B., Naito, S. and Ishikawa, M. (2003) *Arabidopsis TOBAMOVIRUS MULTIPLICATION (TOM) 2* locus encodes a transmembrane protein that interacts with TOM1. *The EMBO Journal* 22, 335–343.

Waigmann, E., Ueki, S., Trutnyeva, K. and Citovsky, V. (2004) The ins and out of nondestructive cell-to-cell and systemic movement of plant viruses. *Critical Reviews in Plant Sciences* 23, 195–250.

Wang, H., Hao, L., Shung, C.Y., Sunter, G. and Bisaro, D.M. (2003) Adenosine kinase is inactivated by geminivirus AL2 and L2 proteins. *The Plant Cell* 15, 3020–3032.

Watson, J.M., Fusaro, A.F., Wang, M. and Waterhouse, P.M. (2005) RNA silencing platforms in plants. *FEBS Letters* 579, 5982–5987.

Whitham, S.A. and Wang, Y. (2004) Roles for host factors in plant viral pathogenicity. *Current Opinion in Plant Biology* 7, 365–371.

Wittmann, S., Chatel, H., Fortin, M.G. and Laliberte, J.F. (1997) Interaction of the viral protein genome linked of turnip mosaic potyvirus with the translational eukaryotic initiation factor (iso) 4E of *Arabidopsis thaliana* using the yeast two-hybrid system. *Virology* 234, 84–92.

Xie, Q., Sanz-Burgos, A.P., Guo, H., Garcia, J.A. and Gutierrez, C. (1999) GRAB proteins, novel members of the NAC domain family, isolated by their interaction with a geminivirus protein. *Plant Molecular Biology* 39, 647–656.

Yamanaka, T., Ohta, T., Takahashi, M., Meshi, T., Schmidt, R., Dean, C., Naito, S. and Ishikawa, M. (2000) *TOM1*, an *Arabidopsis* gene required for efficient multiplication of a tobamovirus, encodes a putative transmembrane protein. *Proceedings of the National Academy of Sciences of the United States of America* 97, 10107–10112.

Yamanaka, T., Imai, T., Satoh, R., Kawashima, A., Takahashi, M., Tomita, K., Kubota, K., Meshi, T., Naito, S. and Ishikawa, M. (2002) Complete inhibition of tobamovirus multiplication by simultaneous mutations in two homologous host genes. *Journal of Virology* 76, 2491–2497.

Yoshii, M., Nishikiori, M., Tomita, K., Yoshioka, N., Kozuka, R., Naito, S. and Ishikawa, M. (2004) The *Arabidopsis cucumovirus multiplication 1* and *2* loci encode translation initiation factors 4E and 4G. *Journal of Virology* 78, 6102–6111.

5 Mechanisms of Plant Virus Evolution and Identification of Genetic Bottlenecks: Impact on Disease Management

M.J. Roossinck and A. Ali

Abstract

The majority of characterized plant viruses have RNA genomes. Genetic variability is a fundamental feature of RNA viruses. High mutation rates, recombination and reassortment are the three basic mechanisms that are responsible for the enormous genetic polymorphism and rapid evolution of RNA viruses. Mutations are most frequently introduced into the viral genome during the replication process due to the low fidelity of RNA-dependent RNA polymerases. Recombination is a widespread phenomenon described in many plant viruses with both RNA and DNA genomes and is responsible for more profound changes within the viral genome (sequence deletion or insertion or strand exchange). Reassortment is also an important mechanism responsible for swapping or introducing a whole genomic segment of the viral genome, but is limited only to the segmented viruses. However, these three mechanisms are counterbalanced by selection and genetic bottlenecks which reduce the genetic variation of plant viruses in nature. Recently, genetic bottlenecks have been identified experimentally in plant virus populations during the systemic movement within the plant and horizontal transmission from plant to plant by aphid vectors.

Introduction

Plant viruses are economically important plant pathogens and are responsible for billions of dollars of economic loss every year (Gray and Banerjee, 1999). Virus infections can cause significant reductions in both quality and quantity of crops, severely reduce food production and thereby affect animal and human health. Virtually all plants are affected by at least one virus. Depending on factors such as plant species, geographic location and growing season, virus diseases can preclude the ability to grow specific crops in certain locations (Falk and Hull, 2004). Significant resources are invested in efforts to control plant virus diseases. The only direct means of controlling virus-induced diseases of plants is by vector control or by the use of genetic resistance. Successful genetic resistance can be compromised by evolution of virus populations to overcome resistance. Vector control frequently involves the use of costly and environmentally harmful chemicals.

About 80% of the approximately 1000 currently recognized plant-infecting viruses have RNA as their genetic material. The remaining 20% constitute only three families (*Caulimoviridae*, *Geminiviridae* and *Nanoviridae*) of plant viruses with DNA genomes (Fauquet *et al.*, 2005). Plant viruses with RNA genomes have a large potential for variation in their genomes that provide them with increased adaptability, allowing rapid response to many environmental challenges, including those posed by host resistance responses.

The RNA genomes of plant viruses have genetically diverse populations that are sometimes referred to as quasispecies. The level of genetic diversity in individual quasispecies varies among closely related viruses (Schneider and Roossinck, 2000) and among hosts infected with the same virus (Schneider and Roossinck, 2001). In addition, all RNA viruses do not generate detectable levels of variation in their populations, indicating that quasispecies are more complex than simple accumulation of mutations. However, factors that control virus population diversity are still unknown (Roossinck and Schneider, 2005).

With the increasingly widespread use of rapid nucleotide sequencing methods and particularly since the advent of the polymerase chain reaction (PCR), extensive genetic data has been obtained for many plant viruses, advancing our understanding of sequence variability among plant viruses. In the last decade, the number of publications in the area of phylogenetic analysis and comparative studies of genetic variability of plant viruses has increased and some reviews have been devoted to these topics (Karasev, 2000; García-Arenal *et al.*, 2001, 2003).

This chapter focuses on the mechanisms of plant virus evolution that are responsible for the genetic diversity of viruses, how the genetic structures of plant virus populations are shaped and their implications for disease management.

Background

Mutation, recombination and reassortment are the primary forces generating the variability in plant virus populations. These populations are shaped by selection that specifically limits variation, and genetic bottlenecks that randomly limit variation. Mutations provide the starting material upon which Darwinian natural selection acts, while recombination or reassortment between different viruses may result in evolutionary leaps through a process of symbiogenesis (Roossinck, 2005). The nature of viral populations differs from conventional populations, and the theoretical framework of quasispecies has been used to reflect these differences (Eigen, 1993; Domingo, 2002; Wilke, 2005).

Driving forces

Mutation is a change in the nucleotide sequence of an organism and is a fundamental source of genetic variation. Point mutations are most often generated by polymerase error, and change a single nucleotide base that is either synonymous or non-synonymous, i.e. they either maintain or change the amino acid coding sequence. Deletions and insertions may include single or several nucleotides, and often result in frame-shift mutations that cause global changes. Most mutations are either neutral or deleterious, but single nucleotide changes can result in substantial changes in a virus, such as altered aphid transmission (Perry *et al.*, 1998), long distance movement (Koshkina *et al.*, 2003), maintenance of replication and virulence (Masuta *et al.*, 1998; Karasawa *et al.*, 1999) and symptom expression (Suzuki *et al.*, 1995).

Mutation rate refers to nucleotide misincorporation (including insertions and deletions) by either polymerase error, RNA editing or other means like environmental mutagens. Mutation frequency refers to the detectable mutations in the population after natural selection and genetic bottlenecks have acted on the mutant spectrum produced by the underlying mutation rate (Domingo and Holland, 1994).

Evidence of mutations in plant viruses was first reported as early as 1926 (McKinney, 1926) on the basis of symptoms induced by tobacco mosaic virus (TMV). Later, several studies on the genetic variation of other plant viruses were reported on the basis of symptoms (Price, 1934; McKinney, 1935) until the development of molecular techniques in the 1960s when the study of genetic variability of plant viruses began at the nucleic acid level. In most cases, mutations in plant viruses have always been documented for a portion of the genome, including the structural coat protein that may also be involved in vector transmission or cell to cell movement. Very few reports are available for other parts of plant viral genomes.

Recombination is the process by which segments of genes are swapped between different genetic variants or strains during the process of replication. Recombination for a plant virus was first proposed and reported in 1955 (Best and Gallus, 1955). When two strains of tomato spotted wilt virus (TSWV) were inoculated together onto a plant, new strains were produced that were different from each of the original parental strains but combined some of the character determinants of each. Later, a similar phenomena was described for two potyviruses, potato virus Y and potato virus C (Watson, 1960). Since then, recombination has been reported for numerous RNA and DNA plant viruses (Aaziz and Tepfer, 1999).

Two types of molecular events result in recombination. Homologous recombination (also known as legitimate recombination) occurs between two nearly identical RNAs (or within nearly identical regions). It can be subdivided into two types: (i) precise recombination occurs when the recombinant junction sites are located accurately at the corresponding nucleotides of the RNAs and (ii) imprecise recombination occurs when the junction sites occupy different positions within recombining RNAs. Imprecise recombination produces RNA in which some sequences are duplicated (inserted) or deleted.

Non-homologous recombination (also referred as illegitimate recombination) occurs between unrelated RNAs or dissimilar regions at non-corresponding sites. The resultant recombinants differ significantly from the parental RNAs. Recombination can be intermolecular or intramolecular, resulting in the insertion of unrelated sequences as well as in exchange, duplication or deletion of existing viral sequence elements.

Homologous and non-homologous recombinations were first shown in brome mosaic virus (BMV) (Bujarski and Kaesberg, 1986). Recombination is a general phenomenon and is considered to play a pivotal role in the genetic variability and evolution of plant viruses (Aaziz and Tepfer, 1999). At the population level, recombination may result in dramatic changes in the biological properties of the virus with major epidemiological consequences, including the appearance of virulent strains or the acquisition of broader host ranges (Legg and Thresh, 2000; Monci et al., 2002). On the other hand, RNA recombination can be an efficient tool for viruses to repair viral genomes, thus contributing to virus fitness (Cheng and Nagy, 2003).

Reassortment is the process in which whole genomic segments are exchanged among segmented viral genomes. It was originally described by Botstein (1980) who detected it in the bacteriophages of coliform bacteria. In plant viruses, reassortment was first reported between the RNAs of tobacco rattle virus and pea early browning virus (Robinson et al., 1987). Plant viruses with segmented genomes usually package their genomic components in separate virions that probably facilitate reassortment events by allowing the exchange of independent genomic components during transmission of viruses (Roossinck, 2005). Reassortment in plant DNA viruses was first shown experimentally in the 1980s (Stanley et al., 1985).

Viruses as quasispecies

RNA viruses were first described as quasispecies to reflect the extreme amount of variation that is possible in their populations (Domingo et al., 1996, 1998, 2005; Eigen, 1996; Domingo, 2002). There is evidence that DNA viruses may also be considered as quasispecies (Isnard et al., 1998). Mutations in a quasispecies are generated and lost at a constant rate, so the mutation frequency is stable. The average nucleotide at each position forms the consensus sequence. This may be the same as the master sequence, the sequence of the most common individual in the population. A quasispecies is defined as a single replicating population at equilibrium. This key point has led to a great deal of confusion for two reasons: a single replicating population is not easily defined, and technically may mean viruses in a single cell, or single region of the plant; and populations that are not at equilibrium may not reflect a quasispecies-like nature. There are a few important significant differences between conventional populations and quasispecies. The quasispecies is the unit upon which selection acts, meaning that a viral quasispecies can be thought of as an individual with thousands of alleles. The quasispecies may carry within it many mutant genomes that can provide extended function. This also can lead to much greater adaptability in changing environments. Another important aspect of quasispecies is that the effects of genetic drift are minimal, because if the population passes through a narrow bottleneck, the most fit variant, if lost, will be rapidly regenerated by the highly error-prone polymerase (Manrubia et al., 2005).

Mechanisms and Consequences of Virus Evolution

Mutation rates and frequencies

Replication of genetic information is a key process in all biological entities that is achieved enzymatically through polymerases. Plant RNA virus and pararetrovirus replication are directed by virally encoded replicases, while plant DNA viruses use the host DNA-dependent DNA polymerase (DdDp) for their replication. RNA-dependent RNA polymerases (RdRps) are thought to lack proofreading capability, resulting in very high error rates. No direct experimental data is available for the mutation or error rates of any plant viral RdRps, although the error rates of the RdRps of animal RNA viruses have been assessed either in cell culture or in vitro and are about 1 in 10^{-4} (Domingo and Holland, 1994). The mutation rate of TMV was calculated to be similar to other RNA viruses (Malpica et al., 2002). Eukaryotic cellular DdDps have error rates on the order of 10^{-9}, but geminiviruses that use host DdDps for replication are highly variable, and may have found ways to enhance their variability through alternate uses of host DdDps (Roossinck, 1997).

The mutation frequency or genetic diversity of plant viruses is a more directly measurable population parameter, but it requires very careful experimental methods to be accurate. The use of error-prone cDNA cloning methods has clouded the analysis of mutation frequencies in many studies. Mutation frequencies have been reported for a number of plant RNA viruses, either in individual isolates of viruses, or in experimental evolution studies (Table 5.1). The low levels of variation found within individual isolates of some plant viruses show that the high mutability of RNA viruses does not necessarily result in high genetic variation. Other factors such as purifying selection and genetic bottlenecks can reduce the genetic diversity of viruses (Roossinck, 1997; Holland and Domingo, 1998). A set of experiments with three related plant viruses, TMV, cucumber mosaic virus (CMV) and cowpea chlorotic mottle virus (CCMV) and several different host plants, demonstrated that the mutation frequency was constant for a given plant–virus combination, but varied with different viruses in the

Table 5.1. Mutation frequencies of plant viruses.

Virus/isolate	Genome	Gene/RNA[a]	Host	Mutation frequency (10^{-3})[b]	References
Plant RNA viruses					
CCMV	ssRNA	RNA3	*Nicotiana benthamiana*	0.05	Schneider and Roossinck (2000)
CMV-Fny	ssRNA	RNA3	*N. benthamiana*	0.6	Schneider and Roossinck (2000)
		RNA3	Pepper	1.8	Schneider and Roossinck (2000)
		RNA3	Squash	0.7	Schneider and Roossinck (2000)
		RNA3	Tobacco	1.0	Schneider and Roossinck (2000)
		RNA3	Tomato	0.7	Schneider and Roossinck (2000)
		RNA3	Tobacco protoplast	2.5	Schneider and Roossinck (2000)
CYSDV	ssRNA	HSP70	Cucurbit	0.28	Rubio *et al.* (2001a)
		CP	Cucurbit	0.25	Rubio *et al.* (2001a)
KGMMV	ssRNA	Replicase	Cucumber	1.47	Kim *et al.* (2005)
		Replicase	Zucchini	1.61	Kim *et al.* (2005)
		CP	Cucumber	2.70	Kim *et al.* (2005)
		CP	Zucchini	2.01	Kim *et al.* (2005)
TMV-U1	ssRNA	30 kD/CP	*N. benthamiana*	0.4	Schneider and Roossinck (2000)
		30 kD/CP	Pepper	1.5	Schneider and Roossinck (2001)
		30 kD/CP	Tobacco	0.9	Schneider and Roossinck (2001)
		30 kD/CP	Tomato	0.2	Schneider and Roossinck (2001)
		30 kD/CP	Tobacco protoplast	2.0	Schneider and Roossinck (2001)
TMV-U1	ssRNA	CP	Buckwheat	0.3	Kearney *et al.* (1999)
		CP	*Collinsia* spp	0.4	Kearney *et al.* (1999)
		CP	Marigold	0.6	Kearney *et al.* (1999)
		CP	Nightshade	0.4	Kearney *et al.* (1999)

Continued

Table 5.1. Continued

Virus/isolate	Genome	Gene/ RNA[a]	Host	Mutation frequency (10^{-3})[b]	References
		CP	Phacelia	1.0	Kearney et al. (1999)
		CP	Plantain	0.8	Kearney et al. (1999)
WSMV	ssRNA	CP	Barley	0.81	Hall et al. (2001b)
		CP	Corn	0.71	Hall et al. (2001b)
		CP	Wheat	0.58	Hall et al. (2001b)
CTV	ssRNA	P-PRO	Sweet orange	0.13656	Rubio et al. (2001b)
		MTR	Sweet orange	0.05993	Rubio et al. (2001b)
		CP	Sweet orange	0.03792	Rubio et al. (2001b)
		P20	Sweet orange	0.06029	Rubio et al. (2001b)
CLBV	ssRNA	Replicase	Sweet orange	3.1	Vives et al. (2002)
		CP	Sweet orange	2.0	Vives et al. (2002)
BanMMV	ssRNA	RdRP	Banana	18.9	Teycheney et al. (2005)
		CP	Banana	19.3	Teycheney et al. (2005)
Plant DNA viruses					
MSV-SP1	ssDNA	Full genome	Coix lacryma-jobi	0.38	Isnard et al. (1998)
MSV-SP2		Full genome	Maize	1.05	Isnard et al. (1998)
MSV-N2A		Full genome	Maize	0.69	Isnard et al. (1998)

CCMV, cowpea chlorotic mottle virus; CMV, cucumber mosaic virus; CYSDV, cucurbit yellow stunting disorder virus; KGMMV, kyuri green mottle mosaic virus; TMV, tobacco mosaic virus; WSMV, wheat streak mosaic virus; CTV, citrus tristeza virus; CLBV, citrus leaf blotch virus; BanMMV, banana mild mosaic virus; MSV, maize streak virus.
[a]Portion of the genome analysed: HSP70, heat shock protein 70; CP, coat protein; 30 kD, movement protein; P-Pro, papain-like protease; MTR, methyletransferase; P20, P-20 protein; RdRP, RNA-dependent RNA polymerase.
[b]Mutation frequency/10^3 bases, measured by sequence analysis of cDNA clones.

same host, or with the same virus in different hosts. In addition, CCMV did not develop any detectable variation except for a single recombination event. In these studies, there was no evidence of genetic drift even after ten passages in plants (Schneider and Roossinck, 2000, 2001).

Apart from one example of an ssDNA virus, maize streak virus (Table 5.1), little work is available to document the mutation frequency of plant DNA viruses. More work is needed to compare the mutation frequency of various plant DNA viruses in different hosts, and to better understand the dynamics of DNA virus populations.

High levels of variation in viral quasispecies are thought to give viruses much greater adaptability, and be responsible for emergence of new viral diseases. While logical, there is little experimental evidence to substantiate this idea.

Recombination

Several models for RNA recombination have been suggested, but the most popular model is the template switching, or copy choice mechanism, which predicts that viral RdRps switch templates during RNA synthesis (Cheng and Nagy, 2003). Experimental evidence supporting the template-switching model has been obtained with BMV and CMV. The recombination hot spot regions frequently contain AU-rich stretches, form intramolecular or intermolecular secondary structures or are localized within cis-acting elements such as replication enhancers (Cheng et al., 2005). Two main factors are thought to affect RNA recombination: the structure of recombining molecules and the ability of the particular viral RdRp to switch templates. Recently, it was shown that the mechanisms of homologous and non-homologous recombinations are different and depend on the virus mode of replication (Alejska et al., 2005).

Recombination occurs in both plant RNA and DNA viruses (Monci et al., 2002) and has been documented to occur between viral and host genes in potato leaf roll virus (Mayo and Jolly, 1991). High rates of recombination have been reported for positive single-stranded RNA viruses both experimentally and naturally (Table 5.2). Homologous and non-homologous recombinations were observed in the RNAs of BMV (Nagy and Bujarski, 1992), CCMV (Allison et al., 1990) and tomato bushy stunt virus (White and Morris, 1995). The frequency of homologous recombination was tenfold higher than non-homologous recombination. Recombination occurs less frequently in negative-strand RNA viruses, most likely because of the ribonucleoprotein complex that inhibits the RNA polymerase from switching templates (Chare et al., 2003). Recombination has also been reported in plant pararetroviruses (caulimoviruses) and DNA viruses (geminiviruses) (Table 5.2).

Recombination occurs more frequently in some host species than in others (Worobey and Holmes, 1999; Desvoyes and Scholthof, 2002), indicating that host genes may affect the RNA recombination process. Recent work in yeast has identified a number of host genes that can affect the recombination process, including an exoribonuclease (Cheng et al., 2006; Serviene et al., 2006).

Reassortment

Plant viruses with segmented genomes can undergo pseudorecombination or reassortment. When plants are infected with two different strains of a related virus, which is probably very common in natural infections, the progeny virus particles can contain mixtures of RNA segments which are derived from either parental strain, thus providing a ready mechanism for generating new viruses. Phylogenetic analyses of segmented plant virus genomes show different evolutionary histories for different genes, suggesting that reassortment has been an important factor in their evolution

Table 5.2. Recombination in plant viruses.

Virus genus	Virus name	Genome	Genome configuration	References
Plant RNA viruses				
Alfamovirus	AMV	ssRNA (+)	3 segments	Huisman et al. (1989); Kyul et al. (1991)
Benyvirus	BNYVV	ssRNA (+)	5 segments	Bouzoubaa et al. (1991)
Bromovirus	BMV	ssRNA (+)	3 segments	Bujarski and Kaesberg (1986)
	CCMV	ssRNA (+)	1 segment	Allison et al. (1990)
Carmovirus	TCV	ssRNA (+)	1 segment	Cascone et al. (1990)
Cucumovirus	CMV	ssRNA (+)	3 segments	Fernández-Cuartero et al. (1994)
	TAV	ssRNA (+)	3 segments	Fernández-Cuartero et al. (1994)
Hordeivirus	BSMV	ssRNA (+)	3 segments	Edwards et al. (1992)
Polerovirus	PLRV	ssRNA (+)	1 segment	Mayo and Jolly (1991)
Potyvirus	PVY	ssRNA (+)	1 segment	Watson (1960)
	ZYMV	ssRNA (+)	1 segment	Gal-On et al. (1994)
	PPV	ssRNA (+)	1 segment	Varrelmann et al. (2000)
	PVA	ssRNA (+)	1 segment	Paalme et al. (2004)
Tobamovirus	TMV	ssRNA (+)	1 segment	Beck and Dawson (1990)
Tobravirus	PEBV	ssRNA (+)	2 segments	Robinson et al. (1987); Goulden et al. (1991)
Tombusvirus	TBSV	ssRNA (+)	1 segment	Hillman et al. (1987)
	CNV	ssRNA (+)	1 segment	White and Morris (1994)
Tospovirus	TSWV	ssRNA (−)	3 segments	Best and Gallus (1955)
Nepovirus	ToRSV	ssRNA (+)	2 segments	Rott et al. (1991)
	GFLV	ssRNA (+)	2 segments	Vigne et al. (2004)
Plant DNA viruses				
Begomovirus	ACMV	ssDNA	1 segment	Pita et al. (2001)
	AYVV	ssDNA	1 segment	Saunders et al. (2001)
	BGMV	ssDNA	1 segment	Garrido-Ramirez et al. (2000)
	CLCuV	ssDNA	1 segment	Sanz et al. (2000)
	TYLCV	ssDNA	1 segment	Monci et al. (2002)
Reassortment				
Cucumovirus	CMV	ssRNA (+)	3 segments	White et al. (1995); Lin et al. (2004)
	PSV	ssRNA (+)	3 segments	Hu and Ghabrial (1996)
Bromovirus	CYBV	ssRNA (+)	3 segments	Iwahashi et al. (2005)
Phytoreovius	RDV	dsRNA	12 segments	Uyeda et al. (1995)
Tospovirus	TSWV	ssRNA (−)	3 segments	Qiu and Moyer (1999)
Tenuivirus	RGSV	ssRNA (−)	6 segments	Miranda et al. (2000)
Tobravirus	TRV and PEBV	ssRNA (+)	2 segments	Robinson et al. (1987)

AMV, alfalfa mosaic virus; BNYVV, beet necrotic yellow vein virus; BMV, brome mosaic virus; CCMV, cowpea chlorotic mottle virus; TCV, turnip crinkle virus; CMV, cucumber mosaic virus; TAV, tomato aspermy virus; BSMV, barley strip mosaic virus; PLRV, potato leaf roll virus; PVY, potato virus Y; ZYMV, zucchini yellow mosaic virus; PPV, plum pox virus; PVA, potato virus A; TMV, tobacco mosaic virus; PEBV, pea early-browning virus; TBSV, tomato bushy stunt virus; CNV, cucumber necrosis virus; TSWV, tomato spotted wilt virus; ToRSV, tomato ring spot virus; GFLV, grapevine fanleaf virus; ACMV, African cassava mosaic virus; AYVV, ageratum yellow vein virus; BGMV, bean golden mosaic virus; CLCuV, cotton leaf curl virus; TYLCV, tomato yellow leaf curl virus; PSV, peanut stunt virus; CYBV, cassia yellow blotch virus; RDV, rice dwarf virus; RGSV, rice grassy stunt virus; TRV, tobacco rattle virus; PEBV, pea early browning virus

(Roossinck, 2002, 2005). Reassortant viruses have been reported for numerous plant viruses (Table 5.2) with either positive-sense or negative-sense, single-stranded RNA (Hu and Ghabrial, 1998; Qiu and Moyer, 1999; Miranda et al., 2000; Lin et al., 2004) and also for double-stranded RNA viruses (Uyeda et al., 1995).

Reassortment has been reported to occur both experimentally (Garrido-Ramirez et al., 2000; Ramos et al., 2003) and naturally (Sanz et al., 2000; Pita et al., 2001) for a number of plant DNA viruses, mostly belonging to the genus *Begomovirus* of the family *Geminiviridae* (Table 5.2). In at least some cases, it has led to the generation of new viral strains with expanded host ranges (Garrido-Ramirez et al., 2000).

Natural selection, bottlenecks and genetic drift

Natural selection is the process by which the fittest variants in a specific environment increase their frequency in the population (positive selection), while less fit variants decrease their frequency (negative selection). The effect of selection is directional and results in decreased population diversity. If a population undergoes a bottleneck where only a limited number of individuals are passed through, genetic drift occurs. Plant viruses face bottlenecks at different points in their infection cycles. Systemic infection can impose a bottleneck, where limited numbers of viral genomes are able to move systemically from the initially infected leaf, and transmission events can also impose bottlenecks.

Bottlenecks occurring during systemic movement of viruses were estimated by determining the number of marker-bearing mutants of CMV that moved from inoculated leaves to systemically infected leaves (Li and Roossinck, 2004). Estimates have also been made for TMV (Sacristán et al., 2003) and wheat streak mosaic virus (Hall et al., 2001a; French and Stenger, 2003), based on population diversity. In all cases, significant bottlenecks were found during systemic infections of plant viruses. Recently, genetic bottleneck events were also demonstrated during the horizontal transmission of CMV population by aphid vectors (Ali et al., 2006).

Genetic bottlenecks are random events that reduce genetic variation of a population as opposed to the directional changes resulting from selection. Bottlenecks randomly decrease diversity within populations. When a few or even a single genome of a virus population generates a new population, there is a high probability that it carries a mutation relative to the consensus genome of the parental population, resulting in drift. However, even though beneficial adaptations may be eliminated, they can be regenerated as the population re-establishes its mutant swarm. This could explain why populations of CMV do not change their consensus sequences after multiple passages (Schneider and Roossinck, 2000, 2001), even though they undergo significant bottlenecks during systemic infection (Li and Roossinck, 2004).

Development of New Technologies

Implications of virus evolution for disease control

The evolutionary potential of RNA viruses results in a rapid adaptability of viral genomes to new environments. This makes direct control of viruses difficult, because most resistance mechanisms imposed by the host are overcome by newly emerging viral strains. However, it is important to remember that the goal of a virus is simply to replicate. Causing disease is a side effect of virus infection, and may be exacerbated by monoculture practices common in agriculture. Virus disease is much less common in natural settings where plant species are quite diverse. Virus disease may, in fact, be a product of human culture (Palumbi, 2001). There are numerous examples of plant viruses that infect tolerant hosts. Unfortunately, very little work has been done on viruses that do not cause disease, but these may hold the key to effective and long-lasting control of plant virus diseases. If the disease process could be avoided without interfering with virus replication, there would be no selection pressure imposed on the virus to change. In conclusion, we will have to learn to live with viruses, because we will never be able to get rid of them for very long.

Conclusions and Future Directions

Mutation, recombination and reassortment are the three main sources of genetic polymorphism that contribute to the rapid evolution of plant viruses. Error-prone replication of RdRps introduces a wide spectrum of point mutations at the rate of 10^{-4}–10^{-5} per nucleotide per replication cycle into the viral RNA genome. Recombination is a widespread phenomenon described in many groups of plant viruses, whereas reassortment is limited to plant viruses with segmented genomes. Our understanding of plant virus evolution is in its early stages and more knowledge is needed to understand the role of various forces that shape populations. The mutation rates of plant viral RdRps, the replication modes of positive single-stranded, negative single-stranded and double-stranded plant RNA viruses, the mutation frequencies of dsRNA and DNA viruses, the roles of various host genes shaping virus populations, the effects of genetic bottlenecks during the systemic infection of various hosts and horizontal and vertical transmission of plant viruses are the areas that are largely unstudied. Without a complete understanding of virus evolution we are certain to see the cycle of emerging viruses and new viral diseases repeat itself endlessly. In addition, it may be necessary to rearrange our thinking to come up with successful strategies for plant virus disease control and to learn to live with viruses rather than eradicating them.

Acknowledgement

This work was supported by the Samuel Roberts Noble Foundation.

References

Aaziz, R. and Tepfer, M. (1999) Recombination between genomic RNAs of two cucumoviruses under conditions of minimal selection pressure. *Virology* 263, 282–289.

Alejska, M., Figlerowicz, M., Malinowska, N., Urbanowicz, A. and Figlerowicz, M. (2005) A universal BMV-based RNA recombination system – how to search for general rules in RNA recombination. *Nucleic Acids Research* 33, e105.

Ali, A., Li, H., Schneider, W.L., Sherman, D.J., Gray, S., Smith, D. and Roossinck, M.J. (2006) Analysis of genetic bottlenecks during horizontal transmission of *Cucumber mosaic virus*. *Journal of Virology* 80, 8345–8350.

Allison, R., Thompson, G. and Ahlquist, P. (1990) Regeneration of a functional RNA virus genome by recombination between deletion mutants and requirement for cowpea chlorotic mottle virus 3a and coat genes for systemic infection. *Proceedings of the National Academy of Sciences of the United States of America* 87, 1820–1824.

Beck, D.L. and Dawson, W.O. (1990) Deletion of repeated sequences from *Tobacco mosaic virus* mutants with two coat protein genes. *Virology* 177, 432–469.

Best, R.J. and Gallus, H.P.C. (1955) Further evidence for the transfer of character-determinants (recombination) between strains of tomato spotted wilt virus. *Enzymologia* 17, 207–221.

Botstein, D. (1980) A theory of modular evolution for bacteriophages. *Annals of the New York Academy of Science* 354, 484–490.

Bouzoubaa, S., Niesbach-Klösgen, U., Jupin, I., Guilley, H., Richards, K. and Jonard, G. (1991) Shortened forms of beet necrotic yellow vein virus RNA-3 and -4: internal deletions and a subgenomic RNA. *Journal of General Virology* 72, 259–266.

Bujarski, J.J. and Kaesberg, P. (1986) Genetic recombination between RNA components of a multipartite plant virus. *Nature* 321, 528–530.

Cascone, P.J., Carpenter, C.D., Li, X.H. and Simon, A.E. (1990) Recombination between satellite RNAs of turnip crinkle virus. *The EMBO Journal* 9, 1709–1715.

Chare, E., Gould, E. and Holmes, E. (2003) Phylogenetic analysis reveals a low rate of homologous recombination in negative-sense RNA viruses. *Journal of General Virology* 84, 2691–2703.

Cheng, C.-P. and Nagy, P.D. (2003) Mechanism of RNA recombination in carmo- and tombusviruses: evidence for template switching by the RNA-dependent RNA polymerase *in vitro*. *Journal of Virology* 77, 12033–12047.

Cheng, C.-P., Panavas, T., Luo, G. and Nagy, P.D. (2005) Heterologous RNA replication enhancer stimulates *in vitro* RNA synthesis and template-switching by the carmovirus, but not by the tombusvirus, RNA-dependent RNA polymerase: implication for modular evolution of RNA viruses. *Virology* 341, 107–121.

Cheng, C.-P., Serviene, E. and Nagy, P.D. (2006) Suppression of viral RNA recombination by a host exoribonuclease. *Journal of Virology* 80, 2631–2640.

Desvoyes, B. and Scholthof, H.B. (2002) Host-dependent recombination of a *Tomato bushy stunt virus* coat protein mutant yields truncated capsid subunits that form virus-like complexes which benefit systemic spread. *Virology* 304, 434–442.

Domingo, E. (2002) Quasispecies theory in virology. *Journal of Virology* 76, 463–465.

Domingo, E. and Holland, J.J. (1994) Mutation rates and rapid evolution of RNA viruses. In: Morse, S.S. (ed.) *The Evolutionary Biology of Viruses*. Raven Press, New York, pp. 161–184.

Domingo, E., Escarmís, C., Sevilla, N., Moya, A., Elena, S.F., Quer, J., Novella, I. and Holland, J.J. (1996) Basic concepts in RNA virus evolution. *FASEB Journal* 10, 859–864.

Domingo, E., Baranowski, E., Ruiz-Jarabo, C.M., Martin-Hernandez, A.M., Saiz, J.C. and Escarmis, C. (1998) Quasispecies structure and persistence of RNA viruses. *Emerging Infectious Diseases* 4, 521–527.

Domingo, E., Escarmís, C., Lázaro, E. and Manrubia, S.C. (2005) Quasispecies dynamics and RNA virus extinction. *Virus Research* 107, 129–139.

Edwards, M.C., Petty, T.D. and Jackson, A.O. (1992) RNA recombination in the genome of barley stripe mosaic virus. *Virology* 189, 389–392.

Eigen, M. (1993) Viral quasispecies. *Scientific American* 269, 42–49.

Eigen, M. (1996) On the nature of virus quasispecies. *Trends in Microbiology* 4, 216–217.

Falk, B.W. and Hull, R. (2004) Plant RNA virus diseases. In: Goodman, R.M. (ed.) *Encyclopedia of Plant and Crop Science*. Marcel Dekker, New York, pp. 1023–1025.

Fauquet, C.M., Mayo, M.A., Maniloff, J., Desselberger, U. and Ball, L.A. (eds) (2005) *Virus Taxonomy Eighth Report of the International Committee on Taxonomy of Viruses*. Elsevier Academic Press, San Diego, California.

Fernández-Cuartero, B., Burgyán, J., Aranda, M.A., Salánki, K., Moriones, E. and García-Arenal, F. (1994) Increase in the relative fitness of a plant virus RNA associated with its recombinant nature. *Virology* 203, 373–377.

French, R. and Stenger, D.C. (2003) Evolution of wheat streak mosaic virus: dynamics of population growth within plants may explain limited variation. *Annual Review of Phytopathology* 41, 199–214.

Gal-On, A., Kaplan, I., Roossinck, M.J. and Palukaitis, P. (1994) The kinetics of infection of zucchini squash by cucumber mosaic virus indicate a function for RNA 1 in virus movement. *Virology* 205, 280–289.

García-Arenal, F., Fraile, A. and Malpica, J.M. (2001) Variability and genetic structure of plant virus populations. *Annual Review of Phytopathology* 39, 157–186.

García-Arenal, F., Fraile, A. and Malpica, J.M. (2003) Variation and evolution of plant virus populations. *International Microbiology* 6, 225–232.

Garrido-Ramirez, E.R., Sudarshana, M.R. and Gilbertson, R.L. (2000) *Bean golden yellow mosaic virus* from Chiapas, Mexico: characterization, pseudorecombination with other bean-infecting geminiviruses and germ plasm screening. *Virology* 90, 1224–1232.

Goulden, M.G., Lomonssoff, G.P., Wood, K.R. and Davies, J.W. (1991) A model for the generation of tobacco rattle virus (TRV) anomalous isolates: pea early browning virus RNA-2 acquires TRV sequences from both RNA-1 and RNA-2. *Journal of General Virology* 72, 1751–1754.

Gray, S.M. and Banerjee, N. (1999) Mechanisms of arthropod transmission of plant and animal viruses. *Microbiology and Molecular Biological Reviews* 63, 128–148.

Hall, J.S., French, R., Hein, G.L., Morris, T.J. and Stenger, D.C. (2001a) Three distinct mechanisms facilitate genetic isolation of sympatric wheat streak mosaic virus lineages. *Virology* 282, 230–236.

Hall, J.S., French, R., Morris, T.J. and Stenger, D.C. (2001b) Structure and temporal dynamics of populations within wheat streak mosaic virus isolates. *Journal of Virology* 75, 10231–10243.

Hillman, B.I., Carrington, J.C. and Morris, T.J. (1987) A defective interfering RNA that contains a mosaic of a plant virus genome. *Cell* 51, 427–433.

Holland, J. and Domingo, E. (1998) Origin and evolution of viruses. *Virus Gene* 16, 13–21.

Hu, C.C. and Ghabrial, S.A. (1996) Molecular characterization of peanut stunt virus (PSV) strain BV-15, a natural reassortant between subgroups I and II of PSV strains. *Phytopathology* 86, S17.

Hu, C.C. and Ghabrial, S.A. (1998) Molecular evidence that strain BV-15 of peanut stunt cucumovirus is a reassortant between subgroup I and II strains. *Phytopathology* 88, 92–97.

Huisman, M.J., Cornelissen, B.J.C., Groenendijk, C.F.M., Bol, J.F. and Vloten-Doting, L.V. (1989) Alfalfa mosaic virus temperature-sensitive mutants V. The nucleotide sequence of TBTS 7 RNA 3 shows limited nucleotide changes and evidence for heterologous recombination. *Virology* 171, 409–416.

Isnard, M., Granier, M., Frutos, R., Reynaud, B. and Peterschmitt, M. (1998) Quasispecies nature of three maize streak virus isolates obtained through different modes of selection from a population used to assess response to infection of maize cultivars. *Journal of General Virology* 79, 3091–3099.

Iwahashi, F., Fujisaki, K., Kaido, M., Okuno, T. and Mise, K. (2005) Synthesis of infectious *in vitro* transcripts from *Cassia yellow blotch bromovirus* cDNA clones and a reassortment analysis with other bromoviruses in protoplasts. *Archives of Virology* 150, 1301–1314.

Karasawa, A., Okada, I., Akashi, K., Chida, Y., Hase, S., Nakazawa-Nasu, Y., Ito, A. and Ehara, Y. (1999) One amino acid change in cucumber mosaic virus RNA polymerase determines virulent/avirulent phenotypes on cowpea. *Virology* 89, 1186–1192.

Karasev, A.V. (2000) Genetic diversity and evolution of closteroviruses. *Annual Review of Phytopathology* 38, 293–324.

Kearney, C.M., Thomson, M.J. and Roland, K.W. (1999) Genome evolution of *Tobacco mosaic virus* populations during long-term passaging in a diverse range of hosts. *Archives of Virology* 144, 1513–1526.

Kim, T., Youn, M.Y., Min, B.E., Choi, S.H., Kim, M. and Ryu, K.H. (2005) Molecular analysis of quasispecies of *Kyuri green mottle mosaic virus*. *Virus Research* 110, 161–167.

Koshkina, T.E., Baranova, E.N., Usacheva, E.A. and Zavriev, S.K. (2003) A point mutation in the coat protein gene affects long-distance transport of the tobacco mosaic virus. *Molecular Biology* 37, 742–748.

Kyul, A.C.V.D., Neeleman, L. and Bol, J.F. (1991) Complementation and recombination between alfalfa mosaic virus RNA2 mutants in tobacco plants. *Virology* 783, 731–738.

Legg, J.P. and Thresh, J.M. (2000) Cassava mosaic virus disease in East Africa: a dynamic disease in a changing environment. *Virus Research* 71, 135–149.

Li, H. and Roossinck, M.J. (2004) Genetic bottlenecks reduce population variation in an experimental RNA virus population. *Journal of Virology* 78, 10582–10587.

Lin, H.-X., Rubio, L., Smythe, A.B. and Falk, B.W. (2004) Molecular population genetics of *Cucumber mosaic virus* in California: evidence for founder effects and reassortment. *Journal of Virology* 78, 6666–6675.

Malpica, J.M., Fraile, A., Moreno, I., Obies, C.I., Drake, J.W. and García-Arenal, F. (2002) The rate and character of spontaneous mutations in an RNA virus. *Genetics* 162, 1505–1511.

Manrubia, S.C., Escarmís, C., Domingo, E. and Lázaro, E. (2005) High mutation rates, bottlenecks, and robustness of RNA viral quasispecies. *Gene* 347, 273–282.

Masuta, C., Nishimura, M., Morishita, H. and Hataya, T. (1998) A single amino acid change in viral genome-associated protein of potato virus Y correlates with resistance breaking in Virgin A tobacco. *Phytopathology* 89, 118–123.

Mayo, M.A. and Jolly, C.A. (1991) The 5'-terminal sequence of potato leafroll virus RNA: evidence of recombination between virus and host RNA. *Journal of General Virology* 72, 2591–2595.

McKinney, H.H. (1926) Virus mixtures that may not be detected in young tobacco plants. *Phytopathology* 16, 893.

McKinney, H.H. (1935) Evidence of virus mutation in the common mosaic of tobacco. *Journal of Agricultural Research* 51, 951–981.

Miranda, G.J., Azzam, O. and Shirako, Y. (2000) Comparison of nucleotide sequences between Northern and Southern Philippine isolates of rice grassy stunt virus indicates occurrence of natural genetic reassortment. *Virology* 266, 26–32.

Monci, F., Sánchez-Campos, S., Navas-Castillo, J. and Moriones, E. (2002) A natural recombinant between the geminiviruses *Tomato yellow leaf curl, Sardinia virus* and *Tomato yellow leaf curl virus* exhibits a novel pathogenic phenotype and is becoming prevalent in Spanish populations. *Virology* 303, 317–326.

Nagy, P.D. and Bujarski, J.J. (1992) Genetic recombination in brome mosaic virus: effect of sequence and replication of RNA on accumulation of recombinants. *Journal of Virology* 66, 6824–6828.

Paalme, V., Gammelgard, E., Järvekülg, L. and Valkonen, J.P.T. (2004) *In vitro* recombinants of two nearly identical potyviral isolates express novel virulence and symptom phenotypes in plants. *Journal of General Virology* 85, 739–747.

Palumbi, S.R. (2001) Humans as the world's greatest evolutionary force. *Science* 293, 1786–1790.

Perry, K.L., Zhang, L. and Palukaitis, P. (1998) Amino acid changes in the coat protein of cucumber mosaic virus differentially affect transmission by the aphids *Myzus persicae* and *Aphis gossypii*. *Virology* 242, 204–210.

Pita, J.S., Fondong, V.N., Sangaré, A., Otim-Nape, G.W., Ogwal, S. and Fauquet, C.M. (2001) Recombination, pseudorecombination and synergism of geminiviruses are determinant keys to the epidemic of severe cassava mosaic disease in Uganda. *Journal of General Virology* 82, 655–665.

Price, W.C. (1934) Isolation and study of some yellow strains of cucumber mosaic virus. *Phytopathology* 24, 743–761.

Qiu, W. and Moyer, J.W. (1999) Tomato spotted wilt tospovirus adapts to the TSWV N gene-derived resistance by genome reassortment. *Phytopathology* 89, 575–582.

Ramos, P.L., Guevara-González, R.G., Peral, R., Ascencio-Ibañez, J.T., Polston, J.E., Agrüello-Astorga, G.R., Vega-Arreguín, J.C. and Rivera-Bustamante, R.F. (2003) Tomato mottle Taino virus pseudorecombines with PYMV but not with ToMoV: implications for the delimitation of *cis*-and *trans*-acting replication specificity determinants. *Archives of Virology* 148, 1697–1712.

Robinson, D.J., Hamilton, W.D.O., Harrison, B.D. and Baulcombe, D.C. (1987) Two anomalous tobravirus isolates: evidence for RNA recombination in nature. *Journal of General Virology* 68, 2551–2561.

Roossinck, M.J. (1997) Mechanisms of plant virus evolution. *Annual Review of Phytopathology* 35, 191–209.

Roossinck, M.J. (2002) Evolutionary history of *Cucumber mosaic virus* deduced by phylogenetic analyses. *Journal of Virology* 76, 3382–3387.

Roossinck, M.J. (2005) Symbiosis *versus* competition in the evolution of plant RNA viruses. *Nature Reviews Microbiology* 3, 917–924.

Roossinck, M.J. and Schneider, W.L. (2005) Mutant clouds and occupation of sequence space in plant RNA viruses. In: Domingo, E. (ed.) *Quasispecies: Concepts and Implications for Virology*. Springer, Heidelberg, Germany, pp. 337–348.

Rott, M.E., Tremaine, J.H. and Rochon, D.M. (1991) Comparison of the 5' and 3' termini of tomato ringspot virus RNA1 and RNA2: evidence for RNA recombination. *Virology* 185, 468–472.

Rubio, L., Abou-Jawdah, Y., Lin, H.-X. and Falk, B.W. (2001a) Geographically distant isolates of the crinivirus *Cucurbit yellow stunting disorder virus* show very low genetic diversity in the coat protein gene. *Journal of General Virology* 82, 929–933.

Rubio, L., Ayllón, M.A., Kong, P., Fernández, A., Polek, M., Guerri, J., Moreno, P. and Falk, B.W. (2001b) Genetic variation of *Citrus tristeza virus* isolates from California and Spain: evidence for mixed infections and recombination. *Journal of Virology* 75, 8054–8062.

Sacristán, S., Malpica, J., Fraile, A. and García-Arenal, F. (2003) Estimation of population bottlenecks during systemic movement of *Tobacco mosaic virus* in tobacco plants. *Journal of Virology* 77, 9906–9911.

Sanz, A.I., Fraile, A., García-Arenal, F., Zhou, X., Robinson, D.J., Khalid, S., Butt, T. and Harrison, B.D. (2000) Multiple infection, recombination and genome relationships among begomovirus isolates found in cotton and other plants in Pakistan. *Journal of General Virology* 81, 1839–1849.

Saunders, K., Bedford, I.D. and Stanley, J. (2001) Pathogenicity of a natural recombinant associated with Ageratum yellow vein disease: implication for Geminivirus evolution and disease aetiology. *Virology* 282, 38–47.

Schneider, W.L. and Roossinck, M.J. (2000) Evolutionarily related sindbis-like plant viruses maintain different levels of population diversity in a common host. *Journal of Virology* 74, 3130–3134.

Schneider, W.L. and Roossinck, M.J. (2001) Genetic diversity in RNA viral quasispecies is controlled by host–virus interactions. *Journal of Virology* 75, 6566–6571.

Serviene, E., Shapka, N., Cheng, C.-P., Panavas, T., Phuangrat, B., Baker, J. and Nagy, P.D. (2006) Genome-wide screen identifies host genes affecting viral RNA recombination. *Proceedings of the National Academy of Sciences of the United States of America* 102, 10545–10550.

Stanley, J., Townsend, R. and Curson, S.J. (1985) Pseudorecombinants between cloned DNAs of two isolates of cassava latent virus. *Journal of General Virology* 66, 1055–1061.

Suzuki, M., Kuwata, S., Masuta, C. and Takanami, Y. (1995) Point mutations in the coat protein of cucumber mosaic virus affect symptom expression and virion accumulation in tobacco. *Journal of General Virology* 76, 1791–1799.

Teycheney, P.-Y., Laboureau, N., Iskra-Caruana, M.-L. and Candresse, T. (2005) High genetic variability and evidence for plant-to-plant transfer of *Banana mild mosaic virus*. *Journal of General Virology* 86, 3179–3187.

Uyeda, I., Ando, Y., Murao, K. and Kimura, I. (1995) High resolution genome typing and genomic reassortment events of rice dwarf *Phytoreovirus*. *Virology* 212, 724–727.

Varrelmann, M., Palkovics, L. and Maiss, E. (2000) Transgenic or plant expression vector-mediated recombination of *Plum pox virus*. *Journal of Virology* 74, 7462–7469.

Vigne, E., Bergdoll, M., Guyader, S. and Fuchs, M. (2004) Population structure and genetic variability within isolates of *Grapevine fanleaf virus* from a naturally infected vineyard in France: evidence for mixed infection and recombination. *Journal of General Virology* 85, 2435–2445.

Vives, M.C., Rubio, L., Galipienso, L., Navarro, L., Moreno, P. and Guerri, J. (2002) Low genetic variation between isolates of *Citrus leaf blotch virus* from different host species and of different geographical origins. *Journal of General Virology* 83, 2587–2591.

Watson, M.A. (1960) Evidence for interaction or genetic recombination between potato viruses Y and C in infected plants. *Virology* 10, 211–232.

White, K.A. and Morris, T.J. (1994) Recombination between defective tombusvirus RNAs generates functional hybrid genomes. *Proceedings of the National Academy of Sciences of the United States of America* 91, 3642–3646.

White, K.A. and Morris, T.J. (1995) RNA determinants of junction site selection in RNA virus recombinants and defective interfering RNAs. *RNA* 1, 1029–1040.

White, P.S., Morales, F.J. and Roossinck, M.J. (1995) Interspecific reassortment in the evolution of a cucumovirus. *Virology* 207, 334–337.

Wilke, C.O. (2005) Quasispecies theory in the context of populations genetics. *BMC Evolutionary Biology* 5, 44.

Worobey, M. and Holmes, E.C. (1999) Evolutionary aspects of recombination in RNA viruses. *Journal of General Virology* 80, 2535–2543.

6 Molecular Understanding of Viroid Replication Cycles and Identification of Targets for Disease Management

R.A. Owens

Abstract
The replicative strategies of viroids differ in several fundamental aspects from those of RNA plant viruses. Rather than replicating in the cytoplasm, like most viruses, viroids must first be transported into either the nucleus (pospiviroids) or the chloroplast (avsunviroids) before the beginning of replication. The RNA polymerases involved in viroid replication – DNA-dependent RNA polymerase II or a nuclear-encoded chloroplastic RNA polymerase – are entirely host-encoded, and the absence of viroid-encoded polypeptides strongly suggests that the subsequent cell-to-cell and long-distance movement of progeny is also completely dependent on normal host cell pathways. Recent studies with potato spindle tuber viroid (PSTVd) have identified specific structural features that are involved in: (i) cleavage or ligation of replicative intermediates; and (ii) transport of viroid progeny across tissue boundaries. Significant progress has also been made in unraveling the molecular mechanism(s) of viroid pathogenesis, in particular, the possible role of viroid-induced RNA silencing in disrupting host gene expression. The availability of detailed information regarding host contributions to the disease process has provided new opportunities to target disruption of specific events in the viroid replication cycle. Plants that are resistant or immune to infection can be used to augment the diagnostic testing and seed or nursery stock certification schemes currently used to control viroid diseases.

Introduction

Ever since their discovery in 1971, viroids have been the subject of intense interest. Viroids are the smallest known agents of infectious disease – small (246–401 nucleotides), highly structured, circular, single-stranded RNAs lacking detectable messenger RNA activity. Viruses supply some or most of the genetic information required for their replication, whereas viroids are regarded as 'obligate parasites of the cell's transcriptional machinery'. Over the last 35 years, much has been learned about the molecular biology of viroids and viroid–host interaction, but the precise nature of the molecular signals that allow these agents to replicate autonomously and induce disease in many of their plant hosts remains elusive. A series of questions first posed by Diener, their discoverer, summarize many gaps in our current understanding of viroids:

1. What are the molecular signals that induce certain host DNA-dependent RNA polymerases to accept viroids as templates for the synthesis of complementary RNAs?
2. Are the molecular mechanisms that are responsible for viroid replication operative in uninfected cells? If so, what are their functions?
3. How do viroids induce disease? In the absence of viroid-specified proteins, disease must arise from the direct interaction of host cell constituents with either viroids themselves or viroid-derived RNAs. What role does RNA silencing play in the disease process?
4. What determines viroid host range? In the broadest terms, are viroids restricted to higher plants, or do they have counterparts in animals?

The impact of efforts designed to answer these questions extends far beyond the immediate areas of virology and plant pathology. For example, the discoveries of hammerhead ribozymes (Hutchins et al., 1986) and RNA-dependent DNA methylation (Wassenegger et al., 1996) as well as the characterization of the RNA-dependent RNA polymerase involved in RNA silencing (Schiebel et al., 1998) are landmarks in plant molecular biology and biotechnology. In the area of plant disease management, the first application of nucleic acid-based diagnostics involved the diagnosis of potato spindle tuber viroid (PSTVd) infections (Owens and Diener, 1981). Subsequently, PSTVd has been virtually eliminated from both potato breeding programmes and commercial production in Europe and North America, thereby illustrating the value of combining a rapid and sensitive diagnostic method with a rigorous clean stock or seed certification scheme. In contrast, attempts to create plants that are resistant or immune to viroid infection using knowledge gained from fundamental studies have been only partially successful so far. In summarizing our current understanding of the viroid replication cycle, this chapter will attempt to identify potential 'weak points' where future biotechnological interventions may lead to useful levels of resistance.

Background

A review by Diener (2003) provides a comprehensive personal perspective on the discovery of viroids in the late 1960s to early 1970s. At the same time that characterization of PSTVd was underway at Beltsville, Maryland, USA, the causal agents of two other viroid diseases – chrysanthemum stunt and citrus exocortis – were also under study by independent groups. Thus, many plant virologists quickly recognized the potential significance of viroids as a novel class of infectious agents, but broader acceptance of the 'viroid concept' required their physical recognition as ultraviolet (UV) light-absorbing RNA molecules in polyacrylamide gels or small circular molecules visible in the electron microscope. This pioneering phase of viroid research ended in 1978 with determination of the complete nucleotide sequence of PSTVd (Gross et al., 1978).

Shortly thereafter, sequences of several other viroids, including those of citrus exocortis (CEVd), chrysanthemum stunt (CSVd) and avocado sunblotch (ASBVd) appeared, and the knowledge of their molecular properties began to expand at a rapid rate. A key event in this rapid expansion was the development of cloned viroid cDNAs using recombinant DNA techniques. In addition to greatly facilitating nucleotide sequence determination, cloned viroid cDNAs found immediate use as hybridization probes. In rapid succession, cloned viroid cDNA probes were used to demonstrate that: (i) PSTVd replication proceeds through an asymmetric rolling circle mechanism (Branch and Robertson, 1984) and (ii) the rapid and sensitive detection of PSTVd using dot-blot hybridization (Owens and Diener, 1981). Comparison of the complete nucleotide sequences of several viroids led Symons and colleagues (Keese and Symons, 1985) to propose that PSTVd and related viroids contain five structural domains. Sequence differences between naturally occurring mild and severe strains of PSTVd and CEVd were seen to cluster in the 'pathogenicity domain', an observation that has had a major influence on the course of viroid research. Sensitivity to low levels of α-amanitin implicated DNA-dependent RNA polymerase II as the host enzyme responsible for the replication of PSTVd and related viroids (Schindler and Muhlbach, 1992).

During this time, ASBVd RNAs of both polarities were shown to undergo spontaneous self-cleavage *in vitro* (Hutchins et al., 1986), revealing for the first time the existence of hammerhead ribozymes. Characterization of ASBVd replicative intermediates revealed that this (and presumably other) ribozyme-containing viroids replicate through a symmetric rolling circle mechanism (Daròs et al., 1994). Yet another key event was the demonstration in 1983 that inoculation of susceptible host plants with greater-than-full-length PSTVd cDNAs resulted in systemic infection (Cress et al., 1983), thereby allowing the application of reverse genetics to the study of viroid–host interaction. Characterization of novel viroid chimeras assembled using recombinant DNA techniques revealed that symptom expression is regulated by multiple sequence or structural elements, some of which are located outside the pathogenicity domain (Sano et al., 1992).

By the mid-1990s, most of the 29 currently recognized viroid species had been discovered and the broad outlines of their two contrasting replicative mechanisms were established. Over the past decade, molecular studies of viroid–host interaction have focused on three broad areas. First, sequence elements that act as promoters for either RNA polymerase II (PSTVd and other pospiviroids replicating in the nucleus) or a nuclear-encoded RNA polymerase found in the chloroplast (ASBVd and other avsunviroids replicating in chloroplasts) have been identified (Navarro and Flores, 2000; Kolonko et al., 2006), and the series of molecular rearrangements necessary for multimeric PSTVd RNAs to undergo cleavage or ligation by host enzymes has been defined (Baumstark et al., 1997; Schrader et al., 2003). The picture of viroids that emerges is one of a group of highly dynamic molecules in which alternative secondary structures and tertiary interactions play a crucial role in almost every phase of viroid–host interaction. Second, separate groups of studies have focused on the ability of viroids to move systemically in their hosts without the aid of viroid-encoded proteins. Of the many different techniques used to study viroid movement – from the cytoplasm into the nucleus or chloroplasts prior to replication, from cell to cell through the plasmodesmata and long distance in the vascular system – *in situ* hybridization has proven particularly useful in identifying potential control points in the infection process (Ding et al., 2005). Third, significant progress has been made in assessing the effects of viroid infection on host gene expression. Macroarray analysis has revealed that PSTVd infection triggers complex changes in host gene expression (Itaya et al., 2002), and at least one component of a potential signal transduction cascade (i.e. a viroid-induced protein kinase) has been identified (Hammond and Zhao, 2000). Viroid infection also induces RNA silencing (Itaya et al., 2001; Papaefthimiou et al., 2001; Martinez de Alba et al., 2002), and the possible role of this phenomenon in regulating viroid pathogenicity is currently under investigation in several laboratories (Markarian et al., 2004; Wang et al., 2004).

Molecular biology of viroid replication

In focusing on aspects of viroid replication that currently seem most likely to involve potential disease management targets, it has been necessary to lay aside several related topics such as the origin and evolution of viroids. Fortunately, an up-to-date monograph (Hadidi et al., 2003) and three recent reviews (Tabler and Tsagris, 2004; Ding et al., 2005; Flores et al., 2005) are available for those desiring additional information on other aspects of viroid molecular biology.

As shown in Table 6.1, the 29 officially recognized species of viroids are divided into two families (i.e. the *Pospiviroidae* and the *Avsunviroidae*) that contain a total of seven genera. All 25 species (and one provisional species) in the family *Pospiviroidae* have a rod-like secondary structure that contains five structural or functional domains (Keese and Symons, 1985) and replicate in the nucleus. Three of the four members of the *Avsunviroidae*, in contrast, have a branched secondary structure, and all replicate or accumulate in the chloroplast. All members of the *Avsunviroidae* contain hammerhead ribozymes in both the infectious (+)strand and complementary (−)strand RNAs. With the possible exception of PLMVd, viroids appear to contain no modified nucleotides or unusual phosphodiester bonds.

Viroids differ dramatically in host range. Some like PSTVd and HSVd have comparatively broad host ranges that include diverse herbaceous and woody species; others are restricted to either a single species (CChMVd) or a closely related group of species (ASBVd). Most detailed studies of viroid rep-

Table 6.1. Officially recognized viroid species (Eighth Report, ICTV).

Genus[a]	Species	Reported variants[b]	Length (nt)
Family *Pospiviroidae*			
Pospiviroid	Potato spindle tuber (PSTVd)	109	341–364
	Chrysanthemum stunt (CSVd)	19	348–356
	Citrus exocortis (CEVd)	86	366–475
	Columnea latent (CLVd)	17	359–456
	Iresine (IrVd)	3	370
	Mexican papita (MPVd)	6	359–360
	Tomato apical stunt (TASVd)	5	360–363
	Tomato chlorotic dwarf (TCDVd)	2	360
	Tomato planta macho (TPMVd)	2	360
Hostuviroid	Hop stunt (HSVd)	144	294–303
Cocadviroid	Coconut cadang-cadang (CCCVd)	8	246–301
	Coconut tinangaja (CTiVd)	2	254
	Citrus bark cracking (CBCVd)	6	284–286
	Hop latent (HLVd)	10	255–256
Apscaviroid	Apple scar skin (AASVd)	8	329–333
	Apple dimple fruit (ADFVd)	2	306
	Apple fruit crinkle (AFCVd)[c]	29	368–372
	Australian grapevine (AGVd)	1	369
	Citrus bent leaf (CBLVd)	24	315–329
	Citrus dwarfing (CDVd)	53	291–297
	Grapevine yellow speckle 1 (GYSVd-1)	49	365–368
	Grapevine yellow speckle 2 (GYSVd-2)	1	363
	Pear blister canker (PBCVd)	18	314–316
Coleviroid	*Coleus blumei*-1 (CbVd-1)	9	248–251
	Coleus blumei-2 (CbVd-2)	2	295–301
	Coleus blumei-3 (CbVd-3)	3	361–364
Family *Avsunviroidae*			
Avsunviroid	Avocado sun blotch (ASBVd)	83	239–251
Pelamoviroid	Chrysanthemum chlorotic mottle (CChMVd)	21	397–401
	Peach latent mosaic (PLMVd)	168	335–351
Elaviroid	Eggplant latent (ELVd)	9	332–335

[a]Names of viroid genera are derived from those of the respective type species (listed first).
[b]Sequences available online from the Subviral RNA Database (http://subviral.med.uottawa.ca).
[c]Provisional species (not officially recognized).

lication have been carried out using either PSTVd- or ASBVd-type members of the nuclear and the chloroplast families of viroids. Organizing this information as a schematic diagram of the infected cell (see Fig. 6.1) highlights several points in the infection process where it may be possible to disrupt the viroid–host interactions necessary for systemic spread and disease induction.

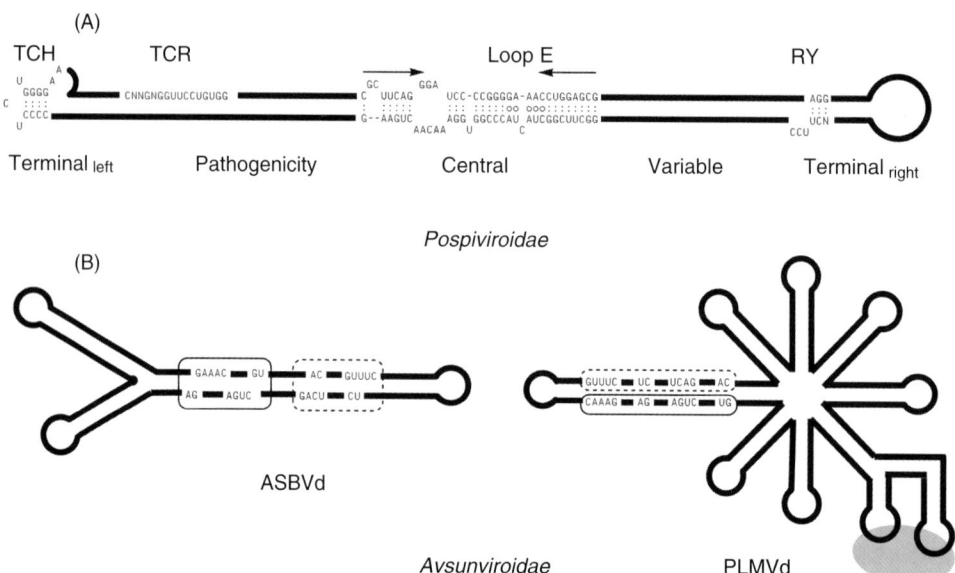

Fig. 6.1. Structural features of viroids. (A) The rod-like secondary structures of PSTVd and other members of the family *Pospiviroidae* contain five structural or functional domains; i.e. terminal$_{left}$, pathogenicity, central, variable and terminal$_{right}$. Members of the genera *Pospiviroid* and *Apscaviroid* and the two largest members of the genus *Coleviroid* also contain a terminal conserved region (TCR). Members of the genera *Hostuviroid* and *Cocadviroid* contain a terminal conserved hairpin (TCH). Arrows indicate a pair of inverted repeats that defines the limits of the central conserved region with its loop E motif containing an array of non-Watson–Crick base pairs (open circles). Members of the genus *Pospiviroid* also contain 1–2 copies of a purine–pyrimidine-rich (RY) motif in their T_R domain. (B) Quasi-rod-like (ASBVd) and branched secondary structures of other members of the family *Avsunviroidae*. Sequences conserved in most (+)strand and (−)strand hammerhead ribozymes are enclosed by solid or dashed boxes, respectively. Shaded oval denotes a kissing loop interaction in PLMVd. (Modified from Fig. 1 of Flores *et al.*, 2005, with permission from the publisher.)

Intracellular Transport to the Nucleus or the Chloroplast

Viroids enter the host cell in one of the two ways: through microscopic wounds in epidermal cells (mechanical inoculation) or through the plasmodesmata connecting most vascular and non-vascular cells (grafting or slash inoculation). Once in the cytoplasm, they must be transported to either the nucleus or the chloroplast before replication begins.

Two different experimental strategies have been used to study the movement of PSTVd into the nucleus. Addition of full-length, fluorescently labelled PSTVd RNA transcripts to a suspension of permeabilized tobacco protoplasts is followed by accumulation in the nucleus. Through the use of various inhibitors, nuclear import was shown to be a cytoskeleton-independent process mediated by a specific and saturable receptor and independent of the Ran GTPase cycle (Woo *et al.*, 1999). In the second system, green fluorescent protein (GFP) expression from a *potato virus X* gene vector replicating in the cytoplasm was completely blocked by inserting an intron into the coding sequence of GFP. When a full-length copy of PSTVd was inserted into this intron, however, the resulting mRNA was transported into the nucleus, the intron was removed, and the perfectly rejoined mRNA was returned to the cytoplasm for translation into protein (Zhao *et al.*, 2001). Subsequent experiments have shown that nuclear import of PSTVd requires only the presence of sequences derived from the upper portion of the central conserved region (R.W. Hammond, Beltsville, Maryland, 2007, personal communication). Although it is likely that PSTVd is transported to the nucleus as a ribonucleoprotein complex, the host proteins involved in transport remain to be identified. The central domain of PSTVd and related viroids contains a loop E motif (Branch *et al.*, 1985), so ribosomal protein L5 and T(ranscription) F(actor) IIIa could be involved. Another possibility is VIRP1, a bromodomain-containing protein from tomato that binds specifically within the right terminal domain of PSTVd and contains a nuclear localization signal (Martínez de Alba *et al.*, 2003; Maniataki *et al.*, 2003).

The entry or exit of ASBVd and other avsunviroids in the chloroplast is unknown. The outer chloroplast membrane contains no structures corresponding to the nuclear pore complex. Protein import into the chloroplast (followed in some cases by insertion into the thylakoid membranes) depends on the presence of N-terminal signal sequences (Jarvis and Robinson, 2004). Flores and colleagues (Daròs and Flores, 2002) have identified two chloroplast proteins that behave like RNA chaperones and facilitate the hammerhead-mediated self-cleavage of ASBVd. These proteins are encoded by the nuclear (rather than the chloroplast) genome, and thus they could also play a role in viroid movement into the chloroplast. As no viral or cellular RNAs are known to move from the cytoplasm to the chloroplast, the pathway by which these few viroids enter the chloroplast is completely unknown.

Rolling circle replication

As shown schematically in Fig. 6.2, nucleic acid extracts from infected leaf tissue contain a variety of viroid-related RNAs of both polarities. Some of these molecules (especially those having a complementary or '(−)strand' polarity) are considerably longer than the infectious circular viroid (+)strand. Characterization by Northern analysis using strand-specific probes and/or primer extension has shown that these molecules represent the intermediates expected for a 'rolling circle' mechanism of replication.

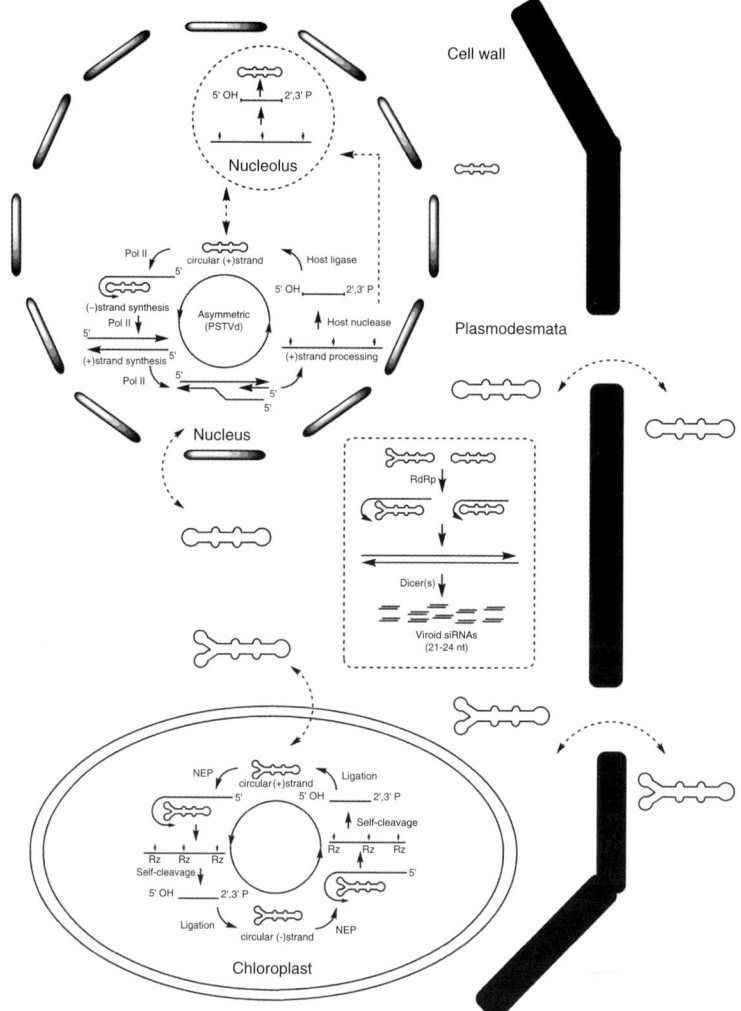

Fig. 6.2. Rolling circle replication schemes for *Pospiviroids* and *Avsunviroids* together with potential pathways for generation of viroid-specific siRNAs. Following transport into the nucleus, the circular genomic RNA of *Pospiviroids* is transcribed by DNA-dependent RNA polymerase II into an oligomeric complementary RNA that is designated as (−)strand even though the genomic (i.e. (+)strand) RNA has no mRNA activity. Transcription of this oligomeric (−)strand RNA by DNA-dependent RNA polymerase II results in the synthesis of oligomeric (+)strand RNA that is subsequently 'processed' into monomeric circular viroid progeny. The (+)strand progeny accumulates in the nucleolus, but the precise location of the cleavage or ligation reaction (i.e. nucleoplasm or nucleolus) is not yet clear. *Avsunviroids* replicate in the chloroplast where transcription appears to involve a tagetitoxin-insensitive RNA polymerase encoded by the nuclear genome. Both the (+)strand and (−)strand RNAs contain hammerhead ribozymes and are able to cleave spontaneously. Following cleavage, the monomeric (−)strand RNA is circularized, thereby resulting in a replicative pathway that is 'symmetrical'. Replication of viroids from either family results in the accumulation of viroid-specific siRNAs. Because the subcellular location (nucleus or cytoplasm) of key steps in the siRNA pathway(s) is not yet clear, only a single pathway is shown in the dashed box. Dashed arrows denote intracellular and intercellular transport pathways. (Modified from Fig. 2 of Tabler and Tsagris, 2004, with permission from the publisher.)

Analysis of ASBVd-infected leaf tissue reveals the presence of monomeric circular RNAs of both polarities (Daròs et al., 1994); thus, ASBVd (and presumably other avsunviroids) replicate through a symmetric rolling circle mechanism. Replication of PSTVd, in contrast, proceeds through an asymmetric rolling circle mechanism in which progeny (+)strands are synthesized on a multimeric linear (−)strand template (Branch and Robertson, 1984). The presence of hammerhead ribozymes in both strands allows multimeric ASBVd RNA to cleave spontaneously, thereby releasing the corresponding linear monomers. Processing of longer-than-unit-length PSTVd (+)strand RNA requires the central conserved region to fold into a multihelix junction containing at least one GNRA tetraloop hairpin followed by cleavage by an as-yet-unidentified host nuclease (Baumstark et al., 1997). Although evidence has been presented suggesting that monomeric linear PLMVd molecules can spontaneously circularize with the formation of a 2',5'-phosphodiester linkage (Cote et al., 2001), circularization of most viroids appears to require the action of a host RNA ligase.

A central question about viroid replication concerns the identity of the polymerase(s) involved. Inhibition of (+)strand and (−)strand PSTVd RNA syntheses by α-amanitin exhibits exactly the same dose–response effect in nuclear run-off experiments as does host mRNA synthesis (Schindler and Muhlbach, 1992), thereby implicating host DNA-dependent RNA polymerase II as the enzyme responsible for pospiviroid replication. Actinomycin D (a widely used inhibitor of rRNA synthesis) had no effect. Direct evidence for an association between RNA polymerase II and CEVd has been presented by Warrilow and Symons (1999), who showed that addition of a monoclonal antibody directed against the C-terminal domain of the largest subunit of RNA pol II results in immunoprecipitation of a nucleoprotein complex containing both (+)strand and (−)strand CEVd RNAs. Resistance of ASBVd RNA synthesis in permeabilized chloroplasts to tagetitoxin inhibition (Navarro et al., 2000) suggests that a nuclear-encoded RNA chloroplastic polymerase (and *not* the eubacterial-like RNA polymerase encoded by the plastid genome) is responsible for ASBVd strand elongation.

Initiation of both ASBVd and PSTVd RNA syntheses appears to be promoter-driven. For ASBVd, both (+)strand and (−)strand syntheses initiate within AU-rich regions located in the terminal hairpin loops of the rod-like native structure. Interestingly, the nucleotide sequences around the ASBVd start sites bear a striking resemblance to the promoter sequences of certain chloroplast genes transcribed by the same nuclear-encoded RNA polymerase believed to be responsible for ASBVd replication (Navarro and Flores, 2000). A recent publication from the Riesner laboratory (Kolonko et al., 2006) has shown that transcription of PSTVd (+)strand template by RNA polymerase II starts at either position 359 or 1 in the left terminal loop. Exactly how ASBVd or PSTVd redirect the respective host DNA-dependent RNA polymerases to accept its quasi-double-stranded RNA genome as template remains to be determined.

Movement from the nucleoplasm to the nucleolus

The involvement of RNA polymerase II implies that PSTVd replication occurs in the nucleoplasm of infected cells, and two studies (Harders et al., 1989; Qi and Ding, 2003a) have examined the relative distribution of (+)strand and (−)strand PSTVd RNAs between the nucleoplasm and the nucleolus in some detail. The picture that emerges indicates that: (i) synthesis of both (−)strand and (+)strand PSTVd RNAs occurs in the nucleoplasm; (ii) (−)strand PSTVd is somehow 'anchored' in the nucleoplasm, while (+)strand PSTVd is selectively transported into the nucleolus; and (iii) a small amount of (+)strand PSTVd moves back into the nucleoplasm and eventually returns to the cytoplasm before spreading to adjacent cells through the plasmodesmata. Because the *in situ* hybridization techniques used cannot distinguish multimeric from monomeric PSTVd RNAs, it is not clear whether the cleavage or ligation of nascent (+)strand PSTVd multimers occurs in the nucleoplasm (i.e. the site of synthesis) or in the nucleolus (site of accumulation). Following movement to the nucleolus, PSTVd colocalizes around the periphery with small nucleolar RNAs (snoRNAs) U3 and U14. Unlike hepatitis delta virus where two different host RNA polymerases may required to complete the replication cycle (Lai, 2005; for opposing view, see Taylor, 2006), there is currently no evidence that PSTVd replication involves any polymerase other than RNA pol II.

Export from the nucleus and formation of siRNA

How PSTVd and related viroids leave the nucleus is currently unclear. In addition to its role in ribosome biosynthesis, the nucleolus plays an important role in many other important cellular processes involving RNA and protein trafficking (Kim et al., 2004). The presence of a loop E motif in their central domain suggests that, like 5S rRNA, pospiviroids may interact with ribosomal protein L5 and transcription factor III A. In *Arabidopsis*, TFIIIA is concentrated at several nuclear foci, including the nucleolus, but is absent from the cytoplasm. Ribosomal protein L5 also accumulates in the nucleus and the nucleolus, and is also present in the cytoplasm (Mathieu et al., 2003). It is possible that viroid transport to the cytoplasm involves the pathway used by host ribosomal RNAs.

A review by Baulcombe (2004) summarizes evidence for the existence of at least three RNA silencing pathways in plants. Silencing signals can be amplified and transmitted between cells, and the biological roles of these pathways – defence against virus infection, regulation of gene expression and formation of heterochromatin – are diverse. Over the past several years, several reports have appeared which documented the presence of viroid-specific silencing-induced RNAs in plants infected with PSTVd and ASBVd (reviewed in Tabler and Tsagris, 2004). As shown in Fig. 6.2, it is not yet clear where and how these viroid-specific siRNAs are produced. Their site(s) of action is also unclear. Because Dicer (an RNase III-like enzyme) is not known to be present in the chloroplast, formation of ASBVd (and presumably other avsunviroid) siRNAs probably takes place in the cytoplasm. PSTVd-related siRNAs may or may not originate in the nucleus, but biochemical analyses have shown that they accumulate in the cytoplasm (Denti et al., 2004). Whether Dicer acts on single-stranded viroid RNAs themselves and/or double-stranded by-products of the replication is also unknown. Wheat germ extracts contain a Dicer activity that is able to cleave one of the long hairpin stems in PLMVd (Landry and Perrault, 2005), but human dicer is unable to cleave either ASBVd or PSTVd (Chang et al., 2003).

What, if any, role does RNA silencing play in regulating viroid replication and pathogenicity? Two observations – first, an inverse correlation between avsunviroid titre and siRNA concentration (Martinez de Alba et al., 2002); second, the accumulation of PSTVd siRNA preceding recovery from severe disease (Sano and Matsura, 2004) – indicate that RNA silencing can suppress viroid replication. On the other hand, the fact that plants infected by mild and severe strains of PSTVd or CChMVd contain very similar levels of siRNA suggests that siRNA concentration alone cannot explain the often dramatic differences in viroid symptom expression. The possible role of viroid-induced RNA silencing in regulating host gene expression is considered in more detail below (see the section on evolving concepts of viroid pathogenesis).

Cell-to-cell movement via the plasmodesmata

Plasmodesmata function as a supracellular control network in plants, allowing proteins and RNA to move from cell to cell and affect developmental programmes in a non-cell-autonomous manner (Lucas and Lee, 2004). Following microinjection into symplastically connected leaf mesophyll cells, fluorescently labelled PSTVd moves rapidly from cell to cell where it accumulates in the nucleus (Ding et al., 1997). Further evidence that viroids contain specific sequence or structural motifs for plasmodesmatal transport comes from the ability of otherwise non-mobile RNAs to move from cell to cell following their fusion to PSTVd. Movement of many viral genomes (both RNA and DNA) through plasmodesmata requires specific viral-encoded 'movement proteins' (Lucas, 2006). Viroid movement from cell to cell presumably involves interaction with one or more host proteins, but their identity remains to be determined.

Phloem-mediated long-distance movement

Viroids, like most plant viruses, move systemically in the host phloem. Dot-blot hybridization analysis of tomato seedlings infected with wild-type PSTVd (Palukaitis, 1987) has shown that overall viroid movement follows the flow of photoassimilates; i.e. from photosynthetic source (mature leaves) to metabolic sinks (shoot apex, young leaves and roots).

Upon closer examination using *in situ* hybridization, however, a much more nuanced picture of viroid–host interaction emerges. PSTVd movement in the phloem is tightly regulated by developmental and cellular factors and probably sustained by replication in phloem-associated cells (Zhu *et al.*, 2001). For example, viroid could not be detected in shoot apical meristems of PSTVd-infected tomato or *Nicotiana benthamiana* plants, even though it was present in the underlying procambium and protophloem. Similarly, PSTVd appears unable to enter developing flowers of *N. benthamiana*, but is found in the sepals (although not other portions) of mature flowers (Zhu *et al.*, 2002). These types of restrictions on viroid trafficking are not absolute, however, because PSTVd (as well as several other viroids) is seed-transmitted in certain hosts.

Viroid movement in the phloem almost certainly involves interaction with host proteins and formation of a ribonucleoprotein complex. Two studies have shown that the most abundant protein in cucumber phloem exudate, a dimeric lectin known as phloem protein 2 (PP2), binds non-specifically to HSVd RNA *in vitro* (Gomez and Pallas, 2001; Owens *et al.*, 2001). A follow-up (Gomez and Pallas, 2004) revealed that cucumber PP2 contains a dsRNA-binding motif and is able to move from an HSVd-infected host rootstock into a non-host (i.e. pumpkin) scion. These interactions may be part of a 'systemic small RNA signalling system' that involves a variety of phloem proteins and plays a key role in regulating plant development and defence against pathogens (Yoo *et al.*, 2004).

Although much remains to be learned about the factors that control the ability of viroids to enter and exit host vascular tissue, a study by Ding and colleagues (Qi *et al.*, 2004) has demonstrated a direct role for a bipartite sequence motif in mediating the directional movement of PSTVd across a specific cellular boundary separating the bundle sheath from the leaf mesophyll. A single C/U substitution at position 259 within the loop E motif of PSTVd strain KF440-2 was known to confer upon this molecule the ability to replicate in tobacco (Wassenegger *et al.*, 1996), and further passage in tobacco resulted in a several-fold increase in viroid titre and the appearance of five additional sequence changes; i.e. four changes at positions 47, 309, 313 and 315 in pathogenicity domain on the left side of the molecule and one change at position 201 in the right terminal loop. Analysis of the effects of individual mutations revealed that four of these five changes (i.e. all but a U/C change at position 315) were required to allow PSTVd to leave the bundle sheath and move into the mesophyll. The presence or absence of this bipartite motif had no effect on movement in the opposite direction (Qi *et al.*, 2004; see also Qi and Ding, 2002).

Evolving concepts of viroid pathogenesis

Shortly after the appearance of the first complete viroid nucleotide sequence in 1978, RNA fingerprinting studies revealed the presence of only minor sequence differences between mild and severe strains of

PSTVd (Dickson et al., 1979). Nearly all of these changes later proved to be located in the 'pathogenicity domain' of PSTVd, and much effort has been expended over the years to determine exactly how mutations affecting only 1–2 positions can have such dramatic biological consequences.

Initial results suggested that PSTVd symptom severity was inversely correlated with the structural stability of a 'virulence modulating' region located within the pathogenicity domain. Nucleotides in the virulence-modulating region were proposed to interact directly with one or more unidentified host factors, the strength of this interaction thereby regulating viroid pathogenicity (Schnölzer et al., 1985). However, several later studies showed this model to be overly simplistic. For example, characterization of a series of novel viroid chimeras containing sequences derived from TASVd and CEVd revealed that pospiviroid pathogenicity is regulated by determinants located in multiple structural domains – not just the pathogenicity domain (Sano et al., 1992). Within the pathogenicity domain itself, three-dimensional conformation proved to be a much better predictor of PSTVd pathogenicity than structural stability (Owens et al., 1996). Finally, single mutations in a loop E motif located in the central domain of PSTVd can have dramatic effects on symptom expression (Qi and Ding, 2003b) and host range (Wassenegger et al., 1996). Similar analyses have identified pathogenicity determinants in HSVd (Reanwarakorn and Semancik, 1998), CChMVd (De la Pena and Flores, 2002) and PLMVd (Malfitano et al., 2003).

At present, much less is known regarding the host contribution to disease development. As described earlier, the interactions of several host proteins with viroids have been characterized in some detail; e.g. those involving tomato VirP1 (Martinez de Alba et al., 2002), cucumber phloem lectin PP2 (Gomez and Pallas, 2004) and two RNA chaperones from avocado (Daròs and Flores, 2002). Differential activation of a mammalian interferon-induced, dsRNA-activated protein kinase by PSTVd strains of varying pathogenicity (Diener et al., 1991) as well as the isolation of a viroid-induced serine–threonine protein kinase from tomato (Hammond and Zhao, 2000) have also been reported. Viroid infection results in the accumulation of p(athogenesis)-r(elated) proteins and low-molecular weight metabolites such as genistic acid involved in systemic signalling (Bellés et al., 2006), and macroarrays of tomato cDNAs have been used to assess the broader effects of PSTVd infection on host gene expression (Itaya et al., 2002).

An often unspoken assumption in these studies is that disease results from the direct interaction of host cell components with either the viroid genomic RNA or perhaps the less abundant, complementary (–)strand RNA synthesized during replication. The recent discovery that viroid-infected plants also contain a variety of small viroid-related siRNAs (see earlier) highlights the many gaps in current concepts of viroid–host interaction. Filling these gaps will require a shift in emphasis from genomics of the pathogen to that of their hosts. Unfortunately, although *Arabidopsis thaliana* has been shown to contain the enzymatic machinery necessary to replicate representative species of the *Pospiviroidae*, replication rates are low, and systemic movement is impaired (Daròs and Flores, 2004).

Development of new technologies

Table 6.2 summarizes the economic effects, distribution and measures currently used to control the viroid diseases affecting a variety of economically important crops. Under most conditions, seed and insect transmission plays only a minor role in disease spread; the primary means of viroid spread are vegetative propagation and mechanical transmission. Measures currently used to control viroid diseases are extensions of those used for virus diseases and include: (i) elimination of viroids from planting material (certified stock programmes); (ii) control of viroid spread in the field (eradication); and (iii) quarantine exclusion of new infection. Proper sanitation, including sterilization of tools and equipment between each plant (for certified stock) or between each bed, row or section (to prevent spread in the field), is critical. Many quarantine and certification programmes are currently in place to prevent viroid introduction from germplasm collections. Results are uneven, but in those cases where control has been achieved (e.g. PSTVd elimination from certified seed potato production in North America and Western Europe, Singh, 1988), three factors have proven critical: first, the availability of rapid and sensitive diagnostic test(s) (see Chapter 12, this volume); second, adoption of either a one-pass production system or a clean stock programme where mother plants maintained under protected conditions are repeatedly tested for their infection status; and third, strict adherence to a zero-tolerance policy.

To date, no useful sources of genetic resistance to viroid infection have been identified in any plant species. For example, many tomato and potato varieties respond to PSTVd infection with very mild or no detectable symptoms, but none has been shown to be resistant or immune to infection (Singh and O'Brien, 1970). Several efforts to create transgenic plants that are resistant to infection by PSTVd or related pospiviroids have been reported over the past decade. RNA-based strategies (Matousek *et al.*, 1994; Atkins *et al.*, 1995; Yang *et al.*, 1997) have involved the constitutive expression of antisense RNA – either antisense RNA alone or coupled to a *trans*-acting ribozyme. An alternative, protein-based (and sequence-independent) strategy relies on the constitutive expression of *pacI*, a dsRNA-specific RNase derived from *Schizosaccharomyces pombe* (Sano *et al.*, 1997). Although viroid accumulation and symptom expression were often reduced, useful levels of resistance have not been achieved. Many of the inhibitory effects observed are probably due to RNA silencing.

As information concerning the structural determinants regulating viroid replication and movement across tissue boundaries continues to accumulate, a third strategy to create resistance becomes increasingly feasible; i.e. the design of RNA decoys that are able to out-compete viroids for essential host factors. A study involving hepatitis C virus (Zhang *et al.*,

Table 6.2. Viroid diseases: economic effects, distribution and current control measures. (Modified from Table 1.1 in Hadidi et al., 2003, with permission from the publisher.)

Viroid(s)	Economic host	Distribution	Damage	Seed transmission	Control strategy
ASBVd	Avocado	Widespread	Severe	Yes	Eradication/certified stock
CCCVd/CTiVd	Coconut/oil palm	Limited	Lethal	Low	Quarantine/replant
CSVd	Chrysanthemum	Widespread	Severe		Eradication/certified stock
Citrus viroids	Citrus	Widespread	Variable		Eradication/certified stock
Grapevine viroids	Grapevine	Widespread	Minor	Yes	Eradication/certified stock
HSVd	Hops	Limited	Moderate		Eradicate/replant
	Cucumber	Limited	Severe		Eradication
Apscaviroids	Pome/stone fruits	Widespread	Mild		Eradication/certified stock/quarantine
PLMVd	Peach/*Prunus* spp.	Widespread	Mild		Quarantine/certified stock
PSTVd	Potato	Limited	Moderate	Yes	Eradication/certified stock/quarantine
	Tomato	Sporadic	Severe		Eradication
TASVd/CLVd	Tomato	Sporadic	Severe		Eradication
TPMVd	Tomato	Limited	Severe		Eradication

2005) illustrates the potential value of this approach. In this study, adenovirus-mediated expression of small RNA molecules containing stem-loop structures that act as *cis*-acting replication elements for hepatitis C virus reduced virus accumulation by 35- to 38-fold – presumably by preventing the binding of the viral replicase to hepatitis C virus genomic RNA.

Conclusions and New Directions

The tools needed to control or eliminate many viroid diseases – a sensitive and reliable diagnostic test as well as appropriate zero-tolerance schemes for the production of certified seed or nursery stock – have been available for some time. The virtual eradication of PSTVd from commercial potato production in North America and Europe demonstrates the efficacy of existing methods. Nevertheless, efforts continue to reduce both the cost per test and the expertise required; i.e. equipment as well as training. None of the existing hybridization- or PCR-based diagnostic tests provides results in 'real time' or under field conditions. Genetic resistance to viroid infection remains an important goal, especially for fruit trees like citrus and other woody perennials where susceptible varieties are maintained for years under unprotected field conditions. Results from previous attempts to create resistance to viroid infection using strategies developed for use against viruses have been disappointing, possibly because these strategies do not take into account significant differences between the structure and replicative strategies of viroids and RNA viruses. In this respect, two recent developments offer reason for cautious optimism.

First, several specific structural features such as the loop E motif of PSTVd and an alternative multihelix structure have been shown to play crucial roles in regulating viroid replication and host range. As described earlier, a specific host protein (i.e. VirP1) has been shown to bind to the RY motif found in the right terminal loop of PSTVd and other pospiviroids, an interaction that may be well required for both intracellular and intercellular movements. As demonstrated by initial characterization of a bipartite sequence motif that regulates PSTVd movement between mesophyll and bundle sheath cells in the leaf, these interactions will undoubtedly prove to be complex. Nevertheless, they provide the first viroid-specific targets against which RNA decoys and other yet-to-be-developed resistance strategies can be tested.

Second, the still-to-be-clarified role(s) of small viroid-related RNAs during the infection process has added a previously unsuspected degree of complexity to the interaction between viroids and their hosts. As in the case of plant viruses, current evidence indicates that RNA silencing plays an important role in the host's attempts to mount a defence against viroid infection. Indeed, it has recently been proposed that the compact, highly base-paired structure of viroids and certain satellite RNAs has evolved to resist the effects of RNA silencing (Wang *et al.*, 2004). This study also raises the possibility that these small viroid-related RNAs may be responsible for many of the characteristic symptoms of viroid infection. Several animal viruses encode small RNAs that target genes involved in the host immune response (Cullen, 2006). Even more intriguingly, replication of hepatitis C virus is dependent on the binding of a specific host microRNA to the 5'-UTR of the genomic RNA (Jopling *et al.*, 2005). Either scenario offers exciting new opportunities to create resistance to viroid diseases.

Critical evaluation of the role of small RNAs in viroid–host interactions requires much more detailed information about likely changes in both host-encoded and viroid-derived small RNA populations during the course of infection. Fortunately, the combination of new sequencing technologies and bioinformatic analysis now allows sampling that is 'deep' enough to distinguish host microRNAs from short-interfering RNAs (Lu *et al.*, 2005). These types of studies also require detailed knowledge of the sequence of the host genome, and it is here that the inability of presently known viroids to replicate and move freely in *Arabidopsis* (Daròs and Flores, 2004) has created what will hopefully be only a temporary bottleneck. Sequencing of the tomato genome is now approximately 20% complete, and a microarray containing approximately 10,000 tomato cDNAs is freely available (http://sgn.cornell.edu). The next 2–3 years should see a much clearer picture of:

(i) changes in host gene expression associated with viroid infection and (ii) molecular mechanisms responsible for these changes. This information can then be used to design and test currently unimaginable strategies to render plant resistance or immunity to viroid infection.

Acknowledgements

I thank Yan Zhao and Jonathan Shao (Molecular Plant Pathology Laboratory), as well as Til Baumstark (University of the Sciences in Philadelphia) for stimulating discussions during the preparation of this chapter.

References

Atkins, D., Young, M., Uzzell, S., Kelly, L., Fillatti, J. and Gerlach, W.L. (1995) The expression of antisense and ribozyme genes targeting citrus exocortis viroid in transgenic plants. *Journal of General Virology* 76, 1781–1790.

Baulcombe, D. (2004) RNA silencing in plants. *Nature* 431, 356–363.

Baumstark, T., Schröder, A.R.W. and Riesner, D. (1997) Viroid processing: switch from cleavage to ligation is driven by a change from a tetraloop to a loop E conformation. *The EMBO Journal* 16, 599–610.

Bellés, J.M., Garro, R., Pallás, V., Fayos, J., Rodrigo, I. and Conejero, V. (2006) Accumulation of gentisic acid as associated with systemic infections but not with the hypersensitive response in plant–pathogen interactions. *Planta* 223, 500–511.

Branch, A.D. and Robertson, H.D. (1984) A replication cycle for viroids and other small infectious RNA's. *Science* 223, 450–455.

Branch, A.D., Benenfeld, B.J., Baroudy, B.M., Wells, F.V., Gerin, J.L. and Robertson, H.D. (1985) Ultraviolet light-induced cross-linking reveals a unique region of local tertiary structure in potato spindle tuber viroid and HeLa 5S RNA. *Proceedings of the National Academy of Sciences of the United States of America* 82, 6590–6594.

Chang, J., Provost, P. and Taylor, J.M. (2003) Resistance of human hepatitis delta virus RNAs to dicer activity. *Journal of Virology* 77, 11910–11917.

Cote, F., Levesque, D. and Perreault, J.P. (2001) Natural 2',5'-phophodiester bonds found at the ligation sites of peach latent mosaic viroid. *Journal of Virology* 75, 19–25.

Cress, D.E., Kiefer, M.C. and Owens, R.A. (1983) Construction of infectious potato spindle tuber viroid cDNA clones. *Nucleic Acids Research* 11, 6821–6835.

Cullen, B.R. (2006) Viruses and microRNAs. *Nature Genetics* 38 (Suppl. 1), S25–S30.

Daròs, J.A. and Flores, R. (2002) A chloroplast protein binds a viroid RNA *in vivo* and facilitates its hammerhead-mediated self-cleavage. *The EMBO Journal* 21, 749–759.

Daròs, J.A. and Flores, R. (2004) *Arabidopsis thaliana* has the enzymatic machinery for replicating representative viroid species of the family *Pospiviroidae*. *Proceedings of the National Academy of Sciences of the United States of America* 101, 6792–6797.

Daròs, J.A., Marcos, J.F., Hernández, C. and Flores, R. (1994) Replication of avocado sunblotch viroid: evidence for a symmetric pathway with two rolling circles and hammerhead ribozyme processing. *Proceedings of the National Academy of Sciences of the United States of America* 91, 12813–12817.

De la Pena, M. and Flores, R. (2002) Chrysanthemum chlorotic mottle viroid RNA: dissection of the pathogenicity determinant and comparative fitness of symptomatic and non-symptomatic variants. *Journal of Molecular Biology* 321, 411–421.

Denti, M.A., Boutla, A., Tsagris, M. and Tabler, M. (2004) Short interfering RNAs specific for potato spindle tuber viroid are found in the cytoplasm but not in the nucleus. *The Plant Journal* 37, 762–769.

Dickson, E., Robertson, H.D., Niblett, C.L., Horst, R.K. and Zaitlin, M. (1979) Minor differences between nucleotide sequences of mild and severe strains of potato spindle tuber viroid. *Nature* 277, 60–62.

Diener, T.O. (2003) Discovering viroids – a personal perspective. *Nature Reviews, Microbiology* 1, 75–80.

Diener, T.O., Hammond, R.W., Black, T. and Katze, M.G. (1991) Mechanisms of viroid pathogenesis: differential activation of the interferon-induced, double-stranded RNA-activated, M(r) 68,000 protein kinase by viroid strains of varying pathogenicity. *Biochemie* 75, 533–538.

Ding, B., Kwon, M.O., Hammond, R. and Owens, R.A. (1997) Cell-to-cell movement of potato spindle tuber viroid. *The Plant Journal* 12, 931–936.

Ding, B., Itaya, A. and Zhong, X.-H. (2005) Viroid trafficking: a small RNA makes a big move. *Current Opinion in Plant Biology* 8, 606–612.

Flores, R., Hernández, C., Martinez de Alba, A.E., Daròs, J.-A. and Di Serio, F. (2005) Viroids and viroid–host interactions. *Annual Review of Phytopathology* 43, 117–139.

Gomez, G. and Pallas, V. (2001) Identification of an *in vitro* ribonucleoprotein complex between a viroid RNA and a phloem protein from cucumber plants. *Molecular Plant–Microbe Interactions* 14, 910–913.

Gomez, G. and Pallas, V. (2004) A long-distance translocatable phloem protein from cucumber forms a ribonucleoprotein complex *in vivo* with hop stunt viroid RNA. *Journal of Virology* 78, 10104–10110.

Gross, H.J., Domdey, H., Lossow, C., Jank, P., Raba, M., Alberty, H. and Sänger, H.-L. (1978) Nucleotide sequence and secondary structure of potato spindle tuber viroid. *Nature* 273, 203–208.

Hadidi, A., Flores, R., Randles, J.W. and Semancik, J.S. (2003) *Viroids.* CSIRO Publishing, Collingwood, Victoria, Australia.

Hammond, R.W. and Zhao, Y. (2000) Characterization of a tomato protein kinase gene induced by infection with potato spindle tuber viroid. *Molecular Plant–Microbe Interactions* 13, 903–910.

Harders, J., Lukacs, N., Robert-Nicoud, M., Jovin, T.M. and Riesner, D. (1989) Imaging of viroids in nuclei from tomato leaf tissue by *in situ* hybridization and confocal laser scanning microscopy. *The EMBO Journal* 8, 3941–4986.

Hutchins, C.J., Rathjen, P.D., Forster, A.C. and Symons, R.H. (1986) Self-cleavage of plus and minus RNA transcripts of avocado sunblotch viroid. *Nucleic Acids Research* 14, 3627–3640.

Itaya, A., Folimonov, A., Matsuda, Y., Nelson, R.S. and Ding, B. (2001) Potato spindle tuber viroid as inducer of RNA silencing in infected tomato. *Molecular Plant–Microbe Interactions* 14, 1332–1334.

Itaya, A., Matsuda, Y., Gonzales, R.A., Nelson, R.S. and Ding, B. (2002) Potato spindle tuber viroid strains of different pathogenicity induces and suppresses expression of common and unique genes in infected tomato. *Molecular Plant–Microbe Interactions* 15, 990–999.

Jarvis, P. and Robinson, C. (2004). Mechanisms of protein import and routing in chloroplasts. *Current Biology* 14, R1064–R1077.

Jopling, C.L., Yi, M., Lancaster, A.M., Lemon, S.M. and Sarnow, P. (2005) Modulation of hepatitis C virus RNA abundance by a liver-specific microRNA. *Science* 309, 1577–1581.

Keese, P. and Symons, R.H. (1985) Domains in viroids: evidence of intermolecular RNA rearrangements and their contribution to viroid evolution. *Proceedings of the National Academy of Sciences of the United States of America* 82, 4582–4586.

Kim, S.H., Ryabov, E.V., Brown, J.W. and Taliansky, M. (2004) Involvement of the nucleolus in plant virus systemic infection. *Biochemical Society Transactions* 32, 557–560.

Kolonko, N., Bannach, O., Aschermann, K., Hu, K.H., Moors, M., Schmitz, M., Steger, G. and Riesner, D. (2006) Transcription of potato spindle tuber viroid by RNA polymerase II starts in the left terminal loop. *Virology* 347, 392–404.

Lai, M.M. (2005) RNA replication without RNA-dependent RNA polymerase: surprises from Hepatitis delta virus. *Journal of Virology* 79, 7951–7958.

Landry, P. and Perrault, J.-P. (2005) Identification of a peach latent mosaic viroid hairpin able to act as a Dicer-like substrate. *Journal of Virology* 79, 6540–6543.

Lu, C., Tej, S.S, Luo, S., Haudenschild, C.D., Meyers, B.C. and Green, P.J. (2005) Elucidation of the small RNA component of the transcriptome. *Science* 309, 1567–1569.

Lucas, W.J. (2006) Plant viral movement proteins: agents for cell-to-cell trafficking of viral genomes. *Virology* 344, 169–184.

Lucas, W.J. and Lee, J.-Y. (2004) Plasmodesmata as a supracellular control network in plants. *Nature Reviews, Molecular Cell Biology* 5, 712–726.

Malfitano, M., Di Serio, F., Covelli, L., Ragozzino, A., Hernández, C. and Flores, R. (2003) Peach latent mosaic viroid variants inducing peach calico contain a characteristic insertion that is responsible for this symptomatology. *Virology* 313, 492–501.

Maniataki, E., Martínez de Alba, A.E., Sägasser, R., Tabler, M. and Tsagris, M. (2003) Viroid RNA systemic spread may depend on the interaction of a 71-nucleotide bulged hairpin with the host protein VirP1. *RNA* 9, 346–354.

Markarian, N., Li, H.W., Ding, S.W. and Semancik, J.S. (2004) RNA silencing as related to viroid induced symptom expression. *Archives of Virology* 149, 397–406.

Martinez de Alba, A.E., Flores, R. and Hernández, C. (2002) Two chloroplastic viroids induce the accumulation of small RNAs associated with posttranscriptional gene silencing. *Journal of Virology* 76, 13094–13096.

Martínez de Alba, A.E., Sägasser, R., Tabler, M. and Tsagris, M. (2003) A bromodomain-containing protein from tomato specifically binds potato spindle tuber viroid RNA *in vitro* and *in vivo*. *Journal of Virology* 77, 9685–9694.

Mathieu, O., Yukawa, Y., Prieto, J.L., Vaillant, I., Sugiura, M. and Tourmente, S. (2003) Identification and characterization of transcription factor IIIA and ribosomal protein L5 from *Arabidopsis thaliana*. *Nucleic Acids Research* 31, 2424–2433.

Matousek, J., Schroder, A.R., Trnena, L., Reimers, M., Baumstark, T., Dedic, Pl, Vlasak, J., Becker, I., Kreuzaler, Fl, Flading, M. and Riesner, D. (1994) Inhibition of viroid infection by antisense RNA expression in transgenic plants. *Biological Chemistry Hoppe-Seyler* 375, 765–777.

Navarro, J.A. and Flores, R. (2000) Characterization of the initiation sites of both polarity strands of a viroid RNA reveals a motif conserved in sequence and structure. *The EMBO Journal* 19, 2662–2670.

Navarro, J.A., Vera, A. and Flores, R. (2000) A chloroplastic RNA polymerase resistant to tagetitoxin is involved in replication of avocado sunblotch viroid. *Virology* 268, 218–225.

Owens, R.A. and Diener, T.O. (1981) Sensitive and rapid diagnosis of potato spindle tuber viroid disease by nucleic acid hybridization. *Science* 213, 670–672.

Owens, R.A., Steger, G., Hu, Y., Fels, A., Hammond, R.W. and Riesner, D. (1996) RNA structural features responsible for potato spindle tuber viroid pathogenicity. *Virology* 222, 144–158.

Owens, R.A., Blackburn, M. and Ding, B. (2001) Possible involvement of the phloem lectin in long-distance viroid movement. *Molecular Plant–Microbe Interactions* 14, 905–909.

Palukaitis, P. (1987) Potato spindle tuber viroid: investigation of the long-distance, intra-plant transport route. *Virology* 158, 239–241.

Papaefthimiou, I., Hamilton, A.J., Denti, M.A., Baulcombe, D.C., Tsagris, M. and Tabler, M. (2001) Replicating potato spindle tuber viroid RNA is accompanied by short RNA fragments that are characteristic of post-transcriptional gene silencing. *Nucleic Acids Research* 29, 2395–2400.

Qi, Y. and Ding, B. (2002) Replication of potato spindle tuber viroid in cultured cells of tobacco and *Nicotiana benthamiana*: the role of specific nucleotides in determining replication levels for host adaptation. *Virology* 302, 445–456.

Qi, Y. and Ding, B. (2003a) Inhibition of cell growth and shoot development by a specific nucleotide sequence in a noncoding viroid RNA. *The Plant Cell* 15, 1360–1374.

Qi, Y. and Ding, B. (2003b) Differential subnuclear localization of RNA strands of opposite polarity derived from an autonomously replicating viroid. *The Plant Cell* 15, 2566–2577.

Qi, Y., Pelissier, T., Itaya, A., Hunt, E., Wasseneger, M. and Ding, B. (2004). Direct role of a viroid RNA motif in mediating direction of RNA trafficking across a specific cellular boundary. *The Plant Cell* 16, 1741–1752.

Reanwarakorn, K. and Semancik, J.S. (1998) Regulation of pathogenicity in hop stunt viroid-related group II citrus viroids. *Journal of General Virology* 79, 3163–3171.

Sano, T. and Matsura, Y. (2004) Accumulation of short interfering RNAs characteristic of RNA silencing precedes recovery of tomato plants from severe symptoms of potato spindle tuber viroid infection. *Journal of General Plant Pathology* 70, 50–53.

Sano, T., Candresse, T., Hammond, R.W., Diener, T.O. and Owens, R.A. (1992) Identification of multiple structural domains regulating viroid pathogenicity. *Proceedings of the National Academy of Sciences of the United States of America* 89, 10104–10108.

Sano, T., Nagayama, A., Ogawa, T., Ishida, I. and Okada, Y. (1997) Transgenic potato expressing a double-stranded RNA-specific ribonuclease is resistant to potato spindle tuber viroid. *Nature Biotechnology* 15, 1290–1294.

Schiebel, W., Pelissier, T., Riedel, L., Thalmeir, S., Schiebel, R., Kempe, D., Lottspeich, F., Sänger, H.-L. and Wassenegger, M. (1998) Isolation of an RNA-directed RNA polymerase-specific cDNA clone from tomato. *The Plant Cell* 10, 2087–2101.

Schindler, I.-M. and Muhlbach, H.-P. (1992) Involvement of nuclear DNA-dependent RNA polymerases in potato spindle tuber viroid replication: a re-evaluation. *Plant Sciences* 84, 221–229.

Schrader, O., Baumstark, T. and Riesner, D. (2003) A mini-RNA containing the tetraloop, wobble pair, and loop E motifs of the central conserved region of potato spindle tuber viroid is processed into a minicircle. *Nucleic Acids Research* 31, 988–998.

Schnölzer, M., Haas, B., Ramm, K., Hofmann, H. and Sänger, H.L. (1985) Correlation between structure and pathogenicity of potato spindle tuber viroid (PSTV). *The EMBO Journal* 4, 2182–2190.

Singh, R.P. (1988) Occurrence, diagnosis, and eradication of the potato spindle tuber viroid in Canada. In: Kryczynski, S. (ed.) *Viroids of Plants and Their Detection: International Seminar*, August 12–20, 1986. Warsaw Agricultural University Press, Warsaw, Poland, pp. 37–50.

Singh, R.P. and O'Brien, M.J. (1970) Additional indicator plants for potato spindle tuber virus. *American Potato Journal* 47, 367–371.

Tabler, M. and Tsagris, M. (2004) Viroids: petite RNA pathogens with distinguished talents. *Trends in Plant Sciences* 9, 339–348.

Taylor, J.M. (2006) Hepatitis delta virus. *Virology* 344, 71–76.

Wang, M.-B. Bian, X.-Y., Wu, L.-M., Liu, L.-X., Smith, N.A., Isenegger, D., Wu, R.-M., Masuta, C., Vance, V.B., Watson, J.M., Rezaian, A., Dennis, E.S. and Waterhouse, P.M. (2004) On the role of RNA silencing in the pathogenicity and evolution of viroids and viral satellites. *Proceedings of the National Academy of Sciences of the United States of America* 101, 3275–3280.

Warrilow, D. and Symons, R. (1999) Citrus exocortis viroid RNA is associated with the largest subunit of RNA polymerase II in tomato in vivo. *Archives of Virology* 144, 2367–2375.

Wassenegger, M., Spieker, R.L., Thalmeir, S., Gast, F.-U., Riedel, L. and Sänger, H.L. (1996) A single nucleotide substitution converts potato spindle tuber viroid (PSTVd) from a noninfectious to an infectious RNA for *Nicotiana tabacum*. *Virology* 226, 191–197.

Woo, Y.-M., Itaya, A. Owens, R.A., Tang, L., Hammond, R.W., Chou, H.-C., Lai, M.M.C. and Ding, B. (1999) Characterization of nuclear import of potato spindle tuber viroid RNA in permeabilized protoplasts. *The Plant Journal* 17, 627–635.

Yang, X., Yie, Y., Zhu, F., Liu, Y., Kang, L., Wang, X. and Tien, P. (1997) Ribozyme-mediated high resistance against potato spindle tuber viroid in transgenic potatoes. *Proceedings of the National Academy of Sciences of the United States of America* 94, 4861–4865.

Yoo, B.C., Kragler, F., Varkonyi-Gasic, E., Haywood, V., Archer-Evans, S., Lee, Y.M., Lough, T.J. and Lucas, W.J. (2004) A systemic small RNA signaling system in plants. *The Plant Cell* 16, 1979–2000.

Zhang, J., Yamada, O., Sakamoto, T., Yoshida, H., Araki, H., Murata, T. and Shimotohno, K. (2005) Inhibition of hepatitis C virus replication by pol III-directed overexpression of RNA decoys corresponding to stem-loop structures in the NS5B coding region. *Virology* 342, 276–285.

Zhao, Y., Owens, R.A. and Hammond, R.W. (2001) Use of a vector based on potato virus X in a whole plant assay to demonstrate nuclear targeting of potato spindle tuber viroid. *Journal of General Virology* 82, 1491–1497.

Zhu, Y., Green, L., Woo, Y.-M., Owens, R.A. and Ding, B. (2001). Cellular basis of potato spindle tuber viroid systemic movement. *Virology* 279, 69–77.

Zhu, Y., Qi, Y., Xun, Y., Owens, R. and Ding, B. (2002) Movement of potato spindle tuber viroid reveals regulatory points of phloem-mediated RNA traffic. *Plant Physiology* 130, 138–146.

7 Molecular Diagnostics of Soilborne Fungal Pathogens

C.A. LÉVESQUE

Abstract

Microbes underpin most of the soil ecosystem functions and studying plant pathogens amongst the incredible soil biodiversity is a challenge. Soilborne plant pathogens are, most often, within taxonomically difficult genera, but fortunately, this also means that they have been included in many comprehensive phylogenetic studies of fungi, including oomycetes. Such databases are being used for the development of molecular diagnostic tools that can provide qualitative or quantitative data on specific taxa. More than two decades after its invention, polymerase chain reaction (PCR) and its many derived applications remain the core technology for soil molecular studies. There are two general kinds of approaches to study soil with PCR. Universal primers of various taxonomic resolution are used to generate a broad mixture of amplicons that are analysed further through a range of techniques. Alternatively, species-specific PCR reactions are performed to detect and/or quantify the target species. Both approaches can be amenable to the detection of a range of pathogens and beneficials. Molecular techniques can resolve many of the taxonomic difficulties that more traditional approaches need to confront, but the crucial issue of soil sampling in ecological studies remains. Molecular detection uncovered and compounded another problem. Many fungi that are being characterized by direct DNA processing of soil are supposedly unculturable, whereas molecular detection is reducing the need for isolation of the culturable ones. There is a need to try to find ways to grow and describe unculturable species and maintain reference collections of live cultures. The stability of DNA in dead organisms and a bias towards detecting non-living organisms does not appear to be a broad issue in microbiologically active soil. Testing for regulatory purposes is an important niche for DNA-based soil assays, but routine usage of these techniques on a commercial scale has not yet been achieved. The availability of on-site testing devices that can provide a resolution at the pathotype or race level of the pathogens and detect the beneficial species that reduce the impact of pathogens would greatly encourage the wide adoption of DNA-base techniques for soilborne disease management. The biosecurity initiatives supporting the sequencing of many genomes of soilborne pathogens and the DNA barcode initiatives targeting the sequencing of the largest possible number of species will greatly help to achieve this goal.

Introduction

Fungi, including oomycetes, are major components of agricultural ecosystems and limitations in isolating and identifying them using traditional methods make it difficult to improve the management of soilborne diseases. This is compounded by the fact that the most common soiborne genera, such as *Fusarium, Colletotrichum, Trichoderma, Penicillium, Phytophthora* or *Pythium*, also happen to be within the most challenging groups for taxonomists. In wheat field soils, it has been estimated that half of the estimated 2500 kg biomass/ha is of fungal origin (Brookes et al., 1985). A commonly cited statistic is the expected presence of 10,000 microbial species/g of soil, which would include saprophytes, plant pathogens, plant growth promoters, zoophagous fungi and hyperparasites (Torsvik et al., 1996). There is a gap between the taxonomists who study fungi and the plant pathologists or ecologists trying to understand the dynamics of plant diseases. The overwhelming diversity of soil fungi, combined with the technical challenges and knowledge needed in identifying them, leads to a 'black box' syndrome when trying to understand epidemiology of soilborne diseases. Biotechnology can narrow the gap between taxonomy and plant pathology and shed light into this by combining classical taxonomy with modern molecular techniques, new methods for isolating or detecting microbes that are difficult to grow and the design of rapid diagnostic assays that can simultaneously detect and identify hundreds or thousands of soilborne organisms.

Background

The molecular diagnostics era started with the utilization of polyclonal and monoclonal antibodies. In plant pathology, antibodies are used extensively for virus and bacteria testing of plant samples and have been successfully commercialized. The use of antibody-based assays for fungi and for soil or even root testing is limited. Fungal antibodies tend to be less specific than the ones developed for viruses or bacteria and the infected roots or infested soil have higher concentrations of organisms that can cause cross-reactions with antibody-based assays for fungi. Substrate utilization and fatty acid profiles can also be used to characterize soil microbes (see Mazzola [2004] for review).

Over two decades after its invention (Mullis and Faloona, 1987), polymerase chain reaction (PCR) as a method for amplification of DNA or RNA is still going strong and has not been replaced by alternatives that are as robust, cost-effective and flexible. PCR for environmental samples is applied in a number of different ways that can be grouped under two general categories. Species-specific primers are used to detect single species or even multiple species if several species-specific primer sets are used for a given sample. Alternatively, more universal primers can be used and the mixture of PCR products then can be sorted out through different techniques, such as cloning and sequencing individual molecules, hybridization to specific oligonucleotides or electrophoresis techniques, thereby resolving the PCR mixture. Within these kinds of applications, there have been many technological advances to PCR or to the processing of PCR products that helped maintain its popularity and expand potential applications. For example, more robust and efficient Taq polymerases keep increasing its efficiency and robustness and the advent of real-time PCR has reduced time and labour. Specific applications of PCR to plant pathology and studies of soil fungi have been reviewed extensively (Martin *et al.*, 2000; Schaad and Frederick, 2002; Filion *et al.*, 2003; Kageyama *et al.*, 2003; Schaad *et al.*, 2003; Lebuhn *et al.*, 2004; Paplomatas, 2004; Schena *et al.*, 2004; Bonants *et al.*, 2005; Lubeck and Lubeck, 2005; Okubara *et al.*, 2005). Despite all the research being done on PCR, only 15–20% of clinical laboratories use it for medical diagnostics (Anonymous, 2006). Interestingly, PCR is being adopted more rapidly in animal diagnostics, especially for the replacement of enzyme-linked immunosorbent assay (ELISA), which suggests some regulatory hindrance for its more widespread adoption in the medical field (Salisbury, 2006). Nevertheless, some technical issues remain with PCR and one always needs to be aware of them when using it for diagnostics. The presence and effect of inhibitors in the DNA extract always needs to be tested with proper controls. The sensitivity of PCR is one of its main advantages but also one of its potential pitfalls. When the same fragment is repeatedly amplified for detection, the laboratory becomes contaminated with this product, increasing drastically the likelihood of false positives. As such, false-positive reactions can be sporadic, a single negative control reaction cannot always detect the problem reliably. Nested PCR, i.e. running two consecutive PCR reactions to increase sensitivity, compounds the problem of potential false-positive results.

Phylogenetic and genome studies are showing unequivocally that some parts of the genomes are highly conserved across large taxonomic groups (e.g. small subunit [SSU] ribosomal DNA), whereas others are hypervariable. Because of this fundamental nature of the genomes and the rapidly expanding DNA sequence databases, there is an endless source of data to design DNA-based assays with different specificity for different taxonomic groups. This review will focus primarily on DNA-based assays.

Sampling and DNA Extraction

Sampling is probably one of the main challenges in molecular detection of pathogens from soil. Most of the sampling issues for molecular tests are the same as the ones for more traditional microbiological techniques developed to study soilborne fungi. It is hard to process more than a few grams of soil sample for dilution plating or DNA extraction; therefore, the sample is an infinitesimal portion of the field or the plot. If the organism has a rather uniform distribution, this is less of a problem, but for organisms with patchy or clumped distribution, this is a major issue. Keeping the sampling size small per plot reduces the power of an experiment, i.e. a reduction in the probability of detecting a difference between treatments and/or an increase in the size of the minimum difference that can be detected, whereas increasing the sample size will increase power but make the experiment more costly. The right balance always needs to be achieved.

DNA extraction kits vary in the amount of soil or roots they can handle. Being able to extract DNA from grams instead of milligrams would be a major improvement. It was found that soil samples larger than 1 g showed much less variation when assessed for bacterial and fungal community structure than samples of less than 1 g (Ranjard et al., 2003). A simple and cheap extraction technique capable of handling approximately 5 g was developed by Reeleder et al. (2003). The pooling of soil cores is often used to reduce processing cost, while maintaining proper sampling representation of species with uneven distribution. It appears that if large soil samples (1 kg) are thoroughly mixed before DNA extraction, it is possible to achieve homogeneity in DNA assay results for bacterial communities with 250 mg of soil for DNA extraction (Kang and Mills, 2006). The comparisons done so far were made with techniques that provide characterization of communities at a high taxonomic level, probably hiding some of the distribution characteristics. Assays that would provide resolutions at the species level should be assessed with various sampling, pooling and DNA extraction schemes if the distribution needs to be known.

Concentrating the propagules before DNA extraction is a way to process a larger sample without the technical challenge of extracting DNA from a large amount of soil. Inoculum concentration is not new to plant pathology. For example, microsclerotia of *Verticillium albo-atrum* have been concentrated through a combination of sieving and sucrose gradient separation (Huisman and Ashworth, 1974), whereas sporangia of *Synchytrium endobioticum* can be concentrated by a combination of sieving and chloroform flotation followed by centrifugation (Pratt, 1976). The latter is the basis for European and Mediterranean Plant Protection Organization (EPPO)-recommended protocols for testing and descheduling of soils (van Leeuwen et al., 2005). Instead of processing the concentrate by microscopy or plating on semi-selective media, it is possible to extract its DNA in order to run a PCR assay. An extraction procedure for nematode eggs was recently adapted to concentrate sporangia of *S. endobioticum* before PCR, providing a sensitivity of 10 sporangia/100 g of soil (van den Boogert et al., 2005).

The sampling situation is somewhat easier if one wants to study soilborne pathogens in roots. The rhizosphere is a complex environment, but in a monoculture, it is much less variable that the soil itself. Root sampling is also similar to baiting in some ways. The root system and baiting will attract the pathogen in the surrounding soil, using the pathogen chemotaxis towards a suitable substrate, increasing the likelihood of detecting the pathogen and the sensitivity of the overall assay. Baiting was an order of magnitude more sensitive than sieving for detecting *Phytophthora cinnamomi* (Greenhalgh, 1978). However, different hosts or baits will attract different colonists. Arcate *et al.* (2006) and Tambong *et al.* (2006) showed that the oomycete community profile obtained through baiting was different from the one obtained by direct PCR detection from the rhizosphere or soil, respectively. If one wants to assess the community of pathogens and beneficial microorganisms, baiting or root sampling will be of more limited value.

Multiplexing

Especially in soil, many pathogens act synergistically or are affected by beneficial organisms and it becomes increasingly important to detect a wide range of organisms simultaneously. The detection of some pathogens, e.g. pathogens that are regulated or a concern for crop biosecurity, means that immediate action must be taken. The detection of some others does not necessarily mean that the plant or yield will be affected. *Pythium irregulare* or *Pythium dissotocum* were isolated from healthy and diseased strawberry roots at a similar frequency (Leandro *et al.*, 2005). The physical environment and the predisposition of the hosts can be an important factor for such an outcome, but the presence of antagonists or plant growth-promoting organisms also could be a significant factor in reducing the probability of a pathogen species being detected. In ecology, the denaturing gradient gel electrophoresis is commonly used to compare fungal diversity (Marshall *et al.*, 2003; Kowalchuk and Smit, 2004). The power of this technique lies in its ability to characterize the diversity with limited background information about the fungal group that is being affected by treatments. However, when the disease potential needs to be assessed quickly, this technique cannot provide the required species resolution in a timely fashion. DNA array-based technologies are being developed to provide such resolution in a multiplex fashion.

Functional Genomics Versus Molecular Detection from the Environment

The development of molecular tools for the detection of pathogens has been closely linked to advances in genomics, but more recently, there has been a parallel development between functional genomics and molecular detection of organisms from the environment.

Large-scale sequencing

Whole-genome sequencing has been instrumental to the field of functional genomics. Once all the genes of a species are sequenced, they must be annotated. If very few genes have known functions, the data provided by assays to determine which genes are upregulated or downregulated have more limited value. In DNA barcode or molecular phylogeny initiatives, scientists are attempting to sequence one or a few genes across as many species as possible. This 'horizontal genomics' approach provides data about many species and eventually includes all known species, for a very limited set of ubiquitous genes. If the species are not identified properly, the use of these sequences in ecological work to determine which species are fluctuating also has limited value. The resources put towards the sequencing of numerous genomes have had, and continue to have, a tremendous impact on functional genomics. Unfortunately, initiatives towards sequencing one or a few genes for a broad swath of the global diversity have not received the same level of support as of yet (see DNA barcode section below). Without the comprehensive databases that barcode or tree of life initiatives can provide, the molecular detection of organisms will always be limited to economically important pathogens and/or rife with false-positive results because of poor sampling within the genera.

Arrays and hybridization

In functional genomics, many and often all the genes of one species are represented on a microarray or a gene chip. The mRNA is made into fluorescently labelled cDNA before hybridization to the microarray. By comparing the hybridization patterns between different treatments, it is possible to identify the genes that are upregulated or downregulated between these treatments. In phylogenetic oligonucleotide arrays, all the species of a genus (e.g. Tambong et al., 2006; see Fig. 7.1) or different pathogens of a crop (e.g. Sholberg et al., 2005) can be represented on an array. By comparing the hybridization patterns between treatments, one can determine the 'upregulated and downregulated' species instead of genes. The preparation and labelling of the probing material for such a phylogenetic array is most often done by PCR instead of reverse transcriptase.

Confirmation and quantification

In functional genomics, two different dyes representing each treatment are typically used for the same hybridization, allowing for direct comparison of hybridization signals on the same microarray. A significant difference in hybridization intensity between the dyes or treatments is seen as a good indication of upregulation or downregulation. However, the regulation of the most important genes identified by microarray hybridization often needs to be confirmed and quantified more accurately by reverse transcriptase PCR. This is currently the fastest growing market for PCR applications (Anonymous, 2006). Similarly, the hybridization signal on a phylogenetic array processed with an environmental sample is semi-quantitative (Lievens et al., 2005), but the results for the most important

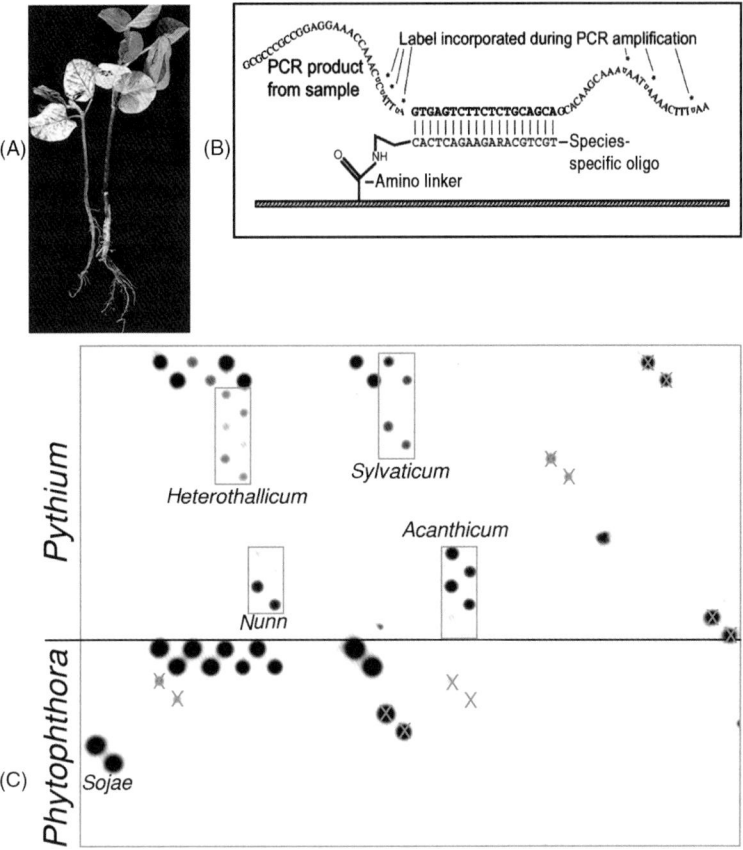

Fig. 7.1. (A) Root rot symptoms on soybean plants from the *Phytophthora* nursery at the Central Experimental Farm, Ottawa. (B) Diagrammatic representation of positive hybridization reaction between a labelled PCR product and a perfectly matching specific and immobilized oligonucleotide. (C) Hybridization reactions on a DNA array after the processing of roots shown in (A). The *Pythium* array is from Tambong et al. (2006) and the *Phytophthora* array is unpublished. The main species in the arrays were also isolated in the nursery. The Xs denote oligonucleotides that are known to cross-react with the species present. The rows or oligonucleotides at the upper right corners of both the *Pythium* and the *Phytophthora* arrays are positive controls or oligonucleotides with genus specificity.

species influenced by treatments should be confirmed and their accuracy should be improved by quantitative PCR. Conversely, it is possible to perform quantitative PCR assays for several species and use DNA array hybridization to confirm that the main species have been tested. This was done to detect and quantify *Pythium* spp. in soils of Washington State (Schroeder et al., 2006). Some companies are developing gene-expression PCR platforms of such high throughput that there would be enough individual PCR reactions to bypass the array hybridization step for many applications (Anonymous, 2006).

Array Development for Pathogen Detection

Most genes in a functional genomics array differ markedly in sequence composition, whereas phylogenetic arrays for molecular detection of fungal species need to exploit very small differences between species to maintain species specificity. This is a fundamental difference in the design of arrays between these two applications. Full cDNAs spotted onto microarrays can work in functional genomics, whereas arrays based on immobilized large fragments (>250 bp) cannot differentiate closely related species (Lévesque et al., 1998). However, signal intensity on microarrays increases as the length of the immobilized probe increases (He et al., 2005a). The current trend in functional genomics of using 70-mer oligonucleotides to increase sensitivity will probably not be applicable in situations where closely related species need to be differentiated using the same DNA region (He et al., 2005b). There is a need for at least a 10% difference in sequence to avoid false positives with long oligonucleotides (see Sessitsch et al., 2006 for review). The side effect of using shorter oligonucleotides is a reduction in sensitivity; therefore, this needs to be compensated by arraying substrate and hybridization detection technologies that maximize sensitivity. Membrane-based arrays are many thousandfold (10^5) more sensitive than planar microarrays (Cho and Tiedje, 2002). The loading limit of the substrate and the amount of immobilized DNA are given as the main explanation for the difference in sensitivity. A porous substrate can 'stack' more immobilized molecules over the same surface area and provide a much stronger signal per unit area. It is also possible that a porous solid substrate enhances hybridization kinetics compared to a planar surface to which bound oligonucleotides must become hybridized to the labelled DNA in aqueous solution. Membrane arrays cannot have the density of planar glass slides; however, developments are being made to make microarrays more porous (e.g. Wu et al., 2004). Gentry et al. (2006) provided an extensive review of all the existing types of arrays for microbial ecology studies. Oligonucleotide suspension microarray is a new technology where the hybridization events are detected in a solution of microspheres with different oligonucleotides bound to them (e.g. Deregt et al., 2006). It will be interesting to compare the sensitivity of this technology with planar microarrays and macroarrays.

Genetic Markers

So far, most of the DNA-based fungal pathogen detection assays have been based on ubiquitous genes or spacers such as intergenic transcribed spacers (ITS). As more genomes are being sequenced, the choice of novel genes will become easier. It is important to have enough polymorphisms in the selected genes or spacers to avoid cross-reactions and false positives; the more polymorphisms, the better. The presence of insertions or deletions (indels) greatly facilitates the design of specific PCR primers or hybridization oligonucleotides. These indels ensure that no cross-reactions will occur. It is also very important to have sequences of the most closely related species to design highly specific assays. Since it is better to have sequences that can be easily aligned for phylogenetic studies, good genes for phylogenies are not necessarily the best to design detection assays. Genes for high-level phylogenies (e.g. SSU) are rarely useful for species detection. When there is a high level of polymorphism in genes for robust species differentiation, it is likely that species-specific markers can be developed. If single nucleotide polymorphisms are evenly scattered along the sequences and cannot be found in clusters within stretches of 20–25 nucleotides, phylogenetic resolution at species level could be achieved, but difficulties in generating hybridization oligonucleotides should be expected. If one wants to detect a single species, phylogenetic studies that identify the most closely related species become invaluable. The first widely used assay for *Phytophthora ramorum* was cross-reacting with the most closely related species known at the time because the ITS was providing few polymorphisms between these two species (Hayden *et al.*, 2004). Assays with genes that had more polymorphisms showed better specificity (Martin et al., 2004; Bilodeau *et al.*, 2006). With the genome of *P. ramorum* now sequenced (Tyler *et al.*, 2006) and with a better sampling of species within the genus (e.g. Brasier *et al.*, 2005), the development of detection assays for *P. ramorum* with highly specific genes, potentially involved even in pathogenicity or avirulence, will be easier. However, the development for multiplex assays targeting all the species of this genus will require genes or spacers found in all species.

Living Versus Non-living

The issue of the viability of the propagules detected through DNA-based molecular techniques is often raised. DNA is a food source and it is very likely that in a microbiologically active substrate, such as moist soil, it will be quickly metabolized by other organisms (Lebuhn *et al.*, 2004). To evaluate this phenomenon, we irradiated spores and used them to inoculate greenhouse potting mix under different treatments. The amount of spores as estimated through quantitative PCR was the same across all treatments (~30,000/g) immediately after inoculation with the same number of irradiated or non-irradiated spores (Fig. 7.2). DNA in the dead spores was stable in a sterile environment, but not if there was microbial activity. One week after inoculation, the number of irradiated spores detected in non-sterile soil dropped to 50 (three logs or 99.8%) and further dropped down to 10 in the wet potting mix when reassessed after another week. The detection of DNA from *Penicillium bilaiae* remained constant over the same period when irradiated spores were mixed with sterile soil. The detection of viable spores remained unchanged except for some slight fluctuation in non-sterile soil. This would confirm that DNA within dead spores is used as food source. One would expect more micro-

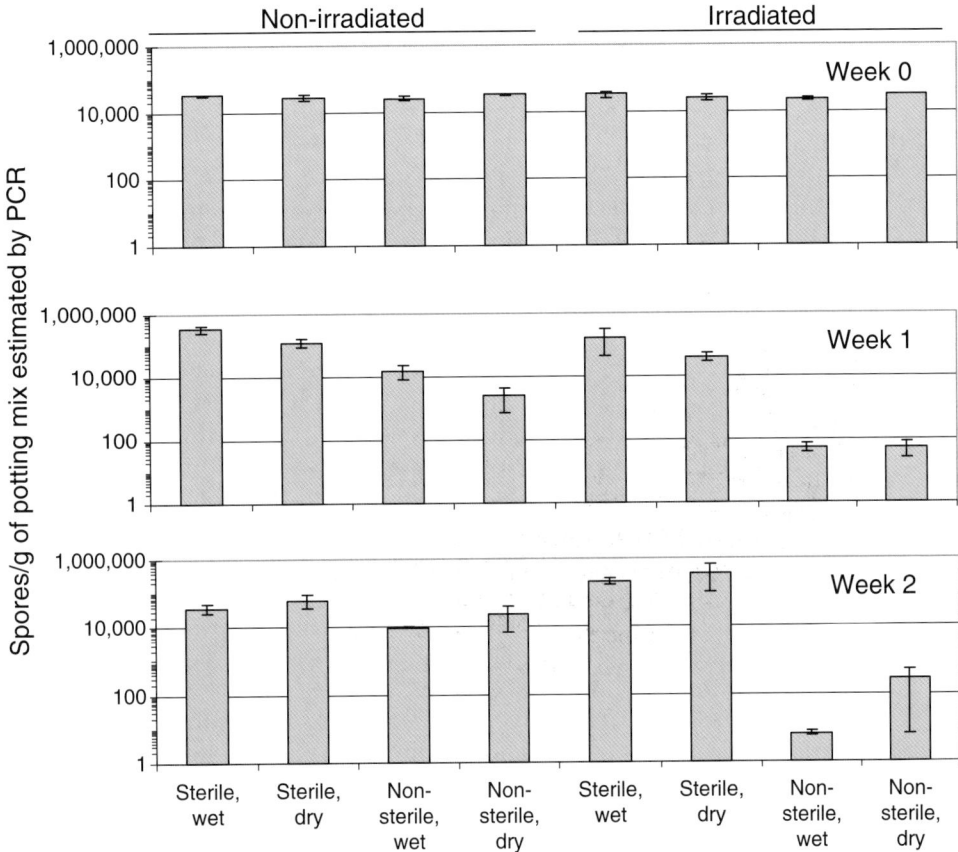

Fig. 7.2. Effect of irradiation of spores, moisture, autoclave sterilization of potting substrate and time after inoculation on PCR-based detection of *Penicillium bilaiae* using DNA extracted from potting mix. The numbers of spores were extrapolated from a standard curve made with DNA extracted from live spores. The D_{10}, i.e. the exposure to kill 90% of the spores, was independently established to be 13 min for *P. bilaiae* (0.05 Mrad). The spores were exposed to a source of 3612 rads/min for 100 min which caused a drop of 6 logs in spore viability. The wet potting soil was kept at 0.1 bar (300% w/w moisture), whereas the dry potting soil was at 3 bars (130% w/w moisture).

bial activity in most natural soils compared to dry greenhouse potting mix; therefore, this is probably a conservative estimate about the rapid degradation of DNA in dead spores in a natural soil environment.

It is believed that RNA-based tests would give a more accurate estimate of viable cells than DNA-based tests that might detect dead cells. A recent medical study found that the rRNA test was actually more sensitive than the DNA test, i.e. more subjects were found to be infected with *Chlamydia trachomatis* through the rRNA test (Yang *et al.*, 2006). One hypothesis is that more rRNA template is present than DNA template. Amplifying mRNA versus ribosomal DNA can give slightly different results; however, they caution that transcriptome may change rapidly, making quantification of cells more difficult (Bodrossy et al., 2006).

Validation

This is one of the most challenging issues with the development of detection systems for microorganisms. The assays must have been tested against a wide rage of strains of the target species and of the species that can cause false positives. They must have been tested with both pure cultures (well-characterized samples with high inoculum for obligates) and field samples that have been assessed for the presence or absence of the target organisms. When developing multiplex assays with over 100 oligonucleotides (e.g. Sanguin et al., 2006; Tambong et al., 2006), this becomes even more challenging as permutations of possible combinations of pathogens at different concentrations are practically infinite. Comparisons of the results among the DNA-based array, traditional microbiological methods and/or the cloning and sequencing of PCR product were used for validation in these studies. When a pathogen is of high economic importance, it is common to have different assays that have been developed to detect it. Recently, different assays for *P. ramorum* were compared in a blind ring trial that included hundreds of DNA samples provided by the World *Phytophthora* Collection, University of California, Riverside (Martin et al., 2006). It appears that redundancy in the assays, i.e. having more than one marker, is a good approach to reduce ambiguous results. This was probably the first example of such an extensive comparison of different assays for the same plant pathogen.

DNA Barcode

One reason why validation of assays for pathogen detection can be so difficult to achieve is the incomplete sampling of species within a genus in the database used to design an assay. Many functional genomics microarrays and their associated PCR assays for confirmation are designed with the full genome already sequenced. This would be equivalent to having all the species of a large taxonomic group sampled and sequenced for at least one marker with enough variation. This is the ultimate goal of the international Consortium for the Barcode of Life (CBOL) (Hebert et al., 2003). In the animal kingdom, the database of the 5' end of the cytochrome oxidase I (COI) gene is expanding rapidly (e.g. Hajibabaei et al., 2006). It would be advantageous to have the same sequence region as barcode for all eukaryotic life forms. One could assess the biodiversity of all eukaryotes in a soil sample by amplifying a single gene region. In many situations, it is difficult to pinpoint the etiology of a disease to a single species, and disease-causing agents appear to be working synergistically (e.g. nematodes, and fungi). The ability to detect simultaneously a wide range of causal agents, their antagonists, their hosts and a wide range of other organisms within the community by the amplification of a single gene would be an elegant way to do this. In fungi and oomycetes, there is already an extensive ITS sequence database and the potential of COI to discriminate species is good (e.g. Martin and Tooley, 2003), but the technique must be tested across a broad range of genera. In *Eumycota*, the frequency and practical limits of introns need to be evaluated for the adoption of COI as barcode. Knowing which plant species are likely to occur in soil samples (e.g. weed versus crop root fragments) is valuable information in some situations, although it appears that COI will not work well as barcode for plants (Chase et al., 2005). Unfortunately, one will need to amplify more than one gene to characterize the different eukaryotic kingdoms in soil. Given that other genes will need to be amplified for bacteria and virus characterization, characterization of all living forms and infectious agents will not be a trivial task, but it is far from being insurmountable.

Detection of Pathogens and Disease Management

Quarantine pathogens and biosecurity

Regulated pathogens have been the prime targets for the development of molecular detection tools. The production of pathogen-free seeds and the early detection of quarantine pathogens remain the best lines of defence. Increased concerns about crop biosecurity (e.g. National Academy of Sciences, 2002) and the recent development of programmes such as the National Plant Diagnostic Network in the USA (Stack et al., 2006) have increased the awareness and the needs for rapid diagnostic tools. As a result, more diagnostic laboratories have been equipped and personnel have been trained in the use of molecular technologies to detect plant pathogens.

Use in research

Molecular diagnostics are part of a research continuum. Taxonomy and molecular phylogeny should be the basis for the development of molecular diagnostics. At the receiving end, there are many research endeavours that are now routinely using molecular diagnostic tools. DNA sequencing and molecular detection techniques are the commonly used methods of identification for research projects in etiology, epidemiology, disease management, plant breeding and soil ecology. Researchers studying root and soil biology are likely to remain important users of molecular diagnostics.

Commercial applications

Compulsory requirement for certification can be a significant impetus for increasing the demand for a particular molecular test. Seed and clonal propagation material are routinely tested and support the commercial viability of the diagnostics industry. In regions where quarantine soilborne pathogens such as *S. endobioticum* are present, there are needs to certify or 'deschedule' fields. It is harder for the commercial diagnostics industry to develop expertise and markets for testing services that are only done when symptoms appear. Moreover, it is often too late to take remedial action once the disease symptoms are present. Therefore, routine testing for early detection of pathogens and to be able to make disease management decisions in a timely fashion is probably where the potential for growth in diagnostic services is the highest. The pathogen detection in hydroponic solutions for greenhouse vegetables is an example where such routine testing services exist. Detection of pathogens on spore traps is another example where routine testing to decide on fungicide sprays could be done for integrated pest management (IPM), following what the entomologists have been doing with pheromone traps since the 1970s. For soilborne pathogens in the field, the applications where routine testing outside a regulatory framework or research could be commercially successful are less obvious. Soil testing before seeding could be done to determine the likelihood of development of certain root diseases for different crops in order to optimize decisions about rotation or seed treatments. The 'soil health' could be determined to make decisions on soil organic amendments. If the testing services can provide information about pathogen races or pathotypes, the cultivar could be selected more judiciously.

Importance of having specimens or live cultures

Molecular studies have shown that a high proportion of soil microorganisms are 'non-culturable' (e.g. Smit *et al.*, 1999); however, it appears that for at least some of them, it is the inability to compete on nutrient-rich media that makes them hard to isolate and grow. Dilution to extinction (Wise *et al.*, 1999) and low nutrient media (Aagot *et al.*, 2001) can help to culture some of these organisms that were thought to be unculturable, an important step to better characterize the unknown species.

The advent of antibody and DNA-based molecular detection tools has reduced the need for the isolation of the pathogens for routine diagnostics. This situation is already causing a side effect in the medical field: it is becoming harder to obtain live strains that have been recently isolated, as it is diagnostically simpler to detect specific genetic information associated with a taxon group in a clinical specimen. This practice has been done for years for essentially unculturable, slowly progressing bacterial pathogens (e.g. *Tropheryma whippelii*, cause of Whipple's disease), but nucleic acid-based tests have replaced culturing as the gold standard diagnostic test for many common human pathogens that can be easily isolated (Nolte and Caliendo, 2003). Therefore, researchers who want to document and study the process of antibiotic resistance acquisition or for presence of other virulence factors in the future, will not have access to a range of live cultures isolated from clinical material. This is compounded by the fact that the infrastructure and technological advances in microbial culture collections have not kept pace with the biotechnology boom. This problem will not be different in plant pathology. If diagnostic laboratories no longer isolate pathogens and culture collections do not get new accessions regularly, it will become increasingly difficult to conduct certain studies in plant pathology. When there is any doubt that an unusual strain might be involved in a plant disease outbreak, isolation should be attempted and a specimen or live culture should be deposited as voucher. Koch's postulate will remain a fundamental protocol in plant pathology whenever a new disease is suspected and it cannot be done without isolation and reisolation after inoculation.

Development of New Technologies

The development of diagnostic kits that growers and IPM consultants can use on-site to diagnose diseases from leaves or seed samples was a breakthrough for antibody-based systems. They were developed mainly for viruses and bacteria. It has been harder to design such system for direct detection of any pathogens from roots or soil. PCR instruments, including real-time ones, are becoming smaller and can run on battery power. Miniaturized and automated PCR systems that can constantly run samples from extraction to real-time reading exist and are being used routinely. The anthrax detection system that easily fits under a small bench is now operational in hundreds of US postal offices (Knight, 2002). The handheld and point-of-care PCR instrument that can perform everything from DNA extraction to automatic reading of results is the holy grail and intensive research continues to design such device (e.g. Higgins *et al.*, 2003; Lee *et al.*, 2006).

Conclusions and Future Directions

There is no doubt that PCR was a revolutionary innovation for molecular detection and diagnostic. However, it is sobering to see how long it is taking for its widespread application in laboratory processing of field samples on a large scale. Maybe, this is just a good example of how long it takes to bring an invention into widespread field applications. PCR is fairly unique because one can pinpoint precisely when it was invented – it is not a technology like many ecological field techniques that was developed by a series of incremental steps over a long period of time. One reason for its slow implementation in the medical field is the rigid regulatory framework under which medical diagnostic laboratories operate, while the more rapid acceptance of PCR in veterinary laboratories is probably because of less-stringent rules (Salisbury, 2006). Then, why the plant diagnostic laboratories have not applied PCR-based tests as rapidly since their regulatory framework is less rigid too? The most probable explanation is the cost of processing samples, i.e. the animal diagnostic market can sustain a higher cost per sample than the plant pathogen detection market. The market for crop diagnostics and the research grants for soil ecology research can probably sustain the cost of licensing the latest technologies only for a limited number of customers or applications. Access to low-cost technologies becomes an essential requirement for commercialization or usage in plant pathology laboratories that are required to process a large number of samples.

Earlier, I made the parallel between phylogeny-based arrays and arrays for functional genomics. The new generation of parallel sequencers will allow the sequencing of a full genome rapidly and cost-effectively. With the expanding number of soilborne fungal genomes that have been or are in the process of being fully sequenced, the combination of the power of functional genomics and phylogeny-based array will provide new insight. Metagenomics can provide entire genomes for a broad range of species in a given community (Venter et al., 2004). New sequencing technology will allow a massive amount of sequencing for environmental samples. The dominant species of a community could be identified rapidly with a phylogeny array at the same time as the main genes being expressed could be characterized by a functional genomics array. Ecological data will be tied to function and taxa if the functional genomic array can have species specificity for the most important genes being expressed.

Combining expression of genes involved in pathogenicity or avirulence with species detection will be a powerful tool for disease management. Complete validation of such systems will be an almost impossible task, as seen in gene expression microarray assays, but running such assays through carefully planned controlled experiments will provide the basis for disease management applications under field conditions. Probably, the best way to avoid questionable results will be through redundancy of markers on the array, i.e. targeting several genes, having several oligonucleotides per gene per species, and by creating artificially mismatched oligonucleotides that can confirm the true matches within a sample (Liu et al., 2005). The amount of data regarding communities and genes being expressed will be phenomenal and a major statistical challenge; however, such data will allow us to shed some light in that soil 'black box' and help us define and predict soil 'health'.

Acknowledgements

I thank Daniel O'Gorman (AAFC, Summerland) and Mary Leggett (Philom Bios Inc., Saskatoon) for designing and conducting the experiment on DNA degradation, Scott Redhead for reviewing this manuscript before submission, Tharcisse Barasubiye for the work on soybean and Kathy Bernard (National Microbiology Laboratory – Public Health Agency of Canada) for fruitful discussions and for making me realize the importance of having microbial collections that maintain and provide access to strains from clinical diagnostics laboratories. My current research is supported by AAFC, CRTI (04-0045RD) and the Canadian Barcode of Life Network.

References

Aagot, N., Nybroe, O., Nielsen, P. and Johnsen, K. (2001) An altered *Pseudomonas* diversity is recovered from soil by using nutrient-poor *Pseudomonas*-selective soil extract media. *Applied and Environmental Microbiology* 67, 5233–5239.

Anonymous (2006) PCR applications continue to expand. *Genetic Engineering News* 26, 1-(22,24)-26.

Arcate, J.M., Karp, M.A. and Nelson, E.B. (2006) Diversity of peronosporomycete (oomycete) communities associated with the rhizosphere of different plant species. *Microbial Ecology* 51, 36–50.

Bilodeau, G.J., Lévesque, C.A., de Cock, A.W.A.M., Duchaine, C., Brière, S., Uribe, P., Martin, F.N. and Hamelin, R.C. (2007) Molecular detection of *Phytophthora ramorum* by Real-Time Polymerase Chain Reaction using TaqMan, ^{SYBR}Green and Molecular Beacons. *Phytopathology* 97, 632–642.

Bodrossy, L., Stralis-Pavese, N., Konrad-Koszler, M., Weilharter, A., Reichenauer, T.G., Schofer, D. and Sessitsch, A. (2006) mRNA-based parallel detection of active methanotroph populations by use of a diagnostic microarray. *Applied and Environmental Microbiology* 72, 1672–1676.

Bonants, P.J.M., Schoen, C.D., Szemes, M., Speksnijder, A., Klerks, M.M., Boogert, P.H.J.F.v.-d., Waalwijk, C., Wolf, J.M.v.-d. and Zijlstra, C. (2005) From single to multiplex detection of plant pathogens: pUMA, a new concept of multiplex detection using microarrays. *Phytopathologia Polonica* 35, 29–47.

Brasier, C.M., Beales, P.A., Kirk, S.A., Denman, S. and Rose, J. (2005) *Phytophthora kernoviae* sp. nov., an invasive pathogen causing bleeding stem lesions on forest trees and foliar necrosis of ornamentals in the UK. *Mycological Research* 109, 853–859.

Brookes, P.C., Powelson, D.S. and Jenkinson, D.S. (1985) The microbial biomass in soil. In: Fitter, A.H. (ed.) *Ecological Interactions in Soil: Plants, Microbes and Animals*. Blackwell, Oxford, pp. 123–125.

Chase, M., Salamin, N., Wilkinson, M., Dunwell, J., Kesanakurthi, R., Haidar, N. and Savolainen, V. (2005) Land plants and DNA barcodes: short-term and long-term goals. *Philosophical Transactions of the Royal Society B: Biological Sciences* 360, 1889–1895.

Cho, J.C. and Tiedje, J.M. (2002) Quantitative detection of microbial genes by using DNA microarrays. *Applied and Environmental Microbiology* 68, 1425–1430.

Deregt, D., Gilbert, S.A., Dudas, S., Pasick, J., Baxi, S., Burton, K.M. and Baxi, M.K. (2006) A multiplex DNA suspension microarray for simultaneous detection and differentiation of classical swine fever virus and other pestiviruses. *Journal of Virological Methods* 136, 17–23.

Filion, M., St-Arnaud, M. and Jabaji-Hare, S.H. (2003) Direct quantification of fungal DNA from soil substrate using real-time PCR. *Journal of Microbiological Methods* 53, 67–76.

Gentry, T.J., Wickham, G.S., Schadt, C.W., He, Z. and Zhou, J. (2006) Microarray applications in microbial ecology research. *Microbial Ecology* 52, 159–175.

Greenhalgh, F.C. (1978) Evaluation of techniques for quantitative detection of *Phytophthora cinnamomi*. *Soil Biology and Biochemistry* 10, 257–259.

Hajibabaei, M., Janzen, D.H., Burns, J.M., Hallwachs, W. and Hebert, P.D. (2006) DNA barcodes distinguish species of tropical Lepidoptera. *Proceedings of the National Academy of Sciences of the United States of America* 103, 968–971.

Hayden, K.J., Rizzo, D., Tse, J. and Garbelotto, M. (2004) Detection of *Phytophthora ramorum* from California forests using a real time polymerase chain reaction assay. *Phytopathology* 94, 1075–1083.

He, Z., Wu, L., Fields, M.W. and Zhou, J. (2005a) Use of microarrays with different probe sizes for monitoring gene expression. *Applied and Environmental Microbiology* 71, 5154–5162.

He, Z., Wu, L., Li, X., Fields, M.W. and Zhou, J. (2005b) Empirical establishment of oligonucleotide probe design criteria. *Applied and Environmental Microbiology* 71, 3753–3760.

Hebert, P.D., Cywinska, A., Ball, S.L. and deWaard, J.R. (2003) Biological identifications through DNA barcodes. *Proceedings of the Royal Society of London. Series B: Biological Sciences* 270, 313–321.

Higgins, J.A., Nasarabadi, S., Karns, J.S., Shelton, D.R., Cooper, M., Gbakima, A. and Koopman, R.P. (2003) A handheld real time thermal cycler for bacterial pathogen detection. *Biosensors and Bioelectronics* 18, 1115–1123.

Huisman, O.C. and Ashworth, L.J. Jr (1974) *Verticillium albo-atrum*: quantitative isolation of microsclerotia from field soils. *Phytopathology* 64, 1159–1163.

Kageyama, K., Komatsu, T. and Suga, H. (2003) Refined PCR protocol for detection of plant pathogens in soil. *Journal of General Plant Pathology* 69, 153–160.

Kang, S. and Mills, A.L. (2006) The effect of sample size in studies of soil microbial community structure. *Journal of Microbiological Methods* 66, 242–250.

Knight, J. (2002) US postal service puts anthrax detectors to the test. *Nature* 417, 579.

Kowalchuk, G.A. and Smit, E. (2004) Fungal community analysis using PCR-denaturing gradient gel electrophoresis (DGGE). In: Kowalchuk, G.A., de Bruijn, F.J., Head, I.M., Akkermans, A.D.L. and Dirk van Elsas, J. (eds) *Molecular Microbial Ecology Manual*. Springer, New York, pp. 771–788.

Leandro, L., Ferguson, L., Fernandez, G. and Louws, F. (2005) Population dynamics of fungi associated with strawberry roots in different soil management systems. *Phytopathology* 95, S57 (abstract).

Lebuhn, M., Effenberger, M., Garces, G., Gronauer, A. and Wilderer, P.A. (2004) Evaluating real-time PCR for the quantification of distinct pathogens and indicator organisms in environmental samples. *Water Science and Technology* 50, 263–270.

Lee, J.G., Cheong, K.H., Huh, N., Kim, S., Choi, J.W. and Ko, C. (2006) Microchip-based one step DNA extraction and real-time PCR in one chamber for rapid pathogen identification. *Lab Chip* 6, 886–895.

Lévesque, C.A., Harlton, C.E. and de Cock, A.W.A.M. (1998) Identification of some oomycetes by reverse dot blot hybridization. *Phytopathology* 88, 213–222.

Lievens, B., Brouwer, M., Vanachter, A.C.R.C., Lévesque, C.A., Cammue, B.P.A. and Thomma, B.P.H.J. (2005) Quantitative assessment of phytopathogenic fungi in various substrates using a DNA macroarray. *Environmental Microbiology* 7, 1698–1710.

Liu, S., Li, Y., Fu, X., Qiu, M., Jiang, B., Wu, H., Li, R., Mao, Y. and Xie, Y. (2005) Analysis of the factors affecting the accuracy of detection for single base alterations by oligonucleotide microarray. *Experimental and Molecular Medicine* 37, 71–77.

Lubeck, M. and Lubeck, P.S. (2005) Universally primed PCR (UP-PCR) and its applications in mycology. In: Deshmukh, S.K. and Rai, M. (eds) *Biodiversity of Fungi: Their Role in Human Life*. Science Publishers, New Hampshire, pp. 409–438.

Marshall, M.N., Cocolin, L., Mills, D.A. and VanderGheynst, J.S. (2003) Evaluation of PCR primers for denaturing gradient gel electrophoresis analysis of fungal communities in compost. *Journal of Applied Microbiology* 95, 934–948.

Martin, F.N., Coffey, M., Berger, P., Hamelin, R., Tooley, P., Garbelotto, M., Hughes, K. and Kubisiak, T. (2006) Validation of molecular markers for *Phytophthora ramorum* detection and identification; evaluation of specificity using a standardized library of isolates. *Phytopathology* 96, S74 (abstract).

Martin, F.N. and Tooley, P.W. (2003) Phylogenetic relationships among *Phytophthora* species inferred from sequence analysis of mitochondrially encoded cytochrome oxidase I and II genes. *Mycologia* 95, 269–284.

Martin, F.N., Tooley, P.W. and Blomquist, C. (2004) Molecular detection of *Phytophthora ramorum*, the causal agent of sudden oak death in California, and two additional species commonly recovered from diseased plant material. *Phytopathology* 94, 621–631.

Martin, R.R., James, D. and Lévesque, C.A. (2000) Impacts of molecular diagnostic technologies on plant disease management. *Annual Review of Phytopathology* 38, 207–239.

Mazzola, M. (2004) Assessment and management of soil microbial community structure for disease suppression. *Annual Review of Phytopathology*, 35–59.

Mullis, K.B. and Faloona, F.A. (1987) Specific synthesis of DNA *in vitro* via a polymerase-catalyzed chain reaction. *Methods in Enzymology* 155, 335–350.

National Academy of Sciences, USA (2002) *Countering Agricultural Bioterrorism*. Washington, DC, pp. 174 (classified appendix E).

Nolte, F.S. and Caliendo, A.M. (2003) Molecular detection and identification of microorganisms. In: Murray, P.R., Baron, E.J., Jorgensen, J.H., Pfaller, M.A. and Yolken, R.H. (eds) *Manual of Clinical Microbiology*. ASM Press, Washington, DC, pp. 234–256.

Okubara, P.A., Schroeder, K.L. and Paulitz, T.C. (2005) Real-time polymerase chain reaction: applications to studies on soilborne pathogens. *Canadian Journal of Plant Pathology* 27, 300–313.

Paplomatas, E.J. (2004) Molecular diagnostics for soilborne fungal pathogens. *Phytopathologia Mediterranea* 43, 213–220.

Pratt, M.A. (1976) A wet-sieving and flotation technique for the detection of resting sporangia of *Synchytrium endobioticum* in soil. *Annals of Applied Biology* 82, 21–29.

Ranjard, L., Lejon, D.P.H., Mougel, C., Schehrer, L., Merdinoglu, D. and Chaussod, R. (2003) Sampling strategy in molecular microbial ecology: influence of soil sample size on DNA fingerprinting analysis of fungal and bacterial communities. *Environmental Microbiology* 5, 1111–1120.

Reeleder, R.D., Capell, B.B., Tomlinson, L.D. and Hickey, W.J. (2003) The extraction of fungal DNA from multiple large soil samples. *Canadian Journal of Plant Pathology* 25, 182–191.

Salisbury, M.W. (2006). PCR: the conquests continue. *Genome Technology*, vol. 62 (June) 20–22.

Sanguin, H., Herrera, A., Oger-Desfeux, C., Dechesne, A., Simonet, P., Navarro, E., Vogel, T.M., Moenne-Loccoz, Y., Nesme, X. and Grundmann, G.L. (2006) Development and validation of a prototype 16S rRNA-based taxonomic microarray for Alphaproteobacteria. *Environmental Microbiology* 8, 289–307.

Schaad, N.W. and Frederick, R.D. (2002) Real-time PCR and its application for rapid plant disease diagnostics. *Canadian Journal of Plant Pathology* 24, 250–258.

Schaad, N.W., Frederick, R.D., Shaw, J., Schneider, W.L., Hickson, R., Petrillo, M.D. and Luster, D.G. (2003) Advances in molecular-based diagnostics in meeting crop biosecurity and phytosanitary issues. *Annual Review of Phytopathology* 41, 305–324.

Schena, L., Nigro, F., Ippolito, A. and Gallitelli, D. (2004) Real-time quantitative PCR: a new technology to detect and study phytopathogenic and antagonistic fungi. *European Journal of Plant Pathology* 110, 893–908.

Schroeder, K.L., Okubara, P.A., Tambong, J.T., Lévesque, C.A. and Paulitz, T.C. (2006) Identification and quantification of pathogenic *Pythium* spp. from soils in eastern Washington using real-time polymerase chain reaction. *Phytopathology* 96, 637–647.

Sessitsch, A., Hackl, E., Wenzl, P., Kilian, A., Kostic, T., Stralis-Pavese, N., Sandjong, B.T. and Bodrossy, L. (2006) Diagnostic microbial microarrays in soil ecology. *The New Phytologist* 171, 719–735.

Sholberg, P., O'Gorman, D. and Bedford, K. (2005) Development of a DNA macroarray for detection and monitoring of economically important apple diseases. *Plant Disease* 89, 1143–1150.

Smit, E., Leeflang, P., Glandorf, B., van Elsas, J.D. and Wernars, K. (1999) Analysis of fungal diversity in the wheat rhizosphere by sequencing of cloned PCR-amplified genes encoding 18S rRNA and temperature gradient gel electrophoresis. *Applied and Environmental Microbiology* 65, 2614–2621.

Stack, J., Cardwell, K., Hammerschmidt, R., Byrne, J., Loria, R., Snover-Clift, K., Baldwin, W., Wisler, G., Beck, H., Bostock, R., Thomas, C. and Luke, E. (2006) The National Plant Diagnostic Network. *Plant Disease* 90, 128–136.

Tambong, J.T., de Cock, A.W., Tinker, N.A. and Lévesque, C.A. (2006) Oligonucleotide array for identification and detection of *Pythium* species. *Applied and Environmental Microbiology* 72, 2691–2706.

Torsvik, V., Sørheim, R. and Goksøyr, J. (1996) Total bacterial diversity in soil and sediment communities – a review. *Journal of Industrial Microbiology and Biotechnology* 17, 170–178.

Tyler, B.M., Tripathy, S., Zhang, X., Dehal, P., Jiang, R.H., Aerts, A., Arredondo, F.D., Baxter, L., Bensasson, D., Beynon, J.L., Chapman, J., Damasceno, C.M., Dorrance, A.E., Dou, D., Dickerman, A.W., Dubchak, I.L., Garbelotto, M., Gijzen, M., Gordon, S.G., Govers, F., Grunwald, N.J., Huang, W., Ivors, K.L., Jones, R.W., Kamoun, S., Krampis, K., Lamour, K.H., Lee, M.K., McDonald, W.H., Medina, M., Meijer, H.J., Nordberg, E.K., Maclean, D.J., Ospina-Giraldo, M.D., Morris, P.F., Phuntumart, V., Putnam, N.H., Rash, S., Rose, J.K., Sakihama, Y., Salamov, A.A., Savidor, A., Scheuring, C.F., Smith, B.M., Sobral, B.W., Terry, A., Torto-Alalibo, T.A., Win, J., Xu, Z., Zhang, H., Grigoriev, I.V., Rokhsar, D.S. and Boore, J.L. (2006) *Phytophthora* genome sequences uncover evolutionary origins and mechanisms of pathogenesis. *Science* 313, 1261–1266.

van den Boogert, P.H.J.F., van Gent-Pelzer, M.P.E., Bonants, P.J.M., De-Boer, S.H., Wander, J.G.N., Lévesque, C.A., van Leeuwen, G.C.M. and Baayen, R.P. (2005) Development of PCR-based detection methods for the quarantine phytopathogen *Synchytrium endobioticum*, causal agent of potato wart disease. *European Journal of Plant Pathology* 113, 47–57.

van Leeuwen, G.C.M., Wander, J.G.N., Lamers, J., Meffert, J.P., van den Boogert, P.H.J.F. and Baayen, R.P. (2005) Direct examination of soil for sporangia of *Synchytrium endobioticum* using chloroform, calcium chloride and zinc sulphate as extraction reagents. *EPPO Bulletin* 35, 25–31.

Venter, J.C., Remington, K., Heidelberg, J.F., Halpern, A.L., Rusch, D., Eisen, J.A., Wu, D., Paulsen, I., Nelson, K.E., Nelson, W., Fouts, D.E., Levy, S., Knap, A.H., Lomas, M.W., Nealson, K., White, O., Peterson, J., Hoffman, J., Parsons, R., Baden-Tillson, H., Pfannkoch, C., Rogers, Y.H. and Smith, H.O. (2004) Environmental genome shotgun sequencing of the Sargasso Sea. *Science* 304, 66–74.

Wise, M.G., McArthur, J.V. and Shimkets, L.J. (1999) Methanotroph diversity in landfill soil: isolation of novel type I and type II methanotrophs whose presence was suggested by culture-independent 16S ribosomal DNA analysis. *Applied and Environmental Microbiology* 65, 4887–4897.

Wu, Y., de Kievit, P., Vahlkamp, L., Pijnenburg, D., Smit, M., Dankers, M., Melchers, D., Stax, M., Boender, P.J., Ingham, C., Bastiaensen, N., de Wijn, R., van Alewijk, D., van Damme, H., Raap, A.K., Chan, A.B. and van Beuningen, R. (2004) Quantitative assessment of a novel flow-through porous microarray for the rapid analysis of gene expression profiles. *Nucleic Acids Research* 32, e123.

Yang, J.L., Schachter, J., Moncada, J., Habte, D., Zerihun, M., House, J.I., Zhou, Z., Hong, K.C., Maxey, K., Gaynor, B.D. and Lietman, T.M. (2006) Comparison of an rRNA-based and DNA-based nucleic acid amplification test for the detection of *Chlamydia trachomatis* in trachoma. *British Journal of Ophthalmology* 91, 293–295.

8 Molecular Detection Strategies for Phytopathogenic Bacteria

S.H. De Boer, J.G. Elphinstone and G.S. Saddler

Abstract
Detection of bacterial plant pathogens on seed and other plant parts used to propagate agricultural and ornamental crops is an important component of disease prevention strategies. Disease prevention or avoidance strategies are particularly important in controlling bacterial diseases since there are few other control options. Such strategies require that the propagation of the agricultural or ornamental crop utilizes seed or vegetative plant parts that are free from pathogenic bacteria. While production methods are important for obtaining pathogen-free plant parts for propagation, indexing and monitoring of such material is required to ensure that specific disease-causing bacteria are not present in or on the planting material. Various molecular methods, based mostly on the polymerase chain reaction (PCR), have been utilized for sensitive and specific detection of bacterial pathogens on seeds and vegetative plant parts. A large number of test methodologies have been explored which target specific DNA regions. These have then been used to develop sensitive tests that can reveal the presence or absence of bacterial species, subspecies or pathovars of concern in consignments of seed and shipments of tubers, bulbs, cuttings, etc. Adequate validation of such molecular diagnostic protocols has been an additional challenge since these methods are being used for certification programmes and trade-related applications.

Introduction

Plant diseases are caused by a number of different etiological agents which include fungi, viruses, nematodes and bacteria. The nature and biology of these agents determines which potential control strategies can be used to prevent the spread and occurrence of plant diseases or mitigate the loss they cause in agricultural production systems.

Several hundred species and subspecies of bacteria have been described as etiological agents of plant disease. Some plant pathogenic bacteria are highly fastidious, such as *Xyllela fastidiosa*, which is exclusively found in association with host plants or its insect vectors. Others such as *Streptomyces* spp. are soil residents and can survive indefinitely in the absence of a plant host. Many bacterial plant diseases, however, are caused by members of genera such as *Clavibacter*, *Curtobacterium*, *Dickeya*, *Erwinia*, *Pectobacterium*, *Pseudomonas* and *Xanthomonas* that may have the capacity to be free-living in the environment, but are principally associated with host plants. For bacterial pathogens that are associated with plant parts used for propagation, whether consisting of sexually derived seed or asexually derived tubers, bulbs, cuttings, etc., an important control strategy is avoidance. Avoidance strategies in plant disease control are characterized by methods to prevent the pathogenic bacteria from becoming associated with the host in the first place. Identification of 'clean' planting material is essential for those plant diseases whose principal inoculum source is infected or contaminated seed or vegetative plant parts.

Seed and vegetative plant propagation units that are exported and imported around the world must be free from bacteria that are of quarantine significance or causal agents of regulated plant diseases. The concept of avoidance as a plant disease control strategy is not only applicable to individual crops, but also to the maintenance of pest-free areas and places of production. The 20th century has seen a shift to a global marketplace, with agricultural commodities exported to locations far away from the place of production. Globalization of trade engenders globalization of plant pests. Hence, determining whether phytobacterial pathogens are present or absent within a given production area has become an important risk mitigation activity required prior to export.

It is often difficult to determine whether pathogenic bacteria are associated with seed lots or various vegetative plant propagation units because usually no symptoms or signs of such infections or associations are visible. Population densities of the bacteria may be variable and the incidence of infected seeds or plant parts is often sporadic. While in some cases bacterial pathogens may be present at high levels in latently infected plants, e.g. bacterial ring rot of potato, in others, they are present at very low levels. Systemically infected cultivars that are considered to be resistant, but not immune, can harbour high densities of pathogenic bacteria without expression of symptoms. Moreover, an etiologic agent of disease can be difficult to distinguish from innocuous bacteria in plant tissues that are typically contaminated by high numbers of saprophytic microorganisms.

Detection methodologies for bacteria have benefited greatly from modern molecular technologies. Molecular methods involving specific amplification of target DNA fragments have been particularly useful for confirming the presence of specific bacteria in plant tissue extracts. Detection methods need to be sensitive, specific and robust. Consequently, various strategies have been utilized to attain appropriate test specificity by targeting different regions of the bacterial genome for amplification. Different amplification strategies and reporter molecules have been explored to achieve adequate test sensitivity. Nevertheless, a number of challenges remain to adequately validate these methodologies and to prevent the occurrence of false-positive and false-negative results.

At present, most molecular tests are applied in combination or alongside other methodologies for disease diagnosis or for the purpose of pathogen detection. A significant question to be addressed is whether adequate criteria can be established so that molecular tests can be used as the sole criterion to measure the bacterial pathogen status of a consignment of seed, a lot of seed potatoes, a shipment of ornamental cuttings, etc. In this chapter, the potential of biotechnology to contribute to this aspect of control of bacterial plant diseases is explored.

Dissemination of Bacteria on Seed and Vegetative Plant Propagation Units

The fact that bacteria cause diseases of plants was already firmly established by the end of the 19th century. Yet, it has taken fully another half a century, and in some cases even longer, to confirm and authenticate the bacterial etiology of some plant disorders. New bacterial plant diseases are still being described today. It has taken at least as long to understand the ecological and epidemiological parameters that characterize bacterial plant diseases and to fully comprehend the impact of these parameters on their control. Moreover, in recent years, the use of molecular technology in systematics is increasing the number of distinct taxa of etiologically significant bacteria. Closely related, but phylogenetically distinct, bacteria are often not distinguished by phenotypic markers (e.g. antigens and enzymes) used in conventional diagnostic methods. Molecular diagnostics will therefore be increasingly required to provide the degree of discrimination needed to detect and monitor bacterial plant diseases due to evolving or newly introduced pathogens.

Bacterial plant diseases are difficult to control. With few exceptions, chemical control options are not available. Agronomic practices in some cases can mitigate disease losses but the major means by which bacterial plant diseases can be controlled is by avoiding them altogether. Avoidance is more practical for some diseases than others, the main consideration being the source of inoculum. If the disease-causing bacteria are present in the environment, i.e. in soil, in insect vectors or on foliage of weed plants, transmission to and subsequent infection into host crop plants is difficult to avoid. Nevertheless, there are many bacterial plant diseases for which the main inoculum source is the planting material itself. Hence, the use of pathogen-free seed or vegetative propagation units is an important disease control strategy. Appropriate, sensitive and robust detection technologies are vital to the implementation of such strategies. There are many examples where detection methodologies represent the only disease control option available. Some examples of bacterial plant diseases that are borne on seed and vegetative plant propagation units (both herbaceous and woody) are described.

Seedborne bacterial pathogens

The black rot disease of crucifers, caused by *Xanthomonas campestris* pv. *campestris*, is typical of seedborne bacterial diseases. Testing seed lots for the presence of the pathogen is essential because the initial inoculum which is carried by infected seeds is a critical factor determining the severity of disease in the subsequent crop (Park *et al.*, 2004). The industry standard for determining whether seed lots are contaminated by *X. campestris* pv. *campestris* involves testing 30,000 seeds from each lot, as three subsamples of 10,000 (Franken *et al.*, 1991). However, successful seed testing depends on the availability of an efficient and reliable method to detect the pathogen.

Similarly, the use of disease-free seed is essential for cost-effective field production of tomato. *Clavibacter michiganensis* subsp. *michiganensis*, the causal agent of bacterial canker of tomato, is a quarantine organism in various countries, including those in the European Union. It is considered to be the most important bacterial disease of tomato causing substantial economic losses worldwide. Infected seed is the primary inoculum source and the major cause of outbreaks of bacterial canker. Even a low (0.01%) transmission rate from seed to seedling can initiate a serious epidemic in field tomatoes (Tsiantos, 1987). Another example of a seedborne pathogen on tomato is the bacterial speck pathogen, *Pseudomonas syringae* pv. *tomato*, which causes significant economic losses to the industry under certain disease-conducive conditions. Indexing of seed for the canker and bacterial speck pathogens is an important component of disease control strategies in tomato production.

The use of certified seed is necessary for effective control of common blight (*X. campestris* pv. *phaseoli*), halo blight (*P. syringae* pv. *phaseolicola*) and bacterial wilt (*Curtobacterium flaccumfaciens* pv. *flaccumfaciens*) on bean. These diseases are widespread and destructive on common bean, a major food crop in Africa and other bean-growing regions. The presence of extremely low levels of primary inoculum can initiate severe epidemics under favourable conditions. Long latency periods prior to the development of disease symptoms and the endophytic nature of pathogens such as the wilt bacterium and its occurrence in low numbers have made pathogen detection particularly difficult in seed certification programmes for bean as well as for quarantine inspection of imports (Guimaraés *et al.*, 2001). Consequently, certification of seed as free of these pathogens requires the use of highly sensitive and specific methods. Similarly, *P. syringae* pv. *pisi*, the etiological agent of pea blight, is primarily seedborne. It has near worldwide distribution and was first recorded in the UK in 1985 in a crop of protein peas grown from infected, imported seed (Stead and Pemberton, 1987). There is strong evidence linking the effectiveness of control with the degree of seed testing undertaken.

Stewart's bacterial wilt and leaf blight of sweetcorn and maize is caused by *Pantoea stewartii* subsp. *stewartii* (syn. *Erwinia stewartii*) and is responsible for serious crop losses. While the pathogen is spread by the corn flea beetle (*Chaetocnema pulicularia*), seed transmission is also a significant component in the epidemiology of the disease (Michener *et al.*, 2002). The potential risk of seed transmission is considered to be so important that more than 50 countries ban the importation of maize seed unless it is certified to be free of *P. stewartii* subsp. *stewartii*. Likewise, rice seeds contaminated with *Acidovorax avenae* subsp. *avenae*, the cause of bacterial stripe and various diseases on other monocotyledonous crop plants, are important sources of primary inoculum and a means of disseminating the pathogen to new areas.

Bacterial dissemination on herbaceous plant propagation units

The impact of molecular test methods has been very significant in the seed potato industry. Potato tubers used for planting (i.e. seed potatoes) are the principle inoculum source for brown rot (*Ralstonia solanacearum*), bacterial ring rot (*C. michiganensis* subsp. *sepedonicus*) and to a lesser extent blackleg and soft rot (*Pectobacterium atrosepticum* [synonym *Erwinia carotovora* subsp. *atroseptica*], *P. carotovorum* subsp. *carotovorum* [syn. *E. carotovora* subsp. *carotovora*] and *Dickeya* spp. [syn. *E. chrysanthemi*]). Asymptomatic, infected potato tubers are a major factor in the dissemination of *R. solanacearum* both locally and internationally. Because pathogen-free seed tubers are required to control pathogen dissemination, assays to detect *R. solanacearum* in tubers are most important. Race 3, biovar 2 of *R. solanacearum* is of particular concern to the potato industry. Although this bacterium was, at one time, considered to be restricted in its distribution to tropical and subtropical regions, its appearance in Sweden in 1976 and intermittent outbreaks across Western Europe since the 1990s has changed that perspective. *R. solanacearum* also infects geraniums and it was detected in cuttings imported from Central America and Africa (Williamson *et al.*, 2002). The testing for the pathogen in geranium cuttings has become an important practice for mitigating further distribution of the bacterium.

Even more than brown rot, bacterial ring rot spread is associated with latent tuber infection. Detection of latent infections of *C. michiganensis* subsp. *sepedonicus* in potato seed lots is of prime importance in controlling the disease. Indexing tubers for the ring rot pathogen is done routinely in some domestic seed potato certification programmes as well as to satisfy international trade requirements. In contrast, some of the soft rot pectobacteria are prevalent in the environment and their presence on seed tubers is of incidental significance (Pérombelon, 2002). Nevertheless, other strains such as *P. atrosepticum*, which cause blackleg, appear to be more restricted to potato and the identification of seed lots free of the bacterium is a useful disease control practice (De Boer, 2002).

Xanthomonas axonopodis pv. *manihotis* causes bacterial blight of cassava. The pathogen is both seedborne and carried on symptomless planting material. In the former case, the bacterium is borne either on the seed coat or in the embryo, and as such, it is of international concern during exchange of germplasm as true seed. At the regional level, cassava is propagated vegetatively from pieces of stem and unless the use of contaminated planting stock is avoided, the pathogen is readily spread into new fields (Lozano, 1986). Similarly, leaf scald disease of sugar cane (*Xanthomonas albilineans*) is controlled by planting only healthy stalks. Latent infection with the pathogen is very common and such stalks are undoubtedly the most important means of pathogen dissemination (Destefano et al., 2003).

Since many ornamental and flower crops are also multiplied vegetatively, the use of pathogen-free mother plants is essential for controlling bacterial diseases. For example, production of anthurium, the second largest tropical flower crop in the world, valued at US$20 million in 2002, is threatened by bacterial blight, caused by *X. axonopodis* pv. *dieffenbachiae* (Robene-Soustrade et al., 2006). Latent infections, which are thought to be involved in the spread of the pathogen within and between countries, may be present for more than 1 year in anthurium propagative material. Disease control consists principally of prevention, sanitation and the use of axenically propagated plants. Also, *Xanthomonas hyacinthi*, causal agent of the yellows disease in hyacinth, is easily spread by wounding of bulbs during mechanical sorting in the presence of diseased bulbs. The development of a rapid and specific test to ascertain whether symptoms are caused by this yellow-pigmented bacterium is of utmost importance to hyacinth growers (van Doorn et al., 2001).

Bacterial dissemination on woody plant propagation units

Most fruit tree production depends on grafting appropriate scion cultivars onto specific rootstocks. Both the scion and the rootstock can be a source of bacterial infections. Apple propagation on dwarf rootstocks in Japan is limited by the occurrence of the crown gall disease, caused by *Agrobacterium tumefaciens* and *Agrobacterium rhizogenes*. Because the spread of the pathogen is enhanced by the use of contaminated scions and rootstocks, the use of pathogen-free plant parts for grafting is a high priority to prevent spread of the disease (Suzaki et al., 2004). Another disease on apple, fireblight (*Erwinia amylovora*), although transmitted mostly by wind, rain and insects, requires that apple fruit and breeding stock be tested for the presence of *E. amylovora* prior to export to areas where the disease is absent (Norelli et al., 2003).

Crown gall, caused by *A. tumefaciens*, is an important pathogen of grapevine in addition to being a pathogen of apple and other woody fruit-producing crops. Economic losses due to this pathogen are not only from intrinsic plant damage, but also from restrictions in trade of potentially infected plants. Again, accurate detection methods are required to identify clean planting material (Cubero et al., 1999). Similarly, bacterial necrosis and canker (*Xylophilus ampelinus*) are serious diseases of table grape cultivars in the Mediterranean and in some regions within South Africa. Up to

80% loss in production occurs in infected plants. The bacterium survives in the vascular tissues and is transmitted by the use of infected propagating material. Because the pathogen is present in symptomless plants, it is crucial that growers screen propagating material prior to use (Panagopoulos, 1987).

X. axonopodis pv. *citri* and *Xylella fastidiosa* cause citrus canker and citrus variegated chlorosis, respectively. Because of the economic and quarantine importance of citrus canker, the accurate detection of the *X. axonopodis* pv. *citri* in seedlings, budwood and asymptomatic plants is critical and is the best approach to prevent further dissemination of the pathogen (Coletta-Filho *et al.*, 2006). While *X. fastidiosa* is primarily transmitted by leafhoppers, it can also be transmitted to seedlings from seed (Li *et al.*, 2003) and can be spread through natural root grafts (He *et al.*, 2000).

Other examples of bacterial diseases on orchard crops for which primary control strategies involve pathogen detection include olive knot, caused by *Pseudomonas savastanoi* and bacterial canker of hazelnut caused by *Pseudomonas avellanae*. Spread of both these pathogens is only limited by early detection of low bacterial levels in propagation material (Scortichini and Lazzari, 1996; Penyalver *et al.*, 2000).

Methods to Detect Bacteria

Detection of pathogenic bacteria in seed and other plant tissues (particularly in latent infections) is challenging because the target bacteria are often irregularly distributed and present as a small component of a much larger bacterial population. Moreover, it is often difficult to distinguish and identify pathogenic bacteria from all the soil-associated and other saprophytic bacteria normally present on plant surfaces. In addition to epiphytic and casual surface contaminants, non-detrimental or beneficial endophytic bacteria may also be present.

Traditional methods

Traditional techniques to detect the presence of pathogenic bacteria involve field inspection for symptoms and signs of disease as well as laboratory tests. Laboratory procedures for detection of the bacteria may involve grow-out assays, serological tests such as enzyme-linked immunosorbent assays (ELISA) and immunofluorescence microscopy. In addition, isolation of the bacteria on selective or semi-selective media is also done. Following isolation, strains need to be characterized by physiological, biochemical and pathogenicity tests. Use of traditional methods are reliable and efficient for some of the bacterial plant pathogens, but for many others they lack adequate sensitivity and specificity. Another major disadvantage is the long times required for grow-out assays, bacterial isolation and pathogenicity tests.

Molecular-based methods

Extraction and purification of bacterial DNA from infected plant material
Since the introduction of the polymerase chain reaction (PCR) in the 1980s, a continuing challenge has been to isolate high-quality amplifiable nucleic acid from plant extracts (Mumford et al., 2006). The basic steps for isolation of nucleic acids involve tissue disruption followed by removal of proteins and washing of the nucleic acids to remove common plant substances which bind to DNA and/or are inhibitory to DNA polymerase (including proteins, polyphenols and acidic polysaccharides). Purification procedures generally use detergents and solvents. Detergents release nucleic acids from protein complexes into the aqueous phase of a phenol or chloroform extraction system. The most commonly used detergent is cetyltrimethyl ammonium bromide (CTAB) (Doyle and Doyle, 1987). In the presence of a high level of salt, CTAB dissociates and precipitates proteins and polysaccharides. Proteins released from the initial dissociation step are then precipitated using deproteinizing agents, generally phenol and chloroform. Nucleic acids can then be precipitated from the aqueous layer using salt and alcohol. Such methods often involve multiple steps and hazardous reagents. For high-throughput routine diagnostic use, an increasing range of non-hazardous automatable alternatives are available, including:

1. Immunocapture-PCR (IC-PCR) methods (e.g. van der Wolf et al., 1996; Khoodoo et al., 2005; Mulholland, 2005). These methods are simple, but are not generic since they require specific antibodies and associated protocols for each target pathogen.
2. Chelex extraction methods involve tissue disruption by heating (98–100°C) in the presence of chelex resin which binds to heavy metals and other cellular components which may inhibit nucleic acid analyses. DNA extracted in this way is single-stranded and is suitable for PCR (Singer-Sam et al., 1989; Boonham et al., 2002) but not for restriction fragment length polymorphism (RFLP) analyses.
3. Silica-based extraction methods rely on the lysing and nuclease-inactivating properties of guanidinium thiocyanate together with the nucleic acid-binding properties of silica particles. Proteins, salts or residual phenol and chloroform do not bind to silica and therefore can be eliminated during washes. Silica is available in different formats, including spin-columns (e.g. Boom et al., 1990) and silica-coated magnetic beads (e.g. Ward et al., 2004; Tomlinson et al., 2005) and a wide range of DNA purification kits are commercially available, some being in automatable and high-throughput formats.
4. ChargeSwitch technology (DNA Research Innovation LTD (DRI), Sittingbourne, UK) is a new non-silica-based technology which avoids the used of ethanol, concentrated salts and organic solvents. Nucleic acids bind to the surface of coated particles when they are positively charged at pH < 6.5, but are eluted into buffers when the charge is neutralized at pH > 8.5. Proteins and other contaminants are not bound to the bead surface and are washed away in the extraction process.

5. Enzymatic protein digestion using proteinase-K, a serine protease with very broad cleavage specificity, for general digestion of protein in biological samples including plant material and bacterial cells, to leave undigested DNA (Mahuku, 2004). A new development of this technology uses a thermophilic protease, which while inactive at 37°C is fully active at 75°C, and is irreversibly denatured at 95°C (Zygem Corp. Ltd., Hamilton, New Zealand). Thus, multiple enzyme reactions can be controlled with temperature by the addition of a mesophilic enzyme (such as lysozyme) and the two activities can be used sequentially with a simple temperature shift from 37°C to 75°C. A final high-temperature cycle at 95°C can be used to remove all residual enzyme activity prior to DNA analysis. This enables single-tube, high-throughput systems without the risk of cross-contamination or misidentification of samples.

Assay formats for amplifying pathogen-specific nucleic acids

Molecular assays which involve amplification of DNA fragments to detect specific pathogen DNA sequences have a number of advantages over traditional methods. Various formats of such assays are available and although each has particular advantages and disadvantages, all have the potential for high specificity, sensitivity and speed.

Conventional polymerase chain reaction
PCR is a technique in which a specific nucleotide sequence is exponentially amplified *in vitro*. The amplification reaction is driven by temperature cycling to promote repeated template denaturation, primer annealing and extension of the annealed primers by a DNA polymerase until there is sufficient product for analysis. Specificity is attained by the design of primers that scan the genomic template for complementary target during the annealing step. Fidelity of primer hybridization depends on the stringency of the reaction, which in part, is a function of temperature and magnesium ion concentration. When PCR is used to detect low levels of bacteria in plant tissues, actual template is very dilute in comparison with the total amount of DNA in the sample. Dilute template presents a challenge for efficiency of the reaction, since collision frequency of primer and template is markedly rare; hence, the first few cycles of the reaction are most sensitive to optimization of conditions, whereas in later cycles, the primer to template ratio becomes considerably more favourable.

DNA fragments (amplicons) generated during PCR reactions are detected by gel electrophoresis in the conventional procedure. Bands of DNA are visualized by staining the gel with ethidium bromide and visualizing under ultraviolet (UV) light. Bands of DNA can be provisionally identified by their degree of migration compared to amplicons generated in control PCR reaction, and to commercially available standards of DNA mixtures of varying sizes (DNA ladders). However, amplicon size alone can be misleading and additional verification of amplicon identity is required either by hybridization with a labelled probe, restriction analysis or sequencing of the amplicon.

Immunocapture-PCR
This method combines two diagnostic tools in series, exploiting high-affinity binding antibodies to selectively capture the target organism, generally from within a complex matrix, prior to detection by conventional PCR amplification (Mulholland, 2005). The method has the advantage that the initial purification step does not have to be specific, and frequently, genus-specific antibodies can be employed. In addition, it also has the advantage of removing the target organisms from substances which may inhibit or compete in the PCR amplification step. IC-PCR has been successfully used to detect *X. axonopodis* pv. *dieffenbachiae*, the cause of bacterial blight of anthurium, in vegetative propagation material (Khoodoo et al., 2005). Similarly, *P. atrosepticum* was detected by IC-PCR in potato peel extracts containing both *P. atrosepticum* and its close relative *P. carotovorum* at low levels (van der Wolf et al., 1996). It was possible to enumerate *P. atrosepticum* to as low as 100 cells/ml, in contrast to conventional PCR alone, in which it was only possible to enumerate cells at 10^5 cells/ml, and only after the extract had been diluted 100 times to reduce the effect of inhibitors (van der Wolf et al., 1996).

PCR–ELISA
An alternate method for detection of PCR products relies on the use of ELISA to detect labels incorporated into the amplicons during PCR amplification. Such protocols may be particularly optimal for automated PCR detection because it avoids the need for gel electrophoretic analysis. A PCR–ELISA was developed for detection of *E. amylovora*, causal agent of fireblight of apple and other roscaceous plants (Merighi et al., 2000). In this protocol, PCR amplification was carried out in the presence of a digoxigenin-labelled nucleotide. Specific hybridization capture of amplicons on streptavidin-coated microtitre plates was achieved by designing a 5' biotinylated probe that could hybridize to an internal region of the PCR amplicon. Detection of the bound amplicons involved binding of anti-digoxigenin-peroxidase-conjugated antibodies and addition of a colorimetric or chemiluminescent substrate that could be read by an automated detector (Merighi et al., 2000).

BIO-PCR
The term BIO-PCR was coined to describe a procedure that involves an enrichment technique prior to extraction and amplification of DNA of the target bacterium (Schaad et al., 1995). The intent of the method is to achieve rapid and specific growth of the target bacterium and suppress the growth of non-target bacteria. The greater proportional population of the target bacterium is then easier to detect with less interference from the non-target microbial flora. Detection of *P. savastanoi* pv. *savastanoi* by PCR was enhanced by pre-enrichment in either non-selective King's B medium or on a semi-selective medium designed for *P. savastanoi* pv. *savatanoi* (Penyalver et al., 2000). Normally, the success of the method is wholly dependent on the availability of a reliable selective medium. Song et al.

(2004) describe the development of such a medium for *A. avenae* subsp. *avenae* on rice seeds. Presumptive utilization of carbon and nitrogen compounds for growth of the bacterium was determined by using Biolog GN Microplates. The most promising carbon and nitrogen compounds were then selected and tested by comparing the specificity and recovery efficiency by dilution-plating techniques. Subsequently, several inhibitors were evaluated to reduce the growth of other rice-associated bacteria without reducing the growth of *A. avenae* subsp. *avenae*. BIO-PCR not only increases the sensitivity of detection, but also avoids the possibility of detecting dead bacterial cells. Moreover, the enrichment step serves to minimize problems with PCR inhibitors because inhibitory molecules are sufficiently diluted out in the enrichment medium. However, quantification of bacterial populations cannot be readily done with BIO-PCR, and if an adequate selective medium is lacking, organisms which compete with or are antagonistic to the target bacterium during enrichment can result in decreased rather than increased sensitivity of detection.

Nested PCR
Sensitivity of molecular detection is increased by nested PCR, which involves the introduction of a second round of amplification using the amplicon of the first PCR reaction as template for the second. However, the manipulation of the previously amplified products vastly increases the risk of cross-contamination in routine analysis. To avoid this problem, nested PCR can be carried out in a single closed tube by designing primers that support amplification at different temperatures. Bertolini *et al.* (2003) developed a nested PCR test for *P. savastanoi* pv. *savastanoi* using external primers that amplified at 62°C and internal primers that amplified at 50°C, but not at 62°C. In one nested PCR test developed to detect *X. axonopodis* pv. *dieffenbachiae* in anthurium, the detection threshold was lowered to approximately 10^3 cfu/ml, which corresponds to one target DNA molecule detected per reaction (Robene-Soustrade *et al.*, 2006). This sensitivity was suitable for detecting the target bacterium in symptomless plants. A nested PCR was also used to detect *X. axonopodis* in cassava seed (Verdier *et al.*, 2001). No DNA amplification was detected using conventional PCR, while with the nested procedure as few as 1.2×10^2 and 8.6×10^2 cfu were detected per reaction. Nested PCR was also used to detect *X. ampelinus* in grapevine cuttings (Botha *et al.*, 2001) where it overcame the problem of excess non-target DNA template compared to target template. The first PCR run generated sufficient copy numbers of the target gene fragment to serve as a template for the internal primer pair. Consequently, low numbers of target cells could be detected when an excess number of saprophytic bacteria were present.

Multiplex PCR
The simultaneous amplification of two or more amplicons in the same reaction mix is called multiplex PCR. This method is used to detect more than one bacterial pathogen at the same time, or to amplify a control DNA

fragment along with target DNA to ascertain the absence of PCR inhibitors when target amplicon is not generated. Multiplex PCR can take on different formats and can be applied to conventional, competitive and real-time PCR assays. Multiplex PCR tests can utilize multiple primer sets, a pair for each amplicon to be generated, or the same primer pair can be used to amplify two different amplicons. Multiplex PCR is particularly useful to demonstrate the competence of amplification when plant extracts are present alongside the target DNA template. Amplification of an internal control indicates both successful DNA isolation and absence of PCR inhibitors within the DNA extract, thus avoiding the possibility of false-negative results.

Pastrik (2000) and Pastrik et al. (2002) developed multiplex PCR tests to detect C. michiganensis subsp. sepedonicus and R. solanacearum in potato tissue, respectively. In addition to amplifying a targeted DNA fragment from the bacterial pathogens, a second set of primers was included to amplify a region of the 18S rRNA gene which is conserved in various eukaryotes, including angiosperms such as potato, tomato and aubergine. A very similar strategy was used to develop a multiplex PCR to detect X. campestris pathovars in Brassica seed (Berg et al., 2005). The amplification of an rRNA gene fragment from Brassica species also served as an internal control to judge the success of DNA extraction and freedom from PCR-inhibiting components. The relative concentration of the primer sets is critical for the sensitivity of the PCR assay. When the concentration of the plant-targeted primer set is too high, amplification of the bacterium-specific DNA fragment decreases, reducing the sensitivity of the test (Pastrik, 2000). In an optimized multiplex PCR, it is possible to detect specific bacteria at a sensitivity equivalent to uniplex PCR. However, optimization can be difficult since the ratio of plant to bacterial DNA is often variable and unknown. To avoid some of the competition between amplification of target and control DNA fragments, it is also possible to construct an artifical control by ligating primer sequences to a heterologous DNA fragment (Smith et al., 2007). Amplification of both target and control DNA is then directed by the same primer set. Subsequent differentiation of amplicons is readily done, depending on the PCR format, by hybridization, sequencing or melt curve analysis.

Competitive PCR
This technique, a form of multiplex PCR, allows the simultaneous detection and quantification of target organisms. Like multiplex PCR, competitive PCR has value in that it can be used to differentiate a true negative reaction from other factors that may result in an unsuccessful amplification, namely the presence of PCR inhibitors. Competitive PCR relies on the construction of a competitor or internal DNA standard which is amplified using the same primers as those designed for the target organism. The competitor DNA is generally cloned into a suitable vector, e.g. *Escherichia coli* and known amounts of whole cells or extracted DNA are introduced into the assay. This method was used to detect and quantify *C. michiganensis*

subsp. *sepedonicus* DNA in infected plant tissues (Hu *et al.*, 1995). The competitive DNA template was generated by the amplification of *Arabidopsis* genomic DNA to yield a 450 bp amplicon distinct from the 250 bp amplicon characteristic of *C. michiganensis* subsp. *sepedonicus*. The ratio of the PCR products amplified in the presence of a constant amount of internal standard DNA template increased linearly and competitively with increased amounts of *C. michiganensis* subsp. *sepedonicus* DNA. Cell numbers estimated by immunofluorescence were consistent with PCR product ratios obtained from cell cultures and inoculated potato plantlets (Hu *et al.*, 1995). Compared to immunofluorescence, competitive PCR was about tenfold more sensitive and could detect as few as 100 cells. A similar approach was used to quantify the potato pathogen, *P. atrosepticum* (Hyman *et al.*, 2000). Predetermined numbers of bacterial cells containing competitor template were added to potato peel extract, either pre-inoculated with *P. atrosepticum* or from naturally contaminated tubers, and pathogen numbers were estimated by comparing the ratio of products generated from *P. atrosepticum* target DNA and competitor template DNA following PCR. Although this method has advantages over conventional PCR in that it can quantify the target and be used to distinguish true negative results, it is difficult to perform routinely and has possibly been superseded by real-time PCR in recent years.

Real-time PCR
This method is so named because it allows the measurement of amplicon generation while the reaction proceeds. It provides further advantages as a one-step reaction over all conventional-based PCR methods, including nested PCR, competitive PCR, etc., which rely on the use of agarose gels to detect the amplicon. Measurement of amplicon accumulation is achieved in the majority of cases using one of two detection chemistries – the 5' nuclease or TaqMan assay, or the use of a DNA-binding dye such as SYBR Green. Although other reporter systems are available (see Wong and Medrano, 2005), all methods are dependent on dedicated instrumentation that is capable of cycling the reaction temperatures and capturing the fluorescent signals. SYBR Green is a double-stranded DNA (dsDNA)-binding dye which is incorporated into amplicons during the PCR. When the dye binds to the minor groove of dsDNA, the intensity of fluorescence emission increases. Hence, measurement of fluorescence during PCR amplification is a function of the amount of dsDNA amplicons in the reaction mix. The advantage of this chemistry for real-time PCR is that no additional probes are required. However, post-amplification analysis is required to identify the amplicon since the method detects all amplified products, including non-specific DNA fragments. A melting-curve analysis is usually the most convenient way to confirm specific amplification. The melting profile should be a single peak.

The 5' nuclease or TaqMan assay requires two primers as in conventional PCR that amplify a DNA fragment in the 50–100 bp size range. In addition, a probe designed to hybridize specifically to the target PCR product is included in the reaction. The probe is labelled with a fluorescent reporter dye and a quencher dye, so that while the reporter dye is in close proximity to the quencher dye, no fluorescence signal escapes from the molecule. The probe is designed to hybridize to the complementary region of the amplicon during each PCR cycle. Subsequently, during the amplification step, the assay exploits the 5' nuclease activity of *Taq* DNA polymerase and the probe is digested by the polymerase, separating the dyes and permitting the escape of the reporter fluorescence signal. Repeated PCR cycles result in exponential amplification of the PCR product and a corresponding increase in fluorescence intensity. Measurement of fluorescence throughout the PCR cycling regime provides a means to analyse the reaction kinetics in real time as well as to quantify amplicon accumulation. The amplicons generated by TaqMan are considerably shorter than the amplicons in conventional PCR, which should be at least 150 bp in length to be readily detected by electrophoretic separation techniques. This increased efficiency is coupled with greater specificity engendered by the requirement of probe hybridization to the amplicons during each PCR cycle. Moreover, the measurement of fluorescence throughout the reaction eliminates the need for post-PCR processing to reveal the presence of amplicon product. Fluorometric readings can be used to simply determine the presence or absence of amplified DNA, corroborating the presence or absence of the target bacterium. Alternatively, the fluorometric readings captured in real time can facilitate quantification of sample DNA. The number of cycles required to reach a predetermined threshold (threshold cycle, C_T) is correlated with the concentration of template DNA originally present in the sample. Comparison of threshold cycles to those of standard preparations of varying DNA concentrations provides a quantitative estimate that relates back to bacterial density in a sample.

Another major advantage of real-time PCR for bacterial detection is the reduced potential for cross-contamination when compared to conventional PCR. Because the real-time method does not require post-amplification analysis, the PCR tubes that contain amplified DNA never need to be opened in the laboratory. The high concentration of specific DNA fragments that are generated by PCR amplification is easily aerosolized by any post-PCR manipulations. Even with the best laboratory practices, cross-contamination of new samples with aerosolized DNA from previous runs is almost inevitable in conventional PCR. The quantitative signals in real-time PCR are measured directly in the test tubes. The assay can be run with high sample numbers and is thus suitable for large-scale monitoring and screening procedures. The advantages of real-time PCR are a quantitative detection of the pathogen, the potential for high sample throughput, and the absence of the need for time-consuming post-PCR analysis.

Real-time PCR detection has been combined with the BIO-PCR strategy to detect *R. solanacearum* race 3, biovar 2, in potato at a high level of sensitivity (Ozakman and Schaad, 2003). In addition, it can be multiplexed to allow the simultaneous detection of two targets as illustrated by an alternative approach for the detection of *R. solanacearum* (Weller *et al.*,

2000). In this assay, primers and probes were designed to allow the detection of specific members of the species (phylotype and sequevar) as the cause of potato brown rot. This assay can be used simultaneously to establish the presence of an infection and determine the identity of the intraspecific group involved.

NASBA

A non-PCR isothermal nucleic acid amplification procedure known as nucleic acid sequence-based amplification (NASBA) has been evaluated for detecting plant pathogenic bacteria. It is a particularly interesting procedure in that it uses RNA rather than DNA as template. The RNA template increases the probability that nucleic acid from only viable bacterial cells is amplified (since the half-life of RNA molecules is very short) and it may also increase sensitivity due to the greater copy number of RNA molecules compared to DNA. NASBA is based on exponential amplification of single-stranded RNA molecules by simultaneous activities of reverse transcriptase and polymerase enzymes (Kievits et al., 1991). The procedure begins with extension of a primer containing a T7 promoter sequence on an RNA template by avian myeloblastosis virus reverse transcriptase (AMV-RT), followed by degradation of the RNA strand of the newly formed RNA/DNA duplex by RNase H. Subsequently, a second primer anneals to the resulting single-stranded DNA and a complementary DNA strand is synthesized by the DNA-dependent DNA polymerase activity of the AMV-RT, yielding a dsDNA molecule including a T7 RNA polymerase promoter sequence. The T7 RNA polymerase then gives a 100- to 1000-fold increase in specific antisense RNA. The sensitivity of NASBA for detection of R. solanacearum in potato tuber extracts was estimated to be 10^4 cfu/ml, equivalent to 100 cfu/reaction (Bentsink et al., 2002). The detection level for purified RNA was estimated to be 10^4 rRNA molecules, which corresponded to less than a bacterial cell. Real-time NASBA protocols have also been devised for detection of C. michiganensis subsp. sepedonicus (van Beckhoven et al., 2002) and R. solanacearum (van der Wolf et al., 2004) in potato by combining RNA amplification with molecular beacons. Molecular beacons are single-stranded oligonucleotides which contain a stem and a loop structure. The loop is complementary to the sequence of the target, and the adjoining stem has a double-stranded structure. Complementary ends of the stem are labelled with a fluorophore and a quencher, respectively. When the loop hybridizes to the target, the stem opens up, causing the fluorophore and quencher to become physically distanced, with resulting fluorescence emission.

Primer or probe selection strategies

Whether molecular detection is by conventional, real-time or other nucleic acid amplification strategy, the success of specific detection is always a function of the uniqueness of the DNA or RNA fragment targeted. In addition, if the targeted nucleic acid sequence is not present in all strains of the bacterial species of interest, complete detection cannot be achieved. Furthermore, if the targeted nucleic acid sequence occurs in other bacteria besides the bacterial species of interest, then false positives will result. Various strategies have been used to identify regions within the bacterial genome that provide the optimal level of target specificity. Targeted areas include the intergenic spacer region of the ribosomal gene cluster, pathogenicity-related genes and non-specified taxon-specific genomic fragments.

Intergenic spacer region
The eubacterial intergenic transcribed spacer (ITS) region between the 16S and 23S rRNA genes, which typically contains one or two tRNA genes, is considered an ideal region for developing PCR primers that are specific to targeted bacterial taxa. This region of the genome contains extensive sequence variation at the genus and species levels, yet retains remarkable conservation of sequences within species, subspecies and pathovars of plant pathogenic bacteria. *C. michiganensis* was one of the first phytopathogenic bacterial species for which the ITS region was targeted for the development of subspecies-specific PCR primers (Li and De Boer, 1995). Sequence similarity of the ITS region, initially amplified by universal primers 1493f and 23r, for the five *C. michiganenesis* subspecies was determined to be in the order of 95%. Yet, sufficient single-base differences, insertions and deletions were present to enable design of primers specific for *C. michiganensis* subsp. *sepedonicus*. These primers have found extensive use in testing for the presence of the bacterial ring rot pathogen in latently infected potato tubers (Li and De Boer, 1995). The two base-pair difference at the 3' end of the forward primer and the one base-pair difference and one insertion near the 3' end of the reverse primer were sufficient to prevent amplification of the other *C. michiganensis* subspecies.

Similarly, ITS-based primers were developed for *P. stewartii* (Coplin et al., 2002) and *X. ampelinus* (Botha et al., 2001) by conventional PCR, and for *A. avenae* subsp. *avenae* (Song et al., 2004) and *Burkholderia glumae* by real-time PCR (Sayler et al., 2006). In these examples, ITS sequences of the target species or subspecies were compared *in silico* with analogous sequences available in GenBank. Subsequently, specific primer sequences were selected using any one of a number of software programmes that are available for primer selection and characterization.

Pathogenicity-related genes
Genes directly involved in plant pathogenicity are logical targets for the development of pathogen-specific DNA primers and probes. Those genes within gene clusters that encode for the hypersensitive reaction and pathogenicity (*hrp*) may contain, in part, the code that determines host specificity as part of the disease-causing process. Furthermore, analysis of the nucleotide sequences of the *hrp* gene clusters in phytopathogenic bacteria such as *Xanthomonas* spp. has revealed that the organization and size of the majority of the *hrp* genes are highly conserved within the genus. The HrpF

protein may play a role in determining the host specificity of pathogenic xanthomonads and the gene showed low homology to sequences of other xanthomonad pathovars. The 3' end of the *hrpF* gene was subsequently used to develop PCR primers to differentiate *X. campestris* from other species (Park *et al.*, 2004; Berg *et al.*, 2005). Similarly, specific primers to *P. syringae* pv. *tomato* were based on a similar *hrp* gene, *hrpZ* (Zaccardelli *et al.*, 2005). *hrpZ* is a chromosomal gene located in the *hrp/hrc* pathogenicity island of *P. syringae* pv. *tomato*, and is essential for symptom production in host plants and the hypersensitive response in non-hosts. The *hrpZ* gene encodes a class of type III secreted proteins which are able to elicit a hypersensitive response in tobacco and trigger systemic acquired resistance in *Arabidopsis*. The *hrpZ* open reading frame is the second gene in an operon which encodes components of the type III secretion apparatus and its physical position is conserved among phytopathogenic pseudomonads, as was demonstrated by multiple alignments of the nucleotide sequences of the *hrpZ* open reading frames of *P. syringae* pathovars (Zaccardelli *et al.*, 2005). Contrary to *avr/hop* genes, *hrpZ* is not known to have homologues in regions of the genome unlinked to the *hrp/hrc* cluster, and, therefore, can be considered a genetically stable trait. A short pathovar-specific sequence with appropriate GC content was selected for potential primer design. Specific primers for *P. avellanae* were selected from the regions of yet another *hrp* gene, *hrpW*, that differed in sequence from the *hrpW* gene of *P. syringae* pv. *tomato*, and *P. syringae* pv. *syringae* (Loreti and Gallelli, 2002).

A conserved region of the *pelADE* cluster (420 bp) specific to *Dickeya* spp. which encodes for three of the five pectate lyase proteins involved in the maceration and soft-rotting of plant tissues has been used successfully to differentiate strains of soft-rotting enterics obtained from different hosts and geographical areas (Nassar *et al.*, 1996). Amplification was obtained with all *Dickeya* spp. while no PCR products were obtained from other soft-rotting pectobacteria or non-pectinolytic relatives.

The *cps* gene region in *P. stewartii* is a second major pathogenicity-related gene cluster in addition to the *hrp* gene cluster of this maize pathogen. While the *hrp* cluster contains genes which encodes a type III secretion system that is necessary for general pathogenicity, the *cps* gene cluster comprises 12 genes that encode for the production of the exopolysaccharide, stewartan. The *cps* gene cluster is similar to the *ams* gene cluster of *E. amylovora* that is responsible for the synthesis of the stewartan-like polysaccharide, amylovoran. However, three glycosyl-transferase genes differ between the *ams* and *cps* clusters. One of these genes, *cpsD*, proved to be useful for developing PCR primers specific for *P. stewartii* (Coplin *et al.*, 2002).

In *X. axonopodis* pv. *citri*, there is a gene cluster that regulates the expression of pathogenicity factors known as *rpf* genes. Analysis of the organization of these genes revealed significant differences from the *rpf* gene cluster in *X. campestris* pv. *campestris* and other plant pathogenic xanthomonads. Primers selected from within the *rpf* gene cluster of *X. axonopodis* pv. *citri* specifically amplified a 581 bp amplicon from the bacterium (Coletta-Filho *et al.*, 2006).

P. savastanoi is uniquely pathogenic to olive, resulting in the production of tumorous outgrowths known as the olive knot disease. Pathogenicity in this bacterium is dependent on the production of the phytohormone, indolacetic acid (IAA), and cytokinins. An *iaaL* gene encodes the conversion of IAA to IAA-lysine. This gene was targeted as a source of external and internal primer sequences for specific nested PCR amplification of the olive knot pathogen (Bertolini *et al.*, 2003).

Pathogenicity of some of the *P. syringae* pathovars is enhanced by the production of phytotoxic compounds. Phaseolotoxin, for instance, is a non-host-specific toxin that induces chlorosis in leaves of several plant species by inhibiting ornithine carbamoyl transferase, a critical enzyme in the urea cycle. Sequences of DNA coding for phaseolotoxin biosynthesis are organized into a large (>30 kb) gene cluster (*tox* cluster). PCR with primers selected for the open reading frame of the phaseolotoxin gene cluster amplified DNA from toxigenic strains of *P. syringae* pv. *phaseolicola*, but not from strains that did not produce the toxin (Marques *et al.*, 2000). However, since non-toxigenic strains of the bacterium also play an important role in the epidemiology of halo blight disease of bean, another set of primers needed to be developed to detect all pathogenic strains. Pathogenic, non-toxigenic strains could be detected by PCR using primers that targeted the avirulence gene, *avrPphF*, that is embedded in a plasmid-borne pathogenicity island (Rico *et al.*, 2003). Although the repertoire of avirulence genes varies with races of the bacterium, DNA from all of the non-toxigenic strains and a few toxigenic strains of *P. syringae* pv. *phaseolotoxin* were PCR-amplified with the avirulence gene primers. Another example in which a toxin gene sequence was successfully employed for development of pathogen-specific PCR primers relates to *P. syringae* pv. *tomato* (Cuppels *et al.*, 2006). In this case, the sequence of a 5.3 kb fragment from the gene cluster controlling production of the phytotoxin coronatine, which was previously used successfully as a DNA detection probe, served as the basis for selecting the primers.

Primers derived from pathogenicity-related plasmid sequences were also useful for PCR detection of *E. amylovora* (Salm and Geider, 2004). All naturally occurring strains of *E. amylovora* contain the non-transmissible plasmid pEA29. Genes encoded by this plasmid greatly enhance pathogenicity of the bacterium although they are not strictly required for the bacterium to cause disease. The real-time TaqMan assay developed from the plasmid sequence amplified a 112 bp fragment. The melting temperature of the primers at 55–60°C was about 10°C lower than the melting temperature of the GC-rich probe. In a similar way, a PCR test with primers based on published sequences of the universal virulence gene, *virC*, encoded by the Ti or Ri plasmids was designed for detection of tumorigenic strains of *Agrobacterium* spp. (Suzaki *et al.*, 2004).

The fact that *X. hyacinthi* can be specifically detected by antibodies to fimbriae suggested that possession of fimbriae is a pathogenesis-related characteristic. van Doorn *et al.* (2001) initially cloned and sequenced the fimbrial gene, by first amplifying it with degenerate primers designed from the amino acid sequence of the fimbriae subunit. Although the *fimA* gene of *X. hyacinthi* was found to have regions of homology with fimbrial and pilin genes of other bacteria, specific primers could be developed from the hypervariable central and C-terminal region of the gene.

Non-specified genomic sequences
Some PCR tests that are very useful for detection of specific bacterial phytopathogens utilize primers that were selected from genomic DNA fragments as being unique to the bacterial pathogen of interest without *a priori* knowledge of the function of the particular gene region. This approach usually involves either a subtraction hybridization step or isolation of a species-specific polymorphic band from a repetitive PCR experiment.

The subtraction hybridization method involves a protocol in which fragmented DNA from the target bacterium is hybridized to DNA from a closely related non-target bacterium on a solid support. Unhybridized DNA from the target bacterium will constitute a fraction that is enriched in fragments which lack complementary sequences in the related bacterium. Genomic fragments unique to the species *P. atrosepticum* and *carotovorum*, for example, were identified from a cloned genomic library by their inability to hybridize with the taxonomically related *E. coli* (Ward and De Boer, 1990). The sequence of one clone that hybridized specifically with strains of *P. atrosepticum* was subsequently used to design species-specific primers (De Boer and Ward, 1995). In another study, DNA fragments unique to *P. atrosepticum* were isolated by using *P. carotovorum* subsp. *carotovorum* as the source of subtractor DNA (Darrasse *et al.*, 1994).

Genomic primers specific for *C. michiganensis* subsp. *sepedonicus* were designed by Mills *et al.* (1997) from clones prepared from DNA of the target bacterium that were not subtracted by hybridization to a DNA cocktail. The cocktail consisted of total DNA extracted and pooled from three related strains, including *C. michiganensis* subsp. *michiganensis* and subsp. *insidiosus*, and *Rhodococcus facians*. Three rounds of subtraction hybridization were carried out and the remaining enriched target DNA fragments were amplified and cloned. Three clones that hybridized only to *C. michiganensis* subsp. *sepedonicus* DNA were sequenced and used to design three different sets of subspecies-specific primers. This unique sequence has also been used to develop real-time BIO-PCR assays for *C. michiganensis* subsp. *sepedonicus* (Schaad *et al.*, 1999).

To identify a unique genomic DNA sequence for *E. amylovora*, five cloned sequences were hybridized to 69 strains of *E. amylovora* and 29 strains of other *Erwinia* spp. (Taylor *et al.*, 2001). One clone that hybridized to DNA from all *E. amylovora* strains, but not to DNA extracted from the other *Erwinias*, was sequenced and used to design PCR primers. Two 30-mer oligonucleotide primers corresponding to sequences near the termini of the cloned insert were shown to direct the synthesis of a 187 bp PCR product exclusively from *E. amylovora*.

Repetitive PCR methods involve amplification of many small DNA fragments with relatively small primers. Subsequent electrophoretic separation of amplification products generates multiple bands. Comparison of banding patterns of the target species with the heterologous species usually reveal polymorphisms useful in the differentiation of taxons. Individual polymorphic bands may constitute genomic sequences that are unique to the source bacterium. To design specific primers for *R. solanacearum*, characteristic banding patterns obtained by random amplification of polymorphic DNA (RAPD) using 15 primers were analysed (Lee and Wang, 2000). One major reproducible fragment of 0.7 kb was cloned and shown to hybridize with a 2.7 kb *Eco*RI fragment of genomic DNA from *R. solanacearum*. The *Eco*RI fragment was excised from the gel, cloned and tested for specificity. After confirming specificity, the fragment was sequenced and used to generate two primers that amplified a 1.1 kb fragment of all *R. solanacearum* strains tested but not other bacteria.

A similar RAPD-based PCR technique was used to identify DNA fragments that were putatively specific to *X. axonopodis* pv. *dieffenbachiae*. One of the random primers resulted in amplification of a major 1.6 kb amplification product from all *X. axonopodis* pv. *dieffenbachiae* strains that was absent in other *Xanthomonas* spp. and subspecies tested. This fragment was specifically hybridized to the target bacterium, and was cloned, sequenced and used as sequence-characterized amplified region (SCAR) markers. One of the SCARs was used to develop a nested PCR protocol (Robene-Soustrade *et al.*, 2006). In the same way, representative isolates of *Pseudomonas corrugata* were subjected to RAPD-PCR with nine primers (Catara *et al.*, 2000). Amplified fragments were labelled and used as hybridization probes to determine specificity. Two fragments were cloned and sequenced. Primers were designed from the fragment sequences and used to develop a PCR test specific for the pathogen causing tomato wilt necrosis.

Insertion elements
These are mobile genetic elements that are only known to encode functions involved in insertion events. A non-coding region of insertion sequence IS*1405* was used to design specific primers to detect race 1 strains of *R. solanacearum* (Lee *et al.*, 2001). IS*1405* is 1174 bp in length and has 18 bp imperfect terminal-inverted repeats and contains a single open reading frame encoding a protein of 321 amino acids. IS*1405* is related to the IS5 subgroup of insertion sequences, and the sequence similarity between members of IS5 subgroup corresponds to the relationship of the host organisms. Comparison of the nucleotide sequence of IS*1405* with that of insertion sequences of the IS5 family indicated that the extent of identity was 51–54%, but only 23–31% in the non-coding region. Therefore, primers could be selected from the more diverse non-coding region to develop a PCR test for *R. solanacearum* race 1.

Sequences within a rare insertion element were used to develop a PCR test for *C. michiganensis* subsp. *insidiosus* (Samac *et al.*, 1998). An element related to IS*1122* that occurs in *C. michiganensis* subsp. *insidiosus* was only found in *C. michiganensis* subsp. *sepedonicus* among more than 40 species of bacteria tested. The two insertion elements, each 1078 bp in size, shared 88% sequence identity and occurred as 30–50 copies within the genomes of the respective pathogens. Regions of significant sequence diversity between the elements permitted development of a subspecies-specific primer set.

Sensitivity and Specificity of Detection

The success of laboratory testing for the presence of bacterial plant pathogens is a function of test specificity and sensitivity. The specificity of a test is evaluated by determining reactivity with organisms which are closely related to the target bacterium, as well as to the organisms generally associated with relevant crop plants. In addition, it is important that the test is applicable to the whole spectrum of representative strains of the pathogen under consideration. It is, of course, a theoretical possibility that entirely unrelated microorganisms, not tested during specificity evaluation, could share identical gene sequences. However, an inability to develop sufficiently specific tests has not yet been reported for any one plant pathogenic bacterium. Furthermore, specificity can be significantly improved by amplifying multiple specific target sequences from different regions of the genome of the target organisms in different PCR reactions.

Test sensitivity is more difficult to establish than specificity. The exquisite sensitivity of PCR itself has been well documented and many PCR assays developed for plant pathogenic bacteria report a sensitivity of less than 10 cells per PCR reaction. However, since high reagent costs dictate that reaction volumes be low (usually 25 µl), the amount of sample DNA analysed in each reaction is also very small (usually 1–2 µl of extracted DNA). The high level of analytical sensitivity also makes molecular techniques susceptible to cross-contamination and carry-over problems, leading to false-positive results. Moreover, the presence of inhibitors can cause false-negative results. After initial validation, test performance needs to be continuously monitored and test results compared to those obtained by other test methods.

The probability of detecting the presence of plant pathogenic bacteria in a plant sample is therefore a function of multiple factors. These include the population size of the target bacterium and the relative density of the target and non-target microbes within the sample, and factors related to the distribution of the target bacterium in the population of seed, tuber or whatever tissue is being sampled, as well as the sampling strategy, sample size, DNA extraction protocol and volume of the extract that can be accommodated in the PCR reaction. Hence, in practical terms, PCR detection of bacteria in plant samples is usually in the range of 10^3–10^5 cells/ml of plant sample homogenate, about 10- to 100-fold less sensitive than when PCR is applied to pure cultures. Measurement of test sensitivity, often conducted by diluting pure cultures in plant sample homogenate, may not accurately reflect the ability to detect bacteria that occur in association with plants or plant parts grown under field conditions. The physiological state of the bacteria, gene copy number, concentration of extraneous DNA and coextracted compounds may affect amplification of target DNA sequences in the final extract used for the PCR reaction.

Validation of Molecular Diagnostic Protocols

While an increasing number of molecular diagnostic methods are becoming available for plant pathogenic bacteria, few have yet to be incorporated into routine testing procedures and fewer, if any, have been adopted as the sole approach for official testing in trade scenarios. At present, molecular assays are mainly used for added confirmation of results obtained through conventional test methods which have been fully validated over many years. Demand is nevertheless increasing, on the grounds of cost-efficiency and ease-of-use, for robust generic molecular tests such as real-time PCR, which permit high-throughput, automated and quantitative testing for multiple target organisms. Only when fully validated will the potential benefits of current molecular diagnostics technology be realized.

Although molecular assays potentially offer increased sensitivity over many conventional methods, this is often not the case in practice, since high costs of molecular reagents tend to limit the sizes of sample which can be tested. In fact, the advantage of molecular over conventional technology in routine testing is often the increased specificity rather than sensitivity. Validation of the specificity of molecular assays is facilitated by the opportunity to perform *in silico* comparisons of target probes and primers against an ever-expanding bacterial DNA sequence database. Nevertheless, sequence databases are far from complete and *in silico* validation does not replace the need to test new assays against a wide range of bacteria, including those related to the target organism as well as those commonly found in test materials. Assay validation should therefore encompass testing of a high number of known positive and negative samples as well as known organisms (related and unrelated) obtained from culture collections. The specificity of molecular assays can be greatly enhanced when detecting more than one specific sequence per target organism, as in real-time PCR assays prescribed for testing potatoes for *C. michiganensis* subsp. *sepedonicus* (Schaad et al., 1999) and *R. solanacearum* (Weller et al., 2000).

There are no standardized rules as to how individual tests should be validated prior to routine use. A number of key principles, examples of good practice, that can be used as a guide are as follows:

- New methods should have been published in peer-reviewed journals with data indicating specificity, sensitivity and reproducibility of the method.
- PCR methods should have internal amplification controls and sufficient other positive and negative controls to allow discrimination of false-positive and false-negative results. At least one positive control should be included such that amplification is detected at or near the limit of detection.

- The performance of new methods should be assessed against an existing method, which serves as a gold standard.
- Data should be obtained by first testing a set of reference strains, followed by a more comprehensive testing phase using both tests in parallel with a large number of known positive (naturally infected) and negative samples.
- Where detection methods require approval for national or supranational testing or monitoring programmes, validation is generally achieved using multi-laboratory trials or ring tests in which the sensitivity, selectivity, ease-of-use and robustness of protocols are determined by supplying methods, blind strains, naturally or artificially infected and non-infected samples and standardized reagents to a number of laboratories and analysing the results obtained by each participant (for a brief discussion of this issue, see Hoorfar et al., 2004).

Such an approach was used to approve conventional, multiplex PCR assays for *C. michiganensis* subsp. *sepedonicus* (Pastrik, 2000) and *R. solanacearum* (Pastrik et al., 2002) for use as a primary screening test in recently revised official European Union test procedures for potatoes (Anonymous, 2006a,b). Initially, the PCR assays were evaluated against a range of strains of each pathogen, close relatives and bacteria which were likely to be found in the host, potato. After this, in the case of *C. michiganensis* subsp. *sepedonicus*, 3500 composite samples of 200 tubers from 143 different potato varieties were tested. All positive findings were confirmed with the established detection methods of immunofluorescence and an aubergine bioassay. Of the 33 positive findings obtained by multiplex PCR, 25 were verified by either immunofluoroscence or bioassay, with 11 of these verified by both methods. A similar approach was employed in the development of the PCR assay for *R. solanacearum* (Pastrik et al., 2002) where 4300 samples of 200 tubers from 143 cultivars of potato were tested. All positive findings were verified by immunofluorescence and a tomato bioassay. Of these, 13 samples tested positive by PCR, immunofluorescence and bioassay, while 12 were positive by immunofluorescence but negative for both PCR and bioassay. These studies, while highlighting good practice, also showed that it is not always possible to verify every result using a completely different test methodology.

Conclusions and Future Directions

Over a relatively short period of time, the molecular detection of plant pathogenic bacteria has become an important component in the development of plant disease management strategies. The value of these methodologies is beyond doubt and there are strong indications that our reliance on them will increase dramatically in the future. At present, molecular detection methods are frequently used in parallel or as confirmatory tests with other methods such as ELISA, immunofluorescence and isolation. However, deploying these methods in combination with alternative approaches is unlikely to continue into the medium to longer term as a combination of improvements in ease-of-use, accuracy and speed continue to be accrued by molecular detection methodologies. The methods and technologies described here highlight trends that may provide clues as to future directions that molecular detection strategies may take. However, the biggest obstacle against large-scale adoption will always be the cost. Molecular detection strategies will be adopted only if they provide significant savings on existing methods of diagnosis and detection. Cost savings are being derived and will continue to do so through automation and the possibility of performing large numbers of tests simultaneously (Mumford et al., 2006). In recent years, multiplexing and quantification have taken diagnostics to new levels (Lievens and Thomma, 2005). Recent developments in high-throughput DNA extraction systems and the introduction of 384 well plate real-time PCR machines are bringing these technologies within the reach of many regulatory laboratories around the world (Lievens and Thomma, 2005). In the immediate future, it is likely that more and more diagnostic laboratories and inspection agencies will embrace these technologies.

Developments in molecular detection strategies for phytobacteria are greatly influenced by technological advances in the medical field and the growing interest in biosecurity. With regards to the latter, a key step in any counterterrorism measure will be the rapid identification of the etiological agent. Suitable methods must meet at least some of the requirements for speed, accuracy and portability, with the capacity to test for a wide range of targets simultaneously. Such requirements also have resonance in the development of diagnostic tests for plant pathogenic bacteria and give indications as to future trends in the development of detection strategies. These include the production of field-based diagnostics and massively parallel testing platforms which can assay for a wide range of pests and pathogens simultaneously.

In terms of field-based diagnostics, possibly the first description of a method for DNA extraction and molecular testing for a plant pathogen, carried out entirely in the field, was described recently for *Phytophthora ramorum*, the causal agent of sudden oak death (Tomlinson et al., 2005). A SmartCycler System (Cepheid, Sunnyvale, California, www.cepheid.com), a portable, real-time PCR machine, was used with reagents stabilized to be transported at ambient temperatures for diagnostic testing at the point of sampling. This system eliminated the time taken to send off samples and receive back test results from a centralized diagnostic laboratory and could allow identification within 2 h from the point of sampling. The inclusion of generic plant cytochrome oxidase-specific primers in the assay provides an internal PCR control. Similarly, Idaho Technology Inc. (Salt Lake City, Utah, www.idahotech.com) markets a product based on real-time PCR technology, called Ruggedized Advanced Pathogen Identification Device (R.A.P.I.D.), for pathogen detection in military field hospitals and other outdoor environments. The portable system includes a rapid thermocycler with concurrent fluorescence monitoring and data analysis. This system has been used successfully to detect *X. citri* and is suitable for use in a mobile field laboratory. The only additional equipment required was a low-speed centrifuge and a vortex mixer (Mavrodieva et al., 2004).

The time is fast approaching when extraction and purification steps can be integrated to allow results to be generated directly from raw materials. The GeneXpert (Cepheid, Sunnyvale, California, www.cepheid.com) is a four-site, self-contained device which integrates automated sample processing and real-time PCR detection of infectious agents (Ulrich et al., 2006). All the steps required for identifying bacterial and viral agents in various biological specimens, including sample preparation, amplification and detection, are combined within a single instrument that provides results in approximately 30–40 min. It is the only platform currently available that is capable of both nucleic acid purification and real-time PCR detection enclosed within a single system, since all sample manipulations are automated, and therefore errors associated with manual processing are eliminated.

Microarrays, initially designed for the study of gene expression, are finding increasing use in diagnostics (Kostrzynska and Bachand, 2006). They consist of a solid support, most commonly glass or silicon, upon which DNA fragments or oligonucleotides are immobilized. Microarrays allow for rapid, simultaneous detection of thousands of specific targets. Generally, it is necessary to perform a PCR amplification step prior to hybridization, although it is possible to detect microorganisms directly with genomic DNA, albeit at a greatly reduced sensitivity. The first use of arrays for the differentiation of plant pathogenic bacteria was performed on pathogens of potato (Fessehaie et al., 2003). Short oligonucleotides of the 3' end of the 16S rDNA gene and the 16S–23S intergenic spacer region from *C. michiganensis* subsp. *sepedonicus*, *R. solanacearum* and a range of pectolytic enterobacteria were spotted onto a nylon array. By using this system and an amplification step prior to hybridization, it was possible to correctly identify pure and mixed cultures recovered from inoculated potato tissue. This approach has been expanded to develop a single microarray for all quarantine pathogens listed by the European Union for potato, including bacteria, fungi, nematodes, viroids and viruses (Mumford et al., 2006; www.diagchip.co.uk). This system provides the very real prospect of screening, within a single test, all the pests and pathogens of a specific host and has enormous potential for screening commodities for the presence of organisms of quarantine significance.

Other novel technologies for rapid detection focus more on the products or effects of plant pathogens. An electronic nose has been used for the early detection of soft-rotting pectobacteria in potato tubers (de Lacy Costello et al., 2000). A prototype device incorporating three sensors to detect compounds known to be evolved by soft-rotting potato tubers has been developed. The device was assessed for its discriminating power under simulated storage conditions and could detect one tuber with soft rot in 100 kg of sound tubers in a simulated storage crate. The device could also detect an inoculated tuber, without visible signs of soft rot, in 10 kg of sound tubers.

In conclusion, it can be envisaged over the longer term that the drive towards taking diagnostics back to the field, silo, store or dockside and utilizing the technologies developed in the areas of medicine and biosecurity, such as or similar to GeneXpert, will provide accurate and sensitive diagnosis performed by the non-specialist at the point of sampling. Research into biosensors or 'lab-on-a-chip' devices may generate further benefits by elimination of sample preparation, enhancements in specificity and reduced analysis times (Deisingh and Thompson, 2004), bringing closer the widespread use of field-based diagnostic tests. The increasing use of microarray technologies will also allow for a wide range of potential pests and pathogens to be screened simultaneously, with a high degree of speed and accuracy. It is therefore highly likely that in many countries, a two-tiered approach to pathogen detection will be adopted using a mix of centralized, highly automated laboratories capable of dealing with high-volume diagnostics for a broad spectrum of pests and pathogens, working in partnership with inspection agencies and extension services using field-based diagnostic devices at the point of sampling. Such improved technologies will undoubtedly contribute significantly to the implementation and success of plant disease control strategies based on the planting of pathogen-tested seed and vegetative propagation units to avoid such diseases altogether.

References

Anonymous (2006a) Commission Directive 2006/56/EC of 12 June 2006 amending the Annexes to Council Directive 93/85/EEC on the control of potato ring rot. *Official Journal of the European Union* L 182, 4 July 2006, 1–43.

Anonymous (2006b) Commission Directive 2006/63/EC of 14 July 2006 amending Annexes II to VII to Council Directive 98/57/EC on the control of *Ralstonia solanacearum* (Smith) Yabuuchi et al. *Official Journal of the European Union* L 206, 27 July 2006, 36–106.

Bentsink, L., Leone, G.O.M., van Beckhoven, J.R.C.M., van Schijndel, H.B., van Gemen, B. and van der Wolf, J.M. (2002) Amplification of RNA by NASBA allows direct detection of viable cells of *Ralstonia solanacearum* in potato. *Journal of Applied Microbiology* 93, 647–655.

Berg, T., Tesoriero, L. and Hailstones, D.L. (2005) PCR-based detection of *Xanthomonas campestris* pathovars in *Brassica* seed. *Plant Pathology* 54, 416–427.

Bertolini, E., Penyalver, R., Garcia, A., Olmos, A., Quesada, J.M., Cambra, M. and Lopez, M.M. (2003) Highly sensitive detection of *Pseudomonas savastanoi* pv. *savastanoi* in asymptomatic olive plants by nested-PCR in a single closed tube. *Journal of Microbiological Methods* 52, 261–266.

Boom, R., Sol, C.J.A., Salimans, M.M.M., Jansen, C.L., Wertheim-van-Dillen, P.M.E. and van der Noordaa, J. (1990) Rapid and simple method for purification of nucleic acids. *Journal of Clinical Microbiology* 28, 495–503.

Boonham, N., Walsh, K., Preston, S., North, J., Smith, P. and Barker, I. (2002) The detection of tuber necrotic isolates of *Potato virus Y*, and the accurate discrimination of PVY^O, PVY^N and PVY^C strains using RT-PCR. *Journal of Virological Methods* 102, 103–112.

Botha, W.J., Serfontein, S., Greyling, M.M. and Berger, D.K. (2001) Detection of *Xylophilus ampelinus* in grapevine cuttings using a nested polymerase chain reaction. *Plant Pathology* 50, 515–526.

Catara, V., Arnold, D., Cirvilleri, G. and Vivian, A. (2000) Specific oligonucleotide primers for the rapid identification and detection of the agent of tomato pith necrosis, *Pseudomonas corrugata*, by PCR amplification: evidence for two distinct genomic groups. *European Journal of Plant Pathology* 106, 753–762.

Coletta-Filho, H.D., Takita, M.A., de Souza, A.A., Neto, J.R., Destefano, S.A.L., Hartung, J.S. and Machado, M.A. (2006) Primers based on the rpf gene region provide improved detection of *Xanthomonas axonopodis* pv. *citri* in naturally and artificially infected citrus plants. *Journal of Applied Microbiology* 100, 279–285.

Coplin, D.L., Majerczak, D.R., Zhang, Y., Kim, W.-S., Jock, S. and Geider, K. (2002) Identification of *Pantoea stewartii* subsp. *stewartii* by PCR and strain differentiation by PFGE. *Plant Disease* 86, 304–311.

Cubero, J., Martinez, M.C., Flop, P. and Lopez, M.M. (1999) A simple and efficient PCR method for the detection of *Agrobacterium tumefaciens* in plant tumors. *Journal of Applied Microbiology* 86, 591–602.

Cuppels, D.A., Louws, F.J. and Ainsworth, T. (2006) Development and evaluation of PCR-based diagnostic assays for the bacterial speck and bacterial spot pathogens of tomato. *Plant Disease* 90, 451–458.

Darrasse, A., Kotoujansky, A. and Bertheau, Y. (1994) Isolation by genomic subtraction of DNA probes specific for *Erwinia carotovora* subsp. *atroseptica*. *Applied and Environmental Microbiology* 60, 298–306.

De Boer, S.H. (2002) Relative incidence of *Erwinia carotovora* subsp. *atroseptica* in stolon end and peridermal tissue of potato tubers in Canada. *Plant Disease* 86, 960–964.

De Boer, S.H. and Ward, L.J. (1995) PCR detection of *Erwinia carotovora* subsp. *atroseptica* associated with potato tissue. *Phytopathology* 85, 854–858.

Deisingh, A.K. and Thompson, M. (2004) Biosensors for the detection of bacteria. *Canadian Journal of Microbiology* 50, 69–77.

de Lacy Costello, B.P.J., Ewen, R.J., Gunson, H.E., Ratcliffe, N.M. and Spencer-Phillips, P.T.N. (2000) The development of a sensor system for the early detection of soft rot in stored potato tubers. *Measurement Science and Technology* 11, 1685–1691.

Destefano, S.A.L., Almeida, I.M.G., Rodrigues Neto, J., Ferreira, M. and Balani, D.M. (2003) Differentiation of *Xanthomonas* species pathogenic to sugarcane by PCR-RFLP analysis. *European Journal of Plant Pathology* 109, 283–288.

Doyle, J.J. and Doyle, J.L. (1987) DNA isolation from small quantities of fresh leaf tissue. *Phytochemical Bulletin* 19, 11–15.

Fessehaie, A., De Boer, S.H. and Levesque, C.A. (2003) An oligonucleotide array for the identification and differentiation of bacteria pathogenic on potato. *Phytopathology* 93, 262–269.

Franken, A.A.J.M., van Zeijl, C., van Bilsen, J.G.P.M., Neuvel, A., Devogel, R, van Wingerden, Y., Birnbaum, Y.E., van Hateren, J. and van der Zouwen, P.S. (1991) Evaluation of a plating assay for *Xanthomonas campestris* pv. *campestris*. *Seed Science and Technology* 19, 215–226.

Guimaraés, P.M., Palmano, S., Smith, J.J., Grossi de Sá, M.F. and Saddler, G.S. (2001) Development of a PCR test for the detection of *Curtobacterium flaccumfaciens* pv. *flaccumfaciens*. *Antonie van Leeuwenhoek* 80, 1–10.

He, C.X., Li, W.B., Ayres, A.J., Hartung, J.S., Viranda, V.S. and Teixeira, D.C. (2000) Distribution of *Xylella fastidiosa* in citrus rootstocks and transmission of citrus variegated chlorosis between sweet orange plants through natural root grafts. *Plant Disease* 84, 622–626.

Hoorfar, J. Wolffs, P. and Rådström, P. (2004) Diagnostic PCR: validation and sample preparation are two sides of the same coin. *Acta Pathologica, Microbiologica et Immunologica Scandinavica* 112, 808–815.

Hu, X., Lai, F.M., Reddy, A.S.N. and Ishimaru C.A. (1995) Quantitative detection of *Clavibacter michiganensis* subsp. *sepedonicus* by competitive polymerase chain reaction. *Phytopathology* 85, 1468–1473.

Hyman, L.J., Birch, P.R.J., Dellagi, A., Avrova, A.O. and Toth, I.K. (2000) A competitive PCR-based method for the detection and quantification of *Erwinia carotovora* subsp. *atroseptica* on potato tubers. *Letters of Applied Microbiology* 30, 330–335.

Kievits, T., van Gemen, B., van Strijp, D., Schukkink, R., Dircks, M., Adriaanse, H., Malek, L., Sooknanan, R. and Lens, P. (1991) NASBA™ isothermal enzymatic in vitro nucleic acid amplification optimized for the diagnosis of HIV-1 infection. *Journal of Virological Methods* 35, 273–286.

Khoodoo, M.H.R., Sahin, F. and Jaufeerally-Fakim, Y. (2005) Sensitive detection of *Xanthomonas axonopodis* pv. *dieffenbachiae* on *Anthurium andreanum* by immunocapture-PCR (IC-PCR) using primers designed from sequence characterized amplified regions (SCAR) of the blight bacterium. *European Journal of Plant Pathology* 112, 379–390.

Kostrzynska, M. and Bachand, A. (2006) Application of DNA microarray technology for detection, identification, and characterization of food-borne pathogens. *Canadian Journal of Microbiology* 52, 1–8.

Lee, Y.-A. and Wang, C.-C. (2000) The design of specific primers for the detection of *Ralstonia solanacearum* in soil samples by polymerase chain reaction. *Botanical Bulletin Academy Sinica* 41, 121–128.

Lee, Y.-A., Fan, S.-C., Chiu, L.Y. and Hsia, K.-C. (2001) Isolation of an insertion sequence from *Ralstonia solanacearum* race 1 and its potential use for strain characterization and detection. *Applied and Environmental Microbiology* 67, 3943–3950.

Li, X. and De Boer, S.H. (1995) Selection of polymerase chain reaction primers from an RNA intergenic spacer region for specific detection of *Clavibacter michiganensis* subsp. *sepedonicus*. *Phytopathology* 85, 837–842.

Li, W.-B., Pria, W.D., Lacava, P.M., Quin, X. and Hartung, J.S. (2003) Presence of *Xylella fastidiosa* in sweet orange fruit and seeds and its transmission to seedlings. *Phytopathology* 93, 953–958.

Lievens, B. and Thomma, B.P.H.J. (2005) Recent developments in pathogen detection arrays: implications for fungal plant pathogens and use in practice. *Phytopathology* 95, 1374–1380.

Loreti, S. and Gallelli, A. (2002) Rapid and specific detection of virulent *Pseudomonas avellanae* strains by PCR amplification. *European Journal of Plant Pathology* 108, 237–244.

Lozano, J.C. (1986) Cassava bacterial blight: a manageable disease. *Plant Disease* 70, 1089–1093.

Mahuku, G.S. (2004) A simple extraction method suitable for PCR-based analysis of plant, fungal, and bacterial DNA. *Plant Molecular Biology Reporter* 22, 71–81.

Marques, A.S.d.A., Corbiere, R., Gardan, L., Tourte, C., Manceau, C., Taylor, H.D. and Samson, R. (2000) Multiphasic approach for the identification of the different classification levels of *Pseudomonas savastanoi* pv. *phaseolicola*. *European Journal of Plant Pathology* 106, 715–734.

Mavrodieva, V., Levy, L. and Gabriel, D.W. (2004) Improved sampling methods for real-time polymerase chain reaction diagnosis of citrus canker from field samples. *Phytopathology* 94, 61–68.

Merighi, M., Sandrini, A., Landini, S., Ghini, S., Girotti, S., Malaguti, S. and Bazzi, C. (2000) Chemiluminescent and colorimetric detection of *Erwinia amylovora* by immunoenzymatic determination of PCR amplicons from plasmid pEA29. *Plant Disease* 84, 49–54.

Michener, P.M., Pataky, J.H.K. and White, D.G. (2002) Rates of transmitting *Erwinia stewartii* from seed to seedlings of a sweet corn hybrid susceptible to Stewart's wilt. *Plant Disease* 86, 1031–1035.

Mills, D., Russell, B.W. and Hanus, J.W. (1997) Specific detection of *Clavibacter michiganensis* subsp. *sepedonicus* by amplification of three unique DNA sequences isolated by subtraction hybridization. *Phytopathology* 87, 853–861.

Mulholland, V. (2005) Immunocapture-polymerase chain reaction. In: Burns, R. (ed.) *Immunochemical Protocols*, 3rd edn. *Methods in Molecular Biology*, Vol. 295. Humana Press, Totowa, New Jersey, pp. 281–290.

Mumford, R., Boonham, N., Tomlinson, J. and Barker, I. (2006) Advances in molecular phytodiagnostics – new solutions for old problems. *European Journal of Plant Pathology* 116, 1–19.

Nassar, A., Darrasse, A., Lemattre, M., Kotoujansky, A., Dervin, C., Vedel, R. and Bertheau, Y. (1996) Characterization of *Erwinia chrysanthemi* by pectinolytic isozyme polymorphism and restriction fragment length polymorphism analysis of PCR-amplified fragments of pel genes. *Applied and Environmental Microbiology* 62, 2228–2235.

Norelli, J.L., Holleran, H., Johnson, W.C., Robinson, I. and Aldwinckle, H.S. (2003) Resistance of Geneva and other apple rootstocks to *Erwinia amylovora*. *Plant Disease* 87, 26–32.

Ozakman, M. and Schaad, N.W. (2003) A real-time BIO-PCR assay for detection of *Ralstonia solanacearum* race 3, biovar 2, in asymptomatic potato tubers. *Canadian Journal of Plant Pathology* 25, 232–239.

Panagopoulos, C.G. (1987) Recent progress on *Xanthomonas ampelina*. *EPPO Bulletin* 17, 225–230.

Park, Y.J., Lee, B.Mo., Ho-Hahn, J., Lee, G.B. and Park, D.S. (2004) Sensitive and specific detection of *Xanthomonas campestris* pv. *campestris* by PCR using species-specific primers based on hrpF gene sequences. *Microbiological Research* 159, 419–423.

Pastrik, K.H. (2000) Detection of *Clavibacter michiganensis* subsp. *sepedonicus* in potato tubers by multiplex PCR with coamplification of host DNA. *European Journal of Plant Pathology* 106, 155–165.

Pastrik, K.H., Elphinstone, J.G. and Pukall, R. (2002) Sequence analysis and detection of *Ralstonia solanacearum* by multiplex PCR amplification of 16S–23S ribosomal intergenic spacer region with internal positive control. *European Journal of Plant Pathology* 108, 831–842.

Penyalver, R., Garcia, A., Ferrer, A., Bertolini, L. and Lopez, M.M. (2000) Detection of *Pseudomonas savastanoi* pv. *savastanoi* in olive plants by enrichment and PCR. *Applied and Environmental Microbiology* 66, 2673–2677.

Pérombelon, M.C.M. (2002) Potato diseases caused by soft rot erwinias: an overview of pathogenesis. *Plant Pathology* 51, 1–12.

Rico, A., Lopez, R., Asensio, C., Aizpun, M.T., Asensio-S.-Manzanera, C. and Murillo, J. (2003) Nontoxigenic strains of *Pseudomonas syringae* pv. *phaseolicola* are a main cause of halo blight of beans in Spain and escape current detection methods. *Phytopathology* 93, 1553–1559.

Robene-Soustrade, I., Laurent, P., Gagnevin, L., Joen, E. and Pruvost, O. (2006) Specific detection of *Xanthomonas axonopodis* pv. *dieffenbachiae* in anthurium (*Anthurium andreanum*) tissues by nested PCR. *Applied and Environmental Microbiology* 72, 1072–1078.

Salm, H. and Geider, K. (2004) Real-time PCR for detection and quantification of *Erwinia amolovora*, the causal agent of fireblight. *Plant Pathology* 53, 602–610.

Samac, D.A., Nix, R.J. and Oleson, A.E. (1998) Transmission frequency of *Clavibacter michiganensis* subsp. *insidiosus* to alfalfa seed and identification of the bacterium by PCR. *Plant Disease* 82, 1362–1367.

Sayler, R.J., Cartwright, R.D. and Yang, Y. (2006) Genetic characterization and real-time PCR detection of *Burkholderia glumae*, a newly emerging bacterial pathogen of rice in the United States. *Plant Disease* 90, 603–610.

Schaad, N.W., Cheong, S.S., Tamaki, S., Hatziloukas, E. and Panopoulos, N.J. (1995) A combined biological and enzymatic amplification (BIO-PCR) technique to detect *Pseudomonas Syringae* PV. *Phaseolicola* in bean seed extracts. *Phytopathology* 85, 243–248.

Schaad, N.W., Berthier-Schaad, Y., Sechler, A. and Knorr, D. (1999) Detection of *Clavibacter michiganensis* subsp. *sepedonicus* in potato tubers by BIO-PCR and an automated real-time fluorescence detection system. *Plant Disease* 83, 1095–1100.

Scortichini, M. and Lazzari, M. (1996) Systemic migration of *Pseudomonas syringae* pv. *avellanae* in twigs and young trees of hazelnut and symptom development. *Journal of Phytopathology* 144, 215–219.

Singer-Sam, J., Tanguay, R.L. and Riggs, A.D. (1989) Use of Chelex to improve the PCR signal from a small number of cells. *Amplifications: A Forum for PCR Users* 3, 11.

Smith, D.S., Gourley, J. and De Boer, S.H. (2007) An artificial positive control for use in the real-time PCR detection of *Clavibacter michiganensis* subsp. *sepedonicus*. *Canadian Journal of Plant Pathology* (abstract) (in press).

Song, W.Y., Kim, H.M., Hwang, C.Y. and Schaad, N.W. (2004) Detection of *Acidovorax avenae* ssp. *avenae* in rice seeds using BIO-PCR. *Journal of Phytopathology* 152, 667–676.

Stead, D.E. and Pemberton, A.W. (1987) Recent problems with *Pseudomonas syringae* pv. *pisi* in UK. *EPPO Bulletin* 17, 291–294.

Suzaki, K., Yoshida, K. and Sawada, H. (2004) Detection of tumorigenic *Agrobacterium* strains from infected apple saplings by colony PCR and improved PCR primers. *Journal of Genetics and Plant Pathology* 70, 342–347.

Taylor, R.K., Guilford, P.J., Clark, R.G., Hale, C.N. and Forster, R.L.S. (2001) Detection of *Erwinia amylovora* in plant material using novel polymerase chain reaction (PCR) primers. *New Zealand Journal of Crop Horticultural Science* 29, 34–43.

Tomlinson, J.A., Boonham, N., Hughes, K.J.D., Griffin, R.L. and Barker, I. (2005) On-site DNA extraction and real-time PCR for detection of *Phytophthora ramorum* in the field. *Applied and Environmental Microbiology* 71, 6702–6710.

Tsiantos, J.H. (1987) Transmission of bacterium *Corynebacterium michiganensis* pv. *michiganensis* by seeds. *Journal of Phytopathology* 119, 142–146.

Ulrich, M.P., Christensen, D.R., Coyne, S.R., Craw, P.D., Henchal, E.A., Sakai, S.H., Swenson, D., Tholath, J., Tsai, J., Weir, A.F. and Norwood, D.A. (2006) Evaluation of the Cepheid GeneXpert® system for detecting *Bacillus anthracis*. *Journal of Applied Microbiology* 100, 1011–1016.

van Beckhoven, J.R.C.M., Stead. D.E. and van der Wolf, J.M. (2002) Detection of *Clavibacter michiganensis* subp. *sepedonicus* by AmpliDet RNA, a new technology based on real time monitoring of NASBA amplicons with a molecular beacon. *Journal of Applied Microbiology* 93, 840–849.

van der Wolf, J.M., Hyman, L.J., Jones, D.A.C., Grevesse, C., van Beckhoven, J.R.C.M., van Vuurde, J.W.L. and Pérombelon, M.C.M. (1996) Immunomagnetic separation of *Erwinia carotovora* subsp *atroseptica* from potato peel extracts to improve detection sensitivity on a crystal violet pectate medium or by PCR. *Journal of Applied Bacteriology* 80, 487–495.

van der Wolf, J.M., van Beckhoven, J.R.C.M., de Haan, E.G., van den Bovenkamp, G.W. and Leone, G.O.M. (2004) Specific detection of *Ralstonia solanacearum* 16S rRNA sequences by AmpliDet RNA. *European Journal of Plant Pathology* 110, 25–33.

van Doorn, J., Hollinger, T.C. and Oudega, B. (2001) Analysis of the type IV fimbrial-subunit gene fimA of *Xanthomonas hyacinthi*: application in PCR-mediated detection of yellow disease in hyacinths. *Applied and Environmental Microbiology* 67, 598–607.

Verdier, V., Ojeda, S. and Mosquera, G. (2001) Methods for detecting the cassava bacterial blight pathogen: a practical approach for managing the disease. *Euphytica* 120, 103–107.

Ward, L.J. and De Boer, S.H. (1990) A DNA probe specific for serologically diverse strains of *Erwinia carotovora*. *Phytopathology* 80, 665–669.

Ward, L.I., Beales, P.A., Barnes, A.V. and Lane, C.R. (2004) A real-time PCR assay based method for routine diagnosis of *Spongospora subterranea* on potato tubers. *Journal of Phytopathology* 152, 633–638.

Weller, S.A., Elphinstone, J.G., Smith, N.C., Boonham, N. and Stead, D.E. (2000) Detection of *Ralstonia solanacearum* strains with a quantitative, multiplex, real-time fluorogenic PCR (TaqMan) assay. *Applied and Environmental Microbiology* 66, 2853–2858.

Williamson, L., Nakaho, K., Hudelson, B. and Allen, C. (2002) *Ralstonia solanacearum* race 3, biovar 2 strains isolated from geranium are pathogenic on potato. *Plant Disease* 86, 987–991.

Wong, M.L. and Medrano, J.F. (2005) Real-time PCR for mRNA quantitation. *Biotechniques* 39, 75–85.

Zaccardelli, M., Spasiano, A., Bazzi, C. and Merighi, M. (2005) Identification and *in planta* detection of *Pseudomonas syringae* pv. *tomato* using PCR amplification of $hrpZ_{Pst}$. *European Journal of Plant Pathology* 111, 85–90.

9 Molecular Diagnostics of Plant-parasitic Nematodes

R.N. Perry, S.A. Subbotin and M. Moens

Abstract

Biochemical and molecular methods of identification provide accurate, reliable diagnostic approaches for the identification of plant-parasitic nematodes. Initially, the techniques were used solely for taxonomic purposes, but increasingly became popular as a component of diagnostic information for farmers, growers and advisors. Diagnostic procedures are now available to differentiate the plant-pathogenic species from related but non-pathogenic species. The microscopic size of plant-parasitic nematodes poses problems and techniques have been developed to enrich samples to obtain qualitative and quantitative information on individual species. In addition, techniques are available to evaluate single nematodes, cysts or eggs of individual species in extracts from soil and plant tissue. Background information on early, pioneering work is presented as a prelude to discussion of diagnostic approaches. These include the use of isoelectric focusing (IEF) and restriction fragment length polymorphisms (RFLPs), progressing to antibody approaches and current polymerase chain reaction (PCR)-based techniques. DNA- or RNA-based techniques are the most widely used approaches for identification, taxonomy and phylogenetic studies, although the development and use of other methods has been, and in some cases still is, important. DNA bar coding and the extraction of DNA from preserved specimens will aid considerably in diagnostic information and these are discussed in the context of the future requirements of accurate and rapid diagnostic protocols.

Introduction

It is essential that the causative organism of plant disease is identified correctly in order to implement effective management strategies. Thus, to differentiate species using diagnostic techniques is a vital component of management of economically important pests, including plant-parasitic nematodes. Conventional methods for nematode identification rely on time-consuming morphological and morphometric analysis of several specimens of the target nematode. The accuracy and reliability of such identification depends largely on the experience and skill of the person making the diagnosis, and the number of such qualified and experienced nematode taxonomists is small and currently in decline. Biochemical and molecular methods of nematode identification provide accurate, alternative diagnostic approaches.

For growers and extension workers, economically important species of plant-parasitic nematodes have to be detected from soil samples prior to planting. Frequently, closely related and morphologically similar species are present as mixtures in soil samples and it is vital that diagnostic procedures are available to differentiate the pathogenic species from related but non-pathogenic species. The microscopic size of plant-parasitic nematodes poses problems and techniques have been developed to enrich samples to obtain qualitative and quantitative information on individual species. To assess damage after plant growth, nematodes have to be extracted from plant tissue and techniques have been developed to overcome difficulties associated with this approach. This chapter will indicate the techniques used to evaluate individual nematodes, cysts or eggs of individual species in extracts from soil and plant tissue. The goal of growers and extension workers is on-site diagnostics, and this is now a realistic objective.

Plant health agencies also need molecular techniques to aid the identification of pathogenic nematodes. The implementation of phytosanitary measures is of major importance in reducing the adverse impact of plant-parasitic nematodes (Hockland et al., 2006). Nematodes that may be of phytosanitary importance are intercepted by plant health inspectors at points of entry of goods into a country. Frequently, these nematodes are species that have the potential to become significant pests if allowed to enter and establish. Internationally standardized diagnostic protocols are needed to limit or avoid the spread of these species, and rapid and accurate molecular techniques are especially important where morphological identification is difficult or where only immature specimens have been intercepted.

Diagnostic techniques that rely on the interaction between or separation of specific molecules include serological, biochemical and DNA- or RNA-based approaches. Molecular diagnostics is a term used more specifically for the characterization of an organism based on information on its DNA or RNA structure. For diagnostic purposes, homologous genes or DNA or RNA fragments, whose sequence differs between species but is similar for all individuals of the same population or species, are selected and compared. DNA- or RNA-based techniques are the most widely used approaches for identification, taxonomy and phylogenetic studies (Perry and Jones, 1998). However, the history of diagnostics of plant-parasitic nematodes shows that the development and use of other methods has been, and in some cases still is, important. Initially, the techniques were used solely for taxonomic purposes, but increasingly became popular as a component of diagnostic information for farmers, growers and advisors.

Background

The work of Webster and Hooper (1968) was among the first papers on differentiation of plant-parasitic nematodes which used serological techniques to examine differences between species of the genera *Heterodera* and *Ditylenchus*. Electrophoretic methods separate proteins based on their size or isoelectric point. Trudgill and Carpenter (1971) used polyacrylamide gel electrophoresis (PAGE) to differentiate pathotypes of potato cyst nematodes (PCN), which paved the way for the subsequent description of a new species, *Globodera pallida*, and its differentiation from *Globodera rostochiensis*. Isoelectric focusing (IEF) was used by Fleming and Marks (1982) to separate and quantify the two PCN species in mixed populations. Bakker and Gommers (1982) used two-dimensional electrophoresis combined with protein analysis as a further refinement for differentiation of PCN species. The two species of PCN are economically important pests, mainly in Europe. Globally, species of the root-knot nematode genus, *Meloidogyne*, are of primary economic importance. In the USA, differentiation of *Meloidogyne* spp. was achieved by Esbenshade and Triantaphyllou (1985) using PAGE combined with staining for esterases, a technique still in routine use. Ibrahim and Perry (1992) showed that this approach could be used with *Meloidogyne* females in galled roots, thus obviating the need to separate nematodes from the host tissue. This technique was used on other genera of plant-parasitic nematodes, such as *Aphelenchoides* (Ibrahim *et al.*, 1994). Monoclonal antibodies (MAbs) were used by Schots *et al.* (1986) as a more sensitive technique to differentiate between PCNs. Robinson (1989) used MAbs in a quantitative assay for *Meloidogyne*. The separation of *Meloidogyne incognita* has been achieved using MAbs raised to non-specific esterases, an approach that avoids the need to separate the esterases by electrophoresis (Ibrahim *et al.*, 1996). Antibodies are being used in a method to enhance detection of nematodes in soil samples (described later).

Isoelectric Focusing

Of the techniques mentioned above, IEF is currently being used extensively in many nematology diagnostics laboratories using the automated electrophoresis equipment, PhastSystem, developed by Pharmacia Corp. IEF is used routinely to differentiate the two PCN species, *G. rostochiensis* and *G. pallida*, based on the presence of species-specific bands (Karssen *et al.*, 1995); however, this technique has not yet been tested against other morphologically similar *Globodera* spp., such as *G. artemisiae*, *G. millefolii* and *G. tabacum*. The IEF method may also be useful to differentiate other cyst nematodes (Sturhan and Rumpenhorst, 1996).

For routine identification of *Meloidogyne* spp., perineal patterns (the cuticular lines and associated structures surrounding the vulval region) are often used as the main diagnostic feature. However, species such as

M. paranaensis, *M. konensis* and *M. mayaguensis* have perineal patterns similar to *M. incognita* or *M. arenaria* and could be misidentified (Carneiro *et al.*, 2000). Carneiro *et al.* (2004) concluded that the perineal pattern can be used only as a complementary tool; enzyme characterization is important for checking the morphological consistency of the identification. Extensive enzymatic studies have demonstrated that *Meloidogyne* spp. can be differentiated by enzyme phenotypes (Esbenshade and Triantaphyllou, 1985; Fargette, 1987). Isozyme phenotypes of adult females, especially esterase and malate dehydrogenase, are species-specific and are useful as reliable characters for *Meloidogyne* identification (Esbenshade and Triantaphyllou, 1990; Carneiro *et al.*, 2004). Progress in electrophoretic procedures has made possible nematode identification from the protein extract of a single female in less than 2 h from the time the females are collected from infected plant roots (Esbenshade and Triantaphyllou, 1990). Descriptions of new species of root-knot nematodes are now usually accompanied by phenotypic profiles for both these isozymes (Castillo *et al.*, 2003; Karssen *et al.*, 2004; Carneiro *et al.*, 2005). One of the main disadvantages of the IEF method is its sensitivity to nematode age. Only young adult females are used for such diagnostics.

Mass Spectrometry

Matrix-assisted laser desorption/ionization time-of-flight mass spectrometry (MALDI-TOFMS) analyses thermolabile, non-volatile organic compounds, especially those of high molecular mass, and is used successfully in biochemistry for the analysis of proteins, peptides, glycoproteins, oligosaccharides and oligonucleotides. Identification is based on the generation of diagnostic MALDI-TOFMS protein profiles of the test organisms. In this technique, positively charged ions are generated and analysed after laser irradiation of the organism in extracts after disruption of cells, or in intact whole cells of the organism in the presence of an ultraviolet (UV)-absorbing matrix. Recently, MALDI-TOFMS was successfully used for nematode diagnostics (Navas *et al.*, 2002; Perera *et al.*, 2005). Perera *et al.* (2005) used two methods, grinding up nematodes and direct analysis of intact nematodes. After standardization, these methods were applied to analyse the seed-gall nematodes, *Anguina tritici* and *Anguina funesta*, and the root-knot nematode, *Meloidogyne javanica*. Typical protein profiles and diagnostic peaks were identified for these nematode species and for mixtures of *Anguina* spp. (Fig. 9.1). Compared with other biochemical methods, MALDI-TOFMS is a promising technique that enables rapid and reliable identification of plant-parasitic nematodes because sample preparation is simple and analysis is rapid. It could be also automated in terms of sample plate loading, generation of spectra (up to 400-place sample plates), and automated peak calling, leading to automation of species identification (Perera *et al.*, 2005).

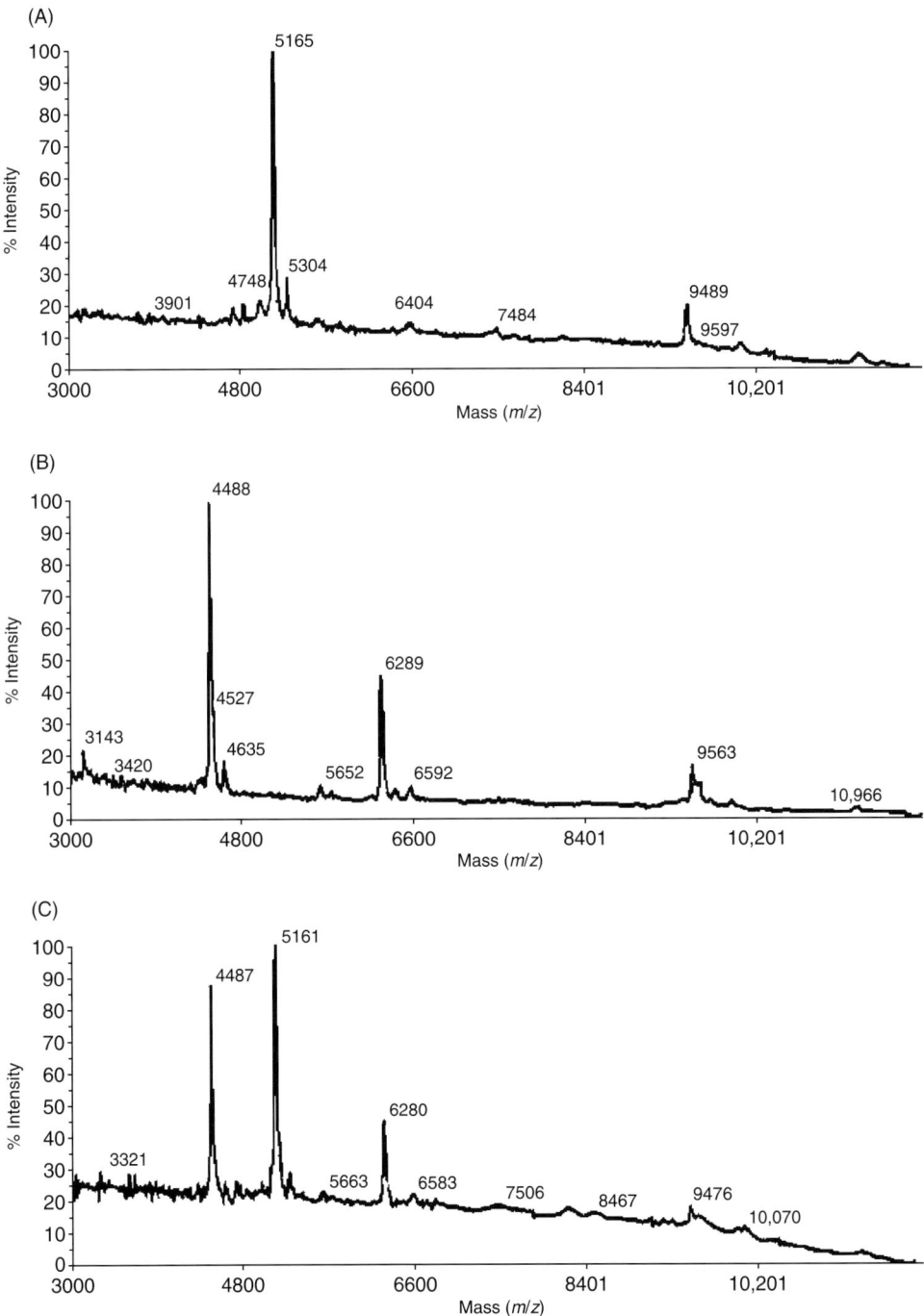

Fig. 9.1. MALDI-TOFMS protein profiles generated by direct analysis of intact nematodes: (A) *Anguina tritici*; (B) *Anguina funesta*; (C) *A. tritici* and *A. funesta* mixture (Perera et al., 2005).

Molecular Diagnostics

The first report of DNA-based techniques to identify plant-parasitic nematodes was published over 20 years ago. This seminal work by Curran et al. (1985) used restriction fragment length polymorphisms (RFLPs) and demonstrated that this technique had greater discriminatory potential than serological and biochemical approaches. Several research groups used DNA probes for identification purposes, including Marshall and Crawford (1987) and Burrows and Perry (1988) who both used probes for PCN, and Palmer et al. (1992) who used probes to identify *Ditylenchus dipsaci*.

Significant progress over the last decade in molecular diagnostics of nematodes has been due to the development and introduction of polymerase chain reaction (PCR). This method enables numerous copies to be obtained from a single or a few molecules of DNA extracted from an organism by chemical synthesis *in vitro*. It has been used extensively to identify species of plant-parasitic nematodes. For example, Ibrahim et al. (1994) used PCR to amplify a fragment of the rDNA array from 12 species and populations of *Aphelenchoides*. RFLPs in the fragment were used successfully to compare and differentiate species and populations. The PCR primers used to amplify the rDNA in this work were based on conserved sequences in the 18S and 26S ribosomal genes of *Caenorhabditis elegans* (Files and Hirsh, 1981) and were first used for work on plant-parasitic nematodes by Vrain et al. (1992), who examined intraspecific rDNA RFLPs in the *Xiphinema americanum* group. The wide application of PCR in diagnostics is a reflection of the advantages of the technique, which is very sensitive, rapid, easy to perform and inexpensive. PCR is used routinely for nematode diagnostics and has been comprehensively reviewed recently by Powers (2004), Blok (2005), Subbotin (2006) and Subbotin and Moens (2006).

Compared with biochemical approaches, molecular diagnostics has several advantages. It does not rely on expressed products and is not influenced by environmental conditions and development stage. Any development stage can be used for diagnosis. It is much more sensitive than any biochemical technique, and can be used with nanograms of DNA extracted from one nematode or even part of a nematode body. It can also be used with various types of samples, such as soil extracts, plant material or formalin-fixed samples.

Ibrahim et al. (2001) conducted an elegant, comparative study estimating the efficiency of detection, identification and quantification of the two PCN species from field soil samples using IEF, ELISA and PCR techniques with standard nematological methods. A greater number of positive results were obtained with PCR with specific primers than with any other method, indicating the greater sensitivity of this method. The results from ELISA did not agree with other methods because of partial cross-reactivity of the two antibodies used. PCR and IEF results can be obtained in 1 day, whereas ELISA results are only available the next day. There were also differences in pricing for sample testing between these methods; the price for IEF testing was significantly higher than that for PCR or ELISA.

Keeping nematodes for molecular studies and DNA extraction

The efficiency of DNA extraction from a sample depends on how nematodes have been prepared and fixed. Various methods of fixation have been proposed and described, but the best approach is to use live nematodes for diagnostics. If the period between nematode extraction and molecular analyses is several days or weeks, nematodes may be kept at low temperatures before use. In some cases, quarantine regulations do not allow live nematodes to be kept and transported, so nematodes should be heated briefly to kill them but leave the DNA undamaged. Often, during long field sampling trips, it is not possible to keep nematodes at low temperatures, so fixation in 75–90% alcohol, glycerol or simply drying the nematodes in a plastic tube are alternative methods to save nematode DNA for further molecular study.

For long-term storage, formalin has been used as a fixative in nematology for many years. It had been assumed that the effects of formalin fixation caused fixed specimens to be unsuitable for DNA analysis, but several methods of DNA extraction from formalin-fixed and glycerine-embedded nematodes stored for days or even years have been tested and have shown promising results (De Giorgi *et al.*, 1994; Thomas *et al.*, 1997; Bhadury *et al.*, 2005; Rubtsova *et al.*, 2005). However, although De Giorgi *et al.* (1994) amplified DNA fragments from fixed nematodes, they reported several artificial mutations in sequences recovered from formalin-fixed nematodes. Artefacts could be the consequence of formalin damaging or cross-linking cytosine nucleotides on either strand, so that DNA polymerase would not recognize them and instead of guanosine incorporate adenosine, thereby creating an artificial C–T or G–A mutation (Williams *et al.*, 1999). In contrast, Thomas *et al.* (1997) and Bhadury *et al.* (2005) did not find ambiguities in sequences obtained from formalin-fixed nematodes after a few days of storage. Rubtsova *et al.* (2005) even reported successful sequencing without ambiguities of a short fragment of the D2 expansion segment of 28S rRNA amplified from *Longidorus* spp. kept in permanent slides for more than 10 years. The development of a successful protocol for DNA extraction and PCR from formalin-fixed and glycerine-embedded nematodes from permanent slides, presently kept in many taxonomic collections in different countries, will provide new opportunities to analyse rare species with limited distribution and, potentially, enable many diagnostic problems in nematology to be solved.

A solution called DESS, which contains dimethyl sulphoxide, disodium ethylene diamine tetraacetic acid (EDTA) and saturated NaCl, has been shown to preserve nematode morphology equally as well as formalin fixation and allowed PCR to be performed on individual nematodes (Yoder *et al.*, 2006). In the future, this may be the solution of choice for nematode preservation.

Antibody approach for sample enrichment

The quantity and quality of DNA is very important for successful diagnosis. In many cases, one specimen, either adult, juvenile or egg, or even part of a nematode might be enough for molecular identification. However, for greater reliability, the use of several specimens of the target nematode is always preferable. Detection of plant-parasitic nematodes in samples is a difficult task due to their microscopic size and uneven dispersal in the soil. A method to enrich nematode extracts from soil samples was proposed and developed by Chen et al. (2001, 2003) using an antibody-based capture system (Fig. 9.2). In this method, an antibody which recognizes the surface of target nematodes is incubated with a nematode suspension extracted from a field sample. Then, magnetic beads coated with the secondary antibody are added and a magnet is used to capture target nematodes while other nematodes are discarded (Chen et al., 2003). The immunomagnetic capture system has been shown to be effective for the enrichment of *Meloidogyne* spp., *X. americanum* and *G. rostochiensis* from total nematode extracts from soil, with up to 80% of the target nematode being recovered (Chen et al., 2001, 2003). The antibody-based capture system is an effective method of detecting specific nematodes in mixed soil samples and results in samples containing target nematodes in large numbers, which are suitable for further diagnostics techniques (Chen et al., 2003).

DNA extraction

Using proteinase K is the most useful, cheap and rapid approach to extract DNA from nematodes (Waeyenberge et al., 2000). It consists of two steps: (i) mechanical destruction of nematode body and tissues in a tube using ultrasonic homogenizer or other tools, or repeatedly freezing samples in liquid nitrogen; and (ii) chemical lyses with proteinase K in a buffer for 1 h or several hours with subsequent brief inactivation of this enzyme at high temperature. Chelex resin protocols can be also successfully applied for DNA extraction from nematodes (Walsh et al., 1991). Various chemical treatments are also applied to remove cell components and purify the DNA. Phenol or phenol with chloroform extractions is often employed to remove proteins and ethanol is then used to precipitate and concentrate the DNA. Stanton et al. (1998) described an efficient method of DNA extraction from nematodes using chemical lyses in alkali solution without prior mechanical breaking of nematode bodies. Effective DNA extraction can be achieved by using commercial kits developed by different companies. These approaches rely on DNA binding to silica in the presence of a high concentration of chaotropic salt (Boom et al., 1990).

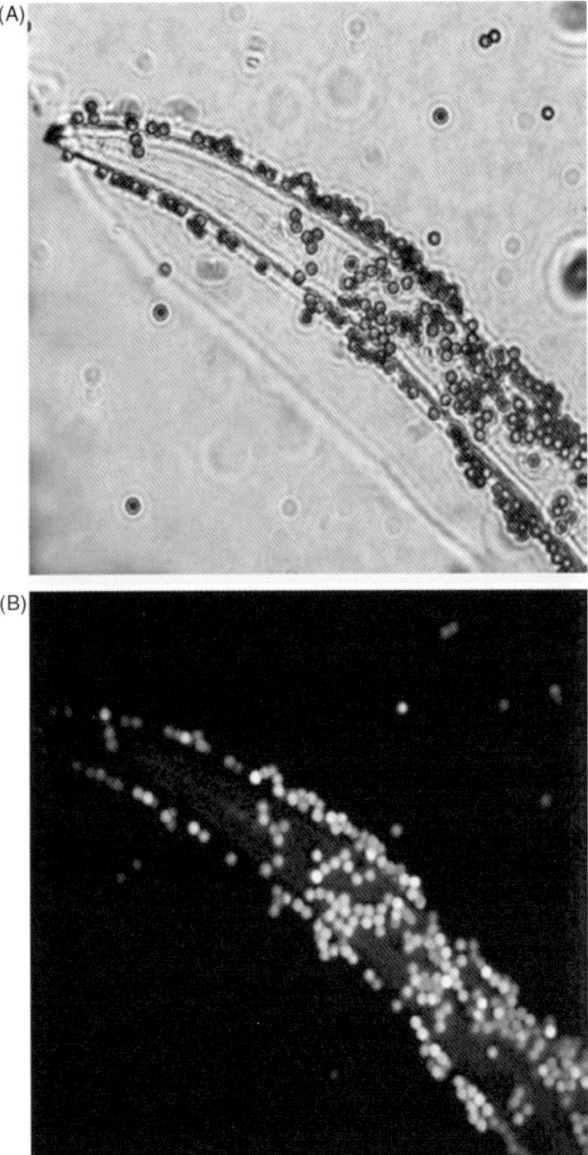

Fig. 9.2. *Meloidogyne arenaria* second-stage juveniles (J2) captured using immunomagnetic capture. J2 are coated with Dynabeads. (A) Bright field image; (B) fluorescence optics (Chen *et al.*, 2001).

Whole genomic amplification

The problem of being able to obtain only limited DNA from nematode samples might also be solved using whole genomic amplification (WGA) approaches. Using these methods, it is possible to generate microgram quantities of DNA starting with as little as a few nanograms of genomic DNA from a single nematode specimen or even part of nematode body. Multiple-displacement amplification (MDA) is a relatively novel technique for WGA. It uses the highly processive Phi29 DNA polymerase and random exonuclease-resistant primers in an isothermal amplification reaction (Dean et al., 2001). Successful amplifications of MDA have been reported using nematode DNA from species of *Bursaphelenchus* and *Meloidogyne*. Application of this method to nematode samples significantly improved efficiency of amplification of ribosomal and protein coding genes (Metge and Burgermeister, 2005; Skantar and Carta, 2005). Skantar and Carta (2005) considered that the potential applications of MDA to nematode identification are far-reaching. Using MDA, it should be possible to archive genetic material from individual nematodes, thereby eliminating the need for more labour-intensive culture methods. MDA could facilitate the development and production of DNA 'type species' that may be shared among scientists, or enable large quantities of genetic material from rare specimens to be archived. Genome amplification is also tolerant of sample degradation and might usefully be applied to formalin-fixed nematode specimens.

DNA sequence targets for diagnostics

There are two main approaches to select target DNA sequences for diagnostic purposes: (i) to use known conserved genes, common to all nematode species, and to explore the specific sequence variation in order to distinguish species; and (ii) to randomly screen the whole genome and find specific DNA fragments that could be used as markers for diagnostics. At present, the first approach is more widely used for nematode diagnostics. The main region targeted for this diagnostic development is nuclear ribosomal RNA genes, especially the internal transcribed spacers 1 (ITS1) and 2 (ITS2), which are situated between 18S and 5.8S, and 5.8S and 28S rRNA genes, respectively. The choice of these genes is partly historical, because they were the first to be characterized in nematodes, and partly due to advantages in methodology, because these genes are present in a cell in many copies and, thus, can be amplified relatively easily from a small sample. Ribosomal genes and their spacers have undergone different mutation rates, and this enables different regions to be used for diagnostics at a higher taxonomic level, such as family and genus, down to species, subspecies or even population levels. Modern diagnostics of nematodes from the genera *Heterodera, Globodera, Bursaphelenchus, Pratylenchus, Anguina, Ditylenchus, Nacobbus* and *Radopholus* are based on nucleotide polymorphisms in sequences of the ITSs. To distinguish most species of root-knot nematodes, the intergenetic spacer (IGS) of nuclear rRNA (Petersen and Vrain, 1996), which is between 28S and 18S rRNA genes, and the intergenic spacer of mitochondrial DNA (Powers and Harris, 1993; Powers et al., 2005), which is between the 5' portion of cytochrome oxidase subunit II and large ribosomal rRNA genes, are used in addition to the ITS-rRNA. Species of root-knot nematodes can be separated based on the length as well as on the nucleotide polymorphism of the amplified fragments, when amplified by PCR primers in the flanking genes (Powers and Harris, 1993; Petersen and Vrain, 1996).

Sequence analyses of the 18S rRNA gene (Floyd *et al.*, 2002) and the D2–D3 expansion segments of the 28S rRNA for many tylenchid nematodes (Subbotin *et al.*, 2006) and longidorids (Rubtsova *et al.*, 2005; He *et al.*, 2005) have revealed that these genes are also reliable diagnostic targets at the species level. Other genes that are increasingly being used for diagnostic purposes include the major sperm protein (Setterquist *et al.*, 1996), heat shock Hsp90 (Skantar and Carta, 2004) and actin (Kovaleva *et al.*, 2005) genes. It is evident that recent progress in nematode genome sequencing and expressed sequence tag (EST) projects (Scholl and Bird, 2005) will give more promising and reliable gene candidates for diagnostic developments.

Targets for development of a diagnostic method also can be identified by screening random regions of the genome to find DNA fragment sequences that are unique for a particular taxon. This can be done using PCR-based techniques, such as random amplified polymorphic DNA (RAPD) or amplified fragment length polymorphism (AFLP), which provide randomly generated fragments from the genome. These fragments are separated by gel electrophoresis and the patterns are compared for different taxa. Potential diagnostic and unique bands are extracted, cloned and sequenced and then used to design specific sequence-characterized amplified region (SCAR) primers (Zijlstra, 2000).

Satellite DNA is present in the genome of almost all eukaryotic organisms and is composed of highly repetitive sequences organized as long arrays of tandemly repeated elements. Satellite DNA sequences have been characterized from a number of plant-parasitic nematodes and have proven to be species-specific, thus constituting useful diagnostic tools for identification of species of agronomic interest (Grenier *et al.*, 1997; Castagnone-Sereno *et al.*, 1999; He *et al.*, 2003).

PCR-based methods

PCR is the most important technique in diagnostics. Currently, there are several PCR-based methods used for nematode diagnostics: PCR-RFLP, PCR with specific primers, PCR–single strand conformation polymorphism (SSCP) and real-time PCR.

PCR-RFLP

This method is used to diagnose monospecific samples and contains two steps: (i) amplification of gene marker region using universal primers (Table 9.1); and (ii) digestion of the resulting amplicon by restriction enzymes. PCR-based techniques have been used for diagnostics of many

Table 9.1. Universal primers useful for nematode diagnostics.

Code	Sequence (5'–3')	Target DNA region	References
C2F3 1108	GGT CAA TGT TCA GAA ATT TGT GG TAC CTT TGA CCA ATC ACG CT	Intergenic region between COII and 16S rRNA gene of mtDNA	Powers and Harris (1993)
18S rDNA1.58S	TTG ATT ACG TCC CTG CCC TTT GCC ACC TAG TGA GCC GCG CA	ITS1 of rRNA	Szalanski et al. (1997)
18S 26S	TTG ATT ACG TCC CTG CCC TTT TTT CAC TCG CCG TTA CTA AGG	ITS1-5.8S-ITS2 of rRNA	Vrain et al. (1992)
F194 F195	CGT AAC AAG GTA GCT GTA G TCC TCC GCT AAA TGA TAT G	ITS1-5.8S-ITS2 of rRNA	Ferris et al. (1993)
D2A D3B	ACA AGT ACC GTG AGG GAA AGT TG TCG GAA GGA ACC AGC TAC TA	D2-D3 region of 28S rRNA	De Ley et al. (1999)
TW81 AB28	GTT TCC GTA GGT GAA CCT GC ATA TGC TTA AGT TCA GCG GGT	ITS1-5.8S-ITS2 of rRNA	Joyce et al. (1994)
SSU18A SSU26R	AAAGATTAAGCCATGCATG CATTCTTGGCAAATGCTTTCG	18S rRNA	Blaxter et al. (1998)
Nem_18S_F Nem_18S_R	CGCGAATRGCTCATTACAACAGC GGGCGGTATCTGATCGCC	18S rRNA	Floyd et al. (2005)
18sl.2 18srb:	GGCGATCAGATACCGCCCTAGTT TACAAAGGGCAGGGACGTAAT	18S rRNA	Powers et al. (2005)

nematode groups (Table 9.2); here, we focus mainly on the applications for diagnosis of root-knot and cyst nematodes.

For diagnostics of root-knot nematodes, several gene markers, including intergenic region of mtDNA, nuclear ITS-rRNA, intergenic region of rRNA, 18S rRNA and D2–D3 expansion segment of 28S rRNA gene, have been developed and used. A detailed description of the discrimination of species of root-knot nematodes using PCR of the intergenic region located between the cytochrome oxidase subunit II (COII) and 16S rRNA (lRNA) of mtDNA has been provided by Powers and Harris (1993). It has been shown that the amplification length and RFLP of this fragment enable the separation of five major *Meloidogyne* spp. Subsequently, this region was studied intensively in several root-knot nematode species using PCR-RFLP and sequencing techniques (Hugall et al., 1994, 1997; Williamson et al., 1997; Orui, 1998; Stanton et al., 1998; Blok et al., 2002; Tigano et al., 2005). Powers et al. (2005) described a modified, detailed procedure for PCR-RFLP identi-

Table 9.2. PCR-RFLP assays used for diagnosis of nematodes.

Nematode genera	Target DNA sequence	References
Heterodera	ITS of rRNA	Bekal *et al.* (1997); Szalanski *et al.* (1997); Orui (1997); Subbotin *et al.* (2000a); Tanha Maafi *et al.* (2003)
Globodera	ITS of rRNA	Fleming and Mowat (1993); Thiéry and Mugniéry (1996); Subbotin *et al.* (1999, 2000b)
Bursaphelenchus	ITS of rRNA	Hoyer *et al.* (1998); Iwahori *et al.* (1998); Braasch *et al.* (2001); Burgermeister *et al.* (2005)
Ditylenchus	ITS of rRNA	Wendt *et al.* (1993)
Aphelenchoides	ITS of rRNA	Ibrahim *et al.* (1994)
Pratylenchus	ITS of rRNA	Orui (1995); Orui and Mizukubo (1999); Waeyenberge *et al.* (2000)
Xiphinema	ITS of rRNA	Vrain *et al.* (1992)
Radopholus	ITS of rRNA	Elbadri *et al.* (2002)
Anguina	ITS of rRNA	Powers *et al.* (2001)
Meloidogyne	MtDNA	Powers and Harris (1993); Orui (1998); Blok *et al.* (2002); Xu *et al.* (2004); Powers *et al.* (2005)
Meloidogyne	ITS of rRNA	Zijlstra *et al.* (1995); Schmitz *et al.* (1998)
Nacobbus	ITS of rRNA	Reid *et al.* (2003)

fication of *Meloidogyne* spp. using two marker regions, a fragment of mitochondrial DNA located between the COII and 16S rRNA (Powers and Harris, 1993) and a partly amplified 18S ribosomal rRNA gene. The mitochondrial gene region provided greater species discrimination and revealed intraspecific variation among many isolates. The samples were amplified by C2F3/1108 primer set (Table 9.1), and the size of individual amplification products was determined in subsequent assays (Fig. 9.3).

Zijlstra *et al.* (1995) were the first to use the amplified ITS region of nuclear rRNA for discrimination of *M. hapla*, *M. chitwoodi* and *M. fallax* from each other and from *M. incognita* and *M. javanica* by using several restriction enzymes. Using another set of primers, Schmitz *et al.* (1998) proposed the simultaneous use of two restriction enzymes, *Hinf*I/*Rsa*I or *Dra*I/*Rsa*I, for identification of several species of root-knot nematodes. However, studies based on these techniques did not reveal diagnostic enzymes that facilitated the separation of *M. incognita* and *M. javanica* (Xue *et al.*, 1993; Zijlstra *et al.*, 1995; Schmitz *et al.*, 1998). Subsequent sequencing of the ITS-rRNA regions of *M. incognita*, *M. javanica* and *M. arenaria* revealed that these sequences were almost identical (Hugall *et al.*, 1999). Similar sequences were found in several other closely related species, including *M. hispanica*, *M. morocciensis* (De Ley *et al.*, 1999) and *M. mayaguensis* (Brito *et al.*, 2004). Thus, unlike mtDNA, using the ITS-rRNA does not enable

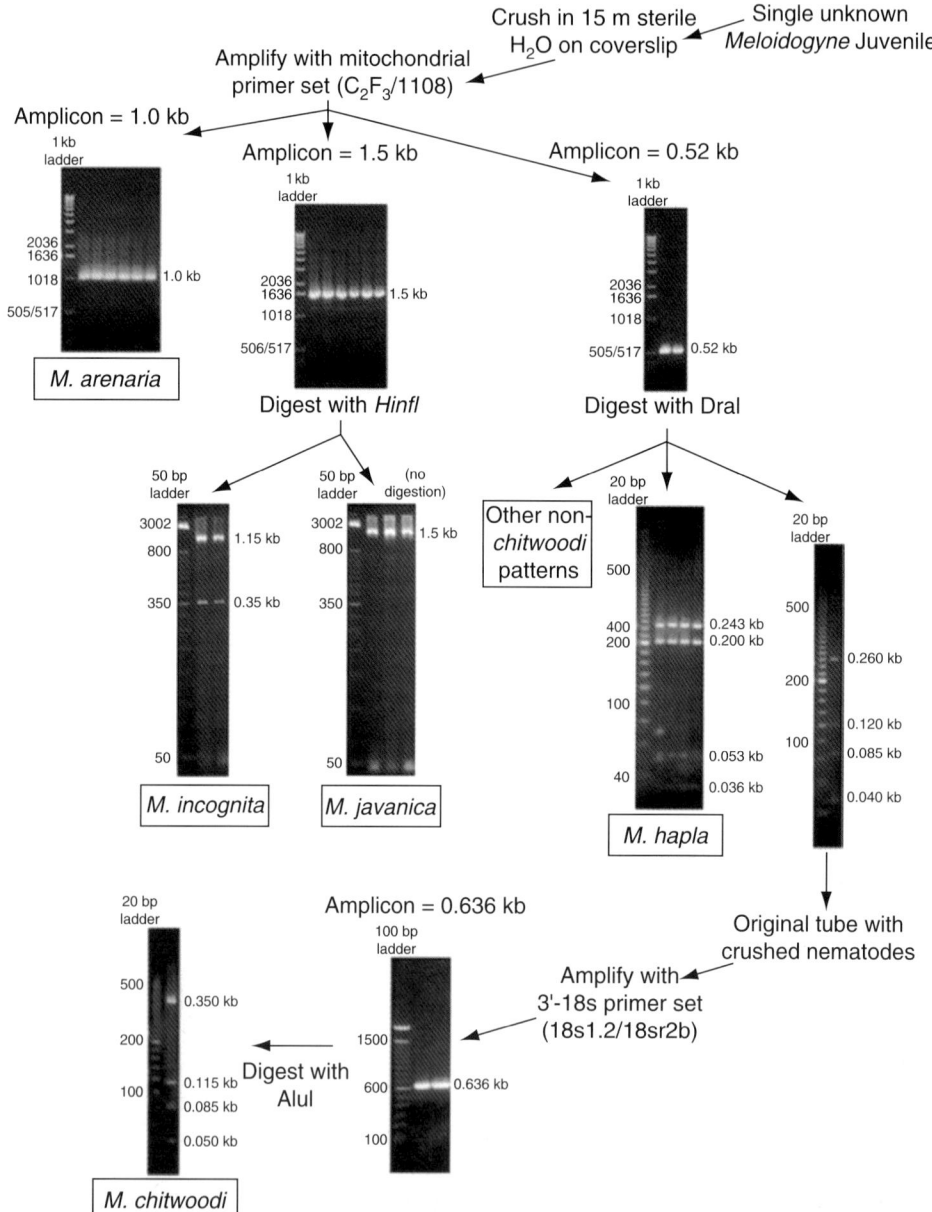

Fig. 9.3. Flow chart depicting the PCR-RFLP steps to identify *Meloidogyne* spp. (Powers et al., 2005).

several agricultural important species of root-knot nematodes to be distinguished.

After sequencing nuclear intergenic rRNA spacer in *M. arenaria*, Vahidi and Honda (1991) found that it contained 5S-rRNA gene dividing the intergenic spacer into two spacers, one of them having nucleotide repeats. In one species, the numbers of such repeats was variable, even within the genome of a single individual, whereas in some other species, the length of intergenic rRNA spacers was constant and could be used for species discrimination. Consequently, Petersen and Vrain (1996) proposed that the length of this spacer could be used to diagnose *M. chitwoodi*, *M. hapla* and *M. fallax*. Diagnostics of cyst nematodes is based on interspecific variation of the ITS-rRNA gene. After sequencing the rRNA gene from several nematodes, Ferris *et al.* (1993, 1994) suggested that the ITS region was conserved for populations of the same species and was sufficiently variable between species to distinguish among species of cyst nematodes. Subsequently, Szalanski *et al.* (1997) successfully applied PCR-RFLP of the ITS1 to differentiate *H. cruciferae*, *H. glycines*, *H. trifolii*, *H. schachtii*, *H. goettingiana* and *H. zeae*. This approach was also used to discriminate species from the Avenae group (Bekal *et al.*, 1997) and other cyst nematodes (Orui, 1997; Fleming *et al.*, 1998).

A comprehensive analysis of PCR-RFLP of the ITS-rRNA of cyst nematodes has been conducted by Subbotin and co-authors and published in a series of articles (Subbotin *et al.*, 1997, 1999, 2000a, 2003; Eroshenko *et al.*, 2001; Amiri *et al.*, 2002; Tanha Maafi *et al.*, 2003; Madani *et al.*, 2004). In these studies, universal TW81 and AB28 primers were used, which generated an amplicon of 1060bp length, although this could vary in different species. Application of the RFLP technique to separate 26 species of cyst nematodes revealed several restriction enzymes, *Cfo*I, *Alu*I, *Bsu*RI, *Bsh*1236I and *Scr*FI, with greater discriminatory powers than other enzymes and which enabled more species of cyst nematodes to be characterized (Fig. 9.4).

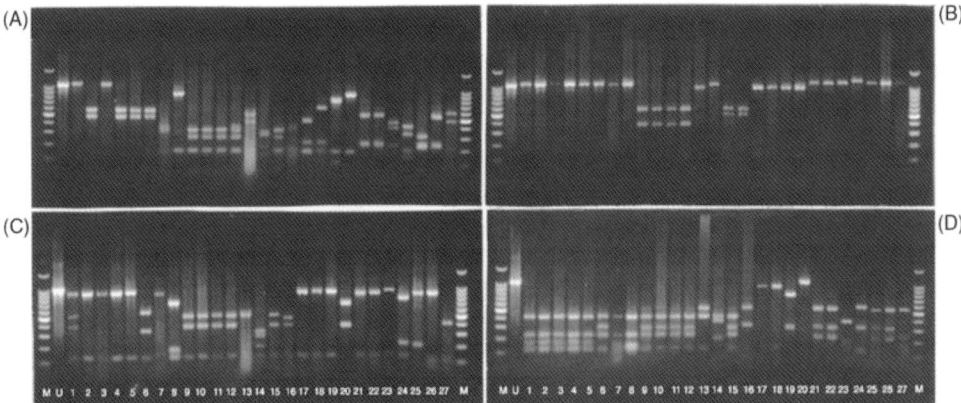

Fig. 9.4. Restriction fragments of amplified ITS regions of cyst-forming nematodes using TW81 and AB28 primers. Restriction enzymes: (A) *Alu*I; (B) *Ava*I; (C) *Bsh*1236I; (D) *Bsi*ZI. Lines: M – 100bp DNA ladder (Promega, USA); U – unrestricted PCR product; 1 and 2 – *Heterodera avenae*, 3 – *H. arenaria*, 4 – *H. filipjevi*, 5 – *H. aucklandica*, 6 – *H. ustinovi*, 7 – *H. latipons*, 8 – *H. hordecalis*, 9 – *H. schachtii*, 10 – *H. trifolii*, 11 – *H. medicaginis*, 12 – *H. ciceri*, 13 – *H. salixophila*, 14 – *H. oryzicola*, 15 – *H. glycines*, 16 – *H. cajani*, 17 – *H. humuli*, 18 – *H. ripae*, 19 – *H. fici*, 20 – *H. litoralis*, 21 – *H. carotae*, 22 – *H. cruciferae*, 23 – *H. cardiolata*, 24 – *H. cyperi*, 25 – *H. goettingiana*, 26 – *H. urticae*, 27 – *Meloidodera alni* (Subbotin *et al.*, 2000a).

For example, *Cfo*I produced 16 polymorphic profiles enabling differentiation of 12 out of 26 species. Some enzymes produced species-specific profiles; for example, *Mva*I produced a specific profile for *H. schachtii*. Restriction of PCR product by seven enzymes usually enabled the most important species of cyst nematodes of the genus *Heterodera* to be distinguished from each other and from their sibling species (Fig. 9.4) (Subbotin *et al.*, 2000a). RFLP of the ITS region has been used to differentiate species of the genus *Globodera* parasitizing solanaceous plants (Fleming and Mowat, 1993; Szalanski *et al.*, 1997; Thiéry and Mugniéry, 1996; Fleming and Powers, 1998). For example, the two species of PCN could be distinguished by at least eight restriction enzymes (Subbotin *et al.*, 2000b).

PCR with specific primers
This type of PCR constitutes a new step in the development of DNA diagnostics and enables the detection of one or several species in a nematode mixture by a single PCR test, thus decreasing diagnostic time and costs. Diagnostics using PCR with specific primers has been developed for a wide range of plant-parasitic nematodes (Table 9.3). In multiplex PCR, several primer sets are utilized for amplification of several target genes. Such an approach has been developed for diagnostics of *H. glycines* (Fig. 9.5) (Subbotin *et al.*, 2001), *H. schachtii* (Amiri *et al.*, 2002) and *Bursaphelenchus xylophilus* (Jiang *et al.*, 2005). Multiplex PCR for detection of these species

Table 9.3. PCR with specific primer assays used for diagnosis of nematodes.

Nematode species and genera	References
Heterodera glycines	Subbotin *et al.* (2001)
Heterodera schachtii	Amiri *et al.* (2002)
Globodera rostochiensis and *G. pallida*	Mulholland *et al.* (1996); Bulman and Marshall (1997); Fullaondo *et al.* (1999)
Bursaphelenchus xylophilus	Liao *et al.* (2001); Matsunaga and Togashi (2004); Kang *et al.* (2004); Jiang *et al.* (2005); Leal *et al.* (2005); Castagnone *et al.* (2005)
Ditylenchus dipsaci	Esquibet *et al.* (2003); Subbotin *et al.* (2005)
Pratylenchus spp.	Uehara *et al.* (1998); Al-Banna *et al.* (2004)
Xiphinema spp.	Hübschen *et al.* (2004a)
Longidorus spp.	Hübschen *et al.* (2004b)
Paratrichodorus and *Trichodorus*	Boutsika *et al.* (2004)
Meloidogyne incognita	Zijlstra (2000); Dong *et al.* (2001); Randing *et al.* (2002)
Meloidogyne hapla	Williamson *et al.* (1997); Zijlstra (1997, 2000); Dong *et al.* (2001)
Meloidogyne chitwoodi	Williamson *et al.* (1997); Petersen *et al.* (1997); Zijlstra (2000)
Meloidogyne javanica	Zijlstra *et al.* (2000); Dong *et al.* (2001)
Meloidogyne arenaria	Zijlstra *et al.* (2000); Dong *et al.* (2001)
Meloidogyne naasi	Zijlstra *et al.* (2004)
Nacobbus spp.	Atkins *et al.* (2005)

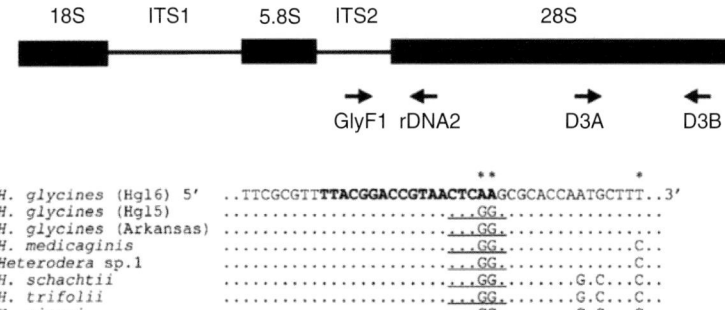

Fig. 9.5. Schematic representation showing positions of four primers in the rRNA genes used in the duplex PCR for identification of *Heterodera glycines* and ITS2 fragment alignment used for development of the species-specific primer (GlyF1), which is indicated in bold (Subbotin et al., 2001).

includes two sets of primers: (i) the first is to amplify an internal control (e.g. universal primers for the expansion regions of 28S rRNA gene) confirming the presence of DNA in the sample and the success of PCR; and (ii) the second, including species-specific primer, is targeted to nematode DNA sequence of interest (Fig. 9.6). Nested PCR in a multiplex assay with specific SCAR primers has been developed for the diagnosis of several species of root-knot nematodes (Zijlstra, 2000).

PCR-SSCP
The PCR-SSCP technique can detect single nucleotide changes in the DNA fragments being studied due to the altered conformation mobility of the single strands of DNA during electrophoresis with a non-denaturing gel. Clapp *et al.* (2000) tested the PCR-SSCP method at two coastal dune locations in the Netherlands to determine the population structure of cyst nematodes. The ITS2 region was sufficiently variable within the taxa investigated to allow species to be separated on the basis of minor sequence variation. The PCR primers used in this study were effective for several species of cyst and root-knot nematodes.

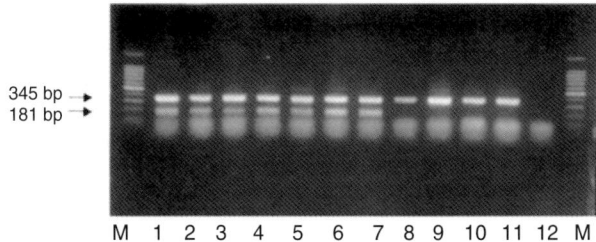

Fig. 9.6. Gel with PCR products (181 and 345 bp) generated by two sets of primers (GluyF1 + rDNA2 and D3A + D3B), respectively. Line: M – 100 bp DNA ladder (Promega, USA); 1–7: *Heterodera glycines*, 8: *H. schachtii*; 9: *H. ciceri*; 10: *H. medicaginis*; 11: *H. cajani*; 12: sample without nematode DNA (Subbotin et al., 2001).

Real-time PCR

The PCR technique also permits the estimation of the number of nematodes in a sample. This has been successful only for nematodes of the same stage, e.g. second-stage juveniles of cyst nematodes and fourth-stage juveniles of *D. dipsaci*. It may be more difficult when trying to quantify nematodes of different stages. This technique indirectly measures the nematode number by assuming that the number of target DNA copies in the sample is proportional to the number of targeted nematodes. The real-time technique allows continuous monitoring of the sample during PCR using hybridization probes (TaqMan, molecular beacons) or dyes such as SYBR Green I. SYBR Green binds only to double-stranded DNA and becomes fluorescent only when bound. This dye has the virtue of being easy to use because it has no sequence specificity and it can be used to detect any PCR product. However, this virtue has a drawback, as the dye binds also to any non-specific product, including primer dimers. To overcome this problem, the melting curve analysis can be employed. The products of PCR reaction are melted by increasing the temperature of the sample. The non-specific product tends to melt at a much lower temperature than the longer specific product. Both the shape and position of the DNA melting curve area are a function of the GC/AT ratio. The length of amplicon can be used to differentiate amplification products separated by less than 2°C in T_m (the melting temperature) (Fig. 9.7).

Bates *et al.* (2002) were the first to use real-time PCR with SYBR Green I with plant-parasitic nematodes, when they used this technique to detect PCN. These authors found relatively large (4°C) differences in T_m between specific PCR products of *G. rostochiensis* and *G. pallida* amplified using the species-specific primers of Bulman and Marshall (1997). These differences ensure that melting peaks of these two products can be clearly distinguished in a multiplex reaction. By calculating the ratio of the melting peak height at T_m of each product and comparing it to the standard run under the same conditions, it was even possible to estimate the proportion of each product in the nematode mixture and finally to determine the ratio of juveniles of these two species in the sample. However, this method has not been applicable for absolute quantification of these nematodes. After testing, Madani *et al.* (2005) improved this approach and described methods for quantification of *G. pallida* and *H. schachtii* juveniles in a sample using the set with universal and species-specific primers. Validation tests showed a high correlation between real numbers of second-stage juveniles in a sample and the expected numbers detected by real-time PCR. A method for quantification using real-time PCR with SYBR Green I was also developed for estimation of the numbers of juveniles of the stem nematode, *D. dipsaci*, in soil samples (Subbotin *et al.*, 2005). All these methods showed a high level of specificity and sensitivity and enabled the detection of a single nematode in a sample. The background of other soil-inhabiting nematodes present in tested samples did not significantly compromise the accuracy of these assays. However, in most cases, accuracy of the quantification decreased with decreasing nematode numbers in the samples.

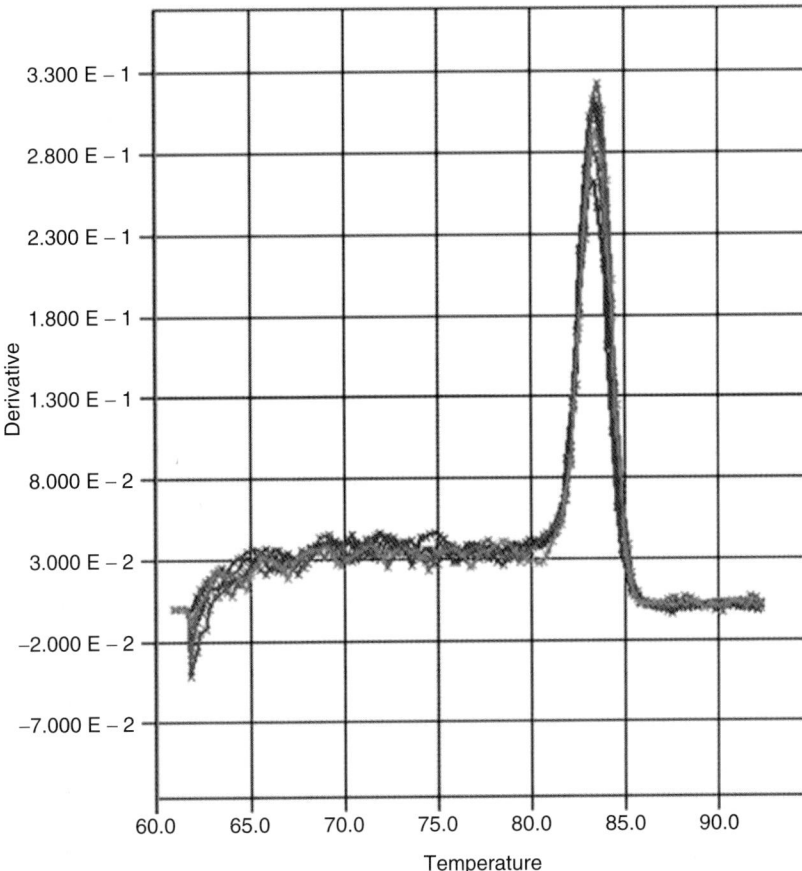

Fig. 9.7. Melting curve (fluorescence vs temperature) of specific amplicon for *Globodera pallida* (Madani et al., 2005).

The disadvantage of using a fluorescent dye is that it binds to any double-stranded DNA and then it cannot be used for quantification of several targets in a multiplex real-time PCR, because it cannot distinguish between different sequences. In this case, sequence-specific fluorescent probes, such as TaqMan probes, are needed. Cao et al. (2005) developed a method for detecting the pinewood nematode, *B. xylophilus*, using TaqMan probes. The PCR assay detected DNA template concentrations as low as 0.01 ng. The Ct values were correlated with the DNA template concentration ($R^2 = 0.996$), indicating the validity of the assay and its potential for quantification of target DNA. The real-time PCR assay also detected DNA from single specimens of *B. xylophilus*. Recently, Holeva et al. (2006) described a novel diagnostic method for two virus–vector nematodes, *Paratrichodorus pachydermus* and *Trichodorus similis*, and associated tobacco rattle virus based on TaqMan probes. They demonstrated the potential of this assay for rapid, accurate and sensitive detection of both

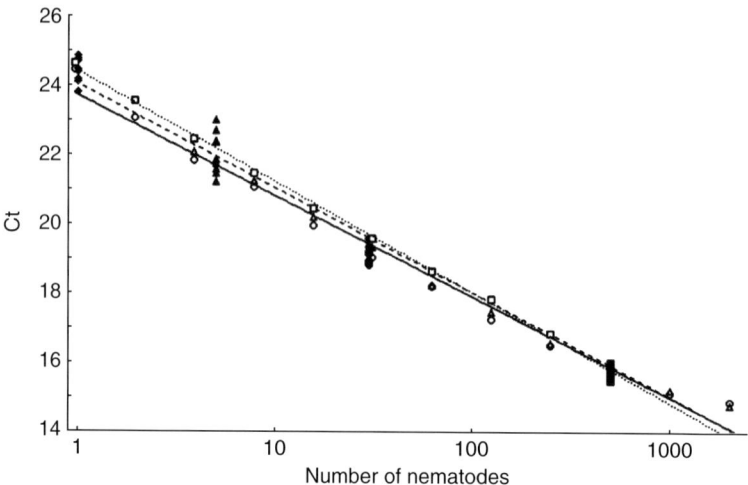

Fig. 9.8. Sensitivity, detection range and specificity of the real-time PCR assay for *Globodera pallida*. The reproducibility of the assay was determined by testing a dilution series of three independent DNA extractions. Black symbols represent the results of the validation test with known numbers of second-stage juveniles (1, 5, 30 or 500 juveniles) (Madani et al., 2005).

trichodorid species and virus from field samples. The real-time PCR method is straightforward, sensitive and reproducible and, compared with conventional PCR methods, it has several advantages (Fig. 9.8). The technique allows a simultaneous faster detection and quantification of target DNA and the automated system overcomes the laborious process of estimating the quality of PCR product after electrophoresis.

Problems with PCR-based techniques

Over many years, the application of various PCR methods with universal and specific primers for diagnostic purposes has revealed several problems. First, PCR amplifies DNA from live and dead specimens. This compromises the use of this method for the estimation of the efficiency of pesticide applications on nematode populations. Use of another reverse transcriptase (RT)-PCR technique can solve this problem. Using this approach, mRNA is extracted from live nematodes and then the RT converts RNA to cDNA, which is subsequently amplified and also could be quantified. The second limitation of this method is the probability of a *false-positive reaction*. Primer design is always based on existing knowledge of DNA sequences for target species and closely related nematodes. However, there is a possibility that similar fragments can be obtained for another previously non-investigated nematode as well as for the target species. The third limitation, which is the opposite of second one, is a probability of a *false-negative reaction*. Although a region with a conserved sequence should be used for primer design, the possibility cannot be excluded that some mutations occurred in this region in some specimens or populations of the target taxa. As a result, they might become non-detectable by the PCR test. The fourth limitation is a probability of sample contamination, which might occur during sample preparation and might give a false-positive reaction due to the great sensitivity of the PCR method. Following strict rules to prevent contamination during preparation of the PCR mixture is imperative for all diagnostics tests.

These limitations indicate that, first, the PCR technique should be used intelligently and the researcher or advisor must be aware of the peculiarities of each method. Second, when there is any doubt, it is always necessary to confirm identification by several methods, including the use of traditional morphological features.

Hybridization DNA arrays

DNA arrays provide a powerful method for the next generation of diagnostics. The distinct advantage of this approach is that it combines DNA amplification with subsequent hybridization to oligonucleotide probes specific for multiple target sequences. DNA arrays can be used to detect many nematode species based on differences in ribosomal RNA gene. For example, nucleic acids can be extracted from a sample and then rRNA gene fragments amplified by PCR using sets of universal primers. The resulting PCR products can be hybridized to an array consisting of many oligonucleotide probes, which are designed to detect nematodes by genus or species and are based on discriminatory sequences.

In general, arrays are described as macroarrays or microarrays, the difference being the size and density of the sample spots, the substrate of hybridization and the type of production. Although the potential of DNA array methods for nematological diagnostics has been recognized (Blok, 2005; Subbotin and Moens, 2006), little progress had been made in their use, and only few research papers have been published on this technique. A reverse dot-blot assay has been developed for identification of seven *Pratylenchus* spp. using oligonucleotides designed from the sequences of the ITS region of rRNA (Uehara *et al.*, 1999). Recently, François *et al.* (2006) were successful in developing a DNA oligonucleotide microarray for identification of *M. chitwoodi* using two types of probes designed from SCAR and satellite DNA sequences.

DNA barcoding

The bar-coding technique is based on the idea that a particular nucleotide sequence from a common gene can serve as a unique identifier for every species, and a single piece of DNA can identify all life forms on earth.

Floyd *et al.* (2002) were the first to develop a 'molecular operation taxonomic unit' approach when they applied a molecular barcode, derived from single specimen PCR and sequencing of the 5' segment of the 18S-rRNA gene, to estimate nematode diversity in Scottish grassland.

The cytochrome c oxidase subunit I (COI) of mtDNA is emerging as the standard barcode for many animals. It is nearly 648 nucleotide pairs long in most groups. Mitochondrial DNA evolves much more quickly and contains more differences than the ribosomal gene or its spacer, making mtDNA more useful for distinguishing closely related species. The COI gene is not well characterized yet for plant-parasitic nematodes, except for a few genera; however, Blouin (2002) found that mtDNA sequence variation among individuals of the same animal-parasitic nematode species averages from a fraction of a percent up to 2%, and the maximum difference observed between a pair of individuals that were clearly members of the same interbreeding population was 6%. MtDNA sequence difference between closely related species is typically in the range of 10–20%, so if two individuals differ by about 10% or more, one might question whether they really are conspecific (Blouin *et al.*, 1998).

A promising approach to standardize nematode identification using DNA bar cording is to characterize not one but two, or even more, gene regions, which must fit three following criteria: (i) show significant species-level genetic variability and divergence; (ii) be an appropriately short sequence length so as to facilitate DNA extraction and amplification; and (iii) contain conserved flanking sites for developing universal primers. Several DNA regions, such as 18S rRNA (Floyd *et al.*, 2002), D2–D3 expansion segments of 28S rRNA (De Ley *et al.*, 2005; Subbotin *et al.*, 2006) or the ITS-RNA gene (De Ley *et al.*, 2005), have been proposed for such a procedure.

The ITS-rRNA region is more precisely characterized for many groups of plant-parasitic nematodes than any other gene fragment. The intraspecific variation for the ITS-rRNA sequence gene varies between nematode genera. For example, it typically does not exceed 1.3% for *D. dipsaci sensu stricto* (Subbotin *et al.*, 2005), 1.4% for *Heterodera* spp. (Tanha Maafi *et al.*, 2003; Madani *et al.*, 2004) or 1.8% for *Globodera* spp. (Subbotin *et al.*, 2000b). Observed differences greater than these values between a sample and a standard cast doubts about its co-specificity. Recent detailed sequence and phylogenetic studies indicated the presence of several sibling species that probably exist within currently defined species of plant-parasitic nematodes, such as *G. pallida* (Subbotin *et al.*, 2000a,b), *H. avenae* (Subbotin *et al.*, 2003), *D. dipsaci* (Subbotin *et al.*, 2005), *B. xylophilus* and *B. mucronatus* (Zheng *et al.*, 2003). However, in some cases, the ITS-rRNA does show differences between sequences and does not distinguish among some closely related and recently diverged species, such as *M. incognita*, *M. javanica* and *M. arenaria* (Hugall *et al.*, 1999) or *H. avenae* and *H. arenaria* (Subbotin *et al.*, 2003).

DNA barcoding still causes a spirited reaction from many scientists. There are some potential limitations to barcoding, which relate to problems such as the presence of groups of organisms with little sequence diversity, a lack of resolution of recently diverged species, identification of hybrids and possible amplification of nuclear pseudogenes. Although all of these problems still exist, exploratory studies have shown that about 96% of eukaryotic species surveyed can be detected with barcoding, although most of these would also be resolvable with traditional means; however, the remaining 4% pose problems and can lead to error rates that are unacceptably high (up to 31% of false attributions) when relying on DNA barcoding alone. A further problem faced by the biologists who are trying to identify nematodes using barcoding is that currently there is insufficient information in databases for extensive nematode species identification based on DNA fragments. However, the increasing deposition of DNA sequences in public databases such as the GenBank and NemATOL will be beneficial for diagnostics (Powers, 2004).

Future Developments and Directions

It is clear from the work over the last 20 years that molecular techniques are powerful tools for nematode diagnosis. By using them, scientists have solved a number of issues; however, an even larger number still remain unsolved. The promising and attractive results have generated increasing demands for applications in new fields and for better performing techniques. Diagnostic results are now expected more rapidly and samples are preferably examined on the spot. Direct diagnosis using soil samples without prior nematode extraction would further reduce the period of time required for diagnosis. Mobile molecular equipment may be particularly useful in quarantine applications, where the detection of single individuals is of paramount importance. Techniques must be able to distinguish between dead and living individuals.

Plant-parasitic nematodes live in communities comprised of variable numbers of species. The expectation of extension services goes beyond the molecular diagnosis of a single species, even if it is the major pathogen of a crop. Knowledge about the composition of the nematode community will help in designing a crop rotation sequence. For the same reason, the molecular quantification of plant-parasitic nematodes is extremely important. Molecular quantification has proven possible for a number of cyst nematodes and for *D. dipsaci*. However, for both of these groups, specimens of the same stage were used (second-stage and fourth-stage juveniles, respectively). Quantification of different stages will be necessary if molecular diagnostics is to be used by extension services. When planning a crop rotation, the identification of pathotypes (e.g. PCN) and races (e.g. *Meloidogyne*) is very important; currently, this identification is time-consuming and may not always be possible if nematode development is interrupted. This issue has been addressed molecularly, but still requires more research before a method can be used routinely.

The future of molecular nematode diagnostics is in the development of nanodiagnostics, which is still primarily in the research stage. Nanotechnology extends the limits of molecular diagnostics to the nanoscale (one billionth of a metre). This scale of sensitivity as applied to diagnostics would include the detection of molecular interaction (Jain, 2003) and it is anticipated that many of the specific nanotechnologies will eventually be applied to the diagnostics of nematodes.

The small dimensions that are detectable using this technology have led to the use of nanochips, which employs the power of an electronic current to separate DNA probes to specific sites on the array, based on charge and size. Each test site on nanochips can be controlled electronically from the system's onboard computer. The use of the electronic-mediated hybridization reduces the time for detection of target DNA sequences to minutes instead of hours required with conventional approaches.

Another application of nanotechnology is the use of magnetic nanoparticles, bound to a suitable antibody, to label specific molecules, structures or microorganisms. Magnetic immunoassay techniques have been developed in which the magnetic field generated by the magnetically labelled targets is detected directly with a sensitive magnetometer (Jain, 2003).

DNA sequencing costs have fallen more than 100-fold over the past decade, fuelled in large part by tools, technologies and process improvements developed as part of the successful effort to sequence the human genome. New technologies open the door to the next generation of sequencing methods, which include, for example, pyrosequencing, sequencing-by-sequencing approach and sequencing using nanopores. There are many opportunities to reduce the cost and increase the throughput of DNA sequencing, which are likely to lead to very different and novel approaches to diagnostics.

References

Al-Banna, L., Ploeg, A.T., Williamson, V.M. and Kaloshian, I. (2004) Discrimination of six *Pratylenchus* species using PCR and species-specific primers. *Journal of Nematology* 36, 142–146.

Amiri, S., Subbotin, S.A. and Moens, M. (2002) Identification of the beet cyst nematode *Heterodera schachtii* by PCR. *European Journal of Plant Pathology* 108, 497–506.

Atkins, S.D., Manzanilla-Lopez, R.H., Franco, J., Peteira, B. and Kerry, B.R. (2005) A molecular diagnostic method for detecting *Nacobbus* in soil and in potato tubers. *Nematology* 7, 193–202.

Bakker, J. and Gommers, F.J. (1982) Differentiation of the potato cyst nematodes, *Globodera rostochiensis* and *G. pallida*, and of two *G. rostochiensis* pathotypes by means of two dimensional electrophoresis. *Proceedings of the Koninklijke Nederlandse Akademie van Wetenschappen, Series C* 85, 309–314.

Bates, J.A., Taylor, E.J.A., Gans, P.T. and Thomas J.E. (2002) Determination of relative proportion of *Globodera* species in mixed populations of potato cyst nematodes using PCR product melting peak analysis. *Molecular Plant Pathology* 3, 153–161.

Bekal, S., Gauthier, J.P. and Rivoal, R. (1997) Genetic diversity among a complex of cereal cyst nematodes inferred from RFLP analysis of the ribosomal internal transcribed spacer region. *Genome* 40, 479–486.

Bhadury, P., Austen, C.M., Bilton, D.T., Lambshead, J.D., Rogers, A.D. and Smerdon, G.R. (2005) Combined morphological and molecular analysis of individual nematodes through short-term preservation in formalin. *Molecular Ecology Notes* 5, 965–968.

Blaxter, M.L., De Ley, P., Garey, J.R., Liu, L.X., Scheldeman, P., Vierstraete, A., Vanfleteren, J.R., Mackey, L.Y., Dorris, M., Frisse, L.M., Vida, J.T. and Thomas, W.K. (1998) A molecular evolutionary framework for the phylum Nematoda. *Nature* 392, 71–75.

Blok, V.C. (2005) Achievements in and future prospects for molecular diagnostics of plant-parasitic nematodes. *Canadian Journal of Plant Pathology* 27, 176–185.

Blok, V.C., Wishart, J., Fargette, M., Berthier, K. and Phillips, M.S. (2002) Mitochondrial DNA differences distinguishing *Meloidogyne mayaguensis* from the major species of tropical root-knot nematodes. *Nematology* 4, 773–781.

Blouin, M.S. (2002) Molecular prospecting for cryptic species of nematodes: mitochondrial DNA versus internal transcribed spacer. *International Journal for Parasitology* 32, 527–531.

Blouin, M.S., Yowell, C.A., Courtney, C.H. and Dame, J.B. (1998) Substitution bias, rapid saturation, and the use of mtDNA for nematode systematics. *Molecular Biology and Evolution* 15, 1719–1727.

Boom, R., Sol, C.J., Salimans, M.M., Jansen, C.L., Wertheim-van Dillen, P.M. and van der Noordaa, J. (1990) Rapid and simple method for purification of nucleic acids. *Journal of Clinical Microbiology* 28, 495–503.

Boutsika, K., Phillips, M.S., MacFarlane, S.A., Brown, D.J.F., Holeva, R.C. and Blok, V.C. (2004) Molecular diagnostics of trichodorid nematodes and their associated *Tobacco rattle virus* (TBV). *Plant Pathology* 53, 110–116.

Braasch, H., Tomiczek, Ch., Metge, K., Hoyer, U., Burgermeister, W., Wulfert, I. and Schönfeld, U. (2001) Records of *Bursaphelenchus* spp. (Nematoda, Parasitaphelenchidae) in coniferous timber imported from the Asian part of Russia. *Forest Pathology* 31, 129–140.

Brito, J., Powers, T.O., Mullin, P.G., Inserra, R.N. and Dickson, D.W. (2004) Morphological and molecular characterization of *Meloidogyne mayaguensis* isolates from Florida. *Journal of Nematology* 36, 232–240.

Bulman, S.R. and Marshall, J.W. (1997) Differentiation of Australasian potato cyst nematode (PCN) populations using the polymerase chain reaction (PCR). *New Zealand Journal of Crop and Horticultural Science* 25, 123–129.

Burgermeister, W., Metge, K., Braasch, H. and Buchbach, E. (2005) ITS-RFLP patterns for differentiation of 26 *Bursaphelenchus* species (Nematoda: Parasitaphelenchidae) and observations on their distribution. *Russian Journal of Nematology* 13, 29–42.

Burrows, P.R. and Perry, R.N. (1988) Two cloned DNA fragments which differentiate *Globodera pallida* from *G. rostochiensis*. *Revue de Nématologie* 11, 453–457.

Cao, A.X., Liu, X.Z., Zhu, S.F. and Lu, B.S. (2005) Detection of the pinewood nematode, *Bursaphelenchus xylophilis*, using a real-time polymerase chain reaction assay. *Phytopathology* 95, 566–571.

Carneiro, R.M.D.G., Almeida, M.R.A. and Quénéherve, P. (2000) Enzyme phenotypes of *Meloidogyne* spp. populations. *Nematology* 2, 645–654.

Carneiro, R.M.D.G., Tigano, M.S., Randing, O., Almeida, M.R. and Sarah, J.-L. (2004) Identification and genetic diversity of *Meloidogyne* spp. (Tylenchida: Meloidogynidae) on coffee from Brazil, Central America and Hawaii. *Nematology* 6, 287–298.

Carneiro, R.M.D.G., Almeida, M.R.A., Gomes, A.C.M.M. and Hernandez, A. (2005) *Meloidogyne izalcoensis* n. sp. (Nematode: Meloidogynidae) a root-knot nematode parasitising coffee in El Salvador. *Nematology* 7, 819–832.

Castagnone, C., Abad, P. and Castagnone-Sereno, P. (2005) Satellite DNA-based species-specific identification of single individuals of the pinewood nematode *Bursaphelenchus xylophilus* (Nematoda: Aphelenchoididae). *European Journal of Plant Pathology* 112, 191–193.

Castagnone-Sereno, P., Leroy, F., Bongiovanni, M., Zijlstra, C. and Abad, P. (1999) Specific diagnosis of two root-knot nematodes, *Meloidogyne chitwoodi* and *M. fallax*, with satellite DNA probes. *Phytopathology* 89, 380–384.

Castillo, P., Vovlas, N., Subbotin, S. and Troccoli, A. (2003) A new root-knot nematode: *Meloidogyne baetica* n. sp. (Nematoda: Heteroderidae) parasitizing wild olive in Southern Spain. *Phytopathology* 93, 1093–1102.

Chen, Q., Robertson, L., Jones, J.T., Blok, V.C., Phillips, M.S. and Brown, D.J.F. (2001) Capture of nematodes using antiserum and lectin-coated magnetized beads. *Nematology* 3, 593–601.

Chen, Q., Brown, D.J.F., Curtis, R.H. and Jones, J.T. (2003) Development of a magnetic capture system for recovery of *Xiphinema americanum*. *Annals of Applied Biology* 143, 283–289.

Clapp, J.P., Stoel, C.D. van der and Putten, W.H. van der (2000) Rapid identification of cyst (*Heterodera* spp., *Globodera* spp.) and root-knot (*Meloidogyne* spp.) nematodes on the basis of ITS2 sequence variation detected by PCR-single-strand conformational polymorphism (PCR-SSCP) in cultures and field samples. *Molecular Ecology* 9, 1223–1232.

Curran, J., Baillie, D.L. and Webster, J.M. (1985) Use of genomic DNA restriction fragment length differences to identify nematode species. *Parasitology* 90, 137–144.

Dean, F.B., Nelson, J.R., Giesler, T.L. and Lasken, R.S. (2001) Rapid amplification of plasmid and phage DNA using Phi 29 DNA polymerase and multiply-primed rolling circle amplification. *Genome Research* 11, 1095–1099.

De Giorgi, C., Sialer, M.F. and Lamberti, F. (1994) Formalin-induced infidelity in PCR-amplified DNA fragments. *Molecular and Cellular Probes* 8, 459–462.

De Ley, I.T., Karssen, G., De Ley, P., Vierstraete, A., Waeyenberge, L., Moens, M. and Vanfleteren, J. (1999) Phylogenetic analyses of internal transcribed spacer region sequences within *Meloidogyne*. *Journal of Nematology* 31, 530–531.

De Ley, P., Tandingan De Ley, I., Morris, K., Abebe, E., Mundo, M., Yoder, M., Heras, J., Waumann, D., Rocha-Olivares, A., Burr, J., Baldwin, J.G. and Thomas, W.K. (2005) An integrated approach to fast and informative morphological vouchering of nematodes for applications in molecular barcoding. *Philosophical Transactions of the Royal Society of London B* 272, 1945–1958.

Dong, K., Dean, R.A., Fortnum, B.A. and Lewis, S.A. (2001) Development of PCR primers to identify species of root-knot nematodes: *Meloidogyne aranaria, M. hapla, M. incognita* and *M. javanica*. *Nematropica* 31, 273–282.

Elbadri, G.A.A., De Ley, P., Wayenberge, L., Vierstraete, A., Moens, M. and Vanfleteren, J. (2002) Interspecific variation in *Radopholus similis* isolates assessed with restriction fragment length polymorphism and DNA sequencing of the internal transcribed spacer region of the ribosomal RNA cistron. *International Journal for Parasitology* 32, 199–205.

Eroshenko, A.S., Subbotin, S.A. and Kazachenko, I.P. (2001) *Heterodera vallicola* sp. n. (Tylenchida: Heteroderidae) from elm trees, *Ulmus japonica* in the Primorsky territory, the Russian Far East, with rDNA identification of closely related species. *Russian Journal of Nematology* 9, 9–17.

Esbenshade, P.R. and Triantaphyllou, A.C. (1985) Use of enzyme phenotypes for identification of *Meloidogyne* species (Nematoda: Tylenchida). *Journal of Nematology* 17, 6–20.

Esbenshade, P.R. and Triantaphyllou, A.C. (1990) Isozyme phenotypes for the identification of *Meloidogyne* species. *Journal of Nematology* 22, 10–15.

Esquibet, M., Grenier, E., Plantard, O., Andaloussi, A. and Caubel, G. (2003) DNA polymorphism in the stem nematode *Ditylenchus dipsaci*: development of diagnostic markers for normal and giant races. *Genome* 46, 1077–1083.

Fargette, M. (1987) Use of the esterase phenotype in the taxonomy of the genus *Meloidogyne*. 2. Esterase phenotypes observed in West African populations and their characterization. *Revue de Nématologie* 10, 45–56.

Ferris, V.R., Ferris, J.M. and Faghihi, J. (1993) Variation in spacer ribosomal DNA in some cyst-forming species of plant parasitic nematodes. *Fundamental and Applied Nematology* 16, 177–184.

Ferris, V.R., Ferris, J.M., Faghihi, J. and Ireholm, A. (1994) Comparisons of isolates of *Heterodera avenae* using 2-D PAGE protein patterns and ribosomal DNA. *Journal of Nematology* 26, 144–151.

Files, J.G. and Hirsh, D. (1981) Ribosomal DNA of *Caenorhaditis elegans*. *Journal of Molecular Biology* 149, 223–240.

Fleming, C.C. and Marks, R.J. (1982) A method for the quantitative estimation of *Globodera rostochiensis* and *Globodera pallida* in mixed species samples. *Ministry of Agriculture, Northern Ireland Record of Agricultural Research* 30, 67–70.

Fleming, C.C. and Mowat, D.J. (1993) Potato cyst nematode diagnostics using the polymerase chain reaction. In: Ebbels, D. (ed.) *Proceedings of Plant Health and the European Single Market*. British Crop Protection Council, Farnham, UK, pp. 349–354.

Fleming, C.C. and Powers, T.O. (1998) Potato cyst nematode diagnostics; morphology, different hosts and biochemical technique. In: Marks, R.J. and Brodie, B.B. (eds) *Potato Cyst Nematode – Biology, Distribution and Control*. CAB International, Wallingford, UK, pp. 91–114.

Fleming, C.C., Turner, S.J., Powers, T.O. and Szalansky, A.L. (1998) Diagnostics of cyst nematodes: use of the polymerase chain reaction to determine species and estimate population level. *Aspect of Applied Biology* 52, 375–382.

Floyd, R., Eyualem, A., Papert, A. and Blaxter, M. (2002) Molecular barcodes for soil nematode identification. *Molecular Ecology* 11, 839–850.

Floyd, R.M., Rogers, A.D., Lambshead, J.D. and Smith, C.R. (2005) Nematode-specific PCR primers for the 18S small subunit rRNA gene. *Molecular Ecology Note* 5, 611–612.

François, C., Kebdani, N., Barker, I., Tomlinson, J., Boonham, N. and Castagnone-Sereno, P. (2006) Towards specific diagnosis of plant-parasitic nematodes using DNA oligonucleotide microarray technology: a case study with quarantine species *Meloidogyne chitwoodi*. *Molecular and Cellular Probes* 20, 64–69.

Fullaondo, A., Barrena, E., Viribay, M., Barrena, I., Salazar, A. and Ritter. E. (1999) Identification of potato cyst nematode species *Globodera rostochiensis* and *G. pallida* by PCR using specific primer combinations. *Nematology* 1, 157–163.

Grenier, E., Castagnone-Sereno, P. and Abad, P. (1997) Satellite DNA sequences as taxonomic markers in nematodes of agronomic interest. *Parasitology Today* 13, 398–401.

He, Y., Li, H., Brown, D.J.F., Lamberti, F. and Moens, M. (2003) Isolation and characterisation of microsatellites for *Xiphinema index* using degenerate oligonucleotide primed PCR. *Nematology* 5, 809–819.

He, Y., Subbotin, S.A., Rubtsova, T.V., Lamberti, F., Brown, D.J.F. and Moens, M. (2005) A molecular phylogenetic approach to Longodoridae (Nematoda: Dorylaimida). *Nematology* 7, 111–124.

Hockland, S., Inserra, R.N., Millar, L. and Lehman, P.S. (2006) International plant health – putting legislation into practice. In: Perry, R.N. and Moens, M. (eds) *Plant Nematology*. CAB International, Wallingford, UK, pp. 327–345.

Holeva, R., Phillips, M.S., Neilson, R., Brown, D.J.F., Young, V., Boutsika, K. and Blok, V.C. (2006) Real-time PCR detection and quantification of vector trichodorid nematodes and *Tobacco rattle virus*. *Molecular and Cellular Probes* 20, 203–211.

Hoyer, U., Burgermeister, W. and Braasch, H. (1998) Identification of *Bursaphelenchus* species (Nematoda, Aphelenchoididae) on the basis of amplified ribosomal DNA (ITS-RFLP). *Nachrichtenblatt des Deutschen Pflanzenschutzdienstes* 50, 273–277.

Hübschen, J., Kling, L., Ipach, U., Zinkernagel., V., Bosselut, N., Esmenjaud, D., Brown, D.J.F. and Neilson, R. (2004a) Validation of the specificity and sensitivity of species-specific primers that provide a reliable molecular diagnostic for *Xiphinema diversicaudatum*, *X. index* and *X. vuittenezi*. *European Journal of Plant Pathology* 110, 779–788.

Hübschen, J., Kling, L., Zinkernagel., V., Brown, D.J.F. and Neilson, R. (2004b) Development and validation of species-specific primers that provide a molecular diagnostic for virus–vector longodorid nematodes and related species in German viticulture. *European Journal of Plant Pathology* 110, 883–891.

Hugall, A., Moritz, C., Stanton, J. and Wolstenholme, D. (1994) Low, but strong structured mitochondrial DNA diversity in root-knot nematodes (*Meloidogyne*). *Genetics* 136, 903–912.

Hugall, A.F., Stanton, C. and Moritz, C. (1997) Evolution of the AT-rich mitochondrial DNA of the root knot nematode, *Meloidogyne hapla*. *Molecular Biology and Evolution* 14, 40–48.

Hugall, A., Stanton, J. and Moritz, C. (1999) Reticulate evolution and the origins of ribosomal internal transcribed spacer diversity in apomictic *Meloidogyne*. *Molecular Biology and Evolution* 16, 157–164.

Ibrahim, S.K. and Perry, R.N. (1992) Use of esterase patterns of females and galled roots for the identification of species of *Meloidogyne*. *Fundamental and Applied Nematology* 16, 187–190.

Ibrahim, S.K., Perry, R.N., Burrows, P.R. and Hooper, D.J. (1994) Differentiation of species and populations of *Aphelenchoides* and of *Ditylenchus angustus* using a fragment of ribosomal DNA. *Journal of Nematology* 26, 412–421.

Ibrahim, S.K., Davies, K.G. and Perry, R.N. (1996) Identification of the root-knot nematode, *Meloidogyne incognita*, using monoclonal antibodies raised to non-specific esterases. *Physiological and Molecular Plant Pathology* 49, 79–88.

Ibrahim, S.K., Minnis, S.T., Barker, A.D.P., Russell, M.D., Haydock, P.P.J., Evans, K., Grove, I.G., Woods, S.R. and Wilcox, A. (2001) Evaluation of PCR, IEF and ELISA techniques for the detection and identification of potato cyst nematodes from field soil samples in England and Wales. *Pest Management Science* 57, 1068–1074.

Iwahori, H., Tsuda, K., Kanzaki, N., Izui, K. and Futai, K. (1998) PCR-RFLP and sequencing analysis of ribosomal DNA of *Bursaphelenchus* nematodes related to pine wilt disease. *Fundamental and Applied Nematology* 21, 655–666.

Jain, K.K. (2003) Nanodiagnostics: application of nanotechnology in molecular diagnostics. *Expert Review of Molecular Diagnostics* 3, 153–161.

Jiang, L.Q., Zheng, J.W., Waeyenberge, L., Subbotin, S.A. and Moens, M. (2005) Duplex PCR based identification of *Bursaphelenchus xylophilus* (Steiner and Buhrer, 1934) Nickle, 1970. *Russian Journal of Nematology* 13, 115–121.

Joyce, S.A., Reid, A., Driver, F. and Curran, J. (1994) Application of polymerase chain reaction (PCR) methods to identification of entomopathogenic nematodes. In: Burnell, A.M., Ehlers, R.-U. and Masson, J.P. (eds) *COST 812 Biotechnology: Genetics of Entomopathogenic Nematode-Bacterium Complexes*. Proceedings of Symposium and Workshop, St. Patrick's College, Maynooth, Co. Kildare, Ireland. Luxembourg, European Commission, DG XII, pp. 178–187.

Kang, J.S., Choi, K.S., Shin, S.C., Moon, I.S., Lee, S.G. and Lee, S.H. (2004) Development of an efficient PCR-based diagnosis protocol for the identification of the pinewood nematode, *Bursaphelenchus xylophilus* (Nematoda: Aphelenchoididae). *Nematology* 6, 279–285.

Karssen, G., van Hoenselaar, T., Verkerk-Bakker, B. and Janssen, R. (1995) Species identification of cyst and root-knot nematodes from potato by electrophoresis of individual females. *Electrophoresis* 16, 105–109.

Karssen, G., Bolk, R.J., van Aelst, A.C., van den Beld, I., Kox, L.F.F., Korthals, G., Molendijk, L., Zijlstra, C., van Hoof, R. and Cook, R. (2004) Description of *Meloidogyne minor* n. sp. (Nematoda: Meloidogynidae), a root-knot nematode associated with yellow patch disease in golf courses. *Nematology* 6, 59–72.

Kovaleva, E.S., Subbotin, S.A., Masler, E.P. and Chitwood, D.J. (2005) Molecular characterization of the actin gene from cyst nematodes in comparison to those from other nematodes. *Comparative Parasitology* 72, 39–49.

Leal, I., Green, M., Allen, E., Humble, L. and Rott, M. (2005) An effective PCR-based diagnostic method for the detection of *Bursaphelenchus xylophilus* (Nematoda: Aphelenchoididae) in wood samples from lodgepole pine. *Nematology* 7, 833–842.

Liao, J.L., Zhang, L.H. and Feng, Z.X. (2001) Reliable identification of *Bursaphelenchus xylophilus* by rDNA amplification, *Nematologia Mediterranea* 19, 131–135.

Madani, M., Vovlas, N., Castillo, P., Subbotin, S.A. and Moens, M. (2004) Molecular characterization of cyst nematode species (*Heterodera* spp.) from the Mediterranean basin using RFLPs and sequences of ITS-rDNA. *Journal of Phytopathology* 152, 229–234.

Madani, M., Subbotin, S.A. and Moens, M. (2005) Quantitative detection of the potato cyst nematode, *Globodera pallida*, and the beet cyst nematode, *Heterodera schachtii*, using Real-Time PCR with SYBR green I dye. *Molecular and Cellular Probes* 19, 81–86.

Marshall, J. and Crawford, A.M. (1987) A cloned DNA fragment that can be used as a sensitive probe to distinguish *Globodera pallida* from *Globodera rostochiensis* and other cyst forming nematodes. *Journal of Nematology* 19, 541 (abstract).

Matsunaga, K. and Togashi, K. (2004) A simple method for discriminating *Bursaphelenchus xylophilus* and *B. mucronatus* by species-specific polymerase chain reaction primer pairs. *Nematology* 6, 273–277.

Metge, K. and Burgermeister, W. (2005) Multiple displacement amplification of DNA for ITS-RFLP analysis of individual juveniles of *Bursaphelenchus*. *Nematology* 7, 253–257.

Mulholland, V., Carde, L., O'Donnell, K.L., Fleming, C.C. and Powers, T.O. (1996) Use of the polymerase chain reaction to discriminate potato cyst nematode at the species level. In: Marshall, C. (ed.) *Proceedings of Diagnostics in Crop Production Symposium*. British Crop Production Council, Farnham, UK, pp. 247–252.

Navas, A., López, J.A., Espárrago, G., Camafeita, E. and Albar, J.P. (2002) Protein viability in *Meloidogyne* spp. (Nematoda: Meloidogynidae) revealed by two-dimensional gel electrophoresis and mass spectrometry. *Journal of Proteome Research* 1, 421–427.

Orui, Y. (1995) Discrimination of the main *Pratylenchus* species (Nematoda: Pratylenchidae) in Japan by PCR-RFLP analysis. *Applied Entomology and Zoology* 31, 505–514.

Orui, Y. (1997) Discrimination of *Globodera rostochiensis* and four *Heterodera* species (Nematoda: Heteroderidae) by PCR-RFLP analysis. *Japanese Journal of Nematology* 27, 67–75.

Orui, Y. (1998) Identification of Japanese species of the genus *Meloidogyne* (Nematoda: Meloidogynidae) by PCR-RFLP analysis. *Applied Entomology and Zoology* 33, 43–51.

Orui, Y. and Mizukubo, T. (1999) Discrimination of seven *Pratylenchus* species (Nematoda: Pratylenchidae) in Japan by PCR-RFLP analysis. *Applied Entomology and Zoology* 34, 205–211.

Palmer, H.M., Atkinson, H.J. and Perry, R.N. (1992) Monoclonal antibodies specific to surface expressed antigens of *Ditylenchus dipsaci*. *Fundamental and Applied Nematology* 15, 511–515.

Perera, M.R., Vanstone, V.A. and Jones, M.G.K. (2005) A novel approach to identify plant parasitic nematodes using matrix-assisted laser desorption/ionization time-of-flight mass spectrometry. *Rapid Communications in Mass Spectrometry* 19, 1454–1460.

Perry, R.N. and Jones, J.T. (1998) The use of molecular biology techniques in Plant Nematology: past, present and future. *Russian Journal of Nematology* 6, 47–56.

Petersen, D.J. and Vrain, T.C. (1996) Rapid identification of *Meloidogyne chitwoodi*, *M. hapla*, and *M. fallax* using PCR primers to amplify their ribosomal intergenic spacer. *Fundamental and Applied Nematology* 19, 601–605.

Petersen, D.J., Zijlstra, C., Wishart, J., Blok, V. and Vrain, T.C. (1997) Specific probes efficiently distinguish root-knot nematode species using signature sequence in the ribosomal intergenetic spacer. *Fundamental and Applied Nematology* 20, 619–626.

Powers, T. (2004) Nematode molecular diagnostics: from bands to barcodes. *Annual Review of Phytopathology* 42, 367–383.

Powers, T.O. and Harris, T.S. (1993) A polymerase chain reaction method for identification of five major *Meloidogyne* species. *Journal of Nematology* 25, 1–6.

Powers, T.O. Szalanski, A.I., Mullin, P.G., Harris, T.S. Bertozzi, T. and Griesbach, J.A. (2001) Identification of seed gall nematodes of agronomic and regular concern with PCR-RFLP of ITS1. *Journal of Nematology* 33, 191–194.

Powers, T.O., Mullin, P.G., Harris, T.S., Sutton, L.A. and Higgins, R.S. (2005) Incorporating molecular identification of *Meloidogyne* spp. into a large-scale regional nematode survey. *Journal of Nematology* 37, 226–235.

Randing, O., Bongiovanni, M., Carneiro, R.M.D.G. and Castagnone-Sereno, P. (2002) Genetic diversity of root-knot nematodes from Brazil and development of SCAR markers specific for the coffee-damaging species. *Genome* 45, 862–870.

Reid, A., Manzanilla-Lopez, R.H. and Hunt, D.J. (2003) *Nacobbus aberrans* (Thorne, 1935) Thorne and Allen, 1944 (Nematode: Pratylenchidae); a nascent species complex revealed by RFLP analysis and sequencing of the ITS-rDNA region. *Nematology* 5, 441–451.

Robinson, M.P. (1989) Quantification of soil and plant populations of *Meloidogyne* using immunoassay techniques. *Journal of Nematology* 21, 583–584.

Rubtsova, T.V., Moens, M. and Subbotin, S.A. (2005) PCR amplification of a rRNA gene fragment from formalin-fixed and glycerine-embedded nematodes from permanent slides. *Russian Journal of Nematology* 13, 137–140.

Schmitz, V.B., Burgermeister, W. and Braasch, H. (1998) Molecular genetic classification of Central European *Meloidogyne chitwoodi* and *M. fallax* populations. *Nachrichtenblatt des Deutschen Pflanzenschutzdienstes* 50, 310–317.

Scholl, E.H. and Bird, D.M. (2005) Resolving tylenchid evolutionary relationships through multiple gene analysis derived from EST data. *Molecular Phylogenetics and Evolution* 36, 536–545.

Schots, A., Bakker, J., Egberts, E. and Gommers, F.J. (1986) Monoclonal antibodies against potato cyst nematodes. *Revue de Nématologie* 9, 309.

Setterquist, R.A., Smith, G.K., Jones, R. and Fox, G.E. (1996) Diagnostic probes targeting the major sperm protein gene that may be useful in the molecular identification of nematodes. *Journal of Nematology* 28, 414–421.

Skantar, A.M. and Carta, L.K. (2004) Molecular characterization and phylogenetic evaluation of the Hsp90 gene from selected nematodes. *Journal of Nematology* 36, 466–480.

Skantar, A.M. and Carta, L.K. (2005) Multiple displacement amplification (MDA) of total genomic DNA from *Meloidogyne* spp. and comparison to crude DNA extracts in PCR of ITS1, 28S D2–D3 rDNA and Hsp90. *Nematology* 7, 285–293.

Stanton, J.M., McNicol, C.D. and Steele, V. (1998) Non-manual lysis of second stage *Meloidogyne* juveniles for identification of pure and mixed samples based on polymerase chain reaction. *Australian Plant Pathology* 27, 112–115.

Sturhan, D. and Rumpenhorst, H.J. (1996) Untersuchungen über den *Heterodera avenae*-Komplex. *Mitteilungen aus der Biologischen Bundesanstalt für Land- und Fortwirtschaft, Berlin-Dahlem* 317, 75–91.

Subbotin, S.A. (2006) Molecular diagnostics of plant parasitic nematodes using the polymerase chain reaction. In: Zinovieva, S.V., Perevertin, K.A., Romanenko, N.D. *et al.* (eds) *Applied Nematology*. Nauka, Russia, pp. 228–252.

Subbotin, S.A. and Moens, M. (2006) Molecular taxonomy and phylogeny. In: Perry, R.N. and Moens, M. (eds) *Plant Nematology*. CAB International, Wallingford, UK, pp. 33–58.

Subbotin, S.A., Sturhan, D., Waeyenberge, L. and Moens, M. (1997) *Heterodera riparia* sp. n. (Tylenchida: Heteroderidae) from common nettle, *Urtica dioica* L., and rDNA-RFLP separation of species from the *H. humuli* group. *Russian Journal of Nematology* 5, 143–157.

Subbotin, S.A., Halford, P.D. and Perry, R.N. (1999) Identification of populations of potato cyst nematodes from Russia using protein electrophoresis, rDNA-RFLPs and RAPDs. *Russian Journal of Nematology* 7, 57–63.

Subbotin, S.A., Waeyenberge, L. and Moens, M. (2000a) Identification of cyst forming nematodes of the genus *Heterodera* (Nematoda: Heteroderidae) based on the ribosomal DNA-RFLPs. *Nematology* 2, 153–164.

Subbotin, S.A., Halford, P.D., Warry, A. and Perry, R.N. (2000b) Variations in ribosomal DNA sequences and phylogeny of *Globodera* parasitising Solanaceae. *Nematology* 2, 591–604.

Subbotin, S.A., Peng, D. and Moens, M. (2001) A rapid method for the identification of the soybean cyst nematode *Heterodera glycines* using duplex PCR. *Nematology* 3, 365–370.

Subbotin, S.A., Sturhan, D., Rumpenhorst, H.J. and Moens, M. (2003) Molecular and morphological characterisation of the *Heterodera avenae* complex species (Tylenchida: Heteroderidae) *Nematology* 5, 515–538.

Subbotin, S.A., Madani, M., Krall, E., Sturhan, D. and Moens, M. (2005) Molecular diagnostics, taxonomy and phylogeny of the stem nematode *Ditylenchus dipsaci* species complex based on the sequences of the ITS-rDNA. *Phytopathology* 95, 1308–1315.

Subbotin, S.A., Sturhan, D., Chizhov, V.N., Vovlas, N. and Baldwin, J.G. (2006) Phylogenetic analysis of Tylenchida Thorne, 1949 as inferred from D2 and D3 expansion fragments of the 28S rRNA gene sequences. *Nematology* 8, 455–474.

Stanton, J.M., McNicol, C.D. and Steele, V. (1998) Non-manual lysis of second stage *Meloidogyne* juveniles for identification of pure and mixed samples based on polymerase chain reaction. *Australian Plant Pathology* 27, 112–115.

Szalanski, A.L., Sui, D.D., Harris, T.S. and Powers, T.O. (1997) Identification of cyst nematodes of agronomic and regulatory concern with PCR-RFLP of ITS1. *Journal of Nematology* 29, 255–267.

Tanha Maafi, Z., Subbotin, S.A. and Moens, M. (2003) Molecular identification of cyst-forming nematodes (Heteroderidae) from Iran and a phylogeny based on the ITS sequences of rDNA. *Nematology* 5, 99–111.

Thiéry, M. and Mugniéry, D. (1996) Interspecific rDNA restriction fragment length polymorphism in *Globodera* species, parasites of Solanaceous plants. *Fundamental and Applied Nematology* 19, 471–479.

Thomas, W.K., Vida, J.T., Frisse, L.M., Mundo, M. and Baldwin, J.G. (1997) DNA sequences from formalin-fixed nematodes: integrating molecular and morphological approaches to taxonomy. *Journal of Nematology* 29, 250–254.

Tigano, M.S., Carneiro, R.M.D.G., Jeyaprakash, A., Dickson, D.W. and Adams, B.J. (2005) Phylogeny of *Meloidogyne* spp. based on 18S rDNA and the intergenic region of mitochondrial DNA sequences. *Nematology* 7, 851–862.

Trudgill, D.L. and Carpenter, J.M. (1971) Disc electrophoresis of proteins of *Heterodera* species and pathotypes of *Heterodera rostochiensis*. *Annals of Applied Biology* 69, 34–41.

Uehara, T., Mizukubo, T., Kushida, A. and Momota, Y. (1998) Identification of *Pratylenchus coffeae* and *P. loosi* using specific primers for PCR amplification of ribosomal DNA. *Nematologica* 44, 357–368.

Uehara, T., Kushida, A. and Momota, Y. (1999) Rapid and sensitive identification of *Pratylenchus* spp. using reverse dot blot hybridization. *Nematology* 1, 549–555.

Vahidi, H. and Honda, B.M. (1991) Repeats and subrepeats in the intergenic spacer of rDNA from the nematode *Meloidogyne arenaria arenaria*. *Molecular and General Genetics* 227, 334–336.

Vrain, T.C., Wakarchuk, D.A., Levesque, A.C. and Hamilton, R.I. (1992) Intraspecific rDNA restriction fragment length polymorphisms in the *Xiphinema americanum* group. *Fundamental and Applied Nematology* 15, 563–574.

Waeyenberge, L., Ryss, A., Moens, M., Pinochet, J. and Vrain, T.C. (2000) Molecular characterization of 18 *Pratylenchus* species using rDNA restriction fragment length polymorphism. *Nematology* 2, 135–142.

Walsh, S.P., Metzger, D.A. and Higuchi, R. (1991) Chelex 100 as medium for simple extraction of DNA for PCR-based typing from forensic material. *Biological Techniques* 10, 506–513.

Webster, J.M. and Hooper, D.J. (1968) Serological and morphological studies on the inter- and intraspecific differences of the plant-parasitic nematodes *Heterodera* and *Ditylenchus*. *Parasitology* 58, 878–891.

Wendt, K.R., Vrain, T.C. and Webster, J.M. (1993) Separation of three species of *Ditylenchus* and some host races of *D. dipsaci* by restriction fragment length polymorphism. *Journal of Nematology* 25, 555–563.

Williams, C., Pohtén, F., Moberg, C., Söderkvist, P., Uhlén, M., Pontén, J., Sitbon, G. and Lundeberg, J. (1999) A high frequency of sequence alterations is due to formalin fixation of archival specimens. *American Journal of Pathology* 155, 1467–1471.

Williamson, V.M., Caswell-Chen, E.P., Westerdahl, B.B., Wu, F.F. and Caryl, G. (1997) A PCR assay to identify and distinguish single juveniles of *Meloidogyne hapla* and *M. chitwoodi*. *Journal of Nematology* 29, 9–15.

Xu, J.H., Liu, P.L., Meng, Q.P. and Long, H. (2004) Characterisation of *Meloidogyne* species from China using isozyme phenotypes and amplified mitochondrial DNA restriction fragment length polymorphism. *European Journal of Plant Pathology* 110, 309–315.

Xue, B., Baillie, D.L. and Webster, J.M. (1993) Amplified fragment length polymorphisms of *Meloidogyne* spp. using oligonucleotide primers. *Fundamental and Applied Nematology* 16, 481–487.

Yoder, M., Tandingan De Ley, I., King, I.W., Mundo-Ocampo, M., Mann, J., Blaxter, M., Poiras, L. and De Ley, P. (2006) DESS: a versatile solution for preserving morphology and extractable DNA of nematodes. *Nematology* 8, 367–376.

Zheng, J., Subbotin, S.A., He, S., Gu, J. and Moens, M. (2003) Molecular characterisation of some Asian isolates of *Bursaphelenchus xylophilus* and *B. mucronatus* using PCR-RFLPs and sequences of ribosomal DNA. *Russian Journal of Nematology* 11, 17–22.

Zijlstra, C. (1997) A fast PCR assay to identify *Meloidogyne hapla*, *M. chitwoodi*, and *M. fallax*, and to sensitively differentiate them from each other and from *M. incognita* in mixtures. *Fundamental and Applied Nematology* 20, 505–511.

Zijlstra, C. (2000) Identification of *Meloidogyne chitwoodi*, *M. fallax* and *M. hapla* based on SCAR-PCR: a powerful way of enabling reliable identification of populations or individuals that share common traits. *European Journal of Plant Pathology* 106, 283–290.

Zijlstra, C., Lever, A.E.M., Uenk, B.J. and Van Silfhout, C.H. (1995) Differences between ITS regions of isolates of root-knot nematodes *Meloidogyne hapla* and *M. chitwoodi*. *Phytopathology* 85, 1231–1237.

Zijlstra, C., Donkers-Venne, D.T.H.M. and Fargette, M. (2000) Identification of *Meloidogyne incognita*, *M. javanica* and *M. arenaria* using sequence characterized amplified region (SCAR) based PCR assays. *Nematology* 2, 847–853.

Zijlstra, C., van Hoof, R. and Donkers-Venne, D. (2004) A PCR test to detect the cereal root-knot nematode *Meloidogyne naasi*. *European Journal of Plant Pathology* 110, 855–860.

10 Molecular Diagnostic Methods for Plant Viruses

A. Olmos, N. Capote, E. Bertolini and M. Cambra

Abstract

Plants are frequently infected by a wide range of viruses that cause important agronomic, economic and social impact. Detection of viruses at early stages of infection is crucial to reduce economic losses. Biological indexing and serological enzyme-linked immunosorbent assay (ELISA) methods are widely used for diagnosis. Nevertheless, molecular techniques have revolutionized plant virus detection and identification. Early molecular hybridization technologies were rapidly supplanted by more powerful nucleic acids amplification methods based on the polymerase chain reaction (PCR). Although molecular methods are highly discriminatory, allowing strain typing, routine testing has been hampered by problems in reproducibility. Continuous efforts have been made to overcome these barriers. Improved systems to prepare plant or insect samples have been developed. Efforts have also been directed at increasing the sensitivity and specificity of detection, which can be limited by the high content of enzyme inhibitors in plant materials. Nested and multiplex PCR offer high sensitivity and the possibility to detect several targets in one assay, respectively. Other technologies allow the amplification of nucleic acids in an isothermal reaction (nucleic acid sequence-based amplification [NASBA] or reverse transcription loop-mediated isothermal amplification [RT-LAMP] procedures). High-throughput testing has been achieved by real-time PCR, which allows the automation of PCR combined with fluorimetry. Real-time PCR simultaneously permits detection and quantitation of targets. The use of integrated protocols combining the specificity of serological techniques with the sensitivity of molecular methods will increase the accuracy and reliability of virus diagnostic. In the future, nucleic acid arrays and biosensors assisted by nanotechnology could revolutionize the detection of plant viruses.

Introduction

Plant viruses are important pathogens in agriculture worldwide. They cause considerable economic damage in different crops, including fruit trees, field crops, vegetables, cereals and ornamentals. As an example, estimated costs associated with sharka disease (caused by *Plum pox virus*, PPV) management in the last 30 years exceed €10,000 million worldwide (Cambra *et al.*, 2006b).

Insufficient knowledge of virus uptake and spread by vectors prevents the development of adequate control measures. However, many systems to control plant virus diseases have been developed. These focus on methods to prevent infection and the use of virus-free seeds and starting planting material. The implementation of such measures requires detection and identification methods of high sensitivity, specificity and reliability. Because viruses may remain latent in susceptible and tolerant hosts, and can be present in very low titre, their detection in seeds, propagative material and other reservoirs is a priority. Latent infections in host tissues can lead to widespread distribution of the virus and large economic losses. In addition, a number of vectors facilitate the spread of viruses from infected plants to healthy ones. The need for rapid and accurate techniques is especially required for quarantine viruses, for which the risk of disease and inoculum spread must be reduced almost to zero. Thus, accurate methods to detect plant viruses in soil, plant materials and vectors are essential to make large-scale diagnosis possible for quarantine purposes, eradication programmes and sanitary control of viruses (López *et al.*, 2003, 2006).

Conventional diagnosis of viral agents in plant pathology is based on biological indexing and serological assays. However, molecular methods are increasingly used for detection and characterization of plant viruses because they provide the advantage of assessing the genes directly, facilitating further identification. In general, sensitive molecular techniques provide a second level of testing, which is particularly suitable for the analysis of those samples in which previous data are not conclusive, or for the rapid screening of critical samples usually related to national or international trade or to the first identification of a virus in a given country.

This chapter provides a brief review of the development of nucleic acid-based molecular detection methods and discusses the state-of-the-art detection methods based on the amplification of nucleic acids. Finally, criteria used to establish internationally accepted diagnostic protocols are also proposed.

Background

Molecular diagnosis began when Morris and Dodds (1979) developed a method for purification of double-stranded RNA (dsRNA) from plant material and fungi. The genome of the majority of plant viruses consists of RNA, and during viral replication, dsRNA is generated. Thus, dsRNAs were used as infection markers, based on the notion that these molecules are of viral origin, even when they are apparently endogenous to plants (Gibbs *et al.*, 2000). The patterns of dsRNA segments resolved in acrylamide gel electrophoresis have been used to differentiate closteroviruses (Dodds and Bar-Joseph, 1983), to characterize *Citrus tristeza virus* (CTV) strains (Moreno *et al.*, 1990), and to discriminate between healthy and infected olive plants (Grieco *et al.*, 2000). However, some of the more important drawbacks of the method are its lack of specificity, low sensitivity and high labour cost.

Another early diagnostic molecular method is nucleic acid hybridization. It was used for the first time in plant pathology to detect potato spindle tuber viroid (Owens and Diener, 1981). Molecular hybridization-based assays were rapidly adapted to virus detection (Hull, 1993). However, although this methodology can provide higher sensitivity than some serological tests, the initial use of radioactive probes, such as ^{32}P-labelled cRNA, prevented the wide application of this technique for routine testing. Labelling of probes with non-radioactive molecules, such as biotin or digoxigenin, although easier to use, are often not very sensitive (Chu *et al.*, 1989). Recent modifications of hybridization approaches may facilitate diagnosis. The imprint hybridization of sections of plant material (Song *et al.*, 1993) has been successfully applied for viroid detection (see Chapter 12, this volume). Today, the most common molecular hybridization format for the detection of viruses is non-isotopic dot-blot hybridization using digoxigenin-labelled probes. The method has been applied for the detection of members of the genus *Nepovirus* and of stone fruit viruses, such as *Apple mosaic virus* (ApMV), *Prunus necrotic ringspot virus* (PNRSV), *Prune dwarf virus* (PDV), PPV and *Apple chlorotic leaf spot virus* (ACLSV) (Más and Pallás, 1995; Herranz *et al.*, 2005). Multiple RNA probes (riboprobes) have been used for the detection of distinct viruses (Ivars *et al.*, 2004). As an alternative to mixing different riboprobes, polyprobes (a unique riboprobe containing partial nucleic acid sequences of different viruses) can be used (Herranz *et al.*, 2005). Conventional and fast flow-through hybridizations have also been used for the specific detection of amplicons generated after amplification techniques, increasing sensitivity and specificity levels (Bertolini *et al.*, 2001) and reducing time (Olmos *et al.*, 2007).

The discovery of the polymerase chain reaction (PCR) (Saiki *et al.*, 1985), dramatically modified the molecular diagnostic scene. Many improvements were adopted early, including coupling reverse transcription (RT) to PCR in a single RT-PCR reaction. This expanded the applications of PCR by allowing the detection of RNA viruses after an initial step that converts single-strand RNA to cDNA (Korschineck *et al.*, 1991; Wetzel *et al.*, 1991; Hadidi *et al.*, 1995). Nevertheless, when conventional PCR or RT-PCR is applied to routine detection, the sensitivity is similar to that obtained by enzyme-linked immunosorbent assay (ELISA) or hybridization techniques. This may be due, at least in part, to the poor quality of samples that contain potential inhibitors of RT-PCR (Wilson, 1997). Thus, sample preparation is critical and is continuously being improved. Different systems for viral target preparation before the amplification have been developed that bypass the need for extensive nucleic acids purification. These include the direct use of raw plant extracts or the spotting of extracts on paper or membranes. In some cases, plant tissues are tested directly without the preparation of an extract by printing or squashing samples on paper (see below).

Modifications to conventional or simple PCR were developed to improve sensitivity and specificity; e.g. heminested, nested RT-PCR and cooperational PCR (Co-PCR). Improvements were also made by coupling serological detection to molecular amplification. Other PCR variants were developed to verify the presence of more than one pathogen in plant material, or to detect related viruses on multiple hosts. Multiplex PCR and polyvalent degenerate oligonucleotide RT-PCR are useful in plant pathology because different viruses frequently infect a single host. Today, real-time PCR and isothermal amplifications are being developed to increase sensitivity and accuracy while maintaining reliability. Real-time PCR combined with user-friendly software packages has made the quantitative detection of plant viruses available to diagnostic laboratories. Other 'thermal cycler free' nucleic acids amplification techniques are also available. Such isothermal amplification has been applied to plant pathology, allowing real-time assays using molecular beacon probes (see below).

The development of automatic sequencing of amplified genome fragments has been instrumental for the improvement of virus detection methods. Sequencing is a very reliable technique for virus identification. In addition, the extensive sequence data available for many viral isolates have permitted the development of strain-specific probes for molecular hybridization assays or primers for amplification-based assays.

Molecular methods facilitate the detection and characterization of viruses at early stages of infection and reduce the impact of viral diseases. However, they require validations to determine their accuracy, sensitivity, specificity, comparative cost and reproducibility before they can be recommended as international standards for routine diagnostic methods used in large-scale indexing.

Preparation of Plant Material Prior to Nucleic Acid Amplification

Analysis of plant material often requires time-consuming preparation of extracts and purification of nucleic acids. Flowers, leaves, bark tissues, seeds or fruits can be used as a source of infected material, but often have to be cut into small pieces and placed into individual plastic bags or containers to avoid contaminations during the grinding process. Published methods, commercial kits and semi-automated systems are available to isolate nucleic acids and to avoid the presence of inhibitory compounds that may compromise the amplification reaction (López et al., 2006). The majority are based on the use of resins or silica.

The homogenization procedure for samples frequently releases polysaccharides and aromatic compounds that inhibit reverse transcriptase and *Taq* polymerase. Plant extract preparation is also time-consuming, requires special buffers and equipment, enhances risks of contamination and entails the possibility of target modification. Grinding limits the number of samples that can be processed, especially when woody plant material is analysed. Several attempts to overcome these problems have been undertaken. Immunocapture prior to amplification was one of the first reported improvements (Wetzel et al., 1992). This method recalls the first steps of ELISA, and allows immobilizing and concentrating viral particles using specific antibodies coated on the walls of tubes in which crude plants homogenates are incubated. Several washing steps are incorporated to eliminate or reduce interfering inhibitors before the amplification cocktail is added. This technique is estimated to increase sensitivity by about 250 times compared to simple RT-PCR (Wetzel et al., 1992). In some cases, virus particles can be captured by direct binding on plastic surfaces without the use of antibodies. Alternatively, amplification can be conducted using microlitres of crude extract immobilized on small pieces of paper or membranes (Olmos et al., 1996; Osman and Rowhani, 2006). Although there is an improvement, these techniques still require plant material homogenization. The preparation of crude extracts in simple buffers and subsequent dilution in sterile water has also been explored.

The need for a grinding step can be circumvented by the use of tissue prints on paper (Whatman 3 MM) (Olmos et al., 1996). The preparation of imprinted membranes is carried out by briefly pressing a fresh section of tender shoots, leaves, flowers or fruits directly on paper or on other solid supports (nylon, nitrocellulose, polyvinylidene fluoride, etc.) to make a print. The usefulness of this system was verified using plant extracts spotted on paper (Osman and Rowhani, 2006). Direct PCR from squashed aphids onto paper has also been reported (Olmos et al., 2005; Cambra et al., 2006a). The squashing procedure is simple: aphids are moved from leaves or flowers with a little brush and carefully deposited onto the paper membrane and then individually squashed with the round bottom of a plastic Eppendorf tube (Olmos et al., 1997). Imprints of plant material, squashed aphids or spotted extracts can be stored in a dry place at room temperature for several months before use without decreasing the sensitivity of detection.

For viral RNA extraction, the piece of membrane harbouring each individual-squashed aphid, tissue print or spotted extract is cut with sterile scissors and introduced in an Eppendorf tube. After addition of 0.5–1% Triton X-100, the extract can be either used directly or submitted to an immunocapture step. However, the major limitation of this method is the reduced amount of sample that can be loaded on a membrane piece, compromising the sensitivity of the assay. Consequently, the most sensitive amplification methods must be used (see below).

Primer and Probe Design

Primer and probe design is of prime importance for successful detection of viral targets by amplification methods. Partial or complete nucleotide sequence of virus genomes can be retrieved using Nucleotide Sequence Search software located in the National Center for Biotechnology Information (NCBI) web page (http://www3.ncbi.nlm.nih.gov/Entrez) (Bethesda, Maryland, USA). Similarity search Advanced BLAST 2.0 (blastn software) (Altschul et al., 1997) can be used to view alignments of sequences drawn from the GenBank, EMBL and DDBJ databases and to analyse sequence identity between virus species or virus isolates, allowing the design of primers and probes in conserved or unique regions of the genome of the target virus.

The size of the amplified product should be as small as possible to increase sensitivity and specificity (Singh and Singh, 1997). In conventional PCR, amplicons are visualized by gel electrophoresis and the size of the PCR product should range from 100 to 1000 bp, although minimum amplicon size is recommended. Nested PCR methods that are based on two consecutive rounds of amplification require the design of four compatible primers. Two external primers amplify a large amplicon that is then used as a target for a second round of amplification using two internal primers. Two critical steps are involved in primer design. First, the four primers should not anneal with each other to avoid unspecific amplifications that could compete with the specific product. Second, the size of the shortest amplicon should be around 200 bp as recommended for conventional PCR.

A method that requires additional considerations for optimization of primer design is the Co-PCR, which is based on the simultaneous annealing of four primers to the same target. In this technique, the largest amplicon is the major product of amplification because internal primers are in very low concentration. After some PCR cycles, the supply of internal primers is depleted and the intermediated amplicons generated are used as additional primers (Olmos et al., 2002). Internal and external primers must be as compatible as in the nested-PCR method, but in addition, they must be designed as close as possible. For instance, if the size of the largest amplicon is around 300 bp, the shortest amplicon should be around 240 bp.

Multiplex reactions are based on mixing different primers specific to several targets. Compatible and non-interfering primers should be designed *in silico* and *in vitro* analysis must be performed to test if the reaction is displaced towards any of the most abundant targets.

Real-time PCR using SYBR Green, TaqMan or molecular beacon probes is designed to amplify small fragments; consequently, specific primers and probes design software have been developed (PrimerExpress, Applied Biosystems). Isothermal amplifications, such as nucleic acid sequence-based amplification (NASBA), AmpliDet or reverse transcription loop-mediated isothermal amplification (RT-LAMP), require the use of different software such as PrimerExplorer (Eiken Chemical Co.) and RNAfold Vienna Package (http://rna.tbi.univie.ac.at/cgi-bin/RNAfold.cgi) and the careful *in vitro* testing of primers and probes designed *in silico*.

PCR-based Amplification Methods

Conventional PCR-based methods (PCR and RT-PCR)

PCR is the 'gold standard' of molecular methods used for sensitive and specific viral detection. Accurate design of primers allows one to fine-tune the specificity level of the reaction. Thus, if primers are targeted to strongly conserved genome regions among members of a genus, this technique will allow the amplification of genomic sequences for many members of a virus genus as reported for *Luteovirus* (Robertson *et al.*, 1991), *Geminivirus* (Deng *et al.*, 1994), *Closterovirus* (Karasev *et al.*, 1994), *Carmovirus* (Morozov *et al.*, 1995), *Potyvirus* (Gibbs and Mackenzie, 1997), *Tospovirus* (Mumford *et al.*, 1996), *Carlavirus* (Badge *et al.*, 1996), *Cucumovirus* (Choi *et al.*, 1999), *Tymovirus* and *Marafivirus* (Sabanadzovic *et al.*, 2000) or *Polerovirus* (Hauser *et al.*, 2000). On the other hand, if primers are targeted to highly specific regions of a viral isolate, the reaction will allow discrimination of viral isolates.

Although theoretically a 10^9-fold amplification expected for a 2–3 h reaction should permit the detection of single molecules, in routine diagnostics, the level of sensitivity is similar to that of ELISA or hybridization techniques (Olmos *et al.*, 2005). This is mainly due to the efficiency of reaction, which is affected by several parameters, including primer specificity, quality of nucleic acid targets, presence of inhibitors in the reaction, enzyme types, buffer composition and stability.

The majority of plant viruses have an RNA genome, which requires an initial step of RT to obtain the suitable cDNA target for amplification. Although these two reactions need to be performed in different tubes for optimal conditions, single-tube RT-PCR approaches have been successfully used (Korschineck *et al.*, 1991; Wetzel *et al.*, 1991), reducing the time needed to complete the reaction.

Heminested and nested PCR

Sensitivity of detection can be improved by using nested PCR-based methods. The reaction consists of two consecutive rounds of amplification: an RT-PCR (for RNA viruses) and a conventional PCR. The RT-PCR is performed using external primers. Usually, the products of this first amplification are transferred to another tube before the nested PCR is carried out using internal primers (Olmos *et al.*, 2003). In the case of heminested PCR, a single nested primer is used along with one external primer. In all cases, the largest amplification product generated in the first reaction is used as template for the second reaction. In addition, the use of internal primers permits further characterization of the amplified product.

Many reports illustrate the potential of nested PCR-based methods. For instance, this method allowed not only an increased sensitivity of detection of PPV, but also clustered isolates into D or M types (Olmos *et al.*, 1997). These methods also permitted an increased level of detection for PDV in peach trees (Helguera et al., 2002) or to detect and identify members of *Tobamovirus* genus (Dovas et al., 2004). However, the use of two rounds of amplification in two separate tubes increases false-positive results as a consequence of amplifying contaminant amplicons. This is especially problematic when the method is used on a large scale. To overcome this, several authors have proposed single-tube nested PCR methodologies. It can be achieved by placing the cocktail for nested amplification in a hanging gel matrix on the top of the reaction tube (Yourno, 1992). Alternatively, primer design can be improved and the relative concentration of outer to inner primers can be adjusted (Mutasa et al., 1996). Finally, the reaction tube can be compartmentalized using the end of a standard pipette tip (Olmos et al., 1999). Coupling nested PCR variants with squashed or printed samples on paper membranes has allowed the detection of RNA targets from several viruses in plant material and in individual insect vectors (Olmos et al., 1997, 1999; Marroquín et al., 2004; Cambra et al., 2006a).

PCR coupled with serological detection (PCR-ELISA and PCR-ELOSA)

PCR products can be captured on a solid phase, directly or indirectly, and subsequently detected by direct or indirect labelling. Proteins such as avidin or streptavidin that bind biotinylated PCR products, or antibodies directed against biotin, can be coated on the plastic to capture DNA targets (PCR-ELISA) (Landgraf et al., 1991). PCR-enzyme-linked oligosorbent assay (PCR-ELOSA) (Mallet et al., 1993) is based on direct fixation of the capture oligonucleotide probes by passive adsorption on the surface of the microtitre wells. Nagata et al. (1985) described the direct binding of DNA targets onto a microtitre plate. In the nucleic acid-based sandwich system, two nucleic acids probes are used as capture and detection probes. Nucleic acids can be coated onto a solid phase covalently or by passive adsorption. Following the initial description of PCR-ELISA and PCR-ELOSA, more than 300 publications have described the development of such techniques in the diagnostic field. These procedures are suitable for routine purposes and lend themselves to automation. A PCR-microplate hybridization method for plant virus detection was initially described for detection of potato virus Y (PVY) (Hataya et al., 1994). The methodology was further simplified and optimized for routine testing and for the detection and characterization of PPV (Olmos et al., 1997). Other approaches have been developed for the simultaneous identification of different viruses (Youssef et al., 2002). The advantages of these methods are the level of sensitivity achieved, which is similar to that obtained by nested PCR-based methods, and the high specificity observed when oligonucleotides are used as capture probes, allowing molecular analysis of mutations.

Cooperational PCR

An interesting variant of PCR is Co-PCR, which is based on the simultaneous annealing of three or four primers to a nucleotide sequence target

(Olmos et al., 2002). The reaction begins with the parallel amplification of different fragments from the same target, with the largest amplicon encompassing the others. Four amplicons are produced from a combination of four primers during the initial steps of the reaction. Because internal primers are in minute quantities, they are depleted early in the reaction and intermediate amplicons are used by the polymerase as new primers. As a result, the largest amplicon becomes the main product of amplification. Small amounts of reagents are used and susceptibility to inhibitors might be a problem to achieve good sensitivity. This can be overcome by initial nucleic acids extraction with available commercial kits. The method has been successfully assayed both in metal block and capillary air thermal cyclers for the detection of plant RNA viruses such as CTV, PPV, *Cucumber mosaic virus* (CMV), *Cherry leaf roll virus* (CLRV) and *Strawberry latent ringspot virus* (SLRSV) (Olmos et al., 2002). The easy adaptation of this technique to other models has permitted it to be rapidly applied to detect DNA from bacteria and phytoplasmas. In contrast to nested PCR, which is performed in two sequential rounds of amplifications and generates the lowest amplicon as main product, Co-PCR requires only one reaction, minimizing manipulation and reducing risk of contamination. This amplification method has been coupled to colorimetric detection to facilitate routine analysis, obtaining a sensitivity level similar to that provided by nested PCR methods.

Multiplex-PCR and polyvalent degenerate oligonucleotide RT-PCR

Multiplex-PCR allows the simultaneous and sensitive detection of different DNA or RNA targets in a single reaction. Multiplex PCR is a useful tool because different viruses frequently infect a single host. Approaches that allow simultaneous detection of multiple plant viruses reduce the number of tests required, reagent usage, time for analysis and consequently the cost (James et al., 2006). However, due to the technical difficulties of designing a reaction involving many compatible primers, there are few examples in which more than three plant viruses have been amplified in a single PCR-based assay (Bariana et al., 1994; Nassuth et al., 2000; Nie and Singh, 2000; Okuda and Hanada, 2001). Two successful examples of multiplex system are the simultaneous detection of the six major characterized viruses described in olive trees (CMV, CLRV, SLRV, *Arabis mosaic virus* (ArMV), *Olive latent virus-1* and *Olive latent virus-2*) (Bertolini et al., 2001) and the simultaneous detection of nine grapevine viruses (ArMV, *Grapevine fanleaf virus, Grapevine virus A, Grapevine virus B, Rupestris stem pitting-associated virus, Grapevine fleck virus, Grapevine leafroll-associated virus-1, -2 and -3*) (Gambino and Gribaudo, 2006).

Multiplex nested RT-PCR method in a single reaction combines the advantages of the multiplex with the sensitivity of the nested PCR. However, an accurate design of compatible primers is required to avoid hairpins and primer–dimer formation. Multiplex nested PCR has been used for the detection of phytoplasmas, fungi and viruses (Dovas and Katis, 2003). To date, there is only one report of multiplex nested PCR performed in a single closed tube (Bertolini *et al.*, 2003). In this work, a multiplex nested RT-PCR was developed for simultaneous detection of CMV, CLRV, SLRSV and ArMV viruses and the bacterium *Pseudomonas savastanoi* pv. *savastanoi* from olive plants using 20 compatible primers in a compartmentalized tube. The sensitivity was increased by at least 100-fold compared to a multiplex RT-PCR, and was similar to the sensitivity reached with monospecific nested PCR for *P. savastanoi* pv. *savastanoi*. This multiplex nested RT-PCR was coupled with colorimetric detection, allowing the discrimination between amplicons of similar size, which would have required additional monospecific analysis if only gel visualization was employed.

Other variations of PCR focused on the detection of several members of a family. The development of a broad-spectrum PCR assay, such as polyvalent nested RT-PCR, has allowed the simultaneous detection of trichoviruses, capilloviruses and foveaviruses of the family *Flexiviridae* (Foissac *et al.*, 2005).

Real-time PCR

Analysis of PCR products requires agarose gel electrophoresis, hybridization or colorimetric detection of the amplicon. Real-time PCR permits the monitoring of reaction during the amplification process, eliminating the need for an analysis step after the reaction and reducing risks of contamination. The PCR reaction is monitored by measuring fluorescent dyes during each cycle, allowing quantitation of the initial number of template copies (Wittwer *et al.*, 1998). Quantitation can be based on non-specific DNA-binding sites, such as SYBR Green, or on specific fluorescent-labelled probes (TaqMan, Molecular Beacons, LUX, FRET or Scorpions). Non-specific fluorescent dyes have the disadvantage that they can bind to any double-stranded DNA. The relative concentration of each primer is critical to avoid non-specific amplifications. In addition, supplementary melting curve analysis must be carried out to analyse the specificity of the reaction products. Probe-based systems produce highly specific real-time PCR products with great sensitivity. However, false-negative results can occur if probes do not recognize their target due to sequence heterogeneity. Real-time PCR is an emerging technology in the diagnosis of plant viruses that is gaining acceptance. It allows a large dynamic range of quantitation of targets. This technology was initially described in 2000 using TaqMan probes for *Potato mop top virus* and *Tobacco rattle virus* (Mumford *et al.*, 2000). Subsequently, the method was successfully applied for the detection of several viruses. More recently, it has been adapted for the detection of *Cauliflower mosaic virus* (Cankar *et al.*, 2005), *Cucumber vein yellowing virus* (Pico *et al.*, 2005) and PPV (Olmos *et al.*, 2005; Varga and James, 2005, 2006a; Capote *et al.*, 2006). In addition, real-time PCR has been successfully used to detect and quantify viral targets in insect vectors, as shown for *Barley yellow dwarf virus* (Fabre *et al.*, 2003) or PPV (Olmos *et al.*, 2005). Multiplex assays have also been developed (Mumford *et al.*, 2000; Varga and James, 2005, 2006a). Real-time PCR has demonstrated great advantages for the detection and quantitation of low levels of viral targets with the possibility of high-throughput testing. In the future, it is likely that this promising technique will be applied to basic studies on virus replication as well as to more applied research concerning epidemiology, breeding for resistance and virus–vector relationships (Mackay *et al.*, 2002; Olmos *et al.*, 2005).

Isothermal Amplification Methods

NASBA and AmpliDet

Nucleic acids sequence-based amplification method (NASBA) is an isothermal amplification process (Compton, 1991) that involves three enzymes: avian myeloblastosis virus reverse transcriptase (AMV-RT), RNase-H and T7 RNA polymerase. Two target sequence-specific primers (one of which bears a bacteriophage T7 promoter sequence appended to its 5' end) are included in the reaction cocktail. The reaction is initiated when the first primer, bearing the T7 sequence tail, is recognized by the AMV-RT to yield an RNA–DNA hybrid. This hybrid is subsequently hydrolysed by RNase-H to release a single-stranded DNA molecule. The second primer anneals to its target sequence and is elongated by AMV-RT to produce a double-stranded DNA molecule. T7 RNA polymerase recognizes its promoter site and begins RNA transcription. Several hybridization methods with specific probes in dot-blot or ELISA formats, or the use of molecular beacon probes in real-time assays named AmpliDet (Leone *et al.*, 1998; Klerks *et al.*, 2001), are required to detect the amplified RNA. The sensitivity of this method has proved to be similar to that obtained by real-time PCR when applied to PPV detection (Olmos *et al.*, 2007). This thermal cycler-free technology is suitable for single-stranded RNA viruses and offers the advantage that no contaminating DNA is amplified.

Loop-mediated isothermal amplification

Loop-mediated isothermal amplification (LAMP) is increasingly being used in the diagnostic field, offering both sensitivity and low cost (Notomi *et al.*, 2000). The method is based on the use of DNA polymerase with strand displacement activity, and four specifically designed primers that recognize six distinct sequences of the target. The LAMP reaction was enhanced by the addition of loop primers (Nagamine *et al.*, 2002), reducing time of reaction and increasing sensitivity. The reaction takes place at 60–65°C during 60 min. Although it was initially developed for DNA amplification, it has been adapted to RNA (RT-LAMP) (Fukuta et al., 2003). The method has been recently applied for the detection of plant viruses such as PPV, with a sensitivity level similar to that obtained with real-time PCR (Varga and James, 2006b).

Selection of a Diagnostic Molecular Method

Beyond sensitivity and specificity

When choosing a diagnostic test, it must be assumed that there are no perfect methods (that never give false-positive or false-negative results). For this reason, it is necessary to know the capacity of each technique. This capacity can be measured by the analysis of sensitivity, specificity, predictive positive and negative values according the prevalence, hit rate and likelihood ratios.

Diagnostic tests are used to classify plants in two groups according to the presence or absence of a specific pathogen. There are two indicators of the operational capacity of a technique: sensitivity and specificity (Altman and Bland, 1994a). Sensitivity means the proportion of true positives that are correctly identified by the test: $a/(a + c)$ (Table 10.1). Specificity is defined by the proportion of true negatives that are correctly identified by the test: $d/(b + d)$ (Table 10.1). These two indicators constitute one approach to evaluate the diagnostic ability of the test. The most sensitive methods must be used to discard the presence of a pathogen and they supply an accurate diagnosis of healthy plants. However, the methods with high sensitivity also yield some false-positive results and consequently, they must not be used as a single technique to accept the presence of a given pathogen. In contrast, the most specific methods can be used to confirm the presence of a pathogen because they give an accurate and reliable diagnosis of true infected plants. However, due to the possibility of false-negative results, they cannot guarantee that a specific pathogen is not present in the sample in minute quantities or in 'subclinical' stages.

Table 10.1. Relationship between the observed diagnostic results obtained by a given method (test results) and the expected results (correct diagnosis).

		Correct diagnosis		
		Infected (+)	Healthy (−)	
Test results	Infected (+)	a	b	a + b
	Healthy (−)	c	d	c + d
		a + c	b + d	N[a]

[a]N: Total number of analysed samples.

Sensitivity and specificity are not the only criteria available to select a diagnostic test. The probability that a diagnostic method makes an accurate diagnosis must be determined by calculating other predictive values (Altman and Bland, 1994b). A positive predictive value represents the proportion of positive samples correctly diagnosed: $[a/(a + b)]$ (Table 10.1). A negative predictive value is the proportion of samples with negative results which are correctly diagnosed. However, the predictive values also depend on the prevalence of infection in the samples tested: $(a + c/N)$ (Table 10.1), and do not apply universally. Prevalence can be interpreted as the probability that the sample is harbouring the pathogen before the test is performed. If the prevalence of the infection is very low, the positive predictive value will not be close to 1 even if both sensitivity and specificity are high. In screening tests, it is expected that many plants giving positive diagnostic results by one specific method are false positives. Positive and negative predictive values can be calculated for any prevalence:

Positive predictive value = sensitivity × prevalence/[sensitivity × prevalence + (1 − specificity) × (1 − prevalence)].

Negative predictive value = specificity × (1 − prevalence)/[(1 − sensitivity) × prevalence + specificity × (1 − prevalence)].

Hit rate is an indicator of a technique and is the total number of samples for which a screening test gave accurate information [true positives + true negatives diagnosed by the technique divided by the entire number of samples: $(a + d/N)$] (Table 10.1). However, hit rate is a misleading statistic and should not be used as an indicator of test accuracy, because in screening tests, there will be far more true negatives than true positives correctly diagnosed by the technique. Consequently, specificity carries excessive weight in the computation of hit rates, which are closer to the specificity index than the sensitivity index and can mask serious flaws in accuracy. Hit rate will only represent the accuracy of a technique when the analysed samples are constituted by 50% of true negatives and 50% of true positives.

Likelihood ratios are independent of the disease prevalence and summarize the diagnostic accuracy. In the case of qualitative tests reporting the presence or absence of a virus, there are two likelihood ratios: the positive or the negative. The positive likelihood ratio is the proportion of true positives correctly identified by the technique $[a/(a + c)]$ (sensitivity) divided by the proportion of false-positive results: $[b/(b + d)]$ (1 − specificity) (Table 10.1). The negative likelihood ratio is the proportion of false negatives $[c/(a + c)]$ (1 − sensitivity) divided by the proportion of true negatives $[d/(b + d)]$ (specificity) (Table 10.1).

A high positive likelihood ratio increases the probability that the disease is present, while a small negative likelihood ratio suggests a low probability for the presence of the disease. In most circumstances, likelihood ratios above 10 and below 0.1 provide strong evidence to rule in or rule out diagnoses, respectively (Jaeschke et al., 2002). Because likelihood ratios are based on a ratio of sensitivity and specificity, they do not vary in different populations or settings and can be used to quantitate the probability of disease for any individual plant, using Bayes theorem (Fagan,

1975; Sackett et al., 2000). Likelihood ratios emphasize that there is never a 100% perfect diagnosis. However, they provide valuable information regarding the accuracy of a diagnostic method.

Agreement between two detection methods and validation of protocols

Cohen's kappa coefficient (Cohen, 1960) constitutes a method for evaluating agreement between two techniques that can be used when a 'gold standard' test is not available. Kappa is an expression of agreement between two methods. This coefficient does not reveal which method is better, but indicates how often the techniques give the same results. The value of kappa ranges between 0 and 1. Landis and Koch (1977) described an interpretation of values in which $k < 0.00$ is poor, $0 < k < 0.20$ is slight, $0.21 < k < 0.40$ is fair, $0.41 < k < 0.60$ is moderate, $0.61 < k < 0.80$ is substantial and $0.81 < k < 1.00$ is in almost perfect agreement.

Kappa coefficient can be calculated according to Table 10.2 as follows: $k =$ (observed agreement − expected agreement)/(1 − expected agreement), where: observed agreement $= (a + d)/N$ and expected agreement $= [(a + b) \times (a + c) + (c + d) \times (b + d)]/N^2$.

An example of application of Kappa coefficient and validation of protocols, reagents and techniques for PPV detection and identification is shown in Table 10.3. Over the years, a large collection of PPV isolates has been analysed by serological and molecular methods in different European laboratories in the frame of research projects, in bilateral actions and in ring tests. Suitable reagents and techniques were selected, validated and included in the OEPP/EPPO protocol (EPPO, 2004). This protocol was tested through a European Union DIAGPRO Project (SMT4-CT98-2252) in 17 laboratories from Austria, Belgium, France, Germany, Great Britain, Greece, Hungary, Italy and Spain in 2002. Eight blind samples were analysed by three techniques for universal detection of PPV (DASI-ELISA using 5B-IVIA monoclonal antibody, IC-PCR using P1/P2 primers, and Co-PCR using P10/P20/P1/P2 primers and a PPV universal probe). In addition, three techniques for identification of PPV-D and PPV-M were used

Table 10.2. Relationship between the results obtained by two different diagnostic methods after the analysis of a given number of samples.

		Method A diagnostic results		
		Infected (+)	Healthy (−)	
Method B diagnostic results	Infected (+)	a	b	a + b
	Healthy (−)	c	d	c + d
		a + c	b + d	N[a]

[a]N: Total number of analysed samples.

Table 10.3. Comparison of serological and molecular techniques and reagents recommended by the European and Mediterranean Plant Protection Organization for *Plum pox virus* detection and identification (EPPO, 2004) (evaluated in a ring test funded by a DIAGPRO project, SMT4-CT98-2252, European Union in 2002).

Techniques and reagents	Characteristics	Parameters											
		N^a	TP^b	TN^c	FP^d	FN^e	Sn^f	Sp^g	PPV^h	NPV^i	Hit rate	$LR+^j$	$LR-^k$	Cohen's kappa coefficient
DASI-ELISA (5B)	Universal detection	128	96	26	0	6	0.92	1.00	1.00	0.81	0.95	∞	0.06	0.86 ± 0.08
DASI-ELISA (4D)	D-specific identification	128	44	64	0	20	0.69	1.00	1.00	0.76	0.84	∞	0.31	0.68 ± 0.08
DASI-ELISA (AL)	M-specific identification	128	52	62	2	12	0.81	0.97	0.96	0.84	0.89	26.04	0.19	0.78 ± 0.08
IC-PCR (P1/P2)	Universal detection	119	70	28	2	19	0.79	0.93	0.97	0.60	0.82	11.79	0.23	0.61 ± 0.08
IC-PCR (P1/PD)	D-specific identification	120	46	58	2	14	0.77	0.97	0.96	0.81	0.87	23.02	0.24	0.73 ± 0.08
IC-PCR (P1/PM)	M-specific identification	119	47	55	5	12	0.80	0.92	0.90	0.82	0.86	9.56	0.22	0.71 ± 0.09
Co-PCR (Universal probe)	Universal detection	120	85	28	2	5	0.94	0.93	0.98	0.85	0.94	14.17	0.06	0.84 ± 0.09
Co-PCR (D probe)	D-specific identification	120	56	54	6	4	0.93	0.90	0.90	0.93	0.92	9.33	0.07	0.83 ± 0.09
Co-PCR (M probe)	M-specific identification	120	50	57	3	10	0.83	0.95	0.94	0.85	0.89	16.67	0.17	0.78 ± 0.09

[a]N: Total samples.
[b]TP: True positives.
[c]TN: True negatives.
[d]FP: False positives.
[e]FN: False negatives.
[f]Sn: Sensitivity.
[g]Sp: Specificity.
[h]PPV: Positive predictive value.
[i]NPV: Negative predictive value.
[j]LR+: Positive likelihood ratio.
[k]LR-: Negative likelihood ratio.

(DASI-ELISA using 4D and AL specific monoclonal antibodies, IC-PCR using P1/PD, P1/PM primers, and Co-PCR using universal primers and D and M specific probes). The results were analysed following the recommendations of the Medical University of South Carolina for diagnostic tests (http://www.musc.edu/dc/icrebm/diagnostictests.html) as indicated earlier. All tested techniques ranged from good to almost perfect agreement as shown in Table 10.3. Reagents used for universal detection were more accurate and reliable than strain-specific reagents. The highest hit rate in routine detection was obtained by DASI-ELISA using 5B-IVIA-specific monoclonal antibody (95%, $k = 0.86 \pm 0.08$) followed by Co-PCR using the PPV universal probe (94%, $k = 0.84 \pm 0.09$). The highest hit rate in identification or characterization of PPV-D and PPV-M was achieved by Co-PCR using specific D and M probes. Although none of the tested techniques reached a 100% accuracy, all techniques are considered to have excellent performance for PPV detection compared with standards used in medical diagnostic.

Other detection techniques not included in this series of tests (i.e. real-time PCR) have now been shown to be reliable enough for universal detection of PPV. In addition, real-time PCR using a universal TaqMan probe has proved to be the most sensitive, rapid and accurate technique for PPV detection when compared with ELISA and conventional PCR (Olmos et al., 2005). Real-time PCR is a valuable technique for PPV-D and PPV-M identification (Varga and James, 2005; Capote et al., 2006). NASBA is a suitable technique for PPV detection in winter and spring seasons (Olmos et al., 2007). Such evaluations must be performed before the adoption of any new diagnostic method.

Comparison of molecular diagnostics methods for plant viruses

A hypothetical comparison of the molecular methods most frequently used in viral diagnosis is reviewed in Table 10.4, showing sensitivity, specificity, strain typing, rapidity and feasibility for routine analysis. Real-time PCR has been considered as the most appropriate method for molecular diagnosis. However, for each particular virus, an evaluation of the techniques using likelihood ratios must be performed. Thus, variations from Table 10.4 could be observed when applied to other viruses and/or hosts.

Development of New Technologies

Conventional ELISA involving specific monoclonal or polyclonal antibodies is still the most widely used method for routine plant virus detection. However, highly sensitive molecular detection methods are rapidly being applied to plant virus diagnosis. Efforts are focused on improving and simplifying the extraction of targets, reducing associated costs, and

Table 10.4. Theoretical comparisons of the sensitivity, specificity, strain-typing abilities, rapidity and feasibility of various molecular techniques for routine virus detection.

Technique	Sensitivity[a]	Specificity[b]	Strain typing[c]	Rapidity	Feasibility[d]
Molecular hybridization	+[e]	+++	+	+	+
Conventional PCR[f]	++	+++	+++	+++	++
Immunocapture-PCR	+++	+++	+++	+++	+++
Heminested and nested PCR (two tubes)	++++	++	+++	++	++
Nested PCR (single tube)	++++	+++	+	++	+++
PCR-ELISA[g] and PCR-ELOSA[h]	++++	++	+++	+	+
Multiplex PCR	++	+++	+++	+++	++
Multiplex nested PCR	++++	+++	+	++	++
Cooperative-PCR	++++	+++	+++	++	+++
Real-time PCR	+++++	+++++	+++++	+++++	+++++
NASBA[i]	++++	+++	+++	++++	+++
LAMP[j]	++++	+++	+++	++++	+++
Macroarray/microarray	+	++++	+++++	+	++

[a]Sensitivity: proportion of true positives correctly identified by the test.
[b]Specificity: proportion of true negatives correctly identified by the test.
[c]Strain typing: ability to discriminate between viral strains.
[d]Feasibility: practicability in routine analysis, execution and interpretation.
[e]For each criterion, methods are rated from acceptable (+) to optimum (+ + + + +).
[f]PCR: Polymerase chain reaction or reverse transcriptase-PCR (RT-PCR).
[g]PCR-ELISA: PCR-enzyme-linked immunosorbent assay.
[h]PCR-ELOSA: PCR-enzyme-linked oligosorbent assay.
[i]NASBA: Nucleic acids sequence-based amplification.
[j]LAMP: Loop-mediated isothermal amplification.

adapting methods for large-scale testing. Some methods are still being developed or in the research phase; i.e. flow-through hybridization (Olmos et al., 2007). The potential power of these methods for plant virus detection and identification is enormous. A promising technology for the development of multipathogen detection system is based on macroarrays and microarrays. A large number of diagnostic nucleic acids probes can be spotted onto a platform (e.g. membrane or glass slide) and hybridized with labelled targets (i.e. nucleic acids extracted from samples and chemically modified with a fluorescent or other reporter). An amplification step prior to the hybridization phase is often needed to achieve satisfactory sensitivity (Agindotan and Perry, 2007). Unfortunately, this increases the cost of the procedure and prevents its use in routine assays. However, there are some indications that at least under certain conditions, such a preliminary amplification could be bypassed without sacrificing too much sensitivity (Boonham et al., 2003). In the near future, when these drawbacks are overcome, a simple hybridization assay may ultimately allow the simultaneous detection of several viruses in an infected plant sample. The development of higher sensitive scanner technology would make these approaches attractive for multiplex analysis.

Another promising but still novel method is the quartz crystal microbalance (QCM)-based DNA biosensor. This technique is a mass-measuring device consisting of a quartz crystal wafer sandwiched between two metal electrodes, which are connected to an external oscillator circuit that records the resonant frequency. This technology has been applied for plant virus detection (*Cymbidium mosaic virus*, *Orchid ringspot virus* and *Tobacco mosaic virus*) based on DNA–RNA hybridization (Eun et al., 2002; Dickert et al., 2004). QCM DNA biosensors could be combined with a flow injection system, thus providing a real-time high-throughput and continuous assays with high sensitivity and specificity.

Conclusions and Future Directions

The greatest challenge for a virology diagnostic laboratory is to carry out reliable analysis to detect and characterize specific virus isolates, to cluster epidemiologically related strains and to discriminate them from unrelated strains. The criteria for evaluation of a molecular method must be a balance between performance (efficacy) and convenience (efficiency). Efficacy includes the capacity of typing strains, the reproducibility between laboratories, the discriminatory power and the agreement between typing techniques. Efficiency includes the ease of execution and interpretation of results, the cost and the availability of reagents and equipment. Therefore, the use of molecular methods for virus detection and characterization still depends on the needs, skill level and resources of a diagnostic laboratory. Robust protocols that consider the extreme sensitivity of the amplification assays, predictive positive and negative values of such methods along with a standardization of the methodology are necessary for high-throughput testing.

International attempts to develop and standardize diagnostic protocols for regulated pests including viruses are coordinated by the European and Mediterranean Plant Protection Organization (OEPP/EPPO) and by the International Plant Protection Convention (IPPC) governed by the Interim Commission on Phytosanitary Measures and hosted by Food and Agriculture Organization (FAO). Diagnostic protocols are published by these organizations and include a combination of approaches that integrate biological, serological and molecular methods previously validated in ring tests. In addition, in the near future, nucleic acid biosensor technologies could revolutionize the diagnostic field with the assistance of nanotechnology.

Acknowledgements

This review was made possible because of the accumulated experience in virus diagnostic supported by Spanish (INIA, IVIA, Generalidad Valenciana and CICYT) and international (EU) grants and bilateral actions (COST-EU).

Dr E. Bertolini is a recipient of a Juan de La Cierva contract of the Spanish Ministry of Educación y Ciencia.

References

Agindotan, B. and Perry, K.L. (2007) Macroarray detection of plant RNA viruses using randomly primed and amplified complementary DNAs from infected plants. *Phytopathology* 97, 119–127.
Altman, D.G. and Bland, J.M. (1994a) Diagnostic tests 1: sensitivity and specificity. *British Medical Journal* 308, 1552.
Altman, D.G. and Bland, J.M. (1994b) Diagnostic tests 2: predictive values. *British Medical Journal* 309, 102.
Altschul, S.F., Madden, T.L., Schaffer, A.A., Zhang, J.H., Zhang, Z., Miller, W. and Lipman, D.J. (1997) Gapped BLAST and PSI-BLAST: a new generation of protein database search programs. *Nucleic Acids Research* 25, 3389–3402.
Badge, J., Brunt, A., Carson, R., Dagless, E., Karamagioli, M., Phillips, S., Seal, S., Turner, R. and Foster, G.D. (1996) A carlavirus-specific PCR primer and partial nucleotide sequence provides further evidence for the recognition of *Cowpea mild mottle virus* as a whitefly-transmitted carlavirus. *European Journal of Plant Pathology* 102, 305–310.
Bariana, H.S., Shannon, A.L., Chu, P.W.G. and Waterhouse, P.M. (1994) Detection of five seedborne legume viruses in one sensitive multiplex polymerase chain reaction test. *Phytopathology* 84, 1201–1205.
Bertolini, E., Olmos, A., Martínez, M.C., Gorris, M.T. and Cambra, M. (2001) Single-step multiplex RT-PCR for simultaneous and colourimetric detection of six RNA viruses in olive trees. *Journal of Virological Methods* 96, 33–41.
Bertolini, E., Olmos, A., López, M.M. and Cambra, M. (2003) Multiplex nested reverse transcription polymerase chain reaction in a single tube for sensitive and simultaneous detection of four RNA viruses and *Pseudomonas savastanoi* pv. *savastanoi* in olive trees. *Phytopathology* 93, 286–292.
Boonham, N., Walsh, K., Smith, P., Madagan, K., Graham, I. and Barker, I. (2003) Detection of potato viruses using microarray technology: towards a generic method for plant viral disease diagnosis. *Journal of Virological Methods* 108, 181–187.
Cambra, M., Bertolini, E., Olmos, A. and Capote, N. (2006a) Molecular methods for detection and quantitation of virus in aphids. In: Cooper, J.I., Kuehne, T. and Polischuk, V. (eds) *Virus Diseases and Crop Biosecurity*. NATO Series C: Environmental Security, Springer, Dordrecht, The Netherlands, pp. 81–88.
Cambra, M., Capote, N., Myrta, A. and Llácer, G. (2006b) *Plum pox virus* and the estimated costs associated with sharka disease. *Bulletin OEPP/EPPO Bulletin* 36, 202–204.
Cankar, K., Ravnikar, M., Zel, J., Gruden, K. and Toplak, N. (2005) Real-time polymerase chain reaction detection of *Cauliflower mosaic virus* to complement the 35S screening assay for genetically modified organisms. *Journal of AOAC International* 88, 814–822.
Capote, N., Gorris, M.T., Martínez, M.C., Asensio, M., Olmos, A. and Cambra, M. (2006) Interference between D and M types of *Plum pox virus* in Japanese plum assessed by specific monoclonal antibodies and quantitative real-time reverse transcription-polymerase chain reaction. *Phytopathology* 96, 320–325.
Choi, S.K., Choi, J.K., Park, W.M. and Ryu, K.H. (1999) RT-PCR detection and identification of three species of cucumoviruses with a genus-specific single pair of primers. *Journal of Virological Methods* 83, 67–73.
Chu, P.W.G., Waterhouse, P.M., Martin, R.R. and Gerlach, W.L. (1989) New approaches to the detection of microbial plant-pathogens. *Biotechnology and Genetic Engineering Reviews* 7, 45–111.

Cohen, J. (1960) A coefficient of agreement for nominal scales. *Educational and Psychological Measurement* 20, 37–46.

Compton, J. (1991) Nucleic-acid sequence-based amplification. *Nature* 350, 91–92.

Deng, D., Mcgrath, P.F., Robinson, D.J. and Harrison, B.D. (1994) Detection and differentiation of whitefly-transmitted geminiviruses in plants and vector insects by the polymerase chain-reaction with degenerate primers. *Annals of Applied Biology* 125, 327–336.

Dickert, F.L., Hayden, O., Bindeus, R., Mann, K.J., Blaas, D., Waigmann, E. (2004) Bioimprinted QCM sensors for virus detection – screening of plant sap. *Analytical and Bioanalytical Chemistry* 378, 1929–1934.

Dodds, J.A. and Bar-Joseph, M. (1983) Double-stranded-RNA from plants infected with closteroviruses. *Phytopathology* 73, 419–423.

Dovas, C.I. and Katis, N.I. (2003) A spot multiplex nested RT-PCR for the simultaneous and generic detection of viruses involved in the aetiology of grapevine leafroll and rugose wood of grapevine. *Journal of Virological Methods* 109, 217–226.

Dovas, C.I., Efthimiou, K. and Katis, N.I. (2004) Generic detection and differentiation of tobamoviruses by a spot nested RT-PCR-RFLP using dl-containing primers along with homologous dG-containing primers. *Journal of Virological Methods* 117, 137–144.

EPPO (2004) Diagnostic protocol for regulated pests. *Plum pox potyvirus*. *Bulletin OEPP/EPPO Bulletin* 34, 247–256.

Eun, A.J., Huang, L., Chew, F., Li, S.F. and Wong, S. (2002) Detection of two orchid viruses using quartz crystal microbalance (QCM) immunosensors. *Journal of Virological Methods* 99, 71–79.

Fabre, F., Kervarrec, C., Mieuzet, L., Riault, G., Vialatte, A. and Jacquot, E. (2003) Improvement of *Barley yellow dwarf virus*-PAV detection in single aphids using a fluorescent real time RT-PCR. *Journal of Virological Methods* 110, 51–60.

Fagan, T.J. (1975) Letter: nomogram for Bayes theorem. *New England Journal of Medicine* 293, 257.

Foissac, X., Svanella-Dumas, L., Gentit, P., Dulucq, M.J., Marais, A. and Candresse, T. (2005) Polyvalent degenerate oligonucleotides reverse transcription-polymerase chain reaction: a polyvalent detection and characterization tool for trichoviruses, capilloviruses, and foveaviruses. *Phytopathology* 95, 617–625.

Fukuta, S., Iida, T., Mizukami, Y., Ishida, A., Ueda, J., Kanbe, M. and Ishimoto, Y. (2003) Detection of *Japanese yam mosaic virus* by RT-LAMP. *Archives of Virology* 148, 1713–1720.

Gambino, G. and Gribaudo, I. (2006) Simultaneous detection of nine grapevine viruses by multiplex reverse transcription polymerase chain reaction with coamplification of a plant RNA as internal control. *Phytopathology* 96, 1223–1229.

Gibbs, A. and Mackenzie, A. (1997) A primer pair for amplifying part of the genome of all potyvirids by RT-PCR. *Journal of Virological Methods* 63, 9–16.

Gibbs, M.J., Koga, R., Moriyama, H., Pfeiffer, P. and Fukuhara, T. (2000) Phylogenetic analysis of some large double-stranded RNA replicons from plants suggests they evolved from a defective single-stranded RNA virus. *Journal of General Virology* 81, 227–233.

Grieco, F., Alkowni, R., Saporani, M., Savino, V. and Martelli, G.P. (2000) Molecular detection of olive viruses. *Bulletin OEPP/EPPO Bulletin* 30, 469–473.

Hadidi, A., Levy, L. and Podleckis, E.V. (1995) Polymerase chain reaction technology in plant pathology. In: Singh, R.P. and Singh, U.S. (eds) *Molecular Methods in Plant Pathology*. CRC Press/Lewis Press, Boca Raton, Florida, pp. 167–187.

Hataya, T., Inoue, A.K. and Shikata, E. (1994) A PCR-microplate hybridization method for plant virus detection. *Journal of Virological Methods* 46, 223–236.

Hauser, S., Weber, C., Vetter, G., Stevens, M., Beuve, M. and Lemaire, O. (2000) Improved detection and differentiation of poleroviruses infecting beet or rape by multiplex RT-PCR. *Journal of Virological Methods* 89, 11–21.

Helguera, P.R., Docampo, D.M., Nome, S.F. and Ducasse, D.A. (2002) Enhanced detection of prune dwarf virus in peach leaves by immunocapture-reverse transcription-polymerase chain reaction with nested polymerase chain reaction (IC-RT-PCR nested PCR). *Journal of Phytopathology – Phytopathologische Zeitschrift* 150, 94–96.

Herranz, M.C., Sánchez-Navarro, J.A., Aparicio, F. and Pallás, V. (2005) Simultaneous detection of six stone fruit viruses by non-isotopic molecular hybridization using a unique riboprobe or 'polyprobe'. *Journal of Virological Methods* 124, 49–55.

Hull, R. (1993) Nucleic acid hybridization procedures. In: Matthews, R.E.F. (ed.) *Diagnosis of Plant Virus Diseases*. CRC Press, Boca Raton, Florida, pp. 253–271.

Ivars, P., Alonso, M., Borja, M. and Hernández, C. (2004) Development of a non-radioactive dot-blot hybridisation assay for the detection of *Pelargonium flower break virus* and *Pelargonium line pattern virus*. *European Journal of Plant Pathology* 110, 275–283.

Jaeschke, R., Guyatt, G. and Lijmer, J. (2002) Diagnostic tests. In: Guyatt, G. and Rennie, D. (eds) *Users' Guides to the Medical Literature*. AMA Press, Chicago, pp. 121–140.

James, D., Varga, A., Pallás, V. and Candresse, T. (2006) Strategies for simultaneous detection of multiple plant viruses. *Canadian Journal of Plant Pathology* 28, 16–29.

Karasev, A.V., Nikolaeva, O.V., Koonin, E.V., Gumpf, D.J. and Garnsey, S.M. (1994) Screening of the closterovirus genome by degenerate primer-mediated polymerase chain-reaction. *Journal of General Virology* 75, 1415–1422.

Klerks, M.M., Leone, G.O., Verbeek, M., van den Heuvel, J.F. and Schoen, C.D. (2001) Development of a multiplex AmpliDet RNA for the simultaneous detection of *Potato leafroll virus* and *Potato virus Y* in potato tubers. *Journal of Virological Methods* 93, 115–125.

Korschineck, I., Himmler, G., Sagl, R., Steinkellner, H. and Katinger, H.W.D. (1991) A PCR membrane spot assay for the detection of *Plum pox virus*-RNA in bark of infected trees. *Journal of Virological Methods* 31, 139–146.

Landgraf, A., Reckmann, B. and Pingoud, A. (1991) Direct analysis of polymerase chain-reaction products using enzyme-linked-immunosorbent-assay techniques. *Analytical Biochemistry* 198, 86–91.

Landis, J.R. and Koch, G.G. (1977) The measurement of observer agreement for categorical data. *Biometrics* 33, 159–174.

Leone, G., van Schijndel, H., van Gemen, B., Kramer, F.R. and Schoen, C.D. (1998) Molecular beacon probes combined with amplification by NASBA enable homogeneous, real-time detection of RNA. *Nucleic Acids Research* 26, 2150–2155.

López, M.M., Bertolini, E., Olmos, A., Caruso, P., Gorris, M.T., Llop, P., Penyalver, R. and Cambra, M. (2003) Innovative tools for detection of plant pathogenic viruses and bacteria. *International Microbiology* 6, 233–243.

López, M., Bertolini, E., Marco-Noales, E., Llop, P. and Cambra, M. (2006) Update on molecular tools for detection of plant pathogenic bacteria and viruses. In: Rao, J., Fleming, C. and Moore, J. (eds) *Molecular Diagnostics: Current Technology and Applications*. Horizon Bioscience, Wymondham, UK, pp. 1–46.

Mackay, I.M., Arden, K.E. and Nitsche, A. (2002) Real-time PCR in virology. *Nucleic Acids Research* 30, 1292–1305.

Mallet, F., Hebrard, C., Brand, D., Chapuis, E., Cros, P., Allibert, P., Besnier, J.M., Barin, F. and Mandrand, B. (1993) Enzyme-linked oligosorbent assay for detection of polymerase chain reaction-amplified human immunodeficiency virus type 1. *Journal of Clinical Microbiology* 31, 1444–1449.

Marroquín, C., Olmos, A., Gorris, M.T., Bertolini, E., Martínez, M.C., Carbonell, E.A., Hermoso de Mendoza, A. and Cambra, M. (2004) Estimation of the number of aphids carrying *Citrus tristeza virus* that visit adult citrus trees. *Virus Research* 100, 101–108.

Más, P. and Pallás, V. (1995) Nonisotopic tissue-printing hybridization – a new technique to study long-distance plant-virus movement. *Journal of Virological Methods* 52, 317–326.

Moreno, P., Guerri, J. and Muñoz, N. (1990) Identification of Spanish strains of *Citrus tristeza virus* by analysis of double-stranded-RNA. *Phytopathology* 80, 477–482.

Morozov, S.Y., Ryabov, E.V., Leiser, R.M. and Zavriev, S.K. (1995) Use of highly conserved motifs in plant-virus RNA-polymerases as the tags for specific detection of *Carmovirus*-related RNA-dependent RNA-polymerase genes. *Virology* 207, 312–315.

Morris, T.J. and Dodds, J.A. (1979) Isolation and analysis of double-stranded-RNA from virus-infected plant and fungal tissue. *Phytopathology* 69, 854–858.

Mumford, R.A., Barker, I. and Wood, K.R. (1996) An improved method for the detection of *Tospoviruses* using the polymerase chain reaction. *Journal of Virological Methods* 57, 109–115.

Mumford, R.A., Walsh, K., Barker, I. and Boonham, N. (2000) Detection of *Potato mop top virus* and *Tobacco rattle virus* using a multiplex real-time fluorescent reverse-transcription polymerase chain reaction assay. *Phytopathology* 90, 448–453.

Mutasa, E.S., Chwarszczynska, D.M. and Asher, M.J.C. (1996) Single-tube, nested PCR for the diagnosis of *Polymyxa betae* infection in sugar beet roots and colorimetric analysis of amplified products. *Phytopathology* 86, 493–497.

Nagamine, K., Hase, T. and Notomi, T. (2002) Accelerated reaction by loop-mediated isothermal amplification using loop primers. *Molecular and Cell Probes* 16, 223–229.

Nagata, Y., Yokota, H., Kosuda, O., Yokoo, K., Takemura, K. and Kikuchi, T. (1985) Quantification of picogram levels of specific DNA immobilized in microtiter wells. *FEBS Letters* 183, 379–382.

Nassuth, A., Pollari, E., Helmeczy, K., Stewart, S. and Kofalvi, S.A. (2000) Improved RNA extraction and one-tube RT-PCR assay for simultaneous detection of control plant RNA plus several viruses in plant extracts. *Journal of Virological Methods* 90, 37–49.

Nie, X. and Singh, R.P. (2000) Detection of multiple potato viruses using an oligo(dT) as a common cDNA primer in multiplex RT-PCR. *Journal of Virological Methods* 86, 179–185.

Notomi, T., Okayama, H., Masubuchi, H., Yonekawa, T., Watanabe, K., Amino, N. and Hase, T. (2000) Loop-mediated isothermal amplification of DNA. *Nucleic Acids Research* 28, E63.

Okuda, M. and Hanada, K. (2001) RT-PCR for detecting five distinct Tospovirus species using degenerate primers and dsRNA template. *Journal of Virological Methods* 96, 149–156.

Olmos, A., Dasí, M.A., Candresse, T. and Cambra, M. (1996) Print-capture PCR: a simple and highly sensitive method for the detection of *Plum pox virus* (PPV) in plant tissues. *Nucleic Acids Research* 24, 2192–2193.

Olmos, A., Cambra, M., Dasí, M.A., Candresse, T., Esteban, O., Gorris, M.T. and Asensio, M. (1997) Simultaneous detection and typing of plum pox potyvirus (PPV) isolates by heminested-PCR and PCR-ELISA. *Journal of Virological Methods* 68, 127–137.

Olmos, A., Cambra, M., Esteban, O., Gorris, M.T. and Terrada, E. (1999) New device and method for capture, reverse transcription and nested PCR in a single closed-tube. *Nucleic Acids Research* 27, 1564–1565.

Olmos, A., Bertolini, E. and Cambra, M. (2002) Simultaneous and co-operational amplification (Co-PCR): a new concept for detection of plant viruses. *Journal of Virological Methods* 106, 51–59.

Olmos, A., Esteban, O., Bertolini, E. and Cambra, M. (2003) Nested RT-PCR in a single closed tube. In: Bartlett, J.M.S. and Stirling, D. (eds) *Methods in Molecular Biology: PCR Protocols*, 2nd edn. Humana Press, Totowa, New Jersey, pp. 151–159.

Olmos, A., Bertolini, E., Gil, M. and Cambra, M. (2005) Real-time assay for quantitative detection of non-persistently transmitted *Plum pox virus* RNA targets in single aphids. *Journal of Virological Methods* 128, 151–155.

Olmos, A., Bertolini, E. and Cambra, M. (2007) Isothermal amplification coupled with rapid flow-through hybridisation for sensitive diagnosis of *Plum pox virus*. *Journal of Virological Methods* 139, 111–115.

Osman, F. and Rowhani, A. (2006) Application of a spotting sample preparation technique for the detection of pathogens in woody plants by RT-PCR and real-time PCR (TaqMan). *Journal of Virological Methods* 133, 130–136.

Owens, R.A. and Diener, T.O. (1981) Sensitive and rapid diagnosis of potato spindle tuber viroid disease by nucleic-acid hybridization. *Science* 213, 670–672.

Pico, B., Sifres, A. and Nuez, F. (2005) Quantitative detection of *Cucumber vein yellowing virus* in susceptible and partially resistant plants using real-time PCR. *Journal of Virological Methods* 128, 14–20.

Robertson, N.L., French, R. and Gray, S.M. (1991) Use of group-specific primers and the polymerase chain reaction for the detection and identification of luteoviruses. *Journal of General Virology* 72, 1473–1477.

Sabanadzovic, S., Abou-Ghanem, N., Castellano, M.A., Digiaro, M. and Martelli, G.P. (2000) Grapevine fleck virus-like viruses in *Vitis*. *Archives of Virology* 145, 553–565.

Sackett, D.L., Straus, S., Richardson, W.S., Rosenberg, W. and Haynes, R.B. (2000) *Evidence Based Medicine. How to Practise and Teach EBM*, 2nd edn. Churchill Livingstone, Edinburgh, UK, pp. 67–93.

Saiki, R.K., Scharf, S., Faloona, F., Mullis, K.B., Horn, G.T., Erlich, H.A. and Arnheim, N. (1985) Enzymatic amplification of beta-globin genomic sequences and restriction site analysis for diagnosis of sickle-cell anemia. *Science* 230, 1350–1354.

Singh, M. and Singh, R.P. (1997) Potato virus Y detection: sensitivity of RT-PCR depends on the size of fragment amplified. *Canadian Journal of Plant Pathology* 19, 149–155.

Song, Y.R., Ye, Z.H. and Varner, J.E. (1993) Tissue-print hybridization on membrane for localization of mRNA in plant tissue. *Methods in Enzymology* 218, 671–681.

Varga, A. and James, D. (2005) Detection and differentiation of *Plum pox virus* using real-time multiplex PCR with SYBR Green and melting curve analysis: a rapid method for strain typing. *Journal of Virological Methods* 123, 213–220.

Varga, A. and James, D. (2006a) Real-time RT-PCR and SYBR Green I melting curve analysis for the identification of *Plum pox virus* strains C, EA, and W: effect of amplicon size, melt rate, and dye translocation. *Journal of Virological Methods* 132, 146–153.

Varga, A. and James, D. (2006b) Use of reverse transcription loop-mediated isothermal amplification for the detection of *Plum pox virus*. *Journal of Virological Methods* 138, 184–190.

Wetzel, T., Candresse, T., Ravelonandro, M. and Dunez, J. (1991) A polymerase chain-reaction assay adapted to *Plum pox potyvirus* detection. *Journal of Virological Methods* 33, 355–365.

Wetzel, T., Candresse, T., Macquaire, G., Ravelonandro, M. and Dunez, J. (1992) A highly sensitive immunocapture polymerase chain-reaction method for *Plum pox potyvirus* detection. *Journal of Virological Methods* 39, 27–37.

Wilson, I.G. (1997) Inhibition and facilitation of nucleic acid amplification. *Applied and Environmental Microbiology* 63, 3741–3751.

Wittwer, C.T., Ririe, K. and Rasmussen, R. (1998) Fluorescence monitoring of rapid cycle PCR for quantification. In: Ferre, F. (ed.) *Gene Quantification*. Birkhauser, New York, pp. 129–144.

Yourno, J. (1992) A method for nested PCR with single closed reaction tubes. *PCR Methods and Applications* 2, 60–65.

Youssef, S.A., Shalaby, A.A., Mazyad, H.M. and Hadidi, A. (2002) Detection and identification of *Prune dwarf virus* and *Plum pox virus* by standard and multiplex RT-PCR probe capture hybridization (RT-PCR-ELISA). *Journal of Plant Pathology* 84, 113–119.

11 Molecular Identification and Diversity of Phytoplasmas

G. Firrao, L. Conci and R. Locci

Abstract

Phytoplasmas are a large, diverse group of organisms and causal agents of plant diseases of great economic relevance. Since they cannot be grown in axenic culture, several aspects of phytoplasma biology are still poorly understood. An absolute prerequisite for the control of phytoplasma diseases is the availability of accurate information on the nature of the causal agents. For many years, knowledge of phytoplasmas was restricted to the micromorphological, phytopathological and epidemiological aspects. Over the last 10–15 years, the application of molecular techniques has rapidly improved our knowledge. These non-cultivable organisms have become suitable targets for phylogenetic and genomic studies. A newly acquired ability to positively identify phytoplasmas has had a substantial impact on disease management and the understanding of epidemics. More recently, the decoding of the genome sequence of two 'Candidatus Phytoplasma asteris' strains has helped to understand the nutritional requirements of phytoplasmas and the molecular mechanisms underlying their relationships with plant and insect hosts, providing novel perspectives on cultivation and disease control. The identity, diversity and genomics of these organisms are reviewed in detail with the view to exploiting this knowledge for disease management.

Introduction

Phytoplasmas are wall-less bacteria associated with hundreds of plant diseases. They are peculiar among prokaryotes in being intracellular parasites of organisms belonging to two different kingdoms, *Plantae* and *Animalia* (insects). So far, the phytoplasmas have resisted *in vitro* cultivation, and, therefore, their physiology and interaction with the host remained poorly understood. Molecular biology is now changing the picture. The introduction in the early 1990s of methods for gene amplification (Deng and Hiruki, 1991) and selective phytoplasma DNA preparation (Sears *et al.*, 1989), with the consequent decoding of phytoplasmal genomic features, has opened up new perspectives in the understanding of these prokaryotes and the diseases they cause.

This chapter focuses on two key topics of phytoplasma biology, i.e. molecular diversity and genomics, to demonstrate how biotechnology has contributed and/or is expected to contribute to our knowledge of epidemiology and to the development of control strategies against phytoplasma diseases.

Background

Yellows diseases have been known since the early 1900s. The first report on aster yellows appeared in 1902 (Kunkel, 1926), its causal agent being thought to be a virus. In 1967, some Japanese scientists, upon electron microscopic examination of young leaves with yellows symptoms, failed to find any virus-like particles, but observed pleomorphic bodies in the sieve tubes. The newly discovered plant pathogens were named mycoplasma-like organisms (MLOs) because of their morphological similarity to the cells of *Mycoplasma* spp (Doi *et al.*, 1967). The MLOs were renamed phytoplasmas, when it became clear that they were mollicutes phylogenetically, clearly distinct from *Mycoplasma* spp.

Phytoplasmas are surrounded by a single unit membrane, but lack a rigid cell wall and are sensitive to the antibiotic, tetracycline (Doi *et al.*, 1967). Under the transmission electron microscope, they appear as pleomorphic bodies (roundish to filamentous) with diameter ranging from 200 to 800 nm (Kirkpatrick, 1991). They inhabit the phloem sieve elements of infected plants and the gut, haemolymph, salivary and other organs of sap-sucking insect vectors (Kirkpatrick, 1991).

The genome size is 530–1350 kbp (Marcone *et al.*, 1999) and the G + C content 23–29 mol% (Kollar and Seemüller, 1989). Phytoplasmas typically possess a tRNA^{ILE} in the spacer region between the 16S and the 23S rRNA genes (Smart *et al.*, 1996). Phylogenies based on the 16S rDNA gene have shown that the phytoplasmas constitute a monophyletic group within the *Mollicutes* (Gundersen *et al.*, 1994; Seemüller *et al.*, 1994), *Acholeplasma palmae* being their closest culturable relative.

Phytoplasmas are transmitted from plant to plant by phloem-sucking insects belonging to the order *Hemiptera*, or more specifically to the superfamilies *Membracoidea* (within which all vectors known to date are *Cicadellidae*), *Fulgoromorpha* (mostly *Cixiidae*) and *Psylloidea* (a few genera of the *Psyllidae*). The insects acquire phytoplasmas passively from the phloem of infected plants during feeding. Phytoplasmas are widespread in the gut, haemocele and salivary glands of arthropods and may be introduced into healthy plant tissues, thus causing disease (Kirkpatrick, 1991). Despite several efforts over 40 years, none of the phytoplasmas infecting insects and plants has been cultivated *in vitro* in axenic culture.

Symptom expression

Phytoplasma infections interfere severely with the physiological and biochemical processes of the host plant (Christensen *et al.*, 2005). Consequently, they cause a wide variety of symptoms that include phyllody (i.e. development of floral parts into leaf structures), virescence (development of green flowers and loss of normal flower pigments), hypertrophy of the flower buds, absence of seed production, premature drop of fruit, proliferation of adventitious or axillary shoots resulting in a witches' broom appearance, abnormal elongations of internodes, generalized decline and stunting (small flowers and leaves and shortened internodes). Other symptoms include unseasonal yellowing or reddening of leaves affecting part or the entire foliage, flower and foliage necrosis, leaf curling or cupping, bunchy appearance of growth at the ends of the stems, and yellow-brown discoloration of roots (Fig. 11.1).

Fig. 11.1. Symptoms of some major phytoplasma diseases: (A) yellows in grapevine flavescence dorée; (B) enlarged stipulae in apple proliferation; (C) stunting and witches' broom in lucerne (ArAWB); (D) decline in China tree.

The symptomatology induced in diseased plants and the lengths of the incubation period vary with the phytoplasma, host genotype, stage of infection and physiological state of the host. Progression of the disease up to the possible death of the host can be rapid, as for papaya dieback (PpDB) (4 weeks) (Guthrie *et al.*, 2001) or slow as in China tree (*Melia azedarach* L.) decline (2–3 years; Vázquez *et al.*, 1983). Similar symptoms can be induced by different types of phytoplasmas, whereas different symptoms may be caused by closely related agents (Davis and Sinclair, 1998).

Inside the plant, the infection can cause abnormal deposition of callose in the sieve areas, followed by collapse of the sieve tubes and the companion cells, and excess formation of phloem, resulting in swollen veins and necrosis of the phloem tissues. In plum and apple trees, a significant increase in phenolic compounds, possibly related to defence mechanisms (inhibition of the pathogen inside the phloem vessels), has been observed (Musetti *et al.*, 2000). In general, symptoms induced by phytoplasmal infection have a clear detrimental effect on the host, although some plant species are tolerant or resistant to the infection. These plants may be symptomless or just exhibit mild symptoms. Economic losses caused by phytoplasmal infections range from partial reduction in yield and quality to nearly total crop loss (Lee *et al.*, 2000).

Geographic distribution

Phytoplasmas occur worldwide (McCoy *et al.*, 1989). In the last decade, the development of sensitive, specific assays for their identification and comprehensive classification schemes has greatly advanced the diagnostics of diseases caused by phytoplasmas. Numerous diseases of previously unknown etiologies were found to be caused by phytoplasmas (Lee *et al.*, 2000).

Several phytoplasma disorders seem to be confined to specific geographical regions. For instance, apple proliferation (AP) is restricted to Europe, X-disease of plum and peach to the Americas, peanut witches' broom and rice yellow dwarf to South-east Asia (Lee *et al.*, 2000). However, for some parts of Eastern Europe, including the former Soviet Union, Africa, Central Asia and South America, little is known about the geographical and ecological occurrence of phytoplasmas and the host species infected.

The geographic distribution of some phytoplasmas seems to be correlated to that of their host plants and the vectors that are native in a particular region. For instance, maize bushy stunt is restricted to Central, South and part of North America where the vectors, *Dalbulus maidis* and *Dalbulus elimatus*, live and feed exclusively on maize and teosintes (Nault, 1980).

On the other hand, phytoplasmas associated with native hosts can be dispersed and redistributed throughout geographical regions when foreign germplasm is introduced into a new area. This is possibly the case with the alfalfa witches' broom (ArAWB) in Argentina. The ArAWB was determined to be a subgroup of the 16SrVII group,[1] till now detected only in Argentina (Conci et al., 2005). The high sequence similarity found between the ArAWB and Erigeron witches' broom (from Brazil) supports the hypothesis of an evolutionary divergence occurred in South American phytoplasmas, originating from geographical or ecological isolation (Montano et al., 2000), whereas ArAWB is possibly a natural pathogen infecting native plants of the region. Lucerne was introduced into Argentina during the 16th century and probably in the presence of a new host and that of a vector capable of transmitting the pathogen from native plants, started with the pathogen dispersion in the area of the new crop.

Economic significance

There are several phytoplasma diseases which have a relevant economic and social impact in different parts of the world. Those which have caused major concern in recent years, both for agricultural crops and landscape plants, are described below.

Palm diseases
Lethal yellowing (LY) is a devastating disease that affects more than 38 species of palms throughout the Caribbean region. The effects of the disease are more conspicuous for *Cocos nucifera* L. palms than for other taxa, since this abundant species is typical of the regional landscape (Harrison et al., 1994). Reports of dying coconut palms, exhibiting LY-type symptoms, date back to the 19th century in the Caribbean region (Eden-Green, 1997). Over the last 30 years, LY epidemics have killed millions of palm trees, resulting in serious economic losses for thousands of families whose income depends directly or indirectly on coconuts (Oropeza et al., 2005). In Jamaica, LY is the single most important plant disease affecting the coconut industry. Since 1961, the LY disease has killed almost 90% of the original population of coconut palms. In addition to the economic consequences, there has been a social and cultural impact resulting from the disappearance of a species that symbolized the local vegetation.

Till now, the most effective control is replanting affected areas with resistant germplasm. In general, the highest levels of resistance have been found in domesticated genotypes such as Malayan yellow dwarf (MYD); wild-type materials have shown the lowest levels of resistance. But, unfortunately, more recent unusual high losses of resistant hybrid coconuts have been reported, up to two-thirds of the resistant genotypes (Harrison et al., 2002).

LY phytoplasma is a representative member of the 16SrIV group (Harrison et al., 1994), and the planthopper, *Myndus crudus*, is apparently involved in transmission (Howard and Thomas, 1980). The possibility of seed transmission has also been considered because DNA of the LY phytoplasma has been

---

[1] Here and later in the text, the phylogenetic group to which the phytoplasma belong is reported according to Lee et al. (2000).

detected in embryo tissues from fruits of diseased 'Atlantic tall' coconut palms by polymerase chain reaction (PCR) analysis (Cordova et al., 2003).

Similar diseases known as lethal yellowing-type diseases (LYD) have also been described in several African regions (Eden-Green, 1997). Like LY, LYD have destroyed millions of palms, causing enormous losses particularly in Ghana and Tanzania. The African phytoplasmas are different from the American LY and there are differences between phytoplasmas from the East and the West African coast (Tymon et al., 1997).

Papaya diseases
Phytoplasmas are associated with several papaya (*Carica papaya* L.) diseases in different regions of the world. In Australia, papaya dieback, caused by '*Ca*. P. australiense' (16SrXII) (Liefting et al., 1998), is responsible for annual plant losses of 10–100% in Central and Southern Queensland plantations (Glennie and Chapman, 1976). Symptom progression and death of the plant is quite fast, often occurring within 4 weeks from the appearance of the first symptoms. The incidence of papaya yellow crinkle (PpYC, 16SrII) is usually low and sporadic, although outbreaks can also occur. Papaya mosaic (PpM, 16SrII) is generally of minor importance and hardly distinguishable from PpYC (De La Rue et al., 1999). The phytoplasmas associated with PpYC and PpM have been named '*Ca*. P. australasia' (White et al., 1998), but the use of this name is discouraged, as its differentiation from '*Ca*. P. aurantifolia' does not conform to the rules for the definition of '*Ca*. Phytoplasma' spp. (Firrao et al., 2005).

In Cuba, papaya is affected by a different disease, called papaya bunchy top, transmitted by *Empoasca papayae* and associated with another 16SrXII phytoplasma, recently described as '*Ca*. P. caricae' (Arocha et al., 2005).

China tree yellows disease
The China tree is a fast-growing species from South-east Asia that was introduced into South America as an ornamental shade tree and for commercial purposes. In spite of its insecticidal and antifungal properties, the plant is severely affected by yellows diseases. Symptoms include leaf yellowing and size reduction, shoot proliferation and progressive decline. Phytoplasma infections have been reported in several South American countries, such as Argentina, Paraguay, Brazil and Bolivia (Gomez et al., 1996; Harrison et al., 2003). China tree decline caused by phytoplasmas is a serious problem in Argentina, both in the north-east, where this species is mainly grown for furniture, and in the rest of the country, where it is used as an ornamental shade tree. By molecular characterization, it was possible to assign the isolates from Argentina, Bolivia and Paraguay to the 16SrIII and 16SrXIII groups, often occurring in mixed infections (Harrison et al., 2003; Galdeano et al., 2004; Arneodo et al., 2005).

Almond witches' broom
Stone fruits (mostly almond) are major fruit crops in Lebanon, and the almond witches' broom (AlmWB) has devastated almond production. Stunting, rosetting and dieback are the main symptoms (Abou-Jawdah

et al., 2002). The associated phytoplasma (16SrIX) was designated '*Ca.* P. phoeniceum' (Verdin *et al.*, 2003) and appears to also attack peach and nectarine seedlings in areas where almond witches' broom prevails. Thousands of young and old almond trees have died over the last 15 years due to the rapid spread of the disease by an unknown vector.

European fruit trees phytoplasmas
In Europe, three closely related phytoplasmas ('*Ca.* P. mali', '*Ca.* P. pyri' and '*Ca.* P. prunorum') cause diseases of economic concern, named AP, pear decline (PD) and European stone fruit yellows (ESFY) (Seemüller and Schneider, 2004). All three pathogens are transmitted by psyllids, a peculiarity among phytoplasmas. AP is a major economic threat in some areas where apple trees are intensely cultivated; in the Trentino region, a mean incidence of 7.1% was reported, with infected plants reaching up to 50% of the total in some cultivars such as 'Golden Delicious' and 'Renette du Canada' (Branz, 2003). Symptoms of AP include elongated stipules, leaf rosettes, fruit size reduction, witches' broom appearance and proliferation of dormant auxiliary buds in summer.

Grapevine yellows
A cause of great concern in Europe is grapevine yellows. The disease may be caused by different phytoplasmas, but the most widespread ones belong to the 16SrV (flavescence dorée (FD)-associated phytoplasma) and the 16SrXII (bois noir (BN) and Vergilbungskrankheit-associated phytoplasmas) groups. The symptoms of grapevine yellows (leaf discoloration, rolling down of laminae, flower withering and berry shrivelling, reduction in quality and quantity of the crop) are similar despite the diversity of the causative agents. Epidemiology and severity of the disease are also significantly different. Since its discovery in 1957 (Caudwell, 1957), FD is considered a major problem in the wine industry, particularly in France and Italy.

Unveiling Molecular Identity and Diversity of Phytoplasmas

Early methods

According to the accepted standards (ICSB, 1995), the classification of culturable mollicutes is based on both phenotypic and genotypic characteristics. As the phytoplasmas have not been grown *in vitro*, their phenotypic characters are inadequate for formal classification.

Before the introduction of molecular methods, the identity of phytoplasmas was determined by evaluating their biological properties (e.g. plant-induced symptomatology, range of plant hosts and specificity of transmission by vectors) and the geographic area where the associated diseases occurred (e.g. Italian alfalfa witches' broom, Yucatan coconut lethal decline and ESFY) (Seemüller *et al.*, 1998). A classification of phytoplasmas maintained on periwinkle (*Catharanthus roseus*) by experimental transmission was attempted by Marwitz (1990); his classifications, based on the symptoms that developed, show some similarities to the current genetic classification, although in several cases, genetically different phytoplasmas were grouped together, whereas others were considered to be different, and are indeed genetically closely related (Seemüller et al., 1998). Although the symptomatology remains an important clue for preliminary characterization of phytoplasmal diseases, in recent years, the analysis of the 16S rDNA genes obtained by PCR amplification showed that symptomatically similar diseases may be caused by very distinct organisms. For example, clover is a known host of at least seven different phytoplasmas, and four of them induce the same symptoms. As a further complication, even in a single plot, individual clover plants hosting several different phytoplasmal agents could be detected (Firrao et al., 1996a).

The introduction of immunological methods provided the first tools for the prompt detection and molecular differentiation of phytoplasmas. Serological detection depends on the availability of specific antisera obtained from phytoplasma proteins purified from infected plants. So far, both polyclonal and monoclonal antibodies against several phytoplasmas have been produced, including those associated with major diseases such as aster yellows, western X-disease, AP, sweet potato witches' broom, grapevine FD, stolbur, faba bean phyllody, rice yellow dwarf, China tree decline and more (refer to Firrao et al. (2007), for a comprehensive list of references). Generic antisera for all phytoplasmas have never been developed, but various antibodies showed different degrees of specificity.

Immunological assays are relatively simple, cost-efficient and can be used for large-scale tests. However, employing polyclonal antisera for phytoplasma detection is only moderately successful because of the high reactivity against plant proteins due to the difficulty in obtaining highly purified phytoplasmal antigens. This raises the background values in healthy controls, and low titre infections may be difficult to detect (Jiang and Chen, 1987; Saeed et al., 1993). Monoclonal antibodies are ideal for tissue immunomicroscopy and tissue blotting experiments to localize phytoplasmas in plants and insects, but for routine field diagnostic application, their monospecific antigen recognition characteristics should be considered in relation to the relatively high strain to strain variability in phytoplasmas.

DNA-based Techniques

Specific detection of numerous phytoplasmas in plant and/or insect hosts has been achieved by hybridization assays using cloned DNA. Several phytoplasma genomic groups have been identified in the past and their relatedness evaluated using reciprocal dot hybridizations with a number of probes (Kirkpatrick et al., 1987; Lee and Davis, 1992). Later, PCR with oligonucleotides derived from DNA of phytoplasma rRNA or anonymous DNA fragments has increased detection sensitivity compared to previously employed DNA probes.

PCR using phytoplasma group-specific or universal primers derived from conserved 16S rRNA gene sequences has provided the most sensitive and versatile tool for the preliminary diagnosis of a broad array of phytoplasmas from plants or vectors, even with very low pathogen titres (Ahrens and Seemüller, 1992). However, 16S rRNA sequences of related phytoplasmas are very similar, thus making it difficult or impossible to design PCR primers capable of specifically identifying a particular phytoplasma. Nevertheless, primer pairs could be developed that efficiently provide selective amplification of rDNA only from strains belonging to certain, phylogenetically well-defined, strain clusters (Lee et al., 1994). The shorter, but less-conserved, intergenic spacer region between 16S and 23S rDNA has also proven useful for the development of specific phytoplasma PCR primers for some groups, such as elm yellows, aster yellows, ash yellows and AP clusters (Smart et al., 1996). An additional advantage of the 16S/23S spacer for diagnostic purposes is that its length varies from group to group of phytoplasmas. Hence, phytoplasmas may be differentiated by the length polymorphism of their undigested amplification product obtained with general primers (Palmano and Firrao, 2000).

Restriction fragment length polymorphism (RFLP) analysis of PCR-amplified 16S rDNA using a number of restriction endonucleases is routinely used for identification and preliminary classification of phytoplasmas. Unknown phytoplasmas can be identified by comparing their patterns with those of known phytoplasmas. Classification by RFLP analysis is a simple and rapid method and can be used for large numbers of unknown phytoplasmas (Lee et al., 1998). The groups delineated by PCR-RFLP of 16S rDNA, comprising about 15 groups and 50 subgroups, are consistent with phylogenetic studies based on analysis of full-length 16S rRNA gene sequences from representative phytoplasmas (Lee et al., 2000). The groups are also in agreement with those delineated by methods based on ribosomal protein genes, elongation factor Tu gene (*tuf*) or 16S/23S intergenic spacer region sequences (Kirkpatrick et al., 1994; Schneider et al., 1997; Lee et al., 1998). Only in a few instance, analyses using PCR-RFLP were not consistent with groups based on 16S rRNA gene sequences (Lee et al., 1998). The combined analyses of both 16S rDNA and a ribosomal protein gene fragment (or of a selected chromosomal fragment) sequences are presently regarded as the gold standard to ascertain phytoplasma identity (Lee et al., 1998).

Nested PCR assay, designed to increase sensitivity, is used for the amplification of phytoplasmas from samples with unusually low titres or when inhibitors that interfere with PCR efficiency are present.

Taxonomy and the genus '*Candidatus* Phytoplasma'

By taking advantage of the ability of PCR to amplify target DNAs even in the presence of excess extraneous nucleic acids, more than 200 phytoplasmas have been characterized during the last 15 years by nearly full-length sequencing of their 16S rRNA gene. Reliable protocols for the amplification of 16S rDNA permitted the systematic study of phytoplasma diversity and helped to clarify the phylogenetic structure of these organisms. The phylogenetic information, however, is not sufficient for a formal taxonomic description according to the minimal standards required for new bacteria (ICSB, 1995), and probably such a simplification would not even be desirable. A merely phylogenetic classification would not be able to take into consideration the aspects of coevolution with the host that are dramatically important in organisms such as the intracellular pathogenic phytoplasmas.

In 1994, Murray and Schleifer proposed the category *Candidatus* 'to provide a proper record of sequence-based potential new taxa'. The *Candidatus* concept is a significant starting point in the nomenclature of bacteria, since it introduces the possibility of referring to a particular organism without having to culture it on an artificial substrate. The idea has had a major impact on the nomenclature of the phytoplasmas, although its implementation has not been so immediate, since care has been taken to preserve both the phylogenetic structure of the group and the potentially significant biological information implied by the strict and sometime highly specific association of the phytoplasmas with their hosts.

A solution to the problem was the proposal to separate levels of similarity, i.e. introducing percentage thresholds, while considering existing ecological separations. Accordingly, the *Candidatus* spp. name should refer to a single, unique 16S rRNA gene sequence, and a strain sharing more than 97.5% of its 16S rRNA gene sequence with an already described species should not be described as a new species, unless it is demonstrated that the organism belongs to an ecologically distinct population. The Phytoplasma/Spiroplasma Working Team – Phytoplasma taxonomy group of the International Research Programme for Comparative Mycoplasmology (IRPCM, 2004) has implemented the rules in a paper, which lists about 30 described or suggested '*Candidatus* Phytoplasma' spp. The concept of distinguishing *Candidatus* spp. even when the 16S rDNA sequences are very similar, on ecobiological grounds, not only is in agreement with the current trends of the taxonomy that suggest an elastic definition of species (Rossello-Mora and Amann, 2001), but also has been supported by the genomic data. This will be discussed later in more detail and shows the relevance of the fraction of genes that are devoted to the relationship with the host cellular environment.

Since phytoplasmas lack a formal taxonomic description, obvious type strains are not defined. However, in most cases, '*Candidatus* Phytoplasma' spp. descriptions include the definition of a 'reference' strain useful for comparative purposes.

Figure 11.2 shows the presently recognized 24 '*Ca.* Phytoplasma' spp., arranged according to their phylogenetic position. The figure also includes the phylogenetic additional phytoplasmas that have been proposed as novel species, but that still await formal proposal (IRPCM, 2004). For more details on individual species, the reader is referred to a recent review (Firrao *et al.*, 2005).

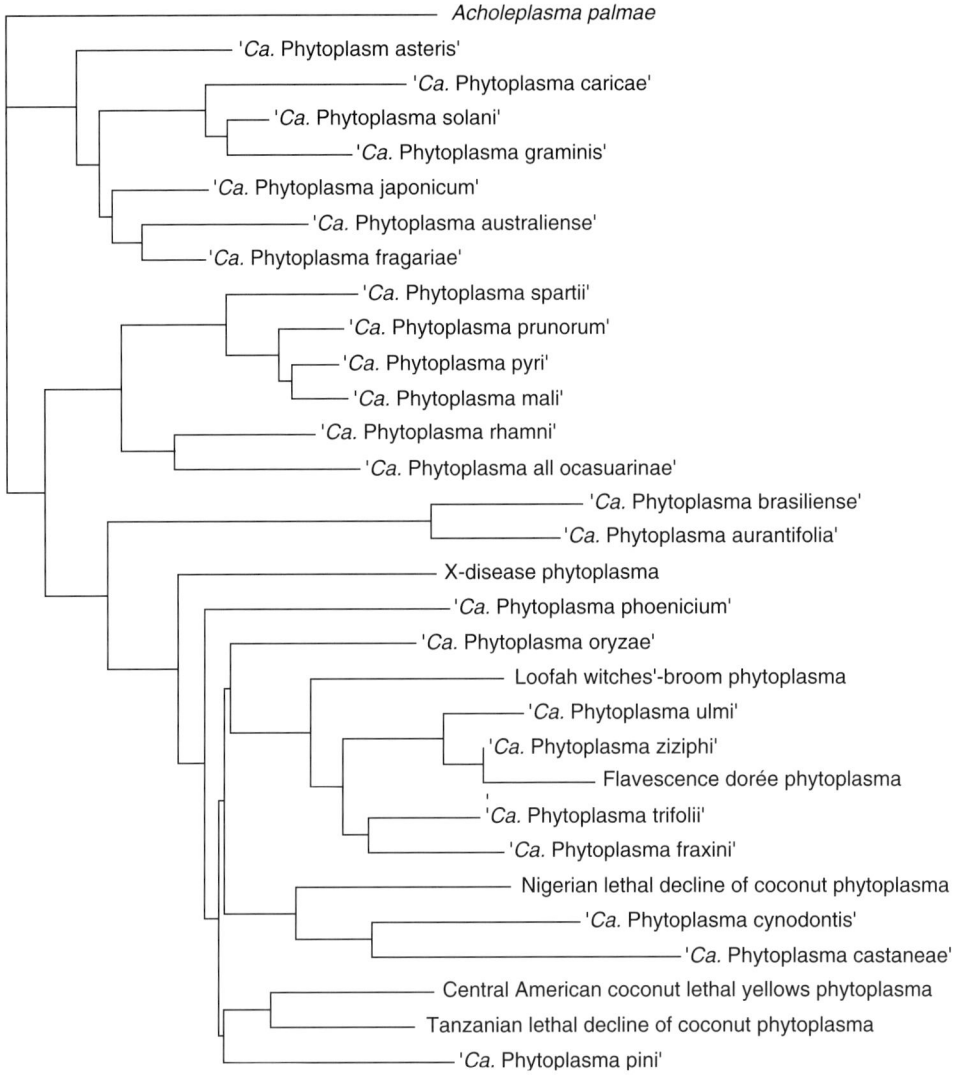

Fig. 11.2. 'Candidatus Phytoplasma' spp. phylogenetic tree based on 16S rDNA sequences. *Acholeplasma palmae* is included as an outgroup to root the tree.

Exploiting molecular diversity

Control of phytoplasma diseases consists essentially of adopting measures capable of avoiding, or at least reducing, plant-to-plant transmission by phloem-feeding insects. In actual practice, this is carried out by a series of phytosanitary procedures, such as spraying insecticides, planting healthy propagation material, eradicating diseased plants that may cause outbreaks and so on. These measures are part of good cultivation practices; however, their efficiency depends strongly on correct diagnostic tools which have only recently become available thanks to progress in the molecular biology of phytoplasmas, as exemplified by the following case studies.

Grapevine yellows
In Europe, the major causes of grapevine yellows are FD and BN phytoplasmas. The agents cause very similar, if not indistinguishable, symptoms on grapevine, although belonging to two phylogenetically clearly distinct groups, i.e. 16SrV and 16SrXII, respectively (Davis et al., 1993). The ability to distinguish between the two phytoplasmas using 16S Rdna-based molecular tools has elucidated their very different epidemiological properties. The FD agent is transmitted from grapevine to grapevine by the leafhopper *Scaphoideus titanus* and may cause severe outbreaks and significant yield losses on many cultivars. Conversely, the BN agent has several herbaceous hosts, which are transmitted by *Hyalesthes obsoletus* (*Cixiidae*), and infects almost exclusively the cultivar 'Chardonnay', causing far less economic losses than FD. Control of BN by means of eradication is unrealistic, since the disease is not transmitted from grapevine to grapevine, but rather from weeds to grapevine. Pesticide treatments are not effective either, as the potential insect vectors are present on grasses most of the time.

In contrast, eradication and insecticide sprays effectively control FD in the Italian region of Friuli Venezia Giulia (FVG). Its introduction was delayed for several years and has since been kept at very low levels by a large-scale diagnostic programmes based on tests, such as PCR analysis of the 16S rDNA, capable of discriminating between BN and FD. While BN, introduced into FVG in the 1980s, is endemic to the region, the areas where FD was recently introduced could be promptly identified and subjected to programmes of eradication and regular spraying with insecticides. As a result, the reported yield losses due to grapevine yellows in the triennium 2001–2003 were notably lower in FVG than in other regions of northern Italy. According to recent estimates (Frausin and Osler, 2004), the economic losses were about €46/ha in FVG, compared to €490/ha in Piedmont during the same period.

Stone fruit phytoplasmas
Stone fruits have been recognized as hosts of phytoplasmas for many years. Among the diseases reported in the literature are X-disease of cherry and peach, peach yellow leaf roll (PYLR), peach rosette, moliere disease of cherry, apricot chlorotic leaf roll, almond witches' broom, plum leptonecrosis and others. Ascertainment of their phylogenetic identity has shown that the disease agents are grouped on a geographical basis. Phytoplasmas belonging to 16SIII and 16SrX groups cause diseases in the USA and Canada, '*Ca.* P. phoenicium' in the Middle East, whereas all diseases of stone fruit trees in Europe are always associated with '*Ca.* P. prunorum' (Lorenz et al., 1994). It should be remembered that not all stone fruits are equally susceptible to the phytoplasmas. For example, '*Ca.* P. prunorum' causes symptoms on the Japanese plum (*Prunus salicina*), but not on the European plum (*Prunus domestica*) (Carraro et al., 1998a). Nevertheless, testing by PCR showed that the European plum is often infected and, like other wild species, may contribute significantly to the disease epidemics in *P. salicina* (Carraro et al., 2002).

Insect vectors

Identification of the vectors of phytoplasmas is an important task in the control of phytoplasma diseases, as insecticidal sprays may significantly reduce the rate of plant-to-plant transmission. Thus, the search for the insect vector of fruit tree phytoplasmas has attracted the attention of researchers for a long time. In California, during the 1960s, the vector of PD was identified as a psyllid (Jensen et al., 1964). It was later shown (Purcell and Suslow, 1984) that psyllids could also transmit PYLR. Later, research on the phylogeny of phytoplasmas confirmed that the causative agents of PD and PYLR are closely related. Similarly, the phytoplasmas causing AP and ESFY are related to the above. Consequently, the search for the vectors, initially focused on leafhoppers, was redirected to psyllids and finally led to the identification of *Cacopsylla costalis* (Frisinghelli et al., 2000) and *C. melanoneura* (Tedeschi et al., 2002) as vectors of AP and *C. pruni* as a vector of ESFY (Carraro et al., 1998b).

The Phytoplasma Genome

Genome sequencing projects

While the unveiling of phytoplasma diversity by molecular tools has been valuable in disease control, genomic data are expected to have an even greater impact. Decoding the genomic information is the most obvious way of clarifying the biology of unculturable organisms. The genome of the phytoplasmas resembles that of the animal mycoplasmas, being 0.5–1.3 Mbp in size (Marcone et al., 1999). The small size facilitates sequence determination as well as understanding and reconstruction of the energy and metabolite fluxes within and in or out of the cell. Until recently, only very limited sequence information on the phytoplasma genome was available. As mentioned earlier, low concentrations of the agents in plants meant that host nucleic acids were a major source of contamination in DNA preparation. It was by exploiting different molar guanine-and-cytosine percentage of the host versus pathogen DNAs that the two nucleic acids could be separated. In the mid-1980s, the repeated CsCl buoyant density gradient centrifugation of total nucleic acids from phytoplasma-infected insects or plants led to the first cloning of phytoplasma DNA (Kirkpatrick et al., 1987). In general, the procedure is not completely capable of removing host nucleic acid contamination, and its efficiency varies greatly in different plant and pathogen combinations. Only very recently the complete genome sequence of '*Ca.* P. australiense' has been obtained using DNA purified by CsCl buoyant density gradient centrifugation, exploiting the high G + C mol% of the natural host strawberry (Liefting et al., 2006).

An alternative technique, leading to extremely pure DNA, has been developed first using pulsed field gel electrophoresis (PFGE) for the visualization (Neimark and Kirkpatrick, 1993) and then the purification and analysis (Firrao et al., 1996b; Liefting and Kirkpatrick, 2003) of intact or digested chromosomes have been prepared in agarose plugs. Using this technique, the size of about 70 different phytoplasmas was calculated (Marcone et al., 1999) and the physical maps of the chromosomes of the western X-disease (Firrao et al., 1996b), AP (Lauer and Seemüller, 2000), sweet potato little leaf (Padovan et al., 2000) and ESFY (Marcone and Seemüller, 2001) phytoplasmas were determined.

Owing to great variability, the genome size can no longer be used as a taxonomic key for differentiating taxa in the mollicute class; however, it must be emphasized that acholeplasmas and spiroplasmas, which are considered phylogenetically 'early' mollicutes, have larger chromosomes than the more 'recent' mycoplasmas and phytoplasmas. This is in agreement with the notion that *Mollicutes* have evolved by reductive evolution, accompanied by significant losses of genomic sequences (Razin et al., 1998; Oshima et al., 2004). The bermudagrass white leaf phytoplasma has the smallest sized genome (530 kbp) found so far in phytoplasmas and the smallest genome known in any living cell (Razin et al., 1998).

Some phytoplasmas reach relatively high concentrations in their host, and purified DNA could be prepared in sufficient amounts for the construction of genomic libraries. Thus, access to the complete genome sequence of two phytoplasmas has been granted in 2006, while several other projects are in progress. New information on the phytoplasma genome that can be used to construct strategies for disease control is now available.

The genome of 'Ca. P. asteris' strain OY is about 861 kbp and contains 754 open reading frames (ORFs), corresponding to 73% coding capacity. The genome of 'Ca. P. asteris' strain AY-WB is smaller, comprising 671 ORFs in about 707 kbp, but the difference between the genomes is only the result of a lower number of multicopy genes and non-coding DNA in the latter strain. Thus, the amount of sequence-coding unique ORFs is very similar in both strains, 433,226 bp in strain OY and 432,553 in strain AY-WB (Oshima et al., 2004; Bai et al., 2006). As noted for other mollicutes, the reductive evolution of the phytoplasma genome did not result in a gene density higher than in other prokaryotes. Like other mollicutes, the phytoplasmas lack genes for the biosynthesis of amino and fatty acids, tricarboxylic acid (TCA) cycle and oxidative phosphorylation (Oshima et al., 2004). While the reduction of biosynthetic genes appears to be the rule in all mollicutes, genes for pathways of intermediate metabolism have been reduced to a different extent in the various genera. Figure 11.3 shows the number of genes in some mollicutes, whose genome has recently been sequenced, their different functional classes and the peculiarity of phytoplasmas. Core functions related to information processing and storage are not significantly different in number among the various chromosomes. The mollicutes have a similar number of genes for translation, transcription and replication, notwithstanding genome size differences. Phytoplasmas are no exception. Conversely, phytoplasmas have reduced gene pools that are thought to be essential for life but not strictly connected with information processing and compartment delimitation. Compartmentalization is a focal point in phytoplasma evolution. Around 30 genes have been identified as related to membrane traffic, about twice the number found in *Mycoplasma genitalium*. Moreover, phytoplasmas have a vast portion of orphan genes that give no similarity hit in the database and whose function is unassigned. A large part of these orphan genes are probably involved in membrane traffic. In the genome of the OY phytoplasma (Oshima et al., 2004), 255 of the 754 recognized ORFs were not assigned. Among them, 91 have at least one predicted (Letunic et al., 2006) transmembrane domain. In addition, 90 of the 255 unassigned ORFs begin with a membrane export signal. We believe that two major driving forces led the phytoplasmas along uncommon tracks, where membrane traffic plays a relevant role: the genome reduction, typical of the class, on one hand, and the strict relationship with hosts belonging to two different kingdoms. Being intracellular parasites, the phytoplasmas have access to a highly elaborate, yet specific, set of nutrients in both the plant and insect environments. As shown by genomic studies, and also by taxonomy, the evolution of the phytoplasmas as organisms, including a very large number of their genes, is determined by their association with the host. This has resulted in profound genetic (gene loss), and morphological (lack of cell walls) and physiological (metabolism integration) changes in the organisms involved.

The most impressive feature of the phytoplasma genome is the absence of genes for adenosine triphosphate (ATP) synthesis, while all other mollicutes whose genome has so far been sequenced have clearly identifiable pathways for energy production. All the mollicutes examined so far have truncated respiratory systems, lack a complete TCA cycle, quinones and cytochromes and may be subdivided into fermentative and non-fermentative. The non-fermentative mycoplasmas may produce ATP through an F-type ATP synthase, which uses the transmembrane potential by the arginine dihydrolase pathway or by oxidizing organic acids such as lactate and pyruvate to acetate and CO_2 (Razin et al., 1998).

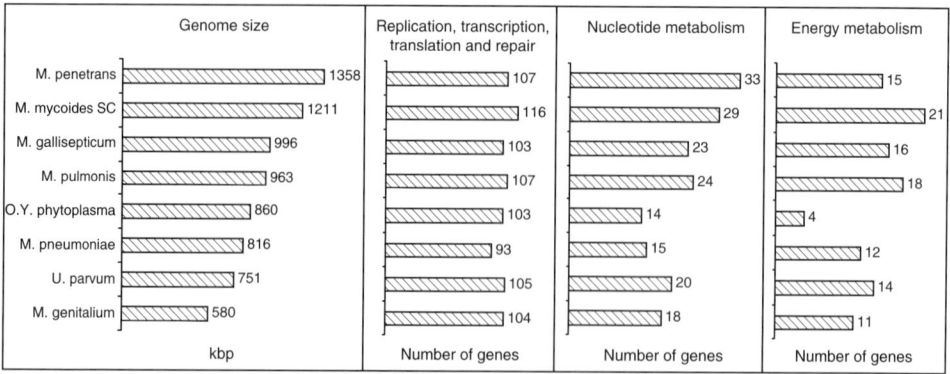

Fig. 11.3. Comparison of the number of genes for different functions in some of the *mollicutes* whose genomes have been sequenced. (From the Molligen database at http://cbi.labri.fr/outils/molligen/).

According to genome sequence data in the phytoplasmas, a complete gene set for the functional phosphoenolpyruvate/sugar phosphotransferase systems for the import of sugars, essential for glycolysis, could not be found; in addition, the pentose phosphate pathway, the arginine dehydrolase pathway and all ATP synthase systems are missing (Oshima et al., 2004; Bai et al., 2006). Due to the high conservation of genome sequences, it is unlikely that functional analogues could be misplaced in the group of orphan genes. Therefore, probably the phytoplasmas import ATP or ATP-containing molecules from the environment for their energy supply.

Other phytoplasma genes

Although complete genome information is, at present, restricted to two chromosomes and a handful of plasmid sequences, other genomic information regarding a large number of phytoplasmas has been gathered in recent years. It is mostly sequence information obtained by sequencing PCR products of reactions carried out using primer sequences deduced from highly conserved gene regions in the prokaryotes or from sequences of purified phytoplasma proteins. This approach has enabled the amplification of homologous gene sequences in a variety of different strains. In addition, several other genes have been detected by random cloning. Therefore, a significant amount of data is now available on genetic characteristics of strains whose entire chromosomal sequence has not yet been decoded.

By far, the best characterized gene is the 16S rDNA. Phytoplasmas share sequence similarity from 88% to 99% among themselves, and from 87% to 88.5% with the closely related *Acholeplasma* spp. (Seemüller et al., 1998; Lee et al., 2000). Certain 16S rDNA oligonucleotide sequences are unique for phytoplasmas, and distinguish them from *Acholeplasma* and other members of the class *Mollicutes* (IRPCM, 2004). PCR primers designed on the basis of these unique sequences from the 16S rRNA gene analysis are the basis for all current phytoplasma detection, classification and phylogenic analyses. Phytoplasmas have two copies of the 16S rDNA operon and heterogeneity between them has been demonstrated in some cases (Schneider and Seemüller, 1994; Liefting et al., 1996). The organization of these rRNA genes is preserved in the same structure and order as in other eubacteria: 16S/23S/5S (Smart et al., 1996; Ho et al., 2001; Oshima et al., 2004). In the 16S/23S spacer region of all phytoplasmal 16S rDNA operons, there is a tRNA gene for isoleucine. This does not occur in animal mycoplasmas, and acholeplasmas have at least two tRNA genes in that intergenic region (Lim and Sears, 1989; Schneider and Seemüller, 1994).

Some ribosomal protein genes from different phytoplasmas have been sequenced and analysed (*rps3*, *rps19* and *rpl22*). These are less conserved than the 16S rRNA gene, with similarities ranging from 60% to 79% within phytoplasmas and from 50% to 57% with the acholeplasmas (Lim and Sears, 1991; Gundersen et al., 1994). Protein ribosomal gene analysis has been used in phytoplasma classification in order to enhance separation within groups of strains with similar 16S rDNA (Marcone et al., 2000; Lee et al., 2004a,b). The elongation factor Tu gene (*tuf*) has also been sequenced in order to evaluate its usefulness for phytoplasma characterization and classification. The similarity values observed among phytoplasmas vary from 87.7% to 97% for different and same groups, respectively (Schneider et al., 1997), indicating that the *tuf* gene sequences are as conserved as those of the 16S rRNA gene in phytoplasmas.

Genes for major antigenic membrane proteins have been cloned and analysed in sweet potato witches' broom (Yu et al., 1998), clover phyllody (Barbara et al., 2002), AP (Berg et al., 1999), western X-disease (Blomquist et al., 2001) and aster yellows (Barbara et al., 2002) phytoplasmas. Similar characteristic domains were found, since they are all cell membrane proteins, but the proteins show significant differences, being serologically variable in different phytoplasmas. These proteins play an important role in pathogens bound to host cells and could be involved in aspects of specific transmission in host–pathogen interactions (Kakizawa et al., 2006).

The *Sec* translocation system allows protein secretion through the plasma membrane and it has been amply studied and characterized in *Escherichia coli*. Genes for SecA and SecY proteins from OY phytoplasma, which correspond to essential components of *Sec* translocation system, have been cloned and analysed. High levels of similarity have been found between OY-SecA, OY-SecY and *Bacillus subtilis* SecA and SecY. This suggests the presence of a functional *Sec* system in phytoplasmas. The transport system, due to the fact that phytoplasmas lack cell walls, might secrete their proteins directly into hosts cells (Kakizawa et al., 2001). Knowing the secretion pathway and related processes could be the first step in understanding and explaining the mechanisms of pathogenicity in phytoplasmas.

Extrachromosomal DNA

The presence of extrachromosomal DNA molecules (EC-DNA) in the cytoplasm of phytoplasmas had been postulated since early electron microscopy (EM) observations showed virus-like particles in the phytoplasmas. However, the preliminary reports were not confirmed. Therefore, phytoplasma-associated DNA molecules of low molecular weight are generally considered plasmidic in nature (Schneider et al., 1992), but their identity and functions have not yet been completely clarified.

A significant amount of sequence data has now been obtained from EC-DNAs of different phytoplasmas, which is congruent with their subdivision into two classes (Firrao et al., 2007). Type I EC-DNAs are true plasmids, as they possess a typically plasmidic replication initiator protein (Rap), similar to that used for rolling circle replication (RCR) by the pLS1 plasmid family, commonly found in Gram-positive bacteria and particularly in members of the *Clostridium* and the *Bacillus* phylogenetic clade. However, the replication initiator protein gene of the phytoplasma plasmids code for an extra C-terminal 100 a.a., whose sequence resembles a virus-like helicase domain, most similar to that of circoviruses, but it is not typical of replication proteins from RCR-type plasmids. The fully sequenced pOYW plasmid from the wild-type strain of the OY phytoplasma belongs to this class (Oshima et al., 2001).

Type II EC-DNAs have the unique property of coding a geminivirus-like replication associated protein (Rep) (Rekab et al., 1999). The type II EC-DNAs show not only prokaryotic features, such as Shine-Dalgarno, promoter sequences and typical bacterial genes, but also eukaryotic polyadenylation signals and TATA boxes, as reported earlier for the geminivirus *abutilon mosaic virus* (Frischmuth et al., 1990).

The origin of type II EC-DNAs is congruent with recombination events between plant viruses and Gram-positive bacterial plasmids within the phytoplasma cell. In these episomes, some genes may have a bacterial or a viral origin. The geminiviruses are single-stranded DNA (ssDNA) viruses which colonize the plant phloem and are vectored by various insects, including leafhoppers, which also transmit phytoplasmas belonging to several phylogenetic groups. The common habitat could have facilitated gene exchange between the viruses and the phytoplasmas, either in the plant or in the insect.

In some OY phytoplasma strains, molecules which are almost the product of recombination of a type I and a type II EC-DNA have been detected. Moreover, ssDNA-binding protein genes, highly similar to those of plasmid pOYW, have been found in several copies in the genomes of both the OY and the AY-WB phytoplasmas (Oshima et al., 2004; Bai et al., 2006).

The high sequence similarity observed among EC-DNAs of phytoplasmas of very different geographical origin and with different plant and insect hosts suggests an ancient common origin, and their possible involvement in pathogenesis, transmission or specificity determination has been considered. Despite a number of reports that preliminarily associated differences in EC-DNA to insect transmissibility (Denes and Sinha, 1992) or mild symptom induction (Kuboyama et al., 1998), the hypothesis that EC-DNA rearrangements could be involved in the modulation of virulence contrasts with the high variability in size and number of EC-DNA molecules that is usually observed among closely related strains or even within a single field population (Rekab et al., 1999; Liefting et al., 2004). Although at present the hypothesis that the EC-DNAs have a role in plant specificity and pathogenicity lacks experimental support, a function for the EC-DNA cannot be ruled out, as these molecules, despite their ancient origin, were not lost during evolution.

Exploiting genomics

Understanding the effects of phytoplasma growth on plant molecular physiology is still at an early stage, but may open the way to disease management.

It is well known that the effects of phytoplasmas vary greatly from species to species and even from cultivar to cultivar (or natural accession of *Arabidopsis thaliana*). It is therefore likely that manipulating the plant physiological regulation with *ad hoc* molecules may dramatically change the interfering effects of phytoplasmas on the host physiology.

The idea that phytoplasmas may cause symptoms on plants by depleting phloem of specific sugar molecules has gained ground in recent years, as information on the pathogenetic mechanisms of the other group of phytopathogenic mollicutes, the cultivable spiroplasmas, has emerged. Using transposon mutagenesis, *Spiroplasma citri* mutants with impaired virulence and insect transmission were identified. Due to an insertion in fructose utilization, operon mutant GMT553 was unable to use fructose, and did not induce symptoms when inoculated into plants (Gaurivaud et al., 2000). Fructose depletion was therefore identified as the cause of yellows symptoms in periwinkle infected by *S. citri*. Lepka et al. (1999) reported the occurrence of changes in the concentration of carbohydrates in the phloem of phytoplasma-infected plants, as compared to the healthy control.

Phytoplasma diseases, however, are characterized by more complex symptoms than those caused by *S. citri*, suggesting that nutrient depletion may not be restricted to fructose, but could include other compounds, depending on the pathogen involved. The genome sequence data, revealing the paucity of energy-yielding pathways in the phytoplasmas and their possible dependence on glycolysis, strongly support this view. Occlusion of the phloem vessels would result in a general reduction of sugar availability and not in the selective depletion of specific molecules. Therefore, it could be postulated that phytoplasmas utilize some sugars in preference to others and that this fact alters pathways that are based on sugar balance in plants.

There is evidence to suggest that the phytoplasmas may obtain the energy needed for their metabolism by glycolysis of host sugars. However, it is worth noting that the genome sequence of both OY and AY-WB does not show any putative IIA enzyme, a component of the phosphotransferase system, so far considered indispensable for glycolysis. The sugar import must therefore occur differently from *S. citri*, where phosphotransferase systems have been detected for fructose, glucose and trehalose (the major sugar in the insect haemolymph). Phytoplasmas have transport systems for maltose, trehalose, sucrose and palatinose and might import these, but they do not have any known gene to make these carbohydrates available for glycolysis (Oshima et al., 2004; Christensen et al., 2005). They may therefore depend on the uptake of phosphorylated hexoses as their carbon source (Christensen et al., 2005). The phytoplasmas may selectively deplete the phloem of specific sugar or sugar-like molecules that have a signalling function. It should be mentioned here that phloem sugars play a prominent role as long-distance regulators of gene expression in plants, and their depletion may produce physiological disorders, as do phytoplasma infections. Glucose and ethylene signal transduction crosstalk have been demonstrated in an *Arabidopsis* glucose-insensitive mutant

(Zhou et al., 1998). Unraveling the phytoplasmal genome is expected to highlight the physiological functions that may be the cause of symptoms in plants.

Conclusions

One may well ask now how phytoplasma taxonomy, genomics and interactions with host will affect plant protection strategies. Discussing species concepts is not just an academic exercise (Ward, 2006). In biology, it is fundamental to define a basic unit potentially of predictive value and useful for understanding the organization of microbial communities, the biology of populations involved in disease outbreaks and the emergence of new diseases. Not being confined by laboratory-based approaches, phytoplasmologists find themselves in a situation not dissimilar from that of eco-microbiologists. They tend to speak of 'ecotypes', to describe molecular populations with a unique distribution inside a community, and of 'geotypes', when dealing with geographically distinct genetic populations (Ward, 2006). At least 'ecotype' and 'geotype' are less controversial than the term 'species'. After all, the IRPCM working team accepts the idea of attributing a taxonomic status to 'ecologically distinct' phytoplasma species (IRPCM, 2004).

Nowadays, it is clear that phytoplasmas are not unique with regard to unculturability. Molecular techniques have shown that just a small portion of existing microorganisms has been studied in the laboratory and this is particularly evident in the field of ecological microbiology. Phytoplasmologists must take advantage of progress in this field where, as will be pointed out later, similar problems are encountered. Having to cope almost completely without phenotypic characterization, unculturability, to make a virtue of necessity, should not be viewed as a disadvantage. At least taxonomic problems can be tackled with an open mind.

Conventional microbiologists had to match rough data on genetic relationship (as deduced from DNA and DNA reassociation studies) with evidence from phenetic studies and sequence-based phylogenetic reconstructions. Ward (2006) gives a sound warning against the temptation to adopt molecular cut-offs to demarcate species, following the experience with culturable bacteria. 'Gold standard' cut-offs, such as 30% variation in DNA–DNA hybridization and 2–3% variation in 16S rRNA sequences, should be adopted with care. The simple reason is that in culturable bacteria 'molecular sequence data have been used to calibrate the amount of sequence divergence of species that were named on phenotypic grounds, hence remaining faithful to the notion that phenotypically defined species are true species' (Ward, 2006). It really looks like a dog chasing its own tail. Conversely, the phytoplasmologists will likely approach taxonomic, genetic, phylogenetic and functional relationships from the genome comparison perspective. At least ten different genome sequences are expected to be completed by 2008. With proper software solutions, genetic relationships equivalent to estimates from DNA and DNA reassociation studies could be obtained. Phylogenies based on multilocus sequence comparison would accurately describe the evolution process, and the understanding of genome functions would comparatively clarify how and which cellular tasks are carried out. Preliminary results of genome comparison among distantly related phytoplasmas (B.C. Kirkpatrick, Cambridge, 2006, personal communication) suggest that there is a common 'core' of gene functions shared by all phytoplasmas, and a set of additional specific genes, probably related to the particular lifestyle and host preferences of a given phytoplasma. In this case, therefore, the genome structure and function will dictate the guidelines defining cut-off values for species and genus definition, and not just based on subjective judgements.

The study of relationships with the hosts will benefit from the genomic data being acquired and the experience gained in recent years in the study of host–microbe interactions. The genomic approach will help to identify the potential pathogen functions which result in alteration of the host physiological behaviour. In endocellular parasites such as the phytoplasmas, the genomic analysis may show up putative import mechanisms that provoke the depletion of regulatory molecules from the host cellular environment, or point out possible effector proteins released by the phytoplasma cell. In any case, known alterations in plant physiology could be corrected by exogenous treatment, thus providing an efficient method of reducing the economic impact of phytoplasma diseases.

To conclude these brief comments, undoubtedly plant pathologists have a knotty problem in trying to clarify phytoplasma biology. On the other hand, having to explore uncharted territories, they can plot their course free from preconceived views, especially with regard to traditional lab-based approaches and take advantage of work carried out in fields facing similar difficulties. If this route is followed, success is likely to occur sooner or later.

References

Abou-Jawdah, Y., Karakashian, A., Sobh, H., Martini, M. and Lee, I.M. (2002) An epidemic of almond witches'-broom in Lebanon: classification and phylogenetic relationships of the associated phytoplasma. *Plant Disease* 86, 477–484.

Ahrens, U. and Seemüller, E. (1992) Detection of DNA of plant pathogenic mycoplasma-like organisms by a polymerase chain reaction that amplifies a sequence of the 16S rRNA gene. *Phytopathology* 82, 828–832.

Arneodo, J.D., Galdeano, E., Orrego, A., Stauffer, A., Nome, S.F. and Conci, L.R. (2005) Identification of two phytoplasmas detected in China-trees with decline symptoms in Paraguay. *Australasian Plant Pathology* 34, 583–585.

Arocha, Y., Lopez, M., Pinol, B., Fernandez, M., Picornell, B., Almeida, R., Palenzuela, I., Wilson, M.R. and Jones, P. (2005) 'Candidatus Phytoplasma graminis' and 'Candidatus Phytoplasma caricae', two novel phytoplasmas associated with diseases of sugarcane, weeds and papaya in Cuba. *International Journal of Systematic and Evolutionary Microbiology* 55, 2451–2463.

Bai, X., Zhang, J., Ewing, A., Miller, S.A., Jancso Radek, A., Shevchenko, D.V., Tsukerman, K., Walunas, T., Lapidus, A., Campbell, J.W. and Hogenhout, S.A. (2006) Living with genome instability: the adaptation of phytoplasmas to diverse environments of their insect and plant hosts. *Journal of Bacteriology* 188, 3682–3696.

Barbara, D.J., Morton, A., Clark, M.F. and Davies, D.L. (2002) Immunodominant membrane proteins from two phytoplasmas in the aster yellows clade (chlorante aster yellows and clover phyllody) are highly divergent in the major hydrophilic region. *Microbiology* 148, 157–167.

Berg, M., Davies, D.L., Clark, M.F., Vetten, H.J., Maier, G., Marcone, C. and Seemüller, E. (1999) Isolation of the gene encoding an immunodominant membrane protein of the apple proliferation phytoplasma, and expression and characterization of the gene product. *Microbiology* 145, 1937–1943.

Blomquist, C.L., Barbara, D.J., Davies, D.L., Clark, M.F. and Kirkpatrick, B.C. (2001) An immunodominant membrane protein gene from the Western X-disease phytoplasma is distinct from those of other phytoplasmas. *Microbiology* 147, 571–580.

Branz, A. (2003) Il monitoraggio sulla diffusione degli scopazzi del melo in Valle di Non e di Sole. Presented at the meeting 'La frutticoltura delle Valli del Noce: 6a giornata tecnica, Feb. 18, Cles, Italy.

Carraro, L., Loi, N., Ermacora, P. and Osler, R. (1998a) High tolerance of European plum varieties to plum leptonecrosis. *European Journal of Plant Pathology* 104, 141–145.

Carraro, L., Osler, R., Loi, N., Ermacora, P. and Refatti, E. (1998b) Transmission of European stone fruit yellows phytoplasma by *Cacopsylla pruni*. *Journal of Plant Pathology* 80, 233–239.

Carraro, L., Ferrini, F., Ermacora, P. and Loi, N. (2002) Role of wild *Prunus* species in the epidemiology of European stone fruit yellows. *Plant Pathology* 51, 513–517.

Caudwell, A. (1957) Deux années d'études sur la Flaescence dorée, nouvelle maladie grave de la vigne. *Annales de l'Amelioration des Plantes* 4, 359–363.

Christensen, N.M., Axelsen, K.B., Nicolaisen, M. and Schulz, A. (2005) Phytoplasmas and their interactions with hosts. *Trends in Plant Science* 10, 526–535.

Conci, L., Meneguzzi, N., Galdeano, E., Torres, L., Nome, C. and Nome, S. (2005) Detection and molecular characterisation of an alfalfa phytoplasma in Argentina that represents a new subgroup in the 16S rDNA Ash Yellows group ('Candidatus Phytoplasma fraxini'). *European Journal of Plant Pathology* 113, 255–265.

Cordova, I., Jones, P., Harrison, N.A. and Oropeza, C. (2003) *In situ* PCR detection of phytoplasma DNA in embryos from coconut palms with lethal yellowing disease. *Molecular Plant Pathology* 4, 99–108.

Davis, R.E. and Sinclair, W.A. (1998) Phytoplasma identity and disease etiology. *Phytopathology* 88, 1372–1376.

Davis, R.E., Dally, E.L., Bertaccini, A., Lee, I.M., Credi, R., Osler, R., Savino, V., Carraro, L., Di Terlizzi, B. and Barba, M. (1993) Restriction fragment length polymorphism analyses and dot hybridisations distinguish mycoplasmalike organisms associated with Flavescence dorée and southern European grapevine yellows disease in Italy. *Phytopathology* 83, 772–776.

De La Rue, S.J., Schneider, B. and Gibb, K.S. (1999) Genetic variability in phytoplasmas associated with papaya yellow crinkle and papaya mosaic diseases in Queensland and the Northern Territory. *Australian Plant Pathology* 28, 108–114.

Denes, A.S. and Sinha, R.C. (1992) Alteration of clover phyllody mycoplasma DNA after *in vitro* culturing of phyllody-diseased clover. *Canadian Journal of Plant Pathology* 14, 189–196.

Deng, S. and Hiruki, C. (1991) Amplification of 16S rRNA genes from culturable and nonculturable mollicutes. *Journal of Microbiological Methods* 14, 53–61.

Doi, Y., Teranaka, M., Yora, K. and Asuyama, R. (1967) Mycoplasma or PLT- group-like organism found in the phloem elements of plants infected with mulberry dwarf, potato witches' broom, aster yellow, or paolownia witches' broom. *Annals of the Phytopathological Society of Japan* 33, 259–266.

Eden-Green, S. (1997) History, distribution and research on coconut lethal yellowing-like diseases of palms. International Workshop on Lethal Yellowing-like Diseases of Coconut, Chatham, UK, pp. 9–25.

Firrao, G., Carraro, L., Gobbi, E. and Locci, R. (1996a) Molecular characterization of a phytoplasma causing phyllody in clover and other herbaceous hosts in northern Italy. *European Journal of Plant Pathology* 102, 817–822.

Firrao, G., Smart, C.D. and Kirkpatrick, B.C. (1996b) Physical map of the western X-disease phytoplasma chromosome. *Journal of Bacteriology* 178, 3985–3988.

Firrao, G., Gibb, K. and Streten, C. (2005) Short taxonomic guide to the genus 'Candidatus Phytoplasma'. *Journal of Plant Pathology* 87, 249–264.

Firrao, G., Garcia-Chapa, M. and Marzachì, C. (2007) Phytoplasmas: genetics, diagnosis and relationships with the plant and insect host. *Frontiers in Biosciences* 12, 1353–1375.

Frausin, C. and Osler, R. (2004) La flavescenza dorata della vite in Friuli Venezia Giulia: le azioni intraprese per contrastarne la diffusione. *Notiziario Ersa* 5–6 (Suppl.), 40–48.

Frischmuth, T., Zimmat, G. and Jeske, P.A. (1990) The nucleotide sequence of abutilon mosaic virus reveals prokaryotic as well as eukaryotic features. *Virology* 178, 461–468.

Frisinghelli, C., Delaiti, L., Grando, M.S., Forti, D. and Vindimian, M.E. (2000) *Cacopsylla costalis* (Flor 1861), as a vector of apple proliferation in Trentino. *Journal of Phytopathology* 148, 425–431.

Galdeano, E., Torres, L.E., Meneguzzi, N., Guzman, F., Gomez, G.G., Docampo, D.M. and Conci, L.R. (2004) Molecular characterization of 16S ribosomal DNA and phylogenetic analysis of two X-disease group phytoplasmas affecting China-tree (*Melia azedarach* L.) and garlic (*Allium sativum* L.) in Argentina. *Journal of Phytopathology* 152, 174–181.

Gaurivaud, P., Danet, J.L., Laigret, F., Garnier, M. and Bové, J.M. (2000) Fructose utilization and pathogenicity of *Spiroplasma citri*. *Molecular Plant-Microbe Interactions* 13, 1145–1155.

Glennie, J.D. and Chapman, K.R. (1976) A review of dieback – a disorder of the papaw (*Carica papaya* L.) in Queensland. *Queensland Journal of Agricultural and Animal Sciences* 33, 177–188.

Gomez, G.G., Conci, L.R., Ducasse, D.A. and Nome, S.F. (1996) Purification of the phytoplasma associated with China-tree (*Melia azedarach* L) decline and the production of a polyclonal antiserum for its detection. *Journal of Phytopathology* 144, 473–477.

Gundersen, D.E., Lee, I.M., Rehner, S.A., Davis, R.E. and Kingsbury, D.E. (1994) Phylogeny of mycoplasmalike organisms (phytoplasmas): a basis for their classification. *Journal of Bacteriology* 176, 5244–5254.

Guthrie, J.N., Walsh, K.B., Scott, P.T. and Rasmussen, T.S. (2001) The phytopathology of Australian papaya dieback: a proposed role for the phytoplasma. *Physiological and Molecular Plant Pathology* 58, 23–30.

Harrison, N.A., Richardson, P.A., Kramer, J.B. and Tsai, J.H. (1994) Detection of the mycoplasmalike organism associated with lethal yellowing disease of palms in Florida by polymerase chain reaction. *Plant Pathology* 43, 998–1008.

Harrison, N.A., Narvaez, M., Almeyda, H., Cordova, I., Carpio, M.L. and Oropeza, C. (2002) First report of group 16SrIV phytoplasmas infecting coconut palms with leaf yellowing symptoms on the Pacific coast of Mexico. *Plant Pathology* 51, 808.

Harrison, N.A., Boa, E. and Carpio, M.L. (2003) Characterization of phytoplasmas detected in Chinaberry trees with symptoms of leaf yellowing and decline in Bolivia. *Plant Pathology* 52, 147–157.

Ho, K.C., Tsai, C.C. and Chung, T.L. (2001) Organization of ribosomal RNA genes from a loofah witches' broom phytoplasma. *DNA and Cell Biology* 20, 115–122.

Howard, F.W. and Thomas, D.L. (1980) Transmission of palm lethal decline to *Veitchia merrillii* by a planthopper *Myndus crudus*. *Journal of Economic Entomology* 73, 715–717.

ICSB, International Committee on Systematic Bacteriology – Subcommittee on the Taxonomy of Mollicutes (1995) Revised minimum standards for descriptions of new species of the class Mollicutes (Division Tenericutes). *International Journal of Systematic Bacteriology* 45, 605–612.

IRPCM, Phytoplasma/Spiroplasma Working Team – Phytoplasma taxonomy group (2004) 'Candidatus Phytoplasma', a taxon for the wall-less, non-helical prokaryotes that colonize plant phloem and insects. *International Journal of Systematic and Evolutionary Microbiology* 54, 1243–1255.

Jensen, D.D., Griggs, W.H., Gonzales, C.Q. and Schneider, H. (1964) Pear decline virus transmission by pear psylla. *Phytopathology* 54, 1346–1351.

Jiang, Y.P. and Chen, T.A. (1987) Purification of mycoplasma-like organisms from lettuce with aster yellows disease. *Phytopathology* 77, 949–953.

Kakizawa, S., Oshima, K., Kuboyama, T., Nishigawa, H., Jung, H.Y., Sawayanagi, T., Tsuchizaki, T., Miyata, S., Ugaki, M. and Namba, S. (2001) Cloning and expression analysis of phytoplasma protein translocation genes. *Molecular Plant-Microbe Interactions* 14, 1043–1050.

Kakizawa, S., Oshima, K. and Namba, S. (2006) Diversity and functional importance of phytoplasma membrane proteins. *Trends in Microbiology* 14, 254–256.

Kirkpatrick, B.C. (1991) Mycoplasma-like organisms: plant and invertebrate pathogens. In: Balows, A., Trüper, H.G., Dworkin, M., Harder, W. and Schliefer, K.H. (eds) *The Prokaryotes*. Springer, New York, pp. 4050–4067.

Kirkpatrick, B.C., Stenger, D.C., Morris, T.J. and Purcell, A.H. (1987) Cloning and detection of DNA from a nonculturable plant pathogenic mycoplamsa-like organism. *Science* 238, 197–199.

Kirkpatrick, B.C., Smart, C.D., Gardner, S., Gao, J.-L., Ahrens, U., Mäurer, R., Schneider, B., Lorenz, K.H., Seemüller, E., Harrison, N., Namba, S. and Daire, X. (1994) Phylogenetic relationships of plant pathogenic MLOs established by 16/23S rDNA spacer sequences. *IOM Letters* 3, 228–229.

Kollar, A. and Seemüller, E. (1989) Base composition of the DNA of mycoplasma-like organisms associated with various plant diseases. *Journal of Phytopathology* 127, 177–186.

Kuboyama, T., Huang, C.C., Lu, X., Sawayanagi, T., Kanazawa, T., Kagami, T., Matsuda, I., Tsuchizak, I.T. and Namba, S. (1998) A plasmid isolated from phytopathogenic onion yellows phytoplasma and its heterogeneity in the pathogenic phytoplasma mutant. *Molecular Plant-Microbe Interactions* 11, 1031–1037.

Kunkel, L.O. (1926) Studies on aster yellows. *American Journal of Botany* 23, 646–705.

Lauer, U. and Seemüller, E. (2000) Physical map of the chromosome of the apple proliferation phytoplasma. *Journal of Bacteriology* 182, 1415–1418.

Lee, I.M. and Davis, R.E. (1992) Mycoplasmas which infect plants and insects. In: Maniloff, J., McElhansey, R.N., Finch, L.R. and Baseman, J.B. (eds) *Mycoplasmas: Molecular Biology and Pathogenesis*. American Society for Microbiology, Washington, DC, pp. 379–390.

Lee, I.M., Gundersen, D.E., Hammond, R.W. and Davis, R.E. (1994) Use of mycoplasmalike organism (MLO) group-specific oligonucleotide primers for nested-PCR assays to detect mixed-MLO infections in a single host-plant. *Phytopathology* 84, 559–566.

Lee, I.M., Gundersen-Rindal, D.E., Davis, R.E. and Bartoszyk, I.M. (1998) Revised classification scheme of phytoplasmas based on RFLP analyses of 16S rRNA and ribosomal protein gene sequences. *International Journal of Systematic Bacteriology* 48, 1153–1169.

Lee, I.M., Davis, R.E. and Gundersen-Rindal, D.E. (2000) Phytoplasma: phytopathogenic mollicutes. *Annual Review of Microbiology* 54, 221–255.

Lee, I.M., Gundersen-Rindal, D.E., Davis, R.E., Bottner, K.D., Marcone, C. and Seemüller, E. (2004a) 'Candidatus Phytoplasma asteris', a novel phytoplasma taxon associated with aster yellows and related diseases. *International Journal of Systematic and Evolutionary Microbiology* 54, 1037–1048.

Lee, I.M., Martini, M., Marcone, C. and Zhu, S.F. (2004b) Classification of phytoplasma strains in the elm yellows group (16SrV) and proposal of 'Candidatus Phytoplasma ulmi' for the phytoplasma associated with elm yellows. *International Journal of Systematic and Evolutionary Microbiology* 54, 337–347.

Lepka, P., Stitt, M., Moll, E. and Seemüller, E. (1999) Effect of phytoplasmal infection on concentration and translocation of carbohydrates and amino acids in periwinkle and tobacco. *Physiological and Molecular Plant Pathology* 55, 59–68.

Letunic, I., Copley, R.R., Pils, B., Pinkert, S., Schultz, J. and Bork, P. (2006) SMART 5: domains in the context of genomes and networks. *Nucleic Acids Research* 34, D257–D260.

Liefting, L.W. and Kirkpatrick, B.C. (2003) Cosmid cloning and sample sequencing of the genome of the uncultivable mollicute, Western X-disease phytoplasma, using DNA purified by pulsed-field gel electrophoresis. *FEMS Microbiology Letters* 221, 203–211.

Liefting, L.W., Andersen, M.T., Beever, R.E., Gardner, R.C. and Forster, R.L.S. (1996) Sequence heterogeneity in the two 16S rRNA genes of *Phormium* yellow leaf phytoplasma. *Applied and Environmental Microbiology* 62, 3133–3139.

Liefting, L.W., Padovan, A.C., Gibb, K.S., Beever, R.E., Andersen, M.T., Newcomb, R.D., Beck, D.L. and Forster, R.L.S. (1998) 'Candidatus Phytoplasma australiense' is the phytoplasma associated with Australian grapevine yellows, papaya dieback and *Phormium* yellow leaf diseases. *European Journal of Plant Pathology* 104, 619–623.

Liefting, L.W., Shaw, M.E. and Kirkpatrick, B.C. (2004) Sequence analysis of two plasmids from the phytoplasma beet leafhopper-transmitted virescence agent. *Microbiology* 150, 1809–1817.

Liefting, L.W., Havukkala, I., Andersen, M.T., Lough, T.J. and Beever, R.E. (2006) The complete genome sequence of 'Candidatus Phytoplasma australiense'. Proceedings of the 16th International Congress of the International Organization for Mycoplasmology, Cambridge, 9–14 July 2006, p. 43.

Lim, P.O. and Sears, B.B. (1989) 16S rRNA sequence indicates that plant-pathogenic mycoplasma-like organisms are evolutionarily distinct from animal mycoplasmas. *Journal of Bacteriology* 171, 5901–5906.

Lim, P.O. and Sears, B.B. (1991) DNA-sequence of the ribosomal-protein genes *rp12* and *rps19* from a plant-pathogenic mycoplasma-like organism. *FEMS Microbiology Letters* 84, 71–73.

Lorenz, K.H., Dosba, F., Poggi Pollini, C., Llacer, G. and Seemüller, E. (1994) Phytoplasma diseases of *Prunus* species in Europe are caused by genetically similar organisms. *Zeitschrift fur Pflanzenkrankheiten und Pflanzenschutz* 101, 567–575.

Marcone, C. and Seemüller, E. (2001) A chromosome map of the European stone fruit yellows phytoplasma. *Microbiology* 147, 1213–1221.

Marcone, C., Neimark, A., Ragozzino, A., Lauer, U. and Seemüller, E. (1999) Chromosome sizes of phytoplasmas composing major phylogenetic groups and subgroups. *Phytopathology* 89, 805–810.

Marcone, C., Lee, I.M., Davis, R.E., Ragozzino, A. and Seemüller, E. (2000) Classification of aster yellows-group phytoplasmas based on combined analyses of rRNA and *tuf* gene sequences. *International Journal of Systematic and Evolutionary Microbiology* 50, 1703–1713.

Marwitz, R. (1990) Diversity of yellows disease agents in plant infections. In: Stanek, G., Cassel, G.H., Tully, J.G. and Whitcomb, R.F. (eds) *Recent Advances in Mycoplasmology*. Fischer Verlag, Stuttgart, Germany, pp. 431–434.

McCoy, R.E., Caudwell, A., Chang, C.J., Chen, T.A., Chiykowski, L.N., Cousin, M.T., Dale, J.L., de Leeuw, G.T.N., Golino, D.A., Hackett, K.J., Kirkpatrick, B.C., Marwitz, R., Petzold, H., Sinha, R.C., Sugiura, M., Whitcomb, R.F., Yong, I.L., Zhu, B.M. and Seemüller, E. (1989) Plant diseases associated with mycoplama-like organisms. In: Whitcomb, R.F. and Tully, J.G. (eds) *The Mycoplasmas*. Academic Press, New York, pp. 545–640.

Montano, H.G., Davis, R.E., Dally, E.L., Pimentel, J.P. and Brioso, P.S.T. (2000) Identification and phylogenetic analysis of a new phytoplasma from diseased chayote in Brazil. *Plant Disease* 84, 429–436.

Murray, R.G. and Schleifer, K.H. (1994) Taxonomic notes: a proposal for recording the properties of putative taxa of procaryotes. *International Journal of Systematic Bacteriology* 44, 174–176.

Musetti, R., Favali, M.A. and Pressacco, L. (2000) Histopathology and polyphenol content in plants infected by phytoplasmas. *Cytobios* 102, 133–147.

Nault, L.R. (1980) Maize bushy stunt and corn stunt: a comparison of disease symptoms, pathogen host ranges and vectors. *Phytopathology* 70, 659–662.

Neimark, H. and Kirkpatrick, B.C. (1993) Isolation and characterization of full-length chromosomes from non-culturable plant-pathogenic mycoplasma-like organisms. *Molecular Microbiology* 7, 21–28.

Oropeza, C., Escamilla, J.A., Mora, G., Zizumbo, D. and Harrison, N. (2005) Coconut lethal yellowing. In: Batugal, P. and Rao, R. (eds) *Status of Coconut Genetic Resources*. IPGRI-APO, Serdang, Malaysia.

Oshima, K., Kakizawa, S., Nishigawa, H., Kuboyama, T., Miyata, S., Ugaki, M. and Namba, S. (2001) A plasmid of phytoplasma encodes a unique replication protein having both plasmid- and virus-like domains: clue to viral ancestry or result of virus/plasmid recombination? *Virology* 285, 270–277.

Oshima, K., Kakizawa, S., Nishigawa, H., Jung, H.Y., Wei, W., Suzuki, S., Arashida, R., Nakata, D., Miyata, S., Ugaki, M. and Namba, S. (2004) Reductive evolution suggested from the complete genome sequence of a plant-pathogenic phytoplasma. *Nature Genetics* 36, 27–29.

Palmano, S. and Firrao, G. (2000) Diversity of phytoplasmas isolated from insects, determined by a DNA heteroduplex mobility assay and a length polymorphism of the 16S–23S rDNA spacer region analysis. *Journal of Applied Microbiology* 89, 744–750.

Padovan, A.C., Firrao, G., Schneider, B. and Gibb, K.S. (2000) Chromosome mapping of the sweet potato little leaf phytoplasma reveals genome heterogeneity within the phytoplasmas. *Microbiology* 146, 893–902.

Purcell, A.H. and Suslow, K.S. (1984) Surveys of leafhoppers (Homoptera: Cicadellidae) and pear psylla (Homoptera: Psyllidae) in pear and peach orchards relative to the spread of peach yellow leaf roll disease. *Journal of Economic Entomology* 77, 1489–1494.

Razin, S., Yogev, D. and Naot, Y. (1998) Molecular biology and pathology of mycoplasmas. *Microbiology and Molecular Biology Reviews* 62, 1094–1156.

Rekab, D., Carraro, L., Schneider, B., Seemüller, E., Chen, J.C., Chang, C.J., Locci, R. and Firrao, G. (1999) Geminivirus-related extrachromosomal DNAs of the X-clade phytoplasmas share high sequence similarity. *Microbiology* 145, 1453–1459.

Rossello-Mora, R. and Amann, R. (2001) The species concept for prokaryotes. *FEMS Microbiology Reviews* 25, 39–67.

Saeed, E.M., Roux, J. and Cousin, M.T. (1993) Studies of polyclonal antibodies for the detection of MLOs associated with faba bean (*Vicia faba* L.) using different ELISA methods and dot-blot. *Journal of Phytopathology* 137, 33–43.

Schneider, B. and Seemüller, E. (1994) Presence of two sets of ribosomal genes in phytopathogenic mollicutes. *Applied and Environmental Microbiology* 141, 173–185.

Schneider, B., Maurer, R., Saillard, C., Kirkpatrick, B.C. and Seemüller, E. (1992) Occurrence and relatedness of extrachromosomal DNAs in plant pathogenic mycoplasmalike organisms. *Molecular Plant-Microbe Interactions* 5, 489–495.

Schneider, B., Gibb, K.S. and Seemüller, E. (1997) Sequence and RFLP analysis of the elongation factor Tu gene used in differentiation and classification of phytoplasmas. *Microbiology* 143, 3381–3389.

Sears, B.B., Lim, P., Holland, N., Kirkpatrick, B.C. and Klomparens, K.L. (1989) Isolation and characterization of DNA from mycoplasmalike organism. *Molecular Plant-Microbe Interactions* 2, 175–180.

Seemüller, E. and Schneider, B. (2004) '*Candidatus* Phytoplasma mali', '*Candidatus* Phytoplasma pyri' and '*Candidatus* Phytoplasma prunorum', the causal agents of apple proliferation, pear decline and European stone fruit yellows, respectively. *International Journal of Systematic and Evolutionary Microbiology* 54, 1217–1226.

Seemüller, E., Schneider, B., Maurer, R., Ahrens, U., Daire, X., Kison, H., Lorenz, K.H., Firrao, G., Avinent, L., Sears, B.B. and Stackebrandt, E. (1994) Phylogenetic classification of phytopathogenic mollicutes by sequence analysis of 16S ribosomal DNA. *International Journal of Systematic Bacteriology* 44, 440–446.

Seemüller, E., Marcone, C., Lauer, U., Ragozzino, A. and Göschl, M. (1998) Current status of molecular classification of the Phytoplasmas. *Journal of Plant Pathology* 80, 3–26.

Smart, C.D., Schneider, B., Blomquist, C.L., Guerra, L.J., Harrison, N.A., Ahrens, U., Lorenz, K.H., Seemüller, E. and Kirkpatrick, B.C. (1996) Phytoplasma-specific PCR primers based on sequences of the 16S–23S rRNA spacer region. *Applied and Environmental Microbiology* 62, 2988–2993.

Tedeschi, R., Bosco, D. and Alma, A. (2002) Population dynamics of *Cacopsylla melanoneura* (Homoptera: Psyllidae), a vector of apple proliferation phytoplasma in northwestern Italy. *Journal of Economic Entomology* 95, 544–551.

Tymon, A.M., Jones, P. and Harrison, N.A. (1997) Detection and differentiation of African coconut phytoplasmas: RFLP analysis of PCR-amplified 16S rDNA and DNA hybridisation. *Annals of Applied Biology* 131, 91–102.

Vázquez, A., Ducasse, D.A., Nome, S.F. and Muñoz, J.O. (1983) Declinamiento del paraíso (*Melia azedarch* L.), síntomas y estudio etiológico de esta nueva enfermedad. *Revista de Investigaciones Agropecuarias* 18, 309–320.

Verdin, E., Salar, P., Danet, J.L., Choueiri, E., Jreijiri, F., El Zammar, S., Gelie, B., Bove, J.M. and Garnier, M. (2003) '*Candidatus* Phytoplasma phoenicium' sp nov., a novel phytoplasma associated with an emerging lethal disease of almond trees in Lebanon and Iran. *International Journal of Systematic and Evolutionary Microbiology* 53, 833–838.

Ward, D.M. (2006) A macrobiological perspective on microbial species. *Microbe* 1, 269–278.

White, D.T., Blackall, L.L., Scott, P.T. and Walsh, B.K. (1998) Phylogenetic positions of phytoplasmas associated with dieback, yellow crinkle and mosaic diseases of papaya, and their proposed inclusion in '*Candidatus* Phytoplasma australiense' and a new taxon, '*Candidatus* Phytoplasma australasia'. *International Journal of Systematic Bacteriology* 48, 941–951.

Yu, Y.L., Yeh, K.W. and Lin, C.P. (1998) An antigenic protein gene of a phytoplasma associated with sweet potato witches' broom. *Microbiology* 144, 1257–1262.

Zhou, L., Jang, J.-C., Jones, T.L., and Sheen, J. (1998) Glucose and ethylene signal transduction crosstalk have been demonstrated in an Arabidopsis glucose-insensitive mutant. *Proceedings of the National Academy of Sciences of the United States of America* 95, 10294–10299.

12 Molecular Detection of Plant Viroids

R.P. Singh

Abstract

Viroids were discovered as novel plant pathogens in the early 1970s, and consist solely of single-stranded RNA covalently closed as a circular molecule. Viroids are the smallest known plant pathogen, ranging from 247 to 463 nucleotides (nt) in length. About 30 viroids infecting agricultural, horticultural and ornamental plants are known. Most of the reported viroids cause disease symptoms and economic losses, but many more are present in plants without exhibiting symptoms. The unique molecular composition and structure of viroids has demanded the development of many innovative detection protocols. The need for routine viroid detection methodology first resulted in the development of polyacrylamide gel electrophoresis (PAGE). Novel innovations in the form of 'return'-PAGE (R-PAGE), 'sequential'-PAGE (S-PAGE), temperature gradient gel electrophoresis (TGGE) and PAGE-'mobile laboratory' increased the efficiency and effectiveness of viroid detection. Viroids were also the subject matter in the first application of both molecular hybridization (MH) (solid support) and reverse transcription–polymerase chain reaction (RT-PCR) for plant pathogen detection. Although reports of new viroid species have diminished, new outbreaks of existing *Pospiviroid* spp. have increased in many countries, largely as a result of symptomless viroid carriers. Detection procedures have played an important role in the eradication of potato spindle tuber viroid (PSTVd) from Canada and it is theorized that new outbreaks of viroids on crop plants can be better explained by understanding the role of ornamental plants in the evolution of viroids. Consequently, procedures to facilitate large-scale survey of symptomless ornamental or crop plants for *Pospiviroid* spp. have recently been developed.

Introduction

Diseases such as potato spindle tuber, chrysanthemum stunt and coconut cadang cadang, existed long before their viroid nature was elucidated (Werner, 1926; Hollings and Stone, 1973; Randles and Rodriguez, 2003). Initially, the diseases were considered to be caused by viruses; however, transmissions of these 'viruses' to indicator plants were relatively difficult, and consequently, the method of biological bioassay remained limited to natural host plants. As a result, not much progress was made to characterize the causal agents of these diseases until the early 1960s (Raymer and O'Brien, 1962). Initially, these diseases were recognized and plants were rouged out solely by diagnostic symptoms in affected plants, in efforts to avert severe economic losses. For example, the yield losses caused by potato spindle tuber viroid (PSTVd) ranged from 17% to 64% depending on the viroid strain and the potato cultivar (Singh et al., 1971), and reached 100% with multiple years of cultivation of infected seeds (Pfannenstiel and Slack, 1980). Similarly, in a perennial tree such as coconut, over 30 million trees had been killed in the Philippines by coconut cadang cadang viroid (CCCVd) by 1982 (Zelany et al., 1982). In some chrysanthemum cultivars, plants become so stunted due to infection by chrysanthemum stunt viroid (CSVd) that the product was not marketable. Similarly, the cone yields of hops could be reduced by 50% or more (Sasaki and Shikata, 1978) due to hop stunt viroid (HSVd).

In contrast, not all viroid infections result in diagnostic symptoms in infected crops, ornamental plants and trees (Singh and Teixeria da Silva, 2006), thus requiring a large number of plant materials to be tested for the absence of viroids in symptomless plants. Failure to ignore symptomless viroid infection in unrelated plant species could have serious consequences. Columnea latent viroid (CLVd) was isolated from three symptomless ornamental host plants without any economic significance (Singh and Teixeria da Silva, 2006). CLVd was not part of any plant certification programmes, despite greenhouse studies that showed the potential to cause severe symptoms in tomato and potato plants. However, its sudden appearance in tomato crops in 1988 in the Netherlands changed the perception of this viroid. Existence of CLVd isolates capable of causing up to 90% infection in tomato crops similarly caused yield reductions of 72–82% upon transfer to potato (Verhoeven et al., 2004), emphasizing the problem of symptomless carriers of viroids accidentally transmitting disease to agriculturally important crop plants.

Strategies for viroid disease management include: (i) exclusion or elimination of viroids from planting material; (ii) exclusion of viroids by crop certification and quarantine; (iii) natural and transgenic resistance to viroids; and (iv) eradication of viroids from a geographical region or country. The above measures are dependent upon viroid diagnostic procedures to ensure their success. Thus, a producer planting a viroid-susceptible crop can mitigate risk of infection only through planting of viroid-free propagating material and by ensuring their freedom from disease throughout the growth period.

Characterization and Classification of Viroids

Viroids, as a novel class of pathogen, were discovered about 35 years ago in an effort to purify the 'virus' causing the potato spindle tuber disease

(Diener and Raymer, 1967; Singh and Bagnall, 1968). The initial determination of the infectious PSTVd was based on its characteristic low-molecular weight and was identified using a combination of sucrose density-gradient centrifugation and polyacrylamide gel electrophoresis (PAGE) (Diener, 1971; Singh and Clark, 1971). PSTVd was identified from both methods by inoculating indicator plants with the infectious fraction. In addition to being small molecules (i.e. less than one-tenth the size of the smallest plant virus genome), viroids were characterized by their single-stranded, covalently closed circular RNA structure (Sänger et al., 1976) ranging from 247 to 463 nucleotides (nt) (Steger and Riesner, 2003). At present, nearly 30 viroids (Flores et al., 2005) have been detected in agricultural, horticultural and ornamental plants throughout the world, with the majority affecting perennial, vegetatively propagated crops. Viroids do not differ significantly from viruses in their pathological effects on plants or in their modes of transmission in fields and orchards (Singh, 1983), but differ radically in their molecular size, configuration, structure and replication (Steger and Riesner, 2003). Unlike viruses, viroids are devoid of a protein coat and they do not contain the machinery for protein synthesis. However, viroids are considered the most efficient and sophisticated 'molecular plant parasites', apparently shedding the need to synthesize structural or functional proteins by manipulating host enzymes for their replication and are capable of destroying trees, including coconut palms (Randles and Rodriguez, 2003). In general, viroids are highly contagious, particularly members of the genus *Pospiviroid*. Viroids have a wide host range and are transmitted readily by mechanical means, including foliage contact, handling during cultivation and contamination of cutting knives or other tools. Foliage contact from plants affected with CSVd was shown to infect 19–28% of adjoining healthy plants (Hollings and Stone, 1973), while 80% of potato plants were infected by PSTVd when leaves were rubbed against each other (Pfannenstiel and Slack, 1980). All viroids are transmissible by grafting, while only a few are transmissible through the seeds and pollen of infected plants (Wallace and Drake, 1962; Singh, 1970; Singh et al., 1991a). Despite the RNA nature of viroids, some viroids can survive in freeze-dried leaves for several years (Singh and Finnie, 1977) or in true potato seed for over 20 years (Singh et al., 1991b). Most cultivated plants have no natural resistance to viroid diseases and transgenic approaches have had limited success (Singh and Teixeria da Silva, 2006). A detection methodology, based on serology, used on viruses cannot be applied to viroids based on their molecular constitution. Therefore, viroid detection methods are often based on viroid RNA properties and the management of viroid diseases requires modified approaches.

An orderly classification of different viroids in families and genera based on their genomic RNA structure and replication feature has been established (Flores et al., 2005). As a result, viroid species have been divided into two families: *Pospiviroidae* and *Avsunviroidae*. The divisions were based on specific sequences within their molecular structure (e.g. presence or absence of a central conserved region [CCR] and a hammerhead ribozyme structure), the replication strategy and the location of the viroid in the host cell (Flores et al., 2005). Viroids in the family *Pospiviroidae* contain a CCR sequence, do not possess a hammerhead ribozyme structure and are located in the nucleus. *Pospiviroidae* members within genera can potentially form rod-like secondary structures with a high degree of base pairing, thereby forming double-stranded segments interspersed with unpaired bases as bulged loops in a closed circular molecule. The rod-like structure has been assigned to five domains correlated with biological functions, and include: the left and right terminus (T_L and T_R); the CCR; the pathogenicity (P); and the variable (V) domains (Keese and Symons, 1985). Genera are distinguished primarily on the sequences that form the CCR and the presence or absence of a conserved motif, such as the terminal conserved region (TCR) (Koltunow and Rezaian, 1988). In contrast, viroids within the family *Avsunvirodae* include those which possess a hammerhead ribozyme structure, but do not contain a CCR and are located in the chloroplast. Viroids in both families replicate through a rolling-circle mechanism involving plus and minus strands of viroid RNA (Branch and Robertson, 1984). Members of *Avsunviroidae* use two rolling circles (symmetric pathway), while those of *Pospiviroidae* use one rolling circle (asymmetric pathway) (Flores et al., 2005) for their replication.

The unique features of circular RNA without a protein coat and the absence of diagnostic symptoms in most infected plants (Singh and Teixeria da Silva, 2006) have required the development of detection procedures based on both the physical and the biological nature of the RNA molecule. Physical identification based on the circular RNA structure and its molecular characteristics in its denatured state can be accomplished through return-PAGE (R-PAGE). Viroid detection based on intrinsic biological properties of the RNA includes infectivity assays, PAGE, molecular hybridization (MH) and reverse transcription–polymerase chain reaction (RT-PCR). Both physical and biological methodologies serve in the identification and certification of viroid-free planting material at different stages of management.

Biological Assays

Initially, when viroid bioassays were confined to natural hosts, ingenious methods were devised to enhance symptom expression in them. Over 90% infection with increased symptom expression was achieved for PSTVd through an innovative method of 'tuber-grafting' (Goss, 1926). An 8–10mm diameter hole is bored through a healthy seed tuber and replaced with the core tissue from an infected tuber obtained using a slightly larger cork borer. The resulting contact of a relatively large exposed tuber surface improved the viroid transmission rate and appearance of PSTVd symptoms in the emerging plants, as well as in new tubers during the first year of growth. The time required for symptom expression of viroid diseases in a natural host could take 9–12 weeks (e.g. PSTVd) and a period of months or years in woody trees for viroids infecting avocado, apple and coconut.

Indicator plants for viroids were not known until the 1960s (Raymer and O'Brien, 1962), although spindle tuber disease was in existence since 1917 (Werner, 1926). Upon understanding the free-RNA nature of viroids, infectivity bioassays utilized phenol-extracted nucleic acid preparations or plant sap prepared in high pH buffers to inhibit ribonuclease action during sap preparations for inoculation, and consequently viroids were transmitted to many plants (Singh, 1973). Biological indexing involves the use of indicator plants, which, upon inoculation, exhibit systemic diagnostic symptoms or local lesions in the inoculated leaves. Symptom development is affected by viroid strain, host species, cultivar, multiple infections, environmental factors and inoculation methods (Singh and Ready, 2003). The symptom expression has been accelerated in some viroid–host combinations by pruning the top of the plant and forcing new growth, as in tomato plants infected with PSTVd (Whitney and Peterson, 1963). Development of systemic necrosis in tomato leaves and lesions in *Scopolia sinensis* (a local lesion host for PSTVd) has been enhanced by the presence of higher than normal levels of Mn^{2+} in the potting media (Singh et al., 1974). The use of a cross-protection approach, where the presence of a mild strain of a viroid is detected despite the absence of severe symptoms after challenge inoculation with a severe strain of the pre-inoculated indicator plant, has been used for PSTVd (Fernow, 1967; Singh et al., 1970), ChCMVd (Horst, 1975) and PLMVd (Desvignes, 1980) to index large numbers of planting material.

With the availability of recombinant DNA technology, bioassays of molecularly cloned viroids utilizing plasmids containing a full-length molecule of viroid cDNA or RNA (transcript prepared with SP6 polymerase-driven *in vitro* transcription) increased the efficiency of viroid detection. In general, biological indexing is not practical for large-scale testing which requires significant investments in greenhouse space, time and labour, and, therefore, has largely been replaced by biochemical and molecular detection methods.

Biochemical and Molecular Methods

Polyacrylamide gel electrophoresis

The highly complex secondary structure of viroids permitted many modifications of PAGE methods specifically designed for the detection of viroid molecules. The first use of PAGE for viroid diagnosis occurred shortly after the initial use of analytical PAGE to isolate viroid RNA from plant tissues (Morris and Smith, 1977). It involved extraction of total nucleic acids from plant tissues using phenol to enrich the viroid fraction and then electrophoresis of the extracted nucleic acid in a non-denaturing 5% polyacrylamide gel to separate the various RNA species by their molecular size. The presence of a distinct band of RNA in known infected samples and its absence in known healthy samples indicate the presence of a viroid.

The rapidity of PAGE analysis relative to bioassays using natural or indicator plants hastened the adaptation of PAGE for the detection of viroids in several plant species worldwide. To discriminate the viroid band from other low-molecular host RNA bands in PAGE, a nomogram was developed to definitively identify the viroid band. The nomogram was based on the relative distance of viroid migration in PAGE relative to the mobility of 5S host RNA, which is always present in plant extract (Mosch et al., 1978). However, in screening by PAGE, the lines of tuber-bearing *Solanum* spp. infected with PSTVd were unclear due to weaker bands of PSTVd (Harris and Miller-Jones, 1981; Gelder and Treur, 1982). The test was modified by introducing a biological viroid amplification step prior to running PAGE. Increased viroid concentration was achieved by first inoculating tomato seedling with extracts from *Solanum* spp., followed by PAGE of tomato extracts 4–5 weeks after inoculation.

Return- and sequential-polyacrylamide gel electrophoresis

In understanding the thermophysical properties of viroids, it became apparent that viroids undergo a conformational transition during denaturation by heat, urea or formamide. During the transition, a highly base-paired rod-like molecule is transformed into a circular single-stranded structure. Circular forms of viroids migrate much slower in the denatured state through PAGE than linear RNA of the same size. This knowledge enabled the development of a two-dimensional (i.e. non-denaturing and denaturing) PAGE protocol for the separation of circular RNA molecules (Schumacher et al., 1983) and a diagnostic version of PAGE (termed Return PAGE or R-PAGE) based on the differential mobility of circular RNA molecules (Schumacher et al., 1986; Singh and Boucher, 1987). In the R-PAGE procedure, the first electrophoresis is carried out under native conditions, but before viroid bands migrate out of the gel, temperatures are increased and the ionic strength of the buffer is lowered to denature the viroid RNA. The direction of RNA migration in the polyacrylamide gel is reversed by changing the polarity of the electric field, thereby permitting two electrophoretic runs within the same gel (Singh, 1991). In another non-denaturing and denaturing gel system (termed sequential or S-PAGE), two separate gels are required (Rivera-Bustamante et al., 1986) to achieve similar separation of viroid bands.

Application of the original R-PAGE (Schumacher et al., 1986) for large-scale surveys of PSTVd in Canada encountered problems in detection of mild strains of PSTVd, which predominate under field conditions (Singh et al., 1970). It was remedied by heating the buffer to 90°C for denaturation of the viroid molecule and subsequently carrying out return direction electrophoresis at 70–71°C (Singh and Boucher, 1987). Using the modified R-PAGE, several large-scale surveys of PSTVd have been carried out (Singh *et al.*, 1988; Avila *et al.*, 1990; De Boer *et al.*, 2002) and the R-PAGE protocol has been the accredited standard method used in potato certification laboratories in Canada for several years and is also recommended as the primary detection method by the European Union (Jefferies and James, 2005).

Temperature gradient gel electrophoresis

Another form of PAGE, termed temperature gradient gel electrophoresis (TGGE), was specifically developed to follow the conversion of the rod-like viroid RNA structure to an open circular conformation (Rosenbaum and Riesner, 1987). TGGE is carried out in a horizontal polyacrylamide gel resting on an electrically insulated metal plate that is heated on one edge and cooled at the other edge. Besides basic molecular studies, TGGE has been used to determine minor variations in the molecular structure of hop latent viroid (HLVd) induced by thermal stress in hops (Matoušek *et al.*, 2001) and in *Nicotiana benthamiana* plants by PSTVd (Matoušek *et al.*, 2004).

Besides the detection of viroids, PAGE, R-PAGE and S-PAGE have been extensively utilized for the characterization of suspected viroids. The techniques have been used to determine whether the causal agent is RNA or DNA, single-stranded or double-stranded, circular or linear. The techniques have also been used for the determination of molecular masses, secondary and tertiary structures and for the monitoring of two strains of PSTVd in cross-protection studies involving viroids (Singh, 1991; Hanold *et al.*, 2003). However, the versatility of PAGE reached its pinnacle in 1988, when a 'mobile laboratory' based on 20% non-denaturing PAGE for field surveys was successfully used to detect CCCVd in coconut plantations in remote areas of the Philippines (Hanold *et al.*, 2003).

Molecular Hybridization

Molecular hybridization entails a specific binding of two strands of nucleic acids, which can be two strands of RNA and two strands of DNA, or one strand of RNA and one strand of DNA. The hybridization includes several steps (e.g. preparation of probes, target samples, denaturation, immobilization onto the solid support, pre-hybridization, hybridization and detection of hybridized products) and they are covered in detail in a review paper (Mühlbach *et al.*, 2003).

Although solution hybridization was in use in viroid research earlier, the first detection of PSTVd using a solid support was developed in 1981 (Owens and Diener, 1981). In viroid research, the use of a nitrocellulose membrane and denaturation by formaldehyde or heating to 100°C for 5 min and then quick cooling are important considerations because they often increase the sensitivity of detection. Commonly, fragments smaller than 200–300 bp bind poorly to nitrocellulose, but viroid RNA is an exception, immobilizing readily on nitrocellulose (Owens and Diener, 1981; Flores, 1986; Candresse et al., 1990; Singh et al., 2006a).

In earlier studies, molecular probes were prepared using radioisotopes ^{32}P and ^{125}I (Owens and Diener, 1981; Barker et al., 1985; Mühlbach et al., 2003). However, this technology was limited to research centres with isotopic facilities that can handle storage and waste disposal of radioactive material. Furthermore, the short lifetime of some isotope probes required frequent handling of highly radioactive material for probe preparation. As a result, use and preparation of non-radioactive molecular probes, mainly for PSTVd, were evaluated (McInnes et al., 1989; Candresse et al., 1990; Welnicki and Hiruki, 1992). Detection with non-radioactive probes using biotin or photobiotin label has been reported to be as sensitive as radioactive probes (McInnes et al., 1989) or it can be 15 or 5 times less sensitive than radioactive cDNA or cRNA probes, respectively (Candresse et al., 1990). Steroid hapten, digoxigenin (DIG) has been used as an alternative to biotin. In this system, after hybridization to viroid molecules, the DIG-labelled deoxyuridine-triphosphate present in cDNA or cRNA probes is detected by enzyme-linked immunosorbent assay. The detection of target molecules is based on the binding of an anti-DIG alkaline phosphatase conjugate to the DIG moiety and hydrolysis of the particular enzyme substrate, which results in the production of a coloured or chemiluminescent product (Welnicki and Hiruki, 1992; Podleckis et al., 1993; Singh et al., 1994). A comparative study of biotin and DIG labels using colourimetric and chemiluminescent analysis has shown that biotinylated cDNA probes detected 20 pg by colourimetric and 2–20 pg by chemiluminescent assay, while the DIG-labelled probes detected only 200 pg (Kanematsu et al., 1991). However, biotinylated probes prepared with PCR, incorporating dUTP or dATP, can detect 0.2–2 pg of PSTVd. This latter technique appears to be an improvement for acquiring a highly sensitive and very specific probe in large quantities within a short period (Kanematsu et al., 1991).

In the initial studies, a monomeric viroid was used as a template for the synthesis of cDNA and cRNA probes (Palukaitis et al., 1985; McInnes et al., 1989). Multimeric cDNA templates (containing concatamers of monomers, dimers, trimers, tetramers and hexamers of viroid RNA) were used mainly in studies with PSTVd to prepare probes. These probes were shown to be more sensitive than the monomeric cDNA probes (Welnicki and Hiruki, 1992; Singh et al., 1994). Since cRNA probes are single-stranded, they preclude self-annealing and increase stability of RNA and RNA duplexes compared to DNA and DNA duplexes. Furthermore, cRNA probes have greater sensitivity, produce less background reactions and have higher specific activity. Non-radioactive riboprobes, particularly DIG-labelled probes, have become increasingly common for routine detection of several viroids. Non-radioactive hybridization has been applied under field conditions to study the incidence of HSVd and PLMVd in several countries of the Mediterranean Region (Mühlbach et al., 2003).

Imprint hybridization

To simplify plant sample manipulations, an 'imprint hybridization' technique was introduced for viroids (Podleckis *et al.*, 1993), patterned after an approach used for viruses termed 'direct tissue blotting immunological assay'. The procedure involves cut stems pieces (inversely or longitudinally) of about 1 cm size pressed onto the surface of a polyvinylidene fluoride (PVDF) or positively charged nylon membrane. Satisfactory detection of DIG-labelled RNA or DNA probes is observed only in chemiluminescent assays and not in chromogenic. 'Tissue blot' of PSTVd-infected potato tubers and apple scar skin viroid (ASSVd)-infected stem, fruits and petioles of apple have been successfully tried (Podleckis *et al.*, 1993) and 'tissue-imprint' hybridization has been applied under field conditions for large-scale indexing of HSVd in apricot trees, for ASSVd in pears and indexing of CSVd. However, application of imprint hybridization methodology to field samples showed that detection of citrus viroids from species other than citron (most susceptible) was erratic owing in large part to the low viroid concentration in field samples. When field samples are biologically amplified in citron for 3 months at 28–32°C or for 7 months at 18–25°C, viroids can be detected with high sensitivity (Palacio *et al.*, 2000). However, use of biological amplification negates the advantage of time saving in the laboratory test and may not be applicable for reliable detection of citrus viroids in large-scale fields and orchards.

Synthetic probes for hybridization

Synthetic probes of 17–87 nt have been evaluated for PSTVd (Sano *et al.*, 1988; Welnicki *et al.*, 1989) in hopes that the diagnostic probes would be available in large amounts, directed to selected regions of viroid molecules and facilitated with ^{32}P end-labelling. The synthetic probes have been found to be less sensitive than full-length probes for both cDNA and cRNA. The sensitivity can be improved by using either longer (Welnicki *et al.*, 1989) or several oligonucleotides from different parts of the viroid molecule (Nakahara *et al.*, 1998). However, there is some indication that oligonucleotides from different parts of a viroid molecule may differ greatly in their sensitivity (Nakahara *et al.*, 1998). Since sensitivity of hybridization is affected significantly by the size of probe length, it is unlikely that synthetic probes of less than the full-length of the viroid would be adequate for viroid detection.

Reverse transcription–polymerase chain reaction

Within 4 years, following the introduction of PCR to amplify DNA fragments, the technique was successfully applied for the amplification of viroids using RT-PCR (Puchta and Sänger, 1989) and for the detection of viroids (Hadidi and Yang, 1990), demonstrating its applicability for plant pathogen detection. The method comprises of four main steps: preparation of nucleic acid templates preferably devoid of enzyme inhibitors; reverse transcription of viroid RNA by reverse transcriptase; PCR amplification of cDNA with a pair of viroid-specific primers; and detection of amplified fragments by electrophoresis on agarose gel or other methods. Preparation of cDNA and the PCR amplification steps are carried out in one step (termed as one-tube RT-PCR) (Ragozzino et al., 2004) or both steps can be performed separately. For the amplification of full-length viroid genomes, primer pairs are designed so that they anneal to the CCR part of the viroid molecule with the 3' ends facing away from each other. Full-length amplification of a number of viroids has been obtained in this manner (Hadidi and Candresse, 2003). In addition to conventional RT-PCR, real-time PCR (using TaqMan® chemistry) has also been developed for viroids (Boonham et al., 2004, 2005).

The cDNA and PCR steps do not differ from those used for viruses or other RNAs and DNAs. However, there are a few modifications for the detection of amplified fragments. Generally, PCR products are detected by electrophoretic analysis on agarose or polyacrylamide gels, followed by staining and destaining with ethidium bromide for both agarose and acrylamide gels and silver nitrate for polyacrylamide gel. To facilitate the detection of amplified fragments, several modifications have been introduced for viroid detection. For example, in the microtitre plate method (Shamloul et al., 2002), a biotinylated DNA probe is hybridized to DIG-labelled RT-PCR amplicons from viroid-infected tissues. The hybridized probe is captured in a streptavidin-coated microtitre plate by avidin–biotin interaction. Finally, the hybridized amplicon is detected with an enzyme-conjugated anti-DIG antibody. Hybridization requires 3 h and the reading of the absorbance requires many additional steps to complete the protocol. Many viroids, including PSTVd, ASSVd, apple dimple fruit viroid (ADFVd), pear blister canker viroid (PBCVd), HSVd and peach latent mosaic viroid (PLMVd) have been detected by this method (Shamloul et al., 2002). However, no large-scale comparison for any viroid of this protocol with conventional agarose gel electrophoresis using field-generated samples has been conducted.

Grape tissues contain high levels of phenolic compounds, polysaccharides and other complex substances that inhibit PCR. Specific protocols have been developed (Staub et al., 1995; Wah and Symons, 1997) that emphasize the removal of polysaccharides, a known inhibitor of DNA polymerases. At least five viroids are present in grape tissues and most of them are carried symptomlessly. These protocols have shown that very high quality RNA preparations (A_{260}/A_{280} 1.9–2.0) can be obtained from grape leaves. In addition to RNA quality, the primer pair was shown to have a decisive effect on the sensitivity of detection (Wah and Symons, 1997). The detection of low-copy number viroid samples was improved by jumping to the extension step in PCR immediately after annealing the primer rather than performing the denaturation step. These modifications have shown that RT-PCR is 25,000-fold more sensitive than the dot-blot hybridization used on the same tissues. Furthermore, the increased sensitivity enabled the detection of low-copy number viroids in supposedly viroid-free shoot apical meristem-cultured plantlets. Modified RT-PCR protocols have been used to demonstrate that all ten commercial grape cultivars were simultaneously infected with five grape viroids (Wah and Symons, 1997).

Multiplex RT-PCR of viroids

Citrus and grapevine are known to harbour several viroids (Wah and Symons, 1997; Ito et al., 2002). A multiplex RT-PCR has been designed to simultaneously detect fragments specific to seven citrus viroids or variants, using 6% PAGE, on the basis of different molecular sizes (Ito et al., 2002). The entire process only takes 2 days and has facilitated the survey of citrus orchards for seven viroids, e.g. CEVd, citrus bent leaf viroid (CBLVd) and its variant CVd-I-LSS, HSVd, citrus viroid III (CVd-III), CVd-IV and citrus viroid OS (CVd-OS). The survey showed a common occurrence of multiple viroid infection by three viroids (Ito et al., 2002). Similarly, a multiplex RT-PCR has been developed to identify six viroids belonging to four genera and in two families, using RT-PCR probe capture hybridization for the identification of specific viroids (Shamloul et al., 2002).

Multiplex RT-PCR of viruses and viroids

Mixed infection of viruses and viroids is known to exist (Singh and Somerville, 1987; Ito et al., 2002). In RT-PCR, where the detection is based on the specificity of primers to RNA, mixed infection of viruses and viroids can be detected from tissue extract. Using random primers (commercially available hexanucleotides) as universal reverse primer for potato viruses and PSTVd, the viroid was detected by RT-PCR along with five potato viruses (Nie and Singh, 2001). Similarly, using specific primer pairs for seven citrus viroids and apple stem grooving virus (ASGV), both viroids and viruses have been detected by RT-PCR (Ito et al., 2002).

Real-Time RT-PCR detection of viroids

A deterrent to large-scale application of the conventional RT-PCR method is the time-consuming nature of certain steps, such as the identification of the viroids after performing the PCR step. A real-time PCR methodology has been developed for PSTVd (Boonham et al., 2004). This technology uses primers and probes in the PCR process. A short, fluorescent probe hybridizes to the region of a viroid molecule between forward and reverse primers. The 5' end of the probe is labelled with a fluorescent dye as a 'reporter' and the 3' end of the probe is non-fluorescent 'quencher'. When the probe is intact, the signal of the reporter is minimal and the signal of quencher is strong. During primer extension, the 5' end of the probe is sequentially cleaved and as a result the fluorescent signals increase, while quencher signals decrease. The increase in fluorescence is monitored in real-time during amplification, using a combined thermal cycler and a fluorescence reader, with data recorded and plotted as a curve during the PCR cycle. At the threshold cycle (Ct), a significant increase in fluorescence occurs; therefore, a Ct value below a certain number (commonly 30–40) indicates a positive result.

In efforts to improve sensitivity, denaturation of samples prior to cDNA synthesis is imperative considering the highly complex secondary structures of viroids (Boonham et al., 2004). Optimization of parameters such as purity of viroid samples, cDNA synthesis temperatures, types and concentration of reverse transcriptase and single verses two-tube RT-PCR, has shown that real-time PCR can be used efficiently for the detection of PSTVd from infected tomato leaves, potato tubers, in vitro plants and true potato seeds (Boonham et al., 2004). The primer and probe set have been shown to detect all of 100 PSTVd sequence variants available from public databases. These have either been tested with live cultures (Boonham et al., 2004) or using synthetic oligonucleotides representing the mismatched sequences of PSTVd isolates (Boonham et al., 2005). Real-time RT-PCR with TaqMan® technology has proven to be a viable option to handle high-throughput testing and has been evaluated for routine testing of PSTVd.

Management of Viroid Diseases

The problems

Unlike cellular pathogens such as bacteria and fungi, viroid diseases are not controlled by chemical applications on infected plants. Post-infection prevention in viroids is more problematic. At present, the majority of economically significant viroid diseases have been detected in perennial or vegetatively propagated plants, further complicating management. Only a few viroids are transmitted through pollen and/or seeds (Wallace and Drake, 1962; Singh, 1970; Singh et al., 1991a) and are largely a threat in the development of new cultivars. To further complicate prevention management, not all viroid infections result in diagnostic symptoms in infected crops, particularly in ornamental plants and trees (Table 12.1), necessitating a large number of plant materials to be tested to prove the absence of viroids in symptomless plants.

Viroids often cause economic losses to the individual producer, but an additional parameter of economic loss from viroid diseases has arisen from the globalization of the modern agricultural production and export system. Governed by national and international marketing and phytosanitary regulations, dicovery of a new viroid pathogen in a country can trigger an avalanche of restrictive export measures and a shutting down of trade in efforts to

Table 12.1. Naturally occurring viroid infection in ornamental host plants.

Plant species	Viroid	Symptoms
Chrysanthemum indicum	CSVd	Stunting
Dendranthema grandiflora	CSVd	Stunting
Ageratum sp.	CSVd	nd
Petunia × hybrida cv. Surfinia	CSVd	Symptomless
Argyranthemum frutescens	CSVd	Stunting
Vinca major	CSVd	Symptomless
C. indicum	CChMVd	Mottle, chlorosis
Double *Impatiens walleriana*	CEVd	Symptomless
Glandularia puchella	CEVd	Symptomless
Trailing *Verbena × hybida*	CEVd	Symptomless
Columnea erythrophae	CLVd	Symptomless
Nematanthus wettsteinii	CLVd-N	Symptomless
Brunfelsia undulata	CLVd-B	Symptomless
Iresine herbstii cv 'Aureoreticulata'	IrVd	Symptomless
Trailing *verbena × hybrida*	IrVd	Symptomless
Alternanthera sessilis	TCDVd	Symptomless
Coleus blumei cv. 'Ruhm von Luxemburg'	CbVd 1-RL	Symptomless
Coleus blumei cv. 'Ruhm von Luxemburg'	CbVd 2-RL	Symptomless
Coleus bluemei cv. 'Bienvenue'	CbVd-Bv	Symptomless
Coleus bluemei cv. 'Bienvenue'	CbVd 1-BvA	Symptomless
Coleus bluemei cv. 'Fairway Ruby'	CbVd 3-FR	Symptomless
Coleus bluemei cv. 'Rainbow Gold'	CbVd 1-RG	Symptomless
Coleus blumei Benth.	CbVd	Symptomless
Mentha spicata	CbVd1	Symptomless
Mentha arvensis var. *piperascens*	CbVd1	Symptomless
Melissa officinalis	CbVd1	Symptomless
Ocimum basilicum	CbVd1	Symptomless
Perrilla magilla	CbVd	Symptomless
Solanum pseudocapsicum cv. 'New Patterson'	TASVd-S	Symptomless

nd = not described.

prevent establishment of a new pathogen in their country. Postponing the shipment of the commodity becomes the principal loss to the producer, rather than as a result of actual crop or fruit yield loss. For example, PSTVd is a listed quarantine organism for many regions, including the European Union (Boonham *et al.*, 2004, 2005; Jefferies and James, 2005). This aspect of income loss could be more prolonged in vegetatively propagated crops like potatoes, where replacement of seed tubers for large areas takes several years.

The need for diagnostic approaches

The planting of viroid-free material minimizes the risk of viroid spread, because most viroids do not spread by aerial vectors in fields and orchards.

However, production of viroid-free planting material requires the availability of diagnostic methods that are sensitive, rapid, reliable and applicable to large-scale testing. This is a mammoth task even for viruses, which can be detected by widely applicable serological tests, as well as infectivity and molecular approaches. Having viroid RNA as the sole molecular material to develop diagnostic protocols presents limited options for developing diagnostic protocols to ensure viroid freedom in the planting material. In spite of this challenge, considerable progress has been made in managing and eradicating viroid diseases from various countries. A case history of PSTVd from Canada illustrates the role of diagnostic and testing procedures in the eradication of this viroid. In the province of New Brunswick, Canada, the incidence of spindle tuber disease in table-stock fields was at the rate of 4% and in some fields as high as 15% during 1969–1970 (Singh et al., 1971). A series of phytosanitary, regulatory and testing procedures brought the incidence of PSTVd to zero by 1979–1980 (Fig. 12.1) (Singh and Crowley, 1985). The main measures to reduce the viroid inoculum were: (i) the development and application of rapid and sensitive detection methods for seed tuber testing; (ii) testing of potato breeding parental material and cultivars prior to their release to growers; (iii) changes in seed certification regulations to limited generations of growing seed tubers; (iv) adoption of a 'zero' tolerance for PSTVd in seed certification by visual inspection; and (v) establishment of an 'Elite' seed potato production farm to supply initial seed potato stock to commercial Elite seed growers free from PSTVd. Provincial disease eradication 'Acts' were enacted to reduce inoculum and maintain proper hygiene around

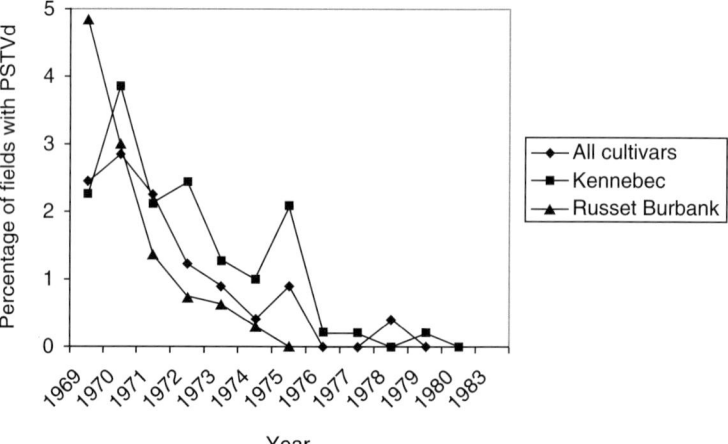

Fig. 12.1. Graphical representation of the decline of potato spindle tuber viroid from 1969 to 1980 in New Brunswick, Canada. The readings are visual inspection records of Potato Seed Certification Inspection Service. 'Russet Burbank' represents a highly susceptible and diagnostic cultivar, while 'Kennebec' is tolerant to PSTVd. 'All cultivars' represent data of ten cultivars grown in New Brunswick during the period.

farms and to regulate the use of cultivators between farms. Most of these changes were common in Canada and as a result, PSTVd has been eliminated from entire seed production areas (Singh, 1988; Singh et al., 1988; De Boer et al., 2002).

Development of New Technologies

To ensure viroid absence in planting material on farms and to be able to test a large number of symptomless plants, particularly various species of ornamental plants as hosts of Pospiviroids (Singh and Teixeria da Silva, 2006), two aspects of the RT-PCR protocol were modified to enhance its suitability for large-scale testing.

Improvements in the nucleic acid extraction procedure

Symptomless hosts serve as reservoirs for viroids and when accidentally transmitted to other plants, the viroids could become destructive pathogens to their new host plants. In most cases, upon encountering a first finding of a diagnostic viroid disease, initial efforts are targeted at its eradication from the initial source. For example, new outbreaks of *Pospiviroid* spp. in tomato and potato crops in recent years have been eradicated in various countries (cited in Singh and Teixeria da Silva, 2006). Considering that viroid incidence could be widely distributed, a large-scale survey would be needed. A procedure for nucleic acid preparation for RT-PCR, which is easy, rapid and reliable, and preferably one without the use of toxic organic solvents such as phenol would be preferable. In earlier studies dealing with PSTVd isolation, where infectivity was the sole criterion for the detection of the viroid, a large amount (150–600g) of leaves and numerous steps in the extraction procedure were used, often requiring several days to complete (Singh and Clark, 1971; Morris and Smith, 1977). Typically, a protocol for nucleic acid extraction from viroid-infected plants could involve many steps, including: (i) homogenization of plant tissues in buffers of high molarity and pH with RNase inhibitors to deter viroid RNA degradation and the use of water-saturated phenol to extract nucleic acids; (ii) centrifugation of homogenates to separate the aqueous phase; (iii) steps to remove polysaccharides from nucleic acid extracts; (iv) digestion of DNA from RNA preparations; (v) enrichment of small RNAs content by chromatography through non-ionic cellulose CF-11 or differential precipitation with lithium chloride; (vi) precipitation of total nucleic acids with ethanol; and (vii) recovery of the nucleic acids by centrifugation and resuspension of the precipitates in an appropriate buffer or water. However, following the introduction of molecular hybridization and PCR, efforts have been made to simplify the viroid isolation procedure and several modifications to the generic protocol in fruit tree tissues have been evaluated. For molecular hybridization, four tissue-homogenizing methods were compared. Two methods involved total nucleic acid extraction with or without organic solvents, followed by formaldehyde or alkaline denaturation. The other two procedures were based on the direct denaturation of crude leaf sap for molecular hybridization. It was observed that alkaline denaturation increased the sensitivity of detection of all four preparations of RNA (Turturo et al., 1998). At the same time, advances have been made to reduce the necessary plant material to only a few milligrams (Nakahara et al., 1999) and methods have become simpler and safer (i.e. protocols not using organic solvents). The use of a commercial material (Levy et al., 1994) can shorten RNA preparation time to 1–2 h from 1 to 3 days and the product is suitable for extracting nucleic acids from both herbaceous and woody plant tissues.

The wide range of plant species found in the ornamental plant industry presents more challenges for generic viroid detection procedures than in crop plants. For example, the variegated foliage of some *Vinca* spp. contain phenolics, while trailing *Verbena* spp. and *Impatiens* spp. contain anthocyanins, phenolics and quinine. These compounds interfere with amplification in RT-PCR, and thereby require removal or reduction prior to RT-PCR. Immobilization of extracts on nitrocellulose membrane and their elution with water has been shown to remove inhibitors of PCR from plant species, which otherwise failed to yield amplified product (Singh et al., 2004).

An inexpensive, rapid and simple method of viroid preparation has been developed based on the alkaline denaturation solution (Turturo et al., 1998). The method involves the binding and elution of the extract onto a nitrocellulose membrane (Singh et al., 2004) as a template suitable for RT-PCR (Singh et al., 2006a). In India, the modified methodology has been used to detect viroids from ornamental plants with considerable success (Singh et al., 2006b). The procedure can be used at four stages of sample preparation for the detection of viroids by RT-PCR. The four formats are diagrammatically presented in Fig. 12.2. Format 1 consists of homogenizing plant tissues in NaOH–EDTA solution with a tissue to solution (50 mM NaOH + 2.5 mM EDTA) ratio of 1:4 (w/v) or higher, incubating at room temperature for 15 min to settle the coarse plant debris and using the supernatant for RT-PCR. Format 2 utilizes the supernatant from format 1 for spotting (10 µl) onto nitrocellulose membrane. The dried spots are cut out and eluted with sterile distilled water and the liquid is used for RT-PCR. Formats 3 and 4 utilize a step of centrifugation instead of sap incubation at room temperature. The centrifuged supernatant sap is directly used for RT-PCR in format 3, while in format 4, the sap is spotted onto nitrocellulose and eluted in sterile distilled water. For the optimum reliability of detection, the spotting is done with the aid of a vacuum device. An extensive evaluation of the four formats has shown that although viroids can be detected in extracts not immobilized on nitrocellulose membrane, the efficiency of detection increases significantly from the membrane-eluted material (Fig. 12.2).

A recent study (Hosokawa et al., 2006) has described another technique of template preparation termed 'microtissue direct RT-PCR method', where a

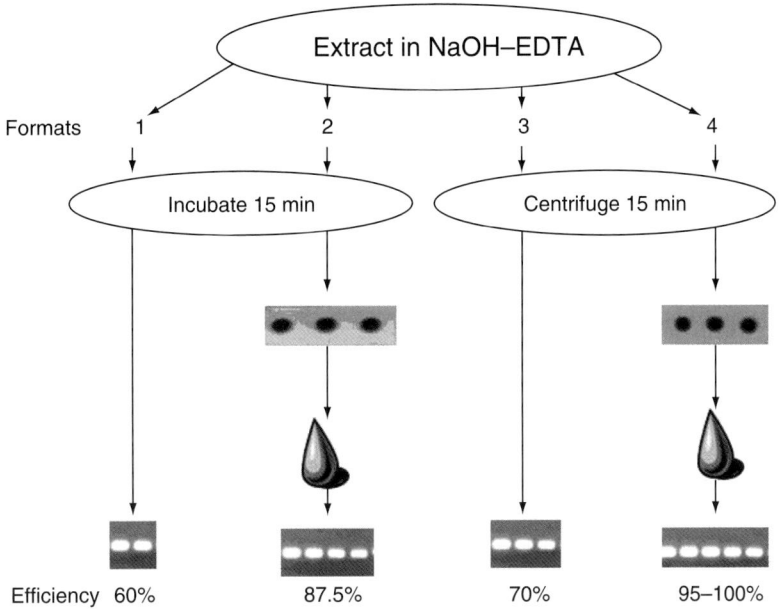

Fig. 12.2. Diagrammatic representation of viroid extraction procedure using sodium hydroxide and ethylenediamine tetraacetic acid. Formats 1–4 represent with or without centrifugation and with or without spotting the extract to nitrocellulose membrane prior to RT-PCR. Efficiency indicates the percentage of detection by each format.

razor or syringe needle is used to pierce the tissue to a depth of 0.1–0.2 mm and the sample is directly transferred to RT mixtures and then amplified by PCR. This method was successfully used to detect CSVd and chrysanthemum chlorotic mottle viroid (CChMVd) from small *in vitro* cultured plantlets. The last two procedures fulfil the requirement of nucleic acid preparations on a large scale for conventional tissues as well as tissue-cultured plantlets.

Pospiviroid universal primer pairs

Most of the reported viroid surveys are specific for a particular group of plants or viroid (e.g. PSTVd strain survey (Singh *et al.*, 1970); ornamental plants from individual nurseries [Bostan *et al.*, 2004]; specific plant species [Singh *et al.*, 1991a; Spieker, 1996]; or fruit trees [Ito *et al.*, 2002]). The symptomless nature of ornamental plants necessitates sampling of all plants without any definite lead for viroid infection. Since most species of the genus *Pospiviroid* (Table 12.1) have been encountered in ornamental and crop plants, a primer pair amplifying the species of the genus *Pospiviroid* has been described and evaluated (Fig. 12.3) (Bostan *et al.*, 2004). The *Pospiviroid* forward primer comprises of 89–108 nt PSTVd in the upper strand of the CCR and the reverse primer includes 259–280 nt from the

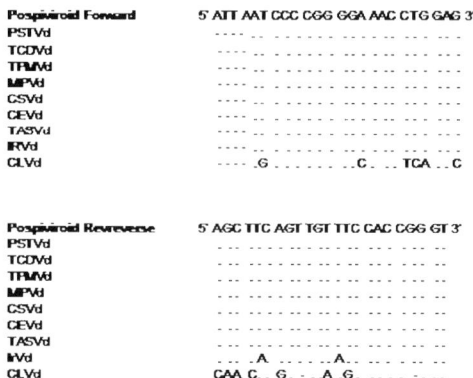

Fig. 12.3. A *Pospiviroid* group-specific primer pair. Forward primer contains bases (ATTA) at the 5' end which are not present in the viroids. The dots represent homology and letters indicate the mismatches from each viroid in both primers segments.

lower strand. The forward primer is fully matched for at least one isolate of eight *Pospiviroid* spp. with the exception of CLVd, and the reverse primer is in full agreement with seven viroids, with two mismatches in IrVd and several mismatches in CLVd (Fig. 12.3). Thus, all *Pospiviroids*, except CLVd, can be detected if present in the plant preparations. Using the generic pospiviroid primer pair, fragment sizes of 196 bp for mexican papita viroid (MPVd) to 228 bp for the largest viroid, IrVd, are expected. CSVd, CEVd and IrVd have been detected from ornamental plants using a generic *Pospiviroid* primer (Bostan *et al.*, 2004; Nie *et al.*, 2005; Singh *et al.*, 2006b). Similarly, two sets of *Pospiviroid* genus-specific primer pairs have been designed which readily amplify CLVd (Verhoeven *et al.*, 2004).

Conclusions and Future Directions

Improvements in molecular biology technologies have been readily adopted for the detection of viroids. The high sensitivity of RT-PCR accompanied by the high quality of RNA preparation has helped in the detection of viroids in low concentrations such as in fruit trees or where viroid distribution may be uneven during different seasons. However, improvements aimed at simplifying viroid RNA preparations and its application to a large number of samples need continuous research. Technologies utilizing microarray chips combined with PCR technology for the detection of all groups of viroids could be envisioned in the near future.

An important aspect of viroid management relies on exclusion of viroids from entering the crop or fruit production chain. It has been shown that long-term natural infections of *Coleus* spp. have resulted in the evolution of a new viroid, by forming a true viroid chimera, consisting of the right half of one existing viroid and the left half of the other (Spieker, 1996). In addition, an artificial 'inverse' chimera has been designed and shown to be infectious in

Coleus plants, demonstrating the ease and potential of the viroid chimera to evolve *in planta* and lead to the emergence of new infectious RNA replicons (Spieker, 1996). Furthermore, considering that *Coleus* viroids are well adapted for rapid dissemination through seed, mechanical and graft transmission to other plants (Singh *et al.*, 1991a), the potential infection to crop plants would generate serious viroid management problems. Fortunately, to date, there is no information about *Coleus* viroids infecting any crops or trees and causing economic damage. However, it is probable that viroids with characteristics similar to *Coleus* viroids may be present in ornamental plants. For example, recent surveys of a few ornamental plants in Canada (Bostan *et al.*, 2004; Nie *et al.*, 2005) and India (Singh *et al.*, 2006b) have shown the existence of viroids in various cultivars of *Alternanthera sessilis*, *Impatiens* and *Verbena* which are carried symptomlessly in these plants. Upon transfer to tomato and potato plants, they cause visible symptoms, severe malformation of potato tubers and total yield loss (Fig. 12.4). The viroids are highly seed transmissible in *Impatiens* and *Verbena*. Both plants are widely grown as aesthetic ground covers, are components of ornamental hanging baskets and part of the general landscape scene. The viroid in question is CEVd, a known viroid of citrus, but the isolates are from these ornamental plants and not from citrus (R.P. Singh, 2006).

Fig. 12.4. (A) Absence of diagnostic symptoms in leaves and flowers of *Impatiens* spp. infected with citrus exocortis viroid. (B) Effect of viroid from *Impatiens* in potato tubers of cultivar 'Jemseg'. Note the smaller size and malformed tubers.

Ornamental plants comprise a rapidly expanding industry, equipped with modern rapid transportation of live cuttings, bulbs and seeds, capable of being imported and exported all over the world without extensive tests being performed for viroids at quarantine centres. Viroids thrive in warmer climates (Singh, 1983), where generally, multinational seed companies and nurseries propagate their plants (Singh *et al.*, 1991a). These plants could be potential sources of crop-infecting viroids and should be surveyed to determine their viroid content. The basic methodology for such a survey has been outlined in this chapter and continuous improvements in detection technology are being made daily.

Acknowledgements

Critical reading and editing by Dr Avinash Singh, AgraPoint International Inc., Nova Scotia, Canada, and the technical assistance of Andrea D. Dilworth are highly appreciated.

References

Avila, A.C., Singh, R.P., Dusi, A.N., Fonseca, M.E.N. and De Castro, L.A.S. (1990) Lack of evidence of the presence of potato spindle tuber viroid in the main potato crop of Brazil. *Fitopatologia Brasileira* 15, 186–189.

Barker, J.M., McInnes, J.L., Murphy, P.J. and Symons, R.H. (1985) Dot-blot procedure with [^{32}P]DNA probes for the sensitive detection of avocado sunblotch and other viroids in plants. *Journal of Virological Methods* 10, 87–98.

Boonham, N., González Pérez, L., Mendez, M.S., Lilia Peralta, E., Blockley, A., Walsh, K., Barker, I. and Mumford, R.A. (2004) Development of a real-time RT-PCR assay for the detection of potato spindle tuber viroid. *Journal of Virological Methods* 116, 139–146.

Boonham, N., Fisher, T. and Mumford, R.A. (2005) Investigating the specificity of real-time PCR assays using synthetic oligonucleotides. *Journal of Virological Methods* 130, 30–35.

Bostan, H., Nie, X. and Singh, R.P. (2004) An RT-PCR primer pair for the detection of *Pospiviroid* and its application in surveying ornamental plants for viroids. *Journal of Virological Methods* 116, 189–193.

Branch, A.D. and Robertson, H.D. (1984) A replication cycle for viroids and other small infectious RNAs. *Science* 223, 450–455.

Candresse, T., Macquaire, G., Brault, V., Monsion, M. and Dunez, J. (1990) ^{32}P- and biotin-labelled in vitro transcribed cRNA probes for the detection of potato spindle tuber viroid and chrysanthemum stunt viroid. *Research in Virology* 141, 97–107.

De Boer, S.H., Xu, H. and DeHaan, T.L. (2002) *Potato spindle tuber viroid* not found in western Canadian provinces. *Canadian Journal of Plant Pathology* 24, 372–375.

Desvignes, J.C. (1980) Different symptoms of the peach latent mosaic. *Acta Phytopathologica Academiae Scientiarum Hungaricae* 15, 183–190.

Diener, T.O. (1971) Potato spindle tuber 'virus' IV. A replicating, low molecular weight RNA. *Virology* 45, 411–428.

Diener, T.O. and Raymer, W.B. (1967) Potato spindle tuber virus: a plant virus with properties of a free nucleic acid. *Science* 158, 378–381.

Fernow, K.H. (1967) Tomato as a test plant for detecting mild strains of potato spindle tuber virus. *Phytopathology* 57, 1347–1352.

Flores, R. (1986) Detection of citrus exocortis viroid in crude extracts by dot-blot hybridization: conditions for reducing spurious hybridization results and for enhancing the sensitivity of the technique. *Journal of Virological Methods* 13, 161–169.

Flores, R., Hernández, C., Martinez de Alba, A.E., Daròs, J.-A. and Di Serio, F. (2005) Viroids and viroid–host interactions. *Annual Review of Phytopathology* 43, 117–139.

Gelder, W.M.J. and Treur, A. (1982) Testing of imported potato genotypes for potato spindle tuber viroid with a tomato-intermediate/electrophoresis combined method. *European Plant Protection Organization Bulletin* 12, 297–305.

Goss, R.W. (1926) A simple method of inoculating potatoes with spindle-tuber disease. *Phytopathology* 16, 233.

Hadidi, A. and Candresse, T. (2003) Polymerase chain reaction. In: Hadidi, A., Flores, R., Randles, J.W. and Semanick, J.S. (eds) *Viroids*, CSIRO Publishing, Collingwood, Australia, pp. 115–122.

Hadidi, A. and Yang, X. (1990) Detection of pome fruit viroids by enzymatic cDNA amplification. *Journal of Virological Methods* 30, 261–270.

Hanold, D., Semancik, J.S. and Owens, R.A. (2003) Polyacrylamide gel electrophoresis. In: Hadidi, A., Flores, R., Randles, J.W. and Semanick, J.S. (eds) *Viroids*, CSIRO Publishing, Collingwood, Australia, pp. 95–102.

Harris, P.S. and Miller-Jones, D.N. (1981) An assessment of the tomato/polyacrylamide gel electrophoresis test for potato spindle tuber viroid in potato. *Potato Research* 27, 399–408.

Hollings, M. and Stone, O.M. (1973) Some properties of chrysanthemum stunt, a virus with the characteristics of an uncoated ribonucleic acid. *Annals of Applied Biology* 74, 333–348.

Horst, R.K. (1975) Detection of a latent infectious agent that protects against infection by chrysanthemum chlorotic mottle viroid. *Phytopathology* 65, 1000–1003.

Hosokawa, M., Matsushita, Y., Uchida, H. and Yazawa, S. (2006) Direct RT-PCR method for detecting two chrysanthemum viroids using minimal amounts of plant tissue. *Journal of Virological Methods* 131, 28–33.

Ito, T., Ieki, H., Ozaki, K., Iwanami, T., Nakahara, K., Hataya, T., Ito, T., Isaka, M. and Kano, T. (2002) Multiple citrus viroids in citrus from Japan and their ability to produce exocortis-like symptoms in citron. *Phytopathology* 92, 542–547.

Jefferies, C. and James, C. (2005) Development of an EU protocol for the detection and diagnosis of *Potato spindle tuber pospiviroid*. *European Plant Protection Organization Bulletin* 35, 125–132.

Kanematsu, S., Hibi, T., Hashimoto, J. and Tsuchizaki, T. (1991) Comparison of nonradioactive cDNA probes for detection of potato spindle tuber viroid by dot-blot hybridization assay. *Journal of Virological Methods* 35, 189–197.

Keese, P. and Symons, R.H. (1985) Domains in viroids: evidence of intermolecular RNA rearrangement and their contribution to viroid evolution. *Proceedings of the National Academy of Sciences of the United States of America* 82, 4582–4586.

Koltunow, A.M. and Rezaian, M.A. (1988) Grapevine yellow speckle viroid. Structural features of a new viroid group. *Nucleic Acids Research* 16, 849–864.

Levy, L., Lee, L.-M. and Hadidi, A. (1994) Simple and rapid preparation of infected plant tissue extracts for PCR amplification of virus, viroid, and MLO nucleic acids. *Journal of Virological Methods* 49, 295–304.

Matoušek, J., Patzak, J., Orctová, L., Schubert, J., Vrba, L., Steger, G. and Riesner, D. (2001) The variability of hop latent viroid as induced upon heat treatment. *Virology* 287, 349–358.

Matoušek, J., Orctová, L., Steger, G., Škopek, J., Moors, M., Dĕdic, P. and Riesner, D. (2004) Analysis of thermal stress-mediated PSTVd variation and biolistic inoculation of progeny of viroid 'thermomutants' to tomato and *Brassica* species. *Virology* 323, 9–13.

McInnes, J.L., Habili, N. and Symons, R.H. (1989) Nonradioactive, photobiotin-labelled DNA probes for routine diagnosis of viroids in plant extracts. *Journal of Virological Methods* 23, 299–312.

Morris, T.J. and Smith, E.M. (1977) Potato spindle tuber disease: procedures for the detection of viroid RNA and certification of disease-free potato tubers. *Phytopathology* 67, 145–150.

Mosch, W.H.M., Huttinga, H., Hakkaart, F.A. and De Bokx, J.A. (1978) Detection of chrysanthemum stunt and potato spindle tuber viroids by polyacrylamide gel electrophoresis. *Netherlands Journal of Plant Pathology* 84, 85–93.

Mühlbach, H.-P., Weber, U., Gómez, G., Pallás, V., Duran-Vila, N. and Hadidi, A. (2003) Molecular hybridization. In: Hadidi, A., Flores, R., Randles, J.W. and Semancik, J.S. (eds) *Viroids.* CSIRO Publishing, Collingwood, Australia, pp. 103–114.

Nakahara, K., Hataya, T., Hayashi, Y., Sugimoto, T., Kimura, I. and Shikata, E. (1998) A mixture of synthetic oligonucleotide probes labeled with biotin for the sensitive detection of potato spindle tuber viroid. *Journal of Virological Methods* 71, 219–227.

Nakahara, K., Hataya, T. and Uyeda, I. (1999) A simple, rapid method of nucleic acid extraction without tissue homogenization for detecting viroids by hybridization and RT-PCR. *Journal of Virological Methods* 77, 47–58.

Nie, X. and Singh, R.P. (2001) A novel usage of random primers for multiplex RT-PCR detection of virus and viroid in aphids, leaves, and tubers. *Journal of Virological Methods* 91, 37–49.

Nie, X., Singh, R.P. and Bostan, H. (2005) Molecular cloning, secondary structure and phylogeny of three pospiviroids from ornamental plants. *Canadian Journal of Plant Pathology* 27, 592–602.

Owens, R.A. and Diener, T.O. (1981) Sensitive and rapid diagnosis of potato spindle tuber viroid disease by nucleic acid spot hybridization. *Science* 213, 113–117.

Palacio, A., Foissac, X. and Duran-Vila, N. (2000) Indexing of citrus viroids by imprint hybridisation. *European Journal of Plant Pathology* 105, 897–903.

Palukaitis, P., Cotts, S. and Zaitlin, M. (1985) Detection of viroids and viral nucleic acids by 'dot-blot' hybridization. *Acta Horticulturae* 164, 109–118.

Pfannenstiel, M.A. and Slack, S.A. (1980) Response of potato cultivars to infection by the potato spindle tuber viroid. *Phytopathology* 70, 922–926.

Podleckis, E.V., Hammond, R.W., Hurtt, S.S. and Hadidi, A. (1993) Chemiluminescent detection of potato and pome fruit viroids by digoxigenin-labelled dot blot and tissue blot hybridization. *Journal of Virological Methods* 43, 147–158.

Puchta, H. and Sänger, H.L. (1989) Sequence analysis of minute amounts of viroid RNA using the polymerase chain reaction (PCR). *Archives of Virology* 106, 335–340.

Ragozzino, E., Faggioli, F. and Barba, M. (2004) Development of a one tube-one step RT-PCR protocol for the detection of seven viroids in four genera: *Apscaviroid Hostuviroid, Pelamoviroid* and *Pospiviroid. Journal of Virological Methods* 121, 25–29.

Randles, J.W. and Rodriguez, M.J.B. (2003) Palm tree viroids. In: Hadidi, A., Flores, R., Randles, J.W. and Semancik, J.S. (eds) *Viroids.* CSIRO Publishing, Collingwood, Australia, pp. 233–245.

Raymer, W.B. and O'Brien, M.J. (1962) Transmission of potato spindle tuber virus to tomato. *American Potato Journal* 39, 401–408.

Rivera-Bustamante, R., Gin, R. and Semanick, J.S. (1986) Enhanced resolution of circular and linear molecular forms of viroid and viroid-like RNA by electrophoresis in a discontinuous-pH system. *Annals of Biochemistry* 156, 91–95.

Rosenbaum, V. and Riesner, D. (1987) Temperature-gradient gel electrophoresis: thermodynamic analysis of nucleic acids and proteins in purified form and in cellular extracts. *Biophysical Chemistry* 26, 235–246.

Sänger, H.L., Klotz, G., Riesner, D., Gross, H.J. and Kleinschmidt, A.K. (1976) Viroids are single-stranded covalently closed circular RNA molecules existing as highly base-paired rod-like structures. *Proceedings of the National Academy of Sciences of the United States of America* 73, 3852–3856.

Sano, T., Kudo, H., Sugimoto, T. and Shikata, E. (1988) Synthetic oligonucleotide hybridization probes diagnose hop stunt viroid strains and citrus exocortis viroid. *Journal of Virological Methods* 19, 109–120.

Sasaki, M. and Shikata, E. (1978) Studies on hop stunt disease I. Host range. *Report of Research Laboratory*, Kirin Brewery Company Limited, Number 21, pp. 27–39.

Schumacher, J., Randles, J.W. and Riesner, D. (1983) A two-dimensional electrophoretic technique for the detection of circular viroids and virusoids. *Annals of Biochemistry* 135, 228–295.

Schumacher, J., Meyer, N., Riesner, D. and Weidemann, J.L. (1986) Diagnostic procedure for detection of viroids and viruses with circular RNAs by 'return' gel electrophoresis. *Journal of Phytopathology* 115, 332–343.

Shamloul, A.M., Faggioli, F., Keith, J.M. and Hadidi, A. (2002) A novel multiplex RT-PCR probe capture hybridization (RT-PCR-ELISA) for simultaneous detection of six viroids in four genera: *Apscaviroid, Hostuviroid, Pelamoviroid,* and *Pospiviroid. Journal of Virological Methods* 105, 115–121.

Singh, R.P. (1970) Seed transmission of potato spindle tuber virus in tomato and potato. *American Potato Journal* 47, 225–227.

Singh, R.P. (1973) Experimental host range of potato spindle tuber virus. *American Potato Journal* 50, 111–123.

Singh, R.P. (1983) Viroids and their potential danger to potatoes in hot climates. *Canadian Plant Disease Survey* 63, 13–18.

Singh, R.P. (1988) Occurrence, diagnosis and eradication of the potato spindle tuber viroid in Canada. In: *Viroids of Plants and Their Detection: International Seminar*, 12–20 August 1986. Warsaw Agricultural University Press, Warsaw, pp. 37–50.

Singh, R.P. (1991) Return-polyacrylamide gel electrophoresis for the detection of viroids. In: Maramorosch, K. (ed.) *Viroids and Satellites: Molecular Parasites at the Frontier of Life*. CRC Press, Boca Raton, Florida, pp. 89–118.

Singh, R.P. and Bagnall, R.H. (1968) Infectious nucleic acid from host tissues infected with the potato spindle tuber virus. *Phytopathology* 58, 696–699.

Singh, R.P. and Boucher, A. (1987) Electrophoretic separation of a severe from mild strains of potato spindle tuber viroid. *Phytopathology* 77, 1588–1591.

Singh, R.P. and Clark, M.C. (1971) Infectious low-molecular weight ribonucleic acid from tomato. *Biochemical Biophysical Research Communications* 44, 1077–1083.

Singh, R.P. and Crowley, C.F. (1985) Successful management of potato spindle tuber viroid in seed potato crop. *Canadian Plant Disease Survey* 65, 9–10.

Singh, R.P. and Finnie, R.E. (1977) Stability of potato spindle tuber viroid in freeze-dried leaf powder. *Phytopathology* 67, 283–286.

Singh, R.P. and Ready, K.F.M. (2003) Biological indexing. In: Hadidi, A., Flores, R., Randles, J.W. and Semancik, J.S. (eds) *Viroids*. CSIRO Publishing, Collingwood, Australia, pp. 89–94.

Singh, R.P. and Somerville, T.H. (1987) New disease symptoms observed on field-grown potato plants with potato spindle tuber viroid and potato virus Y infections. *Potato Research* 30, 127–132.

Singh, R.P. and Teixeria da Silva, J.A. (2006) Ornamental plants: silent carrier of evolving viroids. In: Teixeria da Silva, J.A. (ed.) *Floriculture, Ornamental and Plant Biotechnology*, Vol. III. Global Science Books, London, pp. 531–539.

Singh, R.P., Finnie, R.E. and Bagnall, R.H. (1970) Relative prevalence of mild and severe strains of potato spindle tuber virus in Eastern Canada. *American Potato Journal* 47, 289–293.

Singh, R.P., Finnie, R.E. and Bagnall, R.H. (1971) Losses due to the potato spindle tuber virus. *American Potato Journal* 48, 262–267.

Singh, R.P., Lee, C.R. and Clark, M.C. (1974) Manganese effect on the local lesion symptom of potato spindle tuber 'virus' in *Scopolia sinensis. Phytopathology* 64, 1015–1018.

Singh, R.P., De-Haan, T.-L. and Jaswal, A.S. (1988) A survey of the incidence of potato spindle tuber viroid in Prince Edward Island using two testing methods. *Canadian Journal of Plant Science* 68, 1229–1236.

Singh, R.P., Boucher, A. and Singh, A. (1991a) High incidence of transmission and occurrence of a viroid in commercial seeds of *Coleus* in Canada. *Plant Disease* 75, 184–187.

Singh, R.P., Boucher, A. and Wang, R.G. (1991b) Detection, distribution and long-term persistence of potato spindle tuber viroid in true potato seed from Heilongjiang, China. *American Potato Journal* 68, 65–74.

Singh, R.P., Boucher, A., Lakshman, D.K. and Tavantzis, S.M. (1994) Multimeric non-radioactive cRNA probes improve detection of potato spindle tuber viroid (PSTVd). *Journal of Virological Methods* 49, 221–233.

Singh, R.P., Dilworth, A.D., Singh, M. and McLaren, D.L. (2004) Evaluation of a simple membrane-based nucleic acid preparation protocol for RT-PCR detection of potato viruses from aphid and plant tissues. *Journal of Virological Methods* 121, 163–170.

Singh, R.P., Dilworth, A.D., Singh, M. and Babcock, K.M. (2006a) An alkaline solution simplifies nucleic acid preparation for RT-PCR and infectivity assays of viroids from crude sap and spotted membrane. *Journal of Virological Methods* 132, 204–211.

Singh, R.P., Dilworth, A.D., Baranwal, V.K. and Gupta, K.N. (2006b) Detection of *Citrus exocortis viroid*, *Iresine viroid*, and *Tomato chlorotic dwarf viroid* in new ornamental host plants in India. *Plant Disease* 90, 1457.

Spieker, R.L. (1996) In vitro-generated 'inverse' chimeric *Coleus blumei* viroids evolve *in vivo* into infectious RNA replicons. *Journal of General Virology* 77, 2839–2846.

Staub, U., Polivka, H. and Gross, H.J. (1995) Two rapid microscale procedures for isolation of total RNA from leaves rich in polyphenols and polysaccharides: application for sensitive detection of grapevine viroids. *Journal of Virological Methods* 52, 209–218.

Steger, G. and Riesner, D. (2003) Molecular characteristics. In: Hadidi, A., Flores, R., Randles, J.W. and Semancik, J.S. (eds) *Viroids*. CSIRO Publishing, Collingwood, Australia, pp. 15–29.

Turturo, C., Minafra, A., Ni, H., Wang, G., Di Terlizzi, B. and Savino, V. (1998) Occurrence of peach latent mosaic viroid in China and development of an improved detection method. *Journal of Plant Pathology* 80, 165–169.

Verhoeven, J.Th.J., Jansen, C.C.C., Willemen, T.M., Kox, L.F.F., Owens, R.A. and Roenhorst, J.W. (2004) Natural infections of tomato by *Citrus exocortis viroid*, *Columnea latent viroid*, *Potato spindle tuber viroid* and *Tomato chlorotic dwarf viroid*. *European Journal of Plant Pathology* 110, 823–831.

Wah, Y.F.W.C. and Symons, R.H. (1997) A high sensitivity RT-PCR assay for the diagnosis of grapevine viroids in field and tissue culture samples. *Journal of Virological Methods* 63, 57–69.

Wallace, J.M. and Drake, R.J. (1962) A high rate of seed transmission of avocado sunblotch virus from symptomless trees and origin of such trees. *Phytopathology* 52, 237–241.

Welnicki, M. and Hiruki, C. (1992) Highly sensitive digoxigenin-labelled DNA probe for the detection of potato spindle tuber viroid. *Journal of Virological Methods* 39, 91–99.

Welnicki, M., Skrzeczkowski, J., Soltyñska, A., Joñczyk, P., Markiewicz, W., Kierzek, R., Imiolczyk, B. and Zagórski, W. (1989) Characterisation of synthetic DNA probe detecting potato spindle tuber viroid. *Journal of Virological Methods* 24, 141–152.

Werner, H.O. (1926) The spindle-tuber disease as factor in seed potato production. University of Nebraska, Agricultural Experiment Station, Research Bulletin 32, p. 128.

Whitney, E.D. and Peterson, L.C. (1963) An improved technique for inducing diagnostic symptoms in tomato infected by potato spindle tuber virus. *Phytopathology* 53, 893.

Zelany, B., Randles, J.W., Boccardo, G. and Imperial, J.S. (1982) The viroid nature of the cadang-cadang disease of coconut palm. *Scientia Filipinas* 2, 46–63.

13 Application of Cationic Antimicrobial Peptides for Management of Plant Diseases

S. Misra and A. Bhargava

Abstract

Commercial production of crop plants is often threatened by recurring bacterial, fungal and viral infections. Pesticides and insecticides have commonly been used to contain phytopathogens but their extensive use has contributed to chemical contamination of the environment. Genetic engineering is an effective strategy for developing disease-resistant germplasm that increases yield, reduces loss, and eliminates or reduces the use of pesticides. Antimicrobial peptides are important components of innate disease immunity and have been isolated from a wide variety of organisms. These widespread natural products vary greatly in their properties and spectrum of biological activities. Different peptides and their synthetic derivatives have found applications as antibacterial, antifungal and therapeutic agents. Attempts have been made to bolster plant defences against microorganisms by genetically engineering plants to express cationic peptides. Because the primary target of the cationic peptides is the cell membrane and not a specific receptor or substrate, these peptides confer their activities against a broad spectrum of pathogens and there is only a low probability of resistance arising by changes to metabolic pathways. This chapter highlights salient features of antimicrobial peptides and their applications in plant biotechnology for management of a broad spectrum of diseases.

Introduction

Plant diseases are responsible for enormous losses worldwide ($30–50 billion annually) in cultivated and stored crops, and thus, are a major impediment to effective food production and distribution. In the past, containment of plant pathogens has relied on use of chemical pesticides. Heavy reliance on chemical pesticides has far-reaching implications not only for the environment but also for human health through residual toxicity. In addition, pesticides are becoming less effective because of increasing insecticide resistance in insect populations. Long-term climatic changes leading to changes in vector populations, and concern about the environmental effects of pesticides, as well as consumer concern about pesticide residues in food, have led to increased interest in finding alternative means of controlling phytopathogens.

Traditionally, plant breeding strategies have been successfully used to develop a large number of disease-resistant varieties. However, the increasing intensity of crop management has been accompanied by a growing number of diseases and a large number of pathogenic strains that have outpaced the development of new resistant plant varieties using conventional plant breeding strategies. Unwanted effects such as reduction in yield and fertility are often observed in the transfer of the dominant resistance genes. Transfer of resistance genes into high-yielding crops is a time-consuming process.

The incorporation of specific disease-resistant traits in plants through genetic engineering offers a means to prevent disease-associated losses without damaging the environment. Non-conventional strategies for the production of disease-resistant crop plants have exploited gene transfer technology for molecular resistance breeding (Marcos et al., 1995; Punja, 2001). Such strategies have included: expression of genes of plant defence response pathway components (Cao et al., 1998); expression of genes encoding plant, fungal or bacterial hydrolytic enzymes (Mourgues et al., 1998); and expression of genes encoding elicitors of defence response (Keller et al., 1999) and small peptides (Cary et al., 2000). Expression of antimicrobial peptides in plants, derived not only from plant sources but also from insects and mammals, is a promising strategy that can be exploited to promote disease resistance in plants (Osusky et al., 2000, 2004, 2005).

Antimicrobial Peptides

Cationic antimicrobial peptides have been found in a variety of sources, from prokaryotes to higher eukaryotes (Hancock et al., 1995; Vizioli and Salzet, 2002) (Table 13.1). In the last 25 years, more than 800 cationic, gene-encoded antimicrobial peptides have been described. The majority of peptides (96%) have a net positive charge but some have a net negative charge. In recent years, it has become clear that these endogenous peptide antibiotics constitute part of the first line of host defence (Boman, 1995). Cationic peptides, already present in the first line of defence of living organisms, can be induced and synthesized much more rapidly than immunoglobulin upon infection, before the adaptive immune system is activated, and can function without the high specificity and memory of immunoglobulin or immune cells (Boman, 1995). In mammals, antimicrobial peptides are present at high concentrations in phagocytes (e.g. macrophages, neutrophils, NK cells) and mucosal epithelial cells (e.g. Paneth cells). In lower life forms, such as invertebrates, which have no adaptive immunity, cationic peptides are the major defensive system against infection (Boman, 1995). Insects produce cationic peptides in their fat bodies and hemolymph, where they are induced upon bacterial challenge (Boman, 1995). Cationic peptides also function to keep the natural microflora at a steady state in a variety of different niches such as the skin, mouth and intestine. They are active not only against bacteria but also against fungi, viruses and even parasites (Vizioli and Salzet, 2002) (Table 13.1). The natural cationic peptides of animals and plants are synthesized as precursor peptides, and then processed into their mature forms by cleavage of a signal peptide and a pro-sequence (Hancock, 1997).

The earliest peptide antibiotics used extensively in human medicine were the gramicidins and polymyxins. The lantibiotic, nisin, is currently used as a food preservative. MSI-78, a 22-residue magainin analogue, has

Table 13.1. Cationic peptides with broad-spectrum activity against pathogens and viruses.

Peptide	Source	Net Charge	Activity	Reference
Alloferon	Blow fly	+4	Virus/Tumour cells	Chernysh et al. (2002)
α-Basrubrin	Spinach	+2	Virus/Fungi	Wang and Ng (2004)
Brevinin	Frog	+4	Virus	Yasin et al. (2000)
Caerin	Frog	+3	Virus/Bacteria (+/−)	Goraya et al. (2000)
Cathelicidin	Bovine	+2 to +8	Bacteria(+/−)/ Fungi/ Trypanosomes	Zanetti et al. (1995)
Cecropin	Silk moth	+5	Bacteria (+/−)/Fungi	Boman and Hultmark (1987)
Dermaseptin	Frog	+4	Virus/Bacteria(+/−)/ Fungi	Belaid et al. (2002)
Defensin	Human/ Rabbit Rat/Carrot	+3 to +8	Virus/ Bacteria (+/−)/ Fungi	Daher et al. (1986); Yeaman and Yount (2003)
Esculentin	Frog	+6	Virus/Bacteria(+/−)/ Fungi	Chinchar et al. (2001)
Indolicidin	Bovine	+3	Virus/ Bacteria (+/−)	Selsted et al. (1992)
Lentin	Mushroom	+1	Virus/ Bacteria (+/−)/ Fungi	Ngai and Ng (2003)
Magainin	Frog	+4	Virus/Bacteria (+/−)/ Fungi	Aboudy et al. (1994); Egal et al. (1999)
Melittin	Honey bee	+5	Virus/Bacteria (+/−)/ Fungi	Marcos et al. (1995); Wachinger et al. (1998)
Polyphemusin	Horse shoe crab	+7	Virus/Bacteria (+/−)/ Fungi	Murakami et al. (1991); Nakashima et al. (1992)
Protegrin	Pig	+5	Virus/Bacteria (+/−)/ Fungi	Kokryakov et al. (1993)
Panaxagin	Panax ginseng	+4	Virus/Fungi	Ng and Wang (2001)
Rantuerin	Frog	+4	Virus/Bacteria (+/−)/ Fungi	Chinchar et al. (2001)
Temporin	Frog	+2	Bacteria/Fungi	Harjunpaa et al. (1999)

+ Gram-positive bacteria
− Gram-negative bacteria

completed human Phase III clinical trials, showing equivalent efficacy to oral ofloxacin on polymicrobic infections of individuals with diabetic foot ulcers (Hancock, 1997). IB-367 is a synthetic protegrin-like cationic peptide that has shown efficacy in early clinical trials against oral mucositis and the sterilization of central venous catheters. It is currently proceeding through Phase III clinical trials. In addition, the cationic protein rBPI 21 has recently completed Phase III clinical trials for meningococcemia (Hancock and Diamond, 2000). Another promising prospect for cationic peptides is in plant and fish biotechnology where cationic peptides can be engineered into host organisms to provide enhanced disease resistance (Hancock and Lehrer, 1998). The ability of cationic peptides to bind to lipopolysaccharides and their ability to act synergistically with conventional antibiotics as enhancers are few of the features that make them attractive and potentially novel antibiotics. Furthermore, cationic peptides are gene-coded and synthesized as precursors that undergo post-translational modifications to become active. Their production by genetic engineering is becoming possible and resistance against these antimicrobial peptides does not develop easily.

The discovery and characterization of novel antibacterial, antiviral, antiparasitic and antifungal peptides from natural sources as well as their synthetic and more potent variants is a promising strategy to develop new pharmaceuticals against these microorganisms. However, novel and cost-effective production strategies are needed to facilitate their commercial use in combating diseases.

Structural features and categories

Cationic peptides show significant diversity in size, sequence and structure. They range from 12 to 46 amino acids in length with diverse composition (Hancock, 1997). Despite their diversity, cationic antimicrobial peptides have a net charge of at least +2 at neutral pH, usually because of the presence of arginine or lysine residues in their amino acid sequence (Hancock, 1997). Their secondary structures often contain a hydrophobic domain and a hydrophilic domain. The basicity and amphipathicity of cationic peptides are essential for their antimicrobial activities. The hydrophilic (positively charged) surface facilitates the interaction of the peptides with the negatively charged bacterial surface, e.g. lipopolysaccharide on the outer membrane of Gram-negative bacteria, teichoic acid on the Gram-positive bacteria or negatively charged head groups of the phospholipids in the lipid bilayer (Piers and Hancock, 1994).

Nuclear magnetic resonance (NMR) has emerged as a useful technique for studying structural details of most of the known antimicrobial peptides. Analysis of the three-dimensional structure of these peptides has resulted in a better understanding of their function. Because a majority of these peptides are small in length, their three-dimensional structures can be obtained by NMR methods. Based on the NMR structures of known peptides along with sequence analysis, antimicrobial peptides are broadly classified into five groups: helical, cysteine-rich, sheet, antimicrobial peptides rich in regular amino acids and antimicrobial peptides with rare amino acids.

Helical antimicrobial peptides
Peptides in this category are highly amphipathic helices with hydrophobic and charged cationic surfaces. A well-identified and characterized helical cationic peptide is cecropin-A from the moth, *Hyalophora cecropia*. Magainins, another group of well-characterized helical peptides, isolated from the skin of the African clawed frog, *Xenopus laevis*, are composed of 23 amino acid residues. NMR studies showed that both cecropins and magainins form amphipathic helical structures.

Cysteine-rich antimicrobial peptides
This group consists of peptides that are rich in cysteine residues and are present in a wide variety of organisms. The human neutrophil peptides HNP-1, HNP-2 and HNP-3 were the first cysteine-rich peptides isolated from human neutrophil granules. Most of these molecules harbour a consensus motif of six cysteine residues forming three intramolecular disulfide bonds. Drosomycin, isolated from *Drosophila*, contains four disulfide bonds and three antiparallel strands with a helix between the first two strands (Landon *et al.*, 1997) and represents another example of a cysteine-rich peptide.

Sheet antimicrobial peptides
A few of the known antimicrobial peptides of this class are approximately 20 amino acid residues long and contain one or two disulfide linkages that form a single hairpin structure. Horseshoe crab peptides, tachyplesin and polyphemusin, share a hairpin motif stabilized by two disulfide bonds. NMR studies with thanatin, a 21-residue defence peptide isolated from the hemipteran insect, *Podisus maculiventris*, showed results similar to that of tachyplesin. NMR studies have shown that lactoferricin B, a 25 amino acid proteolytic derivative of lactoferrin, adopts a sheet structure stabilized by a single disulfide bond when in solution (Hwang *et al.*, 1998).

Antimicrobial peptides rich in regular amino acids
Some antimicrobial peptides are composed of a high proportion of regular amino acids. The structural conformation of such peptides differs from the regular helical or sheet peptides. Histatin, a peptide isolated from human saliva, is rich in histidine residues and is active against *Candida albicans* (Xu *et al.*, 1991). Cathelicidins are proline-rich peptides, while indolicidin (Selsted *et al.*, 1992) and tritripticin are tryptophan-rich. Bactenecins-Bac-5 and Bac-7, like cathelicidins, are proline-rich. In contrast, peptide PR-39 is rich in arginine residues.

Antimicrobial peptides with rare modified amino acids
Several peptides are unusual in being composed of rare modified amino acids. Examples of such peptides are those produced by lactic acid bacteria.

Nisin, a lantibiotic, is produced by *Lactococcus lactis* and is composed of rare amino acids like lanthionine, 3-methyllanthionine, dehydroalanine and dehydrobutyrine (de Vos et al., 1993). Another peptide, leucocin A, a 37-residue antimicrobial peptide isolated from *Leuconostoc gelidum*, has been shown to form an amphiphilic conformation (Gibbs et al., 1998).

Mechanism of peptide action

The mode of action of cationic peptides is not completely known. However, specificity with regard to the pathogen as well as with the peptide has been demonstrated. The action of these peptides on bacteria, fungi and viruses is discussed below.

Antibacterial action

Cationic peptides function by disrupting the cytoplasmic membrane of bacteria (Hancock and Lehrer, 1998). This action is proposed to involve three steps: (i) binding to the cell surface; (ii) permeabilization of the outer membrane (in Gram-negative bacteria) and then the cytoplasmic membrane; and (iii) loss of cell viability as a result of cell lysis and DNA damage. Cell lysis is supposed to be initiated by the electrostatic interaction of cationic peptides with the negatively charged cell surface. For Gram-negative bacteria, the positively charged domain of the cationic peptides binds to the divalent cation binding sites of lipopolysaccharide (Piers and Hancock, 1994). The displacement of the native cations Ca^{2+} and Mg^{2+} disrupts the structures of the outer membrane, due to the bulky size of the cationic peptides. This disruption subsequently results in the self-promoted uptake of cationic peptides (Hancock et al., 1995). For Gram-positive bacteria, the cell wall contains covalently bound, negatively charged teichuronic acid and carboxyl groups in the peptidoglycan and these are probably the initial binding sites for the cationic peptides. The interaction between the peptides and the cytoplasmic membrane is thought to be determined by factors such as the anionic lipid composition of the bacterial membrane and the presence of an electrochemical potential across the membrane. After positively charged cationic peptides bind to the negatively charged lipid head groups under the influence of a transmembrane potential (oriented internal negative), the peptides insert into the membrane and undergo conformational changes. They then aggregate to form multimers, which allow them to form channels or pores with their hydrophobic faces positioned towards the membrane and their hydrophilic faces oriented towards the interior of these channels or pores (Shaw et al., 2006). This results in leakage of protons, causing dissipation of the membrane potential and leakage of other small compounds causing cell death. After membrane permeability is altered, the simultaneous loss of the proton motive force, cessation of biosynthesis of macromolecules like DNA, RNA and protein, and leakage of intracellular contents are responsible for eventual cell death (Fidai et al., 1997; Hancock, 1997).

The same factors that are responsible for cell death also seem to regulate the selectivity of cationic peptides for bacterial membranes over eukaryotic cell membranes. The composition of the eukaryotic membrane is quite different from that of bacterial membranes that predominantly contain negatively charged lipids, such as phosphatidylglycerol and cardiolipin, whereas the eukaryotic cell membrane is largely composed of zwitterionic lipids, such as phosphatidylcholine and sphingomyelin. Eukaryotic cell membranes are rich in cholesterol, which may inhibit membrane insertion. Bacterial cells have large, transmembrane potentials of around −140 mV, whereas eukaryotic plasma membranes have membrane potentials of only −20 mV (Yeaman and Yount, 2003). All of these factors contribute to the membrane selectivity of cationic peptides between prokaryotic and eukaryotic cells.

Antifungal action
The modes of action of antifungal peptides have been studied extensively (De Lucca and Walsh, 1999). Peptides, which interact specifically with the lipid components of cell membranes, form pores or ion channels that result in leakage of essential cellular minerals or metabolites or dissipate ion gradients in cell membranes. Other peptides have been shown to inhibit chitin synthase or β-D-glucan synthase. The synthetic peptide D4E1 complexes with ergosterol, a sterol present in the germinating conidia of several fungal species, suggest a lytic mode of action (De Lucca and Walsh, 1999). Research is in progress to elucidate the antifungal action of cationic peptides at the molecular level.

Antiviral action
Not much is known about the mechanisms involved in the antiviral activity of antimicrobial peptides. Direct inactivation of the herpes virus by magainins (Egal et al., 1999), α-defensins (Daher et al., 1986), modelin I (Aboudy et al., 1994) and melittin (Baghian et al., 1997); HIV virus by tachyplesin (Murakami et al., 1991) and indolicidin (Robinson et al., 1998); stomatitis virus by tachyplesin (Murakami et al., 1991); and channel catfish virus (CCV) by esculentin (Chinchar et al., 2001) have been reported. The net cationic charge and ability to form amphipathic structures may enable these peptides to interact with the membranes of the enveloped viruses, which are composed of anionic phospholipids, and disrupt membrane structure. Here, the disruption of membrane integrity occurs because of the interaction between antimicrobial peptides and the virion (Daher et al., 1986). Recently, it was shown that dermaseptin S4 (DS4), which displays a broad spectrum of activity against bacterial, yeast, filamentous fungi and herpes simplex virus I, also inhibits HIV-1 by disrupting virion integrity (Lorin et al., 2005). Antimicrobial peptides like esculentin not only lyse the viral envelope, but also affect the stability of the nucleocapsid (Chinchar et al., 2001). This can also be an effective mechanism for inactivating non-enveloped plant viruses.

Interference with virus and host cell surface interactions is another mode of action adopted by antiviral peptides. Antiviral activity of dermaseptins against herpes simplex virus (Belaid et al., 2002) is an example of this mechanism. DS4 showed an inhibitory effect only when applied to the virus before or during virus adsorption to the target cells, suggesting that the activity of this dermaseptin was exerted at a very early stage of virus proliferation, most likely at the virus–cell interface (Belaid et al., 2002). In enveloped viruses, inhibition of viral-cellular membrane fusion has been observed. Examples include human immunodeficiency virus (HIV) by tachyplesin (Morimoto et al., 1991) and polyphemusin (Nakashima et al., 1992), and herpes simplex virus by apolipoprotein (Srinivas et al., 1990). T_{22}, a tachyplesin synthetic derivative, interferes with the process after HIV binding but before transcription of the HIV genome (Nakashima et al., 1992). In these cases, the antimicrobial peptides exerted a more profound effect on the cell fusion process than on virus penetration as seen by the inhibition of complete cell fusion by peptide treatment *in vitro* (Srinivas et al., 1990). Antiviral activity shown by a number of α-helical synthetic cationic peptides is due to inhibition of virus entry in the cells (Jenssen et al., 2004). Defensins protect cells from herpes simplex virus infection by inhibiting viral adhesion and entry (Yasin et al., 2004).

Inhibition of viral gene expression is an effective mode of action of antimicrobial peptides against both non-enveloped and enveloped viruses. Melittin adopts this mechanism against both the tobacco mosaic virus (TMV) (Marcos et al., 1995) and enveloped HIV virus (Wachinger et al., 1998). Inhibition of HIV by melittin is mediated by the amphipathic α-helical part of the peptide and is a result of intracellular impairment of HIV protein production rather than a membrane effect (Wachinger et al., 1998). With TMV, melittin analogues require elicitation of the peptide along with the virus and binding causes a conformational change in the structure of the RNA (Marcos et al., 1995). Thus, these antimicrobial peptides have effects at the level of gene expression.

There is an enhancement of immunomodulatory properties in response to some peptides (Hancock and Diamond, 2000; Chernysh et al., 2002; Salzet, 2002). Antimicrobial peptides have been reported to be involved in many aspects of innate host defences. They are associated with acute inflammation by acting as chemotoxins for monocytes, recruitment of T-cells through chemotaxis, enhancement of chemokine production and the proliferative response of T-helper cells (Hancock and Diamond, 2000). Synthetic alloferon has been shown to stimulate the natural cytotoxicity of human peripheral blood lymphocytes, induce interferon synthesis in mouse and human models and enhance antiviral and antitumour resistance in mice (Chernysh et al., 2002). Corticostatin acts by competing with the basic amino acid residues of adrenocorticotropic hormone for its binding site (Zhu and Solomon, 1992). Despite the number of successful examples, the molecular basis of protein-mediated virus resistance in most cases is not understood.

Cationic peptides and plants

Synthetic antimicrobial peptides
Most of the antimicrobial peptides are cationic and form an amphipathic secondary structure upon interaction with the surface of the cell membrane, resulting in the formation of ion channels and subsequently cell lysis and death of the pathogen. These two properties have led to the design and synthesis of novel peptides with antimicrobial activity. It was shown that antimicrobial activity could be separated from hemolytic activity through certain nucleotide sequence deletions or substitutions (Blondelle and Houghten, 1991). It has been found that some of the synthetic smaller peptides, in the absence of an amphipathic helical structure, have high levels of antimicrobial activity. Putative cationic amphipathic structures of naturally found proteins have been identified and engineered to display broad-spectrum pathogen activities. The modification of cationic antimicrobial peptides to determine structure–function relationships and/or to produce less toxic molecules with increased activity has been performed primarily on α-helical and β-structured peptides. The important factors in the activity of synthetic antimicrobial cationic peptides are the position and nature of positively charged residues, the formation of specific secondary structures, and the creation of a hydrophobic face on the molecule. These factors are being exploited to design novel and effective drugs. We have developed and successfully used several synthetic cationic peptides in our laboratory for the generation of disease-resistant plants. These include the synthetic variants or chimeras of cecropin, melittin, temporin, dermaseptin, indolicidin, cathelicidins and polyphemusin (Table 13.2).

Cationic antimicrobial peptides from plants
A number of small peptides that display the ability to inhibit the growth of fungi, viruses and bacteria have been isolated from plants (Thomma *et al.*, 2002). Thionins were the first antimicrobial peptides to be isolated from plants (Broekaert *et al.*, 1992). They act on both Gram-positive and Gram-negative bacteria, fungi, yeast and various mammalian cell types. Other antimicrobial peptides were found to be structurally related to insect and mammalian defensins and were named 'plant defensins'. Whereas most antimicrobial peptides from animals and bacteria have high antibacterial activity, plant defensins have high antifungal activity (Broekaert *et al.*, 1992). The plant peptides are 50–100 amino acids in length and have broad-spectrum antimicrobial property (Boman, 1995). As these plant-expressed antimicrobial peptides are a promising source of natural and safe alternatives to antibiotics, there is intense interest in these plant-derived peptides. For a recent and detailed review of plant derived antimicrobial peptides, see the review by Thomma *et al.* (2002).

Cationic peptides and disease resistance in plants
In the past few years, it has become apparent that plants expressing heterologous cationic peptides exhibit broad-spectrum disease resistance,

Table 13.2. Peptide variants developed, tested and expressed in plants in the author's laboratory for broad-spectrum disease resistance.

Peptide variant	Parent peptide	In vitro activity	In planta activity	Plant species	Promoter used	Reference
MsrA1	Cecropin–Melittin	Bacteria/ Fungi	Bacteria/ Fungi	Tobacco, potato	Constitutive	Osusky et al. (2000)
CEMA	Cecropin–Melittin	Bacteria/ Fungi	Bacteria/ Fungi	Tobacco, potato	Wound inducible[a]	Yevtushenko et al. (2005)
MsrA2	Dermaseptin B	Bacteria/ Fungi	Bacteria/ Fungi	Tobacco, potato	Constitutive	Osusky et al. (2005)
MsrA3	Temporin A	Bacteria/ Fungi	Bacteria/ Fungi	Tobacco, potato	Constitutive	Osusky et al. (2004)
PV5	Polyphemusin	Bacteria/ Fungi/ TMV	Bacteria/ Fungi/ TMV	Tobacco	Constitutive	Misra, S. (2005); Bhargava (2005)
PV8	Polyphemusin	Bacteria/ Fungi/ TMV	Bacteria/ Fungi/ TMV	Tobacco	Constitutive	Misra, S. (2005); Bhargava (2005)
10R	Indolicidin	Bacteria/ Fungi/ TMV	Bacteria/ Fungi/ TMV	Tobacco	Constitutive	Misra, S. (2005); Bhargava (2005)
11R	Indolicidin	Bacteria/ Fungi/ TMV	Bacteria/ Fungi/ TMV	Tobacco	Constitutive	Misra, S. (2005); Bhargava (2005)
BMAP-18	Cathelicidin	Bacteria/ fungi/ Trypanosomes	Bacteria/ Fungi/ TMV	Potato	Constitutive, wound inducible[b]	Misra, S. (2005)

[a] win 3.12T promoter from poplar.
[b] BiP promoter from Douglas fir.

including the ability to kill bacteria, fungi, protozoa, and, to some extent, viruses (Florack et al., 1995; Allefs et al., 1996; Osusky et al., 2000, 2004, 2005; Ponti et al., 2003). The antibacterial, antifungal and antiviral activity shown by transgenic plants expressing cationic peptides is discussed below.

Antimicrobial peptides for bacterial resistance in plants

Many different genetic strategies have been proposed to engineer plant resistance to bacterial diseases, including inhibiting bacterial pathogenicity or virulence factors, enhancing natural plant defences, artificially inducing programmed cell death at the site of infection and producing antibacterial proteins of non-plant origin (Mourgues et al., 1998). Genes encoding antibacterial proteins have been cloned and expressed in plants in an attempt to confer resistance to bacterial diseases. Antimicrobial amphipathic peptides like cecropins and their synthetic analogues, Shiva-1 and SB-37, have been expressed in transgenic potato and tobacco plants. Transgenic tobacco plants expressing the Shiva-1 gene showed delayed symptoms and reduced mortality following inoculation with *Ralstonia solanacearum* and *Pseudomonas syringae* (Jaynes et al., 1993). However, as a result of the degradation of cecropins by plant proteases (Florack et al., 1995), no resistance to *R. solanacearum* or *P. syringae* was found in transgenic plants. A stable cecropin analogue (MB39) has been expressed in transgenic tobacco and plants showed no necrosis after leaf infiltration with *P. syringae* (Jaynes et al., 1993). Expression of tachyplesin along with a signal sequence in plants has been shown to confer resistance against *Erwinia* soft rot (Allefs et al., 1996). The attacins, isolated from the giant silk moth, introduced into apple plants, have also shown a reduced susceptibility to *Erwinia amylovora*. Three different lysozyme genes (eggwhite, T4-bacteriophage and human lysozyme) have been expressed in plants. Extracellular extracts from transgenic tobacco plants producing hen-egg-lysozyme inhibited the growth of several species of bacteria, but the susceptibility of the transgenic lines to bacterial diseases has not yet been reported. Greenhouse and *in vitro* experiments indicated a partial resistance to *Erwinia carotovora* in transgenic potato plants producing the T4-bacteriophage lysozyme (During, 1996) and a slight decrease in the symptoms caused by *P. syringae* in tobacco plants producing a human lysozyme. The expression of a human lactoferrin gene in tobacco delayed the onset of symptoms caused by *R. solanacearum* (Zhang et al., 1998). Esculentin from frog skin conferred resistance against *Pseudomonas aeruginosa* when expressed in tobacco (Ponti et al., 2003). The efficiency of these strategies has been improved by the modifications of the peptide genes and construction of synthetic molecules with enhanced expression and stability in plant tissues (Gao et al., 2000; Osusky et al., 2000, 2004, 2005). Synthetic peptide D4E1 (Cary et al., 2000) and a plant defensin, alfAFP from *Medicago sativa* (Gao et al., 2000), have been successfully expressed in tobacco and potato for bacterial and fungal resistance. Similarly, in our laboratory, it was shown that a synthetic cecropin–melittin chimeric peptide and modified temporin provided resistance against *E. carotovora* in potato (Osusky *et al.*, 2000, 2004). Also, indolicidin variants (Bhargava, 2005; Xing *et al.*, 2006) and polyphemusin variants, PV5 and PV8 (Fig. 13.1) showed enhanced resistance to *Erwinia* when expressed in tobacco. Combined expression of several peptide genes, allowing synergistic effects, is a promising strategy for providing broad-spectrum disease resistance in plants.

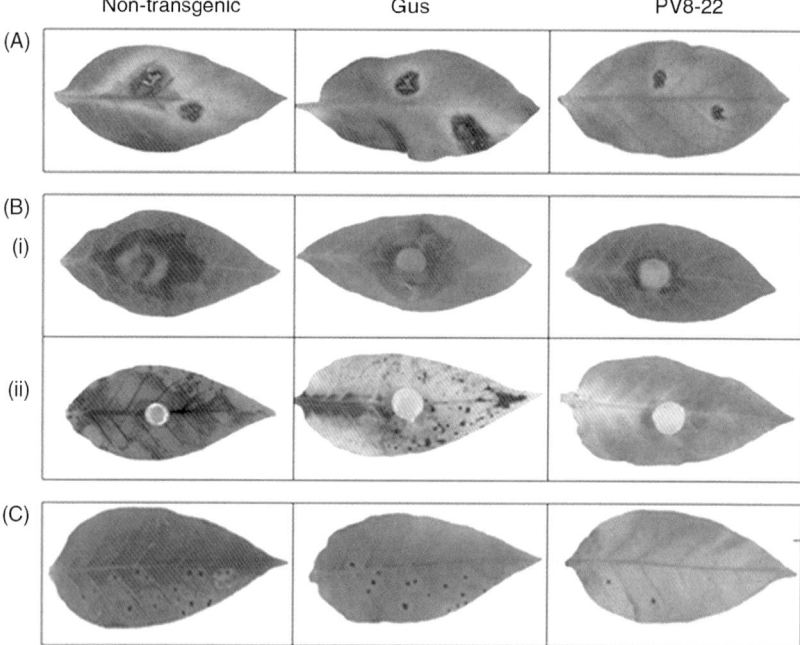

Fig. 13.1. Resistance of transgenic tobacco expressing polyphemusin variant PV8 to phytopathogens. Detached leaves from mature non-transgenic tobacco (control), transgenic GUS and transgenic tobacco expressing PV8 (line#22) were infected with (A) *E. carotovora*; bacterial suspension (1 × 10^5 cells) was applied on to a tobacco leaf surface followed by piercing with needle as a way of inoculating bacteria into the leaf. The inoculated leaves were kept at room temperature and picture was taken after 7 days of incubation. (B) *Botrytis cinerea* (i) and *Verticillium* sp. (ii); Tobacco leaves were detached and placed in Petri dishes with moist filter paper discs. Plugs of inoculum were prepared from the fungal strains, cultured on potato dextrose agar (PDA) and were placed on each leaf. The inoculated leaves were incubated in diffused light at room temperature. The extent of lesion development was recorded at appropriate times and pictures were taken. (C) TMV; 10 µl (5 µg) of TMV was applied to each half of the detached leaves from transgenic and non-transgenic plants. Leaves were rinsed with water and placed on dH_2O soaked filter paper in 15 cm Petri dishes. The samples were then moved to a growth chamber (photoperiod of 12 h, 25°C) where they were monitored daily for appearance of symptoms. Difference in the lesions was observed after 3 days.

Our group has tapped into the potential of the cationic peptide expression in N-terminus-modified, cecropin–melittin cationic peptide chimera (MsrA1), including antiviral peptides (Table 13.2). We expressed a synthetic gene encoding a broad-spectrum antimicrobial activity into potato cultivars. The morphology and yield of transgenic plants and tubers was unaffected. Highly stringent challenges with bacterial or fungal phytopathogens demonstrated powerful resistance. Tubers retained their resistance to infectious challenge for more than a year, and did not appear to be harmful when fed to mice (Osusky et al., 2000). Similarly, temporin A was N-terminally modified (MsrA3) and expressed in potato plants. MsrA3 conveyed strong resistance to late blight and pink rot phytopathogens in addition to the bacterial pathogen E. carotovora. Transgenic tubers remained disease-free during storage for more than 2 years (Osusky et al., 2004).

Antimicrobial peptides for fungal resistance in plants

Genes encoding hydrolytic enzymes, such as chitinase and glucanase, which can degrade fungal cell-wall components, are attractive candidates for the antifungal genetic engineering approach, and are preferentially used for the production of fungal disease-resistant plants (Jach et al., 1995). Cecropin A-derived peptides have been shown to be potent inhibitors of fungal plant pathogens (Cavallarin et al., 1998). The overexpression of defensins and thionins in transgenic plants reduced development of several fungal pathogens, such as *Alternaria* sp., *Fusarium* sp. and *Plasmodiophora* sp., and provided resistance to *Verticillium* sp. under field conditions (Gao et al., 2000). Expression of barley seed ribosome-inactivating proteins reduced development of *Rhizoctonia solani* in transgenic tobacco (Wang et al., 1998). Combined expression of chitinase and ribosome-inactivating proteins in transgenic tobacco had an inhibitory effect on *Rhizoctonia solani* development (Jach et al., 1995). Human lysozyme expression in transgenic carrot and tobacco enhanced resistance to several pathogens, including *Erysiphe* and *Alternaria* sp. (Takaichi and Oeda, 2000). Pokeweed (*Pytolacca americana*) antiviral protein expression in transgenic tobacco reduced *Rhizoctonia solani* infection (Wang et al., 1998). MSI-99, and a magainin analogue, imparted enhanced disease resistance in transgenic tobacco and banana against *Fusarium oxysporum* and *Mycosphaerella musicola* (Chakrabarti et al., 2003). The same magainin analogue expressed via the chloroplast genome to obtain high levels of expression in transgenic tobacco showed inhibition of growth against spores of three fungal species: *Aspergillus flavus*, *Fusarium moniliforme* and *Verticillium dahliae* (DeGray et al., 2001). Heliomicin and drosomycin expressed in transgenic tobacco conferred a minor but statistically significant enhanced resistance to the fungus *Cercospora nicotianae* (Banzet et al., 2002). Cecropin A-melittin hybrid and cecropin A-derived peptides were synthesized and tested for their ability to inhibit growth of *Phytophthora infestans* and other pathogens *in vitro* (Cavallarin *et al.*, 1998). These and other synthetic cationic peptides (e.g. cecropin, melittin, temporin and their modified variants) with *in vitro* broad-spectrum antimicrobial activity expression in transgenic potato and tobacco have provided enhanced resistance against a number of fungal plant pathogens, including *Colletotrichum, Fusarium, Verticillium* sp. and *Phytophthora* sp. (Cavallarin *et al.*, 1998; Osusky *et al.*, 2000, 2004). Figure 13.1 demonstrates the antifungal resistance in transgenic tobacco plants expressing polyphemusin variant PV8.

Dermaseptin B1 is a potent cationic antimicrobial peptide found in skin secretions of the arboreal frog *Phyllomedusa bicolor*. A synthetic derivative of dermaseptin B1, MsrA2, was expressed at low levels in the transgenic potato. Stringent challenges of these transgenic potato plants with a variety of highly virulent fungal phytopathogens – *Alternaria, Cercospora, Fusarium, Phytophthora, Pythium, Rhizoctonia* and *Verticillium* sp. – and with the bacterial pathogen *E. carotovora* demonstrated that the plants had an unusually broad spectrum and powerful resistance to infection (Osusky *et al.*, 2005).

Antimicrobial peptides for viral resistance in plants

A relatively small number of antiviral peptides have been described in an antimicrobial peptide database (Wang and Wang, 2004); however, use of antimicrobial peptides for engineering antiviral resistance in plants is scarce (Bhargava, 2005). Virus resistance in plants has been obtained by expressing specific proteins. Examples of protein-mediated virus resistance mainly include the expression of viral coat proteins but cases of protein-dependent pathogen-derived resistance due to the expression of viral movement proteins or replicases are also known (Tacke *et al.*, 1996). In some instances, resistance is based on the expression of intact, functional proteins; in others, the expression of the intact protein led only to weak resistance or even to enhanced susceptibility. In contrast, expression of a dysfunctional protein may lead to strong resistance (Tacke *et al.*, 1996). A new strategy for engineering virus-resistant plants by transgenic expression of a dominant interfering peptide was shown by Rudolph *et al.* (2003). Transgenic *Nicotiana benthamiana* lines expressing the peptide fused to a carrier protein showed strong resistance against *Tomato spotted wilt virus, Tomato chlorotic spot virus, Groundnut ring spot virus* and *Chrysanthemum stem necrosis virus*. This presents a promising strategy for expressing small peptides in plants. The advantage of using short sequences to engineer resistance is to minimize unpredictable or even deleterious effects observed in several cases after the expression of functional viral proteins (Prins *et al.*, 1997). Expression of only a short peptide or artificial peptides not only minimizes the potential deleterious effect on the plant cell but also prevents other undesirable consequences. One such concern is the evolution of new viruses by recombination with the transgene or by transcapsidation, which can virtually be excluded when an interfering but non-homologous and non-functional molecule is expressed in the plant. Thus, expressing cationic peptides with *in vitro* antiviral activities is a promising strategy to provide broad-spectrum disease resistance in plants.

Indolicidin (10R and 11R) and polyphemusin derivatives (PV5 and PV8) were expressed in tobacco plants and were tested for antibacterial, antifungal and antiviral activity (Bhargava, 2005). These peptides are known to have antiviral activity against animal viruses. *In vitro* tests in our laboratory showed promising results against plant pathogens as well as against TMV. *In planta* resistance was also shown against *Erwinia*, pathogenic fungi and TMV (Bhargava, 2005).

Expression of polyphemusin variant PV8 provided *in planta* resistance against *E. carotovora*, *Botrytis cineria*, *Verticillium* sp. and TMV (Fig. 13.1). We have now expressed these peptides using both constitutive and a wound-inducible promoter from Douglas fir previously characterized in our laboratory (Forward *et al.*, 2002).

Drawbacks of Peptide Expression in Plants

Despite benefits gained from genetic engineering for disease resistance, this technology has some drawbacks and research for solutions is in progress. In addition to transgene integration, disease resistance depends on peptide expression levels in plants, which can, in some cases, be affected by homology-dependent gene silencing. This is a serious problem for strategies using homologous sequences. Furthermore, in the case of proteins/peptides, these must be synthesized, exported from the cell and transported to their desired location without undergoing major modification during the process, and remain stable at their destination, avoiding degradation by plant proteases (Hancock and Lehrer, 1998). *In vitro* testing of the leaf extracts from the plants expressing cationic peptides shows that the expressed peptides are unstable or degraded by proteases (Cavallarin *et al.*, 1998; Hancock and Lehrer, 1998; Cary *et al.*, 2000; Li *et al.*, 2001). Thus, strategies are needed to optimize expression and stability of expressed peptides in transgenic plants. Molecular modelling of peptides and modifications at the N-terminus proved to be an effective method to increase stability of peptides in plants without comprising activity or toxicity (Osusky *et al.*, 2000, 2004, 2005).

Future Directions and Conclusions

The best strategy for providing enhanced and broad-spectrum resistance using cationic peptides is the co-expression of different molecules with complementary modes of action that act at different stages of disease development. There are reports of synergistic action of antimicrobial peptides. Combinations of potential peptides may be a successful strategy for generating broad-spectrum disease resistance, including resistance against viruses in plants. Expression of defensive genes from a promoter that is specifically activated in response to pathogen invasion is highly desirable for engineering disease-resistant plants. Our group had shown earlier that the *win3.12*T promoter from hybrid poplar (*Populus trichocarpa* × *Populus deltoides*) affected a strong systemic activity in aerial parts of potato in response to fungal infection (Yevtushenko *et al.*, 2004). Transcriptional fusion between this pathogen-responsive promoter from poplar and the gene encoding the novel cecropin A-melittin hybrid peptide (CEMA) with strong antimicrobial activity was evaluated in transgenic tobacco for enhanced plant resistance against a highly virulent pathogenic fungus *Fusarium solani*. The antifungal resistance of transgenic plants was strong and accumulation of cationic peptide in transgenic tobacco had no deleterious effect on plant growth and development (Yevtushenko *et al.*, 2005). This was the first report showing the application of a heterologous pathogen-inducible promoter to direct the expression of an antimicrobial peptide in plants, and the feasibility of this approach to provide disease resistance in tobacco and, possibly, other crops. Strategies for a regulated and/or inducible, and tissue-specific expression of peptides, may prove to be effective for better performance in the greenhouse as well as in the field. We are now evaluating several promoters for inducible and tissue-specific expression of engineered peptides.

Ethical concerns have often been expressed about the production of genetically modified plants and animals. However, organisms naturally produce many cationic peptides as part of their innate defences against infection. The results in our laboratory showed low cytotoxicity of synthetic antimicrobial peptides in plants and are promising. The incorporation of cationic peptides into plants through genetic engineering offers a means to prevent disease-associated losses as well as to protect the environment.

Successful applications of a transgenic approach using these peptides to control plant diseases, particularly viruses, will likely help eradicate certain plant diseases, reduce the environmental impact of intensive agriculture – especially use of pesticides, and improve the quality and safety of our food.

References

Aboudy, Y., Mendelson, E., Shalit, I., Bessalle, R. and Fridkin, M. (1994) Activity of two synthetic amphiphilic peptides and magainin-2 against herpes simplex virus types 1 and 2. *International Journal of Peptide Protein Research* 43, 573–582.

Allefs, S.J.H.M., De Jong, E.R., Florack, D.E.A., Hoogendoorn, C. and Stiekema, W.J. (1996) *Erwinia* soft rot resistance of potato cultivars expressing antimicrobial peptide tachyplesin I. *Molecular Breeding* 2, 97–105.

Bhargava, A. (2005) Evaluation and expression of indolicidins and polyphemusin variants in plants for broad-spectrum disease resistance. MSc thesis, University of Victoria, Canada.

Baghian, A., Jaynes, J., Enright, F. and Kousoulas, K.G. (1997) An amphipathic alpha-helical synthetic peptide analogue of melittin inhibits herpes simplex virus-1 (HSV-1)-induced cell fusion and virus spread. *Peptides* 18, 177–183.

Banzet, N., Latorse, M., Bulet, P., Francois, E., Derpierre, C. and Dubald, M. (2002) Expression of insect cysteine-rich antifungal peptides in transgenic tobacco enhances resistance to a fungal disease. *Plant Science* 162, 995–1006.

Belaid, A., Aouni, M., Khelifa, R., Trabelsi, A., Jemmali, M. and Hani, K. (2002) In vitro antiviral activity of dermaseptins against herpes simplex virus type 1. *Journal of Medical Virology* 66, 229–234.

Blondelle, S.E. and Houghten, R.A. (1991) Hemolytic and antimicrobial activities of the twenty-four individual omission analogues of melittin. *Biochemistry* 30, 4671–4678.

Boman, H.G. (1995) Peptide antibiotics and their role in innate immunity. *Annual Review of Immunology* 13, 61–92.

Boman, H.G. and Hultmark, D. (1987) Cell-free immunity in insects. *Annual Review of Microbiology* 41, 103–126.

Broekaert, W.F., Marien, W., Terras, F.R., De Bolle, M.F., Proost, P., Van Damme, J., Dillen, L., Claeys, M., Rees, S.B., Vanderleyden, J. and Cammue, B.P.A. (1992) Antimicrobial peptides from *Amaranthus caudatus* seeds with sequence homology to the cysteine/glycine-rich domain of chitin-binding proteins. *Biochemistry* 31, 4308–4314.

Cao, H., Li, X. and Dong, X. (1998) Generation of broad-spectrum disease resistance by overexpression of an essential regulatory gene in systemic acquired resistance. *Proceedings of the National Academy of Sciences of the United States of America* 95, 6531–6536.

Cary, J.W., Rajasekaran, K., Jaynes, J.M. and Cleveland, T.E. (2000) Transgenic expression of a gene encoding a synthetic antimicrobial peptide results in inhibition of fungal growth *in vitro* and *in planta*. *Plant Science* 154, 171–181.

Cavallarin, L., Andreu, D. and San Segundo, B. (1998) Cecropin A-derived peptides are potent inhibitors of fungal plant pathogens. *Molecular Plant-Microbe Interactions* 11, 218–227.

Chakrabarti, A., Ganapathi, T.R., Mukherjee, P.K. and Bapat, V.A. (2003) MSI-99, a magainin analogue, imparts enhanced disease resistance in transgenic tobacco and banana. *Planta* 216, 587–596.

Chernysh, S., Kim, S.I., Bekker, G., Pleskach, V.A., Filatova, N.A., Anikin, V.B., Platonov, V.G. and Bulet, P. (2002) Antiviral and antitumor peptides from insects. *Proceedings of the National Academy of Sciences of the United States of America* 99, 12628–12632.

Chinchar, V.G., Wang, J., Murti, G., Carey, C. and Rollins-Smith, L. (2001) Inactivation of frog virus 3 and channel catfish virus by esculentin-2P and ranatuerin-2P, two antimicrobial peptides isolated from frog skin. *Virology* 288, 351–357.

Daher, K.A., Selsted, M.E. and Lehrer, R.I. (1986) Direct inactivation of viruses by human granulocyte defensins. *Journal of Virology* 60, 1068–1074.

DeGray, G., Rajasekaran, K., Smith, F., Sanford, J. and Daniell, H. (2001) Expression of an antimicrobial peptide via the chloroplast genome to control phytopathogenic bacteria and fungi. *Plant Physiology* 127, 852–862.

De Lucca, A.J. and Walsh, T.J. (1999) Antifungal peptides: novel therapeutic compounds against emerging pathogens. *Antimicrobial Agents and Chemotherapy* 43, 1–11.

de Vos, W.M., Mulders, J.W., Siezen, R.J., Hugenholtz, J. and Kuipers, O.P. (1993) Properties of nisin Z and distribution of its gene, nisZ, in *Lactococcus lactis*. *Applied and Environmental Microbiology* 59, 213–218.

During, K. (1996) Genetic engineering for resistance to bacteria in transgenic plants by introduction of foreign genes. *Molecular Breeding* 2, 297–305.

Egal, M., Conrad, M., MacDonald, D.L., Maloy, W.L., Motley, M. and Genco, C.A. (1999) Antiviral effects of synthetic membrane-active peptides on herpes simplex virus, type 1. *International Journal of Antimicrobial Agents* 13, 57–60.

Fidai, S., Farmer, S.W. and Hancock, R.E. (1997) Interaction of cationic peptides with bacterial membranes. *Methods in Molecular Biology* 78, 187–204.

Florack, D., Allefs, S., Bollen, R., Bosch, D., Visser, B. and Stiekema, W. (1995) Expression of giant silkmoth cecropin B genes in tobacco. *Transgenic Research* 4, 132–141.

Forward, B.S., Osusky, M. and Misra, S. (2002) The douglas-fir BiP promoter is functional in Arabidopsis and responds to wounding. *Planta* 215, 569–576.

Gao, A.G., Hakimi, S.M., Mittanck, C.A., Wu, Y., Woerner, B.M., Stark, D.M., Shah, D.M., Liang, J. and Rommens, C.M. (2000) Fungal pathogen protection in potato by expression of a plant defensin peptide. *Nature Biotechnology* 18, 1307–1310.

Gibbs, A.C., Kondejewski, L.H., Gronwald, W., Nip, A.M., Hodges, R.S., Sykes, B.D. and Wishart, D.S. (1998) Unusual beta-sheet periodicity in small cyclic peptides. *Nature Structural Biology* 5, 284–288.

Goraya, J., Wang, Y., Li, Z., O'Flaherty, M., Knoop, F.C., Platz, J.E. and Conlon, J.M. (2000) Peptides with antimicrobial activity from four different families isolated from the skins of the North American frogs *Rana luteiventris*, *Rana berlandieri* and *Rana pipiens*. *European Journal of Biochemistry* 267, 894–900.

Hancock, R.E. (1997) Peptide antibiotics. *Lancet* 349, 418–422.

Hancock, R.E. and Diamond, G. (2000) The role of cationic antimicrobial peptides in innate host defences. *Trends in Microbiology* 8, 402–410.

Hancock, R.E. and Lehrer, R. (1998) Cationic peptides: a new source of antibiotics. *Trends in Biotechnology* 16, 82–88.

Hancock, R.E., Falla, T. and Brown, M. (1995) Cationic bactericidal peptides. *Advances in Microbial Physiology* 37, 135–175.

Harjunpaa, I., Kuusela, P., Smoluch, M.T., Silberring, J., Lankinen, H. and Wade, D. (1999) Comparison of synthesis and antibacterial activity of temporin A. *FEBS Letters* 449, 187–190.

Hwang, P.M., Zhou, N., Shan, X., Arrowsmith, C.H. and Vogel, H.J. (1998) Three-dimensional solution structure of lactoferricin B, an antimicrobial peptide derived from bovine lactoferrin. *Biochemistry* 37, 4288–4298.

Jach, G., Gornhardt, B., Mundy, J., Logemann, J., Pinsdorf, E., Leah, R., Schell, J. and Maas, C. (1995) Enhanced quantitative resistance against fungal disease by combinatorial expression of different barley antifungal proteins in transgenic tobacco. *The Plant Journal* 8, 97–109.

Jaynes, J.M., Nagpala, P., Destéfano-Beltrán, L., Huang, J.H., Kim, J.H., Denny, T. and Cetiner, S. (1993) Expression of a cecropin B lytic peptide analog in transgenic tobacco confers enhanced resistance to bacterial wilt caused by *Pseudomonas solanacearum*. *Plant Science* 89, 43–53.

Jenssen, H. Andersen, J.H., Mantzilas, D. and Gutteberg, T.J. (2004) A wide range of medium-sized, highly cationic, alpha-helical peptides show antiviral activity against herpes simplex virus. *Antiviral Research* 64, 119–126.

Keller, H., Pamboukdjian, N., Ponchet, M., Poupet, A., Delon, R., Verrier, J.L., Roby, D. and Ricci, P. (1999) Pathogen-induced elicitin production in transgenic tobacco generates a hypersensitive response and nonspecific disease resistance. *The Plant Cell* 11, 223–235.

Kokryakov, V.N., Harwig, S.S., Panyutich, E.A., Shevchenko, A.A., Aleshina, G.M., Shamova, O.V., Korneva, H.A. and Lehrer, R.I. (1993) Protegrins: leukocyte antimicrobial peptides that combine features of corticostatic defensins and tachyplesins. *FEBS Letters* 327, 231–236.

Landon, C., Sodano, P., Hetru, C., Hoffmann, J. and Ptak, M. (1997) Solution structure of drosomycin, the first inducible antifungal protein from insects. *Protein Science* 6, 1878–1884.

Li, Q., Lawrence, C.B., Xing, H.Y., Babbitt, R.A., Bass, W.T., Maiti, I.B. and Everett, N.P. (2001) Enhanced disease resistance conferred by expression of an antimicrobial magainin analog in transgenic tobacco. *Planta* 212, 635–639.

Lorin, C., Saidi, H., Belaid, A., Zairi, A., Baleux, F., Hocini, H., Belec, L., Hani, K. and Tangy, F. (2005) The antimicrobial peptide dermaseptin S4 inhibits HIV-1 infectivity *in vitro*. *Virology* 334, 264–275.

Marcos, J.F., Beachy, R.N., Houghten, R.A., Blondelle, S.E. and Perez-Paya, E. (1995) Inhibition of a plant virus infection by analogs of melittin. *Proceedings of the National Academy of Sciences of the United States of America* 92, 12466–12469.

Morimoto, M., Mori, H., Otake, T., Ueba, N., Kunita, N., Niwa, M., Murakami, T. and Iwanaga, S. (1991) Inhibitory effect of tachyplesin I on the proliferation of human immunodeficiency virus in vitro. *Chemotherapy* 37, 206–211.

Mourgues, F., Brisset, M.N. and Chevreau, E. (1998) Strategies to improve plant resistance to bacterial diseases through genetic engineering. *Trends in Biotechnology* 16, 203–210.

Murakami, T., Niwa, M., Tokunaga, F., Miyata, T. and Iwanaga, S. (1991) Direct virus inactivation of tachyplesin I and its isopeptides from horseshoe crab hemocytes. *Chemotherapy* 37, 327–334.

Nakashima, H., Masuda, M., Murakami, T., Koyanagi, Y., Matsumoto, A., Fujii, N. and Yamamoto, N. (1992) Anti-human immunodeficiency virus activity of a novel synthetic peptide, T22 ([tyr-5,12, lys-7]polyphemusin II): a possible inhibitor of virus-cell fusion. *Antimicrobial Agents and Chemotherapy* 36, 1249–1255.

Ng, T.B. and Wang, H. (2001) Panaxagin, a new protein from chinese ginseng possesses anti-fungal, anti-viral, translation-inhibiting and ribonuclease activities. *Life Sciences* 68, 739–749.

Ngai, P.H. and Ng, T.B. (2003) Lentin, a novel and potent antifungal protein from shitake mushroom with inhibitory effects on activity of human immunodeficiency virus-1 reverse transcriptase and proliferation of leukemia cells. *Life Sciences* 73, 3363–3374.

Osusky, M., Zhou, G., Osuska, L., Hancock, R.E., Kay, W.W. and Misra, S. (2000) Transgenic plants expressing cationic peptide chimeras exhibit broad-spectrum resistance to phytopathogens. *Nature Biotechnology* 18, 1162–1166.

Osusky, M., Osuska, L., Hancock, R.E., Kay, W.W. and Misra, S. (2004) Transgenic potatoes expressing a novel cationic peptide are resistant to late blight and pink rot. *Transgenic Research* 13, 181–190.

Osusky, M., Osuska, L., Kay, W. and Misra, S. (2005) Genetic modification of potato against microbial diseases: in vitro and in planta activity of a dermaseptin B1 derivative, MsrA2. *Theoretical and Applied Genetics* 111, 711–722.

Piers, K.L. and Hancock, R.E. (1994) The interaction of a recombinant cecropin/melittin hybrid peptide with the outer membrane of *Pseudomonas aeruginosa*. *Molecular Microbiology* 12, 951–958.

Ponti, D., Mangoni, M.L., Mignogna, G., Simmaco, M. and Barra, D. (2003) An amphibian antimicrobial peptide variant expressed in *Nicotiana tabacum* confers resistance to phytopathogens. *Biochemical Journal* 370, 121–127.

Prins, M., Kikkert, M., Ismayadi, C., de Graauw, W., de Haan, P. and Goldbach, R. (1997) Characterization of RNA-mediated resistance to tomato spotted wilt virus in transgenic tobacco plants expressing NS(M) gene sequences. *Plant Molecular Biology* 33, 235–243.

Punja, Z.K. (2001) Genetic engineering of plants to enhance resistance to fungal pathogens – a review of progress and future prospects. *Canadian Journal of Plant Pathology* 23, 216–235.

Robinson, W.E. Jr., McDougall, B., Tran, D. and Selsted, M.E. (1998) Anti-HIV-1 activity of indolicidin, an antimicrobial peptide from neutrophils. *Journal of Leukocyte Biology* 63, 94–100.

Rudolph, C., Schreier, P.H. and Uhrig, J.F. (2003) Peptide-mediated broad-spectrum plant resistance to tospoviruses. *Proceedings of the National Academy of Sciences of the United States of America* 100, 4429–4434.

Salzet, M. (2002) Antimicrobial peptides are signaling molecules. *Trends in Immunology* 23, 283–284.

Selsted, M.E., Novotny, M.J., Morris, W.L., Tang, Y.Q., Smith, W. and Cullor, J.S. (1992) Indolicidin, a novel bactericidal tridecapeptide amide from neutrophils. *Journal of Biological Chemistry* 267, 4292–4295.

Shaw, J.E., Alattia, J.R., Verity, J.E., Prive, G.G. and Yip, C.M. (2006) Mechanisms of antimicrobial peptide action: studies of indolicidin assembly at model membrane interfaces by in situ atomic force microscopy. *Journal of Structural Biology* 154, 42–58.

Srinivas, R.V., Birkedal, B., Owens, R.J., Anantharamaiah, G.M., Segrest, J.P. and Compans, R.W. (1990) Antiviral effects of apolipoprotein A-I and its synthetic amphipathic peptide analogs. *Virology* 176, 48–57.

Tacke, E., Salamini, F. and Rohde, W. (1996) Genetic engineering of potato for broad-spectrum protection against virus infection. *Nature Biotechnology* 14, 1597–1601.

Takaichi, M. and Oeda, K. (2000) Transgenic carrots with enhanced resistance against two major pathogens, *Erysiphe heraclei* and *Alternaria dauci*. *Plant Science* 153, 135–144.

Thomma, B.P., Cammue, B.P. and Thevissen, K. (2002) Plant defensins. *Planta* 216, 193–202.

Vizioli, J. and Salzet, M. (2002) Antimicrobial peptides from animals: focus on invertebrates. *Trends in Pharmacological Sciences* 23, 494–496.

Wachinger, M., Kleinschmidt, A., Winder, D., von Pechmann, N., Ludvigsen, A., Neumann, M., Holle, R., Salmons, B., Erfle, V. and Brack-Werner, R. (1998) Antimicrobial peptides melittin and cecropin inhibit replication of human immunodeficiency virus 1 by suppressing viral gene expression. *Journal of General Virology* 79, 731–740.

Wang, H. and Ng, T.B. (2004) Antifungal peptides, a heat shock protein-like peptide, and a serine-threonine kinase-like protein from Ceylon spinach seeds. *Peptides* 25, 1209–1214.

Wang, P., Zoubenko, O. and Tumer, N.E. (1998) Reduced toxicity and broad spectrum resistance to viral and fungal infection in transgenic plants expressing pokeweed antiviral protein II. *Plant Molecular Biology* 38, 957–964.

Wang, Z. and Wang, G. (2004) APD: the antimicrobial peptide database. *Nucleic Acids Research* 32, 590–592.

Xing, H., Lawrence, C.B., Chambers, O., Davies, H.M., Everett, N.P. and Li, Q.Q. (2006) Increased pathogen resistance and yield in transgenic plants expressing combinations of the modified antimicrobial peptides based on indolicidin and magainin. *Planta* 223, 1024–1032.

Xu, T., Levitz, S.M., Diamond, R.D. and Oppenheim, F.G. (1991) Anticandidal activity of major human salivary histatins. *Infection and Immunity* 59, 2549–2554.

Yasin, B., Pang, M., Turner, J.S., Cho, Y., Dinh, N.N., Waring, A.J., Lehrer, R.I. and Wagar, E.A. (2000) Evaluation of the inactivation of infectious herpes simplex virus by host-defense peptides. *European Journal of Clinical Microbiology and Infectious Disease* 19, 187–194.

Yasin, B., Wang, W., Pang, M., Cheshenko, N., Hong, T., Waring, A.J., Herold, B.C., Wagar, E.A. and Lehrer, R.I. (2004) Theta defensins protect cells from infection by herpes simplex virus by inhibiting viral adhesion and entry. *Journal of Virology* 78, 5147–5156.

Yeaman, M.R. and Yount, N.Y. (2003) Mechanisms of antimicrobial peptide action and resistance. *Pharmacological Review* 55, 27–55.

Yevtushenko, D.P., Sidorov, V.A., Romero, R., Kay, W.W. and Misra, S. (2004) Wound-inducible promoter from poplar is responsive to fungal infection in transgenic potato. *Plant Science* 167, 715–724.

Yevtushenko, D.P., Romero, R., Forward, B.S., Hancock, R.E., Kay, W.W. and Misra, S. (2005) Pathogen-induced expression of a cecropin A-melittin antimicrobial peptide gene confers antifungal resistance in transgenic tobacco. *Journal of Experimental Botany* 56, 1685–1695.

Zanetti, M., Gennaro, R. and Romeo, D. (1995) Cathelicidins: a novel protein family with a common proregion and a variable C-terminal antimicrobial domain. *FEBS Letters* 374, 1–5.

Zhang, Z., Coyne, D.P., Vidaver, A.K. and Mitra, A. (1998) Expression of human lactoferrin cDNA confers resistance to *Ralstonia solanacearum* in transgenic tobacco plants. *Phytopathology* 88, 730–734.

Zhu, Q. and Solomon, S. (1992) Isolation and mode of action of rabbit corticostatic (antiadrenocorticotropin) peptides. *Endocrinology* 130, 1413–1423.

14 Molecular Breeding Approaches for Enhanced Resistance Against Fungal Pathogens

R.E. Knox and F.R. Clarke

Abstract

Marker-assisted selection for fungal plant resistance is the most important tool in molecular breeding at the applied level. Markers for disease resistance have been sought by researchers and breeders since the discovery that genes can be linked to each other. The dearth of visual markers has been the limiting factor in their application, but that has changed with the development of techniques to detect variation in DNA. The differences in DNA are visualized as polymorphisms which currently are predominantly identified as changes in fragment size, made possible through techniques such as polymerase chain reaction, electrophoresis, fluorescent dye detection and the use of restriction enzymes. Because of the many examples of monogenic inheritance of disease resistance genes and the importance of resistance traits, the processes of marker discovery have developed in large part around disease resistance. There are now a vast number of markers for the many resistance genes to fungal diseases in numerous crop species. The integration and use of these markers takes breeding from integrating the technology in marker-assisted selection to the development of breeding strategies around marker use in 'molecular breeding'.

Introduction

Plant breeding is directed towards changes in gene frequencies of plant species to produce superior genotypes for the betterment of humankind. Molecular breeding is the use of molecular genetic tools to assist changes in gene frequency in favourable directions. Like classical breeding, molecular breeding can be viewed to include plant pathology, genetics, cytogenetics, molecular genetics, physiology, biochemistry, quantitative genetics and other disciplines that provide the information to make breeding efforts effective (Fig. 14.1). Molecular breeding is intimately intertwined with traditional breeding, with marker-assisted selection being the predominant tool used. This chapter emphasizes applied molecular breeding with

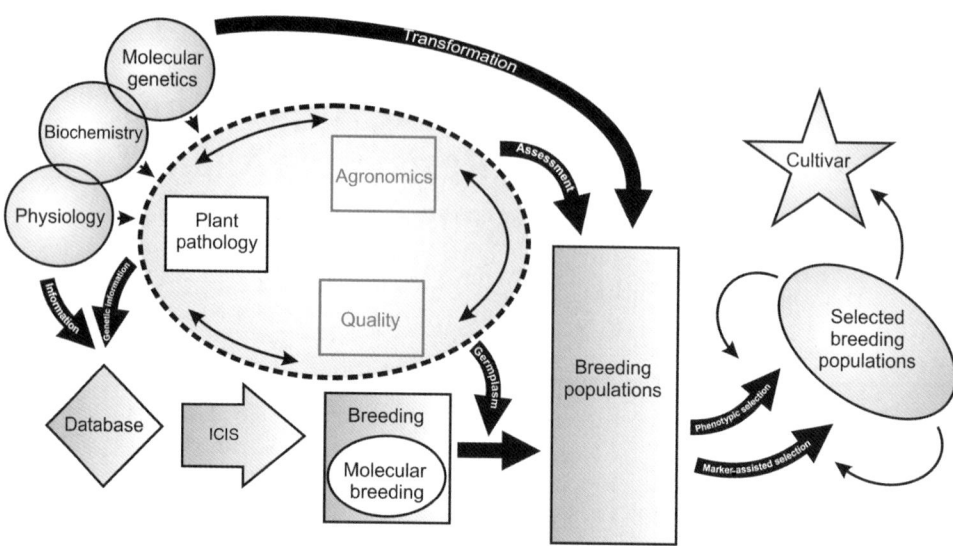

Fig. 14.1. The relationship of molecular breeding with traditional breeding and flow of upstream information and technology from other disciplines into molecular breeding.

reference to the many areas of molecular genetics as they impact crop improvement for resistance to fungal pathogens. Much of this chapter is built on the authors' experience with wheat (*Triticum* spp.) and therefore will be slanted more heavily towards self-pollinated crops.

Fungal pathogens of plants cause the most important diseases of crop plants (Agrios, 1988), and present major challenges to stable crop production (Strange and Scott, 2005). Unlike abiotic stresses, biotic stresses imposed by fungal pathogens are in a rapid state of flux. Although environment can shift through climate change, pathogens not only change but adapt to genotypes used in crop production. The ability of pathogens to thwart the defences of crop plants drives breeding efforts to stay a step ahead. A major tool in this process has been the use of host genetic resistance. Molecular breeding can improve response time to shifts in pathogen populations. Host resistance to fungal pathogens is controlled by genes which are unique not only to the particular species of pathogen, but also in many cases, to particular strains or races of the pathogen (Keen, 1990). Resistance can be monogenic or multigenic. The many pathogens that attack just one crop species impose a large demand on breeding resources in plant pathology and genetics to continually provide a fresh arsenal of resistance genes for disease control.

Among the most significant epidemics of fungal pathogens are stem rust (*Puccinia graminis* Pers. f. sp. *tritici* Eriks. & Henn.) of wheat in the mid-20th century; southern corn leaf blight (*Cochliobolus heterostrophus* (Drechs.) Drechs.) and potato late blight (*Phytophthora infestans* (Mont.) de Bary) of the 19th century; coffee rust (*Hemileia vastatrix* Berk. & Br.); and recently, soybean rust (*Phakopsora pachyrhizi* Syd.) (Agrios, 1988;

Strange and Scott, 2005). In addition to catastrophic loss when the proper conditions of germplasm vulnerability and environment come together with an aggressive strain of pathogen, many other fungal diseases annually erode crop yield through their low but significant presence. Leaf spot diseases and cereal smuts fall into this category (Saari *et al.*, 1996; Ciuffetti and Tuori, 1999). Diseases need not be present in large amounts to cause substantial economic loss, such as those that reduce quality of product. Common bunt (*Tilletia tritici* (Bjerk.) G. Wint. in Rabenh.) is an example of a disease that imparts a foul odour on grain and distasteful flavour to the flour products (Goates, 1996). Pathogens such as *Fusarium* spp. that impart toxins on the grain present health risks (Steiner *et al.*, 2004).

Biotechnology encompasses a vast array of tools and techniques based on biological processes, physical and genetic mapping, transcriptomics with use of expressed genes and microarrays, proteomics and reverse genetics, gene cloning and transformation, comparative genetics and data mining, all of which lead to gene discovery and a greater understanding of gene function. Through the application of biotechnology, we can discover resistance gene locations on chromosomes and, recently, the actual genes for resistance (Huang *et al.*, 2003). The knowledge of resistance gene locations allows development of DNA markers used in breeding as marker-assisted selection. The implications of understanding fungal resistance are enormous in terms of our ability to manage these diseases. For example, gene pyramiding is now a possibility (Van Sanford *et al.*, 2001). Not only does gene pyramiding improve the level of resistance, but also offers the potential of increasing the durability of resistance.

Genetic transformation also falls within the scope of molecular breeding. This technology involves intimate knowledge of the gene, regulatory components and gene functional environment (i.e. the domain where the gene is located – currently optimized through selection of transformation events) (Hashizume *et al.*, 2006). Although transformation has much potential in molecular breeding for disease resistance at the applied level, in the foreseeable future the greatest impact will be in furthering our understanding of gene function. The slow adoption of transformation is primarily due to reluctance of consumers to embrace genetically modified foods, and until recently, few available genes.

Markers of various types, particularly DNA markers, are important in resistance breeding. Markers, in breeding, open many new options for breeders in their battle against fungal pathogens. As a result, marker-assisted selection is a reality in many breeding programmes.

Background

Biotechnology has been utilized to understand the genetics and genes behind disease resistance in important crop species. DNA markers in particular have been an important part of attempting to discover genes involved in resistance as well as research to understand the structure and function of those genes. Perhaps markers will play a less significant role in the future understanding of genes as other strategies of discovery advance and develop, but markers will continue to play an important role in breeding for disease resistance for the foreseeable future.

The concept of markers to identify a trait has been around for a long time (Mohler and Singrun, 2004). One of the difficulties in selecting for resistance is that the pathogen must be present in sufficient quantity plus the environment must be conducive to infection and disease development (Young, 1999). These prerequisites are inconsistently met, making it difficult to retain resistance in breeding material, because lack of selection results in random assortment for resistance alleles. Morphological markers have been sought and used to some extent to identify resistance in the absence of disease pressure from the pathogen. For example, the wheat leaf rust (*P. recondita* f. sp. *tritici* Rob. Ex Desm.) resistance gene *Lr21* is linked to brown chaff (Kerber and Dyck, 1969) and *Lr34* is linked to leaf tip necrosis (Rosewarne et al., 2006). Therefore, the association of resistance with these visual markers offers opportunities to make selection for disease resistance even in the absence of the pathogen and disease. Breeders can select for leaf tip necrosis to improve the frequency of *Lr34* when leaf rust is absent or brown chaff as in the case of *Lr21*, weighing the disadvantage that the expression can be modified by environment or the same phenotype can be produced by other genetic or environmental means. Leaf tip necrosis can also appear in dry windy environments simply by desiccation of the leaf tip and margins, and many genes control chaff colour. Therefore, more definitive markers are required.

As the detection of molecules such as proteins became more cost-effective with procedures such as electrophoresis, opportunities arose to harness variation in such molecules for use as selectable markers. Variation in isozymes and other types of proteins presented opportunities for discovery of molecular markers; however, the resolution and number of protein markers was eclipsed by advances in DNA markers.

Many of the first DNA markers developed were for disease resistance because of the importance of diseases and because many forms of resistance are simply inherited and easily identified (Michelmore et al., 1991; Michelmore, 1995). Initially, correlated markers to simply inherited resistance traits were developed. This strategy is still in use and will continue for the foreseeable future. The ease with which monogenic resistance traits can be characterized and evaluated genetically made such traits a logical first step in marker development. Initially, the major limitations in the development of trait-related markers were the number of markers available.

A variety of DNA marker types exist, some of which are more amenable to high throughput breeding than others. In general, polymerase chain reaction (PCR)-based markers facilitate high throughput, although some are easier and less expensive to use than others. Markers of any type are useful for filling in molecular maps and for initial stages of gene discovery. We will present some of the common marker types used to enhance our understanding of the genetics of fungal resistance.

Restriction fragment length polymorphism (RFLP) (William et al., 2006) are DNA fragments generated from the application of restriction enzymes to genomic DNA that is separated electrophoretically, blotted and hybridized with a labelled DNA probe. RFLP maps were first used to locate resistance genes. However, RFLP markers have limited use in marker-assisted selection because of the large amount of DNA required, the need for expensive enzymes and the involved process to develop the signal.

Amplified fragment length polymorphism (AFLP) (Dieguez et al., 2006) markers are based on a combination of restriction and PCR amplification. These markers have helped to further map crop species particularly where DNA polymorphism is low. AFLPs are difficult to reproduce quickly as needed in high throughput marker screening. Restriction and adaptor ligation, and the requirement for long acrylamide gels or long capillaries with capillary electrophoresis, are too slow for routine screening.

Random amplified polymorphic DNA (RAPD) (Gupta et al., 2006a) markers are useful because of their simplicity. Random decimer primers are used to amplify regions between the primers. However, many are sensitive to PCR conditions, which require that close tolerances in laboratory protocols be met to repeat the desired differential amplification.

Simple sequence repeat (SSR) (Fjellstrom et al., 2006) or microsatellite markers are more robust PCR markers, although polymorphisms based on one or two base-pair differences are difficult to detect. Their amplified products generally are easy to reproduce, contributing to their popularity in marker-assisted selection in many crop species. A limitation of these markers is that primer development is somewhat complex.

Sequence tagged sites (STS) (Guo et al., 2006) are PCR markers developed from sequence polymorphisms, such as expressed sequence tags (ESTs), AFLP or RFLP fragments and the like. STS markers can be based on actual gene sequences. An STS derived from a resistance gene is the best type of marker because it detects the gene rather than being merely linked.

Sequence characterized amplified region (SCAR) (Gupta et al., 2006a) PCR markers are primarily developed from amplified products such as RAPD fragments to improve the specificity of the marker.

Cleaved amplified polymorphic sequence (CAPS) (Helguera et al., 2005) markers involve a restriction step to reveal a polymorphism. The extra requirement for restriction increases the work required to process CAPS markers, plus the extra reagents and steps add to their cost.

Single-nucleotide polymorphism (SNP) (Edwards and Mogg, 2001) is a single base-pair change between alleles. PCR primers are designed around single nucleotide differences, and under an experimentally derived annealing temperature. A DNA polymorphism is detected as the presence or absence of an amplified band. Non-PCR approaches such as the Invader assay can also be used to visualize this type of marker (Edwards and Mogg, 2001). The Invader assay has advantages over PCR-based markers because of reduced reagents and processing time. As we develop new knowledge about sequence differences between alleles for resistance and susceptibility, SNPs will become important as markers of actual alleles.

Diversity Arrays Technology (DArT) (Semagn *et al.*, 2006) uses restriction and a complexity reduction step similar to AFLP on genomic DNA of a representative group of lines. DNA from each line, which has gone through a reduction step, is spotted on a glass slide for hybridization. Polymorphism is determined by hybridization. Sophisticated equipment is required, but the cost per polymorphism is low for this technology.

Instrumentation and biochemical procedures have evolved to improve detection of molecular differences and to increase analytical capacity while at the same time reducing material and reagent quantities and associated costs. RFLP markers using restriction enzymes on large quantities of DNA with fragment separation on horizontal or vertical electrophoretic gels with autoradiographic visualization have given way, in large part, to PCR-based techniques using small quantities of DNA with capillary electrophoresis separation and fluorescent dye detection. Single sample processing has been replaced with multiwell sample plates and multichannel pipettes. In step with the capability to process large numbers of samples are the advances in computer capacity and software capabilities to handle the ever increasing volumes of information. Linked with the Internet, data is quickly shared for everyone's benefit. These technological advances set the scene for rapid developments in molecular breeding.

Our understanding of the genes that feed into molecular breeding programmes starts with fundamental genetics and cytogenetics. Marker development has progressed quickly by using resistance genes and genetic information developed through classical means. Molecular techniques are now used to assist traditional methods of gene evaluation. DNA markers assist geneticists with the identification of genes and cytogeneticists with the introgression of genes from wild relatives of crop species. These genetic stocks are used by biotechnologists to understand genes for improved markers to increase efficiency of molecular breeding. The knowledge of simpler genomes, such as Arabidopsis and rice, has further improved our understanding of genes through comparative mapping, again with a spillover into molecular breeding through new or improved markers.

The practice of molecular breeding continues to evolve as new technologies are brought to bear on the problem. Many of these are developed in medical research, and are eventually adapted to crop research as economies of scale bring costs down. We will now discuss in more detail the discovery of markers, their validation and their application in practical crop breeding programmes.

Detailed review of topic

Marker discovery

Molecular breeding comprises three distinct phases of development: marker discovery, marker validation and marker use (Young, 1999). The

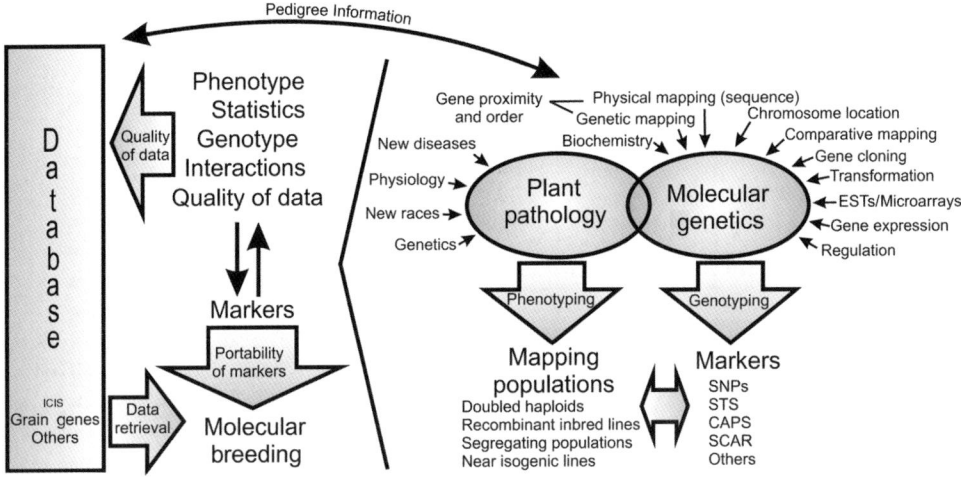

Fig. 14.2. The flow of information and technology in the development of markers and the relationships with molecular breeding.

discovery of a DNA marker linked to a trait is the first step (Van Sanford et al., 2001). Discovery of markers has become relatively routine in major crops, with sufficient markers available to get close to a gene with relatively little effort. Once a marker to the gene has been established, one can locate where it lies on a chromosome map and then use the map information to find markers that are even closer. These markers are run on the population to get a measure of the linkage distance between markers and the resistance genes to identify those which are most closely linked. The improvement of markers depends on a number of strategies in molecular genetics, such as comparative mapping, expression analysis and so on, in combination with phenotypic analysis and mapping (Fig. 14.2). The information generated about a resistance gene through different molecular genetic strategies, when entered into databases, is available to refine molecular breeding.

Many more markers to resistance have been reported than are actually in routine use. However, for markers to be deployed in molecular breeding, a number of steps must be undertaken to understand and validate their use in selection. Questions that create gaps between discovery of markers and use in selection arise (Dubcovsky, 2004). The effort needed to bridge the gap between marker discovery and utilization depends on the particular trait, with greater effort required for more complex traits (Fig. 14.3). Validation of markers for integration in routine use in breeding is the next important step.

Populations and phenotypic characterization
Markers are identified in genetic populations that have been evaluated for the particular disease resistance. With highly expressive and penetrant

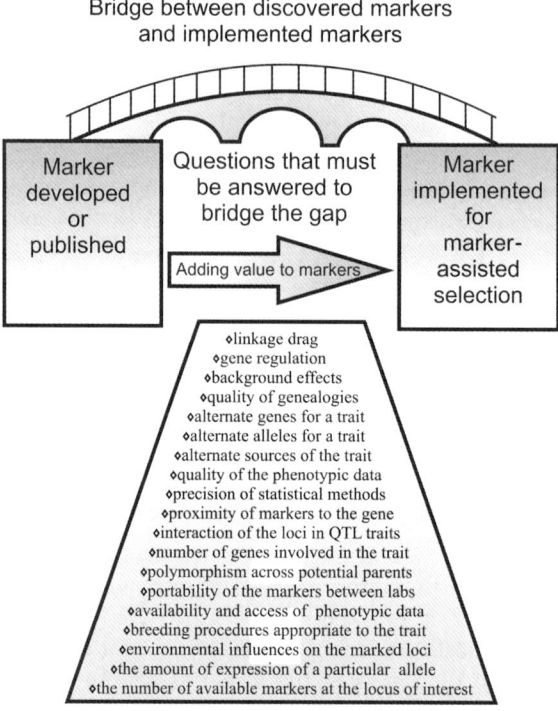

Fig. 14.3. Factors involved between the marker development phase and the implementation phase.

genes, qualitative analysis establishes linkage between markers and the gene. Bulked segregant analysis is useful for quickly finding a marker to a gene for qualitative resistance (Michelmore et al., 1991). With quantitative resistance, quantitative analysis determines the relationship between markers and loci associated with resistance. It is useful to start with populations where resistance is concentrated in one parent and susceptibility in the other. This helps to ensure genes that represent some intermediate level of resistance will not be the same in both parents and thereby render the cross unsuitable for marker discovery.

Different types of genetic populations are available for marker assessment, including traditional F_2 and F_3 families, backcross line populations, near isogenic lines, recombinant inbred line and doubled-haploid populations (Adhikari et al., 2004; Bovill et al., 2006; Lagudah et al., 2006; Cuthbert et al., 2007). Inbred line and doubled-haploid populations are desirable because segregation is minimized, which makes it possible to test them multiple times to improve phenotypic measurement precision of each line, consequently improving determination of gene locations. This is particularly important for quantitative traits such as fusarium head blight (FHB) resistance in wheat (Steiner et al., 2004). Breeding populations undergoing selection can be used to supplement information on putative markers; however, although marker–trait association can be demonstrated, because of selection it is unwise to use such populations to measure degree of proximity.

It is possible to obtain markers that are closer to the desired gene with high resolution maps (Cuthbert *et al.*, 2006). Map distance is based on crossovers, the number of which in a region relates to the number of segregating progeny; so the larger the population, the more crossovers can be sampled at any locus. Greater sample size increases resolution through reduced variance of reasonably close markers and increases the chance to sample close recombinations that are rare but necessary for identification of tightly linked markers. Association mapping is another alternative where linkage disequilibrium can be measured between loci of interest with surrounding markers (Cardon and Bell, 2001). With a large-enough population, the phenotypic variation can be attributed to each of the loci through QTL analysis. If the markers are mapped, information about the identity of the genes is established.

Genotype by environment interaction. Qualitative resistance traits may not vary with environment but the pathogen can vary. A major challenge to breeding for qualitative disease resistance is to have appropriate disease pressure in nurseries used for selection. Often conditions may be appropriate for one disease but not for another, or conditions may be only favourable in certain years, or disease intensity may be too high to discriminate resistance. During marker development, evaluation of the phenotype in multiple environments is necessary to circumvent this problem.

Genotype by environment interaction is an important consideration in phenotypic characterization of populations because expression of some genes is low in certain environments while other genes express across most or all environments (Lynch and Walsh, 1998; Klahr *et al.*, 2007). Breeders prefer the latter. However, with quantitative resistance, even the highest expressing genes may explain only a small portion of the resistance variation. These genes may not be sufficient under some environments to control the disease adequately, and other genes whose function is dependent on environment are needed to supplement the resistance. In environments with high disease pressure, all genes may be necessary whether they are constitutive or not. As our understanding of the genes involved in resistance progresses, we should be able to tailor resistance in certain cultivars to particular environments such as those with high or low disease pressure.

Genes that respond to environmental cues may not be expressed in the environments sampled during population phenotypic evaluation for marker discovery, and as a result, are not found when determining association of the marker with the trait. The solution is to test the population under a variety of environments. FHB resistance in wheat provides an example of this type of problem (Groth *et al.*, 1999; Ma *et al.*, 2006).

Once markers are developed they select for resistance despite environmental conditions and fluctuations in disease intensity.

Genotypic characterization
Genetic considerations. Epistasis occurs when the effect of a gene on a trait is modified by one or several other genes (Falconer, 1989). The same or another trait may be affected. Understanding gene interaction is important in allowing us to understand which combination of genes will maximize resistance. In addition to one resistance gene affecting resistance at another locus, the interaction can be with non-resistance genes (Ma et al., 2006). For example, the wheat height-reducing *rht* can be epistatic to disease resistance (Simon et al., 2004). Usually major genes for resistance are dominantly epistatic, which makes it difficult to develop markers to individual resistance genes where more than one gene is segregating. A gene we are selecting for can also have pleiotropic effects, whereby another phenotypic trait is affected in addition to resistance. Pleiotropy is difficult to demonstrate with classical genetics, but pseudo-black chaff, associated with the *Sr2* stem rust resistance in wheat, may be an example (Kota et al., 2006).

Many sources of disease resistance involve more than one gene with varied levels of phenotypic variation explained by each gene. Common bunt resistance in wheat is an example where genes which confer moderate resistance can be combined to produce highly resistant phenotypes (Knox et al., 1998). Occasionally, resistance genes are simply inherited, highly expressed and explain the majority of phenotypic variation. These major genes are preferred for their simplicity, but they are probably the exception. Even major genes rarely exhibit complete resistance, and resistance can be diminished in genotypes with background effects or modifier genes.

Another consideration with resistance is the breadth of the resistance across pathogen races. Resistance to a disease in a particular region of the world may require a combination of genes. Resistance is effective for as long as virulence in the pathogen does not recombine to match resistance in the host, whereas in another region, virulence may be compatible with the resistance. An example of this is loose smut (*Ustilago tritici* (Pers.) Rostr.) resistance in wheat (Knox et al., 2002).

Numerous sources of resistance are multigenic or quantitative in inheritance, and responsive to different environmental conditions. Markers help us develop an understanding of quantitative resistance. These quantitative trait loci (QTL) present different opportunities and challenges from simply inherited, highly expressive and highly penetrant traits. Quantitative resistance is often controlled by a series of minor genes which through their cumulative effects provide a greater degree of expression of the resistance than any one gene alone (Ma et al., 2006). QTL for resistance are found in many host–pathogen relationships, both biotrophic and necrotrophic (Steiner et al., 2004; Xu et al., 2005). A necrotrophic example is quantitative resistance to FHB in wheat. Multiple loci have been identified which affect the level of this disease, although no locus alone confers full resistance. The combination of a series of loci offers substantially greater resistance than absence of these loci. Other examples are resistance with slow rusting and resistance to leaf spots.

Molecular variants. Molecular variants are different-sized DNA fragments generated from a region identified by a DNA probe applied among lines within a species. The DNA is probed with complementary sequences of DNA, and differences in length of DNA between cultivars, called polymorphisms, are observed through restriction digests or PCR. Visualization is performed through a combination of electrophoresis to sort fragments and fragment detection through autoradiography, fluorescence or staining. Molecular variants are also called molecular alleles; however, the original definition of allele is a 'variant of a gene' and many DNA fragments are neither genes nor parts of genes. The term molecular variant has the advantage of not confusing the functional nature of an allele with the non-functional properties of many DNA fragments.

Polymorphic molecular variants identify regions on a chromosome and are associated with other markers. Resistance-related markers are those molecular variants linked with a locus carrying a resistance gene or genes.

Mapping

Mapping and DNA markers are a great boon to novel gene discovery. Once one gene for resistance is discovered, it can be identified by name. However, subsequent genes must be differentiated from previously identified genes. If the phenotype of genes is not unique, which is often the case with disease resistance in plants, then other information about the genes is required to confirm a difference. A step in gene identification is to demonstrate that the gene is not on the same chromosome as any of the known genes. Markers on a different chromosome are good evidence that the gene is novel and can be assigned a gene designation without having to intercross the gene with all existing genes to see if any are in common. Genome maps assist in determining chromosome location, which helps with identification of genes (Gowda *et al.*, 2006). If named resistance gene locations are known, the appearance in a new location on a map indicates a new resistance gene without having to do tedious gene by gene crosses.

Some plant species have sufficient markers to facilitate marker identification of simply inherited genes. The existing pool of trait-related mapped markers is tested against resistant lines with unknown genes. If, for example, a new line has leaf rust resistance, the next step is to determine if the resistance results from a known gene. By testing resistance gene markers, one can narrow down the relationship of the new gene with existing genes. Specifically, a marker linked to *Lr21* leaf rust resistance in wheat, applied to the new source of resistance and found to be linked, indicates the novel gene or may be *Lr21*. The linkage rules out the gene being any of the genes independent of *Lr21*. As a result, the time and effort required to establish the identity of a gene has improved.

It is important to evaluate a series of markers in the chromosomal region of the trait when assigning location to the genetic factor controlling the trait. We often use markers from consensus maps that are based on linkage distances on relatively small sample sizes. Therefore, the order and proximity of markers may be imprecise. Often, markers are assigned to more than one locus on different chromosomes. This is particularly true in polyploid species. Also, when translocations and inversions take place, loci are moved around in particular cultivar combinations. The larger the region and the greater the diversity of markers in a region that point to a common chromosomal location, the greater the confidence we have that this is the true gene location.

The uncertain proximity of the markers within a QTL to the actual gene is a problem. Higher resolution maps facilitate identification of markers closer to the genetic factors that control resistance. The larger the number of segregating progeny in a population, the greater the number of crossovers sampled at any one locus with reduced map distances between markers (Cuthbert et al., 2006). Greater population size increases resolution through reduced variance associated with sample size (for disease resistance measured phenotypically) and through increased samples of close rarer recombinations necessary for identification of extremely close markers.

QTL analysis also offers opportunities to identify multiple simply inherited dominantly epistatic major resistance genes that are present in a common background. For example, leaf rust resistance in wheat can result from a series of resistance genes within a line. With a large-enough population, phenotypic variation can be attributed to each of the loci through QTL analysis. Mapped markers then provide information to identify the genes.

The synteny between genomes helps gene discovery (Kota et al., 2006; Mateos-Hernandez et al., 2006). The rice (*Oryza sativa* L.) genome has been sequenced and mapped and that information can be used to investigate the genomes of related crops such as barley (*Hordeum vulgare* L.) and wheat which have not yet been sequenced. ESTs in the wheat library can be positioned relative to the rice sequence to enrich the number of markers in a region.

Sequence homology can also be used to develop markers to genes (Guo et al., 2006). For example, if the resistance gene to a particular disease is known in one crop species, that sequence can be used in a related crop species to extract a sequence for resistance to the related disease. Crop-specific polymorphic markers can then be developed as markers to the disease.

Association mapping is an alternative to QTL mapping within biparental cross populations. With association mapping, linkage disequilibrium can be measured between loci of interest with surrounding markers on a diverse group of lines. The hope is that generational crossing over time increases recombination that can be used for the more precise association of markers with genes.

As trait-related markers are discovered, genetic consensus maps provide an aid to selection of markers more closely associated with a gene (Somers et al., 2004). The greater the density of the map, the better is the resolution of association of marker and trait.

Marker validation
Following the discovery of a DNA marker for resistance, marker validation is required (Young, 1999) to understand the number of available

markers at the locus, their proximity to the gene, number of genes involved, alternate genes, alternate alleles, appropriate breeding procedures, portability of the marker between laboratories, interaction of loci in QTL resistance, linkage drag, gene regulation, background effects, polymorphism among potential parents, alternate sources of resistance, quality of genealogies, the amount of expression of a particular allele, quality of phenotypic data, environmental influences, access to available phenotypic and genotypic data, database organization, data retrieval and precision of statistical methods. A lack of understanding of any of these creates a gap between discovery of markers and usage for selection (Fig. 14.2). The amount of understanding required depends on the particular resistance. As we develop markers to more complex resistance, a greater effort is required to understand all of the facets listed above.

Marker development is still a young science because we are still trying to discover the molecular basis for expression of resistance. We have only found coding sequences of genes for relatively few resistance genes, and even less information is available on the sequence information that uniquely defines alleles. Therefore, markers are very often developed from a DNA polymorphism that shows a statistical association with a factor or factors involved in the expression of a trait. Although genes for fungal resistance are being identified with the potential of the perfect or diagnostic marker being developed from the genes, currently the vast majority of markers are only correlated to the resistance loci.

A major challenge to QTL usage in molecular breeding is determining which loci are present in potential parents because of the uncertainty associated with phenotypic characterization of quantitative resistance. When available, markers are used to verify which loci are actually segregating. With the multigenic resistance to FHB in wheat, inheritance of genes and development of markers can be worked out in a particular source of resistance such as the cultivar Sumai 3 (Anderson *et al.*, 2001). However, one cannot assume that descendants with resistance have the same complement of genes. For example, Alsen, a descendant of Sumai 3, has improved FHB resistance among North American spring wheat lines. Sumai 3 has at least three loci controlling FHB resistance (Zhou *et al.*, 2002). The markers cannot be automatically utilized on descendants of Alsen without further genetic analysis to determine which FHB resistance loci Alsen possesses. If only one or a combination of two loci were passed on, selection with markers would be wasted at the loci that did not possess the resistance allele.

In the development of derived lines, other loci that contribute to resistance may have crept in. Moderately susceptible parents may contribute loci to enhance the resistance phenotype. Therefore, markers should be tested in a segregating population of the derived sources of resistance and evaluated against the disease phenotype of the population. Lack of association with a particular marker indicates that the resistance allele was not transferred, and it is pointless to use the marker at that locus for other crosses using the derived parent.

Progeny can be tested to all known markers for FHB resistance to determine if genes from less resistant parents possess previously identified genes that complement those from the main source of resistance. Knowledge of the multiple genes available, their sources and appropriate markers, is important for planning crosses to facilitate pyramiding of complementary genes.

Marker validation is complicated by environmental effects on gene expression, particularly with quantitative traits and traits with low heritability.

Alien genes. Donated chromatin may be integrated to different degrees into the genome of the recipient crop species, depending on the relationship of the two species. Often insertions occur in which crossing over within the host crop species does not take place because of the absence of complementary chromatin. Therefore, dominant markers are often easily identified for the alien insertion as long as the fragment is unique to the insertion. Even in cases where DNA is integrated into the chromosome to allow homeologous pairing, crossing over may be reduced (Paull *et al.*, 1994). In either case, markers may appear to be closer than they really are.

Marker portability. The portability of markers between laboratories is an important consideration. With PCR markers, components of the reaction cocktail, their concentration and PCR temperature stages and cycling must be reproduced carefully for marker reproduction. The requirements of these markers also depend on brand of reagent, the instruments used to produce and detect the marker product, and the specific characteristics of the marker. These factors that affect reproducibility also affect portability. Usually it is not feasible to duplicate instrument and reagent supplies in each laboratory where a marker is used, so implementation of each marker requires resources to work out laboratory-specific protocols. Markers that are robust and tolerant of variation in conditions are more valuable than sensitive markers. Robust markers save time and materials during introduction to a laboratory and during use because there are generally fewer requirements for troubleshooting. Some marker types are more amenable to transfer than others. The robustness of many microsatellite markers makes them appealing.

As previously mentioned, RFLP and AFLP markers do not lend themselves to marker-assisted selection, not only because of the additional steps required, but also because additional supplies and equipment may be required. For example, an incubator is required to provide optimum conditions for the restriction steps. Portability of difficult-to-use markers can be improved by conversion to more amenable marker types, such as STS or SCAR markers (Knox *et al.*, 2002; Gupta *et al.*, 2006a).

Data management

Effective application of marker-assisted selection in breeding programmes requires efficient data management. Marker discovery, development and deployment generate an overwhelming amount of genotypic, phenotypic and genealogy data, all useful for maximizing continued marker application in a breeding programme. Immediate availability of the data is necessary for cross planning to take advantage of the best complementary combination of parents, determining appropriate markers for selection and for marker development.

A database, if organized to allow easy entry and retrieval, is a powerful tool for taking full advantage of disease resistance data generated globally. The International Rice Research Institute (http://cropwiki.irri.org/icis/index.php/Main_Page) is developing a global system that can organize data for global exchange and at the same time address local needs to retrieve data for planning and research purposes.

Databases
Many valuable databases and database interfaces are available for successful molecular breeding for fungal resistance. Two are listed here for the purpose of illustration of databases used in molecular breeding.

The International Crop Information System (ICIS [http://www.icis.cgiar.org]) is a database management system for storage and retrieval of genealogy, phenotypic and genotypic data. The system is being designed to provide researchers with access to public domain crop information from throughout the world, while preserving institutional proprietary information access. This tool is of great value to pathologists developing inventories of disease information about lines and populations, geneticists warehousing genotypic data (e.g. molecular variants of markers, of lines and populations) and for breeders in the application of markers, cross planning and selection planning.

Graingenes (http://wheat.pw.usda.gov/GG2/index.shtml) among other features provides a listing of marker maps and tools to navigate the maps. These features are invaluable for determining alternative markers in a region for a marker that may not be polymorphic in the cross of interest.

Pedigrees
Genealogical information is important for tracing cultivars that might have a particular source of resistance. Gene banks and institutionally archived seeds are used as sources of parents in crosses. It is presumed that the seed is labelled correctly and that any data on that cultivar is correctly associated with the seed, but sometimes problems with labelling occur. A problem arises when a source of seed thought to be a particular cultivar is something different, so selection with the identified markers may not be as effective. However, markers can be used to help sort out mixing, mislabelling or misclassification of entries. Fingerprinting databases can be established that provide a reference to confirm the identity of seed.

Data and analysis
Often research is under way in different laboratories around the world on the same diseases. It is of great value to all researchers to be able to tap into a global database and pull together information about different alleles and/or different genetic loci involved in a particular resistance. Uniformity

of nomenclature, rating systems and symbols used to measure disease facilitate this.

The database of genes for resistance can be conceptualized as a gene warehouse. As the database expands, breeders can select for combinations of not only disease resistance genes, but also other genes that complement each other for optimum disease, agronomic and quality performance.

As noted above, a major challenge to identifying genes, particularly for quantitative resistance, is that a population of adequate size must be characterized under the environmental conditions in which new cultivars will be grown. Often resistance found in one environment is not effective in another environment. Therefore, it is important to have disease data from a wide variety of environments.

Phenotyping is also critical to the identification of QTL for disease resistance. With a larger population of lines, there will be an increase in the number of lines with each combination of genes. Increasing the population size will thus improve the precision and aid in the discovery of loci for resistance. If loci are responsive to environmental cues, then the population must be grown in multiple environments with careful attention to appropriate design and analysis to identify those loci that are effective under different environmental conditions.

With large populations and complex marker maps comes the need for software and hardware that will handle the massive amount of data that can be generated not only in any one laboratory, but also globally. Algorithms for the analysis of phenotypic and genotypic data must continue to develop such that statistics are generated that identify real associations and minimize artefacts. The best data is of no value if it cannot be effectively put to use. Part of handling data is efficient extraction and use. ICIS can assist in the coordinated retrieval of data so that the information content is maximized.

Molecular breeding

Molecular breeding is more than simply applying markers to segregating populations and selecting a particular molecular variant (Fig. 14.4). With the variety of markers available and more being developed, application of markers is complex. Proper integration depends on tying marker strategies together with traditional breeding concepts. Often breeders have streams of germplasm and commercial cultivar development. With germplasm development, recombination reorganizes genes to optimize the background available for a new trait, such as FHB resistance in wheat. The traditional breeding approach of selection through generations of inbreeding takes advantage of the recombination.

Considerations in the development of breeding strategies
DNA markers can be used in a variety of breeding systems and stages of selection to enhance breeding strategies or to pursue outcomes that have

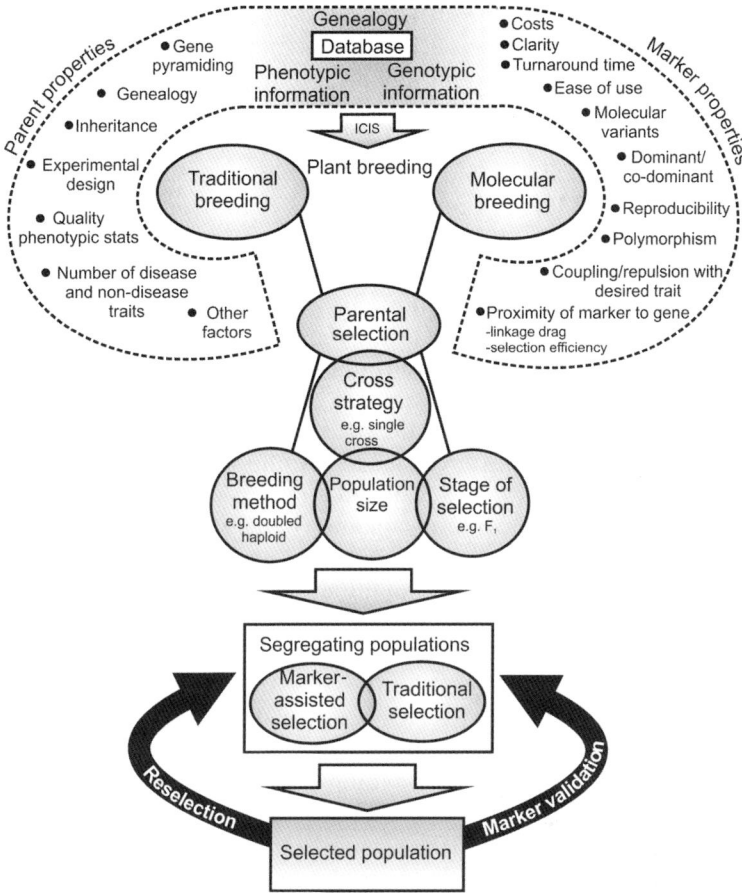

Fig. 14.4. Some of the considerations and decision-making process within the context of molecular and traditional breeding and the population development phase.

been previously unattainable with traditional breeding methods. It is critical to use markers as early as possible for the greatest impact across subsequent generations. However, the cost-effectiveness of selection with markers for a particular resistance at a stage of selection versus selecting the phenotype must be determined (Dekkers and Hospital, 2002). Kuchel et al. (2005) compared resource allocation required for each gene with marker and non-marker strategies. Another costing algorithm incorporates heritability of the resistance and the markers together with the portion of resistance variance explained by the marker (Van Sanford et al., 2001).

Gene pyramiding. Pyramiding or stacking of major genes offers great potential for development of durable resistance to disease. Markers are a useful tool to pyramid genes (Singh et al., 2001). For example, simply inherited, highly expressive and penetrant leaf rust resistance *Lr34* and *Lr21* in wheat, can be pyramided into a single genetic background (Huang et al., 2003; Bossolini et al., 2006). Where genes are masked by dominant epistasis wherein the strongest resistance is usually expressed, it is difficult to pyramid genes with classical methods. Markers for resistance genes at different independent loci can be tracked to pyramid genes with a level of predictability dependent on the proximity of the markers to the genes of interest.

Genes coding for marked quantitative traits can be combined to improve the level of resistance from cumulative or additive effects of the genes on the phenotype. For example, resistance genes to FHB in wheat can be pyramided, whereby the addition of each gene improves the level of resistance expressed to FHB.

Alternate genes. There may be many different qualitative resistance genes for a particular disease. For example, there are 51 named leaf rust resistance genes in wheat. Markers can help us to identify these genes in resistant parental lines where genetic analysis has not been done. With information about the genes present in a background, parental selection is improved as is gene deployment. If a resistant parent is chosen but not polymorphic in a particular cross, an alternate source of resistance can be chosen that is polymorphic. For example, there may not be polymorphism among markers for *Lr21* in a particular cross combination but there may be markers for *Lr42* in a similar cross.

Quantitative sources of resistance, such as slow rusting genes, are thought to be durable. It is possible that similar phenotypes of quantitative forms of resistance can be arrived at from different combinations of genes. However, the genetics of such resistance is not well understood, but markers help us learn which genes are involved and which combinations of genes confer the best resistance. For example, many loci have been discovered for FHB resistance in wheat (Gervais *et al.*, 2003; Yang *et al.*, 2005). Similar levels of resistance have been found as a result of different combinations of resistance loci in different genetic backgrounds (McCartney *et al.*, 2004; Cuthbert *et al.*, 2006).

Allele choices. The alleles utilized must be considered. A particular allele at one locus may work better with a particular allele at another locus, especially with multigenic resistance. When a locus has been characterized to contain a resistance gene, it is important to determine if there are allele differences present. Genotypes identified as having resistance at that locus may possess different resistance alleles, and certain alleles may have more potent resistance than others. Multigenic forms of resistance also likely have allelic series at the various loci that combine to produce resistance. With molecular genetics, we should be able to understand the range of alleles at each locus and determine which alleles combine to produce the most effective forms of resistance. Markers can then be developed around the allele sequence differences to generate perfect markers that can discriminate alleles. With the knowledge of how loci and alleles interact, breeders should be able to select for the combination that gives the best synergy.

Background effects. It is well known that certain genes perform better in certain backgrounds. Background effects are not understood enough to predict which genes function best in which backgrounds. This can only be discovered empirically. This renders marker-assisted selection ineffective in some cases where an introduced gene is poorly expressed compared to the background in which the marker was developed.

Detrimental effects from the incorporation of a gene or associated genes carried by linkage drag may be modified by other genes in the background of the genome. If a resistance gene is important but affects other traits because of linkage drag or pleiotropy, selection in the right background can compensate for the negative effects. For example, the incorporation of the *Lr35/Sr39* leaf rust gene combination of wheat is associated with delayed maturity, but combining them in a background with early maturity genes can compensate for the maturity deficiency (Knox et al., 2000). Parents can be chosen to utilize DNA markers not only to incorporate the *Lr35/Sr39* but also early maturity.

Stage of selection. DNA markers can be applied at different stages of selection with varying degrees of impact. In general, the greatest impact is achieved when markers are applied in early generations to minimize carrying forward lines of no value. Characteristics of the marker determine where marker impact is the greatest relative to the cost of using it.

Segregating F_1 plants derived from complex crosses, such as backcrosses, top crosses or four-way crosses, in combination with marker evaluation offers an opportunity to dramatically shift allele frequencies at the earliest stage. When one dose of the disease-resistant donor parent is used in the first cross of a three-way cross or backcross, the resistance allele will be present in the heterozygous condition in half the derived population. Use of the marker on F_1 progeny will identify the lines with the resistance allele in the heterozygous background. Because the heterozygous F_1 will continue to segregate, further selection will be needed at a later point. Half the population with no potential for resistance will be eliminated prior to population expansion.

If two doses of the resistance are applied in a three-way cross or single backcross, progeny will be half homozygous for the resistant allele and half heterozygous. In this case, a marker can fix the resistance by identifying the heterozygous progeny for removal. No further marker screening would be necessary except for screening under disease pressure at some point to confirm the disease resistance had not been lost through a crossover.

Cross configuration can limit the usage of a dominant marker but not a codominant marker. A dominant marker in coupling with the desired resistance allele is useful when a single dose of resistance is used in only the first cross, but is of no value in testing F_1 progeny if the resistance is included in both the first and second crosses. In the latter case, the dominant marker molecular variant is all that is seen in both the homozygous resistant and heterozygous progeny. If the dominant marker is in repulsion with the resistance allele, the marker will be of value in a three-way cross or single backcross where resistance is two doses. Where the heterozygous progeny can be discarded, the resistance can be fixed in the homozygous resistant progeny. With the dominant marker in repulsion, all top-cross and backcross F_1 progeny will appear the same in a cross of resistant by susceptible parents and the marker cannot discriminate the two segregating genotypes for resistance.

If segregating F_1 plants are grown in greenhouse beds in a grid, the tissue and ultimately the DNA sampled can be tracked by position. Such a system minimizes the time and effort required for labelling. After marker results are generated, the position of the undesirable plants is readily located, and the plant uprooted and discarded.

Maximum efficiency of selection in a doubled-haploid operation is achieved by applying markers to the haploid plants before further resources are used to double them. In the process of developing doubled-haploid plants, they go through a haploid stage. In wide-cross doubled haploids, such as the wheat by maize system, the haploid stage is protracted and to some extent controlled by the investigator by the timing of a colchicine treatment for chromosome doubling. The haploid plant DNA is sampled and DNA markers for resistance are applied. The haploid plants with the marker molecular variant for the susceptible allele are discarded, thereby enriching the remaining pool of haploid plants for resistance. Further investment of doubling the chromosomes and rearing the plants will only be done on those plants with an increased chance of being resistant.

Markers run on doubled-haploid lines produce clean results because there is no segregation. Selection in haploid and doubled-haploid lines fixes the trait at a level consistent with the linkage distance between the marker and trait. With a doubled-haploid programme, segregating F_1 plants can be tested prior to entering the doubled-haploid process to ensure that the plants are carrying the resistance allele.

In an inbred crop, selection might be limited to the F_2 generation or later because the F_1 is not segregating for the resistance (single cross from inbred lines) or the marker configuration does not provide polymorphisms as described above for dominant markers. Whether to evaluate the F_2 with a marker depends on the particular cross and number of markers available relative to the number of traits that can be evaluated in a nursery. For example, if a population is evaluated for a marker to only one resistance gene, the maximum cost of marker assessment is incurred. By putting the population in the field at the F_2 generation and selecting in a disease nursery with multiple diseases, the cost of evaluation per line per disease is much reduced, likely below the cost of a marker determination in the laboratory. If other traits in addition to resistance can be evaluated in the nursery as well, cost of trait evaluation per line is further reduced. Phenotypic selection in the nursery enriches the population for highly heritable traits. Then, the resistance marker can be utilized to select on a much smaller population. Use of multiple markers also reduces costs because, for example, only one DNA extraction is needed.

Marker evaluation may be applied to inbred breeding lines developed through single seed descent or bulk breeding or as a second round of marker selection in which heterozygotes were selected at an earlier generation. The biggest consideration with applying markers to inbred lines is the potential that the association between the resistance gene and the marker will be reduced because of the opportunity for multiple crossovers.

Inbred lines are usually used as parents in crosses in self-pollinated crop species. Often, inbred lines have problems of residual heterozygosity and heterogeneity at unselected loci. It is important when choosing parents for crosses utilized for marker development that sample seed of the parents be preserved. Should variation arise in either the disease phenotype or the molecular genotype, the reference sample is available to determine if the variation is derived from within the population or from outside the population, i.e. outcrossing.

Required population size depends on the number of traits the parents differ for, whether the traits are qualitative or quantitative, level of linkage, type of cross, generation in which the markers are applied, how inbred, desired probability of success and the resources available for testing (Bonnett et al., 2005). Populations required for parents differing for just a few qualitative traits are much smaller than where the parents differ for many qualitative and quantitative traits. Breeders usually select for simply inherited resistance in early filial generations of segregating populations and for more complexly inherited generally quantitative resistance in later generations. Testing for complex resistance is usually more expensive and populations are smaller in later generations. Disease resistance often falls within the simply inherited traits that are evaluated early, and markers are applied with generally predictable results. Quantitative resistance becomes qualitative with application of effective markers, and then can be evaluated in early generations.

Breeding populations usually are large at the F_2 stage because of the high level of attrition due to the combination of agronomic and disease resistance traits evaluated. Sufficient material must be retained for selection of the quantitative traits (disease, quality and agronomic) in later generations. When qualitative traits are being assessed, it makes sense to rank the traits based on cost of assessment and perform the most inexpensive testing first. The evaluation cost per trait per plant is very low in an F_2 disease nursery. For example, in a wheat breeding F_2 nursery, plants are simultaneously assessed for resistance traits such as common bunt, leaf rust, stem rust, a variety of leaf spots including tan spot, septoria tritici blotch, septoria nodorum blotch, spot blotch, crown rot, head blights and stem disorders, as well as for agronomic traits such as straw strength, plant architecture, height, maturity, seed set, grain filling and in some cases plant colour and awnedness. If all 18 of the characteristics were segregating within a population as dominant traits, the probability of one plant being present with all traits is $(3/4)^{18} = 0.0056$, or 1 in 177 plants. To obtain a 95% confidence of selecting the 1 in 177 plants, a population of $(1 - 1/177)^x = 0.05$ must be grown, or about 533 plants (Hanson, 1959). This is a minimum because some of the traits will be recessive and some will be controlled by more than one gene and some loci may be linked, requiring a greater population size to break the linkages. If for example, all 18 traits were recessive, 1 in 69 billion plants would express all these desired traits in the homozygous recessive configuration. Of course, the traits selected for the later generations must be included. Indexes with economic or phenotypic weights could be used with multigenic selection by markers so that lines with potential are not discarded because they lack all markers (Dudley, 1993).

Not all traits will be measured in the nursery; for example, disease may develop poorly, or the genes for resistance to a particular disease are masked by another gene. Evaluation of resistance to any one disease may be unclear if a number of diseases are infecting the same tissue, such as the wide variety of leaf diseases competing for the same infection sites. All things considered, it is still cheaper to evaluate plants in an F_2 field nursery than to process them in a laboratory, even with multiplexing of markers. It is also not likely that markers will be available for all of the traits that are segregating.

An effective approach is to select whatever traits are measurable in the field, and then assess the much-reduced population with markers (Bonnett et al., 2005). However, for every trait not selected in the field (perhaps the disease did not develop), one-quarter to three-quarters of the population will be discarded when a codominant marker is applied. If it is desirable to fix the locus, heterozygotes can be discarded. If resistance genes are being pyramided and multigenic resistance is being selected along with quality and agronomic traits, very large populations will be required so that enough plants remain for effective selection of quantitative traits, such as yield, protein and quality. Starting with 100 plants tested with 10 codominant markers, only 5–6 plants would be left if heterozygotes were retained, not accounting for random variation at each locus. If an attempt were made to fix the traits at this stage, a much larger beginning population size would be required.

Population size must be doubled for every gene to be selected for by markers. For example, if a breeder normally generates 500 top-cross F_1 plants from a cross in wheat and is applying a marker at a particular locus, a population of 1000 plants is required to end up with 500. If applying two markers, 2000 plants will be required to get 500 and so on.

Depending on the marker and the desired outcome, a different number of seeds may need to be sampled from each F_2 plant. For example, if a marker were recessive, to determine if an F_2 plant carried the desired gene as a heterozygote, 10 F_3 seeds would need to be sampled from the F_2 plant to have a 95% confidence of identifying the recessive genotype. If only homozygous recessive F_2-derived lines are to be retained, only three seeds need be tested to have a 95% chance of eliminating heterozygotes and homozygous dominants. In this case, the DNA can be bulked.

With three- and four-way crosses, markers can be applied to F_1 plants to shift gene frequency. For example, if in a backcross with the recipient parent, the donor gene (from the donor parent) can be selected as the heterozygote if a codominant marker or dominant marker in coupling with the donor gene is used. At the same time, codominant markers at other loci can test for homozygotes with favourable alleles provided by the backcross parent.

In all cases, population size must be adjusted upwards to account for selection in the early generations to leave enough segregating lines for selection for other quantitative traits. As more markers become available, resources may need to be shifted from testing of disease in the greenhouse to testing of markers in the laboratory, with verification of marker tested lines in field disease nurseries during the growing season.

Marker-assisted selection

Marker-assisted selection is an indirect selection of one trait by way of another because of linkage of the DNA that codes each trait. A marker can be a protein, a metabolite or phenotype. Marker-assisted selection usually refers to DNA-based markers that are associated with a phenotypic trait. Because DNA markers are the greatest component of molecular breeding, they will be the focus of our discussion on marker-assisted selection (Fig. 14.5).

Sample processing. An important aspect of DNA markers for fungal resistance is the short processing time. Compared to the time required to evaluate resistance in a field disease nursery or greenhouse, where disease expression may not be optimal for resistance evaluation, markers provide a large advantage. Sample throughput is another key to success with molecular breeding and is related to the scale of operation and

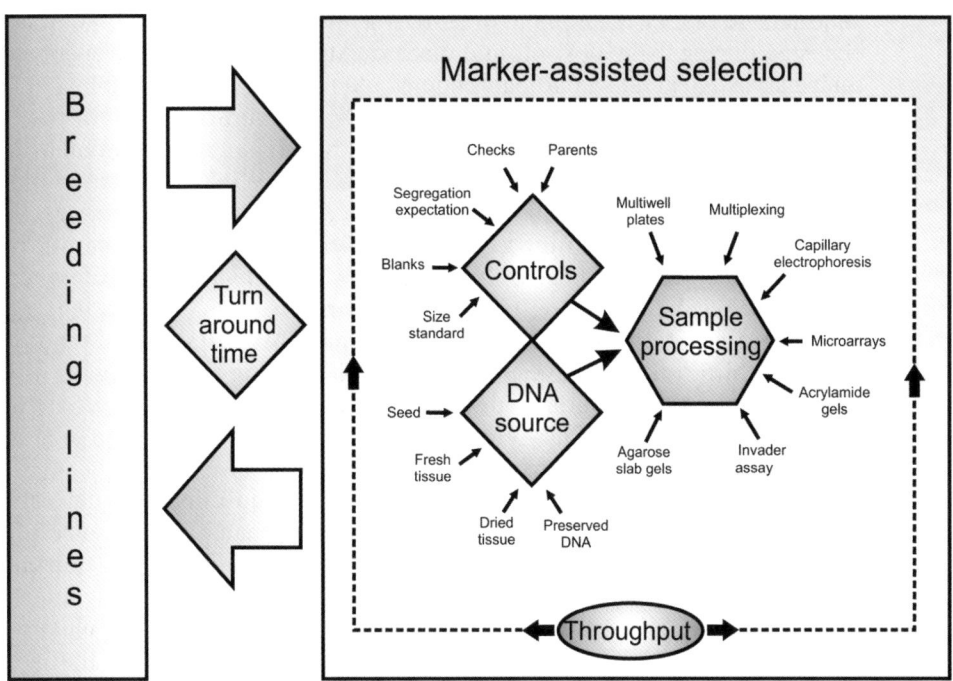

Fig. 14.5. Factors involved in marker-assisted selection.

technology used. High throughput requires a critical mass of determinations to justify investment in the required technology. A range in automation from multichannel pipettes to multichannel capillary electrophoresis systems is available. With increased automation, labour costs tend to drop but capital, material and supply costs tend to increase.

Systematic handling of DNA sampling, extraction and visualization are important to throughput. There are different methods of DNA extractions that can improve throughput by increasing the number of samples that can be processed at one time, or extraction protocols can be modified to a minimum of steps, or both. It is important also to consider adapting markers to technologies that exclude the need to extract the DNA and that do not require PCR, such as the Invader assay (http://www.ngrl.org.uk/wessex/invader.htm).

Selection for disease resistance using DNA markers should be done in conjunction with markers to other pest resistance, agronomic and quality traits so that the cost of DNA extraction is spread over numerous traits. Refined PCR, electrophoretic and data assessment processes, and reduced sample size with a corresponding reduction in reagent costs, also reduces the cost per determination.

Turnaround time. Quick turnaround time is an advantage of markers. Samples collected one day and results provided the next day allow quick assessment of plants for selection. Turnaround time is important for usage of out-of-season nurseries, reducing the number of lines sent to the alternate nursery. Other examples include identifying parents segregating for a resistance gene to reduce the number of crosses with no potential; live plants can be tested during their life cycle and discarded before resources are spent on their harvest. Several technologies contribute to rapid turnaround.

Capillary electrophoresis uses multiple capillaries that contain polymer for separation of DNA fragments. The multichannel operation and sample mulitplexing with detection of different dyes that fluoresce at different wavelengths allows high throughput. Lower amounts of reagents and automated data handling reduce cost. A major advantage of capillary electrophoresis is the high resolution, as low as one base-pair difference in fragment size.

Multiplexing, the amplification of multiple DNA markers that require similar PCR conditions, saves reagents and processing time. By using different fluorescent dyes attached to each set of primers, the PCR products are visualized with capillary electrophoresis even when fragment size is the same between markers in the multiplexed PCR.

With slab gel electrophoresis, agarose slab gel equipment is relatively inexpensive, the technology is less sophisticated than capillary electrophoresis, and less prone to equipment failure, with the further advantage that problem troubleshooting is more straightforward. The disadvantage is the lower resolution than acrylamide, especially on a capillary electrophoresis unit. But high resolution agarose can resolve fragments that are very close in size and gel costs can be reduced by reusing gels. With markers that generate clearly defined molecular variants, sample throughput can be high with slab gels. By using multichannel pipettes and multiple wells spaced not only across the gel but also down the gel, many samples can be evaluated at one time, providing high capacity without capital investment in expensive equipment. A combination of agarose gels and capillary electrophoresis can be a valuable combination. Non-PCR processes like the Invader assay, built around fluorescent signal amplification based on SNP, have the potential to take marker throughput to yet higher levels by reducing steps and complexity.

DNA source. Often in a breeding programme a variety of processes are under way, such as field testing or greenhouse programmes to multiply or evaluate lines, doubled-haploid production, crossing and intergenerational seed storage. Each phase offers opportunities for DNA marker selection. Live plants require careful timing to sample the DNA for marker analysis and discard plants before the next planting cycle. Phases when fresh plant material is available must be set as the highest priority for evaluation. Seed testing is a lower priority that fills in gaps between fresh material sample collections.

DNA can be extracted from seed or seedlings produced from seed stored between generations. Generally, in a breeding programme, diseases are evaluated on crop plants in one field season. Often, lines are increased in an out-of-season nursery, which is grown in a different environment where types of pathogens and races vary from the local field environment. As a result, it is unreliable to select for resistance in an out-of-season nursery. However, while plants are growing for a generational increase in the out-of-season nursery, reserve seed of each line can be tested using DNA markers and only those lines that are carrying the genes of interest are promoted in the breeding programme. Stored seed of new parents may be tested with the battery of markers available to the breeding programme, or seed of parental lines of populations currently in the breeding programme may be tested with new markers.

DNA can be harvested from seedlings. Seed increase and testing for resistance with a pathogen in the greenhouse or growth rooms during the non-growing season provide opportunities to collect DNA from fresh tissue for marker testing. Only lines with the desired resistance as predicted by markers are advanced.

Dried tissue is a useful source of DNA when the number of samples to be processed for DNA extraction overwhelms the available labour, or time or distance between sampling the tissue to extraction in the laboratory is too long to preserve the quality of the tissue. The tissue can be dried and stored until the DNA is extracted.

Preserved DNA is utilized when there is time to extract DNA but insufficient time to process the DNA further. In such cases, the DNA can be stored until there is time for further processing.

Controls. A number of problems can arise during the evaluation of markers, such as establishing the correct molecular variant, lack of amplification of an amplicon, differences in intensity of a band on a gel or level of fluorescence with the capillary electrophoresis unit, mixing of samples, mislabelling of samples and others. Mechanisms to ensure the data is being handled and interpreted correctly are important.

Comparison with the DNA-size standard ensures that the critical molecular variants can be identified. Often, a variety of molecular variants of different sizes exist in the critical size range for a particular marker. Parents should be run with their progeny to display the molecular variants that will be segregating in a cross, and look for potential heterogeneity that would explain all the variation in the progeny. Testing the lines on which the markers were identified assists in determining which DNA fragments are the critical trait-related molecular variants. This is particularly important if a marker is being run for the first time.

Knowledge of the structure of the cross being evaluated with markers allows one to compare the predicted segregation and observed frequency of markers in the population to determine if the population is segregating properly and that the molecular variants are recognized and scored appropriately. This is especially important in populations with a dominant marker. Lines with a null response may either be a null molecular variant or they may be the result of poor amplification. Significant distortion from the expected ratio could indicate that the lines were not scored properly.

Positional blanks can be used to make sure sample clusters are oriented correctly. A blank can be useful to recognize aberrations at various steps in a determination, such as the introduction of contaminated DNA.

Parental selection
Selection of parents is critical to the success of any breeding programme. Molecular breeding has the additional requirement that both phenotypic traits among parents and molecular genotypes must be complementary. A number of factors must be taken into consideration.

Heterogeneity. We often assume inbred lines are a population of homogeneous, homozygous lines. In reality, advanced breeding lines and registered or recognized cultivars are heterogeneous at some level and retain some residual heterozygosity. The heterogeneity arises from residual heterozygosity at the stage of final bulking of a line, genetic aberrations that arise, outcrossing or physical admixtures. This reality must be accommodated both from the perspective of assessing the disease phenotype and the molecular genotype.

Heterogeneity affects interpretation of results during marker development and marker use. For example, the resistance may be present in only a portion of the lines representing a cultivar. If resistance is highly penetrant and expressive or under the control of a single gene, it is easier to detect and control the heterogeneity. However, heterogeneity in resistance with lower penetrance, expressivity, heritability or under multigenic control is often difficult to deduce from the phenotype. This can lead researchers studying the same resistance to come to different conclusions. When heterogeneity arises during marker-assisted selection, the breeder must deal with the uncertainty that a particular molecular variant of the marker is associated with the resistance allele.

To control heterogeneity in a self-pollinated species, the breeder or geneticist can line out and purify the parent or make it a doubled haploid. The selections would be used in crossing. The risk is that the line chosen is not the most representative of the population in general, or worse, may not carry the trait of interest. As noted above, the parents of each cross should be retained as references for unexpected phenotypes and genotypes that may arise as intensive testing progresses.

Phenotypic information. The dilemma breeders and geneticists are faced with is not always knowing the resistance genes that reside in parents they are interested in using for crossing, especially with quantitative traits where it is difficult to assess them in a test cross. The resistance genotype of a line is often surmised by the accumulation of information on the disease phenotype; however, marker genotype information and pedigree provide very important supporting evidence as to whether a line has a particular source of resistance. The phenotype can provide some evidence of the presence of a gene, but often dominant epistasis of one resistance gene over another will mask other genes. In the case of quantitative resistance, different combinations of genes can generate the same phenotype, so the presence of a gene cannot be established based on phenotype alone. If genetic work has been done on an ancestor to indicate the presence of a gene, then the potential of the gene being present is at least known. Knowing both pedigree and phenotype may still not be enough information.

Genotypic information. Using markers along with phenotypic information on a derived line relative to ancestral lines known to possess the resistance increases the certainty of whether a line might have a certain gene. The more marker information associated with the resistance locus, the haplotype, the stronger the evidence the allele of interest is present. A test cross may be performed where the presence of the gene is determined by measuring the disease reaction of the segregating population and assessing whether variation in resistance is associated with markers known to be linked with a gene or genes controlling the resistance. Once the genes have been identified in a particular parent, the markers can be used to select in other progeny that derive from the same parental source of resistance.

During development of a marker or markers, often only one or a few populations are used to establish the markers, and these are usually the first markers tested on breeding material. In a particular cross combination used by the breeder, these markers may be monomorphic. However, the marker choice need not be limited to those markers developed for a few parental combinations. If the markers are mapped, the map can be used to identify other potential markers in the same region as the original markers. The more markers that are available in the region of the resistance locus or loci, the greater the opportunity to find polymorphism between any parental cross combination.

A codominant marker generates a signal, an amplicon in the case of a PCR-based marker, at the locus of interest in each parent. In simple terms, with a dominant marker a signal is present in one parent and not in the other. Codominant markers are superior to dominant/recessive markers because a DNA fragment is associated with both the favourable and unfavourable allele of a trait. Often, when a null allele is present, there is doubt as to whether a determination represents the null molecular variant or simply is the lack of amplification of a line that possesses the identifying DNA fragment. The codominant marker is the marker of choice.

Because most markers are linked, they are free to crossover within the constraints of their proximity to the genetic factor. Genes for disease resistance/susceptibility obviously occur in more than one background and therefore the DNA surrounding the gene differs between cultivars. For example, the stem rust *Sr2* resistance gene in wheat is linked to a 270 bp fragment amplified by the sun2 primers in some cultivars and to the alternate 280 bp molecular variant in other cultivars (Sharp *et al.*, 2001). Likewise, both molecular variants are linked to the stem rust susceptible allele in different cultivars. This can be both advantageous and disadvantageous. The advantage is that polymorphic parents can be selected regardless of the orientation of the marker with the resistance allele. In our example, an adapted parent may be susceptible to stem rust and possess the 280 bp molecular variant of the marker. A resistant line with the 270 bp fragment can be chosen as the other parent. The converse is also true. If the adapted parent is susceptible and possesses the 270 bp fragment, a resistant parent with the 280 bp molecular variant can be used. In situations where the molecular variant is more consistently associated with the particular allele of the trait of interest, in those rare occasions where the alternate type needs to be matched, it may be difficult to identify a parent. An example is with common bunt resistance in wheat. The *Bt10* marker is rarely in repulsion to the 270 bp band. Therefore, if the 270 bp molecular variant is present when the *Bt10* susceptible allele is present, it will be difficult to find a parent polymorphic for the marker with which to cross. Some markers, such as the *Sr38* marker cs1Vrgal3"F, show consistent association of one molecular variant with resistance and the other molecular variant with the susceptible condition and are therefore considered diagnostic (Sharp *et al.*, 2001).

Markers close to resistance loci reduce linkage drag while improving probability of selection. The closest possible polymorphic marker should be chosen. However, some markers are more difficult to use than others, and the extra cost of using the marker, even though close to the gene, must be weighed against the potential misclassification of a more reliable marker further away from the gene of interest. The chance crossing over that separates the marker from the resistance gene requires phenotyping with the pathogen on lines selected with the marker to confirm resistance. The degree of linkage affects the efficiency of selection. A marker within a gene is difficult to obtain because the gene may not be known. Even a marker within a gene is not completely linked to the resistance unless it identifies the region of DNA that describes the expression of the gene as alleles. The polymorphism is commonly in the DNA upstream or downstream from the gene, but markers within the actual gene of interest are being discovered.

Flanking markers improve selection efficiency when the associated markers are not particularly close; however, the disadvantage is linkage drag. Selection for crossover types in the interval defined by those markers is reduced. Rare double crossovers between the markers that flank the gene are required for the undesirable allele to be selected.

Often, wild relatives of crop species are used as sources of disease resistance, and depending on whether the resistance is introgressed as an insertion or a crossover, has implications on the function of markers. In the case of an insertion, it may be difficult to reduce linkage drag with markers to alien chromatin within a recipient species because of little or no crossing over. However, markers developed to the donor species chromatin can be used to assist in selecting recipient lines with the smallest functional piece of chromatin. Where introgression involves crossing over, markers can help reduce linkage drag by selecting markers that are close to or on top of the gene. Reyes-Valdés (2000) discusses further considerations.

One map unit is roughly equivalent to 1 million nucleotide bases. A gene is roughly 5000 bases in length, therefore 200 genes can reside between a gene and a marker that are as close as 1 CM. A crossover will occur in this region in 1 in 100 individuals. The same delinking may be required on the other side of the gene. Therefore, even in large populations, only a few crossover types will occur. If breaking undesirable linkage at multiple loci with a QTL trait is a requirement to get an optimum genetic combination, very large populations will be required. When multiple loci are considered in a cross, the opportunity for selecting away from linkage drag is small, but markers offer the opportunity to break the linkages in directed fashion.

Development of New Technologies

New technologies, such as new markers, methods in data and information handling and sample processing, will continue to enhance molecular breeding for plant disease resistance. Sufficient markers are available for routine marker-assisted selection in breeding programmes of major crop species. With the multiple approaches to marker development, markers are being reported at an increasing rate. One challenge is to stay aware of the available markers and their usage. Access to information through the Internet plus programmes that coordinate this information generation and availability are important for the rapid adoption of new markers. Internet sites that coordinate marker information, such as the 'Wheat Coordinated Agricultural Project' or CAP (http://maswheat.ucdavis.edu) and related sites for other crops are invaluable to the growth of molecular breeding. At this time, half the protocols listed on the Wheat CAP site relate to fungal disease resistance.

Now and in the foreseeable future, the markers that are coming onstream provide the molecular technology with the biggest impact on plant disease management. Somers et al. (2005) demonstrated that crosses could be developed around existing marker information and high throughput genotyping technologies to assemble a number of genes for resistance, together with other important quality and agronomic traits into common backgrounds. This is a model for the integration of molecular technologies in breeding to enhance germplasm for FHB and leaf rust resistance.

Markers to durable resistance genes are much anticipated by breeders and are starting to appear. For example, markers to the wheat *Lr34* leaf rust resistance gene are influencing strategies to breed for resistance to this disease (Bossolini et al., 2006). Such genes are being considered as the basis for gene pyramiding. Ongoing identification of genes provides the basis for gene pyramiding. Furthermore, gene identification allows breeders to deploy genes in a more systematic way. As genes are identified through markers and mapping, other beneficial traits can be associated with those genes particularly through association mapping. Those resistance alleles that confer, or are linked to genes that confer, other beneficial characteristics are being identified and used rather than sources of resistance that have negative effects. An example of this is the linkage of the wheat stripe rust resistance *Yr36* gene to a gene for elevated protein, which makes it a preferred choice over other stripe rust resistance genes (Uauy et al., 2005).

The development of databases and information retrieval tools will continue to be an important area of focus in the near future to offset the volume of information gathered in the course of marker discovery, validation and use in breeding. For example, inbred lines considered as parents are evaluated for multiple markers with a series of molecular variants associated with the phenotypes of various traits that represent several genetic loci, and in the case of quantitative traits measurements are recorded over multiple environments, plus information is generated with each of the interactions of these factors.

Data management and retrieval tools are necessary for the critical steps in cross planning as a foundation for the efficient use of markers. As breeders navigate the new territory of molecular breeding, tools under development, such as ICIS, will need continued refinement to keep pace with new demands. Information handling tools are necessary for markers in breeding to advance from an ad hoc approach to the application of markers to a sophisticated integration of marker technology. Queries are being developed to provide genotypic (marker), pedigree and phenotypic information to assist breeders in cross development and assist each researcher with the retrieval of pertinent information. As molecular information on alleles and phenotypic information on alleles and allele interaction is generated, this too must be readily accessible. With the capacity to evaluate vast numbers of lines for several DNA traits at once, information tools for handling data output are also important to bring phenotypic and genotypic data together for decision making.

Map development will cascade as it feeds into upstream research and in turn upstream research feeds the further refinement of genetic maps. Map information feeds directly into molecular breeding to improve the scope and precision of existing markers for resistance. Genetic maps contribute to discovery of markers for existing resistance genes not yet mapped and to the identification of new genes. Physical maps will continue to help us to compare among crop species the degree of homology of DNA sequences and the synteny of genes (Bossolini *et al.*, 2006). This critical tool in the discovery, verification and isolation of resistance genes along with DNA libraries, expression studies using microarray analysis and transformation for gene knockout and gene silencing, enhance the study of gene function, which further enhances our understanding of how resistance genes interact for the maximum benefit in disease management through resistant cultivars.

Conclusions and Future Directions

Molecular breeding is in its infancy but will grow as breeders adopt new DNA processing and information handling technologies and become familiar with selecting for traits at the gene level with markers. Over the next few years, breeders will continue to wrestle with many of the issues of new markers, new processing technologies, implementation and validation of markers and large volumes of information presented in the previous section. However, we anticipate that computer software tools will be in place for breeders to extract both phenotypic and molecular data on parents and to allow data comparisons for complementary traits with large volumes of detailed data. Breeders will utilize perfect markers to genes as they abandon correlated markers for markers developed around DNA polymorphism within genes.

Selection can be done by measuring any metabolite and keeping the portion of the population within the desirable range, but tracking the gene responsible for the metabolite simplifies the strategy by avoiding downstream complications, such as interactions of molecules and response to the environment. Through proteomics, we can understand which proteins affect production of other molecules, including other proteins, carbohydrates, lipids and other biomolecules and secondary metabolites. Through metabalomics, we can understand which molecules are desirable. These molecules in turn relate to particular enzymes. Understanding proteins through proteomics and metabolomics furthers our understanding of the specific genes involved using reverse genetics (Bender, 2005; Lee *et al.*, 2006).

ESTs in combination with microarray technology provide opportunities to develop markers that relate directly to desired resistance genes (Bossolini *et al.*, 2006; Shen *et al.*, 2006). Through the combined efforts of transcriptomics, metabolomics and proteomics, gene discovery will progress to allow development of SNPs and STSs as perfect markers for resistance genes. As we learn more about genes at the DNA level, we are able to assay for SNP within the genes, and will be able to assess resistance gene alleles at the molecular level to further improve the precision of molecular breeding. Not all sequence differences are associated with trait modification. Being able to track particular resistance alleles will allow the characterization of those alleles on their own and with other alleles. Markers can be developed around sequence differences to generate perfect markers that can discriminate alleles that impact the traits.

As gene discovery progresses, the opportunity to generate markers to a variety of precise alleles will enhance our ability to assign trait variation not only to particular loci but also to the interaction between alleles at particular loci (epistatic effects). Allele combinations can be selected which optimize disease resistance in particular environments.

Discovery of structural genes for resistance only explain part of the picture. Gene regulation must be understood for the effective introduction of resistance into a different genetic background. A gene may be introduced into a background where expression is lost in a portion of the transformation events (Huang et al., 2003). Transformation studies will assist us in understanding the *cis*- and *trans*-acting regulation of resistance genes that can help selection for optimum regulatory sequences.

As we understand the molecular basis of genes involved in resistance, we are able to isolate alleles of those genes and then transform crop plants with those genes. Although transformation has its pitfalls, such as varying levels of expression, differences in expression relative to development and disruption of other genetic loci, these problems are less than those for classical cytogenetic transfers where much larger fragments of chromatin are inserted into a genome. As we have discussed, larger pieces of chromatin often come with detrimental genes. Transformation will also be important for verifying the identity of cloned resistance genes (Huang et al., 2003).

To date, disease complexes such as the leaf spot complex in wheat have made breeding for resistance difficult because selection to one disease is confounded by the presence of another disease competing for the same infection sites and tissues. The development of markers to resistance to the individual diseases in the complex will allow breeders to incorporate the resistance without one disease phenotype masking another.

High throughput technologies are expensive and small breeding programmes will not be able to justify the equipment. Breeding programmes will have to wrestle with issues of testing material in centralized rather than at on-site laboratories. Although the central laboratory provides savings through economies of scale by being able to afford high throughput testing equipment, this is at the expense of flexibility and turnaround time for data use. A balance with capacity split between high throughput laboratories and on-site laboratories could provide a compromise that accommodates a greater range of testing options. The bulk of routine testing that does not require rapid turnaround time can be done at a lower cost per sample determination in central laboratories, whereas some less technologically advanced and thus lower investment cost capacity can be maintained at on-site laboratories. Although higher per sample cost is incurred in the latter, it would be more than offset by the specialized, high impact applications of the procedures. For example, testing segregating parents requires rapid turnaround time, as does taking advantage of breeding materials that may be growing for some other purpose such as disease evaluation.

Although information about the pathogen is not the focus of this chapter, the benefits of research on the pathogen must be acknowledged. Our understanding of pathogen population structure and the nature of virulence at the molecular level will be of value in breeding for resistance to fungal pathogens. As in the past, knowledge of changes in the race structure of a population will aid breeders in selecting the best resistance genes for deployment. Molecular genetics will improve our understanding of changes in virulence and the impact of that virulence. Molecular genetics will also help us to understand the nature of virulence genes and how they interact with resistance, to help us select and build crop genotypes with durable resistance.

We struggle to incorporate disease resistance to a constantly changing multitude of pathogens. Managing fungal disease resistance through molecular breeding is gaining momentum as biotechnology provides robust new tools that are being incorporated into breeding programmes. As breeders, pathologists and molecular geneticists work through the growing pains of molecular breeding, a bright future lies ahead for the development of disease-resistant crop cultivars.

Acknowledgements

We thank Dr John Clarke for his careful review and constructive feedback, Brad Meyer for constructing the figures and Aidan Beaubier for literature search and acquisition.

References

Adhikari, T.B., Yang, X., Cavaletto, J.R., Hu, X., Buechley, G., Ohm, H.W., Shaner, G. and Goodwin, S.B. (2004) Molecular mapping of *Stb1*, a potentially durable gene for resistance to septoria tritici blotch in wheat. *Theoretical and Applied Genetics* 109, 944–953.

Agrios, N.G. (1988) *Plant Pathology*, 3rd edn. Academic Press, San Diego.

Anderson, J.A., Stack, R.W., Liu, S., Waldron, B.L., Fjeld, A.D., Coyne, C., Moreno-Sevilla, B., Fetch, J.M., Song, Q.J., Cregan, P.B. and Frohberg, R.C. (2001) DNA markers for fusarium head blight resistance QTLs in two wheat populations. *Theoretical and Applied Genetics* 102, 1164–1168.

Bender, C.L. (2005) The post-genomic era: new approaches for studying bacterial diseases of plants. *Australasian Plant Pathology* 34, 471–474.

Bonnett, D.G., Rebetzke, G.J. and Spielmeyer, W. (2005) Strategies for efficient implementation of molecular markers in wheat breeding. *Molecular Breeding: New Strategies in Plant Improvement* 15, 75–85.

Bossolini, E., Krattinger, S.G. and Keller, B. (2006) Development of simple sequence repeat markers specific for the *Lr34* resistance region of wheat using sequence information from rice and *Aegilops tauschii*. *Theoretical and Applied Genetics* 113, 1049–1062.

Bovill, W.D., Ma, W., Ritter, K., Collard, B.C.Y., Davis, M., Wildermuth, G.B. and Sutherland, M.W. (2006) Identification of novel QTL for resistance to crown rot in the doubled haploid wheat population 'W21MMT70' × 'Mendos'. *Plant Breeding* 125, 538–543.

Cardon, L.R. and Bell, J.I. (2001) Association study designs for complex diseases. *Nature Reviews Genetics* 2, 91–99.

Ciuffetti, L.M. and Tuori, R.P. (1999) Advances in the characterization of the *Pyrenophora tritici-repentis* – wheat interaction. *Phytopathology* 89, 444–449.

Cuthbert, P.A., Somers, D.J., Thomas, J., Cloutier, S. and Brule-Babel, A. (2006) Fine mapping *Fhb1*, a major gene controlling fusarium head blight resistance in bread wheat (*Triticum aestivum* L.). *Theoretical and Applied Genetics* 112, 1465–1472.

Cuthbert, P.A., Somers, D.J. and Brule-Babel, A. (2007) Mapping of *Fhb2* on chromosome 6BS: a gene controlling fusarium head blight field resistance in bread wheat (*Triticum aestivum* L.). *Theoretical and Applied Genetics* 114, 429–437.

Dekkers, C.M. and Hospital, F. (2002) The use of molecular genetics in the improvement of agricultural populations. *Nature Reviews Genetics* 3, 22–32.

Dieguez, M.J., Altieri, E., Ingala, L.R., Perera, E., Sacco, F. and Naranjo, T. (2006) Physical and genetic mapping of amplified fragment length polymorphisms and the leaf rust resistance *Lr3* gene on chromosome 6BL of wheat. *Theoretical and Applied Genetics* 112, 251–257.

Dubcovsky, J. (2004) Marker-assisted selection in public breeding programs: the wheat experience. *Crop Science* 44, 1895–1898.

Dudley, J.W. (1993) Molecular markers in plant improvement: manipulation of genes affecting quantitative traits. *Crop Science* 33, 660–668.

Edwards, K.J. and Mogg, R. (2001) Plant genotyping by analysis of single nucleotide polymorphisms. In: Henry, R.J. (ed.) *Plant Genotyping: The DNA Fingerprinting of Plants*, CAB International, Wallingford, UK.

Falconer, D.S. (1989) *Introduction to Quantitative Genetics*, 3rd edn. Longman, Burnt Mill, UK.

Fjellstrom, R., McClung, A.M. and Shank, A.R. (2006) SSR markers closely linked to the *Pi-z* locus are useful for selection of blast resistance in a broad array of rice germplasm. *Molecular Breeding: New Strategies in Plant Improvement* 17, 149–157.

Gervais, L., Dedryver, F., Morlais, J.Y., Bodusseau, V., Negre, S., Bilous, M., Groos, C. and Trottet, M. (2003) Mapping of quantitative trait loci for field resistance to fusarium head blight in an European winter wheat. *Theoretical and Applied Genetics* 106, 961–970.

Goates, B.J. (1996) Common bunt and dwarf bunt. In: Wilcoxson, R.D. and Saari, E.E. (eds) *Bunt and Smut Diseases of Wheat: Concepts and Methods of Disease Management*, CIMMYT, Mexico.

Gowda, M., Roy-Barman, S. and Chattoo, B.B. (2006) Molecular mapping of a novel blast resistance gene *Pi38* in rice using SSLP and AFLP markers. *Plant Breeding* 125, 596–599.

Groth, J.V., Ozmon, E.A. and Busch, R.H. (1999) Repeatability and relationship of incidence and severity measures of scab of wheat caused by *Fusarium graminearum* in inoculated nurseries. *Plant Disease* 83, 1033–1038.

Guo, P., Bai, G., Li, R., Shaner, G. and Baum, M. (2006) Resistance gene analogs associated with fusarium head blight resistance in wheat. *Euphytica* 151, 251–261.

Gupta, S.K., Charpe, A., Koul, S., Haque, Q.M.R. and Prabhu, K.V. (2006a) Development and validation of scar markers co-segregating with an *Agropyron elongatum* derived leaf rust resistance gene *Lr24* in wheat. *Euphytica* 150, 233–240.

Gupta, S.K., Charpe, A., Prabhu, K.V. and Haque, Q.M.R. (2006b) Identification and validation of molecular markers linked to the leaf rust resistance gene *Lr19* in wheat. *Theoretical and Applied Genetics* 113, 1027–1036.

Hanson, W.D. (1959) Minimum family sizes for the planning of genetic experiments. *Agronomy Journal* 51, 711–715.

Hashizume, F., Nakazaki, T., Tsuchiya, T. and Matsuda, T. (2006) Effectiveness of genotype-based selection in the production of marker-free and genetically fixed transgenic lineages: ectopic expression of a pistil chitinase gene increases leaf-chitinase activity in transgenic rice plants without hygromycin-resistance gene. *Plant Biotechnology* 23, 349–356.

Helguera, M., Vanzetti, L., Soria, M., Khan, I.A., Kolmer, J. and Dubcovsky, J. (2005) PCR markers for *Triticum speltoides* leaf rust resistance gene *Lr51* and their use to develop isogenic hard red spring wheat lines. *Crop Science* 45, 728–734.

Huang, L., Brooks, S.A., Li, W., Fellers, J.P., Trick, H.N. and Gill, B.G. (2003) Map-based cloning of leaf rust resistance gene *Lr21* from the large and polyploid genome of bread wheat. *Genetics* 164, 655–664.

Keen, N.T. (1990) Gene-for-gene complementarity in plant-pathogen interactions. *Annual Review of Genetics* 24, 447–463.

Kerber, E.R. and Dyck, P.L. (1969) Inheritance in hexaploid wheat of leaf rust resistance and other characters derived from *Aegilops squarrosa*. *Canadian Journal of Genetics and Cytology* 11, 639–647.

Klahr, A., Zimmerman, G., Wenzel, G. and Mohler, V. (2007) Effects of environment, disease progress, plant height and heading date on the detection of QTLs for resistance to fusarium head blight in an European winter wheat cross. *Euphytica* 154, 17–28.

Knox, R.E., Fernandez, M.R., Brule-Babel, A.L. and DePauw, R.M. (1998) Inheritance of common bunt resistance in androgenetically derived doubled haploid and random inbred populations of wheat. *Crop Science* 38, 1119–1124.

Knox, R.E., Campbell, H., DePauw, R.M., Clarke, J.M. and Gold, J.J. (2000) Registration of P8810-B5B3A2A2 white-seeded spring wheat germplasm with *Lr35* leaf and *Sr39* stem rust resistance. *Crop Science* 40, 1512–1513.

Knox, R.E., Menzies, J.G., Howes, N.K., Clarke, J.M., Aung, T. and Penner, G.A. (2002) Genetic analysis of resistance to loose smut and an associated DNA marker in durum wheat doubled haploids. *Canadian Journal of Plant Pathology* 24, 316–322.

Kota, R., Spielmeyer, W., McIntosh, R.A. and Lagudah, E.S. (2006) Fine genetic mapping fails to dissociate durable stem rust resistance gene *Sr2* from pseudo-black chaff in common wheat (*Triticum aestivum* L.). *Theoretical and Applied Genetics* 112, 492–499.

Kuchel, H., Ye, G., Fox, R. and Jefferies, S. (2005) Genetic and economic analysis of a targeted marker-assisted wheat breeding strategy. *Molecular Breeding: New Strategies in Plant Improvement* 16, 67–78.

Lagudah, E.S., McFadden, H., Singh, R.P., Huerta-Espino, J., Bariana, H.S. and Spielmeyer, W. (2006) Molecular genetic characterization of the *Lr34/Yr18* slow rusting resistance gene region in wheat. *Theoretical and Applied Genetics* 114, 21–30.

Lee, J., Bricker, T.M., Lefevre, M., Pinson, S.R.M. and Oard, J.H. (2006) Proteomic and genetic approaches to identifying defence-related proteins in rice challenged with the fungal pathogen *Rhizoctonia solani*. *Molecular Plant Pathology* 7, 405–416.

Lynch, M. and Walsh, B. (1998) *Genetics and Analysis of Quantitative Traits*. Sinauer Associates, Sunderland, Massachusetts.

Ma, H.X., Bai, G.H., Zhang, X. and Lu, W.Z. (2006) Main effects, epistasis, and environmental interactions of quantitative trait loci for fusarium head blight resistance in a recombinant inbred population. *Phytopathology* 96, 534–541.

Mateos-Hernandez, M., Singh, R.P., Hulbert, S.H., Bowden, R.L., Huerta-Espino, J., Gill, B.S. and Brown-Guedira, G. (2006) Targeted mapping of ESTs linked to the adult plant resistance gene *Lr46* in wheat using synteny with rice. *Functional and Integrative Genomics* 6, 122–131.

McCartney, C.A., Somers, D.J., Fedak, G. and Cao, W. (2004) Haplotype diversity at fusarium head blight resistance QTLs in wheat. *Theoretical and Applied Genetics* 109, 261–271.

Michelmore, R.W. (1995) Molecular approaches to manipulation of disease resistance genes. *Annual Review of Phytopathology* 15, 393–427.

Michelmore, R.W., Paran, I. and Kesseli, R.V. (1991) Identification of markers linked to disease-resistance genes by bulked segregant analysis: a rapid method to detect markers in specific genomic regions by using segregating populations. *Proceedings of the National Acadamy of Sciences, USA* 88, 9828–9832.

Mohler, V. and Singrun, C. (2004) General considerations: marker-assisted selection. In: Lorz, H. and Wenzel, G. (eds) *Biotechnology in Agriculture and Forestry*, Springer, Berlin/Heidelberg, Germany.

Paull, J.G., Pallotta, M.A., Langridge, P. and The, T.T. (1994) RFLP markers associated with *Sr22* and recombination between chromosome 7A of bread wheat and the diploid species *Triticum boeoticum. Theoretical and Applied Genetics* 89, 1039–1045.

Reyes-Valdés, M.H. (2000) A model for marker-based selection in gene introgression breeding programs. *Crop Science* 40, 91–98.

Rosewarne, G.M., Singh, R.P., Huerta-Espino, J., William, H.M., Bouchet, S., Cloutier, S., McFadden, H. and Lagudah, E.S. (2006) Leaf tip necrosis, molecular markers and 1-proteasome subunits associated with the slow rusting resistance genes *Lr46/Yr29. Theoretical and Applied Genetics* 112, 500–508.

Saari, E.E., Mamluk, O.F. and Burnett, P.A. (1996) Bunts and smuts of wheat. In: Wilcoxson, R.D. and Saari, E.E. (eds) *Bunt and Smut Diseases of Wheat: Concepts and Methods of Disease Management*, CIMMYT, Mexico.

Semagn, K., Bjornstad, A., Skinnes, H., Maroy, A.G., Tarkegne, Y. and William, M. (2006) Distribution of DArT, AFLP, and SSR markers in a genetic linkage map of a doubled-haploid hexaploid wheat population. *Genome* 49, 545–555.

Sharp, P.J., Johnston, S., Brown, G., McIntosh, R.A., Pallota, M., Carter, M., Bariana, H.S., Khatkar, S., Lagudah, E.S., Singh, R.P., Khairallah, M., Potter, R. and Jones, M.G.K. (2001) Validation of molecular markers for wheat breeding. *Australian Journal of Agricultural Research* 52, 1357–1366.

Shen, X., Francki, M.-G. and Ohm, H.-W. (2006) A resistance-like gene identified by EST mapping and its association with a qtl controlling fusarium head blight infection on wheat chromosome 3BS. *Genome* 49, 631–635.

Simon, M.R., Worland, A.J. and Struik, P.C. (2004) Influence of plant height and heading date on the expression of the resistance to *Septoria tritici* blotch in near isogenic lines of wheat. *Crop Science* 44, 2078–2085.

Singh, S., Sidhu, J.S., Huang, N., Vikal, Y., Li, Z., Brar, D.S., Dhaliwal, H.S. and Khush, G.S. (2001) Pyramiding three bacterial blight resistance genes (xa5, xa13 and Xa21) using marker-assisted selection into indica rice cultivar PR106. *Theoretical and Applied Genetics* 102, 1011–1015.

Somers, D.J., Issac, P. and Edwards, K. (2004) A high-density microsatellite consensus map for bread wheat (*Triticum aestivum* L.). *Theoretical and Applied Genetics* 109, 1105–1114.

Somers, D.J., Thomas, J., De Pauw, R., Fox, S., Humphreys, G. and Fedak, G. (2005). Assembling complex genotypes to resist *Fusarium* in wheat (*Triticum aestivum* L.). *Theoretical and Applied Genetics* 111, 1623–1631.

Steiner, B., Lemmens, M., Griesser, M., Scholz, U., Schondelmaier, J. and Buerstmayr, H. (2004) Molecular mapping of resistance to fusarium head blight in the spring wheat cultivar Frontana. *Theoretical and Applied Genetics* 109, 215–224.

Strange, R.N. and Scott, P.R. (2005) Plant disease: a threat to global food security. *Annual Review of Phytopathology* 43, 83–116.

Uauy, C., Brevis, J.C., Chen, X., Khan, I., Jackson, L., Chicaiza, O., Distelfeld, A., Fahima, T. and Dubcovsky, J. (2005) High-temperature adult-plant (HTAP) stripe rust resistance gene *Yr36* from *Triticum turgidum* ssp. *dicoccoides* is closely linked to the grain protein content locus *Gpc-b1. Theoretical and Applied Genetics* 112, 97–105.

Van Sanford, D., Anderson, J., Campbell, K., Costa, J., Cregan, P., Griffey, C., Hayes, P. and Ward, R. (2001) Discovery and deployment of molecular markers linked to fusarium head blight resistance: an integrated system for wheat and barley. *Crop Science* 41, 638–644.

William, H.M., Singh, R.P., Huerta-Espino, J., Palacios, G. and Suenaga, K. (2006) Characterization of genetic loci conferring adult plant resistance to leaf rust and stripe rust in spring wheat. *Genome* 49, 977–990.

Xu, X.Y., Bai, G.H., Carver, B.F., Shaner, G.E. and Hunger, R.M. (2005) Mapping of QTLs prolonging the latent period of *Puccinia triticina* infection in wheat. *Theoretical and Applied Genetics* 110, 244–251.

Yang, Z., Gilbert, J., Fedak, G. and Somers, D.J. (2005) Genetic characterization of QTL associated with resistance to fusarium head blight in a doubled-haploid spring wheat population. *Genome* 48, 187–196.

Young, N.D. (1999) A cautiously optimistic vision for marker-assisted breeding. *Molecular Breeding: New Strategies in Plant Improvement* 5, 505–510.

Zhou, W.C., Kolb, F.L., Bai, G.H., Shaner, G. and Domier, L.L. (2002) Genetic analysis of scab resistance QTL in wheat with microsatellite and AFLP markers. *Genome* 45, 719–727.

15 Protein-mediated Resistance to Plant Viruses

J.F. Uhrig

Abstract

Viruses are a major threat to global agricultural production. Conventional breeding of virus-resistant crops has been successful to some extent, but selection procedures may be too slow to meet the challenges of fast-evolving strains of pathogenic viruses. Therefore, the development of transgenic approaches might be a promising alternative. Based upon recent progress in understanding the molecular basis of viral infections and plant defence mechanisms, protein-mediated resistance concepts have been developed that might fruitfully complement transgenic resistance based on triggering the RNA silencing machinery. While earlier approaches focused mainly on the utilization of viral proteins, current strategies include the antiviral potential of proteins, peptides and antibodies from a variety of different sources. The expression of proteins or peptides exhibiting dominant antiviral effects *in vivo* emerges as a promising tool to engineer broad-spectrum and potentially durable resistance to plant viruses.

Introduction

Viral diseases cause substantial agricultural yield reduction and crop loss. Although difficult to assess, estimates indicate an overall reduction of worldwide agronomic production by several billion US dollars per year due to viral diseases. The impact of particular virus species has been assessed more specifically, and some of them cause considerable damage. Sometimes, single viral disease outbreaks and epidemics have devastating consequences. The losses in cassava production in 2003, for example, due to *Cassava mosaic virus* infection amounted to more than US$1.9 billion, resulting in famine-related deaths on the African continent (Legg and Fauquet, 2004). *Tomato spotted wilt virus* (TSWV), as a second example, ranges among the most detrimental virus species, regularly causing crop losses of approximately US$1 billion/year (Goldbach and Peters, 1996).

So far, no antiviral drugs for application in plant protection have been developed. Also, the experience with chemical control of the insect vectors which spread viral diseases, such as different thrips species which are known to rapidly develop insecticide resistances, is not very encouraging (Morse and Hoddle, 2006). Currently, control of viral pathogens relies to a great extent upon traditional agricultural measures, such as the avoidance of sources of inoculum, control of insect or nematode vectors, modifications of cultural practices and crop rotation. These preventive steps, however, have sometimes unpredictable effectiveness (Varma, 1993). Therefore, the development of virus-resistant crop varieties by conventional breeding or by genetic engineering is a major goal in agronomic research. Traditionally, breeding of virus resistance has been achieved by selection of resistant varieties or by introgression of resistance traits from wild species into crop plants, a strategy that has proven to be successful in many cases (Lecoq et al., 2004). However, classical breeding procedures are rather time-consuming. Given the rapidly changing challenges by fast-evolving pathogenic viruses, the conventional approach might not be sufficient to deal with the increasing threat of viral diseases and the problem of viruses becoming resistant to established control measures (Garcia-Arenal and McDonald, 2003). Furthermore, traits of interest such as disease resistance are frequently coupled to undesirable traits, such as lower yield or reduced stress tolerance, which are not compatible with the requirements of modern agricultural production (Goldbach et al., 2003; Lecoq et al., 2004). Progress in the development of innovative molecular biological techniques, along with improvements in the genetic transformation of crop plants, has paved the way for the genetic engineering of virus resistance. Furthermore, recent advances in understanding the molecular basis of plant diseases, of viral replication cycles and infection mechanisms, and of plant antiviral defence mechanisms provide the basis for novel, knowledge-based approaches to engineer virus-resistant plants (Chapters 4 and 19, this volume).

Initially, attempts to achieve transgenic resistance to plant viruses were based upon the proposition that it may be possible to interfere with the development of a pathogen by causing the host to synthesize a pathogen gene product, RNA or protein (Sanford and Johnston, 1985). Subsequently, insights into the molecular basis of such 'pathogen-derived resistance' approaches revealed that, in many instances, the viral proteins were not required, but rather, the presence of RNA was sufficient to induce resistance (see Chapter 16, this volume). However, other cases of transgenic resistance did require the accumulation of the viral gene products, either as functional proteins or as dysfunctional polypeptide. RNA-mediated resistance is based on activating the endogenous RNA silencing machinery and has been successfully applied to engineer virus-resistant plants in recent years (Goldbach et al., 2003). While often being very strong, resistance based on RNA silencing relies on exactly matching stretches of nucleotide sequences. Resistance is therefore generally limited to a narrow range of closely related viruses. Protein-mediated resistance, on the other hand, although molecularly much less understood and in its strength often less pronounced, in many cases confers broad-spectrum resistance to a variety of different and even unrelated virus species (Pang et al., 1993; Vaira et al., 1995; Tacke et al., 1996). Going beyond the pathogen-derived resistance concept, with respect to the expression of dominant interfering molecules, a diverse range of promising innovative strategies are being developed, and peptides, antibodies, plant resistance proteins or proteins with antiviral activities from different sources have successfully been used to achieve protein-mediated transgenic virus resistance.

Pathogen-derived Protein-mediated Resistance

Expression of viral proteins in plants can result in different effects, ranging from the induction of disease symptoms and increased susceptibility to virus infections, to effective resistance and even immunity (Abel et al., 1986; Deom et al., 1990; Beachy, 1997; Prins et al., 1997a). This variability in the outcome of protein-mediated resistance has somewhat slowed down the use and further development of this strategy and has led to a relatively limited importance of protein-mediated resistance compared to RNA-mediated resistance (Goldbach et al., 2003). However, novel approaches in devising more knowledge-based concepts might stimulate the use of protein-mediated strategies in the future (Prins, 2003; Rudolph et al., 2003; Uhrig, 2003).

The first example of pathogen-derived resistant plants, the coat protein-mediated resistance utilizing the *Tobacco mosaic virus* (TMV) capsid protein, represents one of the few examples that are mechanistically well investigated. Here, the transgenically expressed TMV capsid protein interferes with the disassembly of TMV particles shortly after entrance/penetration into the cell (Beachy, 1997). Disassembly is an essential early process in positive-sense single-stranded ssRNA viruses, because the viral genomic RNA directly serves as a template for protein translation by the plant ribosomes. Analysis of mutant forms of the TMV coat protein revealed that the mechanism underlying the resistance phenotype relies on the protein–protein interaction properties of the protein (Bendahmane et al., 1997). The wild-type TMV coat protein can induce weak resistance, while a mutant coat protein with decreased ability to form homopolymers did not mediate resistance to TMV. However, expression of a mutant protein exhibiting stronger homotypic interactions than the wild-type protein conferred high-level TMV resistance to the transgenic plants (Bendahmane et al., 1997). This example shows that in protein-mediated resistance, subtle structural and functional properties of the viral proteins might be decisive for success or failure of a transgenic resistance concept. This provides a possible explanation for the very ambiguous and unpredictable success in engineering protein-mediated resistance. Recent results indicate that the TMV coat protein not only interferes with capsid disassembly, but also influences movement protein accumulation and cell-to-cell spread of the infection (Bendahmane *et al.*, 2002). Here again, a mutant of the coat protein had a more pronounced effect than the wild-type protein. Indeed, transgenic expression of wild-type coat protein had a positive effect on the production of TMV movement protein, whereas the mutant form reduced movement protein accumulation and interfered with its function, resulting in high levels of resistance to TMV infection in plants (Bendahmane *et al.*, 2002).

Transgenic expression of different mutants of the TMV coat protein was found not only to confer resistance to TMV infection, but also to interfere with the disease development of other non-related viruses (Anderson *et al.*, 1989). In a recent study, the reciprocal ability of TMV and potato virus X (PVX) coat proteins to confer heterologous resistance has been demonstrated, and it was suggested that the quaternary structure of the TMV coat protein is critical for heterologous coat protein-mediated resistance (Bazzini *et al.*, 2006). The fact that in TMV coat protein-mediated resistance the interaction properties of TMV capsid protein are of central importance might represent a common theme in protein-mediated resistance and potentially, detailed knowledge of the molecular interactions among viral components will be instrumental to devise novel resistance strategies in the future.

In recent years, TSWV has emerged as a model system for negative/ambisense RNA plant viruses and several transgenic strategies have been developed to control TSWV (Pang *et al.*, 1993; Vaira *et al.*, 1995; Rudolph *et al.*, 2003; Prins *et al.*, 2005). In accordance with the general notion of multifunctionality of plant viral capsid proteins, the TSWV nucleocapsid protein (N-protein) has a central position in the coordination of the viral life cycle (Uhrig *et al.*, 1999; Callaway *et al.*, 2001). The TSWV N-protein is a basic protein with sequence non-specific RNA binding properties (Richmond *et al.*, 1998). Specificity for the pseudocircular TSWV genomic RNA might be achieved by a specific affinity to an RNA secondary structure conserved in the 5' regions of tospoviral genomic RNAs (Uhrig, 2005). The TSWV N-protein is able to perform a number of molecular interactions, including binding to the viral RNA, homopolymerization and a physical interaction with NSm, the TSWV movement protein. Furthermore, there are indications that it directly interacts with and regulates the viral RNA-dependent RNA polymerase (RdRp). It may also be in direct or indirect contact with G1/G2, the glycoproteins inserted in the envelope membrane of the virion (Mumford *et al.*, 1996; Richmond *et al.*, 1998; Uhrig *et al.*, 1999; Soellick *et al.*, 2000). A central feature of the N-protein is its ability to form homopolymeric structures that are thought to be the basis for cooperative binding and packaging of the viral tripartite genomic RNA (Uhrig *et al.*, 1999). The minimal infectious TSWV particle consists of the segmented genomic RNA, packaged by multiple copies of the N-protein and associated with a few molecules of the viral RdRp. In addition to packaging and protection of the viral RNA and regulatory functions in transcription, replication and budding, TSWV N-protein directly interacts with the viral movement protein NSm and might therefore have additional functions in the process of intercellular movement (Soellick et al., 2000).

The TSWV N-gene has successfully been used to engineer virus-resistant plants. However, the strength and breadth of resistance was rather variable. Strong resistance was found to be associated with a very narrow range of resistance, being effective solely to TSWV. This strong resistance is based on RNA-mediated mechanisms triggering silencing of the viral N-gene (de Haan et al., 1992; Pang et al., 1993; Vaira et al., 1995). Broad-spectrum N-gene-mediated resistance has also been described, but the strength of resistance was to a great extent variable and rather unpredictable (Pang et al., 1993). However, recent advances in understanding the molecular basis of N-protein function provided insights into the molecular mechanism of N-protein-mediated resistance. In-depth analysis of the protein interaction properties of the TSWV N-protein revealed that homopolymerization was based upon two protein domains localized at the N- and the C-terminus of the protein, respectively. Two evolutionary conserved hydrophobic residues in the C-terminal interaction domain have been identified that contribute mainly to the strength of the homotypic interaction (Uhrig et al., 1999).

There is only limited overall sequence conservation of the N-proteins within the tospovirus family. Nevertheless, the position, function and specificity of the domains mediating homopolymerization are evolutionary conserved. A systematic analysis of interaction specificities revealed two functional classes of N-proteins from tospoviral species. Members of each class are able to interact with one another, while interaction with N-proteins from the other class is not possible (Uhrig, 2006). Constitutive expression of TSWV N-protein in *Nicotiana tabacum* and *Nicotiana benthamiana* resulted in protein-mediated resistance to TSWV. A careful analysis of the degree of resistance to TSWV and a number of different tospovirus species in individual transgenic lines revealed that the strongest and most reliable resistance was observed when the plants were challenged with quite distantly related tospoviruses expressing N-proteins that are not able to interact with the TSWV N-protein (Uhrig, 2006). These findings suggest a dominant negative mechanism based on differential abilities to perform molecular interactions and the competition for RNA- and/or protein-binding sites (Uhrig, 2006). Dominant negative interference as a possible approach to achieve virus resistance in plants has been developed further. Based on this concept, a peptide-based strategy, using peptide 'aptamers' has been employed successfully to engineer broad-spectrum tospovirus resistance in plants (Rudolph et al., 2003; see below).

In addition to viral coat proteins, there are a few examples of successful protein-mediated resistance conferred by the transgenic expression of other viral proteins. Movement proteins seem to be effective as antiviral agents only as dysfunctional molecules, while expression of functional forms increase susceptibility (Cooper et al., 1995). Transgenic *Nicotiana tabacum* plants expressing different mutant forms of the TMV movement protein were partially resistant to TMV and to several other viruses from different taxonomic groups (Lapidot et al., 1993; Malyshenko et al., 1993; Cooper et al., 1995). This property is not restricted to the TMV movement protein. Expression of a mutant form of Potato leaf roll virus (PLRV) movement protein, for example, conferred broad-range protection of potato plants against virus infection. Potato lines accumulating N- or C-terminally extended PLRV movement proteins were resistant to infection by the unrelated Potato virus Y (PYV) and Potato virus X (Tacke et al., 1996). Similarly, transgenic tomato plants expressing wild-type or mutated movement proteins from Bean dwarf mosaic virus (BDMV) resulted in a significant delay in infection by Tomato mottle virus (ToMoV), a related geminivirus (Hou et al., 2000). However, expression of viral movement proteins can have deleterious effects on various aspects of plant growth and development, potentially limiting the use of this class of proteins to engineer protein-mediated resistance in plants (Prins et al., 1997b; Hou et al., 2000).

Peptide-mediated Virus Resistance

Insufficient insight into the molecular mechanisms underlying protein-mediated virus resistance has prevented the rational design of effective antiviral molecules so far. Thus, protein-mediated resistance projects are largely done by trial and error. However, some recent results might be helpful for the development of more knowledge-based approaches. In a number of reported cases of protein-mediated virus resistance, it has been observed that mutated or truncated forms of the respective viral protein expressed in plant cells are far more effective than the native forms (Longstaff et al., 1993; Tacke et al., 1996). One possible general explanation for this phenomenon might be related to the dominant negative concept originally put forward by Ira Herskowitz (Herskowitz, 1987; Longstaff et al., 1993; Rudolph et al., 2003; Uhrig, 2003). The dominant negative concept is based on a structural and functional modularity and multifunctionality of proteins. Essential for the function of most, if not all, proteins are interactions with other cellular components, other proteins, nucleic acids or lipids. Overexpression of mutant polypeptides lacking a functionally essential domain might therefore outcompete endogenous proteins for binding sites and disrupt their activity (Herskowitz, 1987). This idea might lend itself to the development of rationally designed molecules that disrupt vital functions of plant viral proteins in a dominant manner in vivo, thereby producing virus resistance.

Peptides or protein microdomains designed to block protein-binding sites put the dominant negative concept to the extreme. These peptide aptamers have been used to specifically inhibit a wide variety of protein functions in vivo and in vitro (Colas et al., 1996; Norman et al., 1999; Geyer and Brent, 2000). In pharmaceutical research, peptide aptamers are increasingly recognized as a promising tool for drug target identification and their validation *in vivo* (Hoppe-Seyler and Butz, 2000; Burgstaller *et al.*, 2002; Troitskaya and Kodadek, 2004). Applying this rather novel strategy led to the identification of peptide inhibitors of specific protein functions, of signal transduction and of metabolic pathways. Furthermore, antibiotic, anti-inflammatory and antiviral activities of peptide aptamers have been demonstrated (Butz *et al.*, 2000, 2001; Tao *et al.*, 2000). In multicellular organisms, application of peptide aptamers as specific dominant inhibitors of protein functions *in vivo*, has so far been restricted to a 'proof-of-principle' experiment in *Drosophila* and to the recent engineering of virus resistance in tobacco (Kolonin and Finley, 1998; Rudolph *et al.*, 2003).

The yeast two-hybrid system offers a rather straightforward experimental way of selecting peptides with specific binding properties for use as dominant inhibitors of viral protein functions *in vivo* (Colas *et al.*, 1996). In contrast to *in vitro* methods, such as the phage display technology, the yeast two-hybrid method allows their selection in yeast cells, warranting that they can be expressed as functional molecules in the target cells (Colas *et al.*, 1996; Cohen *et al.*, 1998; Norman *et al.*, 1999; Rudolph *et al.*, 2003). Peptide aptamers selected with the yeast two-hybrid system have been found to exhibit very high target specificities and binding affinities with K_D-values between 10^{-6} and 10^{-11} M. Such affinities are comparable to the properties of specific antibodies (Colas *et al.*, 1996; Geyer and Brent, 2000; Troitskaya and Kodadek, 2004).

The aptamer strategy has been applied to engineer virus-resistant plants (Rudolph *et al.*, 2003). As mentioned above, the TSWV N-protein plays a central role in the viral replication cycle (Richmond *et al.*, 1998; Uhrig *et al.*, 1999; Soellick *et al.*, 2000; Kellmann, 2001). Dissection of N-protein domain mediating its homomultimerization allowed the isolation of microdomains of the N-protein that specifically and strongly interact with one of the N-protein homodimerization domains (Uhrig *et al.*, 1999; Rudolph *et al.*, 2003). A minimal peptide exhibiting this binding property is 29 amino acids in length. Semi-quantitative assays revealed that the peptide aptamer binds to the N-protein interaction domain with a higher affinity than the full-length N-protein itself. Furthermore, this peptide aptamer interacts strongly with a number of nucleocapsid proteins from different tospoviruses, indicating that it targets a domain conserved among several tospoviral species (Rudolph *et al.*, 2003). Stable expression of this peptide fused to a carrier protein in transgenic *Nicotiana tabacum* and *N. benthamiana* led to strong resistance not only to TSWV but also to five different tospoviral species, consistent with the interaction specificities of the peptide aptamer (Rudolph *et al.*, 2003). This first example of a peptide-based approach to engineer virus resistance might pave the way for a more directed and knowledge-based development of molecules dominantly interfering with viral infections in plants. Very recently, the identification of peptide aptamers targeting the replication protein of *tomato golden mosaic virus* (TGMV), a member of the geminivirus family, has been reported (Lopez-Ochoa *et al.*, 2006). Using a library of random peptides inserted in a flexible loop of thioredoxin, several binding peptides have been isolated by yeast two-hybrid screening, and these peptides interfered with viral replication in plant cells. Such peptides represent promising tools to develop a resistance strategy to control geminiviruses.

Virus Resistance Based on Intracellular Antibodies

Similar to the rationale behind the peptide aptamer approach, antibodies can be envisioned binding to and inactivating a virtually unlimited variety of viral target proteins. Antibody-based resistance might emerge as a flexible and versatile tool to protect crop plants. This strategy, again, has greatly benefited from the advances in understanding the molecular basis of plant diseases and from the identification of viral proteins essential for infection, replication, systemic movement and pathogenicity.

Intracellular expression of hybridoma-derived single-chain variable antibody fragments (scFv) is successfully being applied in gene therapy to control human viruses, and promising results have recently been obtained with this strategy in plants. Already more than 10 years ago, evidence was provided that it is possible to reduce virus accumulation and to delay disease symptom development by the expression of scFvs (Tavladoraki *et al.*, 1993). However, for a rather long time, only very few examples of partial resistance or suppression of virus symptoms have been reported (Fecker *et al.*, 1996; Schillberg *et al.*, 1999). One possible impediment to the progress of this technology is the failure of most antibodies to properly fold under the reducing conditions in the cytoplasm (Schouten *et al.*, 2002). None the less, several groups succeeded recently in engineering high levels of resistance to different plant viruses by expressing specific single-chain antibodies selected with the phage display technology (Boonrod *et al.*, 2004; Prins *et al.*, 2005; Villani *et al.*, 2005). Single-chain antibodies targeting a conserved domain of a plant viral RdRp inhibited complementary RNA synthesis of different RdRps *in vitro*, and upon transgenic expression in *Nicotiana benthamiana* plants, conferred resistance to four different plant viruses (Boonrod *et al.*, 2004). In the second example, scFvs were selected for specific binding to the TSWV N-protein (Prins *et al.*, 2005). Functional expression of scFvs directed against tospoviral proteins in plant cells has been reported already some time ago, and it was proposed that such an approach could be used to engineer plants resistant to tospoviruses (Franconi *et al.*, 1999). However, it was not until recently that this approach has proven successful to engineer tospovirus resistance. Expression of the scFvs in *N. benthamiana* plants resulted in high levels of resistance against TSWV (Prins *et al.*, 2005).

In another approach to engineer virus resistance, two scFvs specifically targeting the *Cucumber mosaic virus* (CMV) virion have been selected *in vitro* using the phage display technology (Villani *et al.*, 2005). Upon transgenic expression, both scFvs were found to accumulate as soluble proteins in the cytoplasm, and a tomato line fully resistant to CMV infection has been identified. In these plants, the scFv binds the virus in the inoculated leaves and prevents long distance viral movement.

Antibodies targeting whole viruses/virions, nucleocapsid proteins or RdRp can be used to engineer virus-resistant plants, indicating that a variety of different antigens and viral targets may be suitable for this strategy. After a long period of technological developments, antibody-based strategies may now become a promising tool complementing the available antiviral resistance strategies.

Transgenic Expression of Proteins with Antiviral Activity

The approaches to engineer virus-resistant plants discussed so far aim at directly binding to and competing with viral structures. A number of different strategies have been developed that make use of the antiviral properties of proteins that have evolved in plants and animals, facilitating more specific ways to resist virus infection. A number of plant proteins with antiviral activity have been identified that target and inhibit different essential steps of viral replication cycles. A class of proteins commonly termed ribosome-inactivating proteins (RIPs) and protease inhibitors have been exploited using transgenic technologies to engineer resistance in crop plants. Furthermore, the interferon-regulated 2-5A system used in higher vertebrates as a defence against virus infection has successfully been transferred to crop plants (Mitra *et al.*, 1996).

RIPs are produced by a number of different plant species and their antiviral activity is well documented (Wang and Tumer, 2000; Park *et al.*, 2004). Generally, this class of proteins inhibits the translocation step of translation by catalytically removing a specific adenine base from 28S ribosomal RNA (Fong *et al.*, 1991). To prevent damage of the endogenous 28S rRNA, RIPs are synthesized as precursor proteins and targeted to the vacuole. It has been suggested that the RIPs enter the cytoplasm together with the virus and then exert their inhibiting function in the infected cell. A number of genes from different plant species encoding RIPs have been cloned and used to transform crop plants. In several cases, the transgenically expressed RIPs are apparently functional and the transformed plants exhibit broad-spectrum virus resistances (Wang and Tumer, 2000). *Nicotiana benthamiana* plants, for example, expressing PAP, a RIP from pokeweed, were resistant to a wide range of viruses, both mechanically inoculated and transmitted by aphids (Lodge *et al.*, 1993). In another example, Dianthin, a RIP isolated from *Dianthus caryophyllus*, was expressed in *N. benthamiana* from a transactivatable geminivirus promoter, and was found to inhibit virus replication upon inoculation with the respective geminivirus (Hong *et al.*, 1996). Recent results indicate that some RIPs can be applied exogenously and play a role in inducing systemic acquired resistance. External application of purified beetin, a single-chain ribosome-inactivating protein from *Beta vulgaris*, to sugarbeet leaves prevented infection by *Artichoke mottle crinkle virus* (Iglesias *et al.*, 2005).

There is a strong evolutionary pressure for viruses to restrict the size of their genomes. Therefore, many viruses produce polyproteins that are subsequently processed by specific proteases to release the active proteins. Cysteine proteases play an essential role in the replication cycles of many viruses. Consequently, the expression of cysteine protease inhibitors (cystatins) in plants was envisioned to lead to virus resistance. A proof of this concept has been achieved by transgenic expression of a cysteine protease inhibitor gene from rice in tobacco. Transgenic plants were shown to be resistant to *Tobacco etch virus* (TEV) and PVY, two viruses dependent on processing of polyprotein precursers. Consistently, no resistance against TMV infection has been found, for this virus does not rely on polyprotein processing for its propagation (Gutierrez-Campos et al., 1999).

Recent evidence suggests a further activity of cysteine protease inhibitors. Celostatin, a cystatin from *Celosia cristata*, suppressed TMV-induced hypersensitive cell death in *Nicotiana glutinosa*, and it was therefore speculated that this protein could be instrumental in engineering resistance to protease-free viruses (Gholizadeh et al., 2005). However, this needs to be proven experimentally. Furthermore, pleiotropic effects of the overexpression of a cystatin gene in transgenic tobacco could indicate that a general use of cysteine protease inhibitors in plants may be limited (Gutierrez-Campos et al., 2001).

A rather different idea of engineering virus-resistant plants has been followed with the transfer of the 2-5A system, consisting of 2-5A synthetase and the 2-5A-dependent RNAse L, to plants. Higher vertebrates utilize this interferon-regulated RNA degradation system as a defence against virus infection. Co-expression of the human enzymes in transgenic tobacco plants led to the production of functional 2-5A synthetase and activated RNAse L, and these transgenic plants were found to be resistant to three different types of viruses (Mitra et al., 1996; Ogawa et al., 1996). Recently, it has been shown that resistance to CMV in tobacco expressing the 2-5A system may be associated with the establishment of a hypersensitive response (HR), a common early event in natural virus resistance in plants (see below) (Honda et al., 2003).

Transgenic Resistance based on Natural Resistance Genes

Plants succeed in combating infection without adaptive immunity and without a circulating antibody system. A rather complex picture is emerging of how plants recognize pathogens and mobilize their cellular defence machineries. Interestingly, recent advance in unravelling the molecular mechanisms of resistance gene (R-gene) function revealed a striking similarity between the components of the innate immune system in animals and R-proteins with their associated signalling molecules (Holt et al., 2003). The function of R-genes has for a long time attracted much interest, and both conventional breeding efforts and transgenic strategies aim at the utilization of this efficient resistance mechanism (McDowell and

Woffenden, 2003). In fact, most pathogen resistant crop plant varieties obtained by classical breeding strategies, such as selection for specific traits or introgression of the respective characters from wild species, are protected by R-gene-dependent mechanisms (Kang et al., 2005). A large number of R-genes have been identified and characterized, conferring resistance to a diverse range of pathogens, including viruses, bacteria, fungi and even nematodes.

Upon recognition of the pathogen, R-proteins trigger two different responses. The first is a local event called HR. Early processes in the HR include changes in ion fluxes, activation of kinase cascades, the generation and release of reactive oxygen species, and the production of nitric oxide (Nimchuk et al., 2003; Belkhadir et al., 2004). These immediate changes are followed by the recruitment of hormones participating in defence. Often, the HR eventually leads to programmed cell death of the infected cells, phenotypically manifested by necrotic lesions at the site of infection. Viruses are usually confined to the lesions and are thus prevented from spreading into the neighbouring cells. The second R-protein-mediated response to pathogen attack is initiation of the so-called systemic acquired resistance, an inducible defence mechanism mediating increased resistance or even immunity to pathogens in tissues that are distant from the initial infection site.

Over the past 10 years, more than 40 R-genes have been cloned, several of which confer resistance to plant viruses (see Chapter 17, this volume). However, despite the large number of genes available and despite all efforts, there are only a few examples of successful uses of R-genes in transgenic virus resistance. The N-gene, which confers resistance to TMV in tobacco, has been cloned and expressed heterologously in transgenic tomato plants. The transgenic lines exhibited marked resistance against TMV, providing evidence that the N-gene retains its effectiveness to initiate HR in a heterologous system. This was the first example demonstrating the use of R-genes in providing transgenic protection in a crop plant (Whitham et al., 1996). Other examples include the Rx-gene, conferring resistance to PVX in potato that has been used to engineer virus-resistant *N. benthamiana* plants (Bendahmane et al., 1999), and the Tm-2(2) gene from tomato that has been transferred to tobacco, resulting in durable resistance against *Tomato mosaic virus* (Lanfermeijer et al., 2003).

R-gene-mediated resistance appears to be an evolutionary very successful and durable antiviral strategy in the plant kingdom. Potential drawbacks, however, for the application of R-genes in engineering virus resistance may be that there are only a limited number of natural resistance genes available, and that so far, it has been possible to functionally transfer R-genes only between rather closely related plant species.

Conclusions and Future Directions

The major goal in both conventional breeding of virus-resistant crops and transgenic approaches to engineer virus resistance is to achieve broad-spectrum, strong and durable protection against viral infections. Despite the fact that recent advances in the molecular understanding of viral infection cycles and host defences have made possible considerable progress in developing novel strategies and concepts to control viruses, there are still some major obstacles and difficulties limiting the success of these approaches. For example, there are still technical problems in developing suitable transformation protocols for many crop plants. Additionally, there are limitations intrinsic to each antiviral strategy applied. Novel approaches to engineer protein-mediated resistance have greatly benefited from recent advances in understanding the molecular basis of viral infections and plant defence strategies. However, ultimately, combinations of different forms of transgenic resistance might be necessary to succeed in combating fast-evolving pathogens and achieving broad-spectrum, high level and durable resistance of crop plants.

References

Abel, P.P., Nelson, R.S., De, B., Hoffmann, N., Rogers, S.G., Fraley, R.T. and Beachy, R.N. (1986) Delay of disease development in transgenic plants that express the tobacco mosaic virus coat protein gene. *Science* 232, 738–743.

Anderson, E.J., Stark, D.M., Nelson, R.S., Powell, P.A., Tumer, N.E. and Beachy, R.N. (1989) Transgenic plants that express the coat protein genes of tobacco mosaic virus or alfalfa mosaic virus interfere with disease development of some nonrelated viruses. *Phytopathology* 79, 1284–1290.

Bazzini, A.A., Asurmendi, S., Hopp, H.E. and Beachy, R.N. (2006) Tobacco mosaic virus (TMV) and potato virus X (PVX) coat proteins confer heterologous interference to PVX and TMV infection, respectively. *Journal of General Virology* 87, 1005–1012.

Beachy, R.N. (1997) Mechanisms and applications of pathogen-derived resistance in transgenic plants. *Current Opinion in Biotechnology* 8, 215–220.

Belkhadir, Y., Subramaniam, R. and Dangl, J.L. (2004) Plant disease resistance protein signaling: NBS–LRR proteins and their partners. *Current Opinion in Plant Biology* 7, 391–399.

Bendahmane, M., Fitchen, J.H., Zhang, G. and Beachy, R.N. (1997) Studies of coat protein-mediated resistance to tobacco mosaic tobamovirus: correlation between assembly of mutant coat proteins and resistance. *Journal of Virology* 71, 7942–7950.

Bendahmane, A., Kanyuka, K. and Baulcombe, D.C. (1999) The Rx gene from potato controls separate virus resistance and cell death responses. *The Plant Cell* 11, 781–792.

Bendahmane, M., Szecsi, J., Chen, I., Berg, R.H. and Beachy, R.N. (2002) Characterization of mutant tobacco mosaic virus coat protein that interferes with virus cell-to-cell movement. *Proceedings of the National Academy of Sciences of the United States of America* 99, 3645–3650.

Boonrod, K., Galetzka, D., Nagy, P.D., Conrad, U. and Krczal, G. (2004) Single-chain antibodies against a plant viral RNA-dependent RNA polymerase confer virus resistance. *Nature Biotechnology* 22, 856–862.

Burgstaller, P., Girod, A. and Blind, M. (2002) Aptamers as tools for target prioritization and lead identification. *Drug Discovery Today* 7, 1221–1228.

Butz, K., Denk, C., Ullmann, A., Scheffner, M. and Hoppe-Seyler, F. (2000) Induction of apoptosis in human papillomaviruspositive cancer cells by peptide aptamers targeting the viral E6 oncoprotein. *Proceedings of the National Academy of Sciences of the United States of America* 97, 6693–6697.

Butz, K., Denk, C., Fitscher, B., Crnkovic-Mertens, I., Ullmann, A., Schroder, C.H. and Hoppe-Seyler, F. (2001) Peptide aptamers targeting the hepatitis B virus core protein: a new class of molecules with antiviral activity. *Oncogene* 20, 6579–6586.

Callaway, A., Giesman-Cookmeyer, D., Gillock, E.T., Sit, T.L. and Lommel, S.A. (2001) The multifunctional capsid proteins of plant RNA viruses. *Annual Review of Phytopathology* 39, 419–460.

Cohen, B.A., Colas, P. and Brent, R. (1998) An artificial cell-cycle inhibitor isolated from a combinatorial library. *Proceedings of the National Academy of Sciences of the United States of America* 95, 14272–14277.

Colas, P., Cohen, B., Jessen, T., Grishina, I., McCoy, J. and Brent, R. (1996) Genetic selection of peptide aptamers that recognize and inhibit cyclin-dependent kinase 2. *Nature* 380, 548–550.

Cooper, B., Lapidot, M., Heick, J.A., Dodds, J.A. and Beachy, R.N. (1995) A defective movement protein of TMV in transgenic plants confers resistance to multiple viruses whereas the functional analog increases susceptibility. *Virology* 206, 307–313.

de Haan, P., Gielen, J.J., Prins, M., Wijkamp, I.G., van Schepen, A., Peters, D., van Grinsven, M.Q. and Goldbach, R. (1992) Characterization of RNA-mediated resistance to tomato spotted wilt virus in transgenic tobacco plants. *Biotechnology* 10, 1133–1137.

Deom, C.M., Schubert, K.R., Wolf, S., Holt, C.A., Lucas, W.J. and Beachy, R.N. (1990) Molecular characterization and biological function of the movement protein of tobacco mosaic virus in transgenic plants. *Proceedings of the National Academy of Sciences of the United States of America* 87, 3284–3288.

Fecker, L.F., Kaufmann, A., Commandeur, U., Commandeur, J., Koenig, R. and Burgermeister, W. (1996) Expression of single-chain antibody fragments (scFv) specific for beet necrotic yellow vein virus coat protein or 25 kDa protein in *Escherichia coli* and *Nicotiana benthamiana*. *Plant Molecular Biology* 32, 979–986.

Fong, W.P., Wong, R.N., Go, T.T. and Yeung, H.W. (1991) Minireview: enzymatic properties of ribosome-inactivating proteins (RIPs) and related toxins. *Life Science* 49, 1859–1869.

Franconi, R., Roggero, P., Pirazzi, P., Arias, F.J., Desiderio, A., Bitti, O., Pashkoulov, D., Mattei, B., Bracci, L., Masenga, V., Milne, R.G. and Benvenuto, E. (1999) Functional expression in bacteria and plants of an scFv antibody fragment against tospoviruses. *Immunotechnology* 4, 189–201.

Garcia-Arenal, F. and McDonald, B.A. (2003) An analysis of the durability of resistance to plant viruses. *Phytopathology* 93, 941–952.

Geyer, C.R. and Brent, R. (2000) Selection of genetic agents from random peptide aptamer expression libraries. *Methods in Enzymology* 328, 171–208.

Gholizadeh, A., Santha, I.M., Kohnehrouz, B.B., Lodha, M.L. and Kapoor, H.C. (2005) Cystatins may confer viral resistance in plants by inhibition of a virus-induced cell death phenomenon in which cysteine proteinases are active: cloning and molecular characterization of a cDNA encoding cysteine-proteinase inhibitor (celostatin) from *Celosia cristata* (crested cock's comb). *Biotechnology and Applied Biochemistry* 42, 197–204.

Goldbach, R. and Peters, D. (1996) Molecular and biological aspects of tospoviruses. In: Elliot, R.M. (ed.) *The Bunyaviridae*. Plenum, New York, pp. 129–157.

Goldbach, R., Bucher, E. and Prins, M. (2003) Resistance mechanisms to plant viruses: an overview. *Virus Research* 92, 207–212.

Gutierrez-Campos, R., Torres-Acosta, J.A., Saucedo-Arias, L.J. and Gomez-Lim, M.A. (1999) The use of cysteine proteinase inhibitors to engineer resistance against potyviruses in transgenic tobacco plants. *Nature Biotechnology* 17, 1223–1226.

Gutierrez-Campos, R., Torres-Acosta, J., Perez-Martinez, J.D. and Gomez-Lim, M.A. (2001) Pleiotropic effects in transgenic tobacco plants expressing the oryzacystatin I gene. *Hortscience* 36, 118–119.

Herskowitz, I. (1987) Functional inactivation of genes by dominant negative mutations. *Nature* 329, 219–222.

Holt, B.F., Hubert, D.A. and Dangl, J.L. (2003) Resistance gene signaling in plants – complex similarities to animal innate immunity. *Current Opinion in Immunology* 15, 20–25.

Honda, A., Takahashi, H., Toguri, T., Ogawa, T., Hase, S., Ikegami, M. and Ehara, Y. (2003) Activation of defense-related gene expression and systemic acquired resistance in cucumber mosaic virus-infected tobacco plants expressing the mammalian 2' 5' oligoadenylate system – brief report. *Archives of Virology* 148, 1017–1026.

Hong, Y., Saunders, K., Hartley, M.R. and Stanley, J. (1996) Resistance to geminivirus infection by virus-induced expression of dianthin in transgenic plants. *Virology* 220, 119–127.

Hoppe-Seyler, F. and Butz, K. (2000) Peptide aptamers: powerful new tools for molecular medicine. *Journal of Molecular Medicine* 78, 426–430.

Hou, Y.M., Sanders, R., Ursin, V.M. and Gilbertson, R.L. (2000) Transgenic plants expressing geminivirus movement proteins: abnormal phenotypes and delayed infection by tomato mottle virus in transgenic tomatoes expressing the bean dwarf mosaic virus BV1 or BC1 proteins. *Molecular Plant-Microbe Interactions* 13, 297–308.

Iglesias, R., Perez, Y., de Torre, C., Ferreras, J.M., Antolin, P., Jimenez, P., Rojo, M.A., Mendez, E. and Girbes, T. (2005) Molecular characterization and systemic induction of single-chain ribosome-inactivating proteins (RIPs) in sugarbeet (*Beta vulgaris*) leaves. *Journal of Experimental Botany* 56, 1675–1684.

Kang, B.C., Yeam, I. and Jahn, M.M. (2005) Genetics of plant virus resistance. *Annual Review in Phytopathology* 43, 581–621.

Kellmann, J.W. (2001) Identification of plant virus movement–host protein interactions. *Zeitschrift für Naturforschung [C]* 56, 669–679.

Kolonin, M.G. and Finley, R.L., Jr. (1998) Targeting cyclin-dependent kinases in *Drosophila* with peptide aptamers. *Proceedings of the National Academy of Sciences of the United States of America* 95, 14266–14271.

Lanfermeijer, F.C., Dijkhuis, J., Sturre, M.J., de Haan, P. and Hille, J. (2003) Cloning and characterization of the durable tomato mosaic virus resistance gene Tm-2(2) from *Lycopersicon esculentum*. *Plant Molecular Biology* 52, 1037–1049.

Lapidot, M., Gafny, R., Ding, B., Wolf, S., Lucas, W.J. and Beachy, R.N. (1993) A dysfunctional movement protein of tobacco mosaic virus that partially modifies the plasmodesmata and limits virus spread in transgenic plants. *The Plant Journal* 4, 959–970.

Lecoq, H., Moury, B., Desbiez, C., Palloix, A. and Pitrat, M. (2004) Durable virus resistance in plants through conventional approaches: a challenge. *Virus Research* 100, 31–39.

Legg, J.P. and Fauquet, C.M. (2004) Cassava mosaic geminiviruses in Africa. *Plant Molecular Biology* 56, 585–599.

Lodge, J.K., Kaniewski, W.K. and Tumer, N.E. (1993) Broad-spectrum virus resistance in transgenic plants expressing pokeweed antiviral protein. *Proceedings of the National Academy of Sciences of the United States of America* 90, 7089–7093.

Longstaff, M., Brigneti, G., Boccard, F., Chapman, S. and Baulcombe, D. (1993) Extreme resistance to potato virus X infection in plants expressing a modified component of the putative viral replicase. *The EMBO Journal* 12, 379–386.

Lopez-Ochoa, L., Ramirez-Prado, J. and Hanley-Bowdoin, L. (2006) Peptide aptamers that bind to a geminivirus replication protein interfere with viral replication in plant cells. *Journal of Virology* 80, 5841–5853.

Malyshenko, S.I., Kondakova, O.A., Nazarova Ju, V., Kaplan, I.B., Taliansky, M.E. and Atabekov, J.G. (1993) Reduction of tobacco mosaic virus accumulation in transgenic plants producing non-functional viral transport proteins. *Journal of General Virology* 74, 1149–1156.

McDowell, J.M. and Woffenden, B.J. (2003) Plant disease resistance genes: recent insights and potential applications. *Trends in Biotechnology* 21, 178–183.

Mitra, A., Higgins, D.W., Langenberg, W.G., Nie, H., Sengupta, D.N. and Silverman, R.H. (1996) A mammalian 2-5A system functions as an antiviral pathway in transgenic plants. *Proceedings of the National Academy of Sciences of the United States of America* 93, 6780–6785.

Morse, J.G. and Hoddle, M.S. (2006) Invasion biology of thrips. *Annual Review of Entomology* 51, 67–89.

Mumford, R.A., Barker, I. and Wood, K.R. (1996) The biology of the tospoviruses. *Annals of Applied Biology* 128, 159–183.

Nimchuk, Z., Eulgem, T., Holt, B.F., 3rd and Dangl, J.L. (2003) Recognition and response in the plant immune system. *Annual Review of Genetics* 37, 579–609.

Norman, T.C., Smith, D.L., Sorger, P.K., Drees, B.L., O'Rourke, S.M., Hughes, T.R., Roberts, C.J., Friend, S.H., Fields, S. and Murray, A.W. (1999) Genetic selection of peptide inhibitors of biological pathways. *Science* 285, 591–595.

Ogawa, T., Hori, T. and Ishida, I. (1996) Virus-induced cell death in plants expressing the mammalian 2',5' oligoadenylate system. *Nature Biotechnology* 14, 1566–1569.

Pang, S.Z., Slightom, J.L. and Gonsalves, D. (1993) Different mechanisms protect transgenic tobacco against tomato spotted wilt and impatiens necrotic spot tospoviruses. *Biotechnology* 11, 819–824.

Park, S.W., Vepachedu, R., Sharma, N. and Vivanco, J.M. (2004) Ribosome-inactivating proteins in plant biology. *Planta* 219, 1093–1096.

Prins, M. (2003) Broad virus resistance in transgenic plants. *Trends in Biotechnology* 21, 373–375.

Prins, M., Kikkert, M., Ismayadi, C., de Graauw, W., de Haan, P. and Goldbach, R. (1997a) Characterization of RNA-mediated resistance to tomato spotted wilt virus in transgenic tobacco plants expressing NS(M) gene sequences. *Plant Molecular Biology* 33, 235–243.

Prins, M., Storms, M.M.H., Kormelink, R., de Haan, P. and Goldbach, R. (1997b) Transgenic tobacco plants expressing the putative movement protein of tomato spotted wilt tospovirus exhibit aberrations in growth and appearance. *Transgenic Research* 6, 245–251.

Prins, M., Lohuis, D., Schots, A. and Goldbach, R. (2005) Phage display-selected single-chain antibodies confer high levels of resistance against tomato spotted wilt virus. *Journal of General Virology* 86, 2107–2113.

Richmond, K.E., Chenault, K., Sherwood, J.L. and German, T.L. (1998) Characterization of the nucleic acid binding properties of tomato spotted wilt virus nucleocapsid protein. *Virology* 248, 6–11.

Rudolph, C., Schreier, P.H. and Uhrig, J.F. (2003) Peptide-mediated broad-spectrum plant resistance to tospoviruses. *Proceedings of the National Academy of Sciences of the United States of America* 100, 4429–4434.

Sanford, J.C. and Johnston, S.A. (1985) The concept of parasite-derived resistance-deriving resistance genes from the parasites own genome. *Journal of Theoretical Biology* 113, 395–405.

Schillberg, S., Zimmermann, S., Voss, A. and Fischer, R. (1999) Apoplastic and cytosolic expression of full-size antibodies and antibody fragments in *Nicotiana tabacum*. *Transgenic Research* 8, 255–263.

Schouten, A., Roosien, J., Bakker, J. and Schots, A. (2002) Formation of disulfide bridges by a single-chain Fv antibody in the reducing ectopic environment of the plant cytosol. *Journal of Biological Chemistry* 277, 19339–19345.

Soellick, T., Uhrig, J.F., Bucher, G.L., Kellmann, J.W. and Schreier, P.H. (2000) The movement protein NSm of tomato spotted wilt tospovirus (TSWV): RNA binding, interaction with the TSWV N protein, and identification of interacting plant proteins. *Proceedings of the National Academy of Sciences of the United States of America* 97, 2373–2378.

Tacke, E., Salamini, F. and Rohde, W. (1996) Genetic engineering of potato for broad-spectrum protection against virus infection. *Nature Biotechnology* 14, 1597–1601.

Tao, J., Wendler, P., Connelly, G., Lim, A., Zhang, J., King, M., Li, T., Silverman, J.A., Schimmel, P.R. and Tally, F.P. (2000) Drug target validation: lethal infection blocked by inducible peptide. *Proceedings of the National Academy of Sciences of the United States of America* 97, 783–786.

Tavladoraki, P., Benvenuto, E., Trinca, S., De Martinis, D., Cattaneo, A. and Galeffi, P. (1993) Transgenic plants expressing a functional single-chain Fv antibody are specifically protected from virus attack. *Nature* 366, 469–472.

Troitskaya, L.A. and Kodadek, T. (2004) Peptides as modulators of enzymes and regulatory proteins. *Methods* 32, 406–415.

Uhrig, J.F. (2003) Response to Prins: broad virus resistance in transgenic plants. *Trends in Biotechnology* 21, 376–377.

Uhrig, J.F., Soellick, T.R., Minke, C.J., Philipp, C., Kellmann, J.W. and Schreier, P.H. (1999) Homotypic interaction and multimerization of nucleocapsid protein of tomato spotted wilt tospovirus: identification and characterization of two interacting domains. *Proceedings of the National Academy of Sciences of the United States of America* 96, 55–60.

Vaira, A.M., Semeria, L., Crespi, S., Lisa, V., Allavena, A. and Accotto, G.P. (1995) Resistance to tospoviruses in *Nicotiana benthamiana* transformed with the N gene of tomato spotted wilt virus: correlation between transgene expression and protection in primary transformants. *Molecular Plant-Microbe Interactions* 8, 66–73.

Varma, A. (1993) Integrated management of plant viral diseases. *Ciba Foundation Symposia* 177, 140–155.

Villani, M.E., Roggero, P., Bitti, O., Benvenuto, E. and Franconi, R. (2005) Immunomodulation of cucumber mosaic virus infection by intrabodies selected *in vitro* from a stable single-framework phage display library. *Plant Molecular Biology* 58, 305–316.

Wang, P. and Tumer, N.E. (2000) Virus resistance mediated by ribosome inactivating proteins. *Advances in Virus Research* 55, 325–355.

Whitham, S., McCormick, S. and Baker, B. (1996) The N gene of tobacco confers resistance to tobacco mosaic virus in transgenic tomato. *Proceedings of the National Academy of Sciences of the United States of America* 93, 8776–8781.

16 Transgenic Virus Resistance Using Homology-dependent RNA Silencing and the Impact of Mixed Virus Infections

M. Ravelonandro

Abstract

The breeding of crop plants with resistance to plant viruses is an important objective of modern agriculture. Although conventional breeding has proven effective in some cases, in many economically important crop–virus combinations natural resistance is not available. With the development of biotechnology tools, pathogen-derived resistance (PDR) has become the method of choice to engineer resistance to plant viruses. In the majority of highly resistant transgenic lines, PDR was found to be operating at the RNA level and induces homology-dependent RNA silencing, a natural defence mechanism of the plant that causes sequence-specific degradation of the viral RNA. Widely exploited by plant pathologists, RNA silencing represents a highly efficient mean to interfere with viral RNA accumulation. However, mixed viral infections are common in nature and must be considered in terms of stability and durability of virus resistance. Many virus genomes encode specific proteins that promote synergistic interactions between plant viruses by suppressing RNA silencing. These silencing suppressors could potentially affect the engineered virus resistance, especially in the context of mixed infections in which only one of the two virus genomes is targeted by the resistance. In this chapter, the molecular mechanisms of RNA silencing-derived virus resistance will be reviewed and the advantages of these strategies will be briefly outlined. The discovery of plant virus suppressors of the RNA silencing machinery will be discussed with special emphasis on the potyvirus HC-Pro protein. Finally, case studies in which the durability of RNA silencing-derived resistance was tested in the context of mixed infections will be reviewed. In particular, recent studies demonstrating the stability of engineered resistance to *Plum pox virus* in transgenic *Prunus* lines under conditions of mixed infections will be discussed.

Introduction

Plant viruses have an important impact on commercial crops. For example, potyviruses are serious pathogens affecting annual and perennial stone fruit trees (Ravelonandro *et al.*, 1997). Most plant viruses are transmitted by insects, which are usually controlled by pesticides. While preventive measures based on early detection and removal of diseased crops have been used to control the spread of viruses in the field, a more long-term approach is to grow resistant varieties. Yield loss is the most important economic consequence of viral diseases. In addition to yield loss, another important economic impact of plant virus diseases is the restriction of movement of plant material infected with quarantine viruses (e.g. *Plum pox potyvirus* [PPV]). A country affected by PPV cannot export *Prunus* explants, and consequently, any improvement of local stone fruit cultivars requires a certification programme that only permits plantation of virus-free materials (Nemeth, 1992; Boulila *et al.*, 2004). Over the years, strategies to control PPV have included improvement of diagnosis tools and techniques, disease survey and epidemiology control and research for breeding of genetic resistance. Conventional breeding techniques have been used to select for resistance to PPV. However, only tolerant varieties have been identified. Although virus-induced symptoms and yield loss may be reduced in these crops, the virus accumulates to significant levels. Thus, PPV-tolerant varieties do not provide a good source of protection, especially when considering the strict requirements of virus-free certification programmes. Such lack of absolute resistance and the absence of a curative treatment have posed a basic problem to plant pathologists, not only in the case of PPV, but also for many other important plant virus-induced diseases.

The concept of PDR was first introduced in 1985 (Sanford and Johnston, 1985). This concept was applied for the first time in 1986, when the capsid protein (CP) gene of *Tobacco mosaic virus* (TMV) was expressed in transgenic lines, resulting in resistance to the virus (Powell *et al.*, 1986). This significant achievement was followed by a large number of studies aimed at understanding the molecular mechanisms responsible for the virus resistance phenotype (Beachy *et al.*, 1990; Lindbo and Dougherty, 1992; Baulcombe, 1996; Waterhouse *et al.*, 2001; Voinnet, 2005) (see also Chapter 15, this volume). When viruses enter cells, the viral particles are dismantled to release the genome. Thus, overproduction of the CP was predicted to interfere with uncoating of the incoming virus (Beachy *et al.*, 1990). Although in some cases the strength of virus resistance was correlated with the level of expression of the CP, by the beginning of the 1990's, there were also several troubling examples of PDR in which such a correlation was not found. In fact, high levels of virus resistance were often associated with very low levels of expression of the transgene (Lindbo and Dougherty, 1992; Farinelli and Malnoe, 1993; Ravelonandro *et al.*, 1993).

It became apparent that two divergent mechanisms of PDR operate in resistant transgenic lines (Baulcombe, 1996). The first mechanism is protein-mediated and involves protein–protein interactions. The second mechanism is RNA-mediated and requires RNA–RNA interactions. The development of gene constructs that do not express the coat protein was instrumental to distinguish between the two mechanisms (Lindbo and

Dougherty, 1992). In a seminal study, Smith *et al.*, (1994) demonstrated that transcription of the transgene *in planta* plays an important role in the regulation of the threshold level of RNA transgene and also of the homologous virus genomic RNA. Analysis of a large number of resistant lines operating through the RNA-mediated mechanism revealed two distinct phenotypes in the challenging tests. In some cases, virus resistance is effective immediately and virus symptoms do not occur. In other cases, a recovery phenotype is observed in which plants initially display virus symptoms but become healthy at later stages (Ravelonandro *et al.*, 1993).

It is now known that RNA silencing is the molecular mechanism associated with the resistance phenotype. In this mechanism, RNA is targeted and degraded in a sequence-specific manner. As the formation of double-stranded RNA (dsRNA) structures was found to be an essential step of RNA silencing, transgene design was improved to mimic these structures. This led to the elegant design of 'intron-hairpin RNA' construct, in which tight dsRNA structures are formed *in planta*, resulting in a very efficient induction of RNA silencing (Smith *et al.*, 2000). The intron-hairpin RNA transgenes do not encode complete viral genes but are designed to bear highly conserved viral sequences. It is interesting to note that the studies of transgenic lines expressing portions of the viral genome have played a key role in the discovery of RNA silencing and the characterization of the plant enzymes involved in the various RNA silencing pathways (Voinnet, 2001; Waterhouse *et al.*, 2001).

The discoveries that many plant viruses encode suppressors of RNA silencing and that plant viruses are susceptible to RNA silencing in natural viral infections revealed that RNA silencing operates as a natural antiviral defence response (Li and Ding, 2001; Voinnet, 2001, 2005; Roth *et al.*, 2004; Qu and Morris, 2005; Wang and Metzlaff, 2005). The study of synergistic interactions between a potyvirus and a potexvirus provided the first evidence for the existence of viral suppressors of silencing (Vance *et al.*, 1995). As will be discussed below, many other examples followed that have implications on the stability of PDR in the context of mixed virus infections. This chapter will review the following topics: (i) the plant RNA silencing machinery and its implication as a natural plant defence; (ii) the mechanisms for synergistic interactions between viruses in mixed infections; (iii) the recent findings about virus suppressors of RNA silencing; and (iv) the impact of mixed infection on virus resistance in transgenic woody plants.

The Plant Silencing Machinery

As mentioned above, homology-dependent RNA silencing is a natural plant antiviral defence response (Voinnet, 2001; Voinnet, 2005; Wang and Metzlaff, 2005). Induction of RNA silencing is the only strategy that can produce plants with a high level of resistance to viruses. This strategy has been used to engineer high level of resistance to viruses in various herbaceous and woody hosts (Baulcombe, 1996; Waterhouse et al., 1998; Smith et al., 2000; Hily et al., 2005). Thus, RNA silencing is used as a nucleic acid-based antiviral weapon. The specific RNA-mediated inhibition of gene expression is also often referred to as RNA interference (RNAi). Although I focus my discussion on silencing pathways operating in plants, it should be noted that similar pathways have also been elucidated in animals and fungi.

An important feature of RNA silencing is the production of short interfering RNAs (siRNAs) by a ribonuclease III-like enzyme termed Dicer (Voinnet, 2005). The template for siRNA production can be highly structured dsRNA regions of the viral RNA genome, dsRNA replication intermediates produced during viral replication by the viral RNA-dependent RNA polymerase (RdRp) or dsRNA produced from aberrant mRNA transcript by a cellular RdRp. In the case of viral transgene, production of dsRNA forms of the transgene mRNA by a cellular RdRp seems to be a random event that can be triggered by the overproduction of the transgene RNA or by its unusual structure. The use of transgenes that are already in a highly structured dsRNA conformation (intron-hairpin RNA) bypasses this step, resulting in very efficient induction of RNA silencing. The siRNAs control the specificity of RNA silencing in a homology-dependent manner. They are incorporated into the RNA-induced silencing complex (RISC) and guide this complex to inhibit gene expression at the transcriptional or post-transcriptional level. In the case of virus resistance active against RNA viruses, the siRNAs guide the RISC complex to target and degrade homologous sequences present on the incoming virus genome (Hamilton and Baulcombe, 1999; Elbashir et al., 2001). Many argonaute proteins (AGO) have an endonuclease activity and are probably responsible for the degradation of the template RNA. These proteins are essential components of the RISC complex. Although the RNA silencing induced in transgenic lines containing a portion of the viral genome shares many common properties with the natural antiviral defence mechanism, there are some notable differences. Analysis of siRNAs produced in transgenic lines revealed the presence of two classes of siRNAs which play different roles in the silencing process (Hamilton et al., 2002). Short siRNAs (21–24 nucleotides in length) are involved in the sequence-specific degradation of the target, while larger siRNAs (25–27 nucleotides in length) are required for systemic spread of the silencing signal and for RNA-mediated methylation of the transgene, often resulting in transcriptional silencing of the transgene. A recent study comparing siRNAs observed in PPV-resistant transgenic *Prunus* lines and in PPV-infected wild-type *Prunus* plants revealed that while both species of siRNAs are present in the transgenic lines, only the smaller species of siRNAs are present in the PPV-infected *Prunus* plants (Hily et al., 2005). This result suggests that although sharing common properties, the two processes activate different pathways in the plant.

The plant silencing machinery is complex (Brodersen and Voinnet, 2006). Genes coding for four Dicer-like enzymes, six RdRps and ten AGO proteins have been identified in the *Arabidopsis* genome (Bisaro, 2006). These enzymes orchestrate at least three distinct, but interconnected silencing pathways: (i) the post-transcriptional gene silencing (PTGS) pathway, which is principally involved in antiviral defence and transgene-induced RNA silencing; (ii) the miRNA pathway, an endogenous plant pathway that regulates the activity of many plant genes and is involved in plant development; and (iii) the siRNA-directed transcriptional gene silencing (TGS) pathway, which is active during RNA silencing induced by transgene expression and which is also found to operate in the regulation of endogenous silent plant genes. This third pathway probably also plays a role in natural antiviral defence against DNA viruses that replicate in the nucleus. The three pathways have common characteristics in that silencing of target RNAs is directed by small guiding RNA molecules. However, they differ in the structure of the guiding small RNAs, in the specific enzymes that produce these small RNAs and amplify the silencing signal, and in the nature of the silencing mechanism, which can occur at the transcriptional or post-trancriptional level (Bartel, 2004; Xie *et al.*, 2004; Qi and Hannon, 2005; Bisaro, 2006; Brodersen and Voinnet, 2006). In the PTGS pathway, DCL-2 and DCL-4 are the main enzymes responsible for the production of 21–22 nucleotides siRNAs. These siRNAs direct sequence-specific degradation of the target RNA (e.g. viral genomic RNA). Two cellular RdRps (RDR1 and RDR6) have been shown to be associated with the antiviral defence response depending on the virus considered. These RdRps probably play a role in the amplification and systemic spread of the silencing signal. All steps of the PTGS pathway occur in the cytoplasm of the cell. As will be discussed below, many viral-encoded suppressors of silencing have been characterized that interfere with various steps of the PTGS pathway. In the second pathway, DCL-1 produces 21–22 nucleotides miRNAs from larger miRNA precursors which are derived from cellular non-protein-coding genes (Bartel, 2004). The miRNA precursors are produced in the nucleus and transported into the cytoplasm by an exportin. In plants, the perfectly matched miRNAs guide a RISC complex to the target RNAs to promote their degradation. In contrast in animal cells, imperfectly matched miRNAs promote translational repression of the target gene. While the miRNA pathway has been shown to be implicated in antiviral defence in human cells (Lecellier *et al.*, 2005), in plants it is not generally considered a common antiviral defence response and miRNA targets have not been identified on plant viruses so far. Many but not all viral suppressors of silencing interfere with the miRNA pathway in addition to inhibiting the PTGS pathway. The third pathway implicates 24–26 nucleotides siRNAs which are produced by DCL-3. Accumulation of these siRNAs also requires a cellular RdRp (RDR2) and DNA-dependent RNA polymerase (Pol IV). The siRNAs direct methylation and transcriptional silencing of target genes in a sequence-specific manner.

Mixed Virus Infections, Synergistic Interactions and Viral Suppressors of PTGS

Plant viruses cause a set of specific symptoms on their natural hosts and are often named according to these typical symptoms (e.g. *Cucumber mosaic virus* (CMV) causes typical mosaic on cucumber leaves and fruits, *Plum pox virus* causes pox and deformation of plum fruits). However, the simultaneous infection of plants by two or more different viruses is frequent. Such mixed infections may create an antagonistic reaction in which the replication of one virus genome interferes with that of the other virus genome. In other cases, both virus genomes replicate well in the plant without apparently affecting each other's rate of replication. However, in many cases, synergistic interactions between the two viruses result in enhanced replication of one or both virus genomes and exacerbated symptomatology. Among well-known examples of synergistic interactions are the mixed infections of *potexviruses and potyviruses* in tobacco (Damirdagh and Ross, 1967), and the mixed infection of CMV with TMV or *Tobacco ringspot nepovirus* in tobacco (Garces-Orejuela and Pound, 1957).

The biological basis of viral synergistic interactions has been studied in detail in the case of the potexvirus–potyvirus mixed infections. The 5' end region of the potyvirus genome was initially identified as a main determinant of the synergistic interaction (Vance *et al.*, 1995). The ability to express single plant virus genes in transgenic plants allowed the identification of the potyvirus HC-Pro protein as the first known viral suppressor of silencing (Anandalakshmi *et al.*, 1998; Brigneti *et al.*, 1998; Kasschau and Carrington, 1998). It was suggested that the potyvirus HC-Pro assists the replication of potexviruses by inhibiting the plant PTGS machinery. In turn, this led to the hypothesis that PTGS is a natural antiviral defence mechanism. Originally identified in *Tobacco etch virus*, the role of HC-Pro in synergistic interactions and silencing suppression has been confirmed for many other potyviruses (e.g. PPV; Yang and Ravelonandro, 2002). Simultaneously, the CMV 2b protein was also identified as a suppressor of silencing and its role in synergistic interactions potentiated by CMV was also suggested (Beclin *et al.*, 1998; Brigneti *et al.*, 1998). An early survey of the ability of plant viruses to suppress silencing of a green fluorescent reporter gene (GFP) revealed that members of many distinct plant virus families encode strong suppressors of silencing (Voinnet *et al.*, 1999). On the other hand, other plant viruses (e.g. nepoviruses, caulimoviruses and some potexviruses) do not effectively suppress silencing. These viruses either encode weak suppressors of silencing (e.g. *Potato virus X* [PVX]) or escape silencing rather than actively fighting it. These initial studies spurred an intense search for viral suppressors of silencing in many laboratories throughout the world. As a result, an extensive list of characterized viral suppressors of silencing is now available (Table 16.1). The characterization of these suppressors has provided a wealth of information on the mechanisms of viral counter-defence and virulence, as well as on the cellular silencing machinery. Different methods have been used

Table 16.1. Characterized plant virus suppressors of PTGS.

Virus	Genus	Protein	Comments	References
PoLV	*Aureusvirus*	P14	Suppresses virus and transgene-induced silencing; binds long and short ds siRNA	Merai *et al.* (2005)
TGMV	*Begomovirus Geminivirus*	AC2	Pathogenicity factor; transcriptional activator; inactivates adenosine kinase	Wang *et al.* (2005)
TCV	*Carmovirus*	CP	Binds long and short ds siRNA; inhibits Dicer?	Qu *et al.* (2003); Merai *et al.* (2006)
BYV	*Closterovirus*	P21	Binds short ds siRNAs (21 nt); inhibits RISC assembly; crystal structure established	Reed *et al.* (2003); Ye and Patel (2005); Lakatos *et al.* (2006); Merai *et al.* (2006)
CTV	*Closterovirus*	P20	Suppresses intracellular and systemic silencing	Lu *et al.* (2004)
		P23	Suppresses intracellular silencing	
		CP	Suppresses systemic silencing	
SPCSV	*Crinivirus*	RNAse3	Binds siRNAs; endonuclease activity	Kreuze *et al.* (2005)
		P22	Pathogenicity factor; suppresses intracellular and systemic silencing	
CMV, TAV	*Cucumovirus*	2b	Pathogenicity factor; suppresses systemic silencing; inhibits argonaute1; inhibits miRNA pathway	Beclin *et al.* (1998); Brigneti *et al.* (1998); Guo and Ding, (2002); Zhang *et al.* (2006)
BSMV	*Hordeivirus*	γB	Binds short ds siRNA	Yelina *et al.* (2002); Merai *et al.* (2006)
PCV	*Pecluvirus*	P15	Suppresses intracellular and systemic silencing; binds short ds siRNA; inhibits RISC assembly?	Dunoyer *et al.* (2002); Merai *et al.* (2006)
TEV, PVY, TuMV, PVA	*Potyvirus*	HC-Pro	Pathogenicity factor; virus long distance movement factor; aphid transmission factor; proteinase; suppresses intracellular and systemic silencing; inhibits miRNA pathway; binds calmudolin, a cellular suppressor of silencing; binds short ds siRNA	Anandalakshmi *et al.* (1998, 2000); Mallory *et al.* (2002); Kasschau *et al.* (2003); Merai *et al.* (2006)

Continued

Table 16.1. *Continued*

Virus	Genus	Protein	Comments	References
RYMV	*Sobemovirus*	P1	Pathogenicity factor; virus long distance movement factor	Voinnet *et al.* (1999)
TBSV	*Tombusvirus*	P19	Inhibits intracellular and systemic silencing; binds short ds siRNA; inhibits RISC assembly; inhibits miRNA pathway; P19 dimerization required for silencing suppression; crystal structure established	Voinnet *et al.* (1999); Silhavy *et al.* (2002); Baulcombe and Molnar, (2004); Chapman *et al.* (2004); Lakatos *et al.* (2004); Merai *et al.* (2006)
BWYV	*Polerovirus*	P0	Pathogenicity factor	Pfeffer *et al.* (2002)
TYMV	*Tymovirus*	P69	Pathogenicity factor; viral movement protein; inhibits PTGS but enhances miRNA pathway	Chen *et al.* (2004)
GVA	*Vitivirus*	P10	Inhibits intracellular and systemic silencing; binds ss and ds siRNA and miRNA	Zhou *et al.* (2006)
CPMV	*Comovirus*	CP-S	Coat protein; surface-exposed amino acids essential for suppression of PTGS.	Canizares *et al.* (2004); Liu *et al.* (2004)
TMV, ToMV	*Tobamovirus*	130 kDa	Replicase protein	Kubota *et al.* (2003); Ding *et al.* (2004)
TSWV	*Tospovirus*	NSs	Pathogenicity determinant; reverts pre-established PTGS	Bucher *et al.* (2003)
PVX	*Potexvirus*	P25	Viral cell-to-cell movement protein; inhibits systemic silencing	Voinnet *et al.* (2000); Bayne *et al.* (2005)
RCNMV	*Dianthovirus*	?	Inhibits PTGS probably by recruiting dicer in viral replication complex	Takeda *et al.* (2005)

PoLV, *Pothos latent virus*; TGMV, *Tomato golden mosaic virus*; TCV, *Turnip crinkle virus*; BYV, *Beet yellow virus*; CTV, *Citrus tristeza virus*; SPCSV, *Sweet potato chlorotic stunt virus*; CMV, *Cucumber mosaic virus*; TAV, *Tomato aspermy virus*; BSMV, *Barley stripe mosaic virus*; PCV, *Peanut clump virus*; TEV, *Tobacco etch virus*; PVY, *Potato virus Y*; TuMV, *Turnip mosaic virus*; PVA, *Potato virus A*; RYMV, *Rice yellow mosaic virus*; TBSV, *Tomato bushy stunt virus*; BWYV, *Beet western yellow virus*; TYMV, *Turnip yellow mosaic virus*; GVA, *Grapevine vitivirus A*; CPMV, *Cowpea mosaic virus*; TMV, *Tobacco mosaic virus*; ToMV, *Tomato mosaic virus*; TSWV, *Tomato spotted wilt virus*; PVX, *Potato virus X*; RCNMV, *Red clover necrotic mottle virus*.

to express viral suppressors in plants, including agroinfiltration and production of transgenic lines or transfer of individual viral genes in viral vectors derived from the PVX genome (Angell and Baulcombe, 1997). The methods used to monitor the effect of viral suppressors of silencing on the plant machinery, often using plants silenced for the expression of GFP, have been reviewed (Roth et al., 2004; Qu and Morris, 2005).

General characteristics of viral suppressors of silencing

Interestingly, viral suppressors of silencing do not possess consensus sequence motifs or mechanisms of action. Rather, they are very diverse and interfere with various steps of the silencing machinery. Some block the generation of the Dicer cleavage products, while others inhibit further downstream steps, such as the assembly of the RISC complex or the systemic movement of the silencing signal (Roth et al., 2004; Qu and Morris, 2005; Wang and Metzlaff, 2005; Bisaro, 2006; Li and Ding, 2006; Table 16.1). The phenotypes induced after constitutive expression of suppressors of silencing often mimic those observed during natural viral infection and differ from one viral system to another. In some cases, the biological function of viral suppressors has also been shown to differ according to the host plants affected. One common feature of viral suppressors of silencing seems to be their ability to bind small dsRNA (Merai et al., 2006). Most viral suppressors bind small dsRNAs in a size-specific manner while others (*Aureusvirus* P14 and *Carmovirus* coat proteins) bind large precursors as well as the small dsRNAs, suggesting that the ability to bind small dsRNAs may have evolved independently in different viruses.

Finally, it should also be noted that in some cases, viral genomes encode more than one suppressor of silencing that act in concert in the counter-defence mechanism. An example of this is provided by a *Crinivirus*, *Sweet potato chlorotic stunt virus* (SPCSV) which acts synergistically with a *Potyvirus*, *Sweet potato featherly mottle virus* to produce a fatal double infection (Gibson et al., 1998). Two SPCSV proteins, P22 and RNAse3, a putative RNAse3 with dsRNA-specific endonuclease activity, were shown to be involved in silencing suppression (Kreuze et al., 2005). The silencing suppression activity of the P22 protein was enhanced by co-expression of the RNAse3 protein. Interestingly, RNAse3 alone did not have silencing suppression activity although it has the ability to bind dsRNAs. Similarly, *Citrus tristeza virus* has been shown to encode three distinct suppressors of silencing that target different steps of the silencing pathway (Lu et al., 2004).

The potyvirus HC-Pro protein

HC-Pro is a multifunctional protein involved in pathogenicity, long-distance movement of the virus and aphid transmission (Plisson et al., 2003). It is also a proteinase. HC-Pro inhibits PTGS, prevents accumulation of the 21 nt siRNAs and partially inhibits the associated methylation of the transgene, suggesting that it is targeting a maintenance step of the silencing pathway (Llave et al., 2000; Mallory et al., 2001). When expressed in a heterologous mammalian system, HC-Pro was shown to interfere with the RNAi machinery and prevent dsRNA degradation into siRNAs, suggesting that it is targeting a step of the silencing pathway conserved among various organisms (Reavy et al., 2004). It was recently found that HC-Pro interacts with a cellular calmudolin-related protein which also functions as a suppressor of silencing (Anandalakshmi et al., 2000). The recruitment of a cellular suppressor of silencing may contribute to the silencing suppression activity of HC-Pro (Anandalakshmi et al., 2000). HC-Pro has a wide range of action as it also inhibits the miRNA pathway and interferes with plant development, at least partially explaining the symptoms observed in potyvirus infection or after expression of HC-Pro (Kasschau et al., 2003). Several other suppressors of silencing were subsequently also shown to hinder the miRNA pathway (Table 16.1).

Impact of Viral Suppressors of Silencing on RNA Silencing-derived Transgenic Resistance

The initial enthusiasm towards RNA silencing-derived virus resistance has been somewhat tempered by the discovery that many plant viruses have the ability to inhibit silencing. Can plant virus-encoded suppressors of silencing interfere with the strength and durability of engineered virus-resistance? RNA silencing specified by a virus-derived transgene is sequence-specific and has been shown to effectively control the accumulation of the targeted virus, even when this virus encodes a potent suppressor of silencing. Field evaluation of transgenic PPV-resistant *Prunus* lines for an extended period of time (9–10 years, the expected lifespan of fruit crops) revealed that while susceptible *Prunus* lines became readily infected with PPV transmitted by aphids, the resistant lines (C5 lines) remained essentially immune to the virus (Ravelonandro et al., 1997; Scorza et al., 2001; Hily et al., 2004; Malinowski et al., 2006). Thus, aphid transmission of PPV, a virus known to encode a potent suppressor of silencing (HC-Pro), did not result in resistance breaking. It is interesting to note that graft-inoculation of the C5 lines with PPV-infected budwood did result in a low level of PPV infection, although the virus titer was considerably reduced and the symptoms were very mild (Hily et al., 2004). The amount of virus transferred to the trees is likely to be much lower during aphid transmission than during graft-inoculation (Pirone and Thornbury, 1988), raising the possibility that a minimum concentration of a viral protein (probably HC-Pro) is necessary to combat, at least partially, the resistance (Hily et al., 2004). Additionally, the site of inoculation differs during aphid transmission and graft-inoculation. During aphid transmission, the virus is injected in epidermal cells of fully expanded leaves where RNA silencing is known to be very active. On the other hand, during graft-inoculation, the virus is directly in contact with the vascular system, possibly allowing its spread into the plant before initial recognition by the silencing machinery (Hily et al., 2004). These results suggest that RNA silencing-derived virus resistance remains stable under normal field conditions for extended periods of time. A similar example is also provided by the stable resistance to *Papaya ringspot virus* observed during extensive field testing of transgenic papaya trees (Gonsalves, 2002) (Chapter 19, this volume).

Although RNA silencing efficiently controls the targeted virus, it is sequence-specific and does not affect the replication of unrelated viruses. Thus, these unrelated viruses may have the opportunity to express suppressors of silencing and possibly disrupt the engineered resistance. As mentioned above, mixed infection is a common occurrence in the field. It is therefore important to test the durability of PTGS-derived transgenic resistance during mixed virus infection. In herbaceous plants, resistance to *Potato virus A* (PVA) was stable after single inoculation with PVA but was broken during mixed infection with PVA and *Potato virus Y* (PVY), a second potyvirus which is not targeted by the sequence-homology resistance (Savenkov and Valkonen, 2001). Although it is likely that expression of the PVY HC-Pro protein is responsible for the breaking of the resistance, this was not directly tested. Similarly, infection of PPV-resistant *N. benthamiana* plants with CMV was reported to restore transgene expression and break the PPV resistance (Simon-Mateo et al., 2003). In at least one documented case, a virulent isolate of PVY (PVY(n)) was shown to break the resistance targeted to a milder isolate (PVY(o)) even though the degree of sequence homology with the transgene was very high (Maki-Valkama et al., 2000). It is not known whether this resistance-breaking is due to the action of a particularly potent suppressor of silencing encoded by PVY(n). Although resistance-breaking has been reported under conditions of mixed infection of model herbaceous hosts in the laboratory, the relevance of these observations to agronomical crops and to field situations needs to be evaluated.

Mixed Infection in *Prunus*: A Case Study Using PPV-resistant Transgenic *Prunus* Lines

In *Prunus*, PPV is known to coexist with *Prunus necrotic ringspot* (PNRSV) and *Prune dwarf* (PDV) *ilarviruses*; and with *Apple chlorotic leaf spot trichovirus* (ACLSV) (Nemeth, 1992; Boulila et al., 2004). These viruses are transmitted through the pollen or by grafting. The stability of PPV-engineered resistance in the *Prunus* C5 lines was studied in the context of these known mixed viral infections.

As mentioned above, two species of PPV-specific siRNAs (21 and 27 nt in length) are observed in C5 PPV-resistant transgenic lines, which are specifically induced by the PPV CP transgene (Hily et al., 2005). Wild-type susceptible *Prunus* plants infected with PPV contain a single species of PPV-specific siRNA (21 nt in length) which probably originates from the viral genome as a result of the plant antiviral response (Hily *et al.*, 2005). The presence of the 21 nucleotides siRNAs is not sufficient to control the virus and the second species of siRNA (27 nucleotides) in the transgenic C5 lines is associated with the resistance phenotype and with hypermethylation and transcriptional silencing of the transgene (Hily *et al.*, 2005). Hypermethylation and transcriptional silencing of the transgene is also correlated with the stability of the resistance in the field (Hily *et al.*, 2004). C5 lines infected with PDV, PNRSV or ACLSV accumulate the virus as detected by ELISA and contain a single species of siRNAs (21 nt) specific for the corresponding virus, confirming that these viruses replicate in the plants and trigger the natural antiviral response of the plant, a process unrelated to the engineered PPV resistance (M. Ravelonandro, 2006). These results also confirm the notion that RNA silencing is a highly specific response that does not affect the replication of unrelated viruses.

Simultaneous or sequential graft-inoculations of C5 lines with PPV and PDV, PNRSV or ACLSV were conducted to assess the effect of mixed infection on the stability of the PPV resistance. A PPV-susceptible transgenic line (also transformed with the PPV CP gene, line C6) and a control transgenic line (transformed with only reporter genes, line PT23) were inoculated in parallel. To conduct these experiments, transgenic scions were grafted on a PPV-susceptible rootstock (peach GF305). Chip-budding of PPV-infected material at the rootstock ensures successful infection of the rootstock. After the dormancy cycle, the virus moves from the roots to shoots at the bud-breaking stage. As mentioned earlier, in single graft-inoculation with PPV, C5 lines are somewhat susceptible to PPV infection, probably because RNA silencing is only efficient in fully expanded leaves. However, if the virus moves to the scion, it remains around the grafting point. Systemic spread of the virus in the plant was suppressed and upper leaves remained PPV-free as confirmed by RT-PCR or ELISA (Figs 16.1 and 16.2). In contrast, C6 and PT23 lines became systemically infected with PPV under these conditions. To conduct mixed infections, plants were inoculated by double chip-budding, one at the non-transgenic and PPV-susceptible rootstock and one at the shoot section. ELISA or RT-PCR assays were conducted to assess the presence of each virus at different stages of infection. The control C6 and PT23 lines became readily infected by both viruses under all the conditions tested (PPV inoculated first, last or simultaneously with the second test virus). Symptoms observed during mixed infection of PPV with either ACLSV or PNRSV were similar to those observed in single PPV infections. Unexpectedly, typical PPV symptoms were masked by the appearance of chlorotic spots induced during PPV–PDV mixed infection. Thus, symptom evaluation on its own was not a reliable indicator of mixed infection. There was no uniform spread of the two viruses in these plants, as often observed under conditions of mixed infection (M. Ravelonandro, unpublished data). Inoculation of transgenic C-5 scions with ilarviruses or trichoviruses did

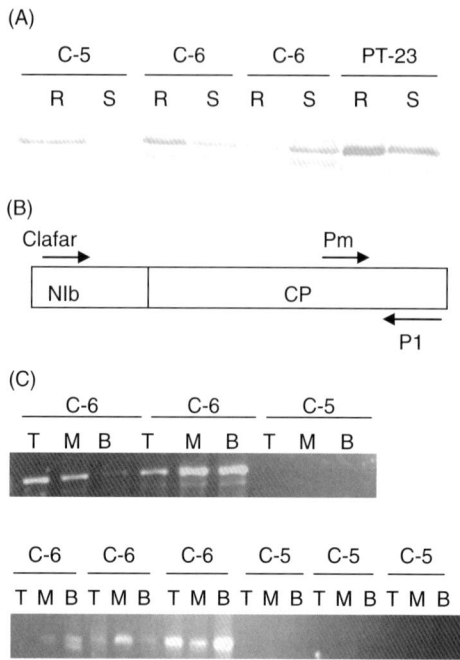

Fig. 16.1. Systemic movement of PPV from a susceptible rootstock does not occur in C-5 transgenic scions. PPV was graft-inoculated to peach GF305 rootstock by chip-budding and systemic invasion of PPV in the transgenic scion was tested. (A) Immunodetection of PPV coat protein. Soluble proteins were extracted from leaves collected either at the PPV-susceptible peach GF305 rootstock (R) or at the transgenic shoots (S) of clones C-5, C-6 and PT-23. The proteins were separated by SDS-polyacrylamide gel electrophoresis and the PPV CP was detected by immunoblotting using PPV-specific antibodies. The arrow indicates the position of PPV CP. (B) Location of primers used for RT-PCR detection of PPV genomic RNA. (C) RT-PCR detection of PPV genomic RNA. Two methods of detection were used: one-step RT-PCR detection and the Clafar-P1 primer pair (upper panel) or heminested-PCR detection and the PM-P1 primer pair (lower panel). Presence of the virus was tested at different leaf positions of the test scions. Samples were collected in the upper leaves (T, tip), the fifth leaf-pair (M, middle) and the lowest leaves (B, bottom).

not affect the PPV resistance. Although initial DAS-ELISA analysis revealed the simultaneous presence of PPV and the challenging virus in a limited number of C-5 scions, the PPV infection remained confined in areas that were in close proximity with the site of the grafting (a result similar to that observed when these plants were inoculated with PPV only) and the upper leaves remained free of PPV (Figs 16.1 and 16.2). The two species of the PPV-specific siRNAs continued to be detected and the methylated status of the PPV CP transgene in clone C-5 remained unchanged in the mixed infections analysed (M. Ravelonandro, 2006). Similarly, preliminary results

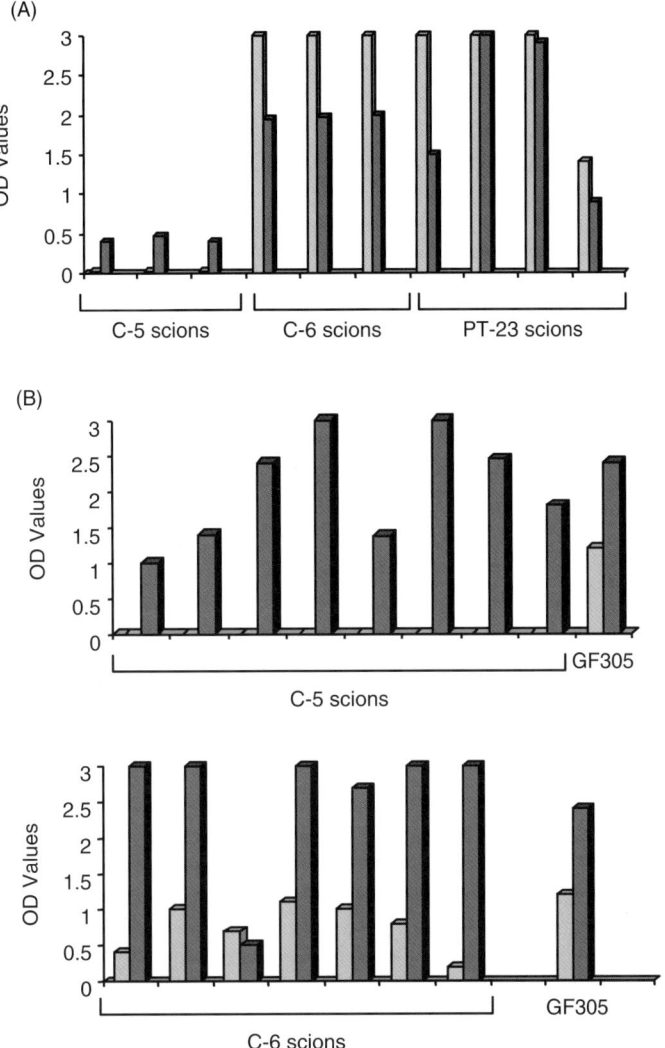

Fig. 16.2. Results of mixed infections of transgenic scions grafted on non-transgenic GF305 peach rootstock. Plants were inoculated with PPV and either PDV (A) or PNRSV (B) by chip-budding of a non-transgenic PPV-susceptible rootstock (GF305). In (A), the plants were simultaneously inoculated with PPV and PDV. In (B), the plants were sequentially inoculated with PNRSV and then with PPV. Relative concentration of each virus in the transgenic scion was tested by ELISA. ELISA values (OD, corresponding to the concentration of PPV) are shown in blue. The relative concentrations of PDV (Panel A) or PNRSV (Panel B) are shown in red. The scions tested were either C5 (transgenic lines transformed with a binary vector allowing the expression of the PPV CP and highly resistant to PPV), C6 (transgenic lines transformed with a binary vector allowing the expression of the PPV CP and susceptible to PPV) or PT23 (control transgenic lines transformed with the empty binary vector). GF305 peach is used as control.

suggest that PPV resistance remains stable under conditions of mixed infection in an open field (Polak et al., 2005).

Although our results suggest that mixed infections with ilarviruses and trichoviruses do not affect the engineered PPV resistance in *Prunus* trees, it remains possible that synergistic interactions between these viruses occur in nature. It has been reported that synergistic interactions between fruit tree viruses in nature differ upon the host considered. Japanese plums are more frequently infected with PPV mixed with ACLSV (Boulila et al., 2004). Mixed infection between PPV, PNRSV and/or PDV is mainly a concern in *P. domestica* plums (Nemeth, 1992). The possible synergistic interactions may also depend on the PPV strain considered. Indeed, six PPV strains have been identified that vary in their virulence and host range (Wang et al., 2006). It is not known whether ilarviruses or trichoviruses encode suppressors of silencing. Thus, it is possible that the stability of the PPV resistance observed in the C5 lines can be explained by the absence of potent suppressors of silencing encoded by these viruses. It would be interesting to test whether infection with PPV, a virus known to encode a potent suppressor of silencing, would affect resistance targeted to ilarviruses or trichoviruses.

Development of New Technologies

The maintenance of resistance to PPV infection in transgenic clone C-5 under conditions of natural mixed infections is an encouraging result for the future application of RNA silencing-derived resistance in the field. However, it should be emphasized that each virus–virus and virus–host combination is likely to differ and that similar studies must be conducted with other virus-resistant crops to confirm stability of the resistance in the field. As breaking of resistance during mixed virus infections has been reported in the laboratory (Maki-Valkama et al., 2000; Savenkov and Valkonen, 2001), it may be useful to improve RNA silencing-derived resistance and target a wider range of viruses. Choosing conserved regions in a virus genome to target divergent isolates is a possible strategy, although not necessarily a guarantee for success (Maki-Valkama et al., 2000). An alternative approach may be to design chimeric transgenes that contain regions from two (or more) different virus genomes and simultaneously target more than one virus. The potential of such approaches has been demonstrated (Jan et al., 2000).

To date, experimental evidence suggests that the siRNA-directed PTGS pathway is the main mechanism involved in the natural antiviral resistance in plants. However, in animals, cellular miRNAs have been identified that target viruses and play a role in the antiviral defence (Bartel, 2004; Lecellier et al., 2005). Interestingly, a chimeric PPV genome, into which miRNA targets were engineered, was shown to be susceptible to miRNA-guided degradation (Simon-Mateo and Garcia, 2006). Also, a recent report demonstrates that the miRNA pathway can be manipulated to control plant virus infection (Niu *et al.*, 2006). An artificial miRNA (amiRNA) was constructed in which viral-specific sequences were inserted in an abundant plant pre-miRNA. Because the miRNA pathway is less temperature-sensitive than the PTGS pathway, the amiRNA approach was shown to effectively control plant viruses at low temperatures (15°C), a significant advantage when considering field control of plant viruses in temperate climates (Niu *et al.*, 2006). Of course, as many virus-encoded suppressors of silencing also affect the miRNA pathway, the possible problems caused by co-infection of plants with unrelated viruses remain. Interestingly, targeting of two viruses with a single amiRNA was shown to be possible (Niu *et al.*, 2006).

Conclusions and Future Directions

The last 10 years has been a very exciting period. From the initial discoveries that most cases of PDR in transgenic plants are RNA-mediated and related to a natural antiviral resistance mechanism, a wealth of detailed studies has followed revealing an unsuspected degree of complexity of the silencing pathways operating in plants. While this has allowed the improvement of strategies for transgenic virus resistance, many questions remain and it is likely that the next 10 years will be equally exciting. The elucidation of the mechanism of actions of viral suppressors, in particular their specific interactions with various hosts, will have practical applications for the engineering of non-herbaceous crops such as fruit trees. A better understanding of the synergistic interactions between viruses is not only important to assess the stability of transgenic virus resistance in the field but also to mitigate economic losses caused by mixed virus infections. In turn, these analyses will also have an impact on quarantine procedures and regulation of import of commercial crops.

Acknowledgements

The author acknowledges the partial sponsoring of this work by EU contract 'TRANSVIR' QLK3-2002-0140, and thanks Dr Hélène Sanfaçon for her valuable comments and helpful discussion about this chapter.

References

Anandalakshmi, R., Pruss, G.J., Ge, X., Marathe, R., Mallory, A.C., Smith, T.H. and Vance, V.B. (1998) A viral suppressor of gene silencing in plants. *Proceedings of the National Academy of Sciences of the United States of America* 95, 13079–13084.
Anandalakshmi, R., Marathe, R., Ge, X., Herr, J.M., Jr., Mau, C., Mallory, A., Pruss, G., Bowman, L. and Vance, V.B. (2000) A calmodulin-related protein that suppresses posttranscriptional gene silencing in plants. *Science* 290, 142–144.

Angell, S.M. and Baulcombe, D.C. (1997) Consistent gene silencing in transgenic plants expressing a replicating potato virus X RNA. *The EMBO Journal* 16, 3675–3684.

Bartel, D.P. (2004) MicroRNAs: genomics, biogenesis, mechanism and function. *Cell* 116, 281–297.

Baulcombe, D.C. (1996) Mechanisms of pathogen-derived resistance to viruses in transgenic plants. *The Plant Cell* 8, 1833–1844.

Baulcombe, D.C. and Molnar, A. (2004) Crystal structure of p19 – a universal suppressor of RNA silencing. *Trends in Biochemical Sciences* 29, 279–281.

Bayne, E.H., Rakitina, D.V., Morozov, S.Y. and Baulcombe, D.C. (2005) Cell-to-cell movement of potato potexvirus X is dependent on suppression of RNA silencing. *The Plant Journal* 44, 471–482.

Beachy, R.N., Loesch-Fries, L.S. and Tumer, N. (1990) Coat-protein mediated resistance against virus infection. *Annual Review of Phytopathology* 28, 451–474.

Beclin, C., Berthome, R., Palauqui, J.C., Tepfer, M. and Vaucheret, H. (1998) Infection of tobacco or Arabidopsis plants by CMV counteracts systemic post-transcriptional silencing of nonviral (trans)genes. *Virology* 252, 313–317.

Bisaro, D.M. (2006) Silencing suppression by geminivirus proteins. *Virology* 344, 158–168.

Boulila, M., Briard, P. and Ravelonandro, M. (2004) Outbreak of plum pox virus in Tunisia. *Journal of Plant Pathology* 86, 197–201.

Brigneti, G., Voinnet, O., Li, W.X., Ji, L.H., Ding, S.W. and Baulcombe, D.C. (1998) Viral pathogenicity determinants are suppressors of transgene silencing in *Nicotiana benthamiana*. *The EMBO Journal* 17, 6739–6746.

Brodersen, P. and Voinnet, O. (2006) The diversity of RNA silencing pathways in plants. *Trends in Genetics* 22, 268–280.

Bucher, E., Sijen, T., de Haan, P., Goldbach, R. and Prins, M. (2003) Negative-strand tospoviruses and tenuiviruses carry a gene for a suppressor of gene silencing at analogous genomic positions. *Journal of Virology* 77, 1329–1336.

Canizares, M.C., Taylor, K.M. and Lomonossoff, G.P. (2004) Surface-exposed C-terminal amino acids of the small coat protein of cowpea mosaic virus are required for suppression of silencing. *Journal of General Virology* 85, 3431–3435.

Chapman, E.J., Prokhnevsky, A.I., Gopinath, K., Dolja, V.V. and Carrington, J.C. (2004) Viral RNA silencing suppressors inhibit the microRNA pathway at an intermediate step. *Genes and Development* 18, 1179–1186.

Chen, J., Li, W.X., Xie, D., Peng, J.R. and Ding, S.W. (2004) Viral virulence protein suppresses RNA silencing-mediated defense but upregulates the role of microRNA in host gene expression. *The Plant Cell* 16, 1302–1313.

Damirdagh, I.S. and Ross, A.F. (1967) A marked synergistic interaction of potato viruses X and Y in inoculated leaves of tobacco. *Virology* 31, 296–307.

Ding, X.S., Liu, J., Cheng, N.H., Folimonov, A., Hou, Y.M., Bao, Y., Katagi, C., Carter, S.A. and Nelson, R.S. (2004) The *Tobacco mosaic virus* 126-kDa protein associated with virus replication and movement suppresses RNA silencing. *Molecular Plant-Microbe Interactions* 17, 583–592.

Dunoyer, P., Pfeffer, S., Fritsch, C., Hemmer, O., Voinnet, O. and Richards, K.E. (2002) Identification, subcellular localization and some properties of a cysteine-rich suppressor of gene silencing encoded by peanut clump virus. *The Plant Journal* 29, 555–567.

Elbashir, S.M., Lendeckel, W. and Tuschl, T. (2001) RNA interference is mediated by 21- and 22-nucleotide RNAs. *Genes and Development* 15, 188–200.

Farinelli, L. and Malnoe, P. (1993) Coat protein gene-mediated resistance to potato virus Y in tobacco-examination of the resistance mechanisms. Is the transgenic coat protein required for protection? *Molecular Plant-Microbe Interactions* 6, 284–292.

Garces-Orejuela, C. and Pound, G.S. (1957) The multiplication of tobacco mosaic virus in the presence of cucumber mosaic virus or tobacco ringspot virus in tobacco. *Phytopathology* 47, 232–239.

Gibson, R.W., Mpembe, I., Alicai, T., Carey, E.E., Mwanga, R.O.M., Seal, S.E. and Vetten, H.J. (1998) Symptoms, aetiology and serological analysis of sweet potato virus disease in Uganda. *Plant Pathology* 47, 95–102.

Gonsalves, D. (2002) Coat protein transgenic papaya: 'acquired' immunity for controlling papaya ringspot virus. *Current Topics in Microbiology and Immunology* 266, 73–83.

Guo, H.S. and Ding, S.W. (2002) A viral protein inhibits the long range signaling activity of the gene silencing signal. *The EMBO Journal* 21, 398–407.

Hamilton, A., Voinnet, O., Chappell, L. and Baulcombe, D. (2002) Two classes of short interfering RNA in RNA silencing. *The EMBO Journal* 21, 4671–4679.

Hamilton, A.J. and Baulcombe, D.C. (1999) A species of small antisense RNA in posttranscriptional gene silencing in plants. *Science* 286, 950–952.

Hily, J.M., Scorza, R., Malinowski, T., Zawadzka, B. and Ravelonandro, M. (2004) Stability of gene silencing-based resistance to plum pox virus in transgenic plum (*Prunus domestica* L.) under field conditions. *Transgenic Research* 13, 427–436.

Hily, J.M., Scorza, R., Webb, K. and Ravelonandro, M. (2005) Accumulation of the long class of siRNA is associated with resistance to plum pox virus in a transgenic woody perennial plum tree. *Molecular Plant–Microbe Interactions* 18, 794–799.

Jan, F.J., Fagoaga, C., Pang, S.Z. and Gonsalves, D. (2000) A single chimeric transgene derived from two distinct viruses confers multi-virus resistance in transgenic plants through homology-dependent gene silencing. *Journal of General Virology* 81, 2103–2109.

Kasschau, K.D. and Carrington, J.C. (1998) A counterdefensive strategy of plant viruses: suppression of posttranscriptional gene silencing. *Cell* 95, 461–470.

Kasschau, K.D., Xie, Z., Allen, E., Llave, C., Chapman, E.J., Krizan, K.A. and Carrington, J.C. (2003) P1/HC-Pro, a viral suppressor of RNA silencing, interferes with Arabidopsis development and miRNA function. *Development Cell* 4, 205–217.

Kreuze, J.F., Savenkov, E.I., Cuellar, W., Li, X. and Valkonen, J.P. (2005) Viral class 1 RNAse III involved in suppression of RNA silencing. *Journal of Virology* 79, 7227–7238.

Kubota, K., Tsuda, S., Tamai, A. and Meshi, T. (2003) Tomato mosaic virus replication protein suppresses virus-targeted posttranscriptional gene silencing. *Journal of Virology* 77, 11016–11026.

Lakatos, L., Szittya, G., Silhavy, D. and Burgyan, J. (2004) Molecular mechanism of RNA silencing suppression mediated by p19 protein of tombusviruses. *The EMBO Journal* 23, 876–884.

Lakatos, L., Csorba, T., Pantaleo, V., Chapman, E.J., Carrington, J.C., Liu, Y.P., Dolja, V.V., Calvino, L.F., Lopez-Moya, J.J. and Burgyan, J. (2006) Small RNA binding is a common strategy to suppress RNA silencing by several viral suppressors. *The EMBO Journal* 25, 2768–2780.

Lecellier, C.H., Dunoyer, P., Arar, K., Lehmann-Che, J., Eyquem, S., Himber, C., Saib, A. and Voinnet, O. (2005) A cellular microRNA mediates antiviral defense in human cells. *Science* 308, 557–560.

Li, F. and Ding, S.W. (2006) Virus counterdefense: diverse strategies for evading the RNA-silencing immunity. *Annual Review of Microbiology* 60, 503–531.

Li, W.X. and Ding, S.W. (2001) Viral suppressors of RNA silencing. *Current Opinion in Biotechnology* 12, 150–154.

Lindbo, J.A. and Dougherty, W.G. (1992) Untranslatable transcripts of the tobacco etch virus coat protein gene sequence can interfere with tobacco etch virus replication in transgenic plants and protoplasts. *Virology* 189, 725–733.

Liu, L., Grainger, J., Canizares, M.C., Angell, S.M. and Lomonossoff, G.P. (2004) Cowpea mosaic virus RNA-1 acts as an amplicon whose effects can be counteracted by a RNA-2-encoded suppressor of silencing. *Virology* 323, 37–48.

Llave, C., Kasschau, K.D. and Carrington, J.C. (2000) Virus-encoded suppressor of posttranscriptional gene silencing targets a maintenance step in the silencing pathway. *Proceedings of the National Academy of Sciences of the United States of America* 97, 13401–13406.

Lu, R., Folimonov, A., Shintaku, M., Li, W.X., Falk, B.W., Dawson, W.O. and Ding, S.W. (2004) Three distinct suppressors of RNA silencing encoded by a 20-kb viral RNA genome. *Proceedings of the National Academy of Sciences of the United States of America* 101, 15742–15747.

Maki-Valkama, T., Valkonen, J.P., Kreuze, J.F. and Pehu, E. (2000) Transgenic resistance to PVY(O) associated with post-transcriptional silencing of P1 transgene is overcome by PVY(N) strains that carry highly homologous P1 sequences and recover transgene expression at infection. *Molecular Plant-Microbe Interactions* 13, 366–373.

Malinowski, T., Cambra, M., Capote, N., Zawadzka, B., Gorris, M.T., Scorza, R. and Ravelonandro, M. (2006) Field trials of plum clones transformed with the plum pox virus coat protein (PPV-CP) gene. *Plant Disease* 90, 1012–1018.

Mallory, A.C., Ely, L., Smith, T.H., Marathe, R., Anandalakshmi, R., Fagard, M., Vaucheret, H., Pruss, G., Bowman, L. and Vance, V.B. (2001) HC-Pro suppression of transgene silencing eliminates the small RNAs but not transgene methylation or the mobile signal. *The Plant Cell* 13, 571–583.

Mallory, A.C., Reinhart, B.J., Bartel, D., Vance, V.B. and Bowman, L.H. (2002) A viral suppressor of RNA silencing differentially regulates the accumulation of short interfering RNAs and micro-RNAs in tobacco. *Proceedings of the National Academy of Sciences of the United States of America* 99, 15228–15233.

Merai, Z., Kerenyi, Z., Molnar, A., Barta, E., Valoczi, A., Bisztray, G., Havelda, Z., Burgyan, J. and Silhavy, D. (2005) *Aureusvirus* P14 is an efficient RNA silencing suppressor that binds double-stranded RNAs without size specificity. *Journal of Virology* 79, 7217–7226.

Merai, Z., Kerenyi, Z., Kertesz, S., Magna, M., Lakatos, L. and Silhavy, D. (2006) Double-stranded RNA binding may be a general plant RNA viral strategy to suppress RNA silencing. *Journal of Virology* 80, 5747–5756.

Nemeth, M. (1992) On the distribution and economic significance of fruit tree viruses in Hungary. *Novenydelem* 28, 26–32 (in Hungarian)

Niu, Q.W., Lin, S.S., Reyes, J.L., Chen, K.C., Wu, H.W., Yeh, S.D. and Chua, N.H. (2006) Expression of artificial microRNAs in transgenic *Arabidopsis thaliana* confers virus resistance. *Nature Biotechnology* 24, 1420–1428.

Pfeffer, S., Dunoyer, P., Heim, F., Richards, K.E., Jonard, G. and Ziegler-Graff, V. (2002) P0 of beet western yellows virus is a suppressor of posttranscriptional gene silencing. *Journal of Virology* 76, 6815–6824.

Pirone, T.P. and Thornbury, D.W. (1988) Quantity of virus required for aphid transmission of a potyvirus. *Phytopathology* 78, 104–107.

Plisson, C., Drucker, M., Blanc, S., German-Retana, S., Le Gall, O., Thomas, D. and Bron, P. (2003) Structural characterization of HC-Pro, a plant virus multifunctional protein. *Journal of Biological Chemistry* 278, 23753–23761.

Polak, J., Pivalova, J., Jokes, M., Svoboda, J., Scorza, R. and Ravelonandro, M. (2005) Preliminary results of interactions of plum pox virus (PPV), prune dwarf virus (PDV), apple chlorotic leaf spot virus (ACLSV) with transgenic plants of plum *Prunus domestica*, clone C-5 grown in an open field. *Phytopathologica Polonica* 36, 115–122.

Powell, A.P., Nelson, R.S., De, B., Hoffmann, N., Rogers, S.G., Fraley, R.T. and Beachy, R.N. (1986) Delay of disease development in transgenic plants that express the tobacco mosaic virus coat protein gene. *Science* 232, 738–743.

Qi, Y. and Hannon, G.J. (2005) Uncovering RNAi mechanisms in plants: biochemistry enters the foray. *FEBS Letters* 579, 5899–5903.

Qu, F. and Morris, T.J. (2005) Suppressors of RNA silencing encoded by plant viruses and their role in viral infections. *FEBS Letters* 579, 5958–5964.

Qu, F., Ren, T. and Morris, T.J. (2003) The coat protein of turnip crinkle virus suppresses posttranscriptional gene silencing at an early initiation step. *Journal of Virology* 77, 511–522.

Ravelonandro, M., Monsion, M., Delbos, R. and Dunez, J. (1993) Variable resistance to plum pox virus and potatovirus Y infection in transgenic plants expressing plum pox virus coat protein. *Plant Science* 91, 157–169.

Ravelonandro, M., Scorza, R., Bachelier, J.C., Labonne, G., Levy, L., Damsteegt, V., Callahan, A.M. and Dunez, J. (1997) Resistance of transgenic *Prunus domestica* to plum pox virus infection. *Plant Disease* 81, 1231–1235.

Reavy, B., Dawson, S., Canto, T. and MacFarlane, S.A. (2004) Heterologous expression of plant virus genes that suppress post-transcriptional gene silencing results in suppression of RNA interference in *Drosophila* cells. *BMC Biotechnology* 4, 18.

Reed, J.C., Kasschau, K.D., Prokhnevsky, A.I., Gopinath, K., Pogue, G.P., Carrington, J.C. and Dolja, V.V. (2003) Suppressor of RNA silencing encoded by beet yellows virus. *Virology* 306, 203–209.

Roth, B.M., Pruss, G.J. and Vance, V.B. (2004) Plant viral suppressors of RNA silencing. *Virus Research* 102, 97–108.

Sanford, J.C. and Johnston, S.A. (1985) The concept of pathogen-derived resistance: deriving resistance genes from the parasite's own genome. *Journal of Theoretical Biology* 113, 395–405.

Savenkov, E.I. and Valkonen, J.P. (2001) Coat protein gene-mediated resistance to potato virus A in transgenic plants is suppressed following infection with another potyvirus. *Journal of General Virology* 82, 2275–2278.

Scorza, R., Callahan, A., Levy, L., Damsteegt, V., Webb, K. and Ravelonandro, M. (2001) Post-transcriptional gene silencing in plum pox virus resistant transgenic European plum containing the plum pox potyvirus coat protein gene. *Transgenic Research* 10, 201–209.

Silhavy, D., Molnar, A., Lucioli, A., Szittya, G., Hornyik, C., Tavazza, M. and Burgyan, J. (2002) A viral protein suppresses RNA silencing and binds silencing-generated, 21- to 25-nucleotide double-stranded RNAs. *The EMBO Journal* 21, 3070–3080.

Simon-Mateo, C. and Garcia, J.A. (2006) MicroRNA-guided processing impairs plum pox virus replication, but the virus readily evolves to escape this silencing mechanism. *Journal of Virology* 80, 2429–2436.

Simon-Mateo, C., Lopez-Moya, J.J., Guo, H.S., Gonzalez, E. and Garcia, J.A. (2003) Suppressor activity of potyviral and cucumoviral infections in potyvirus-induced transgene silencing. *Journal of General Virology* 84, 2877–2883.

Smith, H.A., Swaney, S.L., Parks, T.D., Wernsman, E.A. and Dougherty, W.G. (1994) Transgenic plant virus resistance mediated by untranslatable sense RNAs: expression, regulation, and fate of nonessential RNAs. *The Plant Cell* 6, 1441–1453.

Smith, N.A., Singh, S.P., Wang, M.B., Stoutjesdijk, P.A., Green, A.G. and Waterhouse, P.M. (2000) Total silencing by intron-spliced hairpin RNAs. *Nature* 407, 319–320.

Takeda, A., Tsukuda, M., Mizumoto, H., Okamoto, K., Kaido, M., Mise, K. and Okuno, T. (2005) A plant RNA virus suppresses RNA silencing through viral RNA replication. *The EMBO Journal* 24, 3147–3157.

Vance, V.B., Berger, P.H., Carrington, J.C., Hunt, A.G. and Shi, X.M. (1995) 5' proximal potyviral sequences mediate potato virus X/potyviral synergistic disease in transgenic tobacco. *Virology* 206, 583–590.

Voinnet, O. (2001) RNA silencing as a plant immune system against viruses. *Trends in Genetics* 17, 449–459.

Voinnet, O. (2005) Induction and suppression of RNA silencing: insights from viral infections. *Nature Reviews Genetics* 6, 206–220.

Voinnet, O., Pinto, Y.M. and Baulcombe, D.C. (1999) Suppression of gene silencing: a general strategy used by diverse DNA and RNA viruses of plants. *Proceedings of the National Academy of Sciences of the United States of America* 96, 14147–14152.

Voinnet, O., Lederer, C. and Baulcombe, D.C. (2000) A viral movement protein prevents spread of the gene silencing signal in *Nicotiana benthamiana*. *Cell* 103, 157–167.

Wang, A., Sanfacon, H., Stobbs, L.W., James, D., Thompson, D., Svircev, A. and Brown, D.C.W. (2006) Plum pox virus in Canada: progress in research and future prospects for disease control. *Canadian Journal of Plant Pathology* 28, 182–196.

Wang, H., Buckley, K.J., Yang, X., Buchmann, R.C. and Bisaro, D.M. (2005) Adenosine kinase inhibition and suppression of RNA silencing by geminivirus AL2 and L2 proteins. *Journal of Virology* 79, 7410–7418.

Wang, M.B. and Metzlaff, M. (2005) RNA silencing and antiviral defense in plants. *Current Opinion in Plant Biology* 8, 216–222.

Waterhouse, P.M., Graham, M.W. and Wang, M.B. (1998) Virus resistance and gene silencing in plants can be induced by simultaneous expression of sense and antisense RNA. *Proceedings of the National Academy of Sciences of the United States of America* 95, 13959–13964.

Waterhouse, P.M., Wang, M.B. and Lough, T. (2001) Gene silencing as an adaptive defence against viruses. *Nature* 411, 834–842.

Xie, Z., Johansen, L.K., Gustafson, A.M., Kasschau, K.D., Lellis, A.D., Zilberman, D., Jacobsen, S.E. and Carrington, J.C. (2004) Genetic and functional diversification of small RNA pathways in plants. *PLoS Biology* 2, E104.

Yang, S. and Ravelonandro, M. (2002) Molecular studies of the synergistic interactions between plum pox virus HC-Pro protein and potato virus X. *Archive of Virology* 147, 2301–2312.

Ye, K. and Patel, D.J. (2005) RNA silencing suppressor p21 of beet yellows virus forms an RNA binding octameric ring structure. *Structure (Cambridge)* 13, 1375–1384.

Yelina, N.E., Savenkov, E.I., Solovyev, A.G., Morozov, S.Y. and Valkonen, J.P. (2002) Long-distance movement, virulence, and RNA silencing suppression controlled by a single protein in hordei- and potyviruses: complementary functions between virus families. *Journal of Virology* 76, 12981–12991.

Zhang, X., Yuan, Y.R., Pei, Y., Lin, S.S., Tuschl, T., Patel, D.J. and Chua, N.H. (2006) Cucumber mosaic virus-encoded 2b suppressor inhibits Arabidopsis Argonaute1 cleavage activity to counter plant defense. *Genes and Development* 20, 3255–3268.

Zhou, Z., Dell'orco, M., Saldarelli, P., Turturo, C., Minafra, A. and Martelli, G.P. (2006) Identification of an RNA-silencing suppressor in the genome of grapevine virus A. *Journal of General Virology* 87, 2387–2395.

17 Molecular Characterization of Endogenous Plant Virus Resistance Genes

F.C. Lanfermeijer and J. Hille

Abstract

Plants are exposed to a wide variety of threats. One of the most difficult to counteract is that posed by pathogens. Pathogens constantly adapt to invade plants and take advantage of them. Plants have developed a wide array of defence mechanisms to avoid pathogen infection and limit their detrimental effects on growth and development. These defence mechanisms are continuously being updated to keep in line with the pathogens arsenal development. One of the best studied disease models in plants is the *Tobacco mosaic virus*–tobacco interaction. A century ago, this virus was the first to be isolated and structurally characterized. Its pathogenicity on tobacco has been studied for decades. In this chapter, recent developments in the characterization of plant resistance genes that are effective against plant viruses will be discussed using *Tobacco mosaic virus* (TMV) as a prime example.

Introduction

TMV was the first recognized virus. Its study has played a leading role in the development of virology (Creager et al., 1999). TMV is still a major research subject in plant virology and phytopathology, and as such is one of the best known plant viruses. In recent years, three natural resistance genes have been characterized (*N*, *Tm-2* and *Tm-2²*), which target TMV gene products. At the same time, major advances have been made in understanding the TMV replication cycle. In this chapter, we will use the TMV–tobacco interaction as a model system and as a continuous thread to discuss endogenous plant resistance genes effective against viruses. This focus allows the compilation of research advances in virology and phytopathology. These will be discussed in relation to the molecular analysis of natural viral resistances.

Tobamovirus Infection Process: Know Your Enemy!

Economically, TMV and *Tomato mosaic virus* (ToMV) are the most important species of the genus *Tobamovirus*. But also, *Pepper mild mottle virus* and *Odontoglossum ringspot virus* can cause considerable economic damage in pepper cultures and orchid cultures, respectively. Tobamoviruses belong to the alpha-like supergroup of viruses (Goldbach and de Haan, 1993). TMV, with the *vulgare* isolate (TMV-U1) as an example species, consists of a characteristic proteinaceous rod, made up from 2140 copies of the coat protein (CP, also known as capsid protein) that are helically stacked around the RNA genome. The genome consists of a single molecule of single-stranded linear positive-sense RNA, 6395 nucleotides in length. Traditionally, four open reading frames have been recognized in the genome of TMV (Zaitlin, 1999; Fig. 17.1). The first two open reading frames overlap and encode a 126 kDa and a 183 kDa protein. The 183 kDa protein is produced by occasional read-through of an amber termination codon located at the C-terminus of the 126 kDa protein. The functional TMV replicase consists of the 126 and 183 kDa proteins together with additional host proteins. This replication complex is responsible for the synthesis of the viral genomic RNA and subgenomic RNAs. The movement protein (MP) and the CP are translated from individual subgenomic

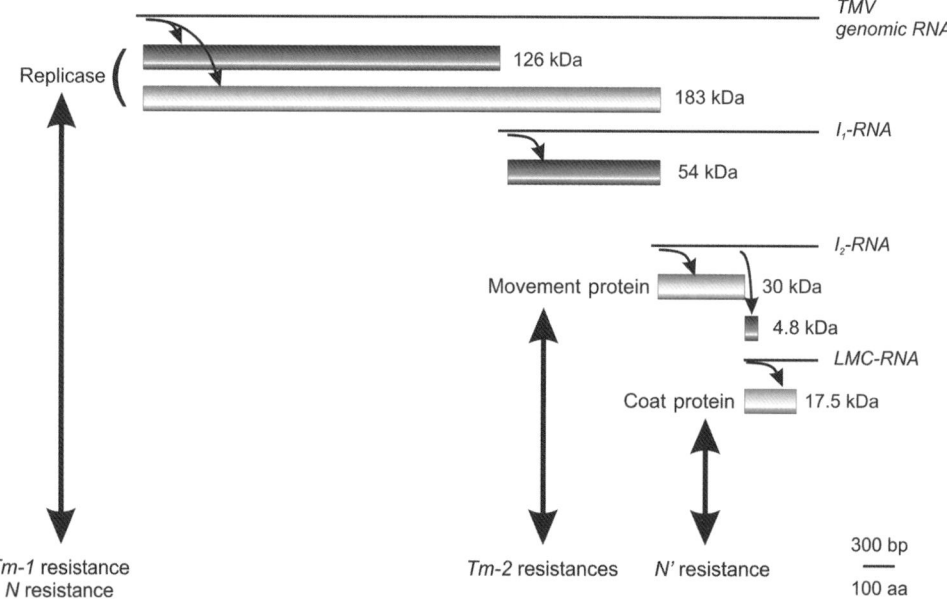

Fig. 17.1. Map of *Tobacco mosaic virus* genome. The lines represent RNAs and the cylinders represent the virus-encoded proteins that are translated from the different viral RNAs. See text for further description. Double-headed arrows indicate the interaction between the viral proteins and the different plant resistance proteins. In addition to the *N'* resistance, the coat protein is also targeted by the *L* resistances, *Hk* resistance and the *S* resistance (see Table 17.1B).

RNAs: I_2 and *LMC*, respectively. An additional subgenomic messenger RNA, I_1, codes for an ambiguous 54 kDa protein, which is not one of the traditionally recognized viral proteins, and which coincides with the read-through 3'-portion of the 183 kDa protein. The function of this protein is not known. A sixth open reading frame has also been recognized in the TMV genome. This open reading frame overlaps the MP and CP open reading frames and encodes a 4.8 kDa protein. This protein is a virulence factor and may be translated from the I_2-subgenomic RNA (Canto *et al.*, 2004).

Production of viral RNA takes place in large replication complexes. These complexes are associated with endoplasmic reticulum (ER) membranes and the cytoskeleton and contain not only the viral replicase proteins, but also the MP, viral RNA and a number of host proteins (Scholthof, 2005). Because of its regulatory role in cell-to-cell movement and RNA transcription, CP was also suggested to interact with the replication complexes (Scholthof, 2005). Another important step of the viral infection cycle is the movement of the virus, first within the cell from the site of replication to the plasmodesmata, second, from cell to cell through the plasmodesmata and third, systemically through the plant. Hence, systemic spread of the virus requires the movement of viral products through tissues with different characteristics (e.g. mesophyll, vascular tissues) and across boundaries between those tissues (e.g. between companion cell and sieve element) (Waigmann *et al.*, 2004: Scholthof, 2005).

Traditionally, individual viral proteins have been implicated with each of these processes. MP was assigned to movement through the cell, cell-to-cell movement and gating of the plasmodesmata and CP was assigned to systemic transport through the phloem of the plants. However, it has become clear that this picture is not as simple as originally thought (Beachy and Heinlein, 2000; Heinlein, 2002; Waigmann *et al.*, 2004; Scholthof, 2005). Both MP and the replicase proteins are able to associate with microfilaments in the host cell, which may function in intracellular movement of the membrane-associated replication complexes towards the plasmodesmata. The association of replication complexes with ER membranes has been attributed to MP, because this protein is a membrane protein with two membrane spanning domains (Scholthof, 2005). Yet, a role of the replicase proteins in this association was also suggested, either indirectly through interaction with membrane-associated host proteins or directly through interaction with the membrane itself (dos Reis Figueira *et al.*, 2002).

Currently, two views exist concerning movement of the virus to neighbouring cells. In the classical view, MP associates with viral RNA and mediates the movement of viral RNA through the plasmodesmata into the neighbouring cells. Subsequently, all viral proteins necessary for replication are produced in the newly infected cell (Beachy and Heinlein, 2000). In the second view, complete viral replication complexes move through the plasmodesmata, a process mediated by MP. This model, which implies the arrival of active replication complexes into the newly infected cell, was proposed to explain the observation that virus replication is rapidly established in secondary infected cells (Scholthof, 2005).

The CP of tobamoviruses has been implicated in the systemic dispersal of the virus through the plant based on mutational and complementation studies (see Waigmann et al., 2004) but also in this case involvement of an additional viral protein, the 126 kDa protein, was suggested. The 126 kDa protein acts either directly by helping virions traversing the mesophyll/phloem parenchyma boundaries or indirectly by suppressing post-transcriptional gene silencing (Scholthof, 2005).

An important conclusion from this short and certainly not complete overview of the tobamovirus replication cycle is that due to the long research history, extensive information on tobamovirus replication cycle and on the function of the viral gene products is available (Creager et al., 1999; Shaw, 1999; Zaitlin and Palukaitis, 2000; Waigmann et al., 2004). More specifically, it is now clear that the viral gene products fulfil their role(s) in cooperation with other viral proteins and host proteins. Hence, this must be considered when designing models for the mode of action of tobamovirus resistance genes.

Protection Against Tobamovirus Infections in Agriculture

In contrast to most other plant viruses, which have specific arthropod or fungal vectors, tobamoviruses are mechanically transmitted, not only by physical handling of the crops but also through opportunistic usage of insects and other animals. The virus particles are extremely stable and highly infectious. They can reside in soil and decaying plant material for a long time and can infect new hosts from this material. Infection takes place probably through small wounds, resulting from herbivory, mechanical stress or human activity. The fact that simple touch with contaminated material allows virus transmission necessitates a high degree of phytosanitary measures and pest prevention. In practice, these measures are not very effective. Therefore, additional means must be deployed. A classical way of preventing damage by virus infection is by cross-protection (Fulton, 1986). Plants are intentionally infected with mild strains of the pathogen in order to prevent infection with a related severe strain. Chemical protection can also be used to prevent the establishment of large populations of insects and other biological vectors. However, the most effective method of prevention of virus infection has been the introduction of natural resistances against plant viruses into susceptible plants through breeding.

A well-known successful example of the introduction of a natural resistance (*R*) gene into a susceptible species is that of the *N* gene of *Nicotiana glutinosa*, which was introduced into *N. tabacum* by classical introgression (Marathe et al., 2002). This led to a large decrease in yield loss resulting from TMV infections. The introgressions of the *Tm* resistances (*Tm-1*, *Tm-2* and *Tm-2²*) from *L. hirsutum*, *L. peruvianum* and *L. peruvianum*, respectively, into *L. esculentum* are other examples of this approach (Pelham, 1966; Hall, 1980). The success rate of the introgression of the three *Tm* resistances differed, which demonstrates that durability of the resistance is one of the major issues with the implementation of natural disease resistances in commercial plant cultures (Hulbert *et al.*, 2001).

Recently, with the development of biotechnology, additional methods to introduce virus resistance in plants have become available. These include: (i) pathogen-derived resistance in which a viral protein or RNA is expressed in transgenic plants (see Chapters 15 and 16, this volume); and (ii) the transfer of natural resistance from one plant species to an unrelated plant species through genetic modification.

Natural Defence Mechanisms Against Viruses in Plants

Plants have developed two major lines of defences against infection by pathogens. First, plants have developed passive and ubiquitously present barriers to prevent entrance of pathogens. These physical barriers consist of cell walls, cuticles and obstacles like hairs and trichomes that repel herbivores and insects, some of which also transmit other pathogens.

The second line of defence is inducible, depends on recognition and detection of the pathogen, and involves active measures towards the pathogen. Within these inducible defences, three types are discriminated: (i) basal and general defence response against all microbes; (ii) pathogen- and host-specific resistance; and (iii) post-transcriptional gene silencing. Basal and general defence (i) depends on general microbial elicitors. Recognition of these elicitors usually takes place outside the cell. This type of resistance has been termed non-host-specific resistance. Pathogen-and host-specific resistance (ii) depends on specific gene products produced by the pathogen (the avirulence protein or Avr protein) and the recognition of these products by specific host resistance proteins (the R protein). This type of resistance is usually associated with a hypersensitive response. Post-transcriptional gene silencing (iii) is also a non-host-specific type of resistance, because no specialized and pathogen-specific plant recognition proteins are necessary. Instead, the specific recognition of pathogenic RNAs is accomplished by the use of small 21–23 nucleotide RNAs, which are derived from pathogenic double-stranded RNAs (Chapter 16, this volume). Gene silencing is only effective against viruses because of the formation of double-stranded RNAs during virus replication. Together, the first two types of inducible resistance are generally called the innate immune response and function against many types of pathogens (bacteria, oomycetes, insects, nematodes, fungi and viruses).

It should be noted that both post-transcriptional gene silencing and basal resistance are unable to completely protect the plant against tobamovirus infections. In the remainder of this chapter, we will concentrate our discussion on host-specific resistance since this type of resistance has been the focus of many breeding attempts to protect plants against tobamoviruses. However, some general remarks considering basal resistance and gene silencing in relation to natural virus resistance and especially to tobamovirus resistance need to be made. Basal resistance is triggered by the perception of general elicitors, which contain specific recognizable molecular patterns that are present in essential and ubiquitous structures from most potential pathogens. The nature of these elicitors ranges from carbohydrates derived from microbe cell walls, other molecules characteristic for bacterial cell walls and peptide epitopes derived from conserved microbe proteins, such as flagellin. These elicitors are generally perceived by receptors located on the plasma membrane of the plant cell. A few of these receptors have been characterized (Gomez-Gomez and Boller, 2002; Jones and Takemoto, 2004; Boller, 2005). Responses triggered by the perception of these elicitors include changes in ion fluxes, changes in internal calcium concentrations, production of reactive oxygen species (ROS), cell-wall reinforcement, phosphorylation, dephosphorylation, callose deposition and induction of general defence-related genes (Gomez-Gomez and Boller, 2002). This response is considered general and independent of the host-specific resistance.

The general non-host response is also triggered by viruses. Exposure of plant tissues to fungal and bacterial elicitors induces the production of ROS in two distinct phases: a rapid phase (within minutes or hours), termed phase I; and a later one (hours or days), termed phase II. While the appearance of the phase II oxidative burst correlates with the presence of specific resistance genes, phase I is a more general response that does not necessitate the presence of host- and pathogen-specific resistance genes (Baker and Orlandi, 1995). Challenging plant tissues with virus also results in rapid ROS production (Allan *et al.*, 2001; Love *et al.*, 2005), which displays similar time kinetics and non-dependency on the presence of resistance genes as that induced by other pathogens (Allan *et al.*, 2001). Other characteristics of ROS production induced by TMV are also similar to general elicitor perception, e.g. dependency on Ca^{2+} flux and (de)phosphorylation (Mittler *et al.*, 1999; Allan *et al.*, 2001). Whether these responses are triggered by perception of specific virus elicitors in analogy with other general elicitors, like flagellin, or whether they are triggered by the entrance process of the viruses into the cells remains to be elucidated. However, it is clear that this perception takes place extracellularly and is independent of the virus-specific intracellular resistance proteins. The role of early responses that are associated with non-host-specific basal resistance needs to be determined: are they necessary for priming host-specific resistance (Allan *et al.*, 2001) or do they have their own defensive capability?

Post-transcriptional gene silencing is deployed against tobamovirus infections and reduces virus accumulation in plants (Xie *et al.*, 2001; Yang *et al.*, 2004). However, tobamoviruses can suppress this silencing, in which the 126 kDa protein has been identified as a key player (Scholthof, 2005).

The reader is referred to several excellent reviews of non-host-specific defence (Gomez-Gomez and Boller, 2002; Jones and Takemoto, 2004; Boller, 2005) and to Chapter 16 for a discussion of post-transcriptional gene silencing.

Resistance Genes: From the Gene-for-Gene Hypothesis to the Guard Hypothesis

Many plant species have developed mechanisms to counteract infection by viruses, which are based on host-specific resistance, and which target viral gene products. In host-specific resistance, the recognition of pathogens by plants has been associated with the 'gene-for-gene' hypothesis (Flor, 1971). The 'gene-for-gene' hypothesis is based on the observation that both the plant and the pathogen need to synthesize a gene product for resistance to take place. As mentioned above, these gene products are the plant R protein and the pathogen Avr protein. If one of those proteins is not produced or is changed, the plant becomes infected. Avr proteins are very diverse in nature and function. On the other hand, most R proteins of plants are members of six families, all with the same function: the detection of the pathogen presence and, subsequently, the activation of the defence mechanism. To fulfil this role, all R proteins are made up from modules which have distinct roles in this process (Hammond-Kosack and Jones, 1997; Martin *et al.*, 2003; Belkhadir *et al.*, 2004; McHale *et al.*, 2006). R proteins contain domains which function in the recognition of the pathogen and domains which function in the activation of the downstream components of the signal transduction network. Several observations suggest that R proteins function in complexes with homologous and heterologous proteins. First, immunoprecipitation experiments showed oligomerization of the N protein in the presence of its Avr protein (Mestre and Baulcombe, 2006). Second, various potential interactors have been identified by yeast two-hybrid experiments (de la Fuente van Bentem *et al.*, 2005).

Originally, the Avr protein and the R protein were thought to interact directly with each other. Due to the limited number of demonstrated direct interactions between host proteins and avirulence proteins, the idea has arisen that R proteins guard host proteins from hijacking and exploitation by pathogens (van der Biezen and Jones, 1998; Ellis *et al.*, 2000; Glazebrook, 2001; Van der Hoorn *et al.*, 2002). In the 'guard' hypothesis, the effects of avirulence proteins on host proteins (the guardees) are monitored by the R proteins, which then trigger the defence reaction. Hence, direct interaction between Avr protein and R protein is not always necessary as long as the effect of the Avr protein on the targeted process or protein is detected. Recently, it was suggested that plant R proteins can function either by directly detecting the corresponding Avr protein, the 'receptor-ligand' model, or by perceiving alterations in plant machines that are targets of Avr protein action in the promotion of pathogen virulence, the 'guard hypothesis' (Dodds *et al.*, 2006). The guard hypothesis might also explain the presence of alleles which do not confer resistance. These alleles may be either non-functional alleles or resistance proteins directed against a distinct pathogen, which is not present in the current plant culture. However, the non-random distribution of mutations in these non-resistance conferring alleles suggests that they are exposed to selective pressure. One possible explanation is that the non-resistance conferring alleles guard the same process or protein as the functional resistance alleles. However, the non-resistance conferring alleles would not detect changes induced by the pathogen, but would rather detect similar changes induced by other phenomena. In this sense, R proteins and their homologues could have a broader function in the biology of the cell (Lanfermeijer et al., 2003): not only discriminating between 'self and non-self' but also between 'self and wrong-self'. From this point of view, it would be interesting to introduce auto-activating mutations (Rathjen and Moffett, 2003; Takken et al., 2006; Tameling et al., 2006) in the proteins encoded by non-resistance conferring alleles to discriminate between signalling-incapable proteins (real non-functional proteins) and functional but recognition-incapable proteins.

The current view is that either upon direct binding of the R protein with the Avr protein or upon binding of the R protein with the host protein, which is affected by the Avr protein, the signal transduction cascade is activated. Considerable progress has been made to unravel the mechanism of signal transduction. Studies with two R proteins, *I-2* conferring resistance to *Fusarium oxysporum* and *Mi-1* conferring resistance to root-knot nematodes and potato aphids, provided the first evidence that ATP hydrolysis is involved in signal transduction via CNL proteins (Tameling et al., 2002, 2006). Studies with various mutants of the protein of the *Rx* resistance of potato against *Potato virus X* demonstrated an important role for intramolecular interactions between the three domains of the R protein and the disruption of these interactions in activation of the resistance (Moffett et al., 2002; Rathjen and Moffett, 2003; Rairdan and Moffett, 2006) Together, the studies on the *I2*, *Mi-1* and *Rx* resistances resulted in a model for intramolecular rearrangement of subdomains as the result of R protein/elicitor interaction (Rathjen and Moffett, 2003; Belkhadir et al., 2004; Albrecht and Takken, 2006; Rairdan and Moffett, 2006; Takken et al., 2006). Recently, it has been shown that the N protein is also able to hydrolyse ATP (Ueda et al., 2006). It has been proposed that the R protein cycles through different conformations, driven by transient interaction with the elicitor and by ATP hydrolysis. The different conformations have different characteristics of interaction with the elicitor and activation of the resistance mechanisms. Subsequently, this iterative mechanism was suggested to result in signal amplification (Rairdan and Moffett, 2006). This mechanism could also explain the absence of experimental evidence for direct interactions between the R proteins and their elicitors because of the transient interaction between the two proteins during the activating cycle.

The Hypersensitive Response

Resistance against plant viruses is usually associated with a hypersensitive response (HR). This HR becomes visible as small necrotic lesions at the infected regions on the plant. These visible lesions are considered to be dead tissue as a result of programmed cell death. This programmed cell death is triggered by the resistance response and is important for resistance as it limits the spread of the pathogen. Within plant virus resistance, however, several different degrees of resistance can be observed. The intensity of the HR and/or its ability to control virus infection varies from case to case. The classical form of resistance, where arrest of infection is associated with necrotic lesions, is observed with the *N* resistance in tobacco (Whitham *et al.*, 1994).

Another case is the extreme resistance observed with the *Rx* resistance against *Potato virus X*. This resistance is characterized by the absence of a visible necrotic HR, although the virus infection is arrested (Bendahmane *et al.*, 1999). Similarly, under normal circumstances, lesions are not visible in the case of the *Tm-2* and *Tm-2^2* resistances. It has been suggested that the rapid HR is so efficient in controlling virus spread that the lesions are too small to be visible (Lanfermeijer *et al.*, 2003). Expression of salicylate hydrolase gene in plants reduces the internal concentration of salicylic acid, a known signalling molecule of HR. When this gene is expressed in *Tm-2^2* plants, visible lesions or even spreading necrosis is observed, suggesting that a slower HR response has occurred (Lanfermeijer *et al.*, 2003). Similar symptom enhancing effects are also observed when the *salicylate hydroxylase* gene is expressed in combination with the *N*-resistance gene (Chivasa and Carr, 1998). Whether the *Rx* extreme resistance has similar characteristics is unknown.

At the other end of the spectrum are cases in which necrotic local lesions are formed with no apparent restriction of virus invasion. For example, the Y-1 resistance, genetically transferred to susceptible potato varieties, displays an HR but is incapable of arresting PVY infection (Vidal *et al.*, 2002). A similar situation occurs when a crucifer- and garlic-infecting TMV is inoculated on tobacco plants in the absence of the *N* gene (Ehrenfeld *et al.*, 2005). These observations point to a complex relation between HR and virus resistance. HR alone is apparently not a prerequisite for resistance. Its role is still unclear and this suggests that plants deploy additional methods to arrest virus infection. However, little is known about these mechanisms.

Classes of Resistance Genes

There are currently five well-defined classes of R proteins (Fig. 17.2; Hammond-Kosack and Jones, 1997; Martin *et al.*, 2003; Belkhadir *et al.*, 2004; McHale *et al.*, 2006) in plants. New forms of resistance proteins are discovered on a regular basis that do not fit in the five established classes of R proteins. The first class is represented by Pto from tomato, which confers resistance to *Pseudomonas syringae* pv. *tomato* (Pedley and Martin, 2003). Pto is unique in its class and no virus resistance genes that encode this type of R proteins have been isolated until now.

The next two classes are plasma membrane-localized proteins, which belong to the related protein families of receptor-like kinase (e.g. Xa21 protein, which is involved in rice bacterial blight disease resistance; Song *et al.*, 1995) or receptor-like proteins (e.g. Cf proteins, which are involved

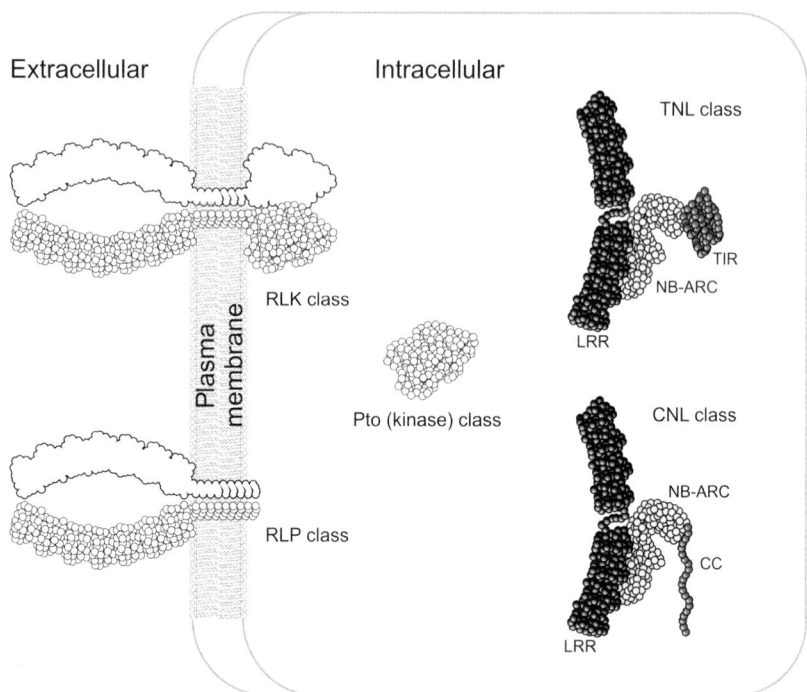

Fig. 17.2. Schematic representation of the five characterized resistance-protein classes. In white circles, the R protein classes which do not include virus resistance proteins; namely, the receptor-like kinase (RLK), the receptor-like protein (RLP) and the Pto classes. Dimerization of RLK and RLP classes is indicated by the presence of a silhouette. The two classes which include virus resistance proteins, the CC-NBS-LRR (CNL) and the TIR-NBS-LRR (TNL) classes, are indicated by structures with domains in different shades of grey. See text for a detailed description of the classes (Hammond-Kosack and Jones, 1997; Martin et al., 2003; Belkhadir et al., 2004; McHale et al., 2006). Domain organization of the CNL and TNL resistance proteins is indicated by shape and shades of grey. The darkest curved structure represents the leucine-rich repeats (LRR) domain; the putative hinch domain of the LRR domain is indicated. The light grey structure with the groove represents the nucleotide binding site domain (NB-ARC). Intramolecular interaction between NB-ARC and LRR domains is schematically shown. The grey helix and the grey diamond shape represent the coiled coil (CC) structure and the Toll and Interleukin 1 Receptor (TIR) domain, respectively.

in *Cladosporium fulvum* resistance in tomato; Kruijt et al., 2005), respectively. Due to their structure, these types of proteins are highly suitable for the extracellular perception of avirulence signals and the subsequent transduction of the signal intracellularly. The completion of the genome sequence of several plant species shows that plants contain a large number of genes encoding these types of proteins (Morillo and Tax, 2006). There are currently no characterized virus resistance genes that are associated with these classes of proteins. This is perhaps not surprising as detection of viral products is likely to occur intracellularly. On the other hand, these types of proteins are prime candidates to function in the extracellular perception of viruses if non-host-specific basal resistance depends on specific receptors (see above).

The last two large classes of resistance genes code for intracellular proteins, which have a C-terminal region of leucine-rich repeats (LRRs) and a central nucleotide binding site (NBS). At the N-terminus of these proteins are either a coiled coil (CC) sequence or a Toll and Interleukin 1 Receptor (TIR) sequence, distinguishing the CC-NBS-LRR (CNL) or TIR-NBS-LRR (TNL) classes of resistance proteins. All characterized virus resistance genes belong to these last two classes (Table 17.1.A). This is in

Table 17.1. Plant virus resistance genes.

A. Isolated and characterized plant virus resistance genes – 2006 status.

Gene	Plant	Virus[a]	Phenotype[b]	Avr protein	Class[c]	References
HRT	A. thaliana	TCV	HR		CNL	Cooley et al. (2000)
N	N. tabacum	TMV	HR	Replicase	TNL	Whitham et al. (1994)
RCY1	A. thaliana	CMV	HR	CP	CNL	Takahashi et al. (2002)
RTM1	A. thaliana	TEV	No HR	Unknown	Jacalin-like[d]	Chisholm et al. (2000)
RTM2	A. thaliana	TEV	No HR	Unknown	HSP[e]-like	Whitham et al. (2000)
Rx1	S. tuberosum	PVX	ER	CP	CNL	Bendahmane et al. (1999)
Rx2	S. tuberosum	PVX	ER	CP	CNL	Bendahmane et al. (2000)
Sw5	L. esculentum	TSWV	HR	Replicase	CNL	Brommonschenkel et al. (2000); Spassova et al. (2001)
Tm-2	L. esculentum	ToMV	HR	MP	CNL	Lanfermeijer et al. (2005)
Tm-2²	L. esculentum	ToMV	HR	MP	CNL	Lanfermeijer et al. (2003)
Y-1	S. tuberosum	PVY	HR	Unknown	TNL	Vidal et al. (2002)

B. Additional tobamovirus resistance genes in *Solanaceae*.

Gene	Plant	Virus[a]	Phenotype[b]	Avr protein	Class[c]	References
N'	N. sylvestris	TMV	HR	CP	Unknown	Knorr and Dawson (1988)
L1/L2/L3/L4	C. annum	TMV	HR	CP	Unknown	Boukema (1980)
S[f]	S. melongena	TMV, ORSV	HR	CP	Unknown	Dardick and Culver (1997)
Hk	C. annum	PaMMV	HR	Replicase	Unknown	Sawada et al. (2005)

[a] CMV, *Cucumber mosaic virus*; ORSV, *Odontoglossum ringspot virus*; PaMMV, *Paprika mild mottle virus*; PVY, *Potato virus Y*; PVX, *Potato virus X*; TCV, *Turnip crinkle virus*; ToMV, *Tomato mosaic virus*; TEV, *Tobacco etch virus*; TMV, *Tobacco mosaic virus*; TSWV, *Tomato spotted wilt virus*.
[b] ER, extreme response; HR, hypersensitive response.
[c] CNL: CC-NBS-LRR R protein; TNL: TIR-NBS-LRR R protein (see text).
[d] jacalin: a D-galactose-specific lectin.
[e] HSP, heath shock protein.
[f] S: *Solanum melongena* resistance.

agreement with the fact that the highest probability of encountering a viral protein is within the cell.

Several *R* genes have been characterized that do not fit the above presented classification. For example, the *RTM* genes allow genome replication and cell-to-cell movement but do not allow long-distance movement of *Tobacco etch virus*, and as such protect the plant against systemic infection. The resistance mechanism does not involve hypersensitive cell death or systemic acquired resistance (Chisholm *et al.*, 2000; Whitham *et al.*, 2000).

Natural Resistance Against Tobamoviruses

Infection of plants by viruses is an opportunistic, not an active process. While fungi and bacteria invest resources into infection systems, viruses depend on biological vectors, wounding or other opportunities to enter the living plant cell. Hence, as long as the primary and basal defences are able to exclude the virus from coming in close contact with living cells, these defences have their function in virus resistance.

The step of entering the living cell by tobamoviruses remains elusive but several mechanisms have been suggested, including among others: entering through small holes in the membrane, membrane attachment followed by membrane passage and endocytosis (Shaw, 1999). No indications exist that plants are able to actively interfere with these processes, although the induction of ROS production by TMV might indicate an attempt by the plant to interfere with this process (see above).

Due to their economical and agricultural importance, *Solanaceae* species have been model organisms in plant physiology, phytopathology and genetics for many years. Extensive breeding programmes have been in place to introduce disease resistances and other traits into commercial varieties, for example, *L. esculentum* and *N. tabacum*. These large breeding efforts led to the production of a large number of different genetic lines, including near-isogenic lines that differ in single traits. This stimulated the isolation of the alleles responsible for these traits. Some of these alleles were the first examples of the *R* gene classes. Several natural resistances against tobamoviruses are or were commercially used in *Solanaceae* (Table 17.1.B) with the most renowned example being the *N* resistance (Marathe *et al.*, 2002).Although resistance-breaking virus isolates have appeared during the last decades, these do not cause major problems within tobacco culture. TMV-Ob is an *N*-resistance breaking isolate, which enabled the identification of the *Avr* gene recognized by the *N* resistance (Padgett *et al.*, 1997). However, the economic damage due to TMV-Ob remains limited due to the fact that the ability of TMV-Ob to overcome the *N* resistance is conditional. At temperatures below 19°C, the *N* resistance recognizes TMV-Ob and triggers an HR.

Temperature sensitivity of the *N*-mediated resistance was used to isolate the *N* gene through transposon tagging. At temperatures below 28°C, plants carrying the *N* gene trigger an HR in response to TMV infection. However, above 28°C, the HR response is not induced and the virus spreads systemically in the plant. When these systemically infected plants are shifted to 21°C, they develop lethal systemic HR. The transposon-tagged and, thus, inactivated *N* gene was isolated from plants that did not display systemic necrosis after the temperature shift (Whitham *et al.*, 1994).

In cultivated tomato, ToMV infections are controlled by the introgressed *Tm-1*, *Tm-2* and *Tm-2²* R genes (Pelham, 1966; Hall, 1980). Among these resistances, the *Tm-2²* resistance has shown to be remarkably durable and, therefore, of practical importance. On the other hand, the *Tm-1* and *Tm-2* were not as successful because these resistances are frequently overcome by naturally occurring ToMV strains (Fraser *et al.*, 1989).

The *Tm-1* resistance gene was introgressed from the wild tomato species *L. hirsutum*. Genetic analysis of ToMV strains capable of overcoming this resistance has shown that two amino acid changes in the replicase proteins are responsible for this phenotype, suggesting that either one of these proteins or both are the matching Avr protein for the *Tm-1* resistance (Meshi *et al.*, 1988). The *Tm*-1 gene has not been isolated and characterized thus far.

Both the *Tm-2* and the *Tm-2²* resistances were introgressed from *L. peruvianum*. These resistances are allelic (Lanfermeijer *et al.*, 2005). Analysis of the nucleotide sequence of ToMV strains, which were capable of overcoming either the *Tm-2* or the *Tm-2²* resistance, revealed that in both cases, MP is the matching Avr protein (Meshi *et al.*, 1989; Calder and Palukaitis, 1992; Weber *et al.*, 1993; Weber and Pfitzner, 1998). The existence of these two alleles, which share the MP as Avr protein but are known to have different degrees of durability, makes the analysis and study of these two resistance genes valuable. The *Tm-2* and *Tm-2²* genes have recently been cloned. Both are members of the CNL resistance protein family (Lanfermeijer *et al.*, 2003, 2005). Like the *N* gene, the *Tm-2²* gene was isolated through transposon tagging. Crosses were made between tomato plants without *Tm-2²* and containing the MP transgene and plants homozygous for the *Tm-2²* allele and containing a linked and activated Ac transposon. The resulting seedlings containing both an intact *Tm-2²* gene and an MP transgene will display systemic necrosis (Weber and Pfitzner, 1998), except when the seedling contains a disrupted and inactive *Tm-2²* gene. From the surviving seedlings, the tagged *Tm-2²* was isolated and characterized (Lanfermeijer *et al.*, 2003). The homologous alleles, *letm-2*, *lptm-2* and *Tm-2*, were isolated using *Tm-2²*-based primers (Lanfermeijer *et al.*, 2003, 2005).

The *N*-resistance Gene

The *N* gene was the first resistance gene isolated that encodes an intracellular TNL resistance protein (Whitham *et al.*, 1994). Due to the long attention this resistance has received, considerable progress has been made in the study on the N protein, i.e. its specific resistance mechanism, its signal transduction and the general concept of disease resistance. It was shown that the transcription of the gene can generate two transcripts, N_S and N_L, as the result of alternative splicing. Both transcripts are required for resistance to TMV (Dinesh-Kumar and Baker, 2000). The N-resistance response is elicited by the 126 kDa TMV protein and more specifically by the C-terminal 50 kDa helicase domain (Padgett et al., 1997). From an agricultural point of view, it is interesting to mention that the tobacco N gene can confer resistance to TMV in heterologous, although phylogenetically closely related, plants, like tomato and N. benthamiana (Whitham et al., 1996; Bendahmane et al., 1999; Liu et al., 2002). Studies on N-signal transduction suggest essential roles of the tobacco genes EDS1, Rar1, NPR1/NIM1 in the N-mediated resistance signalling pathway (Liu et al., 2002). The recent characterization of NRG1 (a CNL-like protein; Peart et al., 2005) shows that CNL proteins function in downstream signalling pathways, in addition to their established role in disease resistance. This involvement of CNL proteins in downstream signalling was also demonstrated for Cf-resistance signal transduction (Gabriels et al., 2006). Even more remarkable is the recent observation that a TNL-like protein plays a key role in the signalling pathway of photomorphogenetic development (Faigon-Soverna et al., 2006), suggesting roles for this type of proteins in immunity and development and offering an additional explanation for the existence of non-resistance conferring alleles of R genes (see above). Other novel observations on N gene are the induction of N protein oligomerization by its Avr protein (Mestre and Baulcombe, 2006) and the observation that the N protein, like the I-2 and Mi proteins (Tameling et al., 2002, 2006), is capable of ATP hydrolysis (Ueda et al., 2006).

The *Tm-2* and *Tm-2²* resistances

The $Tm-2$ and $Tm-2^2$ resistances differ in two features: their durability and the locations of mutations found in the MP protein of virus isolates that can break the resistance. One possible explanation for the durability of a resistance gene is that the loss of the corresponding Avr gene from the pathogen has a significant cost in terms of fitness or pathogenicity (Hulbert et al., 2001). While MP is essential for the tobamovirus replication cycle, the mutations in the different regions of MP known to be essential for the interaction with the Tm-2 and Tm-2² resistance proteins (Calder and Palukaitis, 1992; Weber et al., 1993; Weber and Pfiztner, 1998) have different effects on the biological function of MP. This is reflected in the differences in pathogenicity of the $Tm-2$ and $Tm-2^2$ breaker strains, the $Tm-2$ breaker being as virulent as the wild-type while the $Tm-2^2$ breakers are impaired in their virulence (Fraser et al., 1989), and this explains the different durabilities of the $Tm-2$ and $Tm-2^2$ resistances. Therefore, one might refine the above explanation for R gene durability to state that an R gene may remain durable if amino acid changes in the Avr protein have a significant cost in terms of fitness or pathogenicity. As a result of the importance of the *Tm-2²*-interacting domain for the function of MP and, consequently, for the virulence of the virus, this domain was depicted as the Achilles' heel of the ToMV virus (Lanfermeijer et al., 2003).

Surprisingly, the Tm-2² and the Tm-2 proteins differ by only four amino acids (Lanfermeijer et al., 2005). Of these four amino acid differences, two are located in the NB-ARC domain (Ile257Phe and Met286Ile [*Tm-2²* versus *Tm-2*]) whereas the other two are in the LRR domain and only one amino acid apart (Tyr767Asn and Ser769Thr). It is likely that the difference in virus-specificity of the two alleles is caused by these last two amino acid differences (Lanfermeijer et al., 2005). The differences at positions 257 and 286 are in the Walker B domain and the RNSB-B domain of the NB-ARC domain, respectively (Takken et al., 2006; Tameling et al., 2006) and this domain is presumably not involved in the specificity of the resistance proteins (Lanfermeijer et al., 2005). It is interesting to speculate that the differences between the Tm-2 and Tm-2² proteins may have an effect on the recently established ATP-hydrolytic activity of this type of proteins (Tameling et al., 2002, 2006; Ueda et al., 2006) and, subsequently have an effect on the characteristics of the resistances. In light of the recently proposed mechanism for signal amplification through an iterative mechanism, the characteristics of ATP hydrolysis have been suggested to affect the characteristics of resistance, for instance, extreme resistance (Rairdan and Moffett, 2006). It is possible that the durability of a resistance is related to the speed with which an R protein can cycle through its iterative mechanism and, consequently, can strengthen its signal.

As mentioned above, amino acid changes in MP that are necessary to overcome the *Tm-2* and *Tm-2²* resistances are located in different parts of MP. For the *Tm-2* resistance, those changes are located in the N-terminal half (Meshi et al., 1989; Calder and Palukaitis, 1992), whereas those necessary to overcome the *Tm-2²* resistance are located in the C-terminal half of MP (Calder and Palukaitis, 1992; Weber et al., 1993, 1998). Different regions of MP are also necessary to trigger an HR in the case of the two different resistances (Weber et al., 2004), suggesting that the direct or indirect interaction between MP and the two resistance proteins differed in topology and nature. However, contrary to these observations, the small amino acid difference between the two R genes suggests that the interaction between the Tm-2 and Tm-2² proteins and MP or the MP/guardee complex is highly similar (Lanfermeijer et al., 2005). It is possible that the three-dimensional structure of MP brings the regions in which the mutations are located in close proximity of each other. Alternatively, the mutations while located in different regions may affect similar (direct or indirect) interactions (Lanfermeijer et al., 2005). Further work will be required to explain the different durabilities of the two resistances.

As discussed above, viral proteins do not function isolated. They fulfil their roles in large protein complexes. Additional proteins may also interact with these complexes in a transitory manner and these interactions may affect the direct interaction between MP and the R proteins (e.g. through phosphorylation of essential residues) or one of these proteins is the guardee. However, further experimental evidence is required to determine if host protein(s) are involved in the interaction between MP and either Tm-2 or Tm-2². Some interesting observations have been made that suggest the existence of such proteins. For example, ToMV can replicate in protoplasts from tomato plants containing the *Tm-2²* resistance gene (Motoyoshi and Oshima, 1975) while later it was shown that MP is present in the infected protoplasts (Heinlein, 2002). Although several other explanations are possible, this observation points to a host protein (the guardee?) which is absent in protoplasts and which might function in cell-to-cell movement. Also, it has been observed that phosphorylation of MP plays an important role in its function and a kinase has been shown to interact with MP. Interestingly, one of the residues present in a *Tm-2²* resistance-breaking isolate is a serine, which is a target of phosphorylation (Kawakami *et al.*, 1999).

Conclusions and Future Directions

In this chapter, we have discussed major advances in the field of virus disease resistance. The mechanisms for the signalling of intracellular R proteins (both CNL and TNL proteins) are being deciphered and downstream signalling networks are being mapped. However, several major issues still need to be resolved. For instance, although the guard hypothesis explains the absence of experimental evidence for direct interaction between R proteins and Avr proteins, the nature of the indirect interactions between R proteins, Avr proteins and the guardees needs to be determined, especially in the view of the iterative mechanism of signal transduction. The observation that CNL and TNL proteins also function downstream in signalling pathways places these proteins in a new perspective and might also explain the presence of the non-resistance conferring alleles.

The study of tobamoviruses can supply a wealth of information on the interactions with host proteins, the spatial and temporal behaviour of the viral gene products and supply tools to study the putative interaction between R proteins, Avr proteins and guardees. It can be envisioned that integration of this information in phytopathology will allow the identification of putative guardees.

Recently, a study described the artificial evolution of intracellular Rx-resistance protein resulting in an expanded recognition spectrum (Farnham and Baulcombe, 2006). An *in vitro* random mutagenesis approach provided resistance to a distinct strain of *Potato virus X* and to the distantly related *Poplar mosaic virus*. This seminal study supports many of the observations on resistance, but more importantly hands a new tool in the search for resistances against new pathogens or emerging virulent isolates of controlled pathogens. In combination with the 'Holy Grail' of homologous recombination in plants, this technique could result in 'resistance on demand'.

References

Albrecht, M. and Takken, F.L.W. (2006) Update on the domain architectures of NLRs and R proteins. *Biochemical and Biophysical Research Communications* 339, 459–462.

Allan, A.C., Lapidot, M., Culver, J.N. and Fluhr, R. (2001) An early tobacco mosaic virus-induced oxidative burst in tobacco indicates extracellular perception of the virus coat protein. *Plant Physiology* 126, 97–108.

Baker, C.J. and Orlandi, E.W. (1995) Active oxygen in plant pathogenesis. *Annual Review of Phytopathology* 33, 299–321.

Beachy, R.N. and Heinlein, M. (2000) Role of P30 in replication and spread of TMV. *Traffic* 1, 540–544.

Belkhadir, Y., Subramaniam, R. and Dangl, J.L. (2004) Plant disease resistance protein signaling: NBS–LRR proteins and their partners. *Current Opinion in Plant Biology* 7, 391–399.

Bendahmane, A., Kanyuka, K. and Baulcombe, D.C. (1999) The *Rx* gene from potato controls separate virus resistance and cell death responses. *The Plant Cell* 11, 781–792.

Bendahmane, A., Querci, M., Kanyuka, K. and Baulcombe, D.C. (2000) *Agrobacterium* transient expression system as a tool for the isolation of disease resistance genes: application to the *Rx2* locus in potato. *The Plant Journal* 21, 73–81.

Boller, T. (2005) Peptide signaling in plant development and self/non-self perception. *Current Opinion in Cell Biology* 17, 116–122.

Boukema, I.W. (1980) Allelism of genes-controlling resistance to TMV in *Capsicum* L. *Euphytica* 29, 433–439.

Brommonschenkel, S.H., Frary, A. and Tanksley, S.D. (2000) The broad-spectrum tospovirus resistance gene *Sw-5* of tomato is a homolog of the root-knot nematode resistance gene Mi. *Molecular Plant-Microbe Interactions* 13, 1130–1138.

Calder, V.L. and Palukaitis, P. (1992) Nucleotide sequence analysis of the movement genes of resistance breaking strains of tomato mosaic virus. *Journal of General Virology* 73, 165–168.

Canto, T., MacFarlane, S.A. and Palukaitis, P. (2004) ORF6 of tobacco mosaic virus is a determinant of viral pathogenicity in *Nicotiana benthamiana*. *Journal of General Virology* 85, 3123–3133.

Chisholm, S.T., Mahajan, S.K., Whitham, S.A., Yamamoto, M.L. and Carrington, J.C. (2000) Cloning of the *Arabidopsis RTM1* gene, which controls restriction of long-distance movement of tobacco etch virus. *Proceedings of the National Academy of Sciences of the United States of America* 97, 489–494.

Chivasa, S. and Carr, J.P. (1998) Cyanide restores N gene-mediated resistance to tobacco mosaic virus in transgenic tobacco expressing salicylic acid hydroxylase. *The Plant Cell* 10, 1489–1498.

Cooley, M.B., Pathirana, S., Wu, H.J., Kachroo, P. and Klessig, D.F. (2000) Members of the *Arabidopsis HRT/RPP8* family of resistance genes confer resistance to both viral and oomycete pathogens. *The Plant Cell* 12, 663–676.

Creager, A.N., Scholthof, K.B., Citovsky, V. and Scholthof, H.B. (1999) Tobacco mosaic virus. Pioneering research for a century. *The Plant Cell* 11, 301–308.

Dardick, C.D. and Culver, J.N. (1997) Tobamovirus coat proteins: elicitors of the hypersensitive response in *Solanum melongena* (eggplant). *Molecular Plant-Microbe Interactions* 10, 776–778.

de la Fuente van Bentem, S., Vossen, J.H., Vries, K.J., Wees, S., Tameling, W.I.L., Dekker, H.L., Koster, C.G., Haring, M.A., Takken, F.L.W. and Cornelissen, B.J.C. (2005) Heat shock protein 90 and its co-chaperone protein phosphatase 5 interact with distinct regions of the tomato I-2 disease resistance protein. *The Plant Journal* 43, 284–298.

Dinesh-Kumar, S.P. and Baker, B.J. (2000) Alternatively spliced N resistance gene transcripts: their possible role in tobacco mosaic virus resistance. *Proceedings of the National Academy of Sciences of the United States of America* 97, 1908–1913.

Dodds, P.N., Lawrence, G.J., Catanzariti, A.M., Teh, T., Wang, C.I.A., Ayliffe, M.A., Kobe, B. and Ellis, J.G. (2006) Direct protein interaction underlies gene-for-gene specificity and coevolution of the flax resistance genes and flax rust avirulence genes. *Proceedings of the National Academy of Sciences of the United States of America* 103, 8888–8893.

dos Reis Figueira, A., Golem, S., Goregaoker, S.P. and Culver, J.N. (2002) A nuclear localization signal and a membrane association domain contribute to the cellular localization of the tobacco mosaic virus 126-kDa replicase protein. *Virology* 301, 81–89.

Ehrenfeld, N., Canon, P., Stange, C., Medina, C. and Arce-Johnson, P. (2005) Tobamovirus coat protein CPCg induces an HR-like response in sensitive tobacco plants. *Molecules and Cells* 19, 418–427.

Ellis, J., Dodds, P. and Pryor, T. (2000) The generation of plant disease resistance gene specificities. *Trends in Plant Science* 5, 373–379.

Faigon-Soverna, A., Harmon, F.G., Storani, L., Karayekov, E., Staneloni, R.J., Gassmann, W., Mas, P., Casal, J.J., Kay, S.A. and Yanovsky, M.J. (2006) A constitutive shade-avoidance mutant implicates TIR-NBS-LRR Proteins in *Arabidopsis* photomorphogenic development. *The Plant Cell* 18, 2919–2928.

Farnham, G. and Baulcombe, D.C. (2006) Artificial evolution extends the spectrum of viruses that are targeted by a disease-resistance gene from potato. *Proceedings of the National Academy of Sciences of the United States of America* 103, 18828–18833.

Flor, H.H. (1971) Current status of the gene-for-gene concept. *Annual Review of Phytopathology* 28, 275–296.

Fraser, R.S.S., Gerwitz, A. and Betti, L. (1989) Deployment of resistance genes: implications from studies on resistance-breaking isolates of tobacco mosaic virus. In: *Proceedings of the Fourth International Plant Virus Epidemiology Workshop*. International Society for Plant Pathology, Montpellier, France, pp. 154–155.

Fulton, R.W. (1986) Practices and precautions in the use of cross protection for plant virus disease control. *Annual Review of Phytopathology* 24, 67–81.

Gabriels, S.H., Takken, F.L., Vossen, J.H., de Jong, C.F., Liu, Q., Turk, S.C., Wachowski, L.K., Peters, J., Witsenboer, H.M., De Wit, P.J. and Joosten, M.H. (2006) CDNA-AFLP combined with functional analysis reveals novel genes involved in the hypersensitive response. *Molecular Plant-Microbe Interactions* 19, 567–576.

Glazebrook, J. (2001) Genes controlling expression of defense responses in *Arabidopsis* – 2001 status. *Current Opinion in Plant Biology* 4, 301–308.

Goldbach, R. and de Haan, P.T. (1993) RNA viral supergroups and the evolution of RNA viruses. In: Morse S.S. (ed.) *The Evolutionary Biology of Viruses*. Raven Press, New York, pp. 105–119.

Gomez-Gomez, L. and Boller, T. (2002) Flagellin perception: a paradigm for innate immunity. *Trends in Plant Science* 7, 251–256.

Hall, T.J. (1980) Resistance at the *Tm-2* locus in the tomato to tomato mosaic virus. *Euphytica* 29, 189–197.

Hammond-Kosack, K.E. and Jones, J.D.G. (1997) Plant disease resistance genes. *Annual Review of Plant Physiology and Plant Molecular Biology* 48, 575–607.

Heinlein, M. (2002) The spread of tobacco mosaic virus infection: insights into the cellular mechanism of RNA transport. *Cellular and Molecular Life Sciences* 59, 58–82.

Hulbert, S.H., Webb, C.A., Smith, S.M. and Sun, Q. (2001) Resistance gene complexes: evolution and utilization. *Annual Review of Phytopathology* 39, 285–312.

Jones, D.A. and Takemoto, D. (2004) Plant innate immunity – direct and indirect recognition of general and specific pathogen-associated molecules. *Current Opinion in Immunology* 16, 48–62.

Kawakami, S., Padgett, H.S., Hosokawa, D., Okada, Y., Beachy, R.N. and Watanabe, Y. (1999) Phosphorylation and/or presence of serine 37 in the movement protein of tomato mosaic tobamovirus is essential for intracellular localization and stability *in vivo*. *The Journal of Virology* 73, 6831–6840.

Knorr, D.A. and Dawson, W.O. (1988) A point mutation in the tobacco mosaic virus capsid protein gene induces hypersensitivity in *Nicotiana sylvestris*. *Proceedings of the National Academy of Sciences of the United States of America* 85, 170–174.

Kruijt, M., De Kock, M.J.D. and De Wit, P.J.G.M. (2005) Receptor-like proteins involved in plant disease resistance. *Molecular Plant Pathology* 6, 85–97.

Lanfermeijer, F.C., Dijkhuis, J., Sturre, M.J.G., de Haan, P. and Hille, J. (2003) Cloning and characterization of the durable tomato mosaic virus resistance gene *Tm-2²* from *Lycopersicon esculentum*. *Plant Molecular Biology* 52, 1037–1049.

Lanfermeijer, F.C., Warmink, J. and Hille, J. (2005) The products of the broken *Tm-2* and the durable *Tm-2²* resistance genes from tomato differ in four amino acids. *Journal of Experimental Botany* 56, 2925–2933.

Liu, Y., Schiff, M., Marathe, R. and Dinesh-Kumar, S.P. (2002) Tobacco *Rar1*, *EDS1* and *NPR1/NIM1* like genes are required for *N*-mediated resistance to tobacco mosaic virus. *The Plant Journal* 30, 415–429.

Love, A.J., Yun, B.W., Laval, V., Loake, G.J. and Milner, J.J. (2005) Cauliflower mosaic virus, a compatible pathogen of *Arabidopsis*, engages three distinct defense-signaling pathways and activates rapid systemic generation of reactive oxygen species. *Plant Physiology* 139, 935–948.

Marathe, R., Anandalakshmi, R., Liu, Y. and Dinesh-Kumar, S.P. (2002) The tobacco mosaic virus resistance gene, *N*. *Molecular Plant Pathology* 3, 167–172.

Martin, G.B., Bogdanove, A.J. and Sessa, G. (2003) Understanding the functions of plant disease resistance proteins. *Annual Review of Plant Biology* 54, 23–61.

McHale, L., Tan, X., Koehl, P. and Michelmore, R.W. (2006) Plant NBS–LRR proteins: adaptable guards. *Genome Biology* 7, 212–223.

Meshi, T., Motoyoshi, F., Adachi, A., Watanabe, Y., Takamatsu, N. and Okada, Y. (1988) Two concomitant base substitutions in the putative replicase genes of tobacco mosaic virus confer the ability to overcome the effects of a tomato resistance gene, *Tm-1*. *The EMBO Journal* 7, 1575–1581.

Meshi, T., Motoyoshi, F., Maeda, T., Yoshiwoka, S., Watanabe, H. and Okada, Y. (1989) Mutations in the *Tobacco mosaic virus 30-kD protein* gene overcome *Tm-2* resistance in tomato. *The Plant Cell* 1, 515–522.

Mestre, P. and Baulcombe, D.C. (2006) Elicitor-mediated oligomerization of the tobacco N disease resistance protein. *The Plant Cell* 18, 491–501.

Mittler, R., Lam, E., Shulaev, V. and Cohen, M. (1999) Signals controlling the expression of cytosolic ascorbate peroxidase during pathogen-induced programmed cell death in tobacco. *Plant Molecular Biology* 39, 1025–1035.

Moffett, P., Farnham, G., Peart, J.R. and Baulcombe, D.C. (2002) Interaction between domains of a plant NBS–LRR protein in disease resistance-related cell death. *The EMBO Journal* 21, 4511–4519.

Morillo, S.A. and Tax, F.E. (2006) Functional analysis of receptor-like kinases in monocots and dicots. *Current Opinion in Plant Biology* 9, 460–469.

Motoyoshi, F. and Oshima, N. (1975) Infection with tomato mosaic virus of leaf mesophyll protoplasts from susceptible and resistant lines of tomato. *Journal of General Virology* 29, 81–91.

Padgett, H.S., Watanabe, Y. and Beachy, R.N. (1997) Identification of the TMV replicase sequence that activates the *N* gene-mediated hypersensitive response. *Molecular Plant-Microbe Interactions* 10, 709–715.

Peart, J.R., Mestre, P., Lu, R., Malcuit, I. and Baulcombe, D.C. (2005) NRG1, a CC-NB-LRR protein, together with N, a TIR-NB-LRR protein, mediates resistance against tobacco mosaic virus. *Current Biology* 15, 968–973.

Pedley, K.F. and Martin, G.B. (2003) Molecular basis of *Pto*-mediated resistance to bacterial speck disease in tomato. *Annual Review of Phytopathology* 41, 215–243.

Pelham, J. (1966) Resistance in tomato to tobacco mosaic virus. *Euphytica* 15, 258–267.

Rairdan, G.J. and Moffett, P. (2006) Distinct domains in the ARC region of the potato resistance protein Rx mediate LRR binding and inhibition of activation. *The Plant Cell* 18, 2082–2093.

Rathjen, J.P. and Moffett, P. (2003) Early signal transduction events in specific plant disease resistance. *Current Opinion in Plant Biology* 6, 300–306.

Sawada, H., Takeuchi, S., Matsumoto, K., Hamada, H., Kiba, A., Matsumoto, M., Watanabe, Y., Suzuki, K. and Hikichi, Y. (2005) A new tobamovirus-resistance gene, *Hk*, in *Capsicum annuum*. *Journal of the Japanese Society for Horticultural Science* 74, 289–294.

Scholthof, H.B. (2005) Plant virus transport: motions of functional equivalence. *Trends in Plant Science* 10, 376–382.

Shaw, J.G. (1999) Tobacco mosaic virus and the study of early events in virus infections. *Philosophical Transactions of the Royal Society of London, Series B: Biological Sciences* 354, 603–611.

Song, W.Y., Wang, G.L., Chen, L.L., Kim, H.S., Pi, L.Y., Holsten, T., Gardner, J., Wang, B., Zhai, W.X. and Zhu, L.H. (1995) A receptor kinase-like protein encoded by the rice disease resistance gene, *Xa21*. *Science* 270, 1804–1806.

Spassova, M.I., Prins, T.W., Folkertsma, R.T., Klein-Lankhorst, R.M., Hille, J., Goldbach, R.W. and Prins, M. (2001) The tomato gene *Sw5* is a member of the coiled coil, nucleotide binding, leucine-rich repeat class of plant resistance genes and confers resistance to TSWV in tobacco. *Molecular Breeding* 7, 151–161.

Takahashi, H., Miller, J., Nozaki, Y., Takeda, M., Shah, J., Hase, S., Ikegami, M., Ehara, Y. and Dinesh-Kumar, S.P. (2002) *RCY1*, an *Arabidopsis thaliana RPP8/HRT* family resistance gene, conferring resistance to cucumber mosaic virus requires salicylic acid, ethylene and a novel signal transduction mechanism. *The Plant Journal* 32, 655–667.

Takken, F.L.W., Albrecht, M. and Tameling, W.I.L. (2006) Resistance proteins: molecular switches of plant defence. *Current Opinion in Plant Biology* 9, 383–390.

Tameling, W.I., Elzinga, S.D., Darmin, P.S., Vossen, J.H., Takken, F.L., Haring, M.A. and Cornelissen, B.J. (2002) The tomato *R* gene products I-2 and MI-1 are functional ATP binding proteins with ATPase activity. *The Plant Cell* 14, 2929–2939.

Tameling, W.I., Vossen, J.H., Albrecht, M., Lengauer, T., Berden, J.A., Haring, M.A., Cornelissen, B.J. and Takken, F.L. (2006) Mutations in the NB-ARC domain of I-2 that impair ATP hydrolysis cause autoactivation. *Plant Physiology* 140, 1233–1245.

Ueda, H., Yamaguchi, Y. and Sano, H. (2006) Direct interaction between the tobacco mosaic virus helicase domain and the ATP-bound resistance protein, N factor during the hypersensitive response in tobacco plants. *Plant Molecular Biology* 61, 31–45.

van der Biezen, E.A. and Jones, J.D. (1998) Plant disease-resistance proteins and the gene-for-gene concept. *Trends in Biochemical Sciences* 23, 454–456.

Van der Hoorn, R.A., De Wit, P.J. and Joosten, M.H. (2002) Balancing selection favors guarding resistance proteins. *Trends in Plant Science* 7, 67–71.

Vidal, S., Cabrera, H., Andersson, R.A., Fredriksson, A. and Valkonen, J.P. (2002) Potato gene *Y-1* is an *N* gene homolog that confers cell death upon infection with potato virus Y. *Molecular Plant-Microbe Interactions* 15, 717–727.

Waigmann, E., Ueki, S., Trutnyeva, K. and Citovsky, V. (2004) The ins and outs of nondestructive cell-to-cell and systemic movement of plant viruses. *Critical Reviews in Plant Sciences* 23, 195–250.

Weber, H. and Pfitzner, A.J. (1998) *Tm-2^2* resistance in tomato requires recognition of the carboxy terminus of the movement protein of tomato mosaic virus. *Molecular Plant-Microbe Interactions* 11, 498–503.

Weber, H., Schultze, S. and Pfitzner, A.J. (1993) Two amino acid substitutions in the tomato mosaic virus 30-kilodalton movement protein confer the ability to overcome the *Tm-2^2* resistance gene in the tomato. *The Journal of Virology* 67, 6432–6438.

Weber, H., Ohnesorge, S., Silber, M.V. and Pfitzner, A.J. (2004) The tomato mosaic virus 30 kDa movement protein interacts differentially with the resistance genes *Tm-2* and *Tm-2²*. *Archives of Virology* 149, 1499–1514.

Whitham, S., Dinesh-Kumar, S.P., Choi, D., Hehl, R., Corr, C. and Baker, B. (1994) The product of the tobacco mosaic virus resistance gene *N*: similarity to toll and the interleukin-1 receptor. *Cell* 78, 1101–1115.

Whitham, S., McCormick, S. and Baker, B. (1996) The *N* gene of tobacco confers resistance to tobacco mosaic virus in transgenic tomato. *Proceedings of the National Academy of Sciences of the United States of America* 93, 8776–8781.

Whitham, S.A., Anderberg, R.J., Chisholm, S.T. and Carrington, J.C. (2000) *Arabidopsis RTM2* gene is necessary for specific restriction of tobacco etch virus and encodes an unusual small heat shock-like protein. *The Plant Cell* 12, 569–582.

Xie, Z., Fan, B., Chen, C. and Chen, Z. (2001) An important role of an inducible RNA-dependent RNA polymerase in plant antiviral defense. *Proceedings of the National Academy of Sciences of the United States of America* 98, 6516–6521.

Yang, S.J., Carter, S.A., Cole, A.B., Cheng, N.H. and Nelson, R.S. (2004) A natural variant of a host RNA-dependent RNA polymerase is associated with increased susceptibility to viruses by *Nicotiana benthamiana*. *Proceedings of the National Academy of Sciences of the United States of America* 101, 6297–6302.

Zaitlin, M. (1999) Elucidation of the genome organization of tobacco mosaic virus. *Philosophical Transactions of the Royal Society of London, Series B: Biological Sciences* 354, 587–591.

Zaitlin, M. and Palukaitis, P. (2000) Advances in understanding plant viruses and virus diseases. *Annual Review of Phytopathology* 38, 117–143.

18 Potential for Recombination and Creation of New Viruses in Transgenic Plants Expressing Viral Genes: Real or Perceived Risk?

M. Fuchs

Abstract
The development of virus-resistant transgenic crops has widened the horizons of virus control. A common approach to confer virus resistance relies on the transfer and expression of viral genes in susceptible plants. The successful application of this strategy is illustrated by the commercialization of virus-resistant transgenic squash and papaya in the USA. Since virus-resistant transgenic plants express viral sequences, environmental safety issues have been raised, in particular on the potential for recombination and creation of new viruses. It is conceivable that recombinant viruses can arise from exchange of genetic information as a consequence of RNA recombination between transgene transcripts and the genome of challenging viruses. Resulting chimeric RNA may lead to viable recombinant viruses with identical biological properties as parental lineages or altered biological properties, including increased pathogenicity, expanded host range and changes in vector relationship. The development of recombinant viruses has been extensively documented in transgenic plants expressing viral genes, primarily under conditions of high to moderate selective pressure in confined environments. Under field conditions with limited, if any, selective pressure, no recombinant virus has been found at detectable levels even in transgenic perennial plants established in experimental sites over extended periods of time. So far, although the potential for recombination is real, extensive research indicates that the creation of recombinant viruses in transgenic crops expressing viral genes does not seem to exceed baseline events in conventional plants, thus providing valuable insights into the safe release of virus-resistant transgenic plants and suggesting a reasonable certainty of limited or no hazard.

Introduction

The development of virus-resistant transgenic crops has been a major breakthrough in the successful application of biotechnology to agriculture. Transgenic plants expressing a segment of a plant viral genome display resistance to the cognate virus and to closely related virus strains and viruses (Beachy, 1997; Fuchs and Gonsalves, 2002; Voinnet, 2005). The potential for creating virus-resistant plants was apparent when expression of a viral gene within a plant was shown to provide resistance to the virus from which the gene was derived. This phenomenon was first described with the coat protein (CP) gene of *Tobacco mosaic virus* (TMV) in transgenic *Nicotiana tabacum* cv. Xanthi (Powell Abel et al., 1986). Twenty years after its discovery, pathogen-derived resistance and viral genes continue to be used to confer resistance in susceptible plants and numerous virus-resistant transgenic plants have been developed around the world by using this strategy (Chapter 19, this volume).

Environmental safety issues have been expressed concerning potential risks associated with the release of virus-resistant transgenic crops, since the constitutive expression of viral genes does not naturally occur in conventional crops, except in a few that are infected by pararetroviruses (Harper et al., 2002; Hansen et al., 2005). Of major concern is the possibility of recombination between viral transgene transcripts and the genome from field viruses, which challenge transgenic plants (Hull, 1989; de Zoeten, 1991; Falk and Bruening, 1994; Allison et al., 1996; Rissler and Mellon, 1996; Robinson, 1996; Fuchs and Gonsalves, 1997; Miller et al., 1997; Tepfer and Balazs, 1997; Aaziz and Tepfer, 1999; Hammond et al., 1999; Rubio et al., 1999; Martelli, 2001; Fuchs and Gonsalves, 2002; Tepfer, 2002). Resulting recombinant viruses may have identical biological properties as their parental lineages or new biological properties such as changes in vector specificity, expanded host range and increased pathogenicity. The presence of constitutively expressed viral genes in transgenic plants has even been hypothesized to enhance the rate of plant virus evolution through RNA recombination (Hull, 1989; de Zoeten, 1991; Rissler and Mellon, 1996; Jakab et al., 1997).

RNA recombination is known to play an important role in virus variability, population dynamics and evolution (Garcia-Arenal et al., 2001; Hull, 2002). It is also known to occur in a wide range of plant DNA and RNA viruses upon infection of conventional plants. Therefore, determining baseline rates of recombination in natural virus populations is paramount to determine if transgenic plants expressing viral genes enhance recombination rates over conventional plants. Addressing this issue is needed to examine whether transgenic plants expressing viral genes pose negligible or, in contrast, undue risks with regard to the creation of recombinant viruses.

This chapter discusses the potential for RNA recombination in transgenic plants expressing viral genes and the creation of recombinant viruses. A review of the state-of-the-art knowledge is presented to critically evaluate risks in light of the emergence of new viruses, contrasting differences between risk assessment studies conducted in confined and open environments in relation to the level of selective pressure. Finally, perceived risks will be put into perspective to balance real risks against benefits of the use of transgenic plants expressing viral genes to control virus diseases.

Background

Plant viruses can cause severe damage to crops by substantially reducing vigour, yield and product quality. They can also increase susceptibility to other pathogens and pests, and increase sensitivity to abiotic factors. Detrimental effects of viruses are very costly to agriculture. Losses of over $1.5 billion are reported in rice in South-east Asia, $5.5–9.5 million in potato in the UK and $2 billion in cassava worldwide (Hull, 2002). Also, *Grapevine fanleaf virus* (GFLV) alone is responsible for losses of over $1.0 billion in grapevines in France (M. Fuchs, 2005).

Effective control strategies are needed to mitigate the considerable losses incited by plant viruses. Conventional virus control approaches are designed to limit the introduction of viruses in certain areas, reduce sources of infection and limit vector-mediated virus spread by sanitary and quarantine measures and by cultural practices, i.e. field isolation, removal of weed hosts, rouging, eradication and application of agrochemicals to eradicate vectors. These strategies are costly and not always effective. In addition, the intensive use of agrochemicals to control vector populations is undesirable because of hazardous effects on the environment and concerns for human health.

The use of resistant crop cultivars is the most effective and sustainable strategy to control virus diseases. Virus-resistant plants can be developed by conventional breeding strategies if useful sources of resistance have been identified in appropriate germplasm (Hull, 2002). Major resistance genes have been successfully used in numerous crops to restrict virus infection to inoculation sites, reduce replication to subliminal levels and provide tolerance to virus infection. In many cases, however, host resistance has been difficult to incorporate into agronomically desirable cultivars or no host resistance sources have been found against certain viruses.

The concept of pathogen-derived resistance (Sanford and Johnston, 1985) has opened new avenues for the development of virus-resistant transgenic plants. It was first applied to plant viruses in 1986 when transgenic tobacco plants expressing the CP gene of TMV were shown to be resistant to infection by TMV (Powell Abel *et al.*, 1986). Since this pioneering work, the concept of pathogen-derived resistance has been demonstrated in many transgenic plants as a useful and powerful approach to control viral diseases (Fuchs and Gonsalves, 2002). Applying the concept of pathogen-derived resistance is the most common approach to engineer protection against viruses in plants. Viral genes are engineered in transgenic plants as full-length, truncated or short segment constructs, and in sense, antisense, or invert-repeat orientation. By far, a larger number of transgenic plants engineered for virus resistance express the CP gene than other viral genes.

Extensive research is ongoing not only to develop virus-resistant transgenic crops but also to unravel the mechanisms involved in the engineered protection. Two major mechanisms have been described in association with the engineered protection against viruses in transgenic plants expressing viral genes. The first seems to rely on the expression of a viral gene at the protein level and the second involves regulation of gene expression at the transcript level through RNA silencing. In the case of transgenic plants expressing a TMV CP, an early event after virus entry into cells, maybe blocking of particle disassembly, is involved in resistance (Bendahmane and Beachy, 1999). RNA silencing, in particular homology-dependent post-transcriptional gene silencing (PTGS), is involved in many cases of engineered resistance to viruses in transgenic plants (Voinnet, 2001, 2005; MacDiarmid, 2005). This gene expression regulatory mechanism is triggered by a dsRNA structure derived from the viral transgene, which is processed into 21–24 bp long small interfering RNA (siRNA) duplexes by the RNase III enzyme Dicer. Upon integration of one of the siRNA strands into the multisubunit RNA-induced silencing complex (RISC), cellular RNA that is identical in sequence to the siRNA, i.e. RNA of a homologous challenge virus, is degraded (Voinnet, 2001, 2005; MacDiarmid, 2005).

Environmental safety issues have been expressed on potential risks associated with the release of virus-resistant transgenic plants. Based on the fact that virus-resistant transgenic plants express viral genes, a major issue is RNA recombination and its outcomes (Hull, 1989; de Zoeten, 1991; Falk and Bruening, 1994; Allison et al., 1996; Rissler and Mellon, 1996; Robinson, 1996; Fuchs and Gonsalves, 1997; Miller et al., 1997; Tepfer and Balazs, 1997; Aaziz and Tepfer, 1999; Hammond et al., 1999; Rubio et al., 1999; Martelli, 2001; Fuchs and Gonsalves, 2002; Tepfer, 2002). We have to keep in mind that RNA recombination is not an issue that arose with the development of virus-resistant transgenic plants. RNA recombination is known to be a major source of variation and a major driving force in the evolution of a number of plant viruses. RNA recombination is thought to occur in most, if not all, RNA viruses (Chare and Holmes, 2006) and also in DNA viruses that replicate by reverse transcription of an RNA copy of their genome (Froissart et al., 2005). Furthermore, in the case of *Potato virus Y* (PVY) in grapevines, it was shown to trigger the incorporation of viral sequences into the plant genome (Tanne and Sela, 2005).

RNA recombination refers to the creation of chimeric RNA molecules from distinct segments present in different parental molecules, one donor and one acceptor (Hull, 2002). One mechanism underlying recombination is template switching during RNA replication, in which the synthesis of a nascent RNA strand on a donor RNA molecule is halted. As a consequence, the RNA-dependent RNA polymerase (RdRp) or the nascent RNA strand can interact with the acceptor RNA molecule, leading to template switching and the creation of a chimeric RNA molecule. The sequence or secondary structure of the donor RNA and/or nascent RNA could act as signals for template switching. Another mechanism for recombination is template switching induced by pausing of the RdRp at break points on the RNA template. Recombination can be homologous when it occurs between two RNA molecules that are identical or very similar at the crossover point. It can also be non-homologous when it occurs between two RNA molecules with limited or no obvious homology. Information on natural recombination rates in plant virus populations is becoming more abundant. Recent studies have even indicated that the frequency of recombination can be very high for some viruses (Froissart et al., 2005; Chare and Holmes, 2006).

Transgenic plants can provide unique opportunities for RNA recombination by favouring the exchange of different RNA templates during virus replication. As a result, chimeric RNA molecules consisting of a segment from the viral genome and another segment from viral transgene transcripts can develop. Chimeric RNA molecules can lead to the creation of viable recombinant viruses. Such recombinant viruses may have similar biological properties to their parental lineages or new biological properties, such as changes in vector specificity, expanded host range and increased pathogenicity. RNA recombination in transgenic plants expressing viral genes has been suggested to enhance the rate of virus evolution and facilitate the creation of new virus species (Hull, 1989; de Zoeten, 1991; Rissler and Mellon, 1996; Jakab et al., 1997). Knowing that the CP gene is the preferred viral gene used to engineer resistance, the exchange and incorporation of a CP transgene in a recombinant virus could have profound effects. For example, in the case of numerous viruses, the CP is an important viral determinant of host cell recognition, vector interaction, cell-to-cell movement and pathogenicity among other important steps of the virus multiplication cycle (Callaway et al., 2001). Thus, recombinant viruses with a newly incorporated functional CP gene may express new and eventually undesired properties, such as infection of otherwise non-host plants.

Is RNA recombination in transgenic plants expressing viral genes a real or a perceived risk? Although there is strong evidence for recombination in transgenic plants expressing viral genes, most of the studies reporting on the occurrence of recombination have been carried out under confined environments, i.e. laboratory and greenhouse conditions. These studies are important because they document the occurrence of RNA recombination, the frequency of emergence of recombinant viruses and some of the factors that influence the rate of recombination. They also examine mechanisms underlying the occurrence of RNA recombination. Noteworthy, laboratory studies typically utilize heavy to moderate selective pressure to favour the development of recombinant viruses. For example, a function can be restored if recombination occurs between two defective RNA molecules because there is a strong selection for the recombination event. These conditions are highly unlikely to occur naturally under field conditions where plants are predominantly infected by functional rather than defective viruses.

A more realistic situation to assess the role of transgenic plants expressing viral genes in promoting the emergence of recombinant viruses is to rely on conditions of reduced or no selective pressure. These conditions are prevalent in the field. Therefore, field studies are very important to determine if risks are real and if experiments conducted in confined environments reflect field conditions. Field studies are also important to determine if the creation of recombinant viruses in transgenic plants expressing viral genes is substantially enhanced beyond background recombination events in conventional plants. Also, field data provide valuable insights to balance real risks versus benefits of virus-resistant transgenic plants. This is especially important after the first decade of commercial release of virus-resistant transgenic crops.

The following section reviews the literature on the potential for RNA recombination in transgenic plants expressing viral genes and the creation of new viruses. Emphasis will be placed on recent information from field data in order to critically evaluate our knowledge in light of the emergence of recombinant viruses with harmful properties and to discuss real risks and benefits of the use of virus-resistant transgenic plants for virus disease control.

RNA Recombination and Creation of New Viruses in Transgenic Plants Expressing Viral Genes

Historical aspects

Numerous field experiments have shown the efficacy of pathogen-derived resistance against plant viruses. The first field test was conducted with transgenic tomato plants expressing the CP gene of TMV for resistance to TMV (Nelson *et al.*, 1988). Transgenic plants were highly resistant to mechanical inoculation by TMV, with only 5% of the plants being symptomatic compared to 99% of the control plants. In addition, inoculated transgenic and healthy control plants had identical fruit yield (Nelson *et al.*, 1988), Gonsalves *et al.* (1992), investigated resistance in relation to vector-mediated virus infection and showed that transgenic cucumbers expressing a *Cucumber mosaic virus* (CMV) CP gene construct were resistant to infection by CMV under field conditions with natural challenge inoculations by aphids. These studies opened the way to numerous field experiments with virus-resistant transgenic vegetable, cereal, ornamental, fruit tree and small fruit crops (Fuchs and Gonsalves, 1997, 2002; Tepfer, 2002).

The first virus-resistant transgenic crop, Freedom II, was deregulated in the USA in 1994 and commercialized in the spring of 1995. Freedom II, a summer squash expressing the CP gene of *Zucchini yellow mosaic virus* (ZYMV) and *Watermelon mosaic virus* (WMV), is highly resistant to single and mixed infections by ZYMV and WMV. This was the first vegetable crop and first disease-resistant transgenic crop ever released. Thereafter, transgenic squash expressing the CP gene of ZYMV, WMV and CMV and exhibiting high levels of resistance to these three viruses was deregulated in 1996. Growers in major squash producing areas in the USA have adopted virus-resistant transgenic squash cultivars, which accounted for 2–22% of the total acreage, with the highest adoption in Florida, New

Jersey and Georgia in 2004 (Shankula *et al.*, 2005). Transgenic papaya expressing the CP gene of *Papaya ringspot virus* (PRSV) was the first fruit crop deregulated in the USA in 1997 and subsequently commercialized in Hawaii (Gonsalves, 1998). This release saved the Hawaiian papaya industry from an economic disaster caused by a devastating PRSV epidemic. Also, virus-resistant transgenic tomato and pepper have been commercialized in the People's Republic of China (Huang *et al.*, 2002).

In spite of the great potential of transgenic plants expressing virus-derived genes at controlling virus diseases, several environmental safety issues have been raised following their release. Among potential risks, transgenic plants expressing viral genes may provide increased opportunities for RNA recombination by favouring the exchange of RNA templates during virus replication (Hull, 1989; de Zoeten, 1991; Falk and Bruening, 1994; Allison *et al.*, 1996; Rissler and Mellon, 1996; Robinson, 1996; Fuchs and Gonsalves, 1997, 2002; Miller *et al.*, 1997; Tepfer and Balazs, 1997; Aaziz and Tepfer, 1999; Hammond *et al.*, 1999; Rubio *et al.*, 1999; Teycheney *et al.*, 2000; Martelli, 2001; Tepfer, 2002). The net result of recombination is the development of a recombinant viral RNA derived from two previously distinct RNA molecules. In the case of transgenic plants constitutively expressing viral genes, part or all of viral transgene transcripts can be incorporated into the genome of challenging viruses through RNA recombination.

RNA recombination with challenge viruses in transgenic plants expressing viral genes under confined environments

Lommel and Xiong (1991) were the first to report on the occurrence of recombination in transgenic *Nicotiana benthamiana* expressing a viral gene. An artificial deletion mutant of *Red clover necrotic mosaic virus* (RCNMV) RNA1, i.e. transcripts unable to systemically infect conventional plants, was restored by recombination with a RCNMV RNA2 transgene lacking its 5' leader sequence. RCNMV RNA2 recombinants had 10–17 nucleotides derived from the 5' terminus of RNA1 likely as a result of non-homologous recombination.

Gal *et al.* (1992) detected *Cauliflower mosaic virus* (CaMV) recombinants in transgenic *Brassica napus* expressing the CaMV open reading frame (ORF) VI upon agroinfection with a CaMV isolate lacking ORF VI.

Schoelz and Wintermantel (1993) described the occurrence of recombination in transgenic *N. bigelovii* expressing the CaMV strain D4 ORF VI when challenged with a CaMV strain that is not systemic in solanaceous plants. When transgenic *N. bigelovii* expressing CaMV strain D4 ORF VI was inoculated with a CaMV strain that is systemic in solanaceous plants, recombinant viruses were detected with a competitive advantage over the wild-type virus (Wintermantel and Schoelz, 1996).

Greene and Allison (1994) inoculated transgenic *N. benthamiana* expressing a wild-type *Cowpea chlorotic mottle virus* (CCMV) CP gene, including the 3' non-coding region, with transcripts of an artificial CP deletion mutant and detected recombinants in 3% (4 of 125) of the inoculated transgenic plants. Each recombinant resulted from distinct recombination events. Interestingly, no recombinant was detected in transgenic lines expressing a wild-type CCMV CP for which the 3' non-coding region was deleted (Greene and Allison, 1996). Therefore, the 3' non-coding region of the CP gene may have recruited transgene transcripts to the viral replication complex and triggered the creation of viable recombinant viruses.

Jakab *et al.* (1997) suspected the occurrence of recombination in transgenic *N. tabacum* L. 'Petit Havana' expressing a CP gene and flanking sequences from a PVY strain N when challenge inoculated with transcripts of a CP defective PVY after several passages. Also, PVY recombinants were obtained in four of 12 (33%) PVY-N-resistant transgenic plants inoculated simultaneously with a strain of PVY-N and a strain of PVY-O.

Frischmuth and Stanley (1998) reported recombination between an *African cassava mosaic virus* (ACMV) CP transgene in *N. benthamiana* and an agroinoculated ACMV genomic DNA with a deleted CP gene. Similarly to the CCMV example above, recombination was observed only if viral non-coding sequences were flanking both sides of the CP transgene.

Borja *et al.* (1999) reported on the restoration of a wild-type *Tomato bushy stunt virus* (TBSV) upon inoculation of transgenic *N. benthamiana* expressing a TBSV CP gene construct with TBSV transcripts harbouring a defective CP gene. Recombinants were detected in 13–20% (5–8 of 40) of the inoculated plants of six distinct transgenic lines. Interestingly, the frequency of recombination was significantly reduced if challenge transcripts carried a defective CP gene fused to the chloramphenicol acetyl transferase (CAT) reporter gene (5%, 2 of 40) or a *Cucumber necrosis virus* (CNV) CP deletion mutant (0%, 0 of 160).

Adair and Kearney (2000) detected recombinant TMV RNA but no recombinant TMV virions in 41% (14 of 34) of the transgenic *N. benthamiana* expressing a non-translatable region of the TMV genome encompassing a portion of the RdRp, the movement protein, the CP and a small portion of the 3' non-coding region, which was fused to the green fluorescent protein (GFP), upon challenge inoculation with transcripts of a CP defective TMV mutant. Interestingly, no recombinant RNA or virion was found at detectable levels when transgenic plants were inoculated with transcripts corresponding to a functional TMV.

Varrelmann *et al.* (2000) described recombinant wild-type *Plum pox virus* (PPV) in transgenic *N. benthamiana* expressing a functional PPV CP gene upon inoculation with transcripts of CP defective PPV mutants. As discussed for several other examples above, recombinant viruses were detected only if the transgene construct contained the 3' non-coding region. Also, recombinant PPV was obtained by a double recombination event upon co-inoculation of a CP defective PPV mutant with a functional CP expressed from a plant expression vector.

Teycheney *et al.* (2000) reported on the synthesis of minus-strand RNA molecules from transgene transcripts of the CP gene of *Lettuce mosaic virus* (LMV) or CMV and their 3' non-coding regions in transgenic tobacco upon challenge inoculation with several potyviruses and *Tomato aspermy virus* (TAV), respectively. Interestingly, a complementary RNA strand transcript was produced from the transgene in a manner that was dependent on the presence of the 3' non-coding region. These results indicate that a viral RdRp has the potential to recognize the 3' non-coding region of heterologous viruses expressed in transgenic plants and use them as transcription promoter.

Altogether, studies conducted under confined environments documented the occurrence of RNA recombination between viral transgene transcripts and challenge viruses mainly under conditions of high selective pressure. In most cases, the incorporation of a 3' non-coding region in viral transgenes enhanced the rate of recombination events. This is probably due to the presence of sequence elements that are recognized by the RdRp.

RNA recombination with challenge viruses in transgenic plants expressing viral genes under open field conditions

Thomas *et al.* (1998) analysed more than 65,000 transgenic potato plants expressing various constructs of the CP or the RdRp gene of *Potato leafroll virus* (PLRV) under field conditions of PLRV infection over a six-year period. New viruses or viruses with altered properties, i.e. atypical host range, symptoms and serological features, were not detected in any of the field-grown transgenic plants.

Fuchs *et al.* (1998) investigated the potential of 614 transgenic squash and melon plants expressing the CP gene of an aphid transmissible strain of CMV to facilitate the spread of the aphid non-transmissible strain C of CMV. Transmission of CMV strain C did not occur through recombination from challenge inoculated transgenic plants to uninoculated susceptible non-transgenic plants over two consecutive field seasons.

Fuchs *et al.* (1999) examined 3700 transgenic squash plants expressing the CP gene of an aphid transmissible strain of WMV for their capacity to mediate the spread of the aphid non-transmissible strain MV of ZYMV over two consecutive years. A limited extend of ZYMV strain MV transmission (2%, 77 of 3700) was detected from inoculated transgenic to uninoculated recipient plants as a result of heterologous encapsidation but not recombination. Moreover, no epidemic of ZYMV strain MV developed despite conditions of high disease pressure and the availability of numerous test plants.

Lin *et al.* (2001) assessed the genetic diversity and biological variation of 81 CMV isolates in transgenic squash expressing the CP gene of CMV, conventional squash and a melon cultivar with host resistant to CMV. Single-stranded conformation polymorphism analysis and sequencing indicated no correlation between CMV CP gene variability and host genotype, geographic origin of isolates and capacity to overcome transgenic and/or host resistance. Phylogenetic analyses further inferred no recombination event in CMV isolates from transgenic squash.

Vigne et al. (2004a) used an elegant approach to assess recombination in transgenic grapevines under conditions of heavy disease pressure but low, if any, selective pressure in favour of recombinant viruses. Test plants consisted of conventional scions grafted on to transgenic rootstocks expressing a GFLV CP gene construct. Plants were tested in two sites in a naturally GFLV-infected vineyard. Three of the 18 transgenic lines tested were resistant to GFLV infection over the three-year field trial period while the other lines were as susceptible as non-transgenic controls. Analysis of the CP gene of 347 GFLV isolates (190 from transgenic and 157 from conventional vines) by immunocapture (IC) reverse transcription (RT) polymerase chain reaction (PCR) restriction fragment length polymorphism (RFLP) indicated no detectable recombinants with transgene transcripts. Further analysis of the CP gene of 85 variants representing the different RFLP groups showed GFLV recombinants in three conventional plants that were located outside of the two test sites but not in transgenic plants. A comparative analysis of the nucleotide diversity among subsets of GFLV isolate populations clearly implied a lack of genetic differentiation according to host (transgenic versus conventional) or field site for the majority of the haplotypes tested (Vigne et al., 2004b). Therefore, transgenic grapevines did not assist the creation of viable GFLV recombinants to detectable levels nor did they affect the molecular diversity of GFLV populations during the trial period. Interestingly, and in contrast with studies conducted under high selection pressure, the presence of the 3' non-coding GFLV RNA2 region in the transgene did not have any effect on the emergence of recombinants.

Recently, the potential of transgenic plum trees expressing a PPV CP gene to mediate the emergence of recombinant PPV variants has been investigated under field conditions of high disease pressure but low, if any, selective pressure by IC-RT-PCR (N. Capote and M. Cambra, Valencia, Spain, 2006, personal communication; I. Zagrai, Bistrita, Romania, 2006, personal communication). Transgenic trees were grown in experimental orchards over an 8–10-year period and exposed to PPV infection by graft- and/or aphid-mediated inoculation. Most of the transgenic lines tested were susceptible to PPV infection, except transgenic line C5. Analysis of the CP gene variability of 64 PPV isolates (22 from transgenic and 42 from conventional trees) at the serological and molecular levels revealed no correlation between variation and plant genotype (transgenic versus conventional) in a field trial in Spain. Also, there was no significant difference in the number of aphids, aphid species, nor in the number of viruliferous aphids that visited transgenic and conventional plums (N. Capote and M. Cambra, Valencia, Spain, 2006, personal communication). A similar analysis of the CP gene variability of 82 PPV isolates (15 from transgenic and 67 from conventional trees) from Romania revealed mixed infections by PPV M and D isolates in conventional but not in transgenic plums (I. Zagrai, Bistrita, Romania, 2006, personal communication).

Furthermore, no significant difference in nucleotide variability was found between PPV isolates from transgenic and conventional trees. Interestingly, a few naturally occurring PPV recombinant isolates were found in conventional and in PPV-susceptible but not in PPV-resistant transgenic plums (I. Zagrai, Bistrita, Romania, 2006, personal communication). These recombinants were M type strains for which the recombination event mapped to the RdRp gene but not the CP gene. Therefore, the emergence of these PPV recombinants was not influenced by the PPV CP transgene. In summary, transgenic plums expressing the PPV CP gene did not affect the diversity of viral and aphid populations (N. Capote and M. Cambra, Valencia, Spain, 2006, personal communication; I. Zagrai, Bistrita, Romania, 2006, personal communication).

Altogether, studies conducted under open field conditions did not document the occurrence of RNA recombination between viral transgene transcripts and challenge viruses to detectable levels. This is probably due to the absence of selective pressure in favour of recombinants.

Lessons from RNA recombination studies in transgenic plants expressing viral genes

Numerous studies conducted over the last 15 years or more have provided valuable insights into the potential for RNA recombination in transgenic plants expressing viral genes. RNA recombination has been described for RNA and DNA viruses, primarily in transgenic herbaceous plants grown under confined conditions. A number of factors have been identified to influence the creation and recovery of viable recombinant viruses.

The stringency of selective pressure is an important factor, if not the most prevalent, in the development of recombinant viruses. Across different model systems and viruses, there is clear evidence that high recombination rates can be obtained under conditions of high selective pressure (Lommel and Xiong, 1991; Gal et al., 1992; Schoelz and Wintermantel, 1993; Greene and Allison, 1994; Frischmuth and Stanley, 1998; Borja et al., 1999; Adair and Kearney, 2000; Varrelmann et al., 2000). In contrast, recombination rates are low under conditions of moderate selective pressure (Allison et al., 1996; Wintermantel and Schoelz, 1996) and not detectable under conditions of low, if any, selective pressure (Allison et al., 1996; Fuchs et al., 1998, 1999; Thomas et al., 1998; Adair and Kearney, 2000; Lin et al., 2001; Vigne et al., 2004a,b; N. Capote and M. Cambra, Valencia, Spain, 2006, personal communication; I. Zagrai, Bistrita, Romania, 2006, personal communication).

Most studies carried out under confined conditions rely on high selective pressure (Lommel and Xiong, 1991; Gal et al., 1992; Schoelz and Wintermantel, 1993; Greene and Allison, 1994; Frischmuth and Stanley, 1998; Borja et al., 1999; Adair and Kearney, 2000; Varrelmann et al., 2000). Experiments are designed so that conditions are optimal to recover recombinant RNA and/or recombinant viruses. These conditions are suitable to assess the occurrence of recombination with a limited number of plants and within a short time frame. They also enhance the likelihood of detection of a recombination event that is usually rare and, hence, not easy to identify. For most of these studies, RNA recombination was assessed by the ability of a defective challenge virus to spread systemically in transgenic plants expressing a functional gene. Also, PTGS was likely not active in the transgenic plants tested because most of them were susceptible to challenge infection.

In contrast, no viable recombinant viruses have been detected in transgenic plants expressing viral genes under low or no selective pressure, in particular under field conditions. Interestingly, very similar conclusions were drawn with different test plants exposed to different indigenous vector and viral populations in distinct field environments (Fuchs et al., 1998, 1999; Thomas et al., 1998; Lin et al., 2001; Vigne et al., 2004a,b; N. Capote and M. Cambra, Valencia, Spain, 2006, personal communication; I. Zagrai, Bistrita, Romania, 2006, personal communication). These observations suggest that the likelihood of recombinant viruses to emerge in transgenic plants under field conditions of no selective pressure is extremely low across crops, transgene constructs, challenge viruses, vector species and environments.

The occurrence of RNA recombination depends on the type of viral sequence expressed by transgenic plants. As discussed above, the presence of a 3' non-coding region has been shown to enhance the rate of recombination in many cases (Allison et al., 1996; Greene and Allison, 1996; Varrelmann et al., 2000) and favour the synthesis of minus-strand RNA (Teycheney et al., 2000). Excluding 3' terminal sequences of a viral transgene, and hence removing RdRp binding sites, can reduce the recombination rate (Allison et al., 1996; Greene and Allison, 1996; Varrelmann et al., 2000). It has to be noted, however, that the presence of a viral 3' non-coding region had no impact on the creation of recombinant viruses in some transgenic plants (Adair and Kearney, 2000; Vigne et al., 2004a,b; N. Capote and M. Cambra, Valencia, Spain, 2006, personal communication; I. Zagrai, Bistrita, Romania, 2006, personal communication). Similarly, the degree of relatedness between the viral transgene and the challenge virus is another factor that influences the potential for RNA recombination. Recombination is likely to occur at a higher rate as a result of interactions between related virus species than between unrelated or less related virus species. As a result, RNA recombination can be limited in transgenic plants by reducing the extent of sequence homology between the viral transgene and the challenge virus (Borja et al., 1999).

The challenge virus itself also influences the occurrence of recombination. Recombination rates ranging from 3% to 20% have been obtained with different viruses (ACMV, CaMV, CCMV, PPV, PVY and TBSV) with high propensity for recombination, especially PVY (Chare and Holmes, 2006) and CaMV (Froissart et al., 2005). Restoration of wild-type viruses through RNA recombination required a single recombination event in the case of RCNMV (Lommel and Xiong, 1991) and CCMV (Greene and Allison, 1994, 1996), while a double recombination event was required in the case of TBSV (Borja *et al.*, 1999), PPV (Varrelmann *et al.*, 2000) and TMV (Adair and Kearney, 2000). In the studies with DNA viruses, recombinant viruses likely arose through template switching events during reverse transcription of the viral CaMV RNA (Gal *et al.*, 1992; Schoelz and Wintermantel, 1993; Wintermantel and Schoelz, 1996).

Assessing the occurrence of a rare event and identifying viable recombinant viruses in transgenic plants expressing viral genes can be challenging from a technical point of view. Therefore, robust experimental conditions are necessary to avoid the detection of artefact chimeric RNA molecules. An elegant approach to circumvent some technical difficulties is to target viable recombinant viruses (Fuchs *et al.*, 1998, 1999; Adair and Kearney, 2000; Lin *et al.*, 2001; Vigne *et al.*, 2004a,b; N. Capote and M. Cambra, Valencia, Spain, 2006, personal communication; I. Zagrai, Bistrita, Romania, 2006, personal communication) rather than recombinant RNA molecules (Teycheney *et al.*, 2000; Fernandez-Delmond *et al.*, 2004; Koenig and Büttner, 2004).

The molecular mechanism underlying virus resistance is also likely an important factor for the occurrence of RNA recombination in transgenic plants expressing viral genes. Indeed, if RNA silencing is active, the steady-state accumulation of viral transgene transcripts is low (Voinnet, 2001, 2005; MacDiarmid, 2005) and recombinant RNA molecules arising in silenced transgenic plants would probably be eliminated by the RISC complex. As a consequence, the rate of recombination is expected to be very low in virus-resistant silenced transgenic plants because the amount of transgene transcripts available for template switching may be subliminal. Although direct evidence to substantiate this hypothesis is lacking, it is conceivable that RNA silencing should reduce the potential for RNA recombination in transgenic plants expressing viral genes. Therefore, RNA recombination with viral transgene transcripts is less significant than initially predicted (Hull, 1989; de Zoeten, 1991; Rissler and Mellon, 1996; Jakab *et al.*, 1997).

Also, a cascade of events is necessary for a viable recombinant to develop in transgenic plants expressing viral genes. First, viruliferous vectors need to feed on transgenic plants and consequently transmit virus isolates. Second, the genome of challenging isolates needs to replicate and further interact with transgene transcripts. Third, template switching needs to occur during replication and generate chimeric RNA molecules. Fourth, recombinant RNA molecules need to replicate, be encapsidated, move from cell to cell and subsequently move through the vascular system. There are several constraints attached to each of these steps that will reduce their successful occurrence. It is expected that a viable recombinant virus will develop only if each of these steps has a relatively reasonable probability of fulfillment. The necessity to accommodate a number of successive favourable events substantially reduces the frequency of emergence of recombinant viruses, especially if PTGS is active and degrades both the transgene transcripts and invading viral genome (Voinnet, 2001, 2005; MacDiarmid, 2005).

Is RNA recombination in transgenic plants expressing viral genes a real risk?

The significance of RNA recombination in transgenic plants expressing viral genes must be considered in light of the baseline recombination rate in conventional plants. Such comparative analysis is pertinent to determine if perceived risks are real. Therefore, it is essential to refer to RNA recombination in conventional plants subjected to mixed virus infection or infection with multiple variants as baseline information in order to assess if recombinants in transgenic plants are any different from those in conventional plants. This is especially relevant because hazardous situations have been predicted initially in relation to the potential for RNA recombination in transgenic plants expressing viral genes (Hull, 1989; de Zoeten, 1991; Rissler and Mellon, 1996; Jakab et al., 1997). For example, if a newly developed hybrid virus acquires new transmission characteristics through recombination, it could get spread by new vectors and infect new plant species, thus creating new epidemics (de Zoeten, 1991). A fundamental question relevant to environmental risk assessment is: will transgenic plants expressing viral genes increase the likelihood of development of recombinant virus species beyond that of natural background events?

Abundant evidence from viral nucleotide sequence analysis indicates that RNA recombination is a relatively common phenomenon in some plant RNA viruses, notably potyviruses (Chare and Holmes, 2006) and DNA viruses (Froissart et al., 2005). RNA recombination is a major driving force in plant virus evolution and for the emergence of new variants and new virus species (Garcia-Arenal et al., 2001; Hull, 2002; Tepfer, 2002; Bonnet et al., 2005). Mixed virus infections are frequently described in conventional plants, thus creating ample opportunities for recombinant events to occur between distinct viral RNA molecules. Remarkably, RNA recombination in transgenic plants expressing viral genes has not been shown so far to occur under conditions of high disease pressure and low selective pressure, whether in the field or confined environments. Thus, RNA recombination in transgenic plants does not seem to occur beyond background events in conventional plants.

Relevant to environmental risk assessment is also the outcome of RNA recombination, not so much its occurrence. In other words, it is more critical to assess the creation of viable recombinant viruses rather than the development of chimeric RNA molecules. Field studies are crucial to environmental risk assessment because plants are challenged by functional viruses under conditions of low, if any, selective pressure. Noteworthy, none of the field studies provided evidence for the creation of recombinant viruses at detectable levels (Fuchs et al., 1998, 1999; Thomas et al., 1998; Lin et al., 2001) even when perennial plants have been grown in the field over extended time (Vigne et al., 2004a,b; N. Capote and M. Cambra, Valencia, Spain, 2006, personal communication; I. Zagrai, Bistrita, Romania, 2006, personal communication). These findings are not unexpected because transgenic plants typically express viral transgene transcripts at lower levels than viral RNA or DNA in virus-infected conventional plants. The differential expression level can be orders of different magnitude (Vigne et al., 2004a). Therefore, it is anticipated that the rate of recombination should be substantially lower in transgenic compared to conventional plants.

It is also essential to determine if recombinant viruses have different biological properties, including potentially harmful properties, relative to parental viruses and non-hybrid viruses. This information is critical to evaluate the viability and fitness of recombinant viruses, especially those isolated from transgenic plants and to determine if they outcompete wild-type viruses and show undesirable features. Some recombinant viruses have similar properties compared to non-recombinant viruses (Vigne et al., 2005) while others can exhibit exacerbated properties. For example, an artificial recombinant virus produced in vitro by replacing the 2b gene of CMV by the counterpart of TAV was significantly more virulent than either of its parents (Ding et al., 1996). Another recombinant between CMV and TAV could outcompete the parental viruses in co-infection experiments (Fernandez-Cuartero et al., 1994).

Another consideration for risk assessment is the insertion and expression of viral sequences in plants. This phenomenon is not exclusive to the application of the concept of pathogen-derived resistance. Indeed, the integration of viral sequences, in particular DNA sequences of pararetroviruses and geminiviruses, and RNA of retroviruses, into conventional plants has been described (Harper et al., 2002; Hansen et al., 2005). Integration of viral DNA sequences into host genomes involves recombination, mostly a non-homologous recombination event (Harper et al., 2002). Incorporation of RNA of retroviruses occurs by reverse transcription and the action of a virus-encoded integrase (Hansen et al., 2005). Recently, it has been suggested that RNA recombination between PVY RNA and the RNA of a retrotransposable element in grapevine can lead to the insertion of viral RNA into the plant genome (Tanne and Sela, 2005).

In summary, significant progress has been made over the last 15 years in addressing RNA recombination in transgenic plants expressing viral RNA. Remarkable advances provided valuable insights into the significance of recombination. Extensive research and the commercial release of virus-resistant transgenic squash and papaya over one decade (James, 2005; Shankula et al., 2005) infer a reasonable certainty of limited or no hazard in regard to the occurrence and outcome of RNA recombination. This indicates that transgenic plants expressing viral genes are unlikely to pose undue environmental risk.

Balancing Real Risks of RNA Recombination and Benefits of Virus-resistant Transgenic Plants

Field experiments have highlighted the numerous benefits of virus-resistant transgenic plants to agriculture (Fuchs and Gonsalves, 1997, 2002; Gonsalves, 1998). First, virus-resistant transgenic plants can be developed for crops for which no sources of host resistance is known or can be manipulated by conventional breeding approaches. In that case, pathogen-derived resistance is an effective alternative approach to develop resistance. Second, pyramiding resistance to multiple viruses can be achieved. Third, virus diseases can be managed and their impact mitigated with the use of resistant transgenic cultivars. Fourth, commercial transgenic plants do not serve as primary virus sources for secondary vector-mediated transmission (Klas et al., 2006). Thus, they can reduce the impact of virus epidemics in neighbouring susceptible crops. Fifth, yield is significantly higher for transgenic plants and fruits are of marketable quality even when plants are grown under conditions of high disease pressure. The commercial use of PRSV-resistant transgenic papaya in Hawaii increased production to the level seen prior to the severe PRSV epidemic (Gonsalves, 1998). Also, the net benefit of planting virus-resistant transgenic squash was estimated at $19 million in the USA in 2004 (Shankula et al., 2005). Altogether, benefits related to the use of virus-resistant transgenic plants are of horticultural, epidemiological, environmental, social and economical importance (Fuchs and Gonsalves, 1997, 2002; Gonsalves, 1998).

A comparison of real risks and benefits of the use of virus-resistant transgenic plants should be the basis to decide on the safe deployment of the technology, especially if risks associated with alternative approaches do not outweigh benefits. So far, field evidence does not indicate that transgenic plants expressing viral genes increase the frequency of viral recombinants and create more severe virus variants than mixed infected conventional plants. Similarly, there is little evidence, if any, to infer that transgenic plants expressing viral genes favour the creation of new viruses that could not arise naturally from mixed infected conventional plants (Falk and Bruening, 1994; Fuchs and Gonsalves, 1997, 2002; Miller et al., 1997; Hammond et al., 1999; Rubio et al., 1999; Martelli, 2001). Thus, one can conclude with confidence that the benefits of using virus-resistant transgenic plants by far outweigh the perceived risks of recombination in relation to the deployment of viral transgenes to control virus diseases.

Conclusions and Future Directions

The likelihood of new viruses with undesired biological properties to emerge through recombination in transgenic plants expressing viral genes is extremely low. Also, no adverse effect has been reported in relation to RNA recombination 11 years after the commercialization of virus-resistant transgenic squash and eight years after the commercialization of PRSV-resistant transgenic papaya. Should more research be conducted to address the issue of recombination in transgenic plants expressing viral genes? Providing additional information on the safe use of transgenic plants expressing viral genes could help increase the adoption of the technology.

New research directions could focus on more field research to confirm the current trend, especially with transgenic perennial crops that are exposed continuously to virus infection over extended time. Such studies would be valuable to determine if the likelihood of recombination is higher in perennial than in annual crops or in herbaceous plants. Experiments could also be carried out in different environments by using viruses with different vectors and different transmission modes to assess on a larger scale and across different sites the significance of recombination in transgenic plants.

Limited information is available on the frequency of successful establishment of recombinant viruses after their development, although it appears that only recombinant viruses that are viable and have a selective advantage should outcompete natural populations. Also, little is known on the comparative fitness of recombinant viruses and parental lineages. In general, survival of a viral population, including recombinant variants, depends on the balance between replication fidelity and genomic flexibility. Maintaining a certain complexity of variants, including recombinants, enables a virus population to survive changes in the environment, such as replication in different tissues, systemic spread, adaptation to a new host, etc. Fitness advantages have been described for certain recombinant viruses although it is rare for a viable recombinant to have a selective advantage over parental viruses. Most of these areas of desirable progress, however, are more of academic relevance than of value to risk assessment.

There seem to be an increasing consensus within the scientific community that, while initially perceived as a major concern, RNA recombination in transgenic plants expressing viral genes should not be deemed risky. This is due to the fact that extensive research has provided a reasonable certainty of limited or no hazard.

Acknowledgements

The author is grateful to Dr L.M. Yepes for critical reading of the manuscript and to USDA-APHIS and the European Commission for providing financial support through competitive grants on risk assessment of virus-resistant transgenic crops. Special thanks are due to Drs Nieves Capote, Mariano Cambra and Ioan Zagrai for providing unpublished data, and to Dr Hélène Sanfaçon for helpful comments.

References

Aaziz, R. and Tepfer, M. (1999) Recombination in RNA viruses and in virus-resistant transgenic plants. *Journal of General Virology* 80, 1339–1346.

Adair, T.L. and Kearney, C.M. (2000) Recombination between a 3-kilobase tobacco mosaic virus transgene and a homologous viral construct in the restoration of viral and nonviral genes. *Archives of Virology* 145, 1867–1883.

Allison, R.F., Schneider, W.L. and Greene, A.E. (1996) Recombination in plants expressing viral transgenes. *Seminars in Virology* 7, 417–422.

Beachy, R. (1997) Mechanisms and applications of pathogen-derived resistance in transgenic plants. *Current Opinion in Biotechnology* 8, 215–220.

Bendahmane, M. and Beachy, R.N. (1999) Control of tobamovirus infections via pathogen-derived resistance. *Advances in Virus Research* 53, 369–386.

Bonnet, J., Fraile, A., Sacristan, S., Malpica, J.M. and Garcia-Arenal, F. (2005) Role of recombination in the evolution of natural populations of *Cucumber mosaic virus*, a tripartite RNA plant virus. *Virology* 332, 359–368.

Borja, M., Rubio, T., Scholthof, H.B. and Jackson, A.O. (1999) Restoration of wild-type virus by double recombination of tombusvirus mutants with a host transgene. *Molecular Plant-Microbe Interactions* 12, 153–162.

Callaway, A., Giesman-Cookmeyer, D., Gillock, E.T., Sit, T.L. and Lommel, S.A. (2001) The multifunctional capsid proteins of plant RNA viruses. *Annual Review of Phytopathology* 39, 419–460.

Chare, E.R. and Holmes, E.C. (2006) A phylogenetic survey of recombination frequency in plant RNA viruses. *Archives of Virology* 151, 933–946.

de Zoeten, G.A. (1991) Risk assessment: do we let history repeat itself? *Phytopathology* 81, 585–586.

Ding, S.-W., Shi, B.J., Li, W.X. and Symons, R.H. (1996) An interspecific species hybrid RNA virus is significantly more virulent than either parental virus. *Proceedings of the National Academy of Sciences of the United States of America* 93, 7470–7474.

Falk, B.W. and Bruening, G. (1994) Will transgenic crops generate new viruses and new diseases? *Science* 163, 1395–1396.

Fernandez-Cuartero, B., Burgyan, J., Aranda, M.A., Salanki, K., Moriones, E. and Garcia-Arenal, F. (1994) Increase in the relative fitness of a plant virus RNA associated with its recombinant nature. *Virology* 203, 373–377.

Fernandez-Delmond, I., Pierrugues, O., de Wispelaere, M., Guilbaud, L., Gaubert, S., Diveki, Z., Godon, C., Tepfer, M. and Jacquemond, M. (2004) A novel strategy for creating recombinant infectious RNA virus genomes. *Journal of Virological Methods* 121, 247–257.

Frischmuth, T. and Stanley, J. (1998) Recombination between viral DNA and the transgenic coat protein gene of African cassava mosaic geminivirus. *Journal of General Virology* 79, 1265–1271.

Froissart, R., Roze, D., Uzest, M., Galibert, L., Blanc, S. and Michalakis, Y. (2005) Recombination every day: abundant recombination in a virus during a single multi-cellular host infection. *PLoS Biology* 3, e89.

Fuchs, M. and Gonsalves, D. (1997) Genetic engineering. In: Rechcigl, N.A. and Rechcigl, J.E. (eds) *Environmentally Safe Approaches to Crop Disease Control*. CRC Press, Boca Raton, Florida, pp. 333–368.

Fuchs, M. and Gonsalves, D. (2002) Genetic engineering and resistance to viruses. In: Khachatourians, G.G., McHughen, A., Scorza, R., Nip, W.-K. and Hui, Y.H. (eds) *Transgenic Plants and Crops*. Marcel Dekker, New York, pp. 217–231.

Fuchs, M., Klas, F.E., McFerson, J.R. and Gonsalves, D. (1998) Transgenic melon and squash expressing coat protein genes of aphid-borne viruses do not assist the spread of an aphid non-transmissible strain of cucumber mosaic virus in the field. *Transgenic Research* 7, 449–462.

Fuchs, M., Gal-On, A., Raccah, B. and Gonsalves, D. (1999) Epidemiology of an aphid nontransmissible potyvirus in fields of nontransgenic and coat protein transgenic squash. *Transgenic Research* 8, 429–439.

Gal, S., Pisan, B., Hohn, T., Grimsley, N. and Hohn, B. (1992) Agroinfection of transgenic plants leads to viable *Cauliflower mosaic virus* by intermolecular recombination. *Virology* 187, 525–533.

Garcia-Arenal, F., Fraile, A. and Malpica, J.M. (2001) Variability and genetic structure of plant virus populations. *Annual Review of Phytopathology* 39, 157–186.

Gonsalves, D. (1998) Control of *Papaya ringspot virus* in papaya: a case study. *Annual Review of Phytopathology* 36, 415–437.

Gonsalves, D., Chee, P., Provvidenti, R., Seem, R. and Slightom, J.L. (1992) Comparison of coat protein-mediated and genetically derived resistance in cucumbers to infection by *Cucumber mosaic virus* under field conditions with natural challenge inoculations by vectors. *Bio/Technology* 10, 1562–1570.

Greene, A.E. and Allison, R.F. (1994) Recombination between viral RNA and transgenic plant transcripts. *Science* 263, 1423–1425.

Greene, A.E. and Allison, R.F. (1996) Deletions in the 3' untranslated region of cowpea chlorotic mottle virus transgene reduce recovery of recombinant viruses in transgenic plants. *Virology* 225, 231–234.

Hammond, J., Lecoq, H. and Raccah, B. (1999) Epidemiological risks from mixed virus infections and transgenic plants expressing viral genes. *Advances in Virus Research* 54, 189–314.

Hansen, C.N., Harper, G. and Heslop-Harrison, J.S. (2005) Characterisation of pararetrovirus-like sequences in the genome of potato (*Solanum tuberosum*). *Cytogenetic and Genome Research* 110, 1–4.

Harper, G., Hull., R., Lockhart, B. and Olszewski, N. (2002) Viral sequences integrated into plant genomes. *Annual Review of Phytopathology* 40, 119–136.

Huang, J., Rozelle, S., Pray, C. and Wang, Q. (2002) Plant biotechnology in China. *Science* 295, 674–677.

Hull, R. (1989) Non-conventional resistance to viruses in plants: concepts and risks. In: Gustafson, J.P. (ed.) *Gene Manipulation in Plant Improvement*. Stadler Genetic Symposia Series, Columbia, Michigan, pp. 289–304.

Hull, R. (2002) *Matthew's Plant Virology*, 4th edn. Academic Press, London.

Jakab, G., Vaistij., F.E., Droz, E. and Manoë, P. (1997) Transgenic plants expressing viral sequences create a favourable environmental for recombination between viral sequences. In: Tepfer, M. and Balazs, E. (eds) *Virus-resistant Transgenic Plants: Potential Ecological Impact*, INRA Editions, Springer, Heidelberg, pp. 45–51.

James, C. (2005) Executive summary, brief 34, Global status of commercialized biotech/GM crops: 2005. Available at: http://www.isaaa.org.

Klas, F.E., Fuchs, M. and Gonsalves, D. (2006) Comparative spatial spread over time of *Zucchini yellow mosaic virus* (ZYMV) and *Watermelon mosaic virus* (WMV) in fields of transgenic squash expressing the coat protein genes of ZYMV and WMV, and in fields of nontransgenic squash. *Transgenic Research* 15, 527–541.

Koenig, R. and Büttner, G. (2004) Strategies for the detection of potential *Beet necrotic yellow vein virus* genome recombinants which might arise as a result of growing A type coat protein gene-expressing sugarbeets in soil containing B type virus. *Transgenic Research* 13, 21–28.

Lin, H.-X., Rubio, L., Smythe, A., Jiminez, M. and Falk, B.W. (2001) Genetic diversity and biological variation among California isolates of *Cucumber mosaic virus*. *Journal of General Virology* 84, 249–258.

Lommel, S.A. and Xiong, Z. (1991) Reconstitution of a functional red clover necrotic mosaic virus by recombinational rescue of the cell-to-cell movement gene expressed in a transgenic plant. *Journal of Cell Biochemistry* 15A, 151.

MacDiarmid, R. (2005) RNA silencing in productive virus infections. *Annual Review of Phytopathology* 43, 523–544.

Martelli, G.P. (2001) Transgenic resistance to plant pathogens: benefits and risks. *Journal of Plant Pathology* 83, 37–46.

Miller, W.A., Koev, G. and Mohan, B.R. (1997) Are there risks associated with transgenic resistance in luteoviruses? *Plant Disease* 81, 700–710.

Nelson, R., McCormick, S.M., Delannay, X., Dube, P., Layton, J., Anderson, E.J., Kaniewska, M., Proksch, R.K., Horsch, R.B., Rogers, S.G., Fraley, R.T. and Beachy, R.N. (1988) Virus tolerance, plant growth and field performance of transgenic tomato plants expressing the coat protein from tobacco mosaic virus. *Bio/Technology* 6, 403–409.

Powell Abel, P., Nelson, R.S., De, B., Hoffman, N., Rogers, S.G., Frayley, R.T. and Beachy, R.N. (1986) Delay of disease development in transgenic plants that express the tobacco mosaic coat protein gene. *Science* 232, 738–743.

Rissler, J. and Mellon, M. (1996) *The Ecological Risks of Engineered Crops*. MIT Press, Cambridge, Massachusetts.

Robinson, D.J. (1996) Environmental risk assessment of releases of transgenic plants containing virus-derived inserts. *Transgenic Research* 5, 359–362.

Rubio, T., Borja, M., Scholthof, H.B. and Jackson, A.O. (1999) Recombination with host transgenes and effects on virus evolution: An overview and opinion. *Molecular Plant-Microbe Interactions* 12, 87–92.

Sanford, J.C. and Johnston, S.A. (1985) The concept of parasite-derived resistance – deriving resistance genes from the parasite's own genome. *Journal of Theoretical Biology* 113, 395–405.

Schoelz, J.E. and Wintermantel, W.M. (1993) Expansion of viral host range through complementation and recombination in transgenic plants. *The Plant Cell* 5, 1669–1679.

Shankula, S., Marmon, G. and Blumenthal, E. (2005) Biotechnology-derived crops plants in 2004. Impacts on US Agriculture. Available at: http://www.ncfap.org.

Tanne, E. and Sela, I. (2005) Occurrence of a DNA sequence of a non-retro RNA virus in a host plant genome and its expression: evidence for recombination between viral and host RNAs. *Virology* 332, 614–622.

Tepfer, M. (2002) Risk assessment of virus-resistant transgenic plants. *Annual Review of Phytopathology* 40, 467–491.

Tepfer, M. and Balazs, E. (1997) *Virus-resistant Transgenic Plants: Potential Ecological Impact*. INRA Editions,Springer, Heidelberg.

Teycheney, P.Y., Aaziz, R., Dinant, S., Salanki, K., Tourneur, C., Balazs, E., Jacquemond, M. and Tepfer, M. (2000) Synthesis of (−) strand RNA from the 3′ untranslated region of plant viral genome expressed in transgenic plants upon infection with related viruses. *Journal of General Virology* 81, 1121–1126.

Thomas, P.E., Hassan, S., Kaniewski, W.K., Lawson, E.C. and Zalewski, J.C. (1998) A search for evidence of virus/transgene interactions in potatoes transformed with the potato leafroll virus replicase and coat protein genes. *Molecular Breeding* 4, 407–417.

Varrelmann, M., Palkovics, L. and Maiss, E. (2000) Transgenic or plant expressing vector-mediated recombination of *Plum pox virus*. *Journal of Virology* 74, 7462–7469.

Vigne, E., Komar, V. and Fuchs, M. (2004a) Field safety assessment of recombination in transgenic grapevines expressing the coat protein gene of *Grapevine fanleaf virus*. *Transgenic Research* 13, 165–179.

Vigne, E., Bergdoll, M., Guyader, S. and Fuchs, M. (2004b) Population structure and genetic variability within isolates of *Grapevine fanleaf virus* from a naturally infected vineyard in France: evidence for mixed infection and recombination. *Journal of General Virology* 85, 2435–2445.

Vigne, E., Demangeat, G., Komar, V. and Fuchs, M. (2005) Characterization of a naturally occurring *Grapevine fanleaf virus* recombinant isolate. *Archives of Virology* 150, 2241–2255.

Voinnet, O. (2001) RNA silencing as a plant immune system against viruses. *Trends in Genetics* 17, 449–459.

Voinnet, O. (2005) Induction and suppression of RNA silencing: insights from viral infection. *Nature Reviews Genetics* 6, 206–221.

Wintermantel, W.M. and Schoelz, J.E. (1996) Isolation of recombinant viruses between *Cauliflower mosaic virus* and a viral gene in transgenic plants under conditions of moderate selection pressure. *Virology* 223, 156–164.

19 Virus-resistant Transgenic Papaya: Commercial Development and Regulatory and Environmental Issues

J.Y. SUZUKI, S. TRIPATHI AND D. GONSALVES

Abstract

In Hawaii, transgenic papaya resistant to *Papaya ringspot virus* (PRSV) was developed starting in the 1980s and released commercially in 1998 to combat the widespread destruction of Hawaii's papaya industry. This review describes the proactive development of the transgenic papaya and its impact on stemming the destruction caused by PRSV in Puna, the main papaya producing area in Hawaii. It also focuses on the regulatory issues that were confronted in obtaining approval from the US government's Environmental Protection Agency (EPA), Animal Plant Health Inspection Service (APHIS) and Food and Drug Administration (FDA). The performance of the transgenic papaya is traced over the last 8 years following commercial release, with special observations on the issues of environmental impact and coexistence with non-transgenic papaya. The latter is quite important since a significant part of Hawaii's papaya is exported to Canada and Japan. Canada has approved the transgenic papaya, but efforts to get approval for export of transgenic papaya to Japan are still ongoing.

Introduction

Papaya (*Carica papaya* L.) is an important fruit crop in tropical and subtropical regions due to its economic, nutritional, industrial, pharmaceutical and medicinal values, for local and export markets. *Carica papaya*, a member of the Caricaceae family, probably originated from the Southern part of Mexico and the Northern region of Central America (Badillo, 1993). It is relatively easy to grow from seed. The first mature fruits can be harvested 9 months after sowing, and fruits are produced year-round. The papaya was disseminated into the Asian tropics during the 1600s by seeds taken to the Malay Peninsula, India and Philippines. Documents show a wide distribution in the Pacific Islands by the 1800s (Nakasone, 1975). The Food and Agricultural Organization (FAO) of the United Nations (UN) estimated that about 5.85 million tonnes of fruit were harvested in 2004, almost doubling the 1980 harvest. Brazil (21.6%), Mexico (13.1%), Nigeria (11.6%), Indonesia (11.1%) and India (10.1%) are the largest producers of papaya (FAO, 2004).

A major limiting factor for papaya cultivation worldwide is the disease caused by PRSV. Discovered in 1945, PRSV is the most widespread and damaging papaya virus. The name of the disease, papaya ringspot, is taken from the ringed spots on fruit of infected trees (Jensen, 1949). Trees infected with PRSV develop a range of symptoms: mosaic and chlorosis of leaf lamina, water-soaked oily streaks on the petiole and upper part of the trunk, distortion of young leaves that resembles mite damage, loss of vigour and stunting (Purcifull *et al.*, 1984). Plants infected at the seedling stage or within 2 months after planting do not normally produce mature fruit, while trees infected at later stages produce few fruits of poor quality, due to the presence of ringspots and generally lower sugar concentrations. PRSV is transmitted by numerous species of aphids in a non-persistent manner to a limited host range of cucurbits and papaya, and also produces local lesions on *Chenopodium quinoa* and *C. amaranticolor*. Evidence suggests that PRSV is not seed transmitted, although there has been a report of seed transmission. PRSV is grouped into two types: type P (PRSV-p) infects cucurbits and papaya; whereas type W (PRSV-w) formerly referred to as WMV-1 infects cucurbits but not papaya (Purcifull *et al.*, 1984).

Much progress has been made in the molecular characterization of PRSV. Strains of PRSV-p from Hawaii and Taiwan have been completely sequenced (Yeh *et al.*, 1992; Wang and Yeh, 1997). The genomic RNA consists of 10,326 nucleotides and has the typical array of genes found in potyviruses (Fig. 19.1A). The genome is monocistronic and is expressed via a large polyprotein that is subsequently cleaved to functional proteins. There are two possible cleavage sites, 20 amino acids apart, for the N-terminus of the coat protein (CP). Both of these sites may be functional; the upstream site for producing a functional NIb protein (the RNA-dependent RNA polymerase), and the other, to produce a CP that is capable of functioning in aphid transmission (Quemada *et al.*, 1990; Yeh *et al.*, 1992; Wang *et al.*, 1994). It is impossible to segregate PRSV-p and PRSV-w types by their CP sequences. Within the p types, however, the CP sequences can diverge by as much as 12%.

The Hawaii papaya industry started in the 1940s on the island of Oahu, on about 200 ha (Ferreira *et al.*, 1992). By the 1950s, production on Oahu was affected by PRSV and the industry moved to the island of Hawaii into the area of Puna, which had no PRSV nor commercial papaya production. Growing area increased to 263 ha by 1960 and to 911 ha in 1990. In contrast, the growing area on Oahu fell to less than 20 ha by 1990 (Ferreira *et al.*, 1992). The yellow-fleshed 'Kapoho' was the dominant papaya grown in Puna, distantly followed by the red-fleshed 'Sunrise'. In fact, Kapoho made up 95% of the state of Hawaii's papaya production in 1992.

Remarkably, despite the presence of PRSV in Hilo only 30 km away, Puna remained free of PRSV for over 30 years, thanks to a surrounding

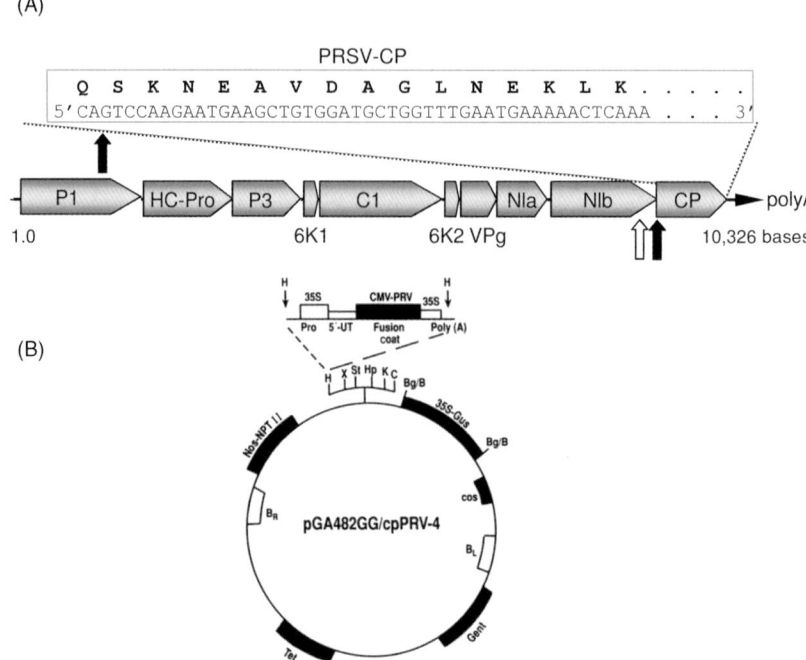

Fig. 19.1. (A) Organization and proteolytic protein products of the 10,326 base monocistonic PRSV genome. The N-terminal sequence of the PRSV HA 5-1 CP is shown at the top. Box arrows represent the proteolytic sites producing the mature CP. (From Tripathi et al., 2006.) (B) Map of the functional genes of the *Agrobacterium* transformation vector pGA482GG/cpPRV-4 used for generating PRSV-resistant papaya. The PRSV *CP* gene cassette consists of the *CP* structural gene of PRSV HA 5-1 translationally fused to the N-terminal end of the cucumber mosaic virus CP (CMV-PRV) including the translation initiation codon, the CMV 5' untranslated sequence (5'UTR) and the Cauliflower Mosaic Virus 35S promoter (35S). The CMV-PRV gene cassette is flanked by selectable and visible marker genes, *nptII* and *uidA* (GUS), respectively. B_R and B_L are the left and right borders of the transformation vector T-DNA sequence. (From Ling et al., 1991.)

area of sparsely populated lava which served as a physical barrier and to diligence by the Hawaii Department of Agriculture (HDOA) in surveying and rogueing infected trees in those nearby communities. However, it was highly probable that PRSV would someday be found in Puna. In 1978, one of the authors (D. Gonsalves) started research towards developing control methods for PRSV. The development of virus-resistant transgenic papaya was initiated in the mid-1980s following unsuccessful attempts at controlling the disease by non-biotechnological means. In 1992, PRSV was indeed discovered in Puna and papaya production there went from 24,045 t in 1992 to 12,134 t in 1998, the year transgenic papaya was commercialized and seeds were first released.

In this chapter, we will cover the development of the transgenic papaya, the environmental risks and regulatory issues involved in its commercialization and acceptance, and its impact on disease management. The readers are also referred to previous publications on the transgenic papaya case (Gonsalves, 1998, 2002, 2006; Gonsalves and Ferreira, 2003; Fermín et al., 2004; Gonsalves and Fermín, 2004; Gonsalves et al., 2006; Tripathi et al., 2006).

Development of Transgenic Papaya

In the mid-1980s, an exciting, yet unproven alternative approach to control viral diseases was introduced. Transgenic tobacco expressing the *CP* gene of *Tobacco mosaic virus* (TMV) showed significant delay in the development of disease symptoms caused by TMV (Powell-Abel et al., 1986). This approach, which provided protection against the detrimental effects of pathogens by expression of genes or sequences of the same or related pathogens, was coined 'Parasite-Derived Resistance' (now referred to as Pathogen-Derived Resistance or PDR) by Sanford and Johnston (Sanford and Johnston, 1985; Baulcombe, 1996; Baulcombe et al., 1996; Beachy, 1997). Of interest, the resistance was genetically inherited, offering a potentially effective and feasible way for controlling PRSV in papaya.

The mild Hawaiian PRSV strain HA 5-1 was used as the source of the *CP* gene for the transgene construct since the goal was to create papaya resistant to Hawaiian strains of the virus (Quemada et al., 1990). The transgene was designed to allow the translation of the *CP* gene, as at that time, it was thought that the CP protein was required for PDR. Since the PRSV CP is produced by post-translational protease cleavage (Fig. 19.1A), there are no native translation signals specific for the *CP* sequence. Therefore, a chimeric gene was designed utilizing the translation signals found in the leader sequence of the *Cucumber mosaic virus* (CMV) *CP* gene fused in frame to the structural sequence of the PRSV *CP* (Fig. 19.1B) (Ling et al., 1991). A key point to transforming a papaya plant is the development of tissue culture conditions, particularly for regeneration. Efforts to develop a papaya regeneration system were unsuccessful until a technique to produce highly embryogenic tissue starting from immature zygotic embryos was developed (Fitch and Manshardt, 1990). The biolistic approach (Sanford et al., 1992) was used to transform papaya with the PRSV *CP* gene construct followed by selection and regeneration of kanamycin-resistant clones (Fitch et al., 1990, 1992). The target cultivars were the red-fleshed Sunrise, Sunset (a sib selection of Sunrise) and the yellow-fleshed Kapoho. A total of 10 papaya plants positive for GUS activity as well as for the *CP* gene by polymerase chain reaction (PCR) amplification were obtained – five Sunset and five Kapoho (Fitch et al., 1990, 1992).

Initial resistance evaluations of R_0 tranformants in the greenhouse and field

To determine the functionality of the PDR-based transgene system in papaya in a timely manner, screening for resistance was initially performed on the original transformants (R_0). Sufficient material for screening was accomplished through micropropagation to produce R_0 clones for greenhouse inoculation tests with the severe Hawaiian PRSV isolate HA. One *CP* gene-positive line tested, a transformed Sunset line designated 55-1, showed excellent resistance to PRSV HA (Fitch *et al.*, 1992).

The research also moved ahead aggressively to determine whether the promising line 55-1 (R_0 material) would be resistant to PRSV and have suitable horticultural characteristics under field conditions (Lius *et al.*, 1997). The experimental samples included micropropagated 55-1 R_0 plants, non-transformed Sunset and a transgenic line lacking the *CP* gene planted in University of Hawaii fields located in Waimanalo, on the island of Oahu. Inoculations were performed either manually or by vector transmission using an isolate found on Oahu. All of the non-transgenic plants became severely infected within 5 months and were completely decimated by the end of the trial, whereas line 55-1 remained symptomless throughout the trial period which lasted from 1992 to 1994.

Comprehensive greenhouse resistance evaluations for progeny (R_1) of promising line

R_1 progeny of 55-1 were created by crossing a 55-1 R_0 plant with non-transgenic papaya. Crossing R_0 55-1 was necessary since it turned out to be a female plant and thus progeny could not be obtained directly from selfing. The 55-1 R_1 seedlings had a 50% segregation ratio for the neomycin phosphotransferase protein, suggesting that the transgene *nptII* insert and presumably the linked *CP* gene was present as a single copy in the parental R_0 line.

R_1 plants were used to screen for resistance to PRSV isolates from Hawaii and other regions under greenhouse conditions. Resistance of the 55-1 R_1 plants was tested against PRSV isolates from Mexico, Florida, Bahamas, Australia, Brazil, China, Okinawa, Ecuador, Guam, Thailand, Jamaica and Hawaii (Tennant *et al.*, 1994). Isolates from Guam, Brazil, Thailand, Ecuador and Okinawa induced severe symptoms on all transgenic plants, although the symptoms were not as severe as those observed on non-transgenic plants. Isolates from Australia, China and Jamaica induced an attenuated phenotype on all transgenic test plants. Excellent resistance was found for Hawaii isolates. A fraction of the plants infected with virus strains from the Bahamas, Mexico and Florida exhibited severe phenotypes whereas others were symptomless. Symptomless plants remained symptomless following reinoculation.

Development of cultivars 'SunUp and Rainbow'

The R_0 transgenic line 55-1 served as a source of germplasm to create SunUp and Rainbow, papaya destined to become commercial cultivars.

As mentioned previously, the 55-1 R_0 plant was a transgenic Sunset, which is a commercial red-fleshed cultivar. The SunUp variety, which is homozygous (*CP/CP*) for the transgene, but is otherwise identical to Sunset, was created as the R_3 generation of the original transformant 55-1. This germplasm held the hope for the development of new resistant varieties since 100% of the progeny from crosses with any other non-transgenic variety would be hemizygous for the *CP* gene (*CP/+*). In Hawaii, the yellow-fleshed Kapoho variety is by far the most popular among farmers and consumers and has a pyriform shape and medium size, which are desirable commercial characteristics for packing and shipping. Thus, in attempt to combine the PRSV resistance and Kapoho characteristics, Rainbow, an F_1 hybrid between SunUp and Kapoho, was created (Manshardt, 1999). The resulting Rainbow cultivar bore pear-shaped fruit with yellow-orange flesh as anticipated and was hemizygous for the transgene (*CP/+*).

Performance of Transgenic Cultivars

Field trial of SunUp and Rainbow in Puna

By 1994, the complete devastation caused by PRSV in Kapoho, a major papaya producing area of Puna, created a critical situation for survival of the Hawaii papaya industry. At the same time, the results from the R_0 field trial on Oahu were quite encouraging. A field trial was conducted in Kapoho to determine if the transgenic papaya could be used to rescue the papaya industry. By late 1994, an application for a field trial was submitted to APHIS. Approval was obtained in early 1995 and the field trial was set up in Kapoho in October 1995 (Ferreira *et al.*, 2002). The field trial was allowed with the stipulation that (i) the field must be sufficiently isolated from commercial orchards to minimize the chance of transgenic pollen escaping to non-transgenic material outside of the field test; (ii) all abandoned papaya trees in the area must be monitored for the introgression of the transgene into fruits of these trees; and (iii) all fruits had to be buried on site.

The results from the field trial (Fig. 19.2) clearly demonstrated the potential value of the transgenic papaya (Gonsalves, 1998; Ferreira *et al.*, 2002). Except for three plants that showed infection at the beginning of the trial, none of the transgenic plants became infected. In contrast, 50% of the non-transgenic control plants were infected within 5 months after transplanting while all were infected by 7 months. Rainbow averaged about 112,082 kg/ha/year of marketable fruit during the trial, a higher yield compared to the average production from non-infected Kapoho, whereas the non-transgenic plants averaged about 5604 kg/ha/year. In addition to evaluating Rainbow for PRSV resistance, it was also critical to analyse its fruit for taste, production, colour, size, and packing and shipping qualities as it was targeted as the alternative variety to Kapoho. The consensus was that Rainbow is a more than adequate substitute for Kapoho even though Rainbow has a slightly larger fruit size.

Fig. 19.2. Aerial view of the transgenic papaya field trial in Puna, Hawaii. At the centre is a block of Rainbow plants surrounded by non-transgenic Sunrise, which are stunted due to PRSV infection. Adjacent to the field at the upper right was a similar block consisting of a non-transgenic version of Rainbow (F_1, Sunset × Kapoho) and a transgenic line similar to Rainbow but with a distinct transgene insertion. The open area in the lower right foreground is the position of the abandoned, PRSV-infected papaya field used as the source of virus inoculum and cleared prior to flowering of the experimental field. (From Tripathi et al., 2006.)

Greenhouse evaluation of Rainbow and SunUp: the effect of transgene copy number, plant development and coat protein homology

As noted above, earlier greenhouse work revealed that the resistance of R_1 plants of line 55-1 was narrow in that they were resistant to PRSV isolates from Hawaii but largely susceptible to isolates outside of Hawaii. In follow up greenhouse studies (Tennant et al., 2001), Rainbow showed similar narrow resistance as R_1 plants of line 55-1 (Table 19.1). In contrast, SunUp showed resistance to Hawaii isolates and to isolates from Jamaica and Brazil. It thus appeared that *CP* gene dosage affected the broadness of resistance since SunUp is homozygous and Rainbow hemizygous for the *CP* gene. Further studies also showed that plant development (age and height) made a difference in that Rainbow at a young age showed variable resistance to Hawaii isolates but complete resistance as plants were older and larger. Likewise, the young SunUp were susceptible to the Thailand isolate but older ones were resistant or showed a long delay in symptom expression.

Table 19.1. *CP* nucleotide sequence homologies of PRSV isolates to PRSV HA 5-1 and summary of reactions of isolates inoculated to Rainbow and SunUp papaya. (From Tripathi *et al.*, 2006.) (Modified from Tennant *et al.*, 2001.)

PRSV isolates	% Homology to transgene *CP*					Reaction to isolates	
	N	core	C	3' ncr	overall	Rainbow	SunUp
Hawaii-HA	99.3	99.8	100	100	99.8	R	R
Hawaii-OA	97.3	98.0	100	95.7	97.9	sR	R
Hawaii-KA	95.3	97.1	98.3	93.6	96.7	sR	R
Hawaii-KE	95.3	97.1	98.3	93.6	96.7	sR	R
Jamaica-JA	89.3	95.0	91.5	69.6	92.5	S	R
Brazil-BR	84.4	93.9	98.3	73.3	91.6	S	R
Thailand-TH	83.7	90.7	91.5	89.4	89.5	S	sR

Rainbow and SunUp are hemizygous (*CP*/+) and homozygous (*CP*/*CP*), respectively, for the PRSV HA 5-1 *CP* transgene. N = 199 nucleotides of the amino terminus, core = 641 nucleotides of the core region, C = 59 nucleotides of the carboxy terminus and 3' ncr = 35 nucleotides of the non-coding regions following the stop codon. R = resistant. sR = susceptible at young stages and resistant at older stages. S = susceptible.

Comparison of the *CP* gene sequences from the various isolates suggested that resistance is affected by *CP* homology to the transgene, with the resistance being strongest against PRSV isolates with the highest homology to the transgene. The *CP* genes of Hawaiian PRSV isolates showed 97–100% homology to the transgene *CP*, while the *CP* genes of other isolates (Jamaica, Brazil, Thailand) showed 89–93% homology. The *CP* gene of the Thai PRSV isolate had the least homology to the transgene *CP*. The above observations are consistent with a resistance mechanism based on homology-dependent post-transcriptional gene silencing (PTGS). Lastly, nuclear run-on experiments confirmed PTGS (Tennant *et al.*, 2001).

Deregulation and Commercialization of SunUp and Rainbow Papaya

In this section, we discuss the steps leading to the deregulation of Rainbow and SunUp in the USA, detailing the roles of the various federal agencies in this process. The timetable of events is shown in Table 19.2. The USA has a coordinated, risked-based system for ensuring that new biotechnology products are safe for the environment based on a policy 'Coordinated Framework for Regulation of Biotechnology', established in 1986 (PEW, 2004). The policy is carried out by three federal agencies: APHIS, the regulatory arm of the US Department of Agriculture (USDA); the EPA; and the FDA. A new web site with coordinated information from the three federal agencies including a database of completed reviews of genetically engineered (GE) crops can be found at http://usbiotechreg.nbii.gov/. The role of the state in regulation of genetically modified organisms (GMO), in

Table 19.2. History of development and deregulation of SunUp and Rainbow in the USA. (Modified from Tripathi et al., 2006.)

Year	Event	Reference
1990	PRSV-resistant papaya R_0 line 55-1 hemizygous for the *CP* transgene is created by biolistic transformation.	Fitch et al. (1992)
1991	APHIS issues permit for field trial of 55-1 in University of Hawaii's experimental farm in Waimanalo.	
1992	Greenhouse evaluation of a R_1 line hemizygous for the CP transgene of 55-1	Tennant et al. (1994)
1992	First field trial of 55-1 transgenic papaya was conducted in Waimanalo on Oahu island. During this time cultivars Rainbow and SunUp hemizygous and homozygous, respectively, for the CP transgene found in 55-1 were developed.	Lius et al. (1994, 1997); Manshardt (1999)
1994	Initial consultation with the FDA	
1995	APHIS issues a permit for a second field trial. Field trial of SunUp and Rainbow began in Puna on Hawaii island	Ferreira et al. (2002)
1996	Transgenic line 55-1 and its derivatives were deregulated by APHIS	Gonsalves (1998); Strating (1996)
1997	Submission of safety and nutritional assessment of 55-1 to FDA	
1997	Exemption from EPA was granted	Gonsalves (1998)
1997	FDA approval was granted for the transgenic lines	Gonsalves (1998)
1998	Bulk seed production of SunUp and Rainbow was completed	Wenslaff and Osgood (1999)
1998	License agreements were obtained from all parties allowing the commercial cultivation of transgenic papaya and its derivatives in Hawaii only. Seeds were released to farmers.	Gonsalves (1998)

general, parallels that of federal regulations although the specifics differ between state to state (Taylor et al., 2004).

APHIS is responsible for most issues relating to environmental safety and regulates GE crops as 'regulated articles', organisms and products known or suspected to be plant pests, or plant pest risks. This regulation is currently administered through the Biotechnology Regulatory Service (BRS) of APHIS. APHIS regulates the import, handling, interstate movement, release into environment including confined experimental use and field trials of GE crops under the Plant Protection Act (PPA) of 2000. Evaluation is based on: potential environmental impact for plant pest risk; disease and pest susceptibilities; the expression of gene products, new enzymes or changes to plant metabolism; weediness and impact on sexually compatible plants; agricultural or cultivation practices; effects on non-target organisms; and the potential for gene transfer to other types of organisms.

In reference to evaluation of papaya line 55-1 from which SunUp and Rainbow were derived, APHIS was largely concerned with the potential risk of heteroencapsidation, recombination, transgene flow to wild relatives and weediness of virus-resistant papaya. These aspects will be discussed in detail later, in the section on Environmental Risk Issues. In November 1996, transgenic line 55-1 and its derivatives were deregulated by a decision document and environmental assessment (EA) document concluding a 'Finding of No Significant Impact' (FONSI) from APHIS (Strating, 1996).

EPA, through its Biopesticides and Pollution Prevention Division of the Office of Pesticide Programmes, regulates the sale, distribution and use of pesticides to protect both human health (food safety) and the environment from pesticides. Under the Federal Insecticide, Fungicide and Rodenticide Act (FIFRA), regulated pesticides include 'pesticidal substances' produced by plants and microbes (products of biotechnology). Pesticides produced by GE plants are referred to as biopesticides or plant-incorporated protectants (PIP) and are regulated by the EPA. However, biopesticides or PIPs produced naturally by non-GE organisms are exempt. EPA regulates the pesticides but not the plant itself. Use permits are issued for field testing. Applicants must register pesticidal products prior to their sale and distribution and EPA may establish conditions for use as part of the registration. The EPA sets tolerance limits for residues of pesticides on and in food and animal feed under the Federal Food, Drug and Cosmetic Act (FFDCA). The EPA, through its Office of Prevention and Toxic Substances Biotechnology Programme, regulates products of biotechnology through its interpretation that organisms are 'chemical substances' under the Toxic Substances Control Act (TSCA). Developers must notify the EPA 90 days prior to manufacture or 60 days prior to field testing of a product regulated by TSCA. According to the EPA, the PRSV *CP* transgene is a pesticide because it confers resistance to plant viruses. Thus, it was subjected to tolerance-levels evaluation in the plant. In the permit application, we petitioned for an exemption from tolerance levels of the CP produced by the transgenic plant. We contended that the pesticide (the *CP* gene) was already present in many fruits consumed by the public, since much of the papaya eaten in the tropics is from PRSV-infected plants. In fact, we had earlier used cross-protection to control PRSV and fruits from these trees were sold to consumers. Furthermore, there is no evidence to date that the CP of PRSV or other plant viruses is allergenic or detrimental to human health in any way. Finally, measured amounts of CP RNA or protein in transgenic plants were much lower than those of infected plants. An exemption from tolerance to lines 55-1 was granted in August 1997.

FDA, through its Centre for Food Safety and Nutrition (CFSN), is responsible for ensuring the safety and proper labelling of all plant-derived foods and feeds, including those developed through bioengineering under the FFDCA. However, the true legal responsibility for food safety falls upon the developer. The FDA is also concerned with possible 'food additives', substances introduced into food that are not pesticides or not generally recognized as safe (GRAS). This agency follows a consultative process whereby the investigators submit an application with data and statements corroborating that the product is not harmful to human health. Evaluation by the FDA is based on the idea of 'substantial equivalence', meaning the nutritional or toxin content of a GE food is within the range normally found in conventional varieties. If the GE food is substantially equivalent, then it is regulated in the same way as non-GE food. GE foods which are nutritionally different, such as canola or soybeans with altered oil content, require labelling. In the evaluation of transgenic papaya 55-1 by the FDA, several aspects were considered: the range of concentration of some important vitamins, including vitamin C; the presence of *uidA* and *nptII* genes; and whether transgenic papaya had abnormally high concentrations of benzyl isothiocyanate (BITC). This latter compound has been reported in papaya (Tang, 1971). FDA approval was granted in September 1997.

Intellectual property rights

In the USA, a transgenic product cannot legally be commercialized unless it is fully deregulated and until licenses are obtained for the use of the intellectual property rights for processes or components that are part of the product or that have been used to develop the product (Gonsalves, 1998). The processes in question were the gene gun and PDR, in particular, CP-mediated protection. The components were translational enhancement leader sequences and genes (*nptII, uidA* and *CP*). This crucial hurdle involved legal and financial considerations beyond our means and expertise. These tasks were taken up by the industry's Papaya Administrative committee (PAC) and its legal counsel, Michael Goldman. The license agreements were obtained from all parties in April 1998, allowing the commercial cultivation of the papaya or its derivatives in Hawaii only. On 1 May, 1998 seeds were distributed free to growers who qualified by watching an educational video and signing an agreement that restricted growing of transgenic papaya only in Hawaii. Fruits can be sold outside Hawaii, provided that the importing state or country allows the importation and sale of transgenic papaya (Gonsalves, 1998). Thus, commercialization of SunUp and Rainbow began in Hawaii in 1998, 8 years after transgenic papaya 55-1 was first created. This followed deregulation by APHIS, which occurred in 1996, 4 years after the first field trials and deregulation by the EPA and FDA in 1997.

Impact of Rainbow on Papaya Production

The impact of the transgenic papaya on the papaya industry can be seen by its rapid rate of adoption in Puna, expressed as the percentage of bearing (actively producing) area of Rainbow and Kapoho (Table 19.3). In 2000,

Table 19.3. Production of transgenic papaya in Puna, Hawaii. (Modified from Tripathi et al., 2006.)

Year	Bearing hectare	% Kapoho	% Rainbow	Production
1998	663.7	100	0	12,134
2000	485.6	38	56	15,399
2001	677.8	45	47	18,275
2002	603.0	53	40	16,275
2003	574.7	45	47	16,209
2004	493.7	35	56	13,606
2005	511.9	22	66	12,206

Bearing area in Puna of non-transgenic Kapoho and transgenic Rainbow in hectares (ha) and the relationship to production ($\times$ 1000 kg) of fresh fruit utilized from the year 2000. Data for 1998 were included for reference. Data were compiled from USDA Statistical Reports of Papaya grown in Hawaii (www.nass.usda.gov/hi).

production in Puna rebounded to 15,399 t from the production low of 11,617 t in 1999 (Table 19.4). This coincided with the first recorded harvests from Rainbow, which comprised 56% of the bearing area that year, while Kapoho comprised 38%. In 2001, Puna papaya production peaked at 18,275 t with a near equal percentage of bearing area of Rainbow and Kapoho. In recent years, the percentage of bearing area of Rainbow has seen an upward trend to the current high of 66% as of 2005. These data indicate that Rainbow has been fully embraced by commercial growers as a popular and profitable cultivar. While the production levels for 2004 and 2005 seem to suggest a downward decline, the actual yields on a per hectare basis have been consistently higher after the introduction of Rainbow compared to yields before its introduction in 1998.

The impact of PRSV on papaya production can be observed by examining the contribution of Puna to Hawaii's total fresh papaya production. In 1992, Puna produced 24,045 t or 95% of the state's 25,310 t of fresh papaya (Table 19.4). Puna's production remained high for 2 years following the discovery of PRSV as a result of massive efforts to control the spread of the virus. In the years of total production decline caused by PRSV from 1995 to 1999, there was a concomitant sharp decrease in the percentage of papaya harvested from Puna. However, since the first recorded harvest of Rainbow in 2000, Puna's contribution to production has steadily and continually climbed, reaching 88% of the state's total in 2005. A substantial portion of this increase as noted above is due to the contribution of Rainbow to the bearing area. The data seem to indicate that acquisition of virus-resistant Rainbow has had a stabilizing effect on papaya production in Puna and subsequently on the industry as a whole. In 2005, when production in Puna reached a recent low of 12,206 t, it still accounted for 88% of Hawaii's total fresh

Table 19.4. Fresh papaya production in the state of Hawaii and in the Puna district from 1992–2005[a]. (Modified from Tripathi et al., 2006.)

Year	Fresh papaya utilization in Hawaii		
	Total (× 1000 kg)	Puna (× 1000 kg)	%
(Virus in Puna) 1992	25,310	24,045	95
1993	26,399	25,079	95
1994	25,492	25,186	99
1995	19,006	17,788	94
1996	17,146	15,511	90
1997	16,193	12,614	78
(Transgenic seeds released) 1998	16,148	12,134	75
1999	17,872	11,617	65
2000	22,793	15,399	68
2001	23,587	18,275	77
2002	19,368	16,275	84
2003	18,507	16,209	87
2004	15,467	13,606	88
2005	13,925	12,206	88

[a]Data were compiled from USDA Statistical Reports of Papaya grown in Hawaii (www.nass.usda.gov/hi).

papaya production, due to low overall productivity in the entire state. This reinforces the important role of Rainbow, since even aside from the negative impact of PRSV on papaya production, other variables such as the weather, rising costs of maintaining healthy orchards and low prices can also profoundly and negatively affect productivity and the health of the papaya industry.

Environmental Risk Issues

Rainbow and SunUp papaya are among the few transgenic crops accepted for commercial production that carry the virus-resistance trait. Due to their potential impact on the environment and human health, the development and release of virus-resistant transgenic plants expressing viral genes continues to raise special concerns beyond general contentions against transgene technology. The major concerns are heteroencapsidation, recombination, transgene flow to wild relatives and potential weediness of virus-resistant plants (Fuchs et al., 1998, 1999; Thomas et al., 1998; Lin et al., 2003; Vigne et al., 2004; Fuchs and Gonsalves, 2007; Chapter 18, this volume). In this section, we will only discuss these concerns as they relate to the transgenic papaya in Hawaii.

Heteroencapsidation

Heteroencapsidation refers to the encapsidation of the genome of a challenge virus by the CP protein subunits expressed in a transgenic plant. Heteroencapsidation has been documented in transgenic herbaceous plants (Osburne et al., 1990; Holt and Beachy, 1991; Candelier-Harvey and Hull, 1993; Lecoq et al., 1993; Hammond and Dienelt, 1997; Fuchs et al., 1998; Fuchs and Gonsalves, 2007).

Regarding the transgenic papaya in Hawaii, heteroencapsidation is of little or no consequence because papayas in Hawaii are infected only by PRSV. There have been reports of the tospovirus *Tomato spotted wilt virus* in Hawaii, but it is not common. The only other major aphid-transmitted potyvirus to infect papaya is *Papaya leaf distortion mosaic virus* (PLDMV), but it does not occur in Hawaii. The PLDMV CP protein is not serologically related to that of PRSV (although they both belong to the same group) which may limit heteroencapsidation in nature. Evidence from our laboratory suggests that the mechanism of resistance in Rainbow and SunUp is via PTGS, with very low expression of both the transgene CP RNA and protein, much lower in fact than that observed upon PRSV infection in a non-transgenic plant. Thus, the likelihood of heteroencapsidation and increased risk beyond that which occurs during mixed infections in nature would presumably be low.

Recombination

Recombination of a viral transgene with an incoming virus can potentially lead to a genetic change which might allow the proliferation of novel recombinants (AIBS, 1995). Recombination is a potential environmental risk issue since it can theoretically result in changes in pathogenicity, such as increased virulence or impact on non-target organisms due to possible changes in host specificity (Fuchs and Gonsalves, 2007). In relation to the transgenic papaya in Hawaii, we have no evidence of recombination occurring under field conditions. A major roadblock to recombination occurring in the transgenic papaya grown in Hawaii is that, so far, none of the PRSV isolates from Hawaii tested have been able to overcome transgenic papaya resistance.

Notwithstanding, the development of a system to produce infectious viral transcripts of PRSV *in vitro* (Chiang and Yeh, 1997) has provided a unique opportunity to begin to functionally identify viral gene segments involved in various functions including pathogenicity. In practice, this was accomplished by construction of *in vitro* transcription templates that consisted of genetically engineered, chimeric PRSV genomes with the normal complement of genes, but composed of segments from two or more parental strains. The recombinant PRSV approach was used for the identification of gene segments involved in pathological differences between YK, a PRSV isolate from Taiwan and the Hawaiian isolate HA on transgenic Rainbow and SunUp papaya. It had been assumed that the YK strain overcame resistance of Rainbow and SunUp mainly due to the low homology (89.9%) of its *CP* gene to the HA 5-1 *CP* transgene. To test this assumption, chimeric constructs were made with the HA virus containing all or portions of its *CP* gene and 3' end non-coding region replaced with that of the corresponding regions of YK (Chiang et al., 2001). Although recombinants with the entire *CP* gene of YK did indeed cause severe symptoms on Rainbow, these were not as severe as the wild type YK genome. Studies utilizing recombinant virus composed of segments from severe and mild strains of PRSV also indicated that the HC-Pro gene plays an important role in viral pathogenicity and acts as suppressor of the gene silencing defence mechanism in papaya (Tripathi et al., 2003, 2004; Yeh et al., 2003; Bau et al., 2004).

These functional experiments show that the CP is not the sole determinant for pathogenicity, but that pathogenic properties of PRSV are governed by the collective contribution of multiple viral genes. In cases such as viral transmission, the *CP* gene functions in this process may require specific interactions with other gene(s) which must already be present in the challenge virus. One interpretation of this data is that should recombination occur, the CP originating from the transgene would only be expected to function in pathogenic processes in the context of closely related or near identical viral genomes which would consequently have similar properties to PRSV.

Transgene flow to wild relatives

One of the major environmental safety issues over virus-resistant transgenic crops is gene flow. Gene flow is not a risk specific to the virus-resistance trait, but its impact on recipient plants could be affected by additional factors, such as plant virus and virus vector prevalence in the environment. Wild relatives of cultivated crops can acquire host genes and/or transgenes through pollen flow and their progeny resulting from gene transfer can exhibit undesired characteristics if the transferred genes provide them with a selective advantage (Fuchs and Gonsalves, 2007).

In Hawaii, there are neither wild relatives nor non-domesticated *Carica papaya*. Even if wild relatives (previously classified in the genus *Carica* but now classified in the genus *Vasconcellea*) were to exist in Hawaii, they are not sexually compatible with *Carica papaya* (Gonsalves et al., 2006). Thus, in Hawaii, there is no risk that gene flow will occur between transgenic papaya and non-domesticated papaya or wild relatives.

Weediness of virus-resistant papaya in Hawaii

PRSV was discovered in Hawaii in the 1940s. In the Territorial records prior to the 1940s, papaya was not listed as a weed. This indicates that even in the absence of disease caused by PRSV, papaya is not nor does it become weedy. Similarly, addition of the virus-resistant trait should not cause papaya to become weedy. Indeed, observations since release of the virus-resistant transgenic papaya in Hawaii confirm that it is not a weed.

Management Issues

In this section, we discuss management issues relating to virus-resistant transgenic crops, citing examples directly from the practices employed in the commercial production of Rainbow and SunUp papaya in Hawaii. The important management issues discussed include understanding, guarding and extending the durability of PRSV resistance of transgenic papaya, factors and measures allowing the production of non-transgenic papaya and maximizing the utility of transgene resistance through cultivar development. The issue of gene flow is discussed here in the context of 'coexistence' in the production of Hawaii's non-transgenic and transgenic papaya as a management issue.

Durable resistance

The breakdown of resistance is of concern for any virus-resistant plant, whether derived from conventional breeding or through transgenics. It is a concern for managing or prolonging the effective period of the transgenic papaya to the point that its economic benefits are maximally realized, particularly in light of the energy expended for its deregulation. As mentioned above, greenhouse inoculations of transgenic 'Rainbow' showed that the transgenic papaya was resistant to only some of the strains of PRSV collected from outside the USA (Tennant et al., 2001). Thus, it is critical to constantly monitor the introduction or emergence of viral strains that could overcome the resistance of the transgenic plants, and accordingly develop a proactive strategy. This practice is extremely important because it takes a long time to develop resistant plants. Since the development of Rainbow, personnel in Hawaii have been continually testing for the breakdown of resistance by routinely challenging Rainbow with locally collected PRSV isolates as well as monitoring Rainbow fields for susceptible plants. In the 8 years since its commercial release, no breakdown in resistance to local isolates has been observed (Ferreira and Gonsalves, 2006).

Performance studies on SunUp and other transgenic papaya have shown that resistance can be broadened to other geographical isolates by increasing the transgene dosage (Tennant et al., 2001). Thus, in recent years, introgression and doubling of the 55-1 transgene into popular local cultivars has been employed as a approach to help sustain disease resistance durability as well as variety in the market (Gonsalves et al., 2006) (see also the section 'Development of new cultivars'). Another important part of our disease management strategy has been to remain proactive and ready for the incidence of resistance breakdown by developing new transgenic lines that are resistant to PRSV strains from outside of Hawaii in addition to local strains (Fermín and Gonsalves, 2003; Gonsalves and Ferreira, 2006).

Production of non-transgenic papaya

One of the major contributions that the transgenic papaya has made to the Hawaiian papaya industry is that it has revived lucrative production of non-transgenic papaya (Gonsalves and Ferreira, 2003). This has occurred in several ways. First, the initial large-scale planting of transgenic papaya in established farms along with the elimination of abandoned virus-infected fields drastically reduced virus inocula and thus allowed for strategic planting of non-transgenic papaya in areas that did not have infection. As early as 1999, the HDOA instituted a plan to ensure the production of non-transgenic papaya in an area known as Kahuawai which is physically isolated from established fields in Puna. Kahuawai was also protected to some extent from aphid vectors carried over from infected fields since the prevailing winds came from the direction of the ocean which bordered the field (Gonsalves and Ferreira, 2003). Growers who followed the recommended practices of monitoring for and rogueing of infected plants were able to economically produce 'Kapoho' without major losses from PRSV. Second, although definitive experiments have not been carried out, it seems that transgenic papaya can provide a buffer zone to protect non-transgenic papayas that are planted within the confines of the buffer. Our reasoning is that viruliferous aphids feeding on transgenic papaya will be purged of virus before travelling to the non-transgenic plantings within the buffer. Thus, growing transgenic and non-transgenic papaya in relatively close proximity may function in management of PRSV infection of non-transgenic papaya. For reasons stated above, production of non-transgenic papaya in Hawaii continues today, and in fact is lucrative and vital, since Japan, which represents a significant share of the Hawaiian papaya export market, has a zero tolerance for transgenic papaya.

An ever-present and continual challenge in maintaining non-transgenic papaya production in Puna is to prevent the significant build-up of virus. This is because PRSV is still around and therefore strict attention is required in planting non-transgenic papaya fields in locations isolated from other non-transgenic fields, and in the timely elimination (rogueing) of infected trees and non-transgenic plantings that are no longer in production to prevent the build-up of virus inoculum. Although important, these simple factors are often not practiced when there are no obvious signs for resurgence of PRSV. It is hoped that people will not forget the tremendous damage that PRSV caused to Hawaii's papaya industry during the period 1992–1998.

Coexistence

Coexistence, the growing of transgenic and non-transgenic papaya in practical proximity to each other such that they can be raised with minimal transfer of genetic characteristics from transgenic to non-transgenic, is in fact, being practiced in the Hawaiian papaya industry today. This situation has been brought about since both transgenic and non-transgenic papayas are necessary for the Hawaiian papaya industry, especially in the growing of organic papaya and in maintaining the Japanese market which at present does not accept transgenic papaya since the latter has not been deregulated in Japan. It should be noted that the USA does not require deregulated crops such as Rainbow and SunUp to be grown in specified locations within the USA.

One means by which practical tools have been introduced to the Hawaiian papaya industry for monitoring and managing non-transgenic papaya production is through adoption of an Identity Preservation Protocol (IPP) (Camp, 2003). This voluntary programme was established by the HDOA at the request of Japanese papaya importers. Documented compliance to the regulations laid out in the IPP allows farmers and shippers to receive a certification letter from the HDOA that accompanies each papaya shipment. The incentive for participation in the programme is that the IPP certification letter allows papaya shipments to be distributed while Japanese officials perform tests for possible contaminating transgenic papaya from samples of the shipment. Shipments without the certification letter must be held until the tests are completed, which may take anywhere from a few days to a week, during which time the fruit may lose quality and marketability.

In order to obtain an IPP certification letter, the non-transgenic papaya must come from papaya orchards approved by the HDOA. Every tree in the orchards in question must be derived from seeds produced in approved, non-GMO fields and each tree must be tested by the applicant and found to be negative for the transgene-linked GUS activity. A papaya-free zone of at least 4.5 m must also separate the non-transgenic orchard. The applicant must subsequently submit detailed records of the transgene detection tests to the HDOA. Prior to final approval of the field by the HDOA, the applicant must in addition randomly test one fruit from 1% of the papaya trees in the field for presence of the transgene. Detailed postharvest protocols for minimizing the chance of contamination of non-transgenic papaya with transgenic papaya, including procedures such as the random testing of papaya prior to packing, must also be submitted to the HDOA and adhered to. If all criteria are met, the certification letter from HDOA will accompany the shipment stating compliance with a properly conducted IPP.

In summary, coexistence is being routinely practiced in Hawaii's papaya industry. The scheme of IPP has proved workable and economical as papaya is still being shipped routinely to Japan without evidence of transgenic fruits.

Development of new cultivars

The successful control of PRSV by Rainbow papaya has spawned the development of new virus-resistant varieties, which have opened new market niches and has enabled the expansion of profitable production of papaya in other regions of Hawaii (Gonsalves et al., 2004). In addition to SunUp and Rainbow, two new cultivars, Poamoho Gold and Laie Gold, were developed primarily for growers on the island of Oahu (Fitch et al., 2002). In contrast, prior to 1998, Hawaii had only one dominant cultivar, the non-transgenic Kapoho and a small areage of Sunrise. Today, Hawaii has SunUp, Rainbow, Kapoho, Sunrise, Laie Gold and Poamoho Gold.

Efforts to Deregulate Transgenic Papaya in Canada and Japan

Although a major constraint to papaya production in Hawaii has been eliminated with the introduction of PRSV-resistant transgenic plants, Hawaii's papaya industry still faces a number of challenges. Some of these challenges have been mentioned previously and include maintaining production of non-transgenic papaya, the durability of the resistance of transgenic papaya, concerns particularly of organic growers that their crops will be contaminated by pollen flow and the general controversy over GMOs. In this section, we address the steps that have been and are being taken to gain market share of transgenic papaya in Canada and Japan.

Canada

Canada accounts for 11% of Hawaii's papaya export market. Canada considers foods derived from GMOs as 'novel foods' and importation requires review and approval by Health Canada (HC), the government organization responsible for food safety. The Canadian Food Inspection Agency (CFIA) and Environment Canada (EC) are two other agencies involved in other aspects of regulation and approval of GMO-derived products. Health Canada approved the import of 'SunUp' and 'Rainbow' transgenic papaya for food purposes only in January 2003. Labelling of the approved transgenic papaya imported into Canada is not required. The data used for the nutritional assessment of the transformant line 55-1 included fruit composition (total soluble solids, carotenoids, vitamin C and minerals), which were within the range found in fruit of non-transgenic cultivars grown in Hawaii. In the toxicology assessment, PRSV CP was not considered a 'novel' protein due to the history of human consumption of PRSV-infected fruit without adverse health effects (http://www.hc-sc.gc.ca/fn-an/gmf-agm/appro/papaya_e.html).

Japan

Currently, Japan accounts for 20% of Hawaii's papaya export market. As mentioned above, the application process for sale of transgenic papaya, specifically derivatives of 55-1, in Japan has not yet been approved, so at present Hawaii exports only non-transgenic papaya to Japan. Obviously, approval for the sale and shipment of transgenic papaya will circumvent much of the concern and consequences of accidental introduction of transgenic papaya into Japan. Recently, government agencies such as the US Foreign Agricultural Service (FAS) have also expressed enthusiasm in support of this goal because of their interests in promoting US biotechnology in other countries. To this end, efforts to allow transgenic papaya into Japan were initiated by the then Papaya Administrative Committee or PAC (later replaced by the present day Hawaii Papaya Industry Association or HPIA) soon after the transgenic papaya was commercialized in Hawaii, with the researchers taking the lead in developing the petition. For the application to allow import of transgenic papaya to Japan, both food for human consumption and environmental safety issues are being evaluated. The petition to the Japanese Ministry of Agriculture, Forestry and Fisheries (MAFF) was approved in 2000, while a petition initially submitted to the Japanese Ministry of Health, Labour and Welfare (MHLW) in 2003 has undergone revision and is in the process of evaluation. Since the initial petitions were filed, additional information has been requested from both Japanese ministries due to subsequent adoption of new policies on the regulation of GMOs.

Currently, advisory committees and expert panelists of the Ministry of Agriculture, Forestry and Fisheries (MAFF) and the Ministry of the Environment (MOE) perform environmental risk assessment and safety evaluations. The environmental safety policies follow the Biosafety Protocol implemented in 2004 by Japan's 'Law Concerning the Conservation and Sustainable Use of Biological Diversity through Regulations on the Use of Living Modified Organisms' or 'Cartagena Law' (Sato, 2006). Japan's laws and policies on environmental risk assessment follow along the lines of a UN agreement called The Cartagena Protocol on Biosafety (CPB) to which it is a member state (http://www.biodiv.org/biosafety/default.aspx). The CPB itself was adopted in 2000 and put into force in 2003. It is a supplementary agreement of the UN Environmental Programme's (UNEP) Convention on Biological Diversity (CBD; http://www.biodiv.org/default.shtml). CBD is an international treaty for the development of national strategies for the conservation and sustainable use of biological diversity (sustainable development) adopted at the UN Conference on Environment and Development (Earth Summit) in Rio de Janeiro in 1992 (http://www.un.org/geninfo/bp/enviro.html). The CPB is based on the 'precautionary principle' and covers regulations dealing with the management or control of risks associated with transfer (particularly across borders), handling, and use of GMOs, termed Living Modified Organisms (LMOs), activities that might adversely affect the environment.

Some of the information and data requested by the MAFF and MOE included alleopathy on plants and soil microbial communities, physiology of papaya, aetiology of PRSV, and pollen biology relating to the potential for transgene contamination by pollination. One of the additional areas we were asked to address was the potential for recombination of transgene encoded sequences with other viruses and its possible impact.

An expert panel and committee of the Food Safety Commission (FSC) are evaluating issues relating to safety of food for human consumption. Their conclusions will be reported to the Ministry of Health, Labour and Welfare (MHLW) under Japan's Food and Sanitation Law. Japan's current policies regarding food safety of GMOs were established in 2004 and draw upon some of the guidelines presented by the Ad Hoc Intergovernmental Task Force on Food Derived from Biotechnology of the Codex Alimentarius 'food code' Commission (http://www.who.int/foodsafety/biotech/codex_taskforce/en/index.html) (http://www.fsc.go.jp/senmon/idensi/gm_kijun_english.pdf). The commission is a subsidiary body of the UN Food and Agriculture Organization (FAO) and the World Health Organization (WHO).

The FSC requested information to determine substantial equivalence between non-transgenic and transgenic papaya for substances including BITC and papain, papaya fruit protein profiles, in addition to allergenicity related studies including PRSV CP heat stability and stability in simulated intestinal and gastric fluid.

Detailed molecular genetic data including Southern hybridization analysis of 55-1 using probes covering the entire transformation plasmid, PCR of the insert border regions and sequence of the insertions and flanking genomic DNA were required and prepared for submission to both the MHLW/FSC and the MAFF/MOE. In addition, open reading frame (ORF) analysis followed by Blast searches to all inclusive and allergen-specific databases were performed on the inserts and flanking genomic DNA to determine the potential expression of toxic or allergenic proteins. These bioinformatic data were submitted to the MHLW/FSC and the MAFF/MOE. Such detailed molecular analysis of the transgene insertion event was not required by the relevant US regulatory agencies.

During review for potential allergenicity of transgene-derived proteins, questions were raised on the potential allergenicity of the PRSV CP. According to the FAO/WHO, 2001 discussion (FAO/WHO, 2001), matches of six amino acids of a protein to known allergens make it a candidate for being an allergen. Using this criterion, Kleter and Peijnenburg (2002) determined that there was a single 6 amino acid match of the PRSV CP to a proposed allergen ABA 1, a protein of the human parasite *Ascaris lumbricoides* or the pig parasite *Ascaris suum*. In response, we claimed that for several reasons, the amino acid homology between PRSV CP and ABA 1 is not relevant with regards to allergenicity; the amino acid sequence is not repeated in the CP sequence like allergenic epitopes usually are; therefore, it would not be expected to trigger the IgE response associated with allergens. The ABA 1 proposed allergenic peptide was found to not be inherently allergenic outside the context of other *Ascaris* proteins (Paterson *et al.*, 2002) and indeed, is not among the officially recognized allergens found at the International Union of Immunological Societies (IUIS) web site (http://www.allergen.org).

If the various agencies approve the Food Safety and Environmental Safety submissions, the form of labelling of the transgenic papaya will then be decided among subcommittee(s) of the MAFF and MHLW followed by notification of the entire approval package to the World Trade Organization (WTO).

Receiving approval for the importation of transgenic papaya into Japan would have huge benefits to Hawaii's papaya industry and would also advance the case for the acceptance and development of transgenic products outside the USA. Following approval, shippers of non-transgenic papaya would likely still have to label their cargo as such, but are also likely to be allowed to continue shipments in cases where errors occur within certain defined tolerance limits, a situation that contrasts with the present day strict zero tolerance policy against transgenic papaya.

The introduction of transgenic papaya fruit to Japan will allow consumers to make a personal choice, serving as a real life example for consumer acceptance of fresh GMO products outside of the USA. Since the transgenic Hawaiian papaya was not developed with support from multinational corporations and is not a major commodity transgenic crop, consumer acceptance should not be clouded by media hype and sentiment against the dominance of multinational corporations or international trade issues. Rather, it is hoped and anticipated that product acceptance will be influenced by factors such as quality, price, advertising, and philosophy of the individual consumer. In this respect, the transgenic papaya will be a ground-breaking biotechnology for the greater, worldwide consumer and governmental acceptance of fresh transgenic products.

Conclusions

In this chapter, we have presented practical accounts on the steps taken to bring about the commercialization of virus-resistant transgenic papaya based on the PDR approach in Hawaii. More than 8 years after its introduction in 1998, the transgenic virus-resistant papaya continues to play a vital role in the Hawaiian papaya industry in the practical and effective management of PRSV, which is essential for the economic production of papaya. Similarly, the more widespread implementation of virus-resistant transgene technology in papaya and other crops, including regional and underrepresented crops, should have a great impact on the management of virus diseases as well as on the economies and health of the local communities not currently enjoying its benefits. Understanding the actual risks and safety issues regarding the implementation of transgene technology under real life conditions and acceptance of the concept are important factors in the development of sensible regulation and the greater adoption of the technology.

References

AIBS (1995) Transgenic virus-resistant plants and new viruses. An American Institute of Biological Sciences Workshop sponsored by the USDA Animal and Plant Health Inspection Service and the Biotechnology Industry Organization. Beltsville, Maryland, 20–21 April. Available at: http://www.aphis.usda.gov/brs/virus/95_virusrept.pdf

Badillo, V. (1993) Caricaceae, segundo esquema. *Revista de la Facultad de Agronomía (Maracay)* 43, 111.

Bau, H.-J., Cheng, Y.-H., Yu, T.-A., Yang, J.-S., Liou, P.-C., Hsiao, C.-H., Lin, C.-Y. and Yeh, S.-D. (2004) Field evaluation of transgenic papaya lines carrying the coat protein gene of papaya ringspot virus in Taiwan. *Plant Disease* 85, 594–599.

Baulcombe, D.C. (1996) Mechanisms of pathogen-derived resistance to viruses in transgenic plants. *The Plant Cell* 8, 1833–1844.

Baulcombe, D.C., English, J., Mueller, E. and Davenport, G. (1996) Gene silencing and virus resistance in transgenic plants. In: Grierson, G.W., Lycett, G.W. and Tucker, G.A. (eds) *Mechanisms and Applications of Gene Silencing*. Nottingham University Press, Nottingham, UK, pp. 127–138.

Beachy, R.N. (1997) Mechanisms and applications of pathogen-derived resistance in transgenic plants. *Current Opinion in Biotechnology* 8, 215–220.

Camp, S.G., III (2003) Identity preservation protocol for non-GMO papayas revised 7 April, 2003. In: Balázs, E. and Gonsalves, D. (eds) *Proceedings of the OECD Co-operative Program Workshop on 'Virus Resistant Transgenic Papaya in Hawaii: A Case Study for Technology Transfer to Lesser Developed Countries'*, Hilo, Hawaii, 22–24 October, pp. 95–100.

Candelier-Harvey, P. and Hull, R. (1993) Cucumber mosaic virus genome is encapsidated in alfalfa mosaic virus coat protein expressed in transgenic plants. *Transgenic Research* 2, 277–285.

Chiang, C.-H. and Yeh, S.-D. (1997) Infectivity assays of *in vitro* and *in vivo* transcripts of papaya ringspot potyvirus. *Botanical Bulletin Academia Sinica* 38, 153–163.

Chiang, C.-H., Wang, J.-J., Jan, F.-J., Yeh, S.-D. and Gonsalves, D. (2001) Comparative reactions of recombinant papaya ringspot viruses with chimeric coat protein (CP) genes and wild-type viruses on CP-transgenic papaya. *Journal of General Virology* 82, 2827–2836.

FAO (2004) Food and Agricultural Organization. FAO statistical database (FAO STAT). Available at: http://faostat.fao.org.

FAO/WHO (2001) Evaluation of allergenicity of genetically modified foods. Report of a joint FAO/WHO expert consultation on allergenicity of foods derived from biotechnology. Available at: http://www.who.int/foodsafety/publications/biotech/en/ec_jan2001.pdf

Fermín, G. and Gonsalves, D. (2003) Papaya: engineering resistance against papaya ringspot virus by native, chimeric and synthetic transgenes. In: Loebenstein, G. and Thottappilly, G. (eds) *Virus and Virus-like Diseases of Major Crops in Developing Countries*. Kluwer Academic Publishers, Dordrecht, The Netherlands, pp. 497–518.

Fermín, G., Tennant, P., Gonsalves, C., Lee, D. and Gonsalves, D. (2004) Comparative development and impact of transgenic papayas in Hawaii, Jamaica, and Venezuela. In: Peña, L. (ed.) *Transgenic Plants: Methods and Protocols*. The Humana Press, Totowa, New Jersey, pp. 399–430.

Ferreira, S.A., Mau, R.F.L., Manshardt, R., Pitz, K.Y. and Gonsalves, D. (1992) Field evaluation of papaya ringspot virus cross protection. In: *Proceedings of the 28th Annual Hawaii Papaya Industry Association Conference*. Honolulu, 29–30 September, pp. 14–19.

Ferreira, S.A., Pitz, K.Y., Manshardt, R., Zee, F., Fitch, M. and Gonsalves, D. (2002) Virus coat protein transgenic papaya provides practical control of papaya ringspot virus in Hawaii. *Plant Disease* 86, 101–105.

Fitch, M. and Manshardt, R.M. (1990) Somatic embryogenesis and plant regeneration from immature zygotic embryos of papaya (*Carica papaya* L.). *Plant Cell Reports* 9, 320–324.

Fitch, M.M.M., Manshardt, R.M., Gonsalves, D., Slightom, J.L. and Sanford, J.C. (1990) Stable transformation of papaya via microprojectile bombardment. *Plant Cell Reports* 9, 189–194.

Fitch, M.M.M., Manshardt, R.M., Gonsalves, D., Slightom, J.L. and Sanford, J.C. (1992) Virus resistant papaya derived from tissues bombarded with the coat protein gene of papaya ringspot virus. *Bio/Technology* 10, 1466–1472.

Fitch, M.M.M., Leong, T., Akashi, L., Yeh, A., White, S., Ferreira, S. and Moore, P. (2002) Papaya ringspot virus resistance genes as a stimulus for developing new cultivars and new production systems. *Acta Horticulturae* 575, 85–91.

Fuchs, M. and Gonsalves, G. (2007) Safety of virus-resistant transgenic plants two decades after their introduction: lessons from realistic field risk assessment studies. *Annual Review of Phytopathology* (in press).

Fuchs, M., Klas, F.E., McFerson, J.R. and Gonsalves, D. (1998) Transgenic melon and squash expressing coat protein genes of aphid-borne viruses do not assist the spread of an aphid non-transmissible strain of cucumber mosaic virus in the field. *Transgenic Research* 7, 449–462.

Fuchs, M., Gal-On, A., Raccah, B. and Gonsalves, D. (1999) Epidemiology of an aphid nontransmissible potyvirus in fields of nontransgenic and coat protein transgenic squash. *Transgenic Research* 8, 429–439.

Gonsalves, D. (1998) Control of papaya ringspot virus in papaya: a case study. *Annual Review of Phytopathology* 36, 415–437.

Gonsalves, D. (2002) Transgenic papaya: a case study on the theoretical and practical application of virus resistance. In: Vasil, I.K. (ed.) *Plant Biotechnology 2002 and Beyond. Proceedings of the 10th International Association of Plant Tissue Culture & Biotechnology Congress*. Kluwer Academic Publishers, Orlando, Florida, pp. 115–118.

Gonsalves, D. (2006) Transgenic papaya: development, release, impact and challenges. In: Thresh, J.M. (ed.) *Plant Virus Epidemiology*. Elsevier, San Diego, California, pp. 318–353.

Gonsalves, D. and Ferreira, S. (2003) Transgenic papaya: a case for managing risks of papaya ringspot virus in Hawaii. Available at: http://plantmanagementnetwork.org/php/elements/sum2.asp?id = 2319.

Gonsalves, D. and Fermín, G. (2004) The use of transgenic papaya to control papaya ringspot virus in Hawaii and transfer of this technology to other countries. In: Christou, P. and Klee, H. (eds) *Handbook of Plant Biotechnology*, Vol. 2. Wiley, London, pp. 1165–1182.

Gonsalves, D., Gonsalves, C., Ferreira, S., Pitz, K., Fitch, M., Manshardt, R. and Slightom, J. (2004) Transgenic virus resistant papaya: from hope to reality for controlling papaya ringspot virus in Hawaii. *APSnet Feature Story for July, 2004*. Available at: http://www.apsnet.org/online/feature/ringspot.

Gonsalves, D., Vegas, A., Prasartsee, V., Drew, R., Suzuki, J.Y. and Tripathi, S. (2006) Developing papaya to control *Papaya ringspot virus* by transgenic resistance, intergeneric hybridization, and tolerance breeding. In: Janick, J. (ed.) *Plant Breeding Reviews*, Vol. 26. Wiley, Hoboken, New Jersey, pp. 35–73.

Hammond, J. and Dienelt, M.M. (1997) Encapsidation of potyviral RNA in various forms of transgene coat protein is not correlated with resistance in transgenic plants. *Molecular Plant-Microbe Interactions* 10, 1023–1027.

Holt, C.A. and Beachy, R.N. (1991) *In vivo* complementation of infectious transcripts from mutant tobacco mosaic virus cDNAs in transgenic plants. *Virology* 181, 109–117.

Jensen, D.D. (1949) Papaya virus diseases with special reference to papaya ringspot. *Phytopathology* 39, 191–211.

Kleter, G.A. and Peijnenburg, A.A. (2002) Screening of transgenic proteins expressed in transgenic food crops for the presence of short amino acid sequences identical to potential, IgE-binding linear epitopes of allergens. *BMC Structural Biology* 2, 1–11.

Lecoq, H., Ravelonandro, M., Wipf-Scheibel, C., Monsion, M., Raccah, B. and Dunez, J. (1993) Aphid transmission of a non-aphid-transmissible strain of zucchini yellow mosaic potyvirus from transgenic plants expressing the capsid protein of plum pox potyvirus. *Molecular Plant-Microbe Interactions* 6, 403–406.

Lin, H.-X., Rubio, L., Smythe, A., Jimenez, M. and Falk, B.W. (2003) Genetic diversity and biological variation among California isolates of *Cucumber mosaic virus*. *Journal of General Virology* 84, 249–258.

Ling, K., Namba, S., Gonsalves, C., Slightom, J.L. and Gonsalves, D. (1991) Protection against detrimental effects of potyvirus infection in transgenic tobacco plants expressing the papaya ringspot virus coat protein gene. *Bio/Technology* 9, 752–758.

Lius, S., Manshardt, R., Gonsalves, D., Fitch, M., Slightom, J. and Sanford, J. (1994) Field test of virus resistance in transgenic papayas. *Hortscience* 29, 483.

Lius, S., Manshardt, R.M., Fitch, M.M.M., Slightom, J.L., Sanford, J.C. and Gonsalves, D. (1997) Pathogen-derived resistance provides papaya with effective protection against papaya ringspot virus. *Molecular Breeding* 3, 161–168.

Manshardt, R.M. (1999) 'UH Rainbow' papaya a high-quality hybrid with genetically engineered disease resistance. *University of Hawaii College of Tropical Agriculture and Human Resources (CTAHR), New Plants for Hawaii (NPH)-1, revised.* Available at: http://www.ctahr.hawaii.edu/ctahr2001/PIO/FreePubs/FreePubs07.asp#NewPlantsForHawaii.

Nakasone, H.Y. (1975) Papaya development in Hawaii. *HortScience* 10, 198.

Osburne, J.K., Sarkar, S. and Wilson, T.M.A. (1990) Complementation of coat protein-defective TMV mutants in transgenic tobacco plants expressing TMV coat protein. *Virology* 179, 921–925.

Paterson, J.C.M., Garside, P., Kennedy, M.W. and Lawrence, C.E. (2002) Modulation of a heterologous immune response by the products of *Ascaris suum*. *Infection and Immunity* 70, 6058–6067.

PEW (2004). Issues in the regulation of genetically engineered plants and animals. Pew Initiative on Food and Biotechnology, Washington, DC, pp. 1–188.

Powell-Abel, P., Nelson, R.S., De, B., Hoffmann, N., Rogers, S.G., Fraley, R.T. and Beachy, R.N. (1986) Delay of disease development in transgenic plants that express the tobacco mosaic virus coat protein gene. *Science* 232, 738–743.

Purcifull, D., Edwardson, J., Hiebert, E. and Gonsalves, D. (1984) Papaya ringspot virus. *CMI/AAB Descriptions of Plant Viruses. No. 292.* (No. 84 Revised, July 1984), pp. 8.

Quemada, H., L'Hostis, B., Gonsalves, D., Reardon, I.M., Heinrikson, R., Hiebert, E.L., Sieu, L.C. and Slightom, J.L. (1990) The nucleotide sequences of the 3'-terminal regions of papaya ringspot virus strains w and p. *Journal of General Virology* 71, 203–210.

Sanford, J.C. and Johnston, S.A. (1985) The concept of parasite-derived resistance – Deriving resistance genes from the parasite's own genome. *Journal of Theoretical Biology* 113, 395–405.

Sanford, J.C., Smith, F.D. and Russell, J.A. (1992) Optimizing the biolistic process for different biological applications. *Methods in Enzymology* 217, 483–509.

Sato, S. (2006) Japan Biotechnology Annual Report. In: Spencer, P. (ed.) USDA Foreign Agricultural Service GAIN Report. US Embassy, Tokyo, Report No. JA6049, pp. 1–26.

Strating, A. (1996) Availability of determination of nonregulated status for papaya lines genetically engineered for virus resistance. *Federal Register* 61, 48663.

Tang, C.S. (1971) Benzyl isothiocyanate of papaya fruit. *Phytochemistry* 10, 117–120.

Taylor, M.R., Tick, J.S. and Sherman, D.M. (2004). Tending the fields: state and federal roles in the oversight of genetically modified crops. A report commissioned by the Pew Initiative on Food and Biotechnology and prepared by Resources for the Future, pp. 1–269. Available at: http://pewagbiotech.org/research/fields/.

Tennant, P.F., Gonsalves, C., Ling, K.-S., Fitch, M., Manshardt, R., Slightom, J,L. and Gonsalves, D. (1994) Differential protection against papaya ringspot virus isolates in coat protein gene transgenic papaya and classically cross-protected papaya. *Phytopathology* 84, 1359–1366.

Tennant, P., Fermín, G., Fitch, M.M., Manshardt, R.M., Slightom, J.L. and Gonsalves, D. (2001) Papaya ringspot virus resistance of transgenic Rainbow and SunUp is affected by gene dosage, plant development, and coat protein homology. *European Journal of Plant Pathology* 107, 645–653.

Thomas, P.E., Hassan, S., Kaniewski, W.K., Lawson, E.C. and Zalewski, J.C. (1998) A search for evidence of virus/transgene interactions in potatoes transformed with *Potato leafroll virus* replicase and coat protein genes. *Molecular Breeding* 4, 407–417.

Tripathi, S., Chen, L.-F., Bau, H.-J. and Yeh, S.-D. (2003) In addition to transgene divergence potyviral *HC-Pro* gene plays an important role in breaking down coat protein gene-mediated transgene resistance. *7th International Congress of Plant Molecular Biology*. Barcelona, Spain, 23–28 June, p. 367 (abstract).

Tripathi, S., Bau, H.-J., Chen, L.-F. and Yeh, S.-D. (2004) The ability of Papaya ringspot virus strains overcoming the transgenic resistance of papaya conferred by the coat protein gene is not correlated with higher degrees of sequence divergence from the transgene. *European Journal of Plant Pathology* 110, 871–882.

Tripathi, S., Suzuki, J. and Gonsalves, D. (2006) Development of genetically engineered resistant papaya for *Papaya ringspot virus* in a timely manner – a comprehensive and successful approach. In: Ronald, P. (ed.) *Plant–Pathogen Interactions: Methods and Protocols*. The Humana Press, New Jersey, pp. 197–240.

Vigne, E., Komar, V. and Fuchs, M. (2004) Field safety assessment of recombination in transgenic grapevines expressing the coat protein of Grapevine fanleaf virus. *Transgenic Research* 13, 165–179.

Wang, C.-H. and Yeh, D.-D. (1997) Divergence and conservation of the genomic RNAs of Taiwan and Hawaii strains of papaya ringspot potyvirus. *Archives of Virology* 142, 271–285.

Wang, C.-H., Bau, H.-J. and Yeh, S.-D. (1994) Comparison of the nuclear inclusion b protein and coat protein genes of five papaya ringspot virus strains distinct in geographic origin and pathogenicity. *Phytopathology* 84, 1205–1210.

Wenslaff, T.F. and Osgood, R.V. (1999) Production of transgenic hybrid papaya seed in Hawaii. *Tropical Fruit Report 1, Hawaii Agricultural Research Center*. Available at: http://www.hawaiiag.org/harc/HARCPB13.htm.

Yeh, S.D., Jan, F.J., Chiang, C.H., Doong, P.J., Chen, M.C., Chung, P.H. and Bau, H.J. (1992) Complete nucleotide sequence and genetic organization of papaya ringspot virus RNA. *Journal of General Virology* 73, 2531–2541.

Yeh, S.-D., Tripathi, S., Bau, H.-J. and Chen, L.-F. (2003) Identification and variability analyses of virus strains capable of breaking transgenic resistance of papaya conferred by the coat protein gene of Papaya ringspot virus. *7th International Congress of Plant Molecular Biology*. Barcelona, Spain, 23–28 June, pp. 367 (abstract).

20 Potential Disease Control Strategies Revealed by Genome Sequencing and Functional Genetics of Plant Pathogenic Bacteria

A.O. CHARKOWSKI

Abstract
Complete genomes of at least one species from nearly every genus of plant pathogenic bacteria have been sequenced, resulting in hundreds of megabases of new genetic information in the last decade. Many new virulence genes, often present in large pathogenicity islands, have been identified in both gram-negative and gram-positive pathogens. The genome sequences, combined with functional genetic analyses, have allowed researchers to identify plant pathogen-specific virulence genes. The products of these virulence genes are attractive targets for disease control measures since methods targeting plant pathogen-specific mechanisms are less likely to endanger human health if bacterial resistance were to develop. Progress towards new disease control strategies depends on obtaining additional genome sequences; on describing the functions of bacterial proteins, RNAs, other polymers and signal molecules during all stages of the bacterial life cycle; and on transfer of technologically innovative pathogen detection methods to use with plant pathogens.

Introduction

Bacteria cause significant agricultural losses each year in both food and ornamental plant production. Bacterial pathogens can cause economic damage at all stages of food production, from seed decay diseases at planting to foliar, fruit and root damage during the growing season; to rots and wilts during fruit, vegetable and ornamental plant transport and storage. Some bacterial diseases, such as fruit russeting or specks and spots on fruit and vegetables, are mainly cosmetic, but are important since they reduce the value of the crop. Others, such as fire blight or bacterial wilt, can cause devastating economic losses and even starvation.

When it comes to control of bacterial diseases, all farmers are essentially organic growers because chemical control options are very limited and growers must rely on integrated pest management. Preventative copper sprays or copper seed treatments are one of the few chemical control methods available. Copper cannot control a bacterial infection once it has been initiated, and many bacteria are resistant to copper. Copper can be also phytotoxic, so its usefulness is limited (Cooksey, 1990). Antibiotics are used in some fruit orchards, but are not used on most crops due to concerns about development and spread of antibiotic resistance among bacterial pathogens (McManus et al., 2002). Plant resistance, pathogen indexing, sanitation, environmental control in greenhouses and use of healthy plant propagules are currently the only other widely used control measures available for bacterial pathogens (Daughtrey and Benson, 2005).

Biocontrol and plant disease resistance activators have been examined for many pathosystems. In fact, the first biocontrol microbe described for a plant pathogen was *Agrobacterium radiobacter*, which produces a nucleotide analogue that is imported by nopaline-producing *Agrobacterium tumefaciens* strains via agrocinopine permease (Farrand and Wang, 1992). *A. radiobacter* has been successfully used commercially to control crown gall for over 25 years, demonstrating that effective biocontrol of bacterial plant pathogens is possible. Among the reasons that this microbe is such an effective biocontrol are that it colonizes the same site as the pathogen, it competes directly with *A. tumefaciens* for nutrients, and the nucleotide analogues that it produces do not affect most bacterial species, so it specifically inhibits the target pathogen. Direct competition for colonization sites also appears to be an important mechanism for how *Pseudomonas fluorescens* 506 (BlightBan506) controls fire blight and russeting on fruit crops (Stockwell and Stack, 2007). Because this biocontrol agent does not kill the target pathogen, it is critical that it be applied at proper times so that it can colonize infection courts, such as newly opened blossoms, before the pathogen. Unfortunately, effective biocontrol methods are not available for most bacterial diseases.

Amazing advances in technology over the past few decades have opened the door to many new possibilities for control of bacterial diseases. In the last 25 years, the first bacterial virulence genes were cloned and sequenced, mutagenesis methods and useful plasmids have been developed for most bacterial pathogens, the epidemiology of most important bacterial pathogens has been elucidated and genome sequences for at least one representative from nearly every genus of plant pathogenic bacteria have been decoded (Table 20.1). Probably the biggest advances in disease control that can be directly traced to genome sequencing are the many new pathogen detection methods available that can be used to contain emerging pathogens and index seed prior to planting to eliminate pathogen-infested seed lots. In addition, multiple strategies for constructing transgenic plants resistant against bacterial pathogens have been devised and tested, ranging from plant expression of insect antimicrobial peptides (Ohshima et al., 1999; Chapter 13, this volume), to plant resistance proteins (Tu et al., 2000), to expression of bacterial enzymes that degrade pathogen signal molecules (Dong et al., 2001). Unfortunately, in

Table 20.1. Sequenced genomes of bacterial plant pathogens.

Pathogen	Disease	Chromosomes	Plasmids	Reference
GRAM-NEGATIVE				
Alphaproteobacteria				
Rhizobiales				
Agrobacterium				
A. tumefaciens C58	Crown gall	2.84 Mb	214 kb	Wood et al. (2001)
		2.08 Mb (linear)	543 kb	Goodner et al. (2001)
A. vitus S4		3.72 Mb	631 kb	http://depts.washington.edu/agro/
		1.28 Mb (linear)	259 kb	
			212 kb	
			130 kb	
			79 kb	
Betaproteobacteria				
Burkholderiales				
Acidovorax				
A. avenae subsp. citrulli		5.3 Mb	No plasmids	http://www.jgi.doe.gov/sequencing/why/CSP2006/Acidovorax.html
Burkholderia				
B. cenocepacia J2315		3.870 Mb	92.7 kb	http://www.sanger.ac.uk/Projects/B_cenocepacia/
		3.217 Mb		
		0.876 Mb		
Ralstonia				
R. solanacearum GMI1000	Tomato wilt	3.7 Mb		Salanoubat et al. (2002)
		2.1 Mb		
R. solanacearum UW551	Brown rot of potato	~6 kb		Gabriel et al. (2006)
R. solanacearum MolK2	Moko disease of banana	~6 kb		http://www.cns.fr/externe/English/Projets/Projet_Y/Y.html
R. solanacearum 1609	Brown rot of potato	~6 kb		
Xylophilus		No genome sequence available		
Gammaproteobacteria				
Enterobacterales				
Dickeya				
D. dadantii 3937	Soft rot and wilt	4.9 Mb	No plasmids	http://www.ahabs.wisc.edu/~pernalab/projects.html
Erwinia				
E. amylovora Ea273	Fire blight	3.8 Mb	71 kb	http://www.sanger.ac.uk/Projects/E_amylovora/
			28 kb	

Continued

Table 20.1. *Continued*

Pathogen	Disease	Genome (circular unless noted)		Reference
		Chromosomes	Plasmids	
Pectobacterium				
P. atrosepticum SCRI1043	Soft rot and blackleg of potato			Bell *et al.* (2004)
P. carotovorum WPP14	Soft rot and wilt	~5 Mb	No plasmids	http://www.ahabs.wisc.edu/~pernalab/projects.html
P. brasiliensis 1692	Soft rot and blackleg of potato	~5 Mb	No plasmids	http://www.ahabs.wisc.edu/~pernalab/projects.html
Pantoea				
P. stewartii	Stewart's wilt of maize	~5 Mb	Numerous plasmids	http://www.hgsc.bcm.tmc.edu/projects/microbial/Pstewartii/
Serratia		No genome sequence		
Pseudomonadales				
Pseudomonas				
P. syringae pv. *tomato* DC3000	Bacterial speck of tomato	6.40 Mb	74 kb 67 kb	Buell *et al.* (2003)
P. syringae pv. *phaseolicola* 1448A	Halo blight of bean	5.93 Mb	132 kb 52 kb	Joardar *et al.* (2005)
P. syringae pv. *syringae* B728a	Bacterial brown spot of bean	6.09 Mb	No plasmids	Feil *et al.* (2005)
Xanthomonadales				
Xanthomonas				
X. axonopodis pv. *citri* 306	Citrus canker	5.17 Mb	33 kb 65 kb	da Silva *et al.* (2002)
X. campestris pv. *campestris* ATCC33913	Black rot of crucifer	5.08 Mb	No plasmids	da Silva *et al.* (2002)
X. campestris pv. *vesicatoria* 85-10	Bacterial spot of pepper and tomato	5.17 Mb	1.8 kb 38 kb 183 kb 19 kb	Thieme *et al.* (2005)
X. oryzae pv. *oryzae* KACC10331	Bacterial blight of rice	4.94 Mb	No plasmids	Lee *et al.* (2005)
Xylella				
X. fastidiosa pv. *citrus* 9a5c	Citrus variegated chlorosis	2.73 Mb	51 kb 1.2 kb	Simpson *et al.* (2000)

Continued

Table 20.1. *Continued*

Pathogen	Disease	Genome (circular unless noted)		Reference
		Chromosomes	Plasmids	
X. fastidiosa pv. *almond* Dixon	Almond leaf scorch	2.43 Mb	51 kb	Bhattacharyya *et al.* (2002b)
X. fastidiosa pv. *oleander* Ann-I	Oleander leaf scorch	2.63 Mb	30 kb	Bhattacharyya *et al.* (2002b)
GRAM-POSITIVE				
Actinobacteria				
Actinomycetales				
Clavibacter				
C. michiganenesis subsp. *Sepedonicus*	Bacterial ring rot of potato		In progress	
C. michiganensis subsp. *Michiganensis*	Bacterial canker		In progress	
Curtobacterium		No genome sequence available		
Leifsonia				
L. xyli subsp. *xyli* CTCB07	Ratoon stunting of sugar cane	2.58 Mb	No plasmids	Monteiro-Vitorello *et al.* (2004)
Rathayibacter		No genome sequence available		
Rhodococcus		No genome sequence available		
Streptomyces				
S. scabies 87.22	Common scab	10.1 Mb (linear)	No plasmids	http://www.sanger.ac.uk/Projects/S_scabies/
Clostridia				
Clostridiales				
Clostridium		No genome sequence available for plant pathogenic *Clostridium*		
Mollicutes				
Acholeplasmatales				
Phytoplasma				
Aster Yellows-Witches Broom	Aster Yellows and Witches Broom	707 kb	4.0 kb	http://www.oardc.ohio-state.edu/phytoplasma/genome.htm
			4.0 kb 5.1 kb 4.3 kb	
Entoplasmatales				
Spiroplasma				
Spiroplasma kunkelii	Corn stunt	In progress		http://www.oardc.ohio-state.edu/spiroplasma/genome.htm

most cases, genetically modified plants resistant against bacterial pathogens have not been widely accepted by consumers. Even if transgenic plants became widely accepted, with current plant transformation technologies, transgenic plants are not a practical option for many areas of agriculture. For example, with ornamental crops, where multiple varieties of a myriad of species are produced, transformation of every variety is not feasible or economical (Daughtrey and Benson, 2005). Conventional plant breeding has also benefited from the genomic revolution and genomic tools have contributed to the development of new disease resistant plant varieties.

Unfortunately, with the exception of molecular tools that aid in plant breeding and the development of new detection methods, very little of the knowledge gained over the last decades has resulted in improved control of bacterial diseases, in great part due to limitations in the use of transgenic plants. However, our knowledge of bacterial genomics is only recently gained and it is likely that great strides in control of bacterial diseases will be made in the coming decades.

Background

Bacteria are the most successful life form on earth, and by their ubiquity, define where life can exist and what biochemistry can do. Bacteria also define variability, with strains considered to be within a single species if they are only 70% genetically identical. Bacterial cell size ranges from 0.3 μm to 0.75 mm in diameter, with cell volumes among species differing by over a million-fold. The genome sizes of free-living bacteria also vary, ranging from 580 kb to over 13 Mb. There are viruses that are larger than some bacteria and protozoa that are smaller than other bacteria. Bacteria can have circular and/or linear chromosomes and plasmids and exchange genetic material across species, allowing them to adapt to many environments.

Many bacteria have symbiotic relationships with plants. The aerial surfaces of plants are harsh environments, laden with antimicrobial compounds and exposed to high daily doses of ultraviolet light, yet this surface can be colonized by multiple bacterial pathogens. Plant roots exist in soils with complex and competitive microbial communities that are dependent on nutrients released by root cells, and some plant pathogens can compete effectively in this niche as well. These pathogenic gram-negative and gram-positive bacterial genera parasitize plants and cause diseases such as wilts, rots, spots, specks and tumours. Related genera cause animal disease, often using homologous virulence genes. The similarity between plant and animal bacterial pathogens is part of the difficulty in controlling plant pathogenic bacteria. Antibiotics work in both animal and plant hosts, but are restricted, in most cases, from use on plant pathogens in order to reduce the spread of antibiotic resistance genes among bacteria (McManus *et al.*, 2002).

Bacterial genome sequencing and functional genetics of plant pathogenic bacteria bring the hope of new disease control methods to assuage losses from bacterial diseases. However, unless control measures are developed for which resistance cannot develop or plant pathogen-specific control measures are found, there will be similar concerns with the spread of resistance among bacterial human pathogens and control measures for bacterial plant pathogens will remain limited.

Currently, control of bacterial plant pathogens is limited to plant resistance, sanitation, pathogen indexing, preventative copper sprays, heat or copper treatment of seeds, and for a limited number of crops, antibiotics. With the exception of transgenic plants, there has been almost no progress in devising new control methods, as clearly illustrated by a review written by Bonde (1937), which describes control methods for bacterial ring rot, a highly contagious and devastating potato disease caused by *Clavibacter michiganensis* subsp. *sepedonicus*. The methods, such as a zero tolerance for this disease in seed potatoes and sanitation, have not changed since the review written almost 70 years ago. Pathogen detection assays are the only addition to the toolbox used to control this pathogen. Unfortunately, the routine use of detection assays is resisted by some growers and regulatory officials in the USA, which produces over $2.4 billion in potatoes annually, since the regulatory consequences of a *C. michiganensis* subsp. *sepedonicus* pathogen detection are costly to farmers.

The deployment and usefulness of plant resistance against bacterial pathogens varies greatly across pathosystems. Plant resistance has proven effective in some cases for foliar pathogens, such as *Pseudomonas* and *Xanthomonas*. In other cases, such as the soft rot pathogens *Pectobacterium* and *Dickeya*, the wilt pathogens *Ralstonia solanacearum* or *Clavibacter michiganensis* subsp. *michiganensis*, no truly resistant cultivars are available. Plant resistance is actually discouraged in the case of the potato pathogen, *C. michiganensis* subsp. *sepedonicus*, which causes bacterial ring rot. Routine inspections of seed potato crops are an important part of controlling this contagious plant pathogen. Seed potato growers found several decades ago that potato cultivars with some resistance to bacterial ring rot could harbour the pathogen asymptomatically and cause it to become widespread in susceptible cultivars grown on the same farm. Therefore, production and certification of cultivars that do not express foliar symptoms of bacterial ring rot, which includes stunting, leaf curling and wilting, is limited in the USA and Canada.

Current control measures for bacterial diseases are not sufficient, particularly in tropical regions where farmers suffer devastating losses from bacterial pathogens, such as *R. solanacearum*, a soil and water borne pathogen that can attack nearly every vegetable and cash crop grown. Even in agricultural systems where good plant resistance is available, such as with *Xanthomonas* on rice (Leach and White, 1996), bacterial pathogens continue to evolve resistance, so a continuous investment in plant breeding to develop new resistant lines is required. Transgenic plants resistant to bacterial diseases have not been accepted by consumers, so are not in widespread use.

Despite the limitations for control of bacterial plant pathogens imposed by concerns over spread of antibiotic resistance among human pathogens and genetically modified plants, advances in genome sequencing have made this a propitious time to discover new disease control measures. In the last decade, representatives from nearly all genera of plant pathogenic bacteria have been sequenced, resulting in hundreds of megabases of new genomic information, providing researchers with new functional genomics methods that have hinted at new possibilities for disease control.

The plethora of genome sequences has allowed progress in functional genomics of plant pathogens. The currently published genome sequences are listed in Table 20.1. This list will grow exponentially in the next few years due to advances in sequencing technologies that allow genome sequences to be decoded in a few hours with relatively little expense (Margulies *et al.*, 2005). Indeed, the challenges of the coming decade are not in obtaining genome sequences, but rather in devising methods to efficiently and accurately annotate genomes and in developing scientific and bioinformatic tools to determine the roles of the many newly discovered genes. An even greater challenge is in determining the functions of these genes in the changing natural environment of microbial communities associated with plants and devising new control methods based on this knowledge.

Development of New Technologies for Disease Control

What will new control methods be?

This chapter focuses on targets for disease control revealed by functional genomics. The actual control mechanisms developed could range from chemical to biological control to transgenic plants, or a combination of these, and other new methods and possible methods are not discussed in depth. In many cases, the use of 'inhibitors' for disease control targets is mentioned; this is meant to refer to anything from a chemical inhibitor to a biological agent interfering with an aspect of bacterial virulence. The most suitable targets were defined as those that are present in plant pathogens and important for virulence, but lacking in animal pathogens and environmental bacteria. If specific inhibitors for these plant pathogen-specific targets can be found, hopefully they will have a lower environmental impact than copper or antibiotic sprays and the many types of broad-spectrum pesticides currently used to control fungi, insects and weeds in agriculture.

These criteria were used because the numerous enzymes required for all bacteria to grow and survive should be reserved for disease control measures that directly impact human disease. Widespread use of new antibiotics or similar new broad-spectrum control measures is likely to result in a faster acquisition of resistance by bacterial pathogens and reduce the usefulness of new drugs for deadly bacterial diseases. Concerns about resistance acquisition are particularly true for plant pathogens since many spend part of their life cycle in soils, which are reservoirs for new resistance mechanisms due to the complex and competitive microbial communities present in soil (D'Costa et al., 2006).

What will the future of bacterial plant disease control look like? Numerous potential targets have been identified by functional genomics, but it is not likely that a chemical silver bullet for bacterial plant pathogens will be found since these pathogens are so diverse. Rather, it seems more likely that a combination of resistant plants developed with the assistance of marker-assisted breeding, plant resistance activators and inhibitors of multiple plant pathogen-specific virulence genes designed for the local bacterial population infecting the crop, will be used. Growers will also probably use sophisticated pathogen detection methods capable of detecting known and emerging pathogens to test seed prior to planting and to describe microbial soil and water communities so that appropriate control measures can be used. Pathogen indexing of fields in combination with GPS information would allow growers to avoid planting or to use targeted biocontrol measures on areas of a field with high or highly virulent pathogen populations. Growers will also still need to take every reasonable sanitation measure with their equipment, perhaps with new methods designed to combat bacterial biofilms.

How will targets be identified?

Few genes in bacterial pathogens are required for pathogenicity, rather most provide subtle and overlapping contributions to fitness on plant hosts. Sometimes the contribution to virulence can only be seen on some hosts or under some conditions. Therefore, it is difficult to identify new bacterial virulence genes through mutagenesis screens coupled with virulence assays since the number of mutants multiplied by the number of hosts and conditions is tremendous. In addition, some bacterial pathogens are difficult to grow and genetically manipulate, so traditional mutagenesis screens are not possible. Many researchers have successfully identified new virulence genes by taking advantage of genomics data that allows identification of virulence genes by location in a pathogenicity island or upregulation in plants or in response to a plant-induced regulator. It is likely that new control targets will be found in similar ways.

Analysis of bacterial genomes has clearly shown that virulence genes, even those with unrelated functions, are often clustered in horizontally transferable pathogenicity islands, and sometimes multiple pathogenicity islands have been inserted into a single genetic locus. These gene islands may be identified by comparing genomes of closely related species to find clusters of genes present in only the species of interest. If no closely related genome sequences are available, putative gene islands can be identified by the many hallmarks of horizontally transferred regions, including differences in GC content, codon usage, adjacent phage, insertion element or plasmid sequences. In addition, in gram-negative bacteria, virulence gene islands are often inserted adjacent to individually encoded tRNAs, making these loci a fruitful place to search for virulence genes.

One of the clearest examples of a mobile pathogenicity island is in the 660 kb island in the plant pathogenic *Streptomyces* sp. (Loria et al., 2006). These pathogens produce a necrosis-inducing toxin, Thaxtomin A, which is required for disease. During a screen designed to find the toxin synthesis genes, a gene encoding a necrosis-inducing peptide, Nec1, was found. Nec1 is only produced by some pathogenic *Streptomyces* sp. and does not appear to play a large role in virulence. However, it is in the midst of a large horizontally transferable pathogenicity island and examination of the genes adjacent to Nec1 allowed identification of several other important virulence genes that may be targets for control of common scab. Similarly, the *hecA* gene, which is an adhesin important for the initial interactions between the soft rot pathogen *Dickeya* and host plants, was first identified because of its insertion adjacent to other *Dickeya* virulence genes in a large pathogenicity island (Rojas et al., 2002).

New control methods identified by high throughput screens

Numerous new drugs and antimicrobials have recently been identified by taking advantage of *in vitro* assays amenable to high throughput screens of compound libraries. Generally, development of new antimicrobials based on genomic data is found by the following process (Mills, 2003): (i) a conserved enzyme target is identified by sequence analysis; (ii) the gene is proven to be essential for *in vitro* or *in vivo* growth or pathogenicity; (iii) an *in vitro* assay with purified protein or a *in vivo* reporter assay is developed; (iv) multiple isozymes of the protein product are tested with numerous compounds; and (v) the active compounds are characterized for potency, mechanism and spectrum. This type of approach has resulted in the identification of new antimicrobials that target essential genes, such as peptide deformylase (Clements et al., 2001), UDP-NAG enolpyruvyl transferase (Baum et al., 2001), 3-ketoacyl acyl carrier protein synthase III (He et al., 2004) and methionyl tRNA synthetase (Jarvest et al., 2002). Two of the main difficulties in converting an enzyme inhibitor into a useful antimicrobial are finding compounds that can pass through the bacterial cell membrane(s) and finding compounds that are not exported out of the cell via an efflux pump (Mills, 2003).

Recently, inhibitors that affect the type III secretion system (T3SS) or its substrates have been identified. For example, the activity of *Pseudomonas aeruginosa* ExoU, a T3SS-secreted phospholipase that kills host cells, can be inhibited by several compounds known to inhibit phospholipases (Phillips et al., 2003) and these compounds also reduced pathogen virulence in animal cell models. Excitingly, acetylated hydrazones that inhibit the *Yersinia pseudotuberculosis* T3SS (Kauppi *et al.*, 2003; Nordfelth *et al.*, 2005) reduced *Yersinia* virulence in animal cell models and also disrupted the life cycle of the obligate animal pathogen *Chlamydia trachomatis* (Muschiol *et al.*, 2006). Thus, it is possible that small molecules discovered to be active against plant pathogens amenable to culturing and high throughput screens may aid in the study of plant pathogens that are difficult or impossible to grow outside of plants.

New chemical inhibitors need not only be synthetic compounds in commercial libraries. Natural compounds produced by cells expressing heterologous genes isolated directly from complex environmental samples or by cells expressing genes modified *in vitro* to produce novel compounds can also be potent antimicrobials (Clardy *et al.*, 2006). The functional genomic revolution has just begun to explore the diversity of natural products useful for disease control and many exciting findings are likely. With either synthetic or natural compounds, uptake into plants and distribution into xylem and other reservoirs of bacterial pathogens in plants, will be critical for the successful use of virulence inhibitors.

The first step: turning off bacterial defences

Plants produce both preformed and induced antimicrobials, which can interfere with the establishment of an initial population of bacteria in host tissues. Consequently, microbial pathogens have evolved mechanisms to counteract the presence of toxic substances, of which multidrug resistance (MDR) efflux pumps are the most important for cell detoxification. Genes encoding MDRs are ubiquitous among bacteria, constituting on average more than 10% of the transporters in an organism. The specificity and importance of most of these MDRs to specific plant pathogens is unknown, although when the individual MDRs have been examined, they have been found to play critical roles in exporting plant antimicrobial compounds (Maggiorani Valecillos *et al.*, 2006).

At least some plants counteract the action of MDRs, which extrude toxins out of cells, by inhibiting the MDRs. For example, the model plant *Berberis fremontii* produces berberine, a cationic alkaloid that can be transported out of bacterial cells, making it essentially ineffective in combating bacterial pathogens (Stermitz *et al.*, 2000). However, this plant species also produces an MDR inhibitor, 5'-methoxyhydnocarpin (5'-MHC). Although 5'-MHC had no antimicrobial activity alone, it potentiated the action of berberine and other efflux pumps substrates, resulting in an accumulation of berberine in cells.

Inhibition of specific bacterial resistance MDRs by small molecules, either sprayed on to plants or produced by plants, may be suitable for control of bacterial pathogens in ornamental and vegetable plant protection, particularly if plant pathogen-specific MDRs can be identified. As more is learned about how naturally produced plant molecules affect bacterial pathogens, the coupling of potent plant-produced MDR inhibitors with plant-produced antimicrobial compounds could prove valuable targets for breeding programmes. MDRs have already proven amenable to high throughput screens for inhibitors with animal pathogens; thus, the strategy of identifying either natural products or new compounds that inhibit efflux of antimicrobial compounds is also likely to work with plant pathogens (Pages *et al.*, 2005). Importantly, combining MDR inhibitors with the strategies discussed below is likely to make many of these strategies more effective as disease control measures.

Deregulation: gram-negative pathogens

Some bacterial species thrive in environments that are considered hostile or competitive, such as a leaf surface exposed to dry conditions and harsh ultraviolet light or a rhizosphere colonized by thousands of diverse species competing for the same resources. To survive in these complex environments, bacteria are able to quickly modulate gene expression in response to changes caused by their hosts, competitor microbes or the environment. Modulation of these responses to cause bacterial pathogens to respond inappropriately to their hosts can be used to control bacterial diseases. Interference with bacterial responses to competitors or environmental stresses could also provide a mechanism to reduce bacterial populations in soil, water or on plant surfaces.

Bacteria have numerous classes of gene regulators that vary in abundance and diversity within a cell, species and genus. Some of these regulatory systems consist of a membrane-bound sensor and a DNA-binding response regulator, such as the two-component systems. Others are cytoplasmic proteins that bind substrates transported into the cell. Some regulators are more commonly associated with laterally transferred gene islands, such as the LysR family, while others are well conserved across species and even genera, such as two-component systems.

Methods to disrupt regulatory cascades depend on the type of regulator targeted and the desired specificity of the effect. Genome analysis has provided useful targets that could be plant pathogen-specific. For example, the HrpXY two-component system is found in most enterobacterial plant pathogens, but is lacking from all related animal pathogens. It controls expression of the T3SS, which encodes a molecular syringe that injects virulence proteins into host cells. This secretion system is required for pathogenicity of some enterobacterial plant pathogens, such as the fire blight pathogen *Erwinia amylovora*, and is therefore a useful target for control measures. Other regulators likely to be important for virulence in some enterobacterial pathogens, such as the LysR homologue adjacent to the *cfa* genes in the blackleg pathogen *Pectobacterium atrosepticum*, are not as widespread and are therefore less useful targets for control measures (Bell *et al.*, 2004). Although the signals that most of the critical virulence regulators respond to are unknown, the ease with which gene expression reporters dependent upon the action of these regulators can be developed into high throughput screens should facilitate identification of virulence regulatory protein inhibitors.

Multiple small molecules are used by bacterial pathogens for cell–cell communication and virulence gene regulation (Visick and Fuqua, 2005). Some, such as the AI-2 system, are widespread in both gram-positive and gram-negative pathogens, and because of this, are probably not useful targets for plant pathogen-specific control measures (Abraham, 2006). Interference with a well-conserved regulatory system that affects expression of virulence genes in many gram-negative pathogens, the acyl-homoserine lactose (AHL) quorum sensing system, has proven effective in controlling bacterial diseases in laboratory assays. Bacteria produce diffusible AHLs as a mechanism to sense bacterial density (Chapter 2, this volume). Once a threshold is reached, pathogens may induce virulence gene expression. For example, the soft rot pathogen *P. carotovorum* induces expression of pectate lyases when AHL concentrations reach a threshold level (Andersson *et al.*, 2000). Transgenic plants that degrade AHLs are resistant to soft rot symptoms caused by *P. carotovorum* (Dong *et al.*, 2001). Plants that overproduce AHLs are also resistant, perhaps because the plants cause the pathogen to inappropriately express virulence genes, triggering plant defences (Mäe *et al.*, 2001).

It is unclear how useful these plants would be in commercial production. Some soft rot pathogens, such as *Dickeya* sp., which is closely related to *P. carotovorum*, also produce AHLs, but only regulate a subset of plant cell wall degrading enzymes in response to AHL concentration, and mutants unable to produce AHLs are not affected in virulence in laboratory assays. Since both *Pectobacterium* and *Dickeya* sp. coexist in many regions of the world, genetically modified plants that interfere with AHL signalling may not significantly reduce the incidence of soft rot disease. Plants that inhibited AHL signalling are also able to be colonized to high levels by *Pectobacterium* in the absence of symptoms (Smadja *et al.*, 2004). If cell wall degrading enzymes can then be triggered by alternate signals, this could lead to devastating crop losses; thus, the usefulness of this strategy is unclear. In addition, since AHL signalling is used by many animal pathogens, and intensive searches are under way for antagonists for use in medicine, this may be an inappropriate target for plant pathogens since resistance mechanisms could affect the efficacy of drugs used to treat human diseases.

Deregulation: gram-positive pathogens

Although the first small signalling molecules, the γ-butyrolactones, were identified in *Streptomyces* sp., very little is known about how these signals affect the virulence of gram-positive pathogens. The structure of these signal molecules is very similar to AHLs, and as with AHLs, γ-butyrolactones, produced by different species differ in side chain length and structure (Takano, 2006). Homologues of the γ-butyrolactone receptor and a zinc metalloprotease involved in the synthesis of an *Enterococcus faecalis* signalling molecule were identified in the *Leifsonia xyli* subsp. *xyli* genome (Monteiro-Vitorello et al., 2004). *L. xyli* subsp. *xyli*, which causes ratoon stunting of sugarcane, has among the smallest genomes of plant pathogens and the fewest virulence genes, yet the genes required for production and/or sensing of signal molecules have been retained. Therefore, although little or nothing is known about signal molecules produced by most gram-negative plant pathogens, it is likely that these or other signalling molecules are produced and are important for virulence.

In the soil bacterium *Streptomyces griseus*, γ-butyrolactones control antibiotic production and sporulation. If this or similar molecules affect production of secondary metabolites or sporulation in the plant pathogenic *Streptomyces*, this could be a target for control of common scab. However, due to the widespread nature of *Streptomyces* in the soil and the important role that it plays in the soil microbial community, disrupting signalling molecules that are not specific to plant pathogenic species may be detrimental to beneficial microbes as well. Considering that the *Streptomyces* genomes are between 8 and 10 Mb, it is likely that many additional signal molecules important for virulence, perhaps specific to plant pathogenic *Streptomyces*, will be identified.

Other small targets: C-di-GMP and iron

Many other small molecules are important for bacterial signalling and nutrition and two important molecules are highlighted here. However, the importance of many of these small molecules and nutrients tends to be widespread among bacterial species. Thus, interfering with the roles or uptake of molecules required for life by most bacteria or which are important as secondary messengers in diverse species is not likely to result in plant pathogen-specific control measures.

Cyclic di-GMP was recently revealed as an important intracellular bacterial signal molecule. Concentration of cyclic di-GMP in cells is increased by proteins with GGDEF domains and reduced by proteins with either EAL or HD-GYP domains (Camilli and Bassler, 2006; Ryan et al., 2006). Proteins may contain combinations of these three domains along with domains likely to sense environmental conditions. It is unclear if and how proteins with both GGDEF and EAL domains regulate, whether production or degradation of cyclic di-GMP is occurring. However, genetic experiments have shown that these proteins regulate important developmental and virulence processes in bacteria, most likely through as yet unknown post-transcriptional processes. Since GGDEF, EAL and HD-GYP proteins are only found in bacteria and not eukaryotes, it may be possible to identify compounds that interfere with the action of these proteins, but that have no effect on eukaryotic host cells. However, because these proteins are also present in environmental bacterial and animal pathogens and it is possible that resistance to inhibitors could develop, inhibitors of these proteins should be reserved for treatment of human diseases.

Iron transport is important for all cells and iron is required for cell function. Iron uptake mechanisms are both heterogeneous within species and redundant within cells. These systems are also necessary for beneficial bacteria and may also play a role in plant health since plant cells may also be able to absorb bacterial iron-siderophore complexes (Wang et al., 1993; Masalha et al., 2000). In fact, production of siderophores by plant growth-promoting bacteria has been suggested as one mechanism by which these beneficial species improve plant health. Therefore, because of the importance of iron acquisition to human pathogens and beneficial microbes and the redundant and heterogeneous iron transport systems used by bacterial pathogens, chemical inhibition of iron transport does not seem likely to lead to useful control measures for plant diseases. However, it is likely to continue to be an important component of biocontrol measures used for bacterial pathogens.

Tell them where to go: interference with taxis

To cause disease, phytopathogenic bacteria must adhere to plants and then grow in or on plant tissue. After a period of growth, and possibly spread throughout the plant, symptoms may develop. Finally, these pathogens must be able to survive in plant debris, insect vectors or elsewhere in the environment until another plant host is encountered. Bacteria that are able to sense and move towards locations conducive to each step of this life cycle are likely to have a survival advantage. These chemosensory (taxis) signals are sensed by methyl-accepting chemotaxis proteins (MCPs) and the signals from these proteins ultimately feed into the flagella apparatus and affect the rotational direction of the bacterial flagella. Since multiple signals are present in the environment, bacteria need to have a balanced response to various attractants so they do not respond by moving toward an attractant present in a location that is poorer for energy generation than the environment that they currently inhabit. For motile pathogens, taxis towards nutrients or away from repellents are likely to be important for transitioning between these life cycle steps and a potential avenue for interference in pathogen life cycles.

One of the most striking differences between gram-negative animal and plant pathogens found through genomics is the abundance of MCPs in plant pathogens. For example, the soft rot pathogen *P. atrosepticum* SCRI1043 has 36 MCPs, compared to 5–12 usually found in animal pathogens. Similarly, the leaf pathogen *Pseudomonas syringae* DC3000 has 49 MCP genes, while the related animal pathogen *P. aeruginosa* has only 20.

It is possible that the multiple MCPs are present not only because they allow recognition of a greater variety of attractants, but because they also allow the plant pathogens to more finely tune taxis responses in different environments. This could be accomplished by encoding multiple MCPs that each respond to the same attractants and repellants, but each of which is also differentially regulated and/or has a different level of transducer activity. This sensitivity to compounds in the environment may also affect how the plant pathogens interact with other bacteria in the environment since they may be attracted or repelled by a wider variety of bacterial and fungal metabolites in microbial communities than related animal pathogens.

Although the MCPs vary greatly among the plant pathogens, if a set of signals important for taxis towards hosts or for virulence could be identified, it may be possible to design antagonists of this signal that would blind bacterial pathogens to the presence of host plants. Since MCPs important for plant pathogenesis are likely to be lacking in animal pathogens, as evidenced by the paucity of MCPs in many animal pathogens, this proposed control mechanism might be plant pathogen-specific. MCPs can also sense repellents, although this has been little studied in plant pathogens. It may also be possible to identify compounds that repel plant pathogens away from host plants, thereby making disease less likely to occur. This strategy is not likely to work with most gram-positive plant pathogens since only a handful of them, such as the soft rot *Clostridium*, are motile via flagella.

Inhibition of structure and function

Most conventional antibiotics and recently discovered chemical inhibitors do not interfere with gene regulation or taxis. Rather, they disrupt the function of structural genes required for cell growth or bacterial pathogenicity. For plant pathogens, inhibitors of structural virulence proteins are attractive targets because many of these virulence proteins are only found in plant pathogens; thus, it may be possible to develop plant pathogen-specific control methods that will not compromise human health. Functional genomics has revealed numerous targets, some of which are discussed below.

Bacterial syringes: type III secretion system inhibitors

The T3SS is a molecular syringe encoded by many gram-negative plant pathogens that delivers effector proteins directly into host cells, where the effector proteins disrupt plant defences and promote bacterial growth (Grant *et al.*, 2006). At least two classes of T3SS important for pathogenicity are present in bacterial plant pathogens (Alfano and Collmer, 1997). These T3SS differ in gene organization, regulation and types of secreted effector proteins. The importance of the T3SS for virulence varies among pathogens. For example, *P. syringae* is dependent on this secretion system for growth in leaves, while in contrast, the T3SS makes only a small contribution to the virulence of soft rot enterobacteria. Animal pathogens also encode T3SSs, but the structure of the outer part of the secretion system differs from those encoded by plant pathogens, and most of the secreted proteins also differ; thus, it should be possible to identify inhibitors of the plant-associated T3SSs and effectors. Some beneficial bacteria also encode T3SSs (Preston *et al.*, 2001), but how widespread these T3SS are and their substrates and functions are unknown. Nor is it clear if and how T3SS inhibitors would affect beneficial bacteria on leaves or in the soil. The function and structure of plant pathogen T3SSs and their substrates are reviewed extensively in Chapter 2 in this book and will not be reviewed here.

Two characteristics of the T3SS make it amenable to high throughput assays for inhibitors. Similar to the flagella secretion system, the T3SS appears to be subject to negative regulation when it is not functioning properly (Preston *et al.*, 1998). Therefore, it might be possible to use gene expression reporters to find structural inhibitors since inhibition of T3SS machinery assembly might result in lower expression of genes encoding either T3SS structural or secreted proteins. Second, in the soft rot pathogen *D. dadantii*, the T3SS is required for biofilm formation (Yap *et al.*, 2005); thus, a simple biofilm assay might be useful as a fast and inexpensive screen for T3SS inhibitors.

Pat-1 inhibitors

Analysis of virulence factors on *C. michiganensis* plasmids led to the identification of Pat-1, a secreted protein with homology to serine proteases. Pat-1 is required by *C. michiganensis* subsp. *michiganensis* for pathogenicity and confers pathogenicity to a non-pathogenic *C. michiganensis* subsp. *michganensis* strain lacking the virulence plasmids (Dreier *et al.*, 1997). Homologous genes are found in other gram-positive and gram-negative plant pathogens, including *L. xyli* subsp. *xyli* (Monteiro-Vitorello *et al.*, 2004) and *Xanthomonas oryzae* pv. *oryzae* (Thieme *et al.*, 2005). In the other plant pathogens, it is not clear if these homologues contribute to virulence. A disrupted homologue is also present in *Xanthomonas campestris* pv. *vesicatoria*, but not in either of the other two sequenced *Xanthomonas* strains (Thieme *et al.*, 2005).

The importance of Pat-1 to pathogenicity in *Clavibacter*, and possibly in other pathogens, makes it an attractive target for control measures. To date, no serine protease activity has been demonstrated for this protein, although a *C. michiganensis* subsp. *michiganensis* strain with a mutation of a conserved motif required for protease activity in related proteins was non-pathogenic. It will be difficult to develop control measures aimed at this virulence factor until an *in vitro* activity assay can be developed for this protein.

No colonization without degradation? Enzyme inhibitors

Genome sequences of plant pathogenic bacteria have revealed that many plant pathogens, including those that do not cause plant maceration, such as *A. tumefaciens* and *P. syringae*, encode numerous plant cell wall degrading enzymes, such as pectinases, cellulases, xylanases and proteases. Of these, proteases are secreted through type I secretion systems and the other cell wall degrading enzymes are secreted through type II secretion systems (T2SS). Most bacteria encode multiple type I secretion systems and one or two T2SSs. Proteins in addition to degradative enzymes are secreted via T2SS and the presence of multiple T2SS is not correlated with pectolytic ability. For example, *P. atrosepticum*, a soft rot pathogen, encodes only one T2SS, while *P. syringae* pv. *phaseolicola*, the bean halo blight pathogen, encodes two T2SS (Bell *et al.*, 2004; Joardar *et al.*, 2005).

The contribution of these enzymes to pathogenicity has been mainly studied in pectolytic pathogens where mutation of T2SS or mutation of multiple genes encoding pectolytic enzymes, generally results in a reduction or elimination of virulence (Ray *et al.*, 2000). Enzyme inhibitors would likely be able to control the soft rot pathogens that rely on these enzymes for pathogenicity. High throughput assays have already been developed for proteases, pectinases, cellulases and xylanases; thus, it should be possible to screen for inhibitors of these enzymes. Inhibition of these enzymes may be a plant pathogen-specific strategy for controlling bacterial diseases since these enzymes are not present in most animal pathogens.

Detoxifying toxins

There is evidence for production of multiple toxins in many plant pathogenic bacteria. For example, *X. oryzae* pv. *oryzae* encodes genes homologous to those required for production of RTX toxin, which may act as inhibitors of ethylene biosynthesis, as well as producing several toxic organic acids (Lee *et al.*, 2005). RTX toxins are also encoded by *Xylella fastidiosa* and *P. atrosepticum* (Bell *et al.*, 2004). *Xanthomonas albilineans* produces the DNA replication inhibitor albicidins, which blocks plastid replication as well as growth of competitor microbes (Birch and Patil, 1985, 1987).

P. syringae produces several broad host range toxins, some of which play important roles in pathogenicity (Bender *et al.*, 1999). These toxins have a wide range of targets. For example, coronatine, a chlorosis-inducing phytotoxin, appears to act as a methyl jasmonate mimic and interferes with plant defences. Tabtoxin, which is produced by *P. syringae* pv. *tabaci*, inhibits glutamine synthetase, which results in a lethal build-up of ammonia in plant cells. Phaseolotoxin acts by irreversibly inhibiting ornithine carbamoylphosphate transferase, a required enzyme in the urea cycle, resulting in an accumulation of ornithine and a lack of arginine and pseudomycin. Peptide and polyketide synthetases with unknown functions are present in many bacterial plant pathogen genomes and some of these genes may produce phytotoxins. Similar toxin-producing genes are rare in animal pathogens; thus, these are possible targets for plant pathogen-specific control methods. Anzai *et al.* (1989) demonstrated that transgenic plants able to detoxify tabtoxin were resistant to symptoms caused by *P. syringae*, but this approach has not been pursued further.

Interference with toxin function or production may also be useful for controlling gram-positive plant pathogens. Plant pathogenic *Streptomyces* spp., which cause common scab on root and tuber crops, produce a toxin (thaxtomin A), which is required for pathogenicity. This toxin has a very broad host range and is able to inhibit cellulose synthesis on all plant species tested (Loria *et al.*, 2006). It also induces necrosis on plant tissue and inhibits seedling growth, probably due to its inhibition of cytokinesis. Because of its broad host range, it seems unlikely that plant resistance to thaxtomin A will be found. Due to its important role in *Streptomyces* pathogenicity, toxin synthesis inhibitors, enzymes that degrade this toxin or other mechanisms of resistance against thaxtomin A would likely provide resistance to common scab.

Primary colours: pigment production inhibitors

Both gram-negative and gram-positive pathogens produce pigments, such as the carotenoids lycopene, spirilloxanthin, β-carotene, cryptoxanthine and canthaxanthin produced by some *C. michiganensis* subsp., which gives these strains yellow, orange and even red hues (Saperstein and Starr, 1954; Saperstein *et al.*, 1954), as well as the yellow xanthomonadins produced by *Xanthomonas* sp. and the blue and red indigiodine pigments produced by *Dickeya* sp. One *C. michiganensis* subsp. is also capable of producing indigoidine (Starr, 1958). These pigments function to protect cells from ultraviolet light and oxidative damage and are important for epiphytic colonization and/or pathogenicity (Jacobs *et al.*, 2005). Interference with pigment production of bacteria on leaf surfaces could reduce bacterial populations, and thus, reduce the probability that foliar symptoms will develop. The importance and diversity of pigments within plant and animal pathogenic bacterial species has been little studied, so it is difficult to predict the efficacy of interfering with pigment production for disease control.

Shedding their coats: biofilm degradation and inhibition

Biofilms are aggregates of bacterial cells held together by an extracellular matrix produced by the bacteria. These aggregates may be adhered to a surface, such as a root or tractor tire or they may be free-floating in irrigation water. Generally, the structural components of these aggregates consist of at least one type of protein filament and at least one type of polysaccharide. Bacterial biofilm formation was only recently widely recognized to be important for plant pathogens, although there were many hints of its importance for decades prior to the widespread use of the term biofilm. Biofilms are likely to be important for bacterial survival on and colonization of plants as well as for survival of pathogens outside of hosts. Since sanitation is an important part of management of bacterial diseases, it is important to determine the degree of resistance of plant pathogens which are protected by biofilm matrices to sanitizers. In at least some cases, such as with *D. dadantii* biofilms, bacterial cells can survive high levels of bleach and would not be killed by conventional sanitation measures (C. Jahn, Wisconsin, 2006, personal communication). Thus, disruption of bacterial biofilms is likely to be an important step to enhance the efficacy of sanitation protocols.

Many gram-negative bacterial plant pathogens produce cellulose as an important part of their extracellular matrix. This was first found in *A. tumefaciens*, where bacterial cellulose production was shown to be important for root infection (Matthysse, 1983). Since then, cellulose synthesis has also been reported in other plant pathogens (Yap *et al.*, 2005), and has even been shown to be important for colonization of plants by human pathogens (J. Barak *et al.*, 2007). For those pathogens that use bacterial cellulose as part of their infection strategy, bacterial cellulose synthesis inhibitors may reduce the ability of these pathogens to adhere to host plants and to form protective biofilms on farm equipment, greenhouse surfaces or plant debris. Cellulases, which can dissolve bacterial biofilms (Yap *et al.*, 2005) could also be used in conjunction with sanitizers to clean agricultural equipment.

Recently, a dramatic example of specific adhesion to plant structures was demonstrated with *Pantoea stewartii*, which colonizes corn xylem (Koutsoudis *et al.*, 2006). In this case, the wild-type cells mainly colonized the annular rings of the xylem. However, a strain that overproduced extracellular polysaccharide (EPS) only diffusely adhered to the xylem and was not localized to the annular rings. It does not seem that this is simply due to EPS masking other adhesin molecules.

Other extracellular carbohydrates, such as colanic acid, alginate and β(1–6) *N*-acetylglucosamine contribute to bacterial biofilm formation in gram-negative and gram-positive bacterial species. Enzymes that degrade these carbohydrates are effective at degrading or even inhibiting biofilm formation (Ramasubbu *et al.*, 2005) and several enzymes have been patented for just this purpose (e.g. see US Patent 6,100,080). However, since individual bacterial strains can form multiple types of genetically and chemically distinct biofilms (Yap *et al.*, 2005), and the differences in biofilm matrices within and across species are still not well understood, more work needs to be done to identify biofilm disruption and dispersion methods appropriate for each pathogen. Even closely related species that cause similar diseases may differ significantly in biofilm formation. For example, although the soft rot enterobacterial pathogen *D. dadantii* produces copious amounts of cellulose in culture, the closely related *P. atrosepticum* does not and examination of its genome revealed that a premature stop codon was present in one of the cellulose synthesis operons in *P. atrosepticum* (Bell *et al.*, 2004).

Protein filaments employed by bacteria for biofilm formation vary widely and can include structures such as fimbriae and hemagglutinens. Some of these adhesins are used to attach to host cells and play important roles in pathogenicity. The roles of adhesins have been more thoroughly examined in animal pathogens, but genome sequences have revealed that adhesion homologues are abundant in plant pathogens, suggesting that these surface appendages also play an important role in plant disease.

Protein adhesins have been shown to be important for plant pathogens with diverse lifestyles. For example, in the opportunistic soft rot pathogen *Dickeya*, a gene encoding a filamentous hemagglutinen homologue, *hecA*, is required for adhesion of *Dickeya* AC4150 to plant tissue and for subsequent disease progression (Rojas *et al.*, 2002). Genome sequencing projects have subsequently revealed that *Dickeya*, as well as many other plant pathogens encode multiple orthologous hemagglutinens (Bell *et al.*, 2004; Joardar *et al.*, 2005; J. Glasner, Wisconsin, 2006, personal communication).

Necrogenic pathogens are also likely to rely on adhesins for initial plant interactions. The four sequenced *Xanthomonas* genomes all encode type IV pili, which vary across these species, perhaps reflecting the different plant hosts of these four *Xanthomonas* sp. *X. campestris* pv. *hyacinthi* type IV pili adhere to stomatal cells, supporting a role for these structures in attachment to and invasion of plants (Van Doorn *et al.*, 1994). Type IV pili are also important for *P. syringae* colonization of leaf surfaces (Roine *et al.*, 1998). Much more needs to be learned about bacterial adhesion to plants before strategies targeted at adhesins can be tested, including which, if any, adhesins are specific to plant pathogens, as well as development of high throughput assays for specific plant pathogen adhesins.

O, NO: inhibition of active oxygen and nitric oxide resistance

Plants generate active oxygen and nitric oxide as a part of their defence response against invading microbes (Klessig *et al.*, 2000) and increasing the production of these molecules by genetic modification of plants can provide resistance against bacterial pathogens (Wu *et al.*, 1995). Multiple antioxidant defences, such as superoxide dismutases, which convert superoxide to hydrogen peroxide, and catalases, which convert hydrogen peroxide to water, are present in many plant pathogen genomes, presumably to protect the cells against active oxygen produced by host cells. In addition, flavohaemoglobins, which catalyse a denitrosylase reaction, can protect cells against nitric oxide (Boccara *et al.*, 2005). These enzymes have been shown to play a role in pathogenicity in multiple pathogens. In addition, protein oxidative damage repair systems are also present and are, at least in some cases, important for pathogenicity (El Hassouni *et al.*, 1999). In addition to enzymes, pigments may also provide defences against active oxygen. For example, the blue pigment indigoidine, which is synthesized by *D. dadantii*, may be a radical scavenger and confers resistance to oxidative stress (Reverchon *et al.*, 2002). The indigoidine synthesis genes are in a large *D. dadantii*-specific gene cluster involved in virulence and are not present in the closely related *Pectobacterium* genomes. Indeed, *Pectobacterium* is not known to produce any pigments, but perhaps one of the many gene clusters with unknown function will be revealed through a functional screen to encode analogous radical scavengers. As discussed above, pigments may be appropriate targets for inhibitors, but enzymes that protect against oxidative damage are widespread in bacteria; thus, they are not ideal targets for plant pathogen-specific control strategies.

Ice nucleation and frost damage

In addition to causing necrotic spots on leaves and fruit and seed decay, some plant pathogens such as *P. syringae* can cause frost damage to plants by nucleating ice on a cell surface protein array. These ice-nucleating strains are probably responsible for frost damage on crops when temperatures are between 0°C and −5°C (Hirano and Upper, 2000). If methods could be found to interfere with bacterial ice nucleation, this could ameliorate crop damage caused by freezing. The ice nucleation protein is a good target for control strategies because it is a surface protein for which the structure is known and for which there is a very simple and inexpensive high throughput assay available, namely freezing water. Many bacteria, plants and animals encode antifreeze proteins and *P. syringae* itself encodes a homologue of a secreted antifreeze protein of unknown activity (Feil *et al.*, 2005). Most bacterial antifreeze proteins reduce the size of ice crystals rather than inhibit freezing, but recently a bacterial antifreeze protein capable of depressing the freezing point of water by over 2°C was discovered in an Antarctic lake-dwelling species (Gilbert *et al.*, 2005), opening up new possibilities for control of frost damage on crops.

Disease control through modification of host biochemistry and physiology

Functional genomics studies with bacterial pathogens have begun to unlock the secrets of how bacteria manipulate plant cells. Induction of plant resistance upon recognition of bacterial pathogens is already an important control measure and the current state of knowledge on how this induction occurs is described elsewhere in this book. In essence, resistance depends on recognition of pathogen molecules or their effects by plant proteins. Some pathogen proteins, such as flagella fragments, are probably recognized when they are on the outside of the host cell. Genomic analysis of bacterial pathogens has allowed us to predict additional secreted proteins produced by pathogens that could be used as targets for engineered resistance genes. However, how to engineer resistance proteins to specifically recognize new bacterial targets is still unknown.

Basic characteristics of plant physiology can also affect the host range of some pathogens. For example, expression of some *Pectobacterium* virulence genes is affected by host pH and a better understanding of how pH affects soft rot pathogen virulence genes could lead to new controls for this broad host range pathogen (Llama-Palacios et al., 2003). In addition, plant nutrition can also affect bacterial diseases. One of the clearest examples of this is the differential expression of a *Pectobacterium* virulence gene in response to calcium concentrations in host leaves (Flego et al., 1997). Calcium content can be altered in plants both genetically and through nutrition, and addition of calcium to soils is currently used as a method to reduce soft rot disease in potato. Bacterial responses to other aspects of plant nutrition are important to understand how plant physiology affects disease progression and how plant nutrition and physiology can be manipulated to better control bacterial diseases.

Phage and predators

Bacteria exist as members of complex communities that include bacterial parasites and predators and both types of organisms have been examined as potential controls for bacterial diseases. Phages were recently tested in combination with systemic acquired resistance inducers and in various formulations for control of bacterial spot on tomato and were found to provide effective control of *X. campestris* on inoculated plants under field and greenhouse conditions (Balogh et al., 2003; Obradovic et al., 2005). Although these results are very promising, much still remains to be done, from determining if phage is as effective on natural infestations with diverse *X. campestris* populations to perfecting the formulations used to prepare and deliver the phage. Because there are numerous types of phage and because bacterial resistance to phage infection can develop, this type of biocontrol is likely to benefit from bacterial genomics. For example, pathogen populations in a region or associated with a seed lot could be typed for phage susceptibility using genomics methods, then appropriate phage mixtures could be applied in conjunction with systemic acquired resistance inducers to control foliar bacterial pathogens.

Bdellovibrio bacteriovorus and related species are obligate predatory bacteria that feast on a broad range of gram-negative bacterial cells. They are ubiquitous in soil and water and have been examined as a biocontrol method for plant diseases (Scherff, 1973; Uematsu, 1980). Unlike some other control measures, predatory bacteria, which are often associated with natural biofilms, can penetrate and kill bacteria which are protected by biofilms (Kadouri and O'Toole, 2005). The *B. bacteriovorus* genome was recently sequenced and found to contain an arsenal of hydrolytic enzymes used to enter prey cells and digest them from within. Interestingly, this predator is unable to synthesize only 11 amino acids; thus, it can probably only produce proteins when it is preying on bacterial cells.

B. bacteriovorus quickly reduces bacterial populations, decimating *P. carotovorum* numbers in culture from 10^8 CFU/ml to less than 10^4 CFU/ml within 20 h (Shemesh and Jurkevitch, 2004). However, over the course of this interaction, a resistant population of enlarged *P. carotovorum* cells remains and this resistant population is able to multiply in the presence of predatory bacteria. The resistance is transient; once the predators are removed, *P. carotovorum* quickly becomes susceptible to *B. bacteriovorus* predation again. Since with some pathogens, such as *P. syringae* and *P. carotovorum*, high numbers of pathogen cells are required to cause plant disease, perhaps *B. bacteriovorus* could be used as a biocontrol agent directly after conditions occur that favor bacterial growth and disease, such as rainfall, to prevent pathogens from reaching high enough levels to cause disease. However, since predatory bacteria are common in the environment, it is possible that the resistance to predation observed in laboratory assays is the normal state for bacteria in nature and that predatory bacteria will not be effective biocontrol organisms.

Detection methods developed from genomics data

Accurate, fast and inexpensive detection of bacterial pathogens is crucial for quality control of planting materials and for excluding quarantine pathogens. Bacterial genomics has made several new pathogen detection methods possible. Many of these methods were first developed to detect animal pathogens in medical samples or food, but all are adaptable to plant pathogens. However, regardless of the ease, sensitivity and accuracy of the pathogen detection method used, detecting quarantine pathogens that are present at a low incidence in large shipments or fields of plants is still a difficult problem, since the labour involved in sampling thousands or millions of plants is cost-prohibitive. Even if every plant is sampled, pathogens may not be evenly distributed throughout the plant; thus, diseased plants may still test negative for pathogens. Therefore, it should be recognized that importing plant materials from locations with quarantine pathogens will eventually result in the import of the pathogen. Abstinence from importation is the only sure method for exclusion of quarantine pathogens; testing is only a mechanism to lengthen the time it takes for importers to spread pathogens to new areas. A few examples of new detection methods are discussed below and in more detail in Chapter 14 in this book. Innovations in detection technologies are occurring rapidly, so these examples are far from an exhaustive review.

PCR-based methods
Ideally, multiple targets are used for PCR-based detection of bacterial pathogens since reliance on single targets has resulted in misidentification of pathogens and in some cases these misidentifications have resulted in costly consequences for growers or importers. In the past, species-specific DNA fragments used for development of PCR-based detection assays were

generally found by subtractive hybridization (Mills *et al.*, 1997). Genome sequence comparisons have provided numerous genus, species and strain-specific targets for development of more accurate multiplex PCR assays for detection and identification of bacterial pathogens.

Array-based methods
Multiplex PCR methods can generally detect, at most, four or five different pathogens and require a substantial amount of knowledge about the pathogen genome sequence to develop useful primers. DNA array methods, although more time-consuming and expensive, have the advantage of being able to detect hundreds of bacterial species at once and of detecting and categorizing new pathogens. For arrays, pathogen DNA is PCR-amplified, then hybridized on to DNA oligos spotted on to nylon or glass supports. Purified, non-amplified DNA or RNA can also be used, but the sensitivity of the assay is greatly reduced. These assays would be most useful for identification of pathogens in samples with unusual symptoms or for identification of emerging pathogens. The power of arrays for identification of an emerging pathogen was demonstrated during the severe acute respiratory syndrome (SARS) outbreak of 2003, when arrays were used to identify the pathogen as a coronavirus (Wang *et al.*, 2003).

Genomics has revealed detection targets that provide layers of identification for bacterial pathogens. For example, rRNA operons and internal transcribed spacer regions, which have highly conserved regions suitable for PCR amplification and variable regions suitable for hybridization to pathogen-specific oligos, allow genus and sometimes species-level identification (Fessehaie *et al.*, 2003). Virulence genes common to many pathogens, such as T3SS genes, can also be used as targets. For finer discrimination among species or subspecies, primer sets can be designed to amplify genus-specific virulence genes. Our laboratory has developed a single PCR primer set that can amplify a small fragment of the T3SS from both enterobacterial and pseudomonad pathogens and a variable region in the middle of this fragment can be used to discriminate among species (R. Reedy, personal communication). These T3SS primers successfully detected a pectolytic *Pseudomonas* sp., for which no previous sequence information was available (R. Reedy, Wisconsin, 2007 personal communication), demonstrating the power of this method for simultaneous identification and virulence gene characterization of new bacterial pathogens.

Immunoassay methods
Enzyme-Linked Immunosorbent Assays (ELISA) and similar immunoassays, such as lateral flow devices, have been the detection method of choice for most plant pathogens in the 1980s since they are simple, inexpensive and in some cases, these assays can be completed in the field. A recent innovation in cell-based sensors based on B lymphocytes, CANARY (cellular analysis and notification of antigen risks and yields) may provide diagnosticians with a very fast method for identification of bacterial pathogens in plant samples (Rider *et al.*, 2003). For this assay, B cells

were constructed to express aequorin, a calcium-sensitive bioluminescent protein and membrane-bound pathogen-specific antibodies. When the antibodies bind to pathogen antigens, the B cell intracellular calcium level elevates, the cells luminesce, and the emitted light is detected by a luminometer. Very low levels of antigen can be detected and generation of light takes only seconds, so this method is both fast, and apart from development of the engineered B cell line specific for the pathogen antigen, is inexpensive. Importantly, CANARY can detect pathogens in samples with very minimal sample preparation, which results in a considerable reduction in the cost and time associated with pathogen detection. Like many other technologically advanced detection methods, the CANARY method has not yet become widely used; thus, it is difficult to assess its strengths and weaknesses for real-world pathogen detection.

Mass spectrometry methods
Recently, matrix-assisted laser desorption ionization-time of flight (MALDI-TOF) mass spectrometry has been developed to identify human pathogens (Claydon et al., 1996; Mazzeo et al., 2006). Briefly, ionized bacterial molecules are separated and detected based on the ratio of molecular weight to charge. A broad molecular weight to charge ratio can be detected; thus, signature ions can be detected when intact bacterial cells are ionized and different bacterial species have different MALDI-TOF fingerprints. Although this method is rapid and promising, it is not yet widely used. One of the main reasons this method is not widespread is that unlike PCR or hybridization methods, MALDI-TOF equipment is expensive and therefore not common in microbiology laboratories. In addition, the reproducibility of this method, particularly for larger ions, is affected by many experimental conditions, ranging from culture medium to sample preparation. Finally, unlike other methods, such as PCR and immunoassays, it has not yet been adapted for detection of pathogens directly from plant samples. However, if a subset of signature ions present in a pathogen under numerous growth conditions are identified, it may be possible to develop this method to quickly detect pathogens in enrichment cultures.

Conclusions and Future Directions

Genus sequencing projects

Currently, genome sequences are not available for all plant pathogenic bacterial species. However, new relatively inexpensive genome sequencing methods should greatly expand the number of plant pathogen genome sequences that are decoded in the next decade (Margulies et al., 2005). Obtaining these sequences is among the highest priority steps in moving forward towards a better understanding of bacterial pathogenesis and eventually, towards new control methods. In order to gain a full understanding of pathogen evolution, the next generation of bacterial genomics needs to include genus sequencing, rather than genome sequencing projects, and sequencing of closely related non-pathogens also needs to be supported.

The rest of the cycle

There are still many conserved genes in bacterial pathogens for which there is no known function and this is clearly one of the largest research needs since some of these genes are likely to lead to new possibilities for control of bacterial pathogens. Many bacterial pathogens spend a substantial time in soil, water or in relative stasis on plant debris or adhered to agricultural equipment. Creative new ways to disrupt these parts of the pathogen life cycle are likely to lead to control measures that are effective against bacterial pathogens. Creativity will also be required to develop assays to both determine what these genes are doing and to find inhibitors for the important processes throughout a pathogen's life cycle. Finally, more work is needed to determine how plant secondary metabolites, potential biocontrol organisms, and variations in plant nutrition and physiology can be used to control bacterial diseases.

Better, cheaper, faster detection

Bacterial genomics has provided a much clearer picture of which bacterial pathogen species are diverse, such as *Pectobacterium* and *Pseudomonas*, and which are not, such as *L. xyli* subsp. *xyli*, and where in the genome this diversity is found. It has also illuminated which genes are conserved within each pathogenic species, highlighting targets for more accurate and less expensive detection assays. For example, comparative analysis of three *X. fastidiosa* subsp. revealed 38 protein clusters present in these *X. fastidiosa* strains that do not have homologues in other bacterial species and which are therefore useful targets for *Xylella* detection assays (Bhattacharyya et al., 2002a).

Rules and regulations

Ideally, work in bacterial functional genomics will lead to a day when a genome sequence can be examined and the response of the bacterial cell to various conditions, such as a host plant cell or insect vector, can be predicted. This would allow us to better gauge the risk of newly introduced pathogens and to better adapt control measures. However, before this distant goal can be reached, we need to be able to build quantitative models of gene regulation so that the expression of various systems can be predicted. Over the last few years, it has become clear through functional genetics that regulatory pathways in bacterial pathogens are a complex and interwoven network, which include multiple layers of post-transcriptional and post-translational modifications with dozens of types of regulatory proteins and RNAs. In most cases, we still do not know which signals the most important global regulators, such as GacS/GacA, are monitoring. To complicate the study of gene regulation further, the quantitative effect of particular regulators on gene expression often differs depending on which type of assay is used, making quantitative and predictive models of gene regulation difficult to build, although some attempts have been made (Sepulchre et al., 2006). Investments in examining the same regulatory cascade with multiple methods to determine which methods best support predictions of cell responses need to be made.

Software for wetlabs

Although it seems that we have been inundated with data in the last few years, it is only the first trickle of an immense flood. To organize and interpret the genome sequence and functional data that we will have access to over the next decades, we need to invest in developing and maintaining freely available genome sequence and functional genomics databases. In particular, consistent annotation of the genomes is necessary for global comparisons, yet this remains a difficult and time-consuming process. The greatest problem is that the software and interfaces for genome projects are almost as diverse as the genomes themselves, making it difficult to compare data across genomes. Unfortunately, genome annotations are often not efficiently updated when new information is revealed through experimentation or by subsequent genome projects. Adding to these problems are the incorrect annotations that have propagated through genomes and the difficulty in determining what, if any, experimental evidence is tied to the proposed function of a gene.

The free web-based database ASAP (A Systematic Annotation Package for Community Analysis of Genomes) solves many of these problems. ASAP was developed so that the *D. dadantii* 3937 research community could annotate this genome as it was being sequenced. This database now includes numerous full and partial annotated genome sequences, including all publicly available enterobacterial genome sequences. Users are allowed various levels of access to ASAP, ranging from guest users, who can view a limited number of annotated genome sequences, to administrators, who can modify all aspects of the database. This has allowed research groups to tailor database access to their projects, which has ranged from the open access and large community annotation of the *D. dadantii* genome to projects where the genome data and annotation was restricted to a small group prior to publication.

ASAP is a curated database, meaning that a research group can assign one or more people to review and accept or reject each annotation added to a genome, thereby reducing the number of annotation errors that propagate through genomes, and also insure that all annotations are linked to appropriate experimental or sequence analysis evidence. ASAP is also capable of storing a variety of data, such as microarray gene expression results, mutant phenotypes, gene ontology (GO) terms and protein motifs associated with each gene.

Each gene in each genome entered into ASAP has an annotation page that can accept annotations ranging from links to journal articles and protein motif definitions to descriptions of mutant phenotypes. Each annotation page can also be linked to the annotation pages of orthologous genes in other genomes, allowing users to quickly determine what is known about orthologues in other species. The annotation pages also have many other tools, including Basic Local Alignment Search Tool (BLAST) searches of nucleotide and protein sequences in the NCBI database as well as within ASAP. Each annotation page in ASAP also links to a map showing the gene context, a feature that allows users to view the likely operon structure of the gene, to compare the context of orthologous genes from different genomes, and to link to annotation pages for surrounding genes. Because each gene has a unique feature ID in addition to its gene name, users can annotate fragmentary draft genomes and transfer the annotations to updated versions of the genome as additional sequence information is acquired and genome gaps are closed. Genomes may be annotated very quickly if a related genome has already been sequenced by automatically transferring annotations from orthologous genes in related species. The linking of annotations for each orthologue, the evidence required for each annotation, the use of GO terms, and the transfer of annotations across genomes both speeds the process of annotation and provides consistency that allows large-scale comparisons. Ideally, as many plant pathogen genome sequences as possible would be entered into ASAP or a similar database and consistently annotated so that we can all more efficiently access and interpret genomics data.

Stepping through the threshold

A little over two decades ago, the first virulence genes were cloned and less than a decade ago, plant pathologists were meeting to discuss which few bacterial genomes were most important to sequence. Even now, scientific societies develop high priority lists yearly for bacterial genomes that should be decoded. With the advent of new genomics technologies, these discussions will quickly become a relic of the past. With creativity, and a bit of luck, this revolution in functional genomics will be followed by a revolution in intelligently designed, environmentally friendly disease control methods that will increase agricultural productivity and reduce hunger.

References

Abraham, W.-R. (2006) Controlling biofilms of gram-positive pathogenic bacteria. *Current Medicinal Chemistry* 13, 1509–1524.

Alfano, J.R. and Collmer, A. (1997) The type III (Hrp) secretion pathway of plant pathogenic bacteria: trafficking harpins, Avr proteins, and death. *Journal of Bacteriology* 179, 5655–5662.

Andersson, R.A., Eriksson, A.R., Heikinheimo, R., Mae, A., Pirhonen, M., Koiv, V., Hyytiainen, H., Tuikkala, A. and Palva, E.T. (2000) Quorum sensing in the plant pathogen *Erwinia carotovora* subsp. *carotovora*: the role of expR(Ecc). *Molecular Plant-Microbe Interactions* 13, 384–393.

Anzai, H., Yoneyama, K. and Yamaguchi, I. (1989) Transgenic tobacco resistant to a bacterial disease by the detoxification of a pathogenic toxin. *Molecular and General Genetics* 219, 492–494.

Balogh, B., Jones, J.B., Momol, M.T., Olson, S.M., Obradovic, A., King, P. and Jackson, L.E. (2003) Improved efficacy of newly formulated bacteriophages for management of bacterial spot of tomato. *Plant Disease* 87, 949–954.

Barak, J.D., Jahn, C.E., Gibson, D.L. and Charkowski, A.O. (2007). The role of cellulose and O-antigen capsule in the colonization of plants by *Salmonella enterica*. *Molecular Plant-Microbe Interactions* (in press).

Baum, E.Z., Montenegro, D.A., Licata, L., Turchi, I., Webb, G.C., Foleno, B.D. and Bush, K. (2001) Identification and characterization of new inhibitors of the *Escherichia coli* MurA enzyme. *Antimicrobial Agents and Chemotherapy* 45, 3182–3188.

Bell, K.S., Sebaihia, M., Pritchard, L., Holden, M.T.G., Hyman, L.J., Holeva, M.C., Thomson, N.R., Bentley, S.D., Churcher, L.J.C., Mungall, K., Atkin, R., Bason, N., Brooks, K., Chillingworth, T., Clark, K., Doggett, J., Fraser, A., Hance, Z., Hauser, H., Jagels, K., Moule, S., Norbertczak, H., Ormond, D., Price, C., Quail, M.A., Sanders, M., Walker, D., Whitehead, S., Salmond, G.P.C., Birch, P.R.J., Parkhill, J. and Toth, I.K. (2004) Genome sequence of the enterobacterial phytopathogen *Erwinia carotovora* subsp *atroseptica* and characterization of virulence factors. *Proceedings of the National Academy of Sciences of the United States of America* 101, 11105–11110.

Bender, C.L., Alarcon-Chaidez, F. and Gross, D.C. (1999) *Pseudomonas syringae* phytotoxins: mode of action, regulation, and biosynthesis by peptide and polyketide synthetases. *Microbiology and Molecular Biology Reviews* 63, 266–292.

Bhattacharyya, A., Stilwagen, S., Ivanova, N., D'Souza, M., Bernal, A., Lykidis, A., Kapatral, V., Anderson, I., Larsen, N., Los, T., Reznik, G., Selkov, E.J., Walunas, T.L., Feil, H., Feil, W.S., Purcell, A., Lassez, J.-L., Hawkins, T.L., Haselkorn, R., Overbeek, R., Predki, P.F. and Kyrpides, N.C. (2002a) Whole-genome comparative analysis of three phytopathogenic *Xylella fastidiosa* strains. *Proceedings of the National Academy of Sciences of the United States of America* 99, 12403–12408.

Bhattacharyya, A., Stilwagen, S., Reznik, G., Feil, H., Feil, W.S., Anderson, I., Bernal, A., D'Souza, M., Ivanova, N., Kapatral, V., Larsen, N., Los, T., Lykidis, A., Selkov, E., Walunas, T.L., Purcell, A., Edwards, R.A., Hawkins, T., Haselkorn, R., Overbeek, R., Kyrpides, N.C. and Predki, P.F. (2002b) Draft sequencing and comparative genomics of *Xylella fastidiosa* strains reveal novel biological insights. *Genome Research* 12, 1556–1563.

Birch, R.G. and Patil, S.S. (1985) Preliminary characterization of an antibiotic produced by *Xanthomonas albilineans* which inhibits DNA synthesis in *Escherichia coli*. *Journal of General Microbiology* 131, 1069–1075.

Birch, R.G. and Patil, S.S. (1987) Evidence that an albicidin-like phytotoxin induces chlorosis in sugarcane leaf scald disease by blocking plastid DNA replication. *Physiological and Molecular Plant Pathology* 30, 207–214.

Boccara, M., Mills, C.E., Zeier, J., Anzi, C., Lamb, C., Poole, R.K. and Delledonne, M. (2005) Flavohaemoglobin HmpX from *Erwinia chrysanthemi* confers nitrosative stress tolerance and affects the plant hypersensitive reaction by intercepting nitric oxide produced by the host. *The Plant Journal* 43, 226–237.

Bonde, R. (1937) A bacterial wilt and soft rot of the potato in Maine. *Phytopathology* 27, 106–108.

Buell, C.R., Joardar, V., Lindeberg, M., Selengut, J., Paulsen, I.T., Gwinn, M.L., Dodson, R.J., Deboy, R.T., Durkin, A.S., Kolonay, J.F., Madupu, R., Daugherty, S., Brinkac, L., Beanan, M.J., Haft, D.H., Nelson, W.C., Davidsen, T., Zafar, N., Zhou, L.W., Liu, J., Yuan, Q.P., Khouri, H., Fedorova, N., Tran, B., Russell, D., Berry, K., Utterback, T., Van Aken, S.E., Feldblyum, T.V., D'Ascenzo, M., Deng, W.L., Ramos, A.R., Alfano, J.R., Cartinhour, S., Chatterjee, A.K., Delaney, T.P., Lazarowitz, S.G., Martin, G.B., Schneider, D.J., Tang, X.Y., Bender, C.L., White, O.,

Fraser, C.M. and Collmer, A. (2003) The complete genome sequence of the *Arabidopsis* and tomato pathogen *Pseudomonas syringae* pv. *tomato* DC3000. *Proceedings of the National Academy of Sciences of the United States of America* 100, 10181–10186.

Camilli, A. and Bassler, B.L. (2006) Bacterial small-molecule signaling pathways. *Science* 311, 1113–1116.

Clardy, J., Fischbach, M.A. and Walsh, C.T. (2006) New antibiotics from bacterial natural products. *Nature Biotechnology* 24, 1541–1550.

Claydon, M.A., Davey, S.N., Edwards-Jones, V. and Gordon, D.B. (1996) The rapid identification of intact microorganisms using mass spectrometry. *Nature Biotechnology* 14, 1584–1586.

Clements, J.M., Beckett, R.P., Brown, A., Catlin, G., Lobell, M., Palan, S., Thomas, W., Whittaker, M., Wood, S., Salama, S., Baker, P.J., Rodgers, H.F., Barynin, V., Rice, D.W. and Hunter, M.G. (2001) Antibiotic activity and characterization of BB-3497, a novel peptide deformylase inhibitor. *Antimicrobial Agents and Chemotherapy* 45, 563–570.

Cooksey, D.A. (1990) Genetics of bactericide resistance in plant pathogenic bacteria. *Annual Review of Phytopathology* 28, 201–214.

da Silva, A.C.R., Ferro, J.A., Reinach, F.C., Farah, C.S., Furlan, L.R., Quaggio, R.B., Monteiro-Vitorello, C.B., Van Sluys, M.A., Almeida, N.F., Alves, L.M.C., do Amaral, A.M., Bertolini, M.C., Camargo, L.E.A., Camarotte, G., Cannavan, F., Cardozo, J., Chambergo, F., Ciapina, L.P., Cicarelli, R.M.B., Coutinho, L.L., Cursino-Santos, J.R., El-Dorry, H., Faria, J.B., Ferreira, A.J.S., Ferreira, R.C.C., Ferro, M.I.T., Formighieri, E.F., Franco, M.C., Greggio, C.C., Gruber, A., Katsuyama, A.M., Kishi, L.T., Leite, R.P., Lemos, E.G.M., Lemos, M.V.F., Locali, E.C., Machado, M.A., Madeira, A.M.B.N., Martinez-Rossi, N.M., Martins, E.C., Meidanis, J., Menck, C.F.M., Miyaki, C.Y., Moon, D.H., Moreira, L.M., Novo, M.T.M., Okura, V.K., Oliveira, M.C., Oliveira, V.R., Pereira, H.A., Rossi, A., Sena, J.A.D., Silva, C., de Souza, R.F., Spinola, L.A.F., Takita, M.A., Tamura, R.E., Teixeira, E.C., Tezza, R.I.D., Trindade dos Santos, M., Truffi, D., Tsai, S.M., White, F.F., Setubal, J.C. and Kitajima, J.P. (2002) Comparison of the genomes of two *Xanthomonas* pathogens with differing host specificities. *Nature* 417, 459–463.

D'Costa, V.M., McGrann, K.M., Hughes, D.W. and Wright, G.D. (2006) Sampling the antibiotic resistome. *Science* 311, 374–377.

Daughtrey, M.L. and Benson, D.M. (2005) Principles of plant health management for ornamental plants. *Annual Review of Phytopathology* 43, 141–169.

Dong, Y.H., Wang, L.H., Xu, J.L., Zhang, H.B., Zhang, X.F. and Zhang, L.H. (2001) Quenching quorum-sensing-dependent bacterial infection by an N-acyl homoserine lactonase. *Nature* 411, 813–817.

Dreier, J., Meletzus, D. and Eichenlaub, R. (1997) Characterization of the plasmid encoded virulence region *pat-1* of phytopathogenic *Clavibacter michiganensis* subsp. *michiganensis*. *Molecular Plant-Microbe Interactions* 10, 195–206.

El Hassouni, M., Chambost, J.P., Expert, D., Van Gijsegem, F. and Barras, F. (1999) The minimal gene set member *msrA*, encoding peptide methionine sulfoxide reductase, is a virulence determinant of the plant pathogen *Erwinia chrysanthemi*. *Proceedings of the National Academy of Sciences of the United States of America* 96, 887–892.

Farrand, S.K. and Wang, C. (1992) Do we really understand crown gall control by *Agrobacterium radiobacter* strain K84? In: Tjamos, E.C., Papavizas, G.C. and Cook, R.J. (eds) *Biological Control of Plant Diseases*. Plenum Press, New York, pp. 287–293.

Feil, H., Feil, W.S., Chain, P., Larimer, F., DiBartolo, G., Copeland, A., Lykidis, A., Trong, S., Nolan, M., Goltsman, E., Thiel, J., Malfatti, S., Loper, J.E., Lapidus, A., Detter, J.C., Land, M., Richardson, P.M., Kyrpides, N.C., Ivanova, N. and Lindow, S.E. (2005) Comparison of the complete genome sequences of *Pseudomonas syringae* pv. *syringae* B728a and pv. *tomato* DC3000. *Proceedings of the National Academy of Sciences of the United States of America* 102, 11064–11069.

Fessehaie, A., De Boer, S.H. and Levesque, C.A. (2003) An oligonucleotide array for the identification and differentiation of bacteria pathogenic on potato. *Phytopathology* 93, 262–269.

Flego, D., Pihoren, M., Saarilahti, H., Palva, T.K. and Palva, E.T. (1997) Control of virulence gene expression by plant calcium in the phytopathogen *Erwinia carotovora*. *Molecular Microbiology* 25, 831–838.

Gabriel, D.W., Allen, C., Schell, M., Denny, T.P., Greenberg, J.T., Duan, Y.P., Flores-Cruz, Z., Huang, Q., Clifford, J.M., Presting, G., Gonzalez, E.T., Reddy, J., Elphinstone, J., Swanson, J., Yao, J., Mulholland, V., Liu, L., Farmerie, W., Patnaikuni, M., Balogh, B., Norman, D., Alvarez, A., Castillo, J.A., Jones, J., Saddler, G., Walunas, T., Zhukov, A. and Mikhailova, N. (2006) Identification of open reading frames unique to a select agent: *Ralstonia solanacearum* race 3 biovar 2. *Molecular Plant-Microbe Interactions* 19, 69–79.

Gilbert, J.A., Davies, P.L. and Laybourn-Parry, J. (2005) A hyperactive, Ca^{2+}-dependent antifreeze protein in an Antarctic bacterium. *FEMS Microbiology Letters* 245, 67–72.

Goodner, B., Hinkle, G., Gattung, S., Miller, N., Blanchard, M., Qurollo, B., Goldman, B.S., Cao, Y.W., Askenazi, M., Halling, C., Mullin, L., Houmiel, K., Gordon, J., Vaudin, M., Iartchouk, O., Epp, A., Liu, F., Wollam, C., Allinger, M., Doughty, D., Scott, C., Lappas, C., Markelz, B., Flanagan, C., Crowell, C., Gurson, J., Lomo, C., Sear, C., Strub, G., Cielo, C. and Slater, S. (2001) Genome sequence of the plant pathogen and biotechnology agent *Agrobacterium tumefaciens* C58. *Science* 294, 2323–2328.

Grant, S.R., Fisher, E.J., Chang, J.H., Mole, B.M. and Dangl, J.L. (2006) Subterfuge and manipulation: Type III effector proteins of phytopathogenic bacteria. *Annual Review of Microbiology* 60, 425–449.

He, X., Reeve, A.M., Desai, U., Kellogg, G.E. and Reynolds, K.A. (2004) 1,2-dithiole-2-ones as potent inhibitors of the bacterial 3-ketoacyl acil carrier protein synthase II (FabH). *Antimicrobial Agents and Chemotherapy* 48, 3093–3102.

Hirano, S.S. and Upper, C.D. (2000) Bacteria in the leaf ecosystem with emphasis on *Pseudomonas syringae* – a pathogen, ice nucleus, and epiphyte. *Microbiology and Molecular Biology Review* 64, 624–653.

Jacobs, J.L., Carroll, T.L. and Sundin, G.W. (2005) The role of pigmentation, ultraviolet radiation tolerance and leaf colonization strategies in the epiphytic survival of phyllosphere bacteria. *Microbiology Ecology* 49, 104–113.

Jarvest, R.L., Berge, J.M., Beryy, V., Boyd, H.F., Brown, M.J., Elder, J.S., Forrest, A.K., Fosberry, A.P., Gentry, D.R., Hibbs, M.J., Jaworski, D.D., O'Hanlon, P.J., Pope, A.J., Rittenhouse, S., Sheppard, R.J., Slater-Radosti, C. and Worby, A. (2002) Nanomolar inhibitors of *Staphlococcus aureus* methionyl tRNA synthetase with potent antibacterial activity against gram-positive pathogens. *Journal of Medicinal Chemistry* 45, 1959–1962.

Joardar, V., Lindeberg, M., Jackson, R.W., Selengut, J., Dodson, R., Brinkac, L.M., Daugherty, S.C., DeBoy, R., Durkin, A.S., Giglio, M.G., Madupu, R., Nelson, W.C., Rosovitz, M.J., Sullivan, S., Crabtree, J., Creasy, T., Davidsen, T., Haft, D.H., Zafar, N., Zhou, L.W., Halpin, R., Holley, T., Khouri, H., Feldblyum, T., White, O., Fraser, C.M., Chatterjee, A.K., Cartinhour, S., Schneider, D.J., Mansfield, J., Collmer, A. and Buell, C.R. (2005) Whole-genome sequence analysis of *Pseudomonas syringae* pv. *phaseolicola* 1448A reveals divergence among pathovars in genes involved in virulence and transposition. *Journal of Bacteriology* 187, 6488–6498.

Kadouri, D. and O'Toole, G.A. (2005) Susceptibility of biofilms to *Bdellovibrio bacteriovorus* attack. *Applied and Environmental Microbiology* 71, 4044–4051.

Kauppi, R., Nordfelth, H., Uvell, H., Wolf-Watz, H. and Elofsson, M. (2003) Targeting bacterial virulence. Inhibitors of type III secretion in *Yersinia*. *Chemistry and Biology* 10, 241–249.

Klessig, D.F., Durner, J., Noad, R., Navarre, D.A., Wendehenne, D., Kumar, D., Zhou, J.M., Shah, J., Zhang, S., Kachroo, P., Trifa, Y., Pontier, D., Lam, E. and Silva, H. (2000) Nitric oxide and salicylic acid signaling in plant defense. *Proceedings of the National Academy of Sciences of the United States of America* 97, 8849–8855.

Koutsoudis, M.D., Tsaltas, D., Minogue, T.D. and Von Bodman, S.B. (2006) Quorum-sensing regulation governs bacterial adhesion, biofilm development and host colonization in *Pantoea stewartii* subspecies *stewartii*. *Proceedings of the National Academy of Sciences of the United States of America* 103, 5983–5988.

Leach, J.E. and White, F.F. (1996) Bacterial avirulence genes. *Annual Review of Phytopathology* 34, 153–179.

Lee, B.M., Park, Y.J., Park, D.S., Kang, H.W., Kim, J.G., Song, E.S., Park, I.C., Yoon, U.H., Hahn, J.H., Koo, B.S., Lee, G.B., Kim, H., Park, H.S., Yoon, K.O., Kim, J.H., Jung, C., Koh, N.H., Seo, J.S. and Go, S.J. (2005) The genome sequence of *Xanthomonas oryzae* pathovar *oryzae* KACC10331, the bacterial blight pathogen of rice. *Nucleic Acids Research* 33, 577–586.

Llama-Palacios, A., Lopez-Solanilla, E., Poza-Carrion, C., Garcia-Olmedo, F. and Rodriguez-Palenzuela, P. (2003) The *Erwinia chrysanthemi* phoP-phoQ operon plays an important role in growth at low pH, virulence and bacterial survival in plant tissue. *Molecular Microbiology* 49, 347–357.

Loria, R., Kers, J. and Joshi, M. (2006) Evolution of plant pathogenicity in *Streptomyces*. *Annual Review of Phytopathology* 44, 469–487.

Mäe, A., Montesano, M., Kõiv, V. and Palva, E.T. (2001) Transgenic plants producing the bacterial pheromone N-acyl-homoserine lactone exhibit enhanced resistance to the bacterial phytopathogen *Erwinia carotovora*. *Molecular Plant-Microbe Interactions* 14, 1035–1042.

Maggiorani Valecillos, A., Rodriguez Palenzuela, P. and Lopez-Solanilla, E. (2006) The role of several multidrug resistance systems in *Erwinia chrysanthemi* pathogenesis. *Molecular Plant-Microbe Interactions* 19, 607–613.

Margulies, M., Egholm, M., Altman, W.E., Attiya, S., Bader, J.S., Bemben, L.A., Berka, J., Braverman, M.S., Chen, Y.J., Chen, Z.T., Dewell, S.B., Du, L., Fierro, J.M., Gomes, X.V., Godwin, B.C., He, W., Helgesen, S., Ho, C.H., Irzyk, G.P., Jando, S.C., Alenquer, M.L.I., Jarvie, T.P., Jirage, K.B., Kim, J.B., Knight, J.R., Lanza, J.R., Leamon, J.H., Lefkowitz, S.M., Lei, M., Li, J., Lohman, K.L., Lu, H., Makhijani, V.B., McDade, K.E., McKenna, M.P., Myers, E.W., Nickerson, E., Nobile, J.R., Plant, R., Puc, B.P., Ronan, M.T., Roth, G.T., Sarkis, G.J., Simons, J.F., Simpson, J.W., Srinivasan, M., Tartaro, K.R., Tomasz, A., Vogt, K.A., Volkmer, G.A., Wang, S.H., Wang, Y., Weiner, M.P., Yu, P.G., Begley, R.F. and Rothberg, J.M. (2005) Genome sequencing in microfabricated high-density picolitre reactors. *Nature* 437, 376–380.

Masalha, J., Kosegarten, H., Elmaci, O. and Mengel, K. (2000) The central role of microbial activity for iron acquisition in maize and sunflower. *Biology and Fertility of Soils* 30, 433–439.

Matthysse, A.G. (1983) Role of bacterial cellulose fibrils in *Agrobacterium tumefaciens* infection. *Journal of Bacteriology* 154, 906–915.

Mazzeo, M.F., Sorrentino, A., Gaita, M., Cacace, G., Di Stasio, M., Facchiano, A., Comi, G., Malorni, A. and Siciliano, R.A. (2006) Matrix-assisted laser desorption ionization-time of flight mass spectrometry for the discrimination of food-borne microorganisms. *Applied and Environmental Microbiology* 72, 1180–1189.

McManus, P.S., Stockwell, V.O., Sundin, G.W. and Jones, A.L. (2002) Antibiotic use in plant agriculture. *Annual Review of Phytopathology* 40, 443–465.

Mills, D., Russell, B.W. and Hanus, J.W. (1997) Specific detection of *Clavibacter michiganensis* subsp. *sepedonicus* by amplification of three unique DNA sequences isolated by subtraction hybridization. *Phytopathology* 87, 853–861.

Mills, S.D. (2003) The role of genomics in antimicrobial discovery. *Journal of Antimicrobial Chemotherapy* 51, 749–752.

Monteiro-Vitorello, C.B., Camargo, L.E.A., Van Sluys, M.A., Kitajima, J.P., Truffi, D., do Amaral, A.M., Harakava, R., de Oliveira, J.C.F., Wood, D., de Oliveira, M.C., Miyaki, C., Takita, M.A., da Silva, A.C.R., Furlan, L.R., Carraro, D.M., Camarotte, G., Almeida, J.N.F., Carrer, H., Coutinho, L.L., El-Dorry, H.A., Ferro, M.I.T., Gagliardi, P.R., Giglioti, E., Goldman, M.H.S., Goldman, G.H., Kimura, E.T., Ferro, E.S., Kuramae, E.E., Lemos, E.G.M., Lemos, M.V.F.,

Mauro, S.M.Z., Machado, M.A., Marino, C.L., Menck, C.F., Nunes, L.R., Oliveira, R.C., Pereira, G.G., Siqueira, W., de Souza, A.A., Tsai, S.M., Zanca, A.S., Simpson, A.J.G., Brumbley, S.M. and Setúbal, J.C. (2004) The genome sequence of the gram-positive sugarcane pathogen *Leifsonia xyli* subsp. *xyli*. *Molecular Plant-Microbe Interactions* 17, 827–836.

Muschiol, S., Bailey, L., Gylfe, A., Sundin, C., Hultenby, K., Bergstrom, S., Elofsson, M., Wolf-Watz, H., Normark, S. and Henriques-Normark, B. (2006) A small-molecule inhibitor of type III secretion inhibits different stages of the infectious cycle of *Chlamydia trachomatis*. *Proceedings of the National Academy of Sciences of the United States of America* 103, 14566–14571.

Nordfelth, R., Kauppi, A.M., Norberg, H.A., Wolf-Watz, H. and Elofsson, M. (2005) Small-molecule inhibitors specifically targeting type III secretion. *Infection and Immunity* 73, 3104–3114.

Obradovic, A., Jones, J.B., Momol, M.T., Olson, S.M., Jackson, L.E., Balogh, B., Guven, K. and Iriarte, F.B. (2005) Integration of biological control agents and systemic acquired resistance inducers against bacterial spot on tomato. *Plant Disease* 89, 712–716.

Ohshima, M., Mitsuhara, I., Okamoto, M., Sawano, S., Nishiyama, K., Kaku, H., Natori, S. and Ohashi, Y. (1999) Enhanced resistance to bacterial diseases of transgenic tobacco plants overexpressing sarcotoxin IA, a bactericidal peptide of insect. *Journal of Biochemistry* 125, 431–435.

Pages, J.M., Masi, M. and Barbe, J. (2005) Inhibitors of efflux pumps in Gram-negative bacteria. *Trends in Molecular Medicine* 11, 382–389.

Phillips, R.M., Six, D.A., Dennis, E.A. and Ghosh, P. (2003) *In vivo* phospholipase activity of the *Pseudomonas aeruginosa* cytotoxin ExoU and protection of mammalian cells with phospholipase A_2 inhibitors. *Journal of Biological Chemistry* 278, 41326–41332.

Preston, G., Deng, W.L., Huang, H.C. and Collmer, A. (1998) Negative regulation of *hrp* genes in *Pseudomonas syringae* by HrpV. *Journal of Bacteriology* 180, 4532–4537.

Preston, G.M., Bertrand, N. and Rainey, P.B. (2001) Type III secretion in plant growth-promoting *Pseudomonas fluorescens* SBW25. *Molecular Microbiology* 41, 999–1014.

Ramasubbu, N., Thomas, L.M., Raqunath, C. and Kaplan, J.B. (2005) Structural analysis of dispersin B, a biofilm-releasing glycoside hydrolase from the periodontopathogen *Actinobacillus actinomycetemcomitans*. *Journal of Molecular Biology* 349, 475–486.

Ray, S.K., Rajeshwari, R. and Sonti, R.V. (2000) Mutants of *Xanthomonas oryzae* deficient in general secretory pathway are virulent deficient and unable to secrete xylase. *Molecular Plant-Microbe Interactions* 13, 394–401.

Reverchon, S., Rouanet, C., Expert, D. and Nasser, W. (2002) Characterization of indigoidine biosynthetic genes in *Erwinia chrysanthemi* and role of this blue pigment in pathogenicity. *Journal of Bacteriology* 184, 654–665.

Rider, T.H., Petrovick, M.S., Nargi, F.E., Harper, J.D., Schwoebel, E.D., Mathews, R.H., Blanchard, D.J., Bortolin, L.T., Young, A.M., Chen, J. and Hollis, M.A. (2003) A B cell-based sensor for rapid identification of pathogens. *Science* 301, 213–215.

Roine, E., Raineri, D.M., Romantschuk, M., Wilson, M. and Nunn, D.N. (1998) Characterization of type IV pilus genes in *Pseudomonas syringae* pv. *tomato* DC3000. *Molecular Plant-Microbe Interactions* 11, 1048–1056.

Rojas, C.M., Ham, J.-H., Deng, W.-L., Doyle, J.J. and Collmer, A. (2002) HecA is a member of a class of adhesins produced by diverse pathogenic bacteria and contributes to the attachment, aggregation, epidermal cell killing, and virulence phenotypes of *Erwinia chrysanthemi* EC16 on *Nicotiana clevelandii* seedlings. *Proceedings of the National Academy of Sciences of the United States of America* 99, 13142–13147.

Ryan, R.P., Fouhy, Y., Lucey, J.F. and Dow, J.M. (2006) Cyclic di-GMP signaling in bacteria: recent advances and new puzzles. *Journal of Bacteriology* 188, 8327–8334.

Salanoubat, M., Genin, S., Artiguenave, F., Gouzy, J., Mangenot, S., Arlat, M., Billault, A., Brottier, P., Camus, J.C., Cattolico, L., Chandler, M., Choisne, N., Claudel-Renard, C., Cunnac, S.,

Demange, N., Gaspin, C., Lavie, M., Moisan, A., Robert, C., Saurin, W., Schiex, T., Siguier, P., Thebault, P., Whalen, M., Winker, P., Levy, M., Wiessenback, J. and Boucher, C.A. (2002) Genome sequence of the plant pathogen *Ralstonia solanacearum*. *Nature* 415, 497–502.

Saperstein, S. and Starr, M.P. (1954) The ketonic carotenoid canthaxanthin isolated from a colour mutant of *Corynebacterium michiganense*. *Biochemistry Journal* 57, 273–275.

Saperstein, S., Starr, M.P. and Filfus, J.A. (1954) Alterations in carotenoid synthesis accompanying a mutation in *Corynebacterium michiganense*. *Journal of Genereal Microbiology* 10, 85–92.

Scherff, R.H. (1973) Control of bacterial blight of soybean by *Bdellovibrio bacteriovorus*. *Phytopathology* 63, 400–402.

Sepulchre, J.-A., Reverchon, S. and Nasser, W. (2006) Modeling the onset of virulence in a pectinolytic bacterium. *Journal of Theoretical Biology* 244, 239–257.

Shemesh, Y. and Jurkevitch, E. (2004) Plastic phenotypic resistance to predation by *Bdellovibrio* and like organisms in bacterial prey. *Environmental Microbiology* 6, 12–18.

Simpson, A.J.G., Reinach, F.C., Arruda, P., Abreu, F.A. plus 111 others. The genome sequence of the plant pathogen *Xylella Fastidiosa*. *Nature* 406, 151–157.

Smadja, B., Latour, X., Fauer, D., Chevalier, S., Dessaux, Y. and Orange, N. (2004) Involvement of N-acylhomoserine lactones throughout plant infection by *Erwinia carotovora* subsp. *atroseptica* (*Pectobacterium atrosepticum*). *Molecular Plant-Microbe Interactions* 17, 1269–1278.

Starr, M.P. (1958) The blue pigment of *Corynebacterium insidiosum*. *Archives of Microbiology* 30, 325–334.

Stermitz, F.R., Lorenz, P., Tawara, J.N., Zenewicz, L.A. and Lewis, K. (2000) Synergy in a medicinal plant: antimicrobial action of berberine potentiated by 5'-methoxyhydnocarpin, a multidrug pump inhibitor. *Proceedings of the National Academy of Sciences of the United States of America* 97, 1433–1437.

Stockwell, V.O. and Stack, J.P. (2007) Using *Pseudomonas* spp. for integrated biological control. *Phytopathology* 97, 244–249.

Takano, E. (2006) γ-Butyrolactones: *Streptomyces* signalling molecules regulating antibiotic production and differentiation. *Current Opinion in Microbiology* 9, 287–294.

Thieme, F., Koebnik, R., Bekel, T., Berger, C., Boch, J., Buttner, D., Caldana, C., Gaigalat, L., Goesmann, A., Kay, S., Kirchner, O., Lanz, C., Linke, B., McHardy, A.C., Meyer, F., Mittenhuber, G., Nies, D.H., Niesback-Klosgen, U., Patschkowski, T., Ruckert, C., Rupp, O., Schneiker, S., Schuster, S.C., Vorholter, F.J., Weber, E., Puhler, A., Bonas, U., Bartels, D. and Kaiser, O. (2005) Insights into genome plasticity and pathogenicity of the plant pathogenic bacterium *Xanthomonas campestris* pv. *vesicatoria* revealed by the complete genome sequence. *Journal of Bacteriology* 187, 7254–7266.

Tu, J., Datta, K., Khush, G.S., Zhang, Q. and Datta, S.K. (2000) Field performance of Xa21 transgenic indica rice (*Oryza sativa* L.), IR72. *Theoretical and Applied Genetics* 101, 15–20.

Uematsu, T. (1980) Ecology of *Bdellovibrio* parasitic to rice bacterial leaf blight pathogen, *Xanthomonas oryzae*. *Review of Plant Protection Research* 13, 12–26.

Van Doorn, J., Boonekamp, P.M. and Oudega, B. (1994) Partial characterization of fimbriae of *Xanthomonas campestris* pv. *hyacinthi*. *Molecular Plant-Microbe Interactions* 7, 334–344.

Visick, K.L. and Fuqua, C. (2005) Decoding microbial chatter: cell–cell communication in bacteria. *Journal of Bacteriology* 187, 5507–5519.

Wang, D., Urisman, A., Liu, Y.T., Springer, M., Ksiazek, T.G., Erdman, D.D., Mardis, E.R., Hickenbotham, M., Magrini, V., Eldred, J., Latreille, J.P., Wilson, R.K., Ganem, D. and DeRisi, J.L. (2003) Viral discovery and sequence recovery using DNA microarrays. *PLOS Biology* 1, 257–260.

Wang, Y., Brown, H.N., Crowly, D.A. and Szaniszlo, P.J. (1993) Evidence for direct utilization of a siderophore, ferrioxamine B, in axenically grown cucumber. *Plant Cell and Environment* 16, 579–585.

Wood, D.W., Setubal, J.C., Kaul, R., Monks, D.E., Kitajima, J.P., Okura, V.K., Zhou, Y., Chen, L., Wood, G.E., Almeida, N.F. Jr., Woo, L., Chen, Y., Paulsen, I.T., Eisen, J.A., Karp, P.D., Bovee, D. Sr., Chapman, P., Clendenning, J., Deatherage, G., Gillet, W., Grant, C., Kutyavin, T., Levy, R., Li, M.J., McClelland, E., Palmieri, A., Raymond, C., Rouse, G., Saenphimmachak, C., Wu, Z., Romero, P., Gordon, D., Zhang, S., Yoo, H., Tao, Y., Biddle, P., Jung, M., Krespan, W., Perry, M., Gordon-Kamm, B., Liao, L., Kim, S., Hendrick, C., Zhao, Z.Y., Dolan, M., Chumley, F., Tingey, S.V., Tomb, J.F., Gordon, M.P., Olson, M.V. and Nester, E.W. (2001) The genome of the natural genetic engineer *Agrobacterium tumefaciens* C58. *Science* 294, 2317–2323.

Wu, G., Shortt, B.J., Lawrence, E.B., Levine, E.B., Fitzsimmons, K.C. and Shah, D.M. (1995) Disease resistance conferred by expression of a gene encoding H_2O_2- generating glucose oxidase in transgenic potato plants. *The Plant Cell* 7, 1357–1368.

Yap, M.-N., Yang, C.H., Barak, J.D., Jahn, C.E. and Charkowski, A.O. (2005) The *Erwinia chrysanthemi* type III secretion system is required for multicellular behavior. *Journal of Bacteriology* 187, 639–648.

21 Molecular Assessment of Soil Microbial Communities with Potential for Plant Disease Suppression

J.D. van Elsas and R. Costa

Abstract
It is known that soil microbial communities, in part, determine the level of disease suppressiveness of soil. However, because of the very high complexity of soil microbial communities, it has been extremely difficult to understand which organisms, or combinations of organisms, cause disease suppression and which mechanisms might be involved in such suppression. In this chapter, currently available modern and advanced molecular techniques that can be used to unravel the structure and function of soil microbial communities are briefly described, and then evaluated for their potential application in research to elucidate and understand the suppressive status of soil. Although these techniques offer clear and indisputable advances over the more traditional methods, there are still caveats and pitfalls that need to be taken into account when interpreting the data. These caveats relate to the level of representativeness of the nucleic acids obtained from soil and to the possible biases of the subsequent amplification, fingerprinting, cloning and sequencing methods. The advanced molecular techniques are thus set in a critical perspective and avenues are explored that may lead to their improvement.

Introduction

Disease suppression

Plant diseases often, although not always, originate from the soil. However, even in the presence of a phytopathogen, soils can be more or less conducive to disease development, leading to the concept of conducive versus suppressive soils. Disease-suppressive soils can be defined as soils where the incidence and severity of certain plant diseases is low or reduced, in spite of the presence of the pathogen(s), susceptible host(s) and climatic conditions that favor disease development (Haas and Défago, 2005; Steinberg *et al.*, 2006). On the other hand, disease-conducive soils are those in which a given plant disease prospers and persists. From a functional viewpoint, these soils are considered to be the opposite of suppressive soils. Plant diseases induced by bacterial and fungal pathogens, as well as by nematodes, have been shown to be regulated by the level of suppressiveness of soils (Weller et al., 2002). Commonly, soil suppressiveness is specific to a given type of disease (Haas and Défago, 2005), but the occurrence of 'general' suppressive soils has also been reported, giving rise to proposals for different classification systems for suppressive soils (Baker and Cook, 1974; Cook and Baker, 1983; Steinberg et al., 2006).

The simple observation that soils lose their disease suppressiveness when subjected to autoclaving or gamma irradiation clearly indicates that the capacity of soils to antagonize the action of phytopathogens is encrypted in the soil microbiota, and may bear a relationship to the structure and function of the soil bacterial and fungal communities (Steinberg et al., 2006). We will address the biological basis for disease suppressiveness in this chapter. Recently, molecular studies of biocontrol *Pseudomonas* spp. have resulted in major breakthroughs in the search for traits, genes and modes of interaction that are involved, in specific cases, in soil suppressiveness and biological control of plant diseases (Keel et al., 1992; Walsh et al., 2001; Haas and Défago, 2005; Mark et al., 2005) (Fig. 21.1). Nevertheless, we still do not have a clear understanding of the types of organisms or genes that are involved in the suppression of most plant diseases, nor of the soil conditions (i.e. mineral composition and physicochemical parameters) that affect disease suppression. Given the fact that the microbiota of most soils is incredibly complex (Torsvik et al., 1990; Gans et al., 2005), it is of primary importance that we develop and apply tools that allow us to unravel the soil microbial communities and at the same time attempt to pinpoint those organisms that are responsible for disease suppression.

Microbial diversity and community structure

The study of soil microbiology and of the role of the soil microbiota in soil functioning is now over 100 years old, and a great wealth of information has been obtained during this time. Using cultivation-based approaches, such as direct dilution plating and enrichment culture, numerous researchers have unraveled the enormous metabolic and phylogenetic diversity that is present in virtually any soil (Van Elsas et al., 2006). However, investigators also discovered that cultivation-based approaches are often prone to so-called cultivation bias, i.e. an apparent distortion of the abundance of specific groups of microorganisms as a result of the cultivation step. Hence, there has been a consistent interest in approaches that would reduce this bias, i.e. that would provide a picture of soil microbial communities closer to the *in situ* actuality. In this context, it has been repeatedly found – for a range of soils as well as of aquatic habitats – that often less than 1% of the extant microbial community is recovered by cultivation on agar plates or

Fig. 21.1. Suppression of the fungal phytopathogen *Rhizoctonia solani* AG1-LB in lettuce provided by the biocontrol agent *Pseudomonas jensenii* strain RU47 in pot experiments. Disease severity caused by *R. solani* (a, treatment containing only the pathogen) is significantly reduced by previous seed inoculation with the biocontrol strain (b, treatment containing both pathogen and biocontrol agent) prior to sowing. Yields observed in (b) are comparable to that of the control (c, non-infested lettuce plants). Photos were taken 7 weeks after sowing. (Courtesy of Modupe Adesina and Kornelia Smalla with permission.)

in liquid culture. This phenomenon has been denoted 'The Great Plate Count Anomaly' by Staley and Konopka (1985). Moreover, to pave the way towards a more comprehensive understanding of key triggers of soil suppressiveness, a major shift of thinking on soil microbiology was necessary, namely, the realization that the rhizosphere (the narrow zone of soil affected by plant roots) (Hiltner, 1904; Lynch, 1984), rather than the bulk soil, is the key microhabitat (a hot spot of microbial abundance and activity) to look at when organisms, metabolites and genes involved in microbe–microbe and plant–microbe interactions are the objects of interest (Haas and Défago, 2005). This is in particular true since a major input of bound carbon enters the soil via the rhizosphere (Sørensen, 1997; Pinton et al., 2001). Recent rhizosphere research – reviewed in a series of papers in *FEMS Microbiology Ecology* (vol. 56, 2006) – has allowed novel insights into the phenomenon of soil suppressiveness which will be addressed in this chapter.

Given the aforementioned Great Plate Count Anomaly, there has been a continuous search for approaches that would overcome the non-culturability of many soil microorganisms and thus unravel the nature of this major part of the microbial community. The idea behind such approaches has been to assess soil microbial communities on the basis of typical cellular constituents (macromolecules) that serve as biomarkers. Two such approaches, either based on nucleic acids or on phospholipid fatty acids (PLFA), have been developed and are currently being applied on a wide scale. In this chapter, we will only address the soil nucleic acid-based methods, as these truly offer the highest resolution. Directly extracted soil DNA as well as RNA lie at the basis of a large suite of detection/analysis methods that have developed into routine methods for many laboratories involved in: (i) fundamental research on soil quality/health; (ii) assessment of soil remediation; (iii) assessment of soil pathogens; and (iv) analysis of soil microbial diversity issues.

In the following sections, we will examine the current status of the soil nucleic acid-based methodologies. We will extrapolate this knowledge to a critical assessment of attempts to link the structure of microbial communities and/or the prevalence of specific genes in soil to disease suppression.

Extraction of Soil Microbial DNA and/or RNA: the Basis for Soil 'Total Community' Analyses

Historical development

Over 25 years have passed since the inception of nucleic acid-based direct approaches to the analysis of soil microbial populations and communities (Torsvik, 1980). However, only following the publication of two key papers in the late 1980s (Ogram et al., 1987; Holben et al., 1988) and with the advent of PCR did nucleic acid-based methods (at that time, DNA) receive due appreciation in soil microbiology. Since the early times, a very broad range of techniques with ever-increasing levels of sophistication have been put forward and employed to unravel the secrets of microbial life underfoot. In particular, the last decade has seen the implementation of a multitude of different techniques that have in common that they all describe microbial communities on the basis of DNA or RNA and in terms of their molecular attributes (Table 21.1).

All currently used 'total community' (TC) nucleic acid extraction methods can be divided into two types of approaches: (i) direct extraction protocols; and (ii) indirect (cell extraction first) extraction protocols. The two approaches find their origin in, respectively, the studies by Holben *et al.* (1988) and Ogram *et al.* (1987), as indicated above. Numerous extraction protocols have been published in the 1990s after these original publications appeared. The sheer number of protocols either mirrors the number of different requirements posed by different soil types or the eagerness of researchers to publish their pet protocols, or both. Recently many authors have based their protocols on that of Zhou *et al.* (1996), even though a protocol of the early 1990s (Smalla *et al.*, 1993) has long been preferred by many workers. At the same time, the development of protocols for the extraction of RNA from soils took place in diverse laboratories (Tsai *et al.*, 1991; Moran

Table 21.1. Molecular methods used to assess soil microbial communities.

Method	Purpose of method	Remarks	References
Clone libraries and sequencing	Inventories of copies of a given gene sampled in the environment	Fine-tuned, in-depth assessment of the diversity of a target gene, often the 16S rRNA gene	Van Elsas *et al.* (2002)
DGGE, T-RFLP, SSCP, (A)RISA, LH-PCR	Molecular fingerprinting of microbial communities	Suitable for comparative analysis of microbial community structure and diversity	Oros-Sichler *et al.* (2006)
Real-time (q)-PCR	Quantification of abundance of target genes in soil	Allows estimation of gene copy numbers in a sample Fundamental for (m-RNA) analysis of gene expression	Garbeva *et al.* (2004c)
Microarrays	Multiplex detection of genes or gene expression	High-throughput hybridization technology. Enormous potential for comparative studies	Wu *et al.* (2001); Loy *et al.* (2002)
FISH and derived methods	Visualization of organisms at cell level on the basis of signal from bound probe	Studies of spatial distribution of organisms in a system.	Schmid *et al.* (2006)

et al., 1993; Felske *et al.*, 1996; Duarte *et al.*, 1998). Van Elsas *et al.* (1999) reviewed the state-of-the-art soil nucleic acid extraction and indicated a trend towards polyphasic approaches, i.e. focusing on DNA as well as RNA at the same time.

As bacterial species often vary their ribosome numbers in accordance with their cellular activities (Wagner, 1994), the investigation of soil microbial communities by RNA-based procedures is generally assumed to be a means of assessing the metabolically active microbial fraction in soils (Duarte *et al.*, 1998; Costa *et al.*, 2004). However, given the fact that ribosome numbers in certain species do not seem to relate to growth or activity rate, this contention has been repeatedly criticized in recent times. In any case, methods for the simultaneous extraction of DNA and RNA from soils were felt to be needed and so several improved methods were published in the first years of the 21st century (Griffiths *et al.*, 2000; Hurt *et al.*, 2001; Gomes *et al.*, 2004).

Current nucleic acid extraction methods

For purely practical reasons, virtually all current soil nucleic acid extraction work is based on methods, available since the turn of the century or so, that utilize easy-to-use commercially available kits. Suppliers of soil DNA extraction kits such as Q Biogene (Carlsbad, California, USA) and MO BIO Laboratories Inc. (Carlsbad, California, USA) provide customers with readily available chemicals, buffers, plasticware and quick protocols that are supposed to enhance the efficiency of DNA recovery and reduce the time spent in the laboratory. Recently, a kit provided by QIAGEN has become available (RNA/DNA Mini Kit – QIAGEN GmbH, Hilden, Germany) that can be used to simultaneously purify DNA and RNA from soil. In general terms, the strategy behind most kit-based protocols for DNA extraction rely on: (i) efficient detachment of microbial cells from soil particles (normally achieved by mechanical means but sonication may also be used); (ii) optimal cell lysis through a combination of chemical (e.g. SDS and/or lysozyme treatment) and mechanical (e.g. bead beating or intense vortexing) means; (iii) removal of PCR inhibitors present in soil such as humic acids; (iv) efficient protein and debris precipitation; (v) binding of DNA to a silica membrane in the presence of high salt concentrations; (vi) washing of the silica-bound DNA with ethanol; (vii) recovery of contaminant-free DNA from the silica membrane by using a salt-free elution buffer.

As can be seen, the handling of hazardous organic solvents, such as phenol and chloroform, very common in previous laboratory protocols, is normally not necessary in currently available 'kit-based protocols', an obvious advantage for users. Also, stepping into the 'kit era' of nucleic acid extraction protocols has led to an unavoidable miniaturization of the whole process, another clear 'plus' in practical terms. Nowadays, DNA extractions can be performed that make use of as little as 0.1 g of soil (wet weight) as the starting point. 'Scaling down', however, might not necessarily be the best choice for some studies. Microbial ecologists, for instance, like to extrapolate their data into major considerations regarding ecosystem structure and functioning. However, cautious consideration must be given to the significance of a 0.25 g soil sample. Another note of caution concerns the possible bias introduced by the kit-based approaches compared to the older and more laborious approaches. Although untested, it is felt that as a result of reliance on quick lysis and column purification methods, the rapid, kit-based approaches are more prone to undersampling of the soil microbial community than the older laboratory-based methods. This potential problem should be borne in mind when considering the data obtained from the nucleic acid-based overall soil analyses.

Irrespective of the approach chosen (i.e. direct or indirect methods; laboratory- or kit-based protocols), robust and reliable procedures designed to extract nucleic acids from soil must ideally provide the researcher with the following features: (i) yield of nucleic acids representative for the microbial community present in soil; (ii) purity and integrity, which allow further reverse transcription (in case of RNA extraction), PCR amplification, community analyses as well as, for instance, metagenomic studies; (iii) rapid processing of a high number of samples; and (iv) application of the method to a broad range of soil types and sediments.

Nucleic Acid-based Tools for Analysing Soil Microbial Communities

Following the preparation of extracts containing the total soil nucleic acids, a range of analysis/detection methods can be used. These are all based on molecular tools, such as amplification by PCR, hybridization, fragment or gene cloning and sequencing and molecular fingerprintings of different kinds (Table 21.1). The second edition of the Molecular Microbial Ecology Manual (Kowalchuk et al., 2004) contains a collection of currently used nucleic acid-based protocols. Furthermore, the use of such techniques has been reviewed by Nybroe et al. (2006), and subsequent analysis methods by fingerprinting tools and/or metagenomics by Oros-Sichler et al. (2006) and Sjoling et al. (2006), respectively. Below, we briefly discuss what can be achieved using some of these methods.

Nucleic acid-based fingerprinting techniques

Fingerprinting techniques of nucleic acids are a means to provide snapshots of the structure and diversity of microbial communities in soil and other environments (Fig. 21.2). Their suitability for quick processing of large sample numbers, a feature essential in most ecological studies, makes comparisons between multiple environments or treatments a fairly achievable task (Costa et al., 2006). Shortly after their inception (Muyzer et al., 1993; Lee et al., 1996; Liu et al., 1997; Schwieger and

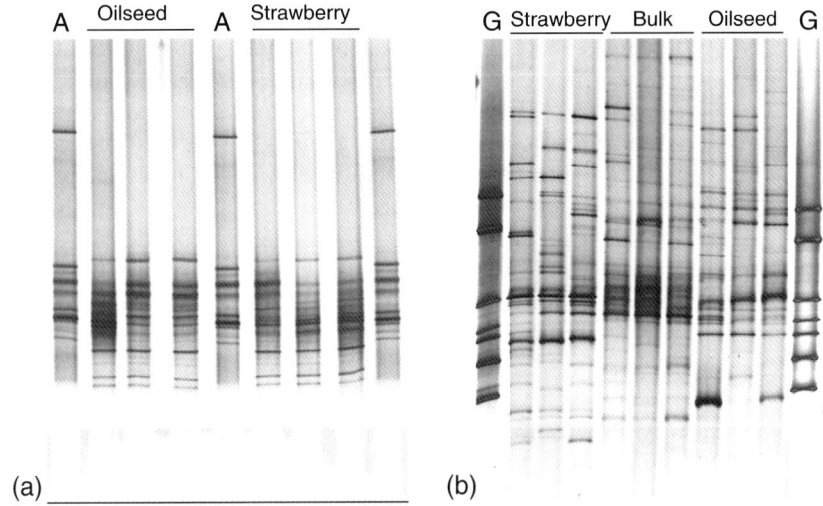

Fig. 21.2. *Pseudomonas*-specific DGGE fingerprints of 16S rRNA (a) and *gacA* (b) gene fragments PCR-amplified from TC-rhizosphere DNA of oilseed rape ('Oilseed') and strawberry ('Strawberry') samples collected at a field site in Germany. Mixtures of 16S rRNA gene and *gacA* gene ('G' lanes in b) fragments amplified from genomic DNA of culturable *Pseudomonas* spp. antagonistic towards the fungal pathogen *Verticillium dahliae* Kleb. are shown ('A' and lanes in picture a and 'G' lanes in picture b, respectively). The diversity of uncultured *Pseudomonas* spp. might be better reflected by *gacA* gene fingerprinting as compared to 16S rRNA gene-based analysis (Costa et al., 2007). (Reproduced with permission from Blackwell Publishing.)

Tebbe, 1998), the wide-scale use of fingerprinting techniques by modern microbial ecologists led to enormous advances in our understanding of the dynamics of complex and diverse microbial communities in soil (Oros-Sichler et al., 2006). Genes most commonly used as targets to fingerprint microbial communities in soil are phylogenetic marker genes, such as those encoding the ribosomal small subunit RNAs, ie. the 16S and 18S ribosomal RNA genes of bacteria and fungi, respectively. Fingerprinting of rRNA genes has been extensively used to assess the 'response' of microbial communities to a plethora of environmental or anthropogenic factors or impacts (Kowalchuk et al., 1997; Peters et al., 2000; Reiter et al., 2002; Schmalenberger and Tebbe, 2002; Garbeva et al., 2003, 2004a,b,c, 2006; Milling et al., 2004; Girvan et al., 2005; Costa et al., 2006). Given the fact that the 16S rRNA operon numbers vary per species (range 1–15), there are recent attempts to use alternative marker genes which occur as single copies, e.g. the *rpoB* gene (Oros-Sichler et al., 2006). Also, as more attention is being paid to soil functioning and to the role microbes play in nature, the assessment of the diversity of 'functional genes' (e.g. the *nifH* gene involved in nitrogen fixation) by fingerprinting methods has increased considerably (Rosado et al., 1997; Avrahami et al., 2002; Galand et al., 2002; Wolsing and Prieme, 2004;

Wertz *et al.*, 2006). One interesting example of a target gene that might be involved in directing soil suppressiveness and whose diversity in soil has been recently assessed by fingerprinting techniques (Costa *et al.*, 2007; Fig. 21.2) is the *gacA* gene of *Pseudomonas* spp, which is not only involved in the regulation of many secondary metabolites produced by beneficial and plant-pathogenic *Pseudomonas* species (Haas and Keel, 2003; Haas and Défago, 2005) but also functions as a suitable phylogenetic marker within this genus (De Souza *et al.*, 2003a). Exploring *gacA* gene variability results in a less biased assessment of the diversity within members of this extremely relevant group of rhizobacteria as compared to 16S rRNA gene-based analysis (Costa *et al.*, 2007; Fig. 21.2).

In principle, fingerprints can be obtained by analysis of the melting curve, folding (terminal) restriction site and/or length difference of DNA molecules generated by PCR amplification. Hence, fingerprinting techniques such as denaturing gradient gel electrophoresis (DGGE) (Fig. 21.2), single-strand conformational polymorphism (SSCP), terminal restriction fragment length polymorphism (T-RFLP), length heterogeneity PCR (LH-PCR) and automated ribosomal intergenic spacer analysis (ARISA) make use of these different potentials to separate a mix of PCR products with divergent sequences that are amplified from whole-community environmental DNA/RNA (Table 21.1). The literature abounds with examples of the successful use of such techniques to describe the microbial community structures in bulk and rhizosphere soils. The use of molecular fingerprinting techniques to characterize soil communities has been recently reviewed by Oros-Sichler *et al.* (2006) and the reader is referred to this compilation for a review of pioneering studies, comparative analysis, applications, technical progress and future perspectives of fingerprinting methods. One should bear in mind, however, that all of these techniques only describe, in a snapshot fashion, the diversity and composition of microbial communities that are locally present, without shedding light on the *in situ* functionalities of the members of these communities.

Quantitative PCR

PCR is a very sensitive method for the detection of specific genes in the soil microbial community. While PCR has for a long time been used only in a qualitative manner, recent work has shown that the technique can be made quantitative by two divergent approaches, i.e. competitive PCR and real-time PCR (Nybroe *et al.*, 2006). The Molecular Microbial Ecology Manual (Kowalchuk *et al.*, 2004) contains up-to-date protocols for PCR-based detection and enumeration of specific genes in soil.

Competitive PCR, which relies on the addition of an internal standard to the PCR mix that is co-amplified with the environmental target DNA sequence, has been used in a limited number of laboratories (Nybroe *et al.*, 2006), but it has now been largely superseded by real-time PCR. Hence, we will further focus on real-time PCR, which truly constitutes the currently preferred protocol for the quantification of specific nucleic acids in soil. Real-time PCR monitors the formation of amplicons during PCR amplification and is not based solely on end point detection. Formation of the product (amplicon) is proportional to a fluorescence signal that is detected during each amplification cycle. In contrast to conventional PCR, real-time PCR requires a specialized thermocycler equipped with a fluorescence detector. The fluorescence can be produced from fluorescent dyes such as SYBR green (Nybroe et al., 2006) which binds to the newly formed amplicons. PCR specificity is crucial when relying on SYBR green detection, as lack of specificity may yield non-target (next to target) products which distorts the quantification.

An approach that is currently widely in use is based on so-called Taqman probes. These are non-extendable dual-labelled fluorogenic hybridization probes, which bind to the target sequence in a manner similar to the primers. A reporter dye (e.g. 6-carboxyfluorescein – FAM) is coupled to one end of the probe and a quencher dye (e.g. 6-carboxy-tetramethyl-rhodamine – TAMRA) to the other end. As long as the probe is intact, the fluorescence of the reporter is quenched by the quencher, so no fluorescence signal is detected. During PCR, the 5'-nuclease activity of the *Taq* polymerase causes probe degradation, releasing (now unquenched) reporter dye into solution, and hence the fluorescence signal increases along with product formation. The signal intensity is automatically recorded by the thermal cycler, and a cycle number at which the amplification curve crosses an arbitrary threshold line (defined as C_T) is used to estimate the number of targets by comparison to a calibration curve.

Real-time PCR was first applied to soil systems in 2001 (Hermansson and Lindgren, 2001) to assess the abundance in soil of ammonia oxidizers, and the method is increasingly being used to study the abundance of different groups of bacteria in soil at both the phylogenetic and functional gene levels. For application to soil, calibration of the signals obtained is based on internal standards consisting of additions of known quantities of cells containing the target gene to soil. In this way, the amplification from the target to be quantified and the internal standard will be subjected to similar PCR background conditions. We discuss below, the use of real-time PCR for the quantification of the *prnD* gene, one of the genes involved in the biosynthesis of pyrrolnitrin, in soils subjected to different treatments.

DNA microarrays

DNA microarrays (also called 'DNA chips' or 'gene arrays') contain nucleic acids as probes attached to solid surfaces (e.g. glass slides) and represent a further development of conventional membrane-based technologies (e.g. Southern hybridization). Microarrays allow the multiplex detection of thousands of genes and represent one of the latest advances in molecular detection technology (Oros-Sichler et al., 2006). The sheer high-throughput power provided by DNA microarray technology allows expression profiling of microbial genomes in a fashion which was unimaginable even two decades ago. Several landmark studies that applied DNA microarrays to assess gene expression of individual microorganisms have emerged in the second half of the 1990s (Schena et al., 1995; DeRisi et al., 1997) and represent a major breakthrough in gene expression research. The simultaneous analysis of thousands of genes which may turn on and off in response to environmental triggers, stress or artificial treatments could, for the first time, be performed on a large scale. Shortly thereafter, the application of simple microarray strategies to complex environmental DNA samples has been attempted (Guschin et al., 1997; Small et al., 2001). Because of their intrinsic features, DNA microarrays are being considered to represent the 'fashionable tools' for ideal high-throughput studies of the sequence diversity of rRNA genes as well as of functional genes in soil or other environmental samples. Recently, so-called functional gene arrays (FGAs; Wu et al., 2001), community gene arrays (CGAs) and phylogenetic oligonucleotide arrays (POAs) have been developed to profile nucleic acids extracted directly from environmental nucleic acid samples (Tiquia et al., 2004). A comprehensive overview of the potential and limitations of such recent technologies is provided by Zhou (2003). As pointed out by this author, while successful for the analysis of gene expression in pure culture, major problems still need to be overcome to adapt microarray hybridization methods to environmental studies. For instance, in the study conducted by Wu et al. (2001), an FGA containing approximately 120 functional genes was used to systematically study the specificity and sensitivity of detection of such genes in DNA isolated from marine sediments. The lack of sensitivity of direct hybridization-based detection was found to be the most limiting factor. The limit of detection of direct microarray hybridization turned out to be at least 10^3–10^4-fold higher than that of detection following PCR amplification. Other sources of technical restrictions also constitute an impediment for microarray hybridization to be fully employed for *in situ* expression profiling of microbial communities, such as specificity of gene detection and construction of comprehensive arrays (Zhou, 2003). An example of the use of POA for environmental DNA samples is the study of Loy et al. (2002), where an array consisting of 132 16S rRNA gene probes specific for all known lineages of sulphate-reducing prokaryotes was successfully tested and employed. In this study, 16S rRNA gene and dissimilatory bisulphite reductase PCR amplicons were fluorescently labelled and used for hybridization. Such an approach holds promise for future applications in analyses of soil microbial communities (Oros-Sichler et al., 2006). At present, there is a need for further development of random (instead of targeted) preamplification technology. Novel approaches, in which random primers of any kind are applied, have shown promise in improving the detectability of targets from soil by microarray hybridization.

Fluorescence *in situ* hybridization

Fluorescence *in situ* hybridization (FISH), also called whole cell hybridization, allows the visualization of specific target bacterial or fungal cells in a natural setting. In this way, a picture of the occurrence and prevalence of target cells can be obtained. The application of FISH to soil microhabitats was recently reviewed by Schmid *et al.* (2006). This review describes 'full-cycle rRNA approach', in which, starting from soil DNA, a picture is obtained of the bacterial community that is locally present. To detect key bacteria in their natural habitat, specific, labelled probes are generated on the basis of the 16S rRNA sequences identified, and applied to whole cells within the system. A range of probes with different levels of specificity have been developed (reviewed in Schmid *et al.*, 2006). Thus, information at the cell level is obtained about the occurrence and dominance of specific organisms.

When applied to rhizospheres or mycospheres of different kinds, FISH has allowed researchers to track specific plant-growth-promoting, biocontrol, fungal-interactive and/or pathogen strains. By localization of specific microorganisms in a system, interactions between microorganisms and interaction with plant roots can be studied (Schmid *et al.*, 2006).

FISH-based detection, however, does not seem to provide a powerful method that would allow all aspects of disease or phytopathogen suppressiveness of a soil to be predicted. It should be used in conjunction with other cultivation-based as well as cultivation-independent methods to shed additional light on microbial abundance, spatial relationships and functioning.

What Do the Nucleic Acid-based Methods Tell Us About Soil Suppressiveness Potential?

Use of microbial diversity assessments

Soils are largely resilient systems that can quickly recover from adverse situations due to: (i) the extreme functional redundancy that exists within its microbial inhabitants and (ii) the intrinsic life strategies of microbes, which allow them to cope with disturbances or environmental changes of any kind by promoting evolutionary novelties that eventually persist in the community as adaptive traits.

Both the functional redundancy in, and adaptability of microbial communities to, changing conditions are intrinsically linked to the issue of diversity. Indeed, a higher microbial diversity may lead to a higher resilience and stability of communities in soils affected by perturbations (Girvan *et al.*, 2005). Soil functioning is, in fact, largely dependent on the activity of microbial communities, as microbes mediate the vast majority of the biogeochemical processes and interactions that take place in this complex habitat. With regard to soil suppressiveness, maintenance of soil microbial diversity would be desirable to guarantee the natural capability of soil to 'regulate' plant disease. Very interestingly, as a relationship between 'above-ground' and 'below-ground' diversity is being reported as a trend in bulk (Garbeva et al., 2006) and rhizosphere (Kowalchuk et al., 2002; Garbeva et al., 2004a, 2006) soils of agroecosystems, organic farmers are investing in agricultural strategies that promote higher 'above-ground diversity' to sustain soil fertility and minimize outbreaks of plant diseases. However, few reports have been published so far which were deliberately designed to link soil community diversity to suppressiveness. One approach to linking molecular data on microbial communities to disease suppression was represented by attempts to correlate microbial diversity measurements on the basis of the 16S rRNA gene with levels of disease suppression. In a series of recent publications, Garbeva et al. (2003, 2004a, 2006) have set out to use DNA-based fingerprinting methods to characterize the microbial communities of soils with different histories and land use, namely: (i) species-rich permanent grassland; (ii) grassland turned into arable land – short-term arable land; and (iii) long-term arable land.

Diversity indices were determined for each soil treatment and linked to data on the strength of suppressiveness of each soil towards the fungal pathogen *Rhizoctonia solani* AG3, causal agent of potato root disease (Garbeva et al., 2004a). The highest suppression of the pathogen was observed for the 'short-term arable land' treatment. In addition, higher numbers of culturable bacteria which antagonized the growth of *R. solani in vitro* were found in this soil and in soil under permanent grassland. As higher microbial diversities were measured for these two soils as compared to long-term arable land, the hypothesis of a positive correlation between microbial diversity and disease suppressiveness of soils was supported by these authors (Garbeva et al., 2004b).

Use of specific (functional) gene detection

One of the clearest relationships between the specific activity of a given group of microbes and the extent of soil suppressiveness is the involvement of 2,4-diacetylphoroglucinol (2,4-DAPG) production by fluorescent *Pseudomonas* spp. in take-all decline (TAD) soils. Studies conducted by Raaijmakers et al. (1997, 1999) and Raaijmakers and Weller (1998) have made use of specific primers and probes targeting the biosynthetic gene locus *phlD*, involved in the production of 2,4-DAPG, to evaluate the occurrence of this gene in natural environments and the role of 2,4-DAPG producers in TAD soils located in the USA. First, it was demonstrated that 2,4-DAPG producers had their densities enriched in TAD soils as compared to conducive soils (Raaijmakers et al., 1997). Second, the elimination of these organisms led to the loss of suppressiveness in TAD soils (Raaijmakers and Weller, 1998). Third, soils conducive to take-all were made suppressive by the introduction of a 2,4-DAPG producing strain via inoculation of wheat, a natural host of *Gaeumannomyces graminis* var.

tritici, the causative agent of take-all (Raaijmakers and Weller, 1999). The observation that suppressiveness is a transferable trait highlights the role of the soil microbiota as a fundamental driving force in suppressiveness (Haas and Défago, 2005). Moreover, a follow-up study by de Souza *et al.* (2003b) with Dutch TAD soils confirmed the previous results obtained for the American TAD soils (American and Dutch TAD soils differ with regard to their physicochemical characteristics), with one additional and crucial 'upgrade'. A mutant of strain SSB17, deficient in 2,4-DAPG production, was unable to control take-all disease of wheat, whereas its corresponding wild type effectively controlled the disease. This demonstrated that 2,4-DAPG was a key factor leading to TAD on its own. Comparisons of wild type and mutant strains are a nice example of how modern genetic engineering technology is helpful in providing clear evidence for the role that microbes play *in situ*. Such studies have relied on robust, cultivation-dependent approaches to sampling target microbes prior to using molecular tools of gene detection and gene silencing. Although such studies represent major advances in our understanding of specific soil suppressiveness, means of directly detecting and quantifying populations and/or genes which drive soil suppressiveness without prior cultivation of microbes are highly desirable and urgently needed. An example of such an approach is the study performed by Garbeva *et al.* (2004c) that consisted of quantifying genes by real-time PCR for production of the antibiotic pyrrolnitrin (i.e. the *prnD* gene locus) and correlating the gene numbers with suppressiveness levels (Fig. 21.3). The working hypothesis in this study was that enhanced numbers of the key gene targeted (*prnD*) would coincide with enhanced levels of microbial populations involved in the suppressiveness and enhanced suppressiveness of the fungal pathogen *Rhizoctonia solani* AG3. It thus took gene numbers to indicate potential

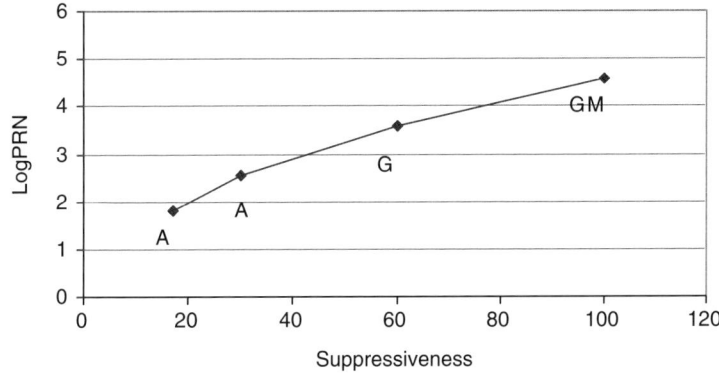

Fig. 21.3. Relationship between *prnD* gene numbers and level of suppression of *Rhizoctonia solani* AG3. (Modified from Steinberg *et al.*, 2006.)

activity. The soil treatments under study were the same as those described previously, i.e. grassland, short-term and long-term arable land. A strongly enhanced abundance of *prnD* genes was detected in the soil treatments of higher levels of suppressiveness aginst *R. solani* AG3, which turned out to be the soils for which the highest microbial diversity values were measured (Fig. 21.3) (Garbeva *et al.*, 2006).

Avenues Towards Future Improved Methods

At present, there is a clear need to define and establish key biological indicators that will allow us to quickly interrogate soil systems as to their status as either disease suppressive or conducive (Steinberg *et al.*, 2006). For the microbially determined suppressiveness, we in principle dispose of the appropriate molecular tools that may allow us to achieve this goal. The *in situ* detection, quantification and assessment of abundance and diversity of genes involved in disease suppression and of their expression can be realistically performed using available molecular tools that microbial ecologists master (Kowalchuk *et al.*, 2004; Nybroe *et al.*, 2006). Nevertheless, the accumulation of meaningful empirical data that can irrefutably link a gene's activity to disease suppression is still lacking for most of the cases in which soil suppressiveness has been reported. One of the underlying reasons for this may be the ephemeral character of gene expression in soil, which may be high at one moment and low or absent just a moment later. As intelligently designed experimental approaches to addressing this question are still scarce, knowledge of their importance in applied biotechnology is being achieved at a low pace, despite the considerable research in this area. For example, in most cases, we are still unable to clearly define which genes or organisms are most likely involved in general and specific disease suppression. Also, an understanding of how the expression of a multitude of genes involved in soil suppressiveness may be triggered/repressed in a multifactorial and spatially constrained world such as the rhizosphere or soil environment is still very limited. In addition, as scientists need to simplify nature to first understand it at a first glance, our comprehension of soil suppressiveness is restricted to studies where bilateral interactions between natural enemies are approached, whereas soil suppressiveness per se may arise as an outcome of the concerted action of microbes in a 'neural' network context, involving multitrophic interactions and other organisms not considered here, such as varied fungi, including mycorrhizae. All constraints indicated above are critical issues when 'ecologically friendly' soil management to improving crop protection is required. Thus, to strike a new path in soil suppressiveness research, we urgently need to develop the targeting systems that will allow us to quickly detect, quantify and assess the abundance and diversity of 'suppressiveness-associated' organisms and their respective genes by cultivation-independent means (Fig. 21.4). We will also need to address their localization in the system by, for instance, FISH-based approaches. This obviously also applies to the study of the

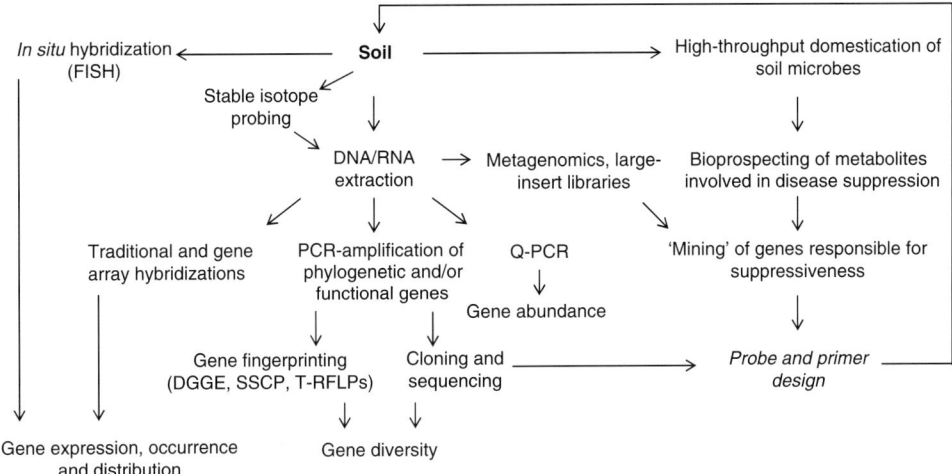

Fig. 21.4. Proposal for an integrated approach to assessing soil suppressiveness by molecular means.

phytopathogenic organisms and their defence mechanisms against inhibitory substances from the plant and their natural enemies in the rhizosphere. Extending our capabilities of 'domesticating' root-associated microbes will provide a means to look into their extraordinary features at both the metabolic and genotypic levels and facilitate the construction of specific gene-based targeting systems. Within this framework, the use of high-throughput technologies such as DNA microarrays holds great promise in unraveling the functioning and diversity of genes and/or organisms most associated with disease suppression in a variety of situations where the expression of a multitude of genes can be simultaneously analysed. In addition to this, genes which play key roles in soil suppressiveness should be specifically addressed. The occurrence (PCR-hybridization), diversity (clone libraries and fingerprints), abundance (quantitative PCR), expression (FISH, mRNA gene-based analysis) and functional relevance (genetic engineering assays in microcosms) of such genes in suppressive soils must be assessed in cutting-edge, robust, polyphasic nucleic acid-based experiments (Fig. 21.4).

References

Avrahami, S., Conrad, R. and Braker, G. (2002) Effect of soil ammonium concentration on N_2O release and on the community structure of ammonia oxidizers and denitrifiers. *Applied and Environmental Microbiology* 68, 5685–5692.

Baker, K.F. and Cook, R.J. (1974) *Biological Control of Plant Pathogens*. W.H. Freeman, San Francisco.

Cook, R.J. and Baker, K.F. (1983) *The Nature and Practice of Biological Control of Plant Pathogens*. The American Phytopathological Society, St Paul, Minnesota, p. 539.

Costa, R., Gomes, N.C.M., Milling, A. and Smalla, K. (2004) An optimized protocol for simultaneous extraction of DNA and RNA from soils. *Brazilian Journal of Microbiology* 35, 230–234.

Costa, R., Götz, M., Mrotzek, N., Lottmann, J., Berg, G. and Smalla, K. (2006) Effects of site and plant species on rhizosphere community structure as revealed by molecular analysis of microbial guilds. *FEMS Microbiology Ecology* 56, 236–249.

Costa, R., Gomes, N.C.M., Opelt, K., Berg, G. and Smalla, K. (2007) *Pseudomonas* community structure and antagonistic potential in the rhizosphere: insights gained by combining phylogenetic and functional gene-based analyses. *Environmental Microbiology* (in press).

DeRisi, J.L., Iyer, V.R. and Brown, P.O. (1997) Exploring the metabolic and genetic control of gene expression on a genomic scale. *Science* 278, 680–686.

De Souza, J.T., Mazzola, M. and Raaijmakers, J.M. (2003a) Conservation of the response regulator gene *gacA* in *Pseudomonas* species. *Environmental Microbiology* 5, 1328–1340.

De Souza, J.T., Weller, D.M. and Raaijmakers, J.M. (2003b) Frequency, diversity, and activity of 2,4-diacetylphloroglucinol-producing fluorescent *Pseudomonas* spp. in Dutch take-all decline soils. *Phytopathology* 93, 54–63.

Duarte, G.F., Rosado, A.S., Seldin, L., Keijzer-Wolters, A.C. and van Elsas, J.D. (1998) Extraction of ribosomal RNA and genomic DNA from soil for studying the diversity of the indigenous microbial community. *Journal of Microbiological Methods* 32, 21–29.

Felske, A., Engelen, B., Nübel, U. and Backhaus, H. (1996) Direct ribosome isolation from soil to extract bacterial rRNA for community analysis. *Applied and Environmental Microbiology* 62, 4162–4167.

Galand, P.E., Saarnio, S., Fritze, H. and Yrjälä, K. (2002) Depth related diversity of methanogen Archaea in Finnish oligotrophic fen. *FEMS Microbiology Ecology* 42, 441–449.

Gans, J., Wolinsky, M. and Dunbar, J. (2005) Computational improvements reveal great bacterial diversity and high metal toxicity in soil. *Science* 309, 1387–1390.

Garbeva, P., van Veen, J.A. and van Elsas, J.D. (2003) Predominant *Bacillus* spp. in agricultural soil under different management regimes detected via PCR-DGGE. *Microbial Ecology* 45, 302–316.

Garbeva, P., van Veen, J.A. and van Elsas, J.D. (2004a) Assessment of the diversity, and antagonism towards *Rhizoctonia solani* AG3, of *Pseudomonas* species in soil from different agricultural regimes. *FEMS Microbiology Ecology* 47, 51–64.

Garbeva, P., van Veen, J.A. and van Elsas, J.D. (2004b) Microbial diversity in soil: selection of microbial populations by plant and soil type and implications for disease suppressiveness. *Annual Review of Phytopathology* 42, 243–270.

Garbeva, P., Voesenek, K. and van Elsas, J.D. (2004c) Quantitative detection and diversity of the pyrrolnitrin biosynthetic locus in soil under different treatments. *Soil Biology and Biochemistry* 36, 1453–1463.

Garbeva, P., Postma, J., van Veen, J.A. and van Elsas, J.D. (2006) Effect of above-ground plant species on soil microbial community structure and its impact on suppression of *Rhizoctonia solani*. *Environmental Microbiology* 8, 233–246.

Girvan, M.S., Campbell, C.D., Killham, K., Prosser, J.I. and Glover, L.A. (2005) Bacterial diversity promotes community stability and functional resilience after perturbation. *Environmental Microbiology* 7, 301–313.

Gomes, N.C.M., Costa, R. and Smalla, K. (2004) Simultaneous extraction of DNA and RNA from bulk and rhizosphere soil. In: Kowalchuk, G.A., de Bruijn, F.J., Head, I.M., Akkermans, A.D.L. and van Elsas, J.D. (eds) *Molecular Microbial Ecology Manual II*. Kluwer Academic Publishers, Dordrecht, The Netherlands, pp. 1743–1761.

Griffiths, R.I., Whiteley, A.S., O'Donnell, A.G. and Bailey, M.J. (2000) Rapid method for coextraction of DNA and RNA from natural environments for analysis of ribosomal DNA- and rRNA-based microbial community composition. *Applied and Environmental Microbiology* 66, 5488–5491.

Guschin, D.Y, Mobarry, B.C., Proudnikov, D., Stahl, D.A., Rittmann, B.E. and Mirzabekov, A.D. (1997) Oligonucleotide microchips as genosensors for determinative and environmental studies in microbiology. *Applied and Environmental Microbiology* 63, 2397–2402.

Haas, D. and Défago, G. (2005) Biological control of soil-borne pathogens by fluorescent pseudomonads. *Nature Reviews Microbiology* 3, 307–319.

Haas, D. and Keel, C. (2003) Regulation of antibiotic production in root-colonizing *Pseudomonas* spp. and relevance for biological control of plant disease. *Annual Review of Phytopathology* 41, 117–153.

Hermansson, A. and Lindgren, P.E. (2001) Quantification of ammonia-oxidizing bacteria in arable soil by real-time PCR. *Applied and Environmental Microbiology* 67, 972–976.

Hiltner, L. (1904) Über neue Erfahrungen und Probleme auf dem Gebiete der Bodenbakteriologie unter besonderer Berücksichtigung der Gründüngung und Brache. *Arbeiten der Deutschen Landwirtschaftgesellschaft* 98, 59–78.

Holben, W.E., Jansson, J.K., Chelm, B.K. and Tiedje, J.M. (1988) DNA probe method for the detection of specific microorganisms in the soil bacterial community. *Applied and Environmental Microbiology* 54, 703–711.

Hurt, R.A., Qiu, X., Wu, L., Roh, Y., Palumbo, A.V., Tiedje, J.M. and Zhou, J. (2001) Simultaneous recovery of RNA and DNA from soils and sediments. *Applied and Environmental Microbiology* 67, 4495–4503.

Keel, C., Schnider, U., Maurhofer, M., Voisard, C., Laville, J., Burger, U., Wirthner, P., Haas, D. and Défago, G. (1992) Suppression of root diseases by *Pseudomonas fluorescens* CHA0: importance of the bacterial secondary metabolite 2,4-diacetylphloroglucinol. *Molecular Plant-Microbe Interactions* 5, 4–13.

Kowalchuk, G.A., Stephens, J.R., de Boer, W., Prosser, J.I., Embley, M.T. and Woldendorp, J.W. (1997) Analysis of ammonia-oxidizing bacteria of the beta subdivision of the class proteobacteria in coastal sand dunes by denaturing gradient gel electrophoresis and sequencing of PCR-amplified 16S ribosomal DNA fragments. *Applied and Environmental Microbiology* 63, 1489–1497.

Kowalchuk, G.A., Buma, D.S., de Boer, W., Klinkhamer, P.G.L. and van Veen, J.A. (2002) Effects of above-ground plant species composition and diversity on the diversity of soil-borne microorganisms. *Antonie van Leeuwenhoek* 81, 509–520.

Kowalchuk, G.A., de Bruijn, F.J., Head, I.M., Akkermans, A.D.L. and van Elsas, J.D. (2004) *Molecular Microbial Ecology Manual II*. Kluwer Academic Publishers, Dordrecht, The Netherlands.

Lee, D.-H., Zo, Y.-G. and Kim, S.-J. (1996) Nonradioactive method to study genetic profiles of natural bacterial communities by PCR-single strand conformation polymorphism. *Applied and Environmental Microbiology* 62, 3112–3120.

Liu, W.-T., Marsh, T.L., Cheng, H. and Forney, L.J. (1997) Characterization of microbial diversity by determining terminal restriction length polymorphisms of genes encoding 16S rDNA. *Applied and Environmental Microbiology* 63, 4516–4522.

Loy, A., Lehner, A., Lee, N., Adamczyk, J., Meier, H., Ernst, J., Schleifer, K-H. and Wagner, M. (2002) Oligonucleotide microarray for 16S rRNA gene-based detection of all recognized lineages of sulfate-reducing prokaryotes in the environment. *Applied and Environmental Microbiology* 68, 5064–5081.

Lynch, J.M. (1984) The rhizosphere – form and function. *Applied Soil Ecology* 1, 193–198.

Mark, G.L., Morrisey, J.P., Higgins, P. and O'Gara, F. (2005) Molecular-based strategies to exploit *Pseudomonas* biocontrol strains for environmental biotechnology applications. *FEMS Microbiology Ecology* 56, 167–177.

Milling, A., Smalla, K., Maidl, F.X., Schloter, M. and Munch, J.C. (2004) Effects of transgenic potatoes with an altered starch composition on the diversity of soil and rhizosphere bacteria and fungi. *Plant and Soil* 266, 23–39.

Moran, M.A., Torsvik, V.L., Torsvik, T. and Hodson, R.E. (1993) Direct extraction and purification of rRNA for ecological studies. *Applied and Environmental Microbiology* 59, 915–918.

Muyzer, G., de Waal, D.C. and Uitterlinden, A.G. (1993) Profiling of complex microbial populations by denaturing gradient gel electrophoresis analysis of polymerase chain reaction-amplified genes coding for 16S rRNA. *Applied and Environmental Microbiology* 59, 695–700.

Nybroe, O., Brandt, K.K., Nicolaisen, M.H. and Sørensen, J. (2006) Methods to detect and quantify bacteria in soil. In: van Elsas, J.D., Jansson, J.K. and Trevors, J.T. (eds) *Modern Soil Microbiology II*. CRC Press, Boca Raton, Florida, pp. 283–316.

Ogram, A., Sayler, G.S. and Barkay, T.J. (1987) DNA extraction and purification from sediments. *Journal of Microbiological Methods* 7, 57–66.

Oros-Sichler, M., Costa, R., Heuer, H. and Smalla, K. (2006) Molecular fingerprinting techniques to analyze soil microbial communities. In: van Elsas, J.D., Jansson, J.K. and Trevors, J.T. (eds) *Modern Soil Microbiology II*. CRC Press, Boca Raton, Florida, pp. 355–386.

Peters, S., Koschinsky, S., Schwieger, F. and Tebbe, C.C. (2000) Succession of microbial communities during hot composting as detected by PCR-single-strand-conformation polymorphism-based genetic profiles of small-subunit rRNA genes. *Applied and Environmental Microbiology* 66, 930–936.

Pinton, R., Varanini, Z. and Nannipieri, P. (2001) The rhizosphere as a site of biochemical interactions among soil components, plants and microorganisms. In: Pinton, R., Varanini, Z. and Nannipieri, P. (eds) *The Rhizosphere – Biochemistry and Organic Substances at Soil–plant Interface*. Marcel Dekker, New York, pp. 1–17.

Raaijmakers, J.M. and Weller, D.M. (1998) Natural plant protection by 2,4-diacetylphloroglucinol-producing *Pseudomonas* spp. in take-all decline soils. *Molecular Plant-Microbe Interactions* 11, 144–152.

Raaijmakers, J.M., Weller, D.M. and Thomashow, L.S. (1997) Frequency of antibiotic-producing *Pseudomonas* spp. in natural environments. *Applied and Environmental Microbiology* 63, 881–887.

Raaijmakers, J.M., Bonsall, R.F. and Weller, D.M. (1999) Effect of population density of *Pseudomonas fluorescens* on production of 2,4-diacetylphloroglucinol in the rhizosphere of wheat. *Phytopathology* 89, 470–475.

Reiter, B., Pfeifer, U., Schwab, H. and Sessitsch, A. (2002) Response of endophytic bacterial communities in potato plants to infection with *Erwinia carotovora* subsp. *atroseptica*, *Applied and Environmental Microbiology* 68, 2261–2268.

Rosado, A.S., Duarte, G.F., Seldin, L. and van Elsas, J.D. (1997) Genetic diversity of *nifH* gene sequences in *Paenibacillus azotofixans* strains and soil samples analyzed by denaturing gradient gel electrophoresis. *Applied and Environmental Microbiology* 64, 2770–2779.

Schena, M., Shalon, D., Davis, R.W. and Brown, P.O. (1995) Quantitative monitoring of gene expression patterns with a complementary DNA microarray. *Science* 270, 467–470.

Schmid, M., Rothballer, M. and Hartmann, A. (2006) Analysis of microbial communities in soil microhabitats using fluorescence *in situ* hybridization. In: van Elsas, J.D., Jansson, J.K. and Trevors, J.T. (eds) *Modern Soil Microbiology II*. CRC Press, Boca Raton, Florida, pp. 317–337.

Schmalenberger, A. and Tebbe, C.C. (2002) Bacterial community composition in the rhizosphere of a transgenic, herbicide-resistant maize (*Zea mays*) and comparison to its non-transgenic cultivar Bosphore. *FEMS Microbiology Ecology* 40, 29–37.

Schwieger, F. and Tebbe, C.C. (1998) A new approach to utilize PCR-single strand conformation polymorphism for 16S rRNA gene-based microbial community analysis. *Applied and Environmental Microbiology* 64, 4870–4876.

Sjoling, S., Stafford, W. and Cowan, D.A. (2006) Soil metagenomics: exploring and exploiting the soil microbial gene pool. In: van Elsas, J.D., Jansson, J.K. and Trevors, J.T. (eds) *Modern Soil Microbiology II*. CRC Press, Boca Raton, Florida, pp. 409–434.

Small, J., Call, D.R., Brockman, F.J., Straub T.M. and Chandler, D.P. (2001) Direct detection of 16S rRNA in soil extracts by using oligonucleotide microarrays. *Applied and Environmental Microbiology* 67, 4708–4716.

Smalla, K., Creswell, N., Mendonça-Hagler, L.C., Wolters, A.C. and van Elsas, J.D. (1993) Rapid DNA extraction protocol from soil for polymerase chain reaction-mediated amplification. *Journal of Applied Bacteriology* 74, 78–85.

Sørensen, J. (1997) The rhizosphere as a habitat for soil microorganisms. In: van Elsas, J.D., Trevors, J.T. and Wellington, E.M.H. (eds) *Modern soil microbiology*. Marcel Dekker, New York, pp. 21–45.

Staley, J.T. and Konopka, A. (1985) Measurement of *in situ* activities of nonphotosynthetic microorganisms in aquatic and terrestrial habitats. *Annual Review of Microbiology* 39, 321–346.

Steinberg, C., Edel-Hermann, V., Alabouvette, C. and Lemanceau, P. (2006) Soil suppressiveness to plant diseases. In: van Elsas, J.D., Jansson, J.K. and Trevors, J.T. (eds) *Modern Soil Microbiology II*. CRC Press, Boca Raton, Florida, pp. 455–478.

Tiquia, S.M., Chong, S.C., Fields, M.W. and Zhou, J. (2004) Oligonucleotide-based functional gene arrays for analysis of microbial communities in the environment. In: Kowalchuk, G.A., de Bruijn, F.J, Head, I.M., Akkermans, A.D.L. and van Elsas, J.D. (eds) *Molecular Microbial Ecology Manual II*. Kluwer Academic Publication, Dordrecht, The Netherlands, pp. 1743–1761.

Torsvik, V.L. (1980) Isolation of bacterial DNA from soil. *Soil Biology and Biochemistry* 12, 18–21.

Torsvik, V., Goksøyr, J. and Daae, F.L. (1990) High diversity in DNA of soil bacteria. *Applied and Environmental Microbiology* 56, 782–787.

Tsai, Y., Park, M.J. and Olson, B.H. (1991) Rapid method for direct electrophoresis and substrate utilization patterns. *Applied and Environmental Microbiology* 57, 765–768.

Van Elsas, J.D., Smalla, K. and Tebbe, C.C. (1999) Extraction and analysis of microbial community nucleic acids from environmental matrices. In: Jansson, J.K., van Elsas, J.D. and Bailey, M.J. (eds) *Tracking Genetically Engineered Microorganisms: Method Development from Microcosms to the Field*. R.G. Landes Co., Georgetown, Texas, pp. 29–51.

Van Elsas, J.D., Jansson, J.K. and Trevors, J.T. (2006) *Modern Soil Microbiology II*. CRC Press, Boca Raton, Florida.

Wagner, R. (1994) The regulation of ribosomal RNA synthesis and bacterial cell growth. *Archives of Microbiology* 161, 100–106.

Walsh, U.F., Morrissey, J.P. and O'Gara, F. (2001) *Pseudomonas* for biocontrol of phytopathogens: from functional genomics to commercial exploitation. *Current Opinion in Biotechnology* 12, 289–295.

Weller, D.M., Raaijmakers, J.M., McSpadden Gardener, B.B. and Thomashow, L.S. (2002) Microbial populations responsible for specific soil suppressiveness to plant pathogens. *Annual Review of Phytopathology* 40, 309–348.

Wertz, S., Degrange, V., Prosser, J.I., Poly, F., Commeaux, C., Freitag, T., Guillaumaud, N. and Le Roux, X. (2006) Maintenance of soil functioning following erosion of microbial diversity. *Environmental Microbiology* 8, 2162–2169.

Wolsing, M. and Prieme, A. (2004) Observation of high seasonal variation in community structure of denitrifying bacteria in arable soil receiving artificial fertilizer and cattle manure by determining T-RFLP of nir gene fragments. *FEMS Microbiology Ecology* 48, 261–271.

Wu, L., Thompson, D.K., Li, G., Hurt, R.A., Tiedje, J.M. and Zhou, J. (2001) Development and evaluation of functional gene arrays for detection of selected genes in the environment. *Applied and Environmental Microbiology* 67, 5780–5790.

Zhou, J. (2003) Microarrays for bacterial detection and microbial community analysis. *Current Opinion in Microbiology* 6, 288–294.

Zhou, J., Bruns, M.A. and Tiedje, J.M. (1996) DNA recovery from soils of diverse composition. *Applied and Environmental Microbiology* 62, 461–468.

22 Enhancing Biological Control Efficacy of Yeasts to Control Fungal Diseases Through Biotechnology

G. Marchand, G. Clément-Mathieu, B. Neveu and R.R. Bélanger

Abstract

Classical biological control of fungal plant diseases by yeasts and yeast-like organisms has been amply described in the literature, yet few commercial products have arisen out of those research efforts. It is generally agreed that a better understanding of: (i) the ecological conditions affecting the activity of a biocontrol agent (BCA); and (ii) the mode(s) of action conferring to a BCA its properties will greatly help improve its efficacy. In this context, the recent advances in biotechnology are providing scientists with new tools to study and improve BCAs. Coincidently, yeasts and yeast-like fungi are particularly amenable to genetic transformation and thus biotechnology techniques. While improving the efficacy of BCAs through techniques affecting gene expression is a promising endeavour, other technologies such as Green Fluorescent Protein (GFP) expression have allowed precise ecological studies of plant–pathogen interactions, and recently, that of BCAs with the host plant and fungal pathogen. These studies can be extremely useful to elucidate the mode of action and determine the ecological fate of the BCAs, as in the case of *Pseudozyma flocculosa*, the registered yeast-like BCA of powdery mildews. Other tools such as reverse genetics approaches based on data from genomic sequencing using RNA interference (RNAi) or gene knockout have the potential to identify key genes involved in biocontrol activity and thus help achieve the improvement of biocontrol efficacy of yeasts through biotechnology.

Introduction

Practical implementation of biological control of plant diseases has lagged behind general expectations and prognostics. Many reasons, including scientific, legal and commercial, can explain this situation (Paulitz and Bélanger, 2001; Fravel, 2005), but in most instances, our inability to properly exploit the intrinsic properties of a BCA accounts for this limited success. In many cases, scientists may have oversimplified or misunderstood the interaction between a plant pathogen and a BCA. In other cases, a BCA may simply not be potent enough to achieve acceptable control of a given plant pathogen in spite of promising performances *in vitro* (Burr *et al.*, 1996; Milus and Rothrock, 1997). These considerations have created a general trend towards deciphering in greater detail how a BCA interacted with a plant pathogen and its environment in order to improve its efficacy. In addition, new technological advances have been developed in recent years (Lorang *et al.*, 2001; Weld *et al.*, 2006) and can be exploited to better understand the ecological behaviour of a BCA and its ensuing impact on both target and non-target organisms. As a result, scientists now have access to a variety of powerful tools, opening the door to applications that were inconceivable only a few years ago. While it is uncertain whether these new technologies will indeed generate a direct increase in the number of BCAs becoming available, they offer unique opportunities to improve their efficacy and facilitate their registration with the proper authorities.

Among BCAs of plant pathogens, yeasts and yeast-like organisms occupy a preponderant place because of their intrinsic properties to interact in the same ecological niche as the one occupied by many parasites (Punja and Utkhede, 2003). At the same time, these organisms are usually amenable to genetic manipulations because of their rather simple structures and life cycle (Fincham, 1989; Ruiz-Díez, 2002). They are thus perfectly suited to be exploited in biotechnology programmes. In this chapter, we will review some of the relevant biotechnology applications that are available for enhancing the properties and our understanding of yeast and yeast-like BCAs, and whenever appropriate, illustrate their benefits through specific examples namely in the context of *P. flocculosa*, a yeast-like epiphyte well known for its antagonistic properties against powdery mildews.

Background

Most BCAs identified to date have been categorized as exerting their activity through the manifestation of one or more of four modes of action: competition, parasitism, antibiosis and induced resistance (Whipps, 2000; Bélanger and Avis, 2002). Genetic enhancement and manipulation of these modes of action require a precise understanding of the gene(s) conferring the specific activity, a feat that has remained elusive in most systems. In fact, most attempts to date to genetically modify biocontrol agents in an effort to improve their efficiency have not yielded the expected levels of success. Many factors can explain these apparent failures but manipulation of 'biocontrol' genes is evidently a more challenging endeavour than originally anticipated. This also probably betrays our ambition to reduce a complex interaction to a simple phenomenon. However, the latest developments and availability of biotechnological tools (Lorang *et al.*, 2001; Tierney and Lamour, 2005) have opened up opportunities to finely dissect the modus operandi of BCAs. These tools have a wide array of applications, extending from studies of a given BCA in a complex ecological interaction, to manipulation of specific genes involved in biocontrol activity.

Ecological Studies

One area of research that has too often been overlooked is the ecological behaviour of a prospective BCA when exposed to fluctuating conditions in a given agricultural system (Huang et al., 2000; Paulitz and Bélanger, 2001; Larkin and Fravel, 2002; Lübeck et al., 2002). It is often believed that our failure to properly understand how a BCA can establish itself in a specific ecological niche is partly responsible for the limited success of biological control on the commercial front. Because biological control of plant diseases typically involves three protagonists – the BCA, the pathogen and the plant itself – the fate of each organism is interdependent. However, very few studies have described the events characterizing this form of tritrophic interaction *in situ* (Lu et al., 2004). In most instances, the interactions involving plant–pathogen or pathogen–BCA have been observed and described by means of electron microscopy (transmission and scanning) and immunocytochemistry techniques (Benhamou and Nicole, 1999; Avis and Bélanger, 2002; Gao et al., 2005; Soylu et al., 2005). Though powerful, these techniques are limited in their scope as they can only offer a punctual appreciation, both in time and space, of the interactions.

In some recent reviews, Errampalli et al. (1999), Lorang et al. (2001), Leveau and Lindow (2002), Jannson (2003) and Larrainzar et al. (2005) have highlighted how studies of plant–pathogen and microbial interactions can be facilitated by using the GFP from *Aequorea victoria* (Tsien, 1998). The great advantage of this reporter protein, once expressed by the microorganism under study, is the immediate visualization under fluorescence microscopy of the latter without any further step. This technique, gaining quickly in popularity, has been particularly exploited to study plant–pathogen interactions (Spellig et al., 1996; Vanden Wymelenberg et al., 1997; Maor et al., 1998; Chen et al., 2003; Si-Ammour et al., 2003; Lu et al., 2004; Skadsen and Hohn, 2004; Bolwerk et al., 2005; Eckert et al., 2005; Spinelli et al., 2005) since it provides an elegant way to examine the progression of the pathogen within plant tissues. So far, plant pathologists have used the GFP transformation technology to study the spatio-temporal behaviour of plant pathogens in systems where fungi are amenable to genetic transformation, such as *Ustilago maydis* in sweet corn (Spellig et al., 1996), *Cochliobolus heterostrophus* in maize (Maor et al., 1998), and *Ophiostoma piceae* in wood (Lee et al., 2002).

As described by Leveau and Lindow (2002) and Larrainzar et al. (2005), GFP technology can offer tremendous insights into microbial function and behaviour in natural environments. Because very few BCAs have been manipulated genetically (Zeilinger et al., 1999; Lübeck et al., 2002; Lu et al., 2004; Bolwerk et al., 2005), this technology has yet to be exploited fully for the study of complex plant–pathogen–BCA interactions.

Mode of Action

Understanding precisely how BCAs kill their target has always been perceived as the key to maximize the efficacy of BCAs in the field. From a theoretical point of view, if the mode of action can indeed be reduced down to a single trait, this greatly facilitates manipulation and over-expression of the trait in key situations for a BCA to eliminate a plant pathogen. As a matter of fact, this strategy has been tried with a few systems with mitigated results. For instance, in cases where parasitism appeared to be the predominant mode of action, several attempts have been made to increase production of lytic enzymes, such as chitinases and glucanases (Kubicek et al., 2001; Lorito et al., 2001). The approach contemplated either the selection of BCA strains with superior ability to produce such enzymes or the direct cloning of relevant genes conferring better substrate degrading properties. While conceptually sound, these initiatives failed to deliver BCAs with notable increased activity. In retrospect, it was generally agreed that parasitism, or at least the enzymes associated with this mode of action, was not the key factor in the biocontrol activity of some fungi, notably in the case of *Trichoderma* spp. (Woo et al., 2006). These results have highlighted the difficulty inherent to ascribing the activity of a BCA to a single factor. In most cases, the genetic analyses employed to study phenotypic variations in an attempt to single out the key factor gene have been termed forward genetics (Tierney and Lamour, 2005). With the advent of rapid advances in genomics and genome sequencing, scientists are now in a position to use a completely different approach, called reverse genetics, where disruption of a gene or gene product is used to study the resulting phenotypes. The objective in reverse genetics is to alter a specific gene and thus infer gene function. Considering the difficulties encountered in identifying 'biocontrol genes', this approach offers new avenues to discover and target with precision those elusive genes involved in the interactions among plant pathogens and BCAs.

Among the many approaches now available, gene silencing via RNAi is being expanded to many organisms since its initial observation in *Caenorhabditis elegans* (Fire et al., 1998). In *C. elegans*, genome-wide RNAi screens against libraries of predicted genes have allowed the study of a variety of biological processes (Kuttenkeuler and Boutros, 2004). Targeted gene disruption by homologous recombination is another strategy being exploited in reverse genetics (Woo et al., 1999; Grevesse et al., 2003). For instance, homologous recombination has been used in embryonic cells in mice and has led to the construction of precise mutations in nearly every gene (Adams et al., 2004). Other approaches being used or developed for the study of specific gene functions include: insertional mutagenesis/transposon-mediated mutagenesis (Vidan and Snyder, 2001), Targeting-Induced Local Lesions in Genomes (TILLING) (McCallum et al., 2000) and Virus-Induced Gene Silencing (VIGS) (Baulcombe, 1999).

While gene function analysis in BCAs is still in its infancy, it is clear that the new biotechnology tools will provide valuable assistance towards that endeavour. Yeasts and yeast-like organisms, having rather small and simple genomes, are likely to be the first fungal BCAs to serve as models for understanding what triggers biocontrol activity.

Studying the Ecology and Mode of Action of BCAs Through Biotechnology: Examples with *Pseudozyma flocculosa*

P. flocculosa (Traquair, L. A. Shaw *et* Jarvis) Boekhout *et* Traquair (syn: *Sporothrix flocculosa* Traquair, Shaw *et* Jarvis) is a yeast-like epiphyte that is found naturally on the surface of leaves (Fig. 22.1). It has been discovered and described by Traquair *et al.* (1988) but its potential as a BCA has generated a lot of research aimed at understanding its ecology and mode of action (Bélanger and Avis, 2002).

From the time of its discovery, *P. flocculosa* has been described almost exclusively as an antagonist of the biotrophic powdery mildews (Fig. 22.2).

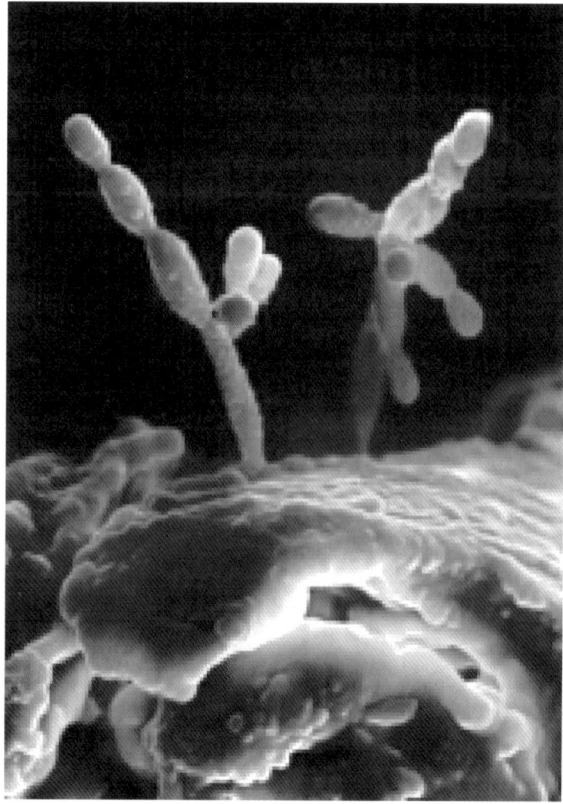

Fig. 22.1. Conidiophores of the biocontrol agent *Pseudozyma flocculosa* on the surface of a cucumber leaf as visualized by scanning electron microscopy. (Reproduced from Bélanger [2006] with permission from the publisher.)

Fig. 22.2. Interaction of the biocontrol agent *Pseudozyma flocculosa* with powdery mildew of cucumber (*Podosphaera xanthii*) on the surface of a cucumber leaf as visualized by scanning electron microscopy. Thin hyphae of *P. flocculosa* cause the collapse of the *P. xanthii* conidial chains.

Several papers have related its efficacy at controlling powdery mildews on different hosts (Hajlaoui and Bélanger, 1993; Bélanger *et al.*, 1994; Dik *et al.*, 1998) but most importantly, efforts at understanding its mode of action have shed some light on the ecology of the organism in relation with optimizing its use as a biocontrol agent. From the very beginning, it was quite clear that *P. flocculosa* was neither a strong competitor in the phyllosphere nor a direct hyperparasite. All observations pointed to a manifestation of antibiosis. The fungus was never seen to penetrate its host but killed it within a few hours of interaction. Microscopic studies confirmed that test fungi exposed to *P. flocculosa* were plasmolyzed very rapidly (Hajlaoui *et al.*, 1992). These results prompted investigations into the active principles that conferred *P. flocculosa* its activity (Avis *et al.*, 2000). In an effort to confirm the role of antibiosis as the predominant mode of action of *P. flocculosa* and to properly understand how it related to its ecology, a genetic transformation system specifically adapted for *P. flocculosa* was developed (Cheng *et al.*, 2001).

Ecological studies with GFP

With the recent strides in genetically manipulating *P. flocculosa*, it became possible to exploit GFP technology for the benefit of this BCA in areas of fungal ecology and tritrophic interactions that were inaccessible a few years ago (Fig. 22.3A and B). Previously, plant pathologists and microbial ecologists have used GFP technology almost exclusively for the study of plant–pathogen interactions in an effort to understand the infection process (Spellig et al., 1996; Maor et al., 1998; Lee et al., 2002). The addition of the BCA component in a foliar system allowed direct observation of the relative influence of one component over the others in a habitat representative of the main biological interacting elements following application of a BCA. Following inundative applications on different plant species or plant–powdery mildew interactions, the growth of a *P. flocculosa* transformed with GFP was closely associated with the presence of powdery mildew colonies regardless of the species or the host plant (Neveu et al., 2007). Development of *P. flocculosa* on control leaves alone was extremely limited 24 h after application and appeared typical of the epiphytic growth characterizing this type of yeast-like fungus. Based on the strong correlation between the colonization pattern of the different powdery mildew species tested and the presence of *P. flocculosa*, as determined by its fluorescence, it seems that growth of the BCA is dependent on the presence of powdery mildews (Fig. 22.3C and D). These results demonstrate that the GFP technology can be used to study plant–pathogen–BCA interactions and fulfil a wide array of purposes, ranging from fundamental observations of the biocontrol behaviour of a BCA to very applied ones serving some of the requirements for the registration of BCAs. In context of requirements for registration of BCAs, one can readily appreciate how GFP technology could provide valuable information regarding, among other specifications, environmental fate. For this purpose, *P. flocculosa*, as the active ingredient of the biofungicide Sporodex, is a good model since it has received temporary registration in Canada and the USA but is still under review in Europe. From the outset, results can alleviate concerns one may have about unrestrained spread of the BCA following mass applications. GFP technology could also help reduce costs of repetitious evaluations on different hosts since multiple host–pathogen combinations can be tested in limited space and time. In addition, corroboration of similar behaviour regardless of the host, as observed, could facilitate requests for data from ecotoxicological studies. Other practical information, for instance residue data and persistence in the environment, could also be obtained in a speedy yet accurate manner, at least initially, by observations of GFP transformants in a specified field. For example, the study of splash dispersal of sporangia of *Phytophthora infestans* generated by raindrop impaction on plants was largely facilitated by the use of GFP expressing transformants (Saint-Jean et al., 2005).

Genes Involved in the Mode of Action

Following the genetic transformation of *P. flocculosa*, Cheng et al. (2003) were able, through insertional mutagenesis, to create mutants that lacked the ability to produce the antibiotics. When the same mutants were tested against other fungi, they were unable to stop their progression in confrontation tests, and more importantly, they had lost all their biocontrol activity against powdery mildews. These results strongly suggested that antibiotic production was an intrinsic property in *P. flocculosa* that played an important part in the life cycle of this epiphyte. However, comparative analysis of the culture filtrates of control and negative mutants revealed that the active ingredient was more complex than originally suspected. The active molecule was in fact a cellobiose lipid, a rather unusual occurrence in the fungal kingdom. In fact, this glycolipid, named flocculosin, was closely related to a particular fungal metabolite, ustilagic acid, produced by *U. maydis* (Lemieux et al., 1951) (Fig. 22.4).

The link between the production of a ustilagic acid by *U. maydis* and the glycolipid flocculosin by *P. flocculosa* is intriguing because the close taxonomic relationship between these two fungi has been verified by molecular biology studies. Boekhout (1995) has shown the closeness of *Pseudozyma* species to yeast-like anamorphs of *U. maydis* based on DNA sequences coding for large subunit of rDNA genes. This suggests that, in the course of their evolution, both fungi have maintained this antibiotic trait because it contributed to their ecological

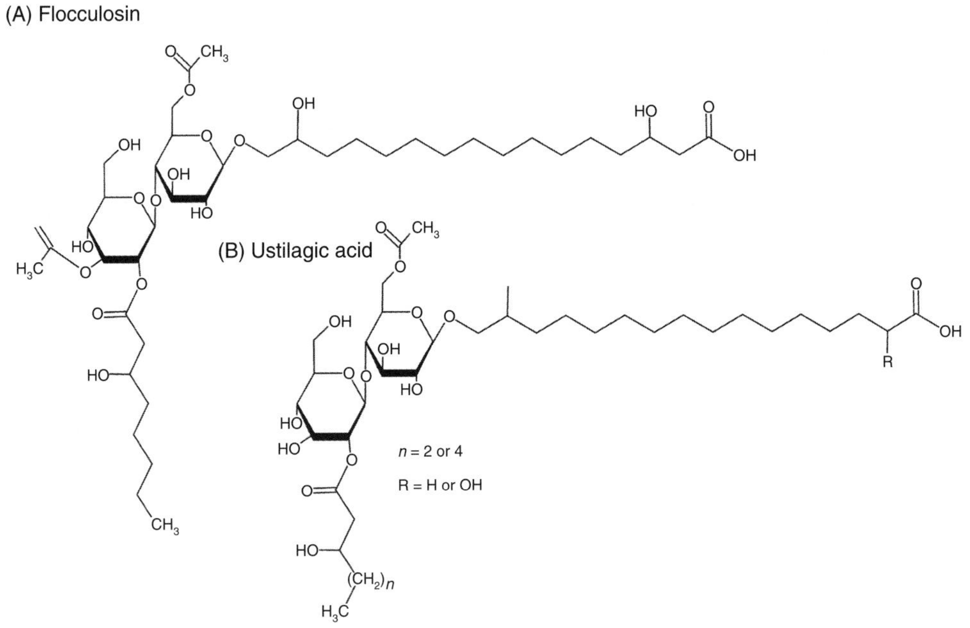

Fig. 22.4. Chemical structure of cellobiose lipids flocculosin (A) produced by the biocontrol agent *Pseudozyma flocculosa*, and ustilagic acid (B) produced by the plant pathogen *Ustilago maydis*.

fitness. From this perspective, it is quite clear that *P. flocculosa* is using this chemical weapon to protect its limited ecological niche and deter potential competitors from invading it (Avis *et al.*, 2000). This is perceived as being rather elegant when this property can be exploited in biological control. However, chemical warfare among microorganisms is not exclusive to 'beneficial' agents and plant pathogens such as *U. maydis* can also find advantages at maintaining their ecological position through this mode of action. For instance, it is well known that *U. maydis* requires plasmogamy between conjugation tubes from two compatible basidiospores to produce dikaryotic mycelium capable of infecting maize. In the process of finding (or waiting for) a compatible mating type, a basidiospore will likely enhance its chance of survival by protecting its immediate surrounding with the release of antifungal compounds.

Using a reverse genetics approach, Hewald *et al.* (2005) recently identified a gene, *cyp*1, which is essential for synthesis of ustilagic acid by *U. maydis*. Deletion of this gene was not lethal and did not appear to alter pathogenicity towards maize but affected pheromone recognition over large distances, abolishing conjugation tube induction and thus preventing mating of compatible strains. Based on these observations, the biosurfactant properties of ustilagic acid were hypothesized to facilitate diffusion of the hydrophobic pheromones that allow the long distance recognition of compatible mating types. It was therefore interesting to determine if and why *Pseudozyma* spp. would retain this trait based on the putative role of ustilagic acid in *U. maydis*. For this purpose, six different *Pseudozyma* species (*P. flocculosa*, *P. antarctica*, *P. rugulosa*, *P. tsukubaensis*, *P. fusiformata* and *P. aphidis*) were screened for homologues of this gene. Results revealed that only two species contained a *U. maydis cyp*1 homologue, *P. flocculosa* and *P. fusiformata* (Marchand *et al.*, 2006). Interestingly, these are the only two *Pseudozyma* species in which production of cellobiose-type lipids such as flocculosin (Mimee *et al.*, 2005) and ustilagic acid (Kulakovskaya *et al.*, 2005), respectively, are reported.

In contrast to *U. maydis*, *P. flocculosa* is believed to lead a strictly asexual anamorphic-like lifestyle. Therefore, if the cellobiose lipid indeed plays a role in recognition of compatible mating types, one can wonder if the *cyp1* gene in *P. flocculosa* is a vestigial trait, since no mating of compatible strains has ever been observed, and strains collected from different parts of the world show remarkably little genetic variation as assessed with the use of molecular markers (Avis *et al.*, 2001). Yet the BCA *P. flocculosa* has conserved the ability to produce flocculosin, and this molecule shows remarkable antifungal properties (Mimee *et al.*, 2005). Did *P. flocculosa* evolve a different role for its cellobiose lipid? On the basis of its antifungal activity, it does seem plausible that flocculosin is involved in protecting the ecological niche and concomitantly in conferring biocontrol activity to *P. flocculosa*. To address these hypotheses, the application of techniques of silencing or knockout to the *P. flocculosa cyp*1 gene would provide useful information about the role of flocculosin and its implication in the mode of action of this BCA. Alternatively, overexpression of this gene could also lead to insights into the apparent specificity of the interaction with powdery mildew fungi, and would represent the bridge from using biotechnology as a tool to further our understanding to potentially improving the biocontrol properties of *P. flocculosa*.

Conclusion

While biotechnology has not yet yielded concrete improvements of the biocontrol efficacy of biocontrol yeasts and fungi to control fungal plant diseases, it has played an important role in improving our fundamental understanding of the BCA–plant–pathogen interaction. For instance, new tools such as GFP expression by BCAs and plant pathogens allow real-time, nondestructive *in situ* observation and ecological studies of complex interactions. *P. flocculosa* is an interesting example of a registered BCA where GFP technology has been exploited to conduct precise ecological studies that complemented the data already obtained through conventional microscopy and molecular approaches. This technology also allows easy tracking of BCAs and thus alleviates concerns over the environmental fate of BCAs in the registration process. Advances in genomic sequencing of BCAs or closely related organisms also provide the opportunity to identify more key genes involved in biocontrol potential. When combined with reverse genetics approaches using techniques such as RNAi or gene knockout to modulate or disrupt the expression of candidate genes, identification of new genes involved in biocontrol activity should yield opportunities to improve the capacity of BCAs to control plant diseases.

Acknowledgements

The authors would like to thank C. Labbé for her help in assembling the figures and also acknowledge the many students, research assistants, colleagues and collaborators who have contributed to this work through their research activities, current and past, and through helpful discussions. The last author is also indebted to NSERC, FQRNT, Plant Products Co. Ltd and the Canada Research Chairs Program for continued financial support.

References

Adams, D.J., Biggs, P.J., Cox, T., Davies, R., van der Weyden, L., Jonkers, J., Smith, J., Plumb, B., Taylor, R., Nishijima, I., Yu, Y.J., Rogers, J. and Bradley, A. (2004) Mutagenic insertion and chromosome engineering resource (MICER). *Nature Genetics* 36, 867–871.

Avis, T.J. and Bélanger, R.R. (2002) Mechanisms and means of detection of biocontrol activity of *Pseudozyma* yeasts against plant-pathogenic fungi. *FEMS Yeast Research* 2, 5–8.

Avis, T.J., Boulanger, R.R. and Bélanger, R.R. (2000) Synthesis and biological characterization of (Z)-9-heptadecenoic and (Z)-6-methyl-9-heptadecenoic acids, fatty acids with antibiotic activity produced by *Pseudozyma flocculosa*. *Journal of Chemical Ecology* 26, 987–1000.

Avis, T.J., Caron, S.J., Boekhout, T., Hamelin, R.C. and Bélanger, R.R. (2001) Molecular and physiological analysis of the powdery mildew antagonist *Pseudozyma flocculosa* and related fungi. *Phytopathology* 91, 249–254.

Baulcombe, D.C. (1999) Fast forward genetics based on virus-induced gene silencing. *Current Opinion in Plant Biology* 2, 109–113.

Bélanger, R.R. (2006) Controlling diseases without fungicides: a new chemical warefare. *Canadian Journal of Plant Pathology* 28, S233–S238.

Bélanger, R.R. and Avis, T.J. (2002) Ecological processes and interactions occurring in leaf surface fungi. In: Lindow, S.E. and Elliott, V.J. (eds) *Ecology of the Phyllosphere*. APS Press, St. Paul, Minnesota.

Bélanger, R.R., Labbé, C. and Jarvis, W.R. (1994) Commercial-scale control of rose powdery mildew with a fungal antagonist. *Plant Disease* 78, 420–424.

Benhamou, N. and Nicole, M. (1999) Cell biology of plant immunization against microbial infection: the potential of induced resistance in controlling plant diseases. *Plant Physiology and Cytochemistry* 37, 703–719.

Boekhout, T. (1995) *Pseudozyma* bandoni emend. Boekhout, a genus for yeast-like anamorphs of Ustilaginales. *Journal of General and Applied Microbiology* 41, 355–366.

Bolwerk, A., Lagopodi, A.L., Lugtenberg, B.J.J. and Bloemberg, G.V. (2005) Visualization of interactions between a pathogenic and a beneficial *Fusarium* strain during biocontrol of tomato foot and root rot. *Molecular Plant–Microbe Interactions* 18, 710–721.

Burr, T.J., Matteson, M.C., Smith, C.A., Corral-Garcia, M.R. and Huang, T.C. (1996) Effectiveness of bacteria and yeasts from apple orchards as biological control agents of apple scab. *Biological Control* 6, 151–157.

Chen, N., Hsiang, T. and Goodwin, P.H. (2003) Use of green fluorescent protein to quantify the growth of *Colletotrichum* during infection of tobacco. *Journal of Microbiological Methods* 53, 113–122.

Cheng, Y.L., Belzile, F., Tanguay, P., Bernier, L. and Bélanger, R.R. (2001) Establishment of a gene transfer system for *Pseudozyma flocculosa*, an antagonistic fungus of powdery mildew fungi. *Molecular Genetics and Genomics* 266, 96–102.

Cheng, Y., McNally, D.J., Labbé, C., Voyer, N., Belzile, F. and Bélanger, R.R. (2003) Insertional mutagenesis of a fungal biocontrol agent led to discovery of a rare cellobiose lipid with antifungal activity. *Applied and Environmental Microbiology* 69, 2595–2602.

Dik, A.J., Verhaar, M.A. and Bélanger, R.R. (1998) Comparison of three biological control agents against cucumber powdery mildew (*Sphaerotheca fuliginea*) in semi-commercial-scale glasshouse trials. *European Journal of Plant Pathology* 104, 413–423.

Eckert, M., Maguire, K., Urban, M., Foster, S., Fitt, B., Lucas, J. and Hammond-Kosack, K. (2005) *Agrobacterium tumefaciens*-mediated transformation of *Leptosphaeria* spp. and *Oeulimacula* spp. with the reef coral gene DsRed and the jellyfish gene gfp. *FEMS Microbiology Letters* 253, 67–74.

Errampalli, D., Leung, K., Cassidy, M.B., Kostrzynska, M., Blears, M., Lee, H. and Trevors, J.T. (1999) Applications of the green fluorescent protein as a molecular marker in environmental microorganisms. *Journal of Microbiological Methods* 35, 187–199.

Fincham, J.R.S. (1989) Transformation in fungi. *Microbiological Reviews* 53, 147–170.

Fire, A., Xu, S., Montgomery, M.M., Kostas, S.A., Driver, S.E. and Mello, C.C. (1998) Potent and specific interference by double-stranded RNA in *Caenorhabditis elegans*. *Nature* 391, 806–811.

Fravel, D.R. (2005) Commercialization and implementation of biocontrol. *Annual Review of Phytopathology* 43, 337–359.

Gao, K., Liu, X., Kang, Z. and Mendgen, K. (2005) Mycoparasitism of *Rhizoctonia solani* by endophytic *Chaetomium spirale* ND35: ultrastructure and cytochemistry of the interaction. *Journal of Phytopathology* 153, 280–290.

Grevesse, C., Lepoivre, P. and Jijakli, M.H. (2003) Characterization of the exoglucanase-encoding gene PaEXG 2 and study of its role in the biocontrol activity of *Pichia anomala* strain K. *Phytopathology* 93, 1145–1152.

Hajlaoui, M.R. and Bélanger, R.R. (1993) Antagonism of the yeast-like phylloplane fungus *Sporothrix flocculosa* against *Erysiphe graminis* var. *tritici*. *Biocontrol Science Technology* 3, 427–434.

Hajlaoui, M.R., Benhamou, N. and Bélanger, R.R. (1992) Cytochemical study of the antagonistic activity of *Sporothrix flocculosa* on rose powdery mildew, *Sphaerotheca pannosa* var. *rosae*. *Phytopathology* 82, 583–589.

Hewald, S., Josephs, K. and Bölker, M. (2005) Genetic analysis of biosurfactant production in *Ustilago maydis*. *Applied and Environmental Microbiology* 71, 3033–3040.

Huang, J.C., Bremer, E., Hynes, R.K. and Erickson, R.S. (2000) Foliar application of fungal biocontrol agents for the control of white mold of dry bean caused by *Sclerotinia sclerotiorum*. *Biological Control* 18, 270–276.

Jannson, J.K. (2003) Marker and reporter genes: illuminating tools for environmental microbiologists. *Current Opinion in Microbiology* 6, 310–316.

Kubicek, C.P., Mach, R.L., Peterbauer, C.K. and Lorito, M. (2001) *Trichoderma*: from genes to biocontrol. *Journal of Plant Pathology* 83, 11–23.

Kulakovskaya, T.V., Shashkov, A.S., Kulakovskaya, E.V. and Golubev, W.I. (2005) Ustilagic acid secretion by *Pseudozyma fusiformata* strains. *FEMS Yeast Research* 5, 919–923.

Kuttenkeuler, D. and Boutros, M. (2004) Genome-wide RNAi as a route to gene function in *Drosophila*. *Briefings in Functional Genomics* 3, 168–176.

Larkin, R.P. and Fravel, D.R. (2002) Effects of varying environmental conditions on biological control of *Fusarium* wilt of tomato by non-pathogenic *Fusarium* spp. *Phytopathology* 92, 1160–1166.

Larrainzar, E., O'Gara, F. and Morrissey, J.P. (2005) Applications of autofluorescent proteins for *in situ* studies in microbial ecology. *Annual Review of Microbiology* 59, 257–277.

Lee, S., Kim, S.H. and Breuil, C. (2002) The use of the green fluorescent protein as a biomarker for sapstain fungi. *Forest Pathology* 32, 153–161.

Lemieux, R.U., Thorn, J.A., Brice, C. and Haskins, R.H. (1951) Biochemistry of the Ustilaginales: II. Isolation and partial characterization of ustilagic acid. *Canadian Journal of Chemistry* 29, 409–414.

Leveau, H.J. and Lindow, S.E. (2002) Bioreporters in microbial ecology. *Current Opinion in Microbiology* 5, 259–265.

Lorang, J.M., Tuori, R.P., Martinez, J.P., Sawyer, T.L., Redman, R.S., Rollins, J.A., Wolpert, T.J., Johnson, K.B., Rodriguez, R.J., Dickman, M.B. and Ciuffetti, L.M. (2001) Green fluorescent protein is lighting up fungal biology. *Applied and Environmental Microbiology* 67, 1987–1994.

Lorito, M., Scala, F., Zoina, A. and Woo, S.L. (2001) Enhancing biocontrol of fungal pests by exploiting the *Trichoderma* genome. In: Gressel, J. and Vurro, M (eds) *Enhancing Biocontrol Agents and Handling Risks*. IOS Press, Amsterdam, The Netherlands.

Lu, Z.X., Tombolini, R., Woo, S., Zeilinger, S., Lorito, M. and Jansson, J.K. (2004) *In vivo* study of *Trichoderma*–pathogen–plant interactions, using a constitutive and inducible green fluorescent protein reporter systems. *Applied and Environmental Microbiology* 70, 3073–3081.

Lübeck, M., Knudsen, I.M.B., Jensen, B., Thrane, U., Janvier, C. and Funck Jensen, D. (2002) GUS and GFP transformation of the biocontrol strain *Clonostachys rosea* IK726 and the use of these marker genes in ecological studies. *Mycological Research* 106, 815–826.

Maor, R., Puyesky, M., Horwitz, B.A. and Sharon, A. (1998) Use of green fluorescent protein (GFP) for studying development and fungal–plant interaction in *Cochliobolus heterostrophus*. *Mycological Research* 102, 491–496.

Marchand, G., Belzile, F. and Bélanger, R.R. (2006) Isolation of a *cyp1* gene homolog in the biocontrol fungus *Pseudozyma flocculosa*. *Phytopathology* 96:S73.

McCallum, C.M., Comai, L., Greene, E.A. and Henikoff, S. (2000) Targeting induced local lesions in genomes (TILLING) for plant functional genomics. *Plant Physiology* 123, 439–442.

Milus, E.A. and Rothrock, C.S. (1997) Efficacy of bacterial seed treatments for controlling *Pythium* root rot of winter wheat. *Plant Disease* 81, 180–184.

Mimee, B., Labbé, C., Pelletier, R. and Bélanger, R.R. (2005) Antifungal activity of flocculosin, a novel glycolipid isolated from *Pseudozyma flocculosa*. *Antimicrobial Agents and Chemotherapy* 49, 1597–1599.

Neveu, B., Labbé, C. and Bélanger, R.R. (2007) GFP technology for the study of biocontrol agents in tritrophic interactions: a case study with *Pseudozyma flocculosa*. *Journal of Microbiological Methods* 68, 275–281.

Paulitz, T.C. and Bélanger R.R. (2001) Biological control in greenhouse systems. *Annual Review of Phytopathology* 39, 103–133.

Punja, Z.K. and Utkhede, R.S. (2003) Using fungi and yeasts to manage vegetable crop diseases. *Trends in Biotechnology* 21, 400–407.

Ruiz-Díez, B. (2002) Strategies for the transformation of filamentous fungi. *Journal of Applied Microbiology* 92, 189–195.

Saint-Jean, S., Testa, A., Kamoun, S. and Madden, L.V. (2005) Use of a green fluorescent protein marker for studying splash dispersal of sporangia of *Phytophthora infestans*. *European Journal of Plant Pathology* 112, 391–394.

Si-Ammour, A., Mauch-Mani, B. and Mauch, F. (2003) Quantification of induced resistance against *Phytophthora* species expressing GFP as a vital marker: beta-aminobutyric acid but not BTH protects potato and *Arabidopsis* from infection. *Molecular Plant Pathology* 4, 237–248.

Skadsen, R.W. and Hohn, T.A. (2004) Use of *Fusarium graminearum* transformed with GFP to follow infection patterns in barley and *Arabidopsis*. *Physiological and Molecular Plant Pathology* 64, 45–54.

Soylu, S., Brown, I. and Mansfield, J.W. (2005) Cellular reactions in *Arabidopsis* following challenge by strains of *Pseudomonas syringae*: from basal resistance to compatibility. *Physiological and Molecular Plant Pathology* 66, 232–243.

Spellig, F., Bottin, A. and Kahmann, R. (1996) Green fluorescent protein (GFP) as a new vital marker in the phytopathogenic fungus *Ustilago maydis*. *Molecular Genetics and Genomics* 252, 503–509.

Spinelli, F., Ciampolini, F., Cresti, M., Geider, K. and Costa, G. (2005) Influence of stigmatic morphology on flower colonization by *Erwinia amylovora* and *Pantoea agglomerans*. *European Journal of Plant Pathology* 113, 395–405.

Tierney, M.B. and Lamour, K.H. (2005) An introduction to reverse genetic tools for investigating gene function. *The Plant Health Instructor* DOI: 10.1094/PHI-A-2005-1025-01.

Traquair, J.A., Shaw, L.A. and Jarvis, W.R. (1988) New species of *Stephanoascus* with *Sporothrix* anamorphs. *Canadian Journal of Botany* 66, 926–933.

Tsien, R.Y. (1998) The green fluorescent protein. *Annual Review of Biochemistry* 67, 509–544.

Vanden Wymelenberg, A.J., Cullen, D., Spear, R.N., Schoenike, B. and Andrews, J.H. (1997) Expression of green fluorescent protein in *Aureobasidium pullulans* and quantification of the fungus on leaf surfaces. *Biotechniques* 23, 686–690.

Vidan, S. and Snyder, M. (2001) Large-scale mutagenesis: yeast genetics in the genome era. *Current Opinion In Biotechnology* 12, 28–34.

Weld, R.J., Plummer, K.M., Carpenter, M.A. and Ridgway, H.J. (2006) Approaches to functional genomics in filamentous fungi. *Cell Research* 16, 31–44.

Whipps, J.M. (2000) Microbial interactions and biocontrol in the rhizosphere. *Journal of Experimental Botany* 52, 487–511.

Woo, S.L., Donzelli, B., Scala, F., Mach, R., Harman, G.E., Kubicek, C.P., Del Sorbo, G. and Lorito, M. (1999) Disruption of the *ech42* (endochitinase-encoding) gene affects biocontrol activity in *Trichoderma harzianum* P1. *Molecular Plant–Microbe Interactions* 12, 419–429.

Woo, S.L., Scala, F., Ruocco, M. and Lorito, M. (2006) Molecular biology of the interactions between *Trichoderma* spp., phytopathogenic fungi and plants. *Phytopathology* 96, 181–185.

Zeilinger, S., Galhaup, C., Payer, K., Woo, S.L., Mach, R.L., Fekete, C., Lorito, M. and Kubicek, C.P. (1999) Chitinase gene expression during mycoparasitic interaction of *Trichoderma harzianum* with its host. *Fungal Genetics and Biology* 26, 131–140.

23 Molecular Insights into Plant Virus–Vector Interactions

D. Rochon

Abstract

Plant-to-plant spread of viruses most often requires a specific vector. Plant viruses are obligate parasites that are usually capable of infecting only specific hosts. The plants they infect are sessile and viruses are incapable of penetrating the plant cell wall. Plant viruses have evolved to overcome these barriers for dissemination and the initiation of infection by developing interactions with specific vectors. The vectors provide the mobility required for plant-to-plant movement, share similar hosts with the virus and are capable of breaching the cell wall. The virus–vector associations observed have been found to be highly specific, implying that viruses have evolved to recognize and attach to specific features of their vectors in a ligand–receptor fashion. This chapter provides a broad overview of the various biological relationships that exist between selected plant viruses and their vectors and recent studies on molecular components of viruses and vectors that mediate these interactions. Basic similarities and differences among the various models for virus–vector interactions are discussed as well as emerging patterns of interaction.

Introduction

Plant viruses are transmitted in a highly specific fashion by a variety of arthropods and by soil-inhabiting nematodes and fungi (Table 23.1). Plant viruses can also be transmitted through seed, pollen and vegetative propagation. A very few viruses can be mechanically transmitted through human and other animal activity, as well as by physical contact between the leaves of adjacent plants in a field (Hull, 2002). Since successful spread of most plant viruses requires specific interactions between the virus and its vector, studies of virus–vector interactions may offer new opportunities for reducing the impact viruses have in the many food, fibre and ornamental crops they infect. This chapter provides a broad overview of vector transmission of plant viruses, focusing on recent research on molecular

Fig. 22.3

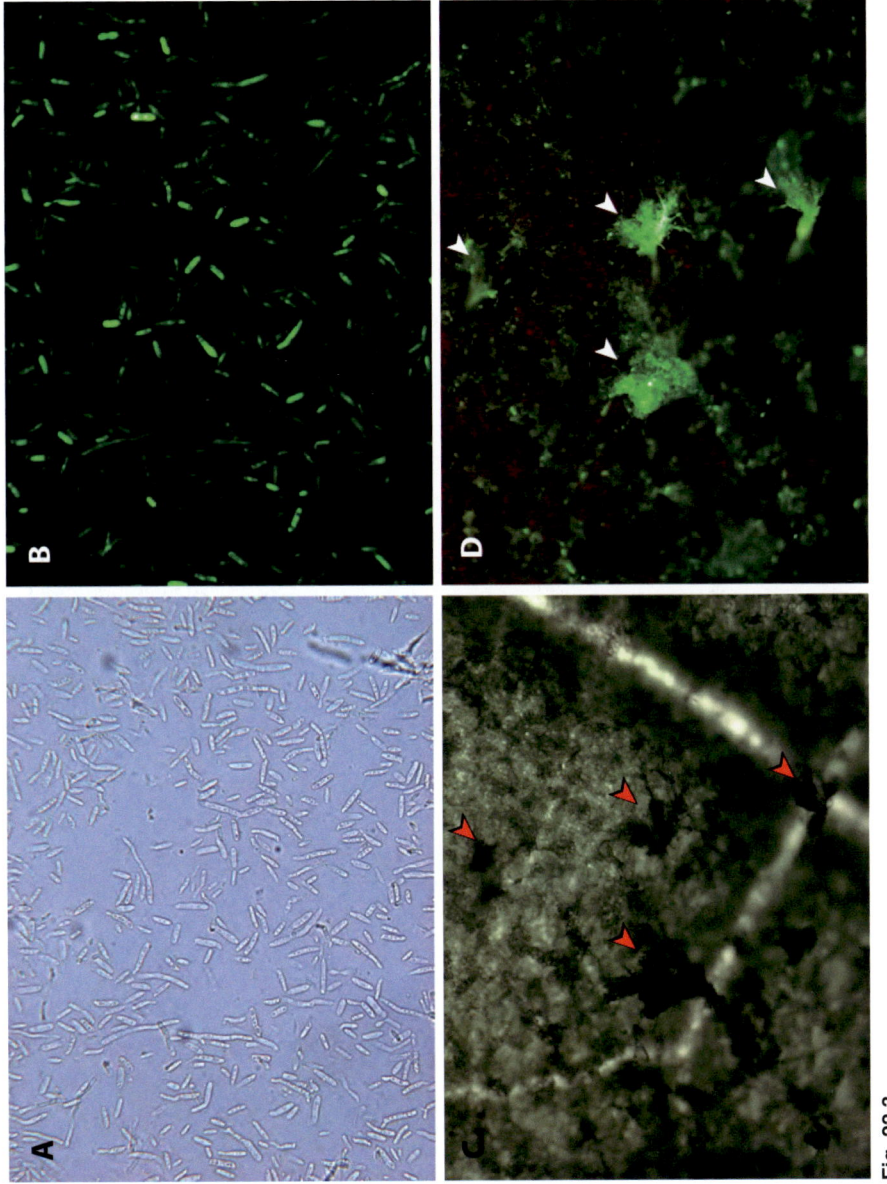

Fig. 22.3
Hyphae and conidia of *Pseudozyma flocculosa* expressing GFP, grown in liquid culture and visualized (A) under visible light and (B) under 488 nm light excitation. Growth of GFP-expressing *P. flocculosa* on the surface of a cucumber leaf, as visualized (C) under visible light and (D) under 488 nm light excitation. Growth of this biocontrol agent correlates with the location of powdery mildew colonies (arrows).

Table 23.1. Summary of vector transmission of plant viruses.

Transmission mode	Characteristics of transmission					Virus–vector examples		Viral-encoded proteins that assist in transmission	
	Acquisition time	Retention time	Virus in haemocoel	Passage across membranes	Replication	Vector	Virus genus/species[a]	Putative and known 'helper proteins'[b]	CP readthrough[c]
Non-persistent	s–min	min–h	No	No	No	Aphids	Potyviruses *Tobacco etch virus*	HC-Pro	No
							Carlaviruses	No	No
							Alfalmovirus	No	No
							Cucumovirus *Cucumber mosaic virus*	No	No
Semi-persistent	min–h	h–days	No	No	No	Whiteflies	Criniviruses *Lettuce infectious yellows virus*	CPm, *HSP70h, p59*	No
						Aphids	Closteroviruses *Beet yellows virus*	CPm, *HSP70h, p64, p20*	No
							Citrus tristeza virus	CPm, *HSP70h, p61*	
							Caulimoviruses *Cauliflower mosaic virus*	p2 (18 kDa), p3 (15 kDa)	No
							Waikaviruses *Anthriscus yellow fleck virus*	Yes (unidentified)	No
						Leafhoppers	Waikaviruses	No	
							Rice tungro virus	Yes (unidentified)	
						Mites	Tritimoviruses *Wheat streak mosaic virus*	HC-Pro	No

Continued

Table 23.1. Continued

Transmission mode	Characteristics of transmission					Virus–vector examples		Viral-encoded proteins that assist in transmission	
	Acquisition time	Retention time	Virus in haemocoel	Passage across membranes	Replication	Vector	Virus genus/species[a]	Putative and known 'helper proteins'[b]	CP readthrough[c]
						Nematodes	Tobraviruses		No
							Pea early browning virus	2b, 2c	
							Tobacco rattle virus	2b	
							Nepoviruses	No	No
							Grapevine fanleaf virus		
Persistent-circulative	h	days–life	Yes	Yes	No	Aphids	Luteoviruses	No	Yes
							Barley yellow dwarf virus		
							Poleroviruses	No	Yes
							Potato leafroll virus		
							Beet western yellows virus		
Persistent-propagative	h	days–life	Yes	Yes	Yes	Thrips	Tospoviruses	No	No
							Tomato spotted wilt virus		
						Leafhoppers, planthoppers	Reoviruses[d]	No	No
						Aphids, leafhoppers, planthoppers	Rhabdoviruses[e]	No	No
In vitro	min	Life of spore	N/A[f]	Possible (see text)	No	Olpidium	Tombusviruses	No	No
							Cucumber necrosis virus		

534   D. Rochon

						Vector	Virus group^a		
In vivo	N/A	Life of spore	N/A	Yes	No	Olpidium	Carmoviruses Aureusviruses Necroviruses Dianthoviruses Varicosaviruses Ophioviruses	No	No
	N/A	Life of spore	N/A	Yes	No	Plasmodiophorids	Benyviruses Furoviruses Pomoviruses Pecluviruses	No	Yes
							Bymoviruses	P2^g	No

^{a}The genus name for the virus group is shown along with examples of members (species) described in the text.
^{b}Refers to proteins encoded by the virus that assist in transmission but that are not a major structural component of purified virus particles. Proteins in italics are putatively involved in transmission.
^{c}Indicates that the virus CP RTD is either known or putatively involved in transmission.
^{d}Refers to members of the *Reoviridae* in general to show the various vectors involved in transmission of the various members.
^{e}Refers to members of the *Rhabdoviridae* in general to show the various vectors involved in transmission of the various members.
^{f}N/A = not applicable.
^{g}Refers to the potential involvement of the 'capsid-like' region of the P2 protein in transmission (see text).

aspects of virus–vector interaction. For details and additional material, readers are referred to several recent comprehensive reviews (Gray and Banerjee, 1999; Hull, 2002; Plumb, 2002; Gray and Gildow, 2003; MacFarlane, 2003; Rush, 2003; Rochon *et al.*, 2004; Whitfield *et al.*, 2005b; Ng and Falk, 2006b). For summaries on virus–vector interactions not covered in this chapter, readers are referred to two additional reviews (Hull, 2002; Plumb, 2002).

Vectors of Plant Viruses

Table 23.1 provides a summary of some common vectors and the viruses they transmit. More comprehensive summaries can be found in several recent reviews (Gray and Banerjee, 1999; Hull, 2002; Plumb, 2002; Gray and Gildow, 2003; MacFarlane, 2003; Rush, 2003; Rochon *et al.*, 2004; Whitfield *et al.*, 2005b; Ng and Falk, 2006b). Approximately 94% of plant virus vectors are arthropods, with insects representing 99% of the arthropods, and aphids representing 55% of the insects. Of the known aphids, nearly all are known to transmit at least one plant virus (Ng and Falk, 2006b). Outside of aphids, the most common vectors include whiteflies and leafhoppers, followed by the less common vectors, thrips, mites, beetles and nematodes. Fungi and plasmodiophorids (protists) represent the remaining least common virus vectors (Rochon *et al.*, 2004; Ng and Falk, 2006b).

The various vectors of plant viruses are known to transmit viruses from diverse taxa and viruses within the same family can be transmitted by different vectors. However, with few exceptions, members of a specific virus genus are transmitted by similar vectors (Table 23.1) (Gray and Banerjee, 1999; Ng and Perry, 2004; Rochon *et al.*, 2004). The latter is most likely related to the fact that a major specificity determinant for vector transmission is the viral coat protein (CP) (Hull, 2002), the sequence and structure of which is a major criterion for virus classification (Van Regenmortel *et al.*, 2000). However, as will be described below, viruses also use one or more additional viral-encoded proteins to assist in transmission.

Modes of Vector Transmission

Plant viruses have evolved diverse strategies for transmission by their vectors (Table 23.1). Nevertheless, each strategy requires three major steps: (i) acquisition of the virus from the host plant; (ii) retention within the vector; and (iii) the ability of the virus to be released from its vector for inoculation of a suitable host. Differences and similarities among these basic criteria have been used to classify the various modes of vector-mediated virus transmission (Gray and Banerjee, 1999; Hull, 2002; Plumb, 2002; Gray and Gildow, 2003; MacFarlane, 2003; Rush, 2003; Rochon *et al.*,

2004; Whitfield *et al.*, 2005b; Ng and Falk, 2006b). These include: the length of time required for acquisition of the virus by its vector, how long viruses can be retained prior to transmission, whether the virus circulates within its vector prior to release and, if so, whether replication of the virus occurs during its circulative pathway (Table 23.1). Some aspects of transmission by fungi and plasmodiophorids have been classified differently (Campbell, 1996; Rochon *et al.*, 2004) due to the distinct biology of the vectors (Table 23.1).

Table 23.1 outlines the various modes of transmission. Viruses that are acquired within a short period of time (i.e. seconds to minutes), and that are transmissible only within a short time following acquisition (minutes to hours), are referred to as non-persistently transmitted viruses. Viruses in which acquisition requires minutes to hours and that are transmissible for hours to days are referred to as being semi-persistently transmitted. Persistently transmitted viruses can require hours for successful acquisition, and are retained by their vectors for days and up to the life of the vector, and circulated through their vector prior to transmission to a suitable host. Viruses that replicate within their hosts during the circulative pathway are referred to as being transmitted in a persistent-propagative manner. Although these general modes of transmission were derived from studies on aphids, thrips and leafhoppers, they apply to transmission by most other arthropod vectors with piercing and sucking mouthparts.

The site of virus acquisition within the vector varies and is closely associated with the length of time required for acquisition. For example, viruses that are non-persistently transmitted are found in association with the stylet, whereas those that are transmitted semi-persistently are usually found associated with the foregut (Gray and Banerjee, 1999; Hull, 2002; MacFarlane, 2003; Ng and Falk, 2006b) (Fig. 23.1). Non-persistently and semi-persistently transmitted viruses have relatively unstable associations with the food canal as they can be lost during feeding probes. Persistently transmitted viruses are usually acquired in the midgut or hindgut and require circulation through the vector. The circulative pathway involves uptake from the lumen of the gut and passage through several membrane barriers (Fig. 23.1B and C). This accounts, in part, for the increased period of time between acquisition and transmission of persistently transmitted viruses. In the case of *Tomato spotted wilt virus* (TSWV), which is transmitted in the persistent-propagative manner, replication occurs in midgut cells and other tissues (Gray and Banerjee, 1999; Hull, 2002; Gray and Gildow, 2003; Whitfield *et al.*, 2005b). Circulative viruses eventually enter the salivary glands and are subsequently released into the salivary canal and egested along with salivary secretions into host cells (Fig. 23.1A).

It is well established that the capsid proteins of plant viruses play essential roles in transmission. However, recently it has become clear that many viruses require additional viral-encoded proteins for transmission. In most cases, these proteins are not major components of the virion and play additional roles in the infection process. As will be described, specific

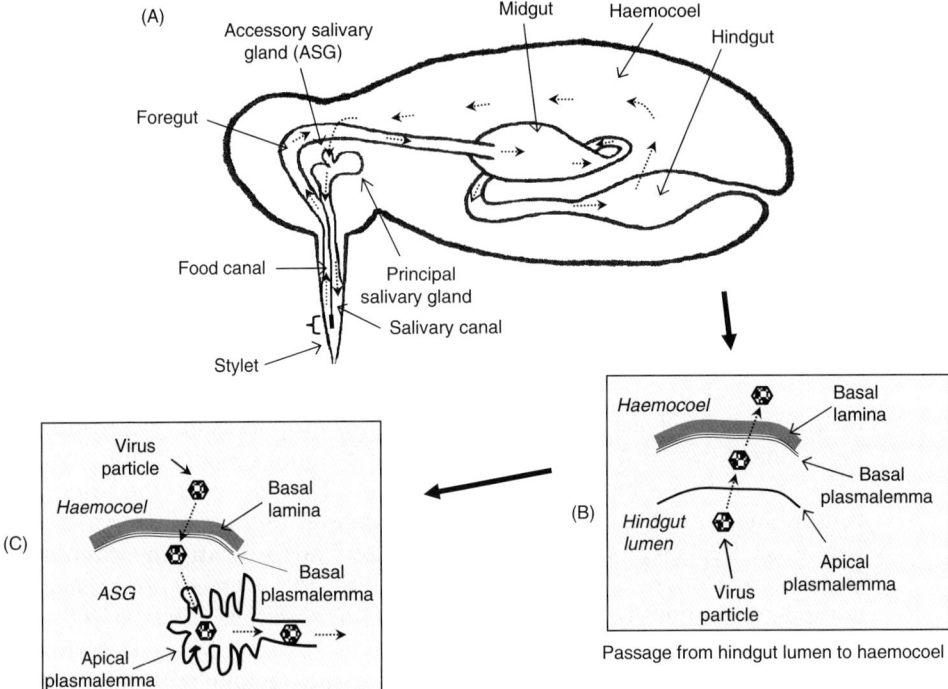

Fig. 23.1. Schematic diagram showing the circulative pathway of a virus through an aphid. (A) The relevant organs are labelled. Dashed arrows correspond to the route taken from the food canal through the aphid and into the salivary canal. The distal portion of the stylet where the food canal fuses with the salivary canal is shown by brackets (see text). Sizes of different insect parts are not drawn to scale. (B) Enlarged basic diagrammatic representation of the passage of a luteovirus particle from the hindgut to the haemocoel [see text and Gray and Gildow (2003) for details]. (C) Enlarged diagrammatic representation of the passage of a luteovirus particle from the haemocoel into the salivary canal [see text and Gray and Gildow (2003) for details].

vector components that associate with virions are being identified. These may represent receptor or receptor-like components that are required for the different stages of the transmission process.

Non-persistent Transmission

Aphids are the only known vectors for non-persistently transmitted viruses (Hull, 2002; Ng and Falk, 2006b). Acquisition of the virus occurs within seconds to minutes during brief probing of the aphid on a plant epidermal cell (Table 23.1). Virus uptake occurs during probing on a host plant but can also occur during probing on non-hosts while the aphid 'searches' for its host. Similarly, acquired virus can be inoculated to a host or non-host plant during subsequent probes by the aphid. During probing, aphids use their stylets to penetrate and access the contents of a cell by puncturing the cell wall and plasma membrane of an epidermal cell. Once a suitable host is found, the aphid will feed for longer times by probing deeper toward phloem sieve tubes. Non-persistently transmitted viruses are most often acquired and transmitted during the brief probing process that occurs while aphids 'search' for a host. In fact, when aphids feed on host plants, they are less likely to transmit a non-persistently transmitted virus since the aphid will likely not move from its host plant in the brief period during which a non-persistently transmitted virus is retained (Ng and Falk, 2006b). The aphid stylet is the site of binding for non-persistently transmitted virus. Although virus is found all along the anterior portion of the alimentary canal (proximal region of the stylet up to the foregut), particles associated with the distal portion of the stylet where it fuses with the salivary canal are the most efficiently transmitted (see bracketed area in Fig. 23.1A). Thus, it is currently believed that efficient transmission (of at least potyviruses) requires that particles become mixed with salivary secretions (rather than simply being regurgitated back through the stylet food canal) (Gray and Banerjee, 1999).

Table 23.1 shows some examples of non-persistently transmitted viruses. At the molecular level, the potyviruses and cucumoviruses have been the most extensively studied of the non-persistently transmitted viruses. Features of their transmission are described below.

Potyviruses

Potyviruses represent one of the most diverse groups of plant viruses and are responsible for significant agronomic losses in food crops throughout the world. The virion is a long flexuous filament that consists of a single CP species that is arranged helically about the monopartite positive-sense RNA genome. The CP subunit consists of two main regions: a highly variable ~30–100 amino acid (aa) N-terminal region, which is exposed on the outside of the particle, and an approximate 220 aa highly conserved C-terminal region that forms the particle core (Fig. 23.2A) (Shukla et al., 1988; Ng and Falk, 2006b). The potyvirus genome is translated into a single long polypeptide that is cleaved into several mature proteins that include the capsid, replication and movement proteins as well as viral polyprotein-specific proteases. One such protein is the helper component-protease (HC-Pro) which plays several roles in the virus infection cycle, including movement and suppression of RNA silencing (Plisson et al., 2003).

Potyvirus transmission requires the presence of particles as well as HC-Pro. The need for an additional viral-encoded protein for transmission of potyviruses was originally suspected when it was found that purified virus is not transmissible. In addition, it was found that a transmissible

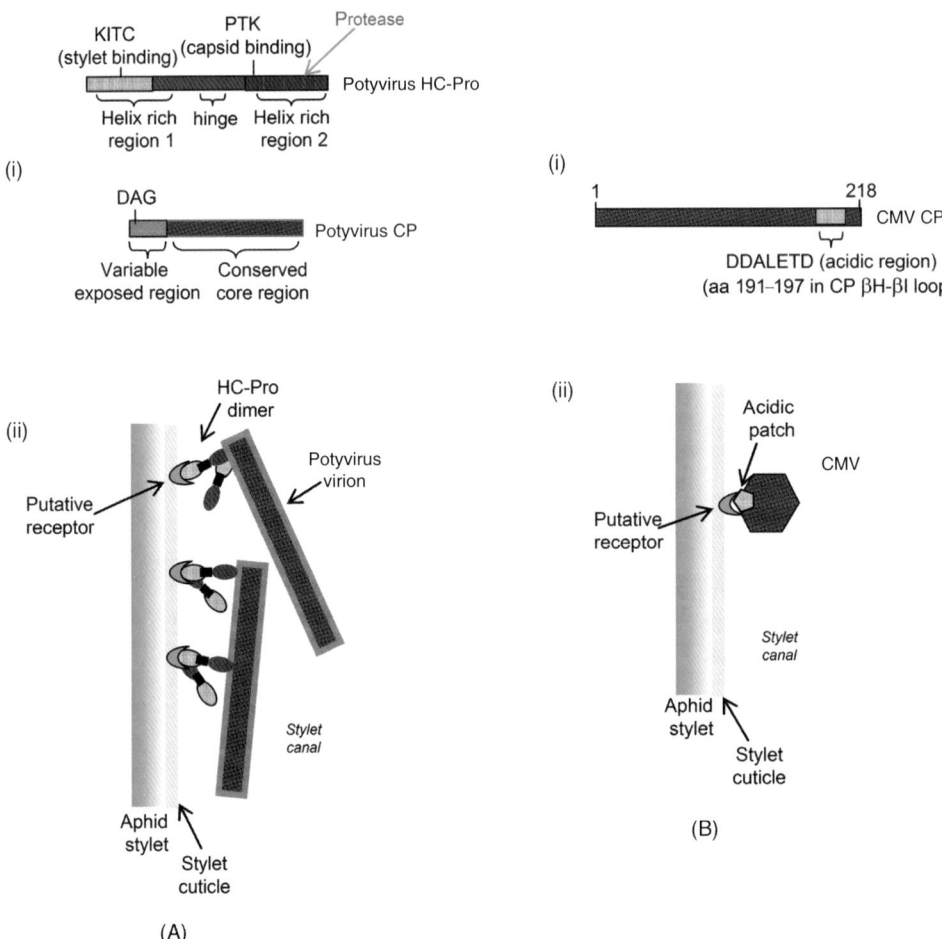

Fig. 23.2. Hypothetical models for interactions of non-persistently transmitted potyviruses (A) and cucumoviruses (B) with the aphid stylet. (A) Part (i) shows structural and functional sites of the potyvirus HC-Pro and CP. Part (ii) shows the hypothetical bridging model for interaction of HC-Pro with the aphid stylet and the N-terminal exposed region of the CP. Pattern schemes used for the structural domains of the CP in parts (i) and (ii) are equivalent. Although a dimer is shown in part (ii) it is possible that the bridging function of HC-Pro may involve other oligomeric forms. Also note that two possible modes of interaction between the HC-Pro dimer with the stylet and CP are shown, as it is not yet clear which parts of the oligomer are involved in the bridging interaction (Ruiz-Ferrer et al., 2005). The model also incorporates the findings (Torrance et al., 2006) that HC-Pro is found at the tip of the potyvirus particle, and the possible role this may have in aphid stylet binding and transmission. See text for additional details.
(B) Hypothetical model for interaction between CMV particles and the lining of the aphid stylet. The CMV CP is shown in part (i) along with the location and amino acid sequence of the acidic region implicated to serve as a site for aphid stylet interaction. Part (ii) shows a model for interaction between the acidic patch associated with pentomers of the particle and a putative receptor on the stylet (Bowman et al., 2002; Liu et al., 2002) (see text for additional references).

potyvirus could assist in transmission of a non-transmissible potyvirus when plants were co-infected. Later, it was found that aphid transmission occurs when HC-Pro is added to purified virus (Pirone and Blanc, 1996). It is now generally accepted that HC-Pro assists in transmission by acting as a bridge between the virion and the cuticular lining of the aphid stylet (Pirone and Blanc, 1996) (Fig. 23.2A). Detailed structural characterization of HC-Pro indicates that it is a homo-oligomer consisting of one or possibly several conjoined dimers. Transmission studies suggest that the oligomeric structure is the form that is involved in transmission (Plisson et al., 2005; Ruiz-Ferrer et al., 2005).

In vivo and in vitro assays have identified specific regions of HC-Pro that are required to bridge the stylet and CP and thereby contribute to transmission (Fig. 23.2A). A PTK (proline-threonine-lysine) tripeptide located in the C-terminal half of HC-Pro has been found to be important for transmission (Peng et al., 1998). The tetrapeptide KITC (lysine-isoleucine-threonine-cysteine), located in the N-terminal region of HC-Pro, contributes to transmission by binding to the aphid stylet (Blanc et al., 1998). Mutational analyses of the CP have shown that a DAG (aspartate-alanine-glycine) tripeptide along with adjacent amino acids within the N-terminal exposed region of the potyvirus CP is required for transmission (Atreya et al., 1995; Pirone and Blanc, 1996). HC-Pro–capsid overlay assays have shown that HC-Pro and CP proteins interact in vitro and that the region surrounding the DAG sequence interacts with the PTK motif in HC-Pro (Blanc et al., 1997) (Fig. 23.2A).

Although the potyvirus particle has been thought to consist of a single type of CP subunit, it has recently been found that HC-Pro is associated with one tip of approximately 10% of purified particles and colocalizes with the 5'-terminal viral genome-linked protein. The basis for this association which is not currently known as HC-Pro is known to play multiple roles including virus movement and transmission (Fig. 23.2A) (Torrance et al., 2006).

Interestingly, it has recently been found that the HC-Pro of the semi-persistently transmitted *Wheat streak mosaic virus* (a tritomovirus in the *Potyviridae*) also acts as a helper factor for mite transmission (Stenger et al., 2005). Thus, HC-Pro activity is not restricted to non-persistent transmission by aphids.

Cucumber mosaic virus

Particles of *Cucumber mosaic virus* (CMV) (a member of the *Cucumovirus* genus) have T=3 icosahedral symmetry, consisting of 180 identical CP subunits (Smith et al., 2000). CMV is transmitted by aphids in a non-persistent manner generally similar to that described for the potyviruses (Hull, 2002; Plumb, 2002; Ng and Perry, 2004; Ng and Falk, 2006b). However, numerous studies have shown that the CP alone contains the determinants for aphid transmission and therefore, unlike the potyviruses, a virus-encoded helper component is not required (Plumb, 2002; Ng and Perry, 2004; Ng and Falk, 2006b).

CMV has a tripartite genome wherein RNA-3 encodes the CP. Plants can be infected by inoculation with synthetic RNA corresponding to each of the three genome components. Synthetic chimeric RNA-3 molecules derived from different CMV strains and co-inoculated with RNA-1 and RNA-2 have been used to assess the role of specific CP regions or amino acids in aphid transmission of different CMV strains. When a series of such experiments were conducted using chimeras of the CP of the aphid-transmissible CMV-Fny and the non-transmissible CMV-M, it was found that 2 amino acid positions (aa 129 and 162) affected efficient transmission by the aphid *Aphis gossypii* and that an additional 3 positions (aa 25, 168 and 214) were required for transmission by *Myzus persicae* (Perry *et al.*, 1998; Ng and Falk, 2006b). Subsequent studies showed that aa 162 and 168 are important for virus particle stability (Ng and Falk, 2006b). Amino acid 162 is not exposed on the surface of the capsids (Smith *et al.*, 2000), suggesting that it may reduce transmission by reducing particle stability or conformation. Substituting the proline at CP aa 129 of CMV-Fny with the leucine that is at this position in CMV-M does not have a pronounced effect on particle stability. However, this change did lead to rapid necrosis of the plant and therefore might decrease transmissibility by affecting the quantity or accessibility of particles (Ng *et al.*, 2005; Ng and Falk, 2006b).

A high resolution crystal structure of CMV has been obtained and, based on this structure, specific mutations have been introduced to determine potential effects on aphid transmissibility (Smith *et al.*, 2000; Liu *et al.*, 2002; Ng and Perry, 2004). A conserved, highly acidic sequence, DDALETD (aspartate-aspartate-alanine-leucine-glutamate-threonine-aspartate),located in the CP βH-βI loop which is exposed on the capsid surface, has been found to have a pronounced effect on transmission without affecting particle stability (Fig. 23.2B). It has been speculated that the acidic patch may mediate electrostatic interactions with components of the lining of the aphid stylet during virus attachment (Smith *et al.*, 2000; Liu *et al.*, 2002). A hypothetical model for interaction between CMV and the aphid stylet is shown in Fig. 23.2B.

Semi-persistent Transmission

As opposed to non-persistent transmission, semi-persistent transmission involves relatively longer acquisition and retention times (Table 23.1) ranging from minutes to hours for the acquisition phase and hours to days for the retention phase. Several viruses from distinct families are known to be transmitted in a semi-persistent manner (Table 23.1). Similarly, semi-persistent transmission occurs among a variety of vectors, including whiteflies, leafhoppers, mites, nematodes and mealybugs (Gray and Banerjee, 1999; Ng and Falk, 2006b).

Cauliflower mosaic virus

A well-studied example of semi-persistent transmission is the aphid-transmitted *Cauliflower mosaic virus* (CaMV). CaMV is a DNA virus with an icosahedral capsid having T = 7 symmetry and consisting of 420 copies of the major CP (p4) (Van Regenmortel et al., 2000). As in non-persistent transmission of potyviruses, CaMV transmission requires virus-encoded helper factors. However, in CaMV, two such helper factors (p2 and p3), in addition to the major CP, have been identified (Fig. 23.3A). Several detailed studies have contributed to an understanding of the complex interaction between CaMV, its helper factors and the aphid stylet (Ng and Falk, 2006b). Recent studies have shown that p3 is loosely anchored within CaMV virions (Plisson et al., 2005). The p3 protein has been shown to interact with p2 and p2 interacts with the aphid foregut. Thus, p2 is functionally similar to HC-Pro in that it acts to bridge an interaction between the aphid mouthpart and virus particles (Fig. 23.3A). However, unlike HC-Pro, p2 interacts in an indirect manner through binding with a minor virion-associated component. Studies have shown that p2 can be found in the stylet without p3:virion complexes. Furthermore, p2:p3:virion complexes are found predominantly in the aphid stylet and are not preformed at a significant level in leaf cells. Thus, it appears that preformed p3:virion complexes interact with p2 following attachment of p2 to the stylet (Leh et al., 2001; Drucker et al., 2002) (see hypothetical model in Fig. 23.3A).

Regions within the N-terminal domain of the CaMV p2 protein have been suggested to contain determinants for attachment to the aphid stylet. A single amino acid change in the N-terminal region can result in loss of transmissibility or affect the efficiency of transmission of CaMV by its various aphid vectors. Moreover, a rapid and spontaneous mutation occurs at this same amino acid position when CaMV is placed under the selective pressure of acquisition by new vectors (Moreno et al., 2005).

Cryo-electron microscopy studies of CaMV particles decorated with p3 have provided insight into the structural arrangement of p3 within the capsid (Plisson et al., 2005). In this model, p3 forms a lateral network of digitations surrounding CaMV capsomeres. The N-terminal region of p3 forms dimeric digitations on the outside of the particle whereas the C-terminal region forms a trimeric anchor that embeds deeply within the shell. The p2 binding domain of p3 has previously been mapped to the N-terminal exposed region of p3 and therefore the structure is compatible with the notion that the p3 N-terminal region is accessible to p2. It is currently not known if p2 binds the p3 dimer or if it binds a monomer. In the latter case, dissociation of the dimer digitations would be required (Plisson et al., 2005). Figure 23.3A provides a hypothetical model for CaMV–stylet interactions.

Criniviruses and closteroviruses

Lettuce infectious yellows virus (LIYV) is a crinivirus in the *Closteroviridae* and is transmitted by whiteflies in a semi-persistent manner (Ng and Falk,

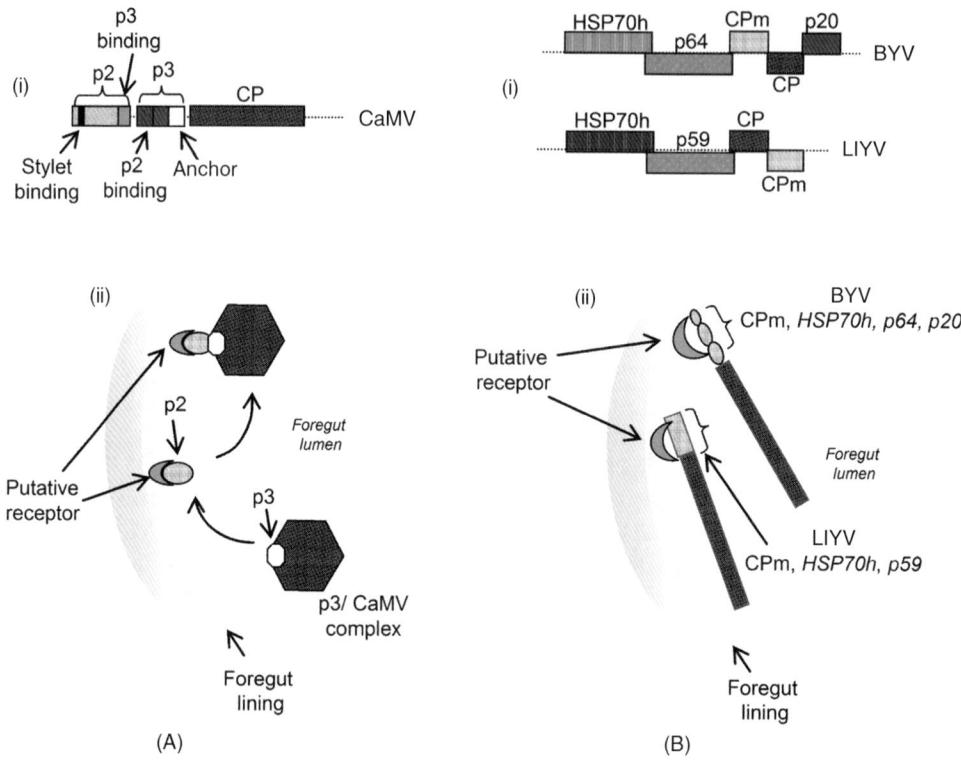

Fig. 23.3. Hypothetical models for interactions of the semi-persistently transmitted viruses, CaMV, BYV and LIYV, with the lining of the aphid stylet and foregut. (A) Organization of the CP, p2 and p3 coding regions in the circular (not shown) CaMV DNA genome is shown in part (i). Regions of p2 involved in binding p3 and the foregut are indicated by shading. Similarly, regions of p3 that bind p2 and the CaMV capsid are also shown. In part (ii) a hypothetical model for interaction between the CaMV particle and the foregut is shown. p3/CaMV complexes, preformed in infected leaves, enter and bind a preformed p2/foregut complex. Detailed shadings corresponding to p2 and p3 binding regions shown in part (i) are not shown. (B) Hypothetical model for interaction between the foregut and particles of BYV and LIYV. Part (i) shows the organization of the BYV and LIYV genome regions encoding the particle-associated proteins. Shading corresponds to proteins sharing sequence identity between BYV and LIYV. Part (ii) shows a model for interaction between the BYV and LIYV particles and the lining of aphid–whitefly foregut. A diagrammatic representation of the structure of the BYV particle and tip is shown. The orientations of the indicated proteins are not known so the entire region is shaded as CPm but indicated as consisting of other proteins. The LIYV particle is shown along with the location of CPm at the virion tip. As for BYV, the location of the particle-associated HSP70 and p59 are currently not known. Thus, only the known CPm regions are shown. Regions of BYV and LIYV that are believed to have roles in transmission are italicized. Italics indicate that the involvement of the particular protein in transmission is not known (see text).

2006b) (Table 23.1). The genome is bipartite, consisting of two long RNA species encapsidated in long filamentous particles (Fig. 23.3B). Studies have shown that highly purified LIYV virions can be successfully transmitted after *in vitro* acquisition, suggesting that transmission does not require an exogenously supplied helper component (Tian et al., 1999). LIYV virions are composed of major and minor CP components (CP and CPm, respectively), wherein CPm is associated with the approximate 5% of one end of the particle and the major CP the remaining 95% (Ng and Falk, 2006b) (Fig. 23.3B). Two additional viral-encoded proteins, HSP70h and p59, have also been found to be present in highly purified LIYV particles (Fig. 23.3B) (Tian et al., 1999).

Particles of the aphid-transmitted closteroviruses, *Beet yellows virus* (BYV) and *Citrus tristeza virus* (CTV), contain similarly organized bipolar virions in which CPm coats approximately 5% of one end of the particle and the major CP the remaining 95%. The HSP70h and p59 homologues of LIYV are also present as minor proteins in BYV and CTV particles. In BYV, particles have been found to contain an additional protein (p20) (Agranovsky et al., 1995; Satyanarayana et al., 2004; Alzhanova et al., 2007) (Fig. 23.3B). The CPm of LIYV (Tian et al., 1999) and both the CPm and CP of BYV appear to be required for transmission (He et al., 1998). In the case of LIYV, incubation of particles with antibodies to CPm prior to feeding aphids inhibited transmission, whereas a similar experiment using antibodies to the major CP did not (Tian et al., 1999). Additionally, LIYV particles that lack an intact CPm are not transmissible (Ng and Falk, 2006a).

It is known that the other minor virion proteins of BYV and CTV (with the exception of p20) are involved in capsid assembly as well as viral movement (Satyanarayana et al., 2004; Alzhanova et al., 2007). However, their roles in the transmission process remain to be determined. A tentative model for interaction of the closteroviruses with the food canal is shown in Fig. 23.3B.

Persistent Non-propagative Transmission

The persistent non-propagative transmission of members of the *Luteoviridae* has been studied extensively for many years. The *Luteoviridae* are responsible for a number of economically important diseases of food crops and have received much attention with regard to their epidemiology, transmission and molecular properties. The family *Luteoviridae* consists of three genera: the *Luteoviruses* (e.g. *Barley yellow dwarf virus* [BYDV]); *Poleroviruses* (e.g. *Potato leafroll virus* [PLRV], *Beet western yellows virus* [BWYV]) and *Enamoviruses* (e.g. *Pea enation mosaic virus* [PEMV]) (Van Regenmortel et al., 2000). Although members of these virus groups have distinctive genome structures, they are similar with respect to the aphid transmission process (Gray and Gildow, 2003). Hereafter, when not specified, *Luteoviridae* members will be referred to as luteovirids.

Luteovirid transmission requires circulation through its aphid vector prior to successful transmission but does not involve replication (Table 23.1). The circulative mode of transmission is responsible for the persistent aspect of luteovirid transmission wherein aphids can remain viruliferous for days and up to the life of the aphid (Table 23.1). Transmission of luteovirids can be described as occurring in distinct stages (Gray and Gildow, 2003): (i) ingestion of virus particles into the food canal during feeding on its host plant; (ii) passage into the insect gut; (iii) passage across the gut epithelial cells and into the haemocoel; and (iv) movement from the haemocoel through the accessory salivary gland (ASG) followed by egestion of virus into plant phloem tissue (Fig. 23.1A). Ingestion of particles occurs when the aphid stylet penetrates through the cell wall and into the cytoplasm of the phloem sieve elements, the site of luteovirid multiplication. One barrier to the specificity of luteovirid acquisition occurs at the stage of transport of ingested virus from the lumen of the gut across the gut epithelial cells and into the haemocoel (Fig. 23.1B). The second barrier is movement from the haemocoel through the aphid salivary gland cells (Fig. 23.1C). The first barrier is relatively non-specific, whereas the ASG barrier plays a much more crucial role in specificity. Additional information on movement across these barriers via transcytosis is provided below and in detail in Gray and Gildow (2003).

Movement of luteovirids from the gut lumen to the haemocoel

The cellular mechanism underlying passage of different luteovirids from the lumen of the alimentary canal across gut cells and into the haemocoel appears to be similar and likely involves a ligand–receptor interaction. A model has been proposed by Gildow (1999) using evidence from ultrastructural studies. In this model, virus present in the lumen of the gut binds to a putative receptor associated with the apical plasmalemma of a gut cell. Invagination of the particle into coated pits occurs followed by complete envelopment of the particle into coated vesicles that later fuse with larger uncoated vesicles. These vesicles become restructured into tubular vesicles containing rows of virus particles. The tubular vesicles then migrate to and fuse with the basal plasmalemma of the gut cell. Fusion with the plasmalemma allows virus to be released from the gut cell and to pass through the basal lamina into the aphid haemocoel. This process is similar to endocytosis–exocytosis mechanisms and strongly suggests the role of specific aphid receptors in luteovirus transmission (Gray and Gildow, 2003).

Movement of luteovirids from the haemocoel into the ASG

Movement of luteovirids into the ASG also appears to involve endocytosis and exocytosis (Gildow, 1999; Gray and Gildow, 2003). Luteovirids are found concentrated near the basal lamina that surrounds the ASG. They are not known to be associated with the principle salivary gland or the basal lamina of other aphid organs or tissues, suggesting that uptake into the ASG is highly specific (Gray and Gildow, 2003). Entry into the ASGs requires passage across the basal lamina as well as the basal plasmalemma that surrounds the secretory cells within the salivary gland. Virus particles are endocytosed at the basal plasmalemma forming virus-containing tubular vesicles. Single particles bud from the tubules into coated vesicles. The coated vesicles then fuse with the apical plasmalemma of the ASG, and then become released into the salivary canal. Virus is then delivered into plant phloem cells via salivary secretions during aphid feeding. Specificity has been shown to function at two ASG barriers, the basal lamina and the plasmalemma, wherein some luteovirid viruses cannot pass through the basal lamina, whereas others can pass this barrier but not that of the plasmalemma (Gildow, 1999; Gray and Gildow, 2003).

Luteovirid-encoded components involved in transmission

Luteovirid particles are small (~25 nm) non-enveloped T = 3 icosahedra that encapsidate a monopartite positive-sense RNA genome that is 5.6–6.0 kb in size. The virus capsid consists of 180 major CP subunits in the range of 22–24 kDa. A minor and variable number of the 180 subunits are readthrough products of CP ORF (Fig. 23.4). The full-length readthrough protein has a predicted size of ~72 kDa but particles also contain smaller such proteins in the range of 55–58 kDa (Van Regenmortel et al., 2000). The CP readthrough domain (RTD) is exposed on the particle surface. Both the CP and the RTD have been implicated in transmission (see below). There is currently no evidence that the luteovirid particle requires a helper component for transmission (Table 23.1) (Plumb, 2002; Ng and Falk, 2006b).

Role of the CP
'CP only' particles can be assembled from luteovirid mutants; however, such particles are not aphid transmissible even though they are capable of establishing systemic infection (Chay et al., 1996; Bruyere et al., 1997; Reinbold et al., 2001). Such mutants can also be ingested by aphids and are found in the hemolymph (Chay et al., 1996; Gildow et al., 2000;

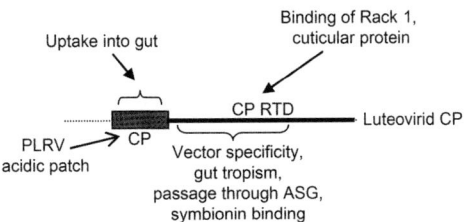

Fig. 23.4. Transmission of luteovirids and involvement of CP and RTD regions. CP structure of luteovirids, highlighting the various roles believed to be associated with the CP and RTD (see text for details).

Reinbold et al., 2001). Therefore, it appears that the CP subunit contains the required determinants for uptake into the gut but may not contain the determinants required for subsequent stages of transmission. PLRV particles lacking the RTD can be produced by baculovirus expression of CP in insect cells. Particles injected into the aphid haemocoel are subsequently found in the ASGs and salivary ducts, suggesting that the CP contains determinants for movement through the ASG (Gildow et al., 2000). However, in similar experiments using baculovirus expressed CP RTD mutants of BWYV, virions were not found in or near the ASG cells (Reinbold et al., 2001).

Structural modelling has identified features of particles common to most luteovirids. In conjunction with results of mutational analyses, an exposed acidic patch found along the perimeter of the particle trimer (pseudo-threefold axis) appears important for aphid transmission (Terradot et al., 2001; Lee et al., 2005).

Role of the CP RT domain
As described above, luteovirid particles that lack the RTD are not transmissible. RTD mutants lacking the C-terminal domain are efficiently transmitted, indicating that this region is not essential for transmission (Bruyere et al., 1997). When the N-terminal region of the luteovirid CP RTD is deleted, the mutated protein does not become incorporated into particles and therefore its role in transmission is difficult to assess in this manner. As will be described below, *in vitro* studies have demonstrated that the N-terminal region of the RTD binds symbionin (van den Heuvel et al., 1997).

A gene exchange strategy was used to assess the roles of the CP and CP RTD in BWYV and a close relative, *Cucurbit aphid-borne yellows virus* (CABYV), in aphid transmission. Both viruses are poleroviruses but they have distinctive aphid specificities as well as gut tropism. BWYV associates with the posterior midgut while CABYV associates with the posterior midgut as well as the hindgut. Hybrid virions containing the CP of BWYV and the RT domain of CABYV and vice versa were produced in plants following inoculation of the respective chimeric cDNA clones. Aphid transmission experiments demonstrated that both vector specificity and aphid gut tropism were determined by the RTD (Brault et al., 2005). Roles of the various regions of the luteovirid CP and RTD are shown in Fig. 23.4.

The BWYV CP and its RT domain have recently been found to be glycosylated in virions (Seddas and Boissinot, 2006). Removal of carbohydrate moieties using periodate oxidation or deglycosylation was found to inhibit aphid transmission. Transmission was also inhibited by lectins specific to alpha-D-galactose when aphids were fed through artificial membranes. The results indicate that glycosylation could play an important role in virus uptake across membrane barriers but it cannot be excluded that the glycoproteins influence particle stability rather than the transmission process per se.

Aphid components involved in transmission

Successful transmission of luteovirids requires the ability to overcome several barriers for entry into the haemocoel and ASG (Gray and Gildow, 2003). With regard to the ASG barriers, two proteins in the heads of transmitting aphids that specifically bind BYDV-MAV virus particles have been identified using overlay assays (Li et al., 2001). An anti-idiotypic antibody raised to a BYDV-MAV-specific monoclonal antibody bound the same two proteins and prevented virus movement from the haemocoel to the salivary glands. These studies suggest that the two identified aphid proteins may be involved in BYDV transmission, possibly by regulating the ability of virus to move across the salivary gland barrier.

Overlay assays using wild-type and mutant BWYV particles and aphid proteins separated by SDS-PAGE or two-dimensional gel electrophoresis have identified additional proteins that may interact with BWYV during circulative transmission (Seddas et al., 2004). Using mass spectrometry, five main proteins were identified: a cuticular protein, symbionin and three membrane-associated proteins: Rack 1, GAPDH3 and actin. The BWYV RTD was implicated in binding to Rack 1 and the cuticular protein. Rack 1 is a highly conserved multifunctional protein known to regulate several cell surface receptors, including integrins, which are components of the basal lamina of invertebrates. Rack 1 is also involved in actin organization in the cytoskeleton. GAPDH3 plays roles in glycolysis but can also bind membranes and regulate endocytosis and exocytosis. It is possible that these proteins are important for passage of luteovirids across aphid membrane barriers in the gut and ASGs (Seddas et al., 2004). Two additional proteins that bound BWYV particles were identified as symbionin.

Luteovirid particles have previously been found to bind symbionin (van den Heuvel et al., 1999) which is a major protein produced by endosymbiotic bacteria harboured by aphids. Symbionin is a homologue of GroEL, a protein with known functions in protein folding and trafficking across membranes. Binding studies have shown that symbionin from several aphid species binds to a variety of luteovirids *in vitro* (van den Heuvel et al., 1999). However, the binding specificities observed *in vitro* do not reflect known luteovirid–aphid specificities, suggesting that if symbionin does play a role in transmission, the effect is likely to be non-specific (van den Heuvel et al., 1997). Binding of symbionin has been suggested to prevent proteolytic digestion of the particle in the hostile environment of the aphid hemolymph. Alternatively, it could promote transfer of particles to the ASG or contribute to a conformational change required for ASG uptake (Gray and Banerjee, 1999; Gray and Gildow, 2003).

Persistent-propagative Transmission

Thrips (*Thysanoptera*) are a highly diverse order of insects known to transmit at least 10 virus species in the *Tospovirus* genus (Whitfield et al.,

2005b). Transmission of tospoviruses occurs in a persistent manner (Table 23.1), involving circulation of the virus from the alimentary canal to the salivary glands. To achieve this, the virus must pass across at least six membrane barriers. Tospoviruses replicate in their vectors and therefore its adaptation to the insect host for successful replication in the vector will affect transmission efficiency. Transmission of tospoviruses is therefore a highly complex process involving several components of both the virus and the vector (Whitfield et al., 2005b).

Tospovirus circulation and replication in thrips

The thrips species that transmit tospoviruses are piercing sucking insects. The mouthparts pierce through epidermal cells and take up the contents of the mesophyll cytoplasm. As stated above, tospoviruses must pass through several barriers in order to be successfully transmitted (Whitfield et al., 2005b). Virus passes from the foregut to the midgut lumen, whereupon it crosses its first membrane barrier, the apical membrane of the microvilli that line the lumen. Virus replication occurs in the midgut cells. Virus then passes through the basement membrane (barrier 2) and then enters the visceral muscle cells (barrier 3) where replication also occurs. The virus then leaves the visceral muscle cells (barrier 4) and enters the primary salivary gland by crossing another basal membrane (barrier 5). Exit across the microvilli (barrier 6) that line the lumen of the salivary canal occurs, whereupon virus present in salivary secretions can be inoculated into its plant host during thrips feeding (for a detailed discussion of this process, see Whitfield et al., 2005b). The finding that tospovirus viral inclusion bodies and non-structural proteins are detected in thrips provides evidence that replication indeed occurs within the vector. TSWV virus proteins have been found in association with secretory vesicles where particles appear to undergo fusion and budding. Thus, exocytosis appears to be involved in transport of TSWV virion and non-structural proteins. The ability of virus to replicate and to pass across these specific barriers likely accounts for the specificity of tospovirus–thrips interactions.

Virus–vector interactions

Tospoviruses are classified along with the animal-infecting *Bunyaviridae* and, as such, share many similarities in genome and particle structure (Van Regenmortel et al., 2000). The genome consists of three ssRNA components that together contain both negative-sense or ambisense coding regions. Viral-encoded nucleoprotein encapsidates the viral RNA and this along with a few copies of the viral RNA-dependent RNA polymerase forms the core of virions. A host-derived double membrane envelops the core. The membrane contains many copies of two viral-encoded glycoproteins (G_N and G_C) that project outward from the membrane as monomers and dimers. Replication of tospoviruses in thrips is likely similar to that of *Bunyaviridae* that infect vertebrates (Whitfield *et al.*, 2005b)

G_N and G_C are logical candidates for interaction with a putative receptor for entry into thrips midgut cells due to their position in the virion membrane. Also, the homologous glycoproteins of other *Bunyaviridae* mediate acquisition by their arthropod vectors and also act as ligands for receptor interaction (Whitfield *et al.*, 2005b). In other *Bunyaviridae*, the membrane glycoproteins initiate fusion with the host membrane, leading to endocytosis and ultimate release of virus particles into the cytoplasm for replication. Anti-idiotypic antibodies to G_C and G_N specifically label midgut cells wherein G_C labels the microvilli lining and G_N the midgut basal membrane (Bandla *et al.*, 1998; Whitfield *et al.*, 2005b). In addition, overlay assays show that the anti-idiotypic antibodies recognize a 50 kDa thrips protein, suggesting that this protein may act as a TSWV entry receptor (Bandla *et al.*, 1998; Medeiros *et al.*, 2000).

Interestingly, the N-terminus of G_N contains an RGD (arginine-glycine-aspartate) motif that is known to be involved in binding integrins. Moreover, this motif has been found to be important in binding of several viral structural proteins to host cells (Akula *et al.*, 2002). In an elegant series of experiments, it was found that the ectodomain of baculovirus-expressed G_N binds specifically to the midgut when fed to thrips, and moreover, competitively inhibited uptake of virus into the midgut. Thus, G_N may specifically bind to the thrips midgut in a ligand–receptor interaction. Further studies have indicated that G_C may serve as a fusion protein that mediates entry into vector cells. TSWV G_C shares a high level of sequence conservation with other *Bunyaviridae* in which the G_C protein was found to be involved in fusion in a pH-dependent manner (Pekosz and Gonzalez-Scarano, 1996; Whitfield *et al.*, 2004). It has recently been found that G_C is cleaved at low pH, consistent with a proposed role in pH-dependent fusion (Whitfield *et al.*, 2005a).

Semi-persistent Transmission by Nematodes

Nematodes are known to be involved in transmission of tobraviruses and nepoviruses (Table 23.1). Nematode vectors belong to distinct species within the families *Longidoridae* and *Trichodoridae*. The longidorids transmit about 30% of known nepoviruses and the trichodorids transmit tobraviruses. These nematodes are ectoparasites that acquire and transmit viruses by feeding on the roots of plants. Acquisition occurs when the extendable stylet of the nematode penetrates cells at or near the root tip and withdraws the contents of the cell. Subsequent feeding on the root of another plant results in inoculation of virus (Brown *et al.*, 1995; Plumb, 2002; MacFarlane, 2003).

Nematodes can retain ingested virus for long periods of time and can also transmit virus to several plants during serial feedings. Studies indicate that ingested virus binds to the lining of the food canal and oesophagus, possibly via carbohydrate moieties present in the lining (Brown *et al.*,

1995). However, virus does not circulate within the nematode. Virus particles are lost during molting of the nematode in which the external cuticle as well as the cuticular lining of the mouthpart are shed. Therefore, nematode transmission is classified as being semi-persistent (Table 23.1) (Brown *et al.*, 1995; Plumb, 2002; MacFarlane, 2003).

Transmission of tobraviruses

The three known tobraviruses ((*Tobacco rattle virus* [TRV], *Pea early browning virus* [PEBV] and *Pepper ringspot virus* [PepRSV]) each have bipartite RNA genomes that are encapsidated into rod-shaped particles. The CP is encoded on RNA-2 along with 2 additional proteins, 2b and 2c. PEBV has an additional ORF that may produce a 9 kDa protein (Fig. 23.5A). The 2b proteins of TRV and PEBV have significant amino acid sequence identity, whereas the 2c proteins have little or no discernable sequence similarity (MacFarlane, 2003).

Infectious clones derived from several tobraviruses have facilitated a more precise assignment of the regions of the CP and other proteins encoded on RNA-2 in nematode transmission. Studies of recombinant tobravirus genomes containing ORFs or partial ORFs of one coding region inserted into another have revealed that the 2b protein encoded by RNA-2 plays a critical role in nematode transmission, since its deletion results in loss of transmission of both TRV and PEBV. The 2b protein has been found to complement transmission of a mutant that lacks the 2b protein when co-inoculated into plants (Vassilakos *et al.*, 2001), suggesting that 2b acts as a helper component similar to that of the potyvirus HC-Pro. Deletion of 15 aa at the C-terminus of the CP of the PEBV and *Tobacco rattle virus* (TRV) isolates also results in loss of transmissibility (MacFarlane *et al.*, 1996). This region of the CP is located on the surface of the particle and has been shown to be relatively unstructured (Blanch *et al.*, 2001), raising the hypothesis that this portion of the CP may interact with other viral or nematode factors to promote transmissibility.

In vitro studies have shown that the TRV 2b protein can interact with the CP, but that removal of the C-terminal domain of the CP results in loss of binding (Visser and Bol, 1999). Further binding studies using a yeast two-hybrid assay have shown that 2b–CP binding interactions are complex. Deletion of a coiled-coil domain of the 2b protein prevents its interaction with CP but removal of the N- and C-terminal regions enhances CP interaction (Holeva and MacFarlane, 2006). Studies showing binding of 2b with the nematode food canal should assist in verifying the role of 2b in bridging an interaction between TRV particles and the nematode food canal.

The 2c protein of PEBV has a significant effect on transmission; however, the 2c protein of TRV does not. It is possible that the difference is due to these viruses being transmitted by nematodes belonging to different tichodorid genera. It may additionally suggest that PEBV requires a third protein for successful transmission as has been observed for caulimoviruses (see

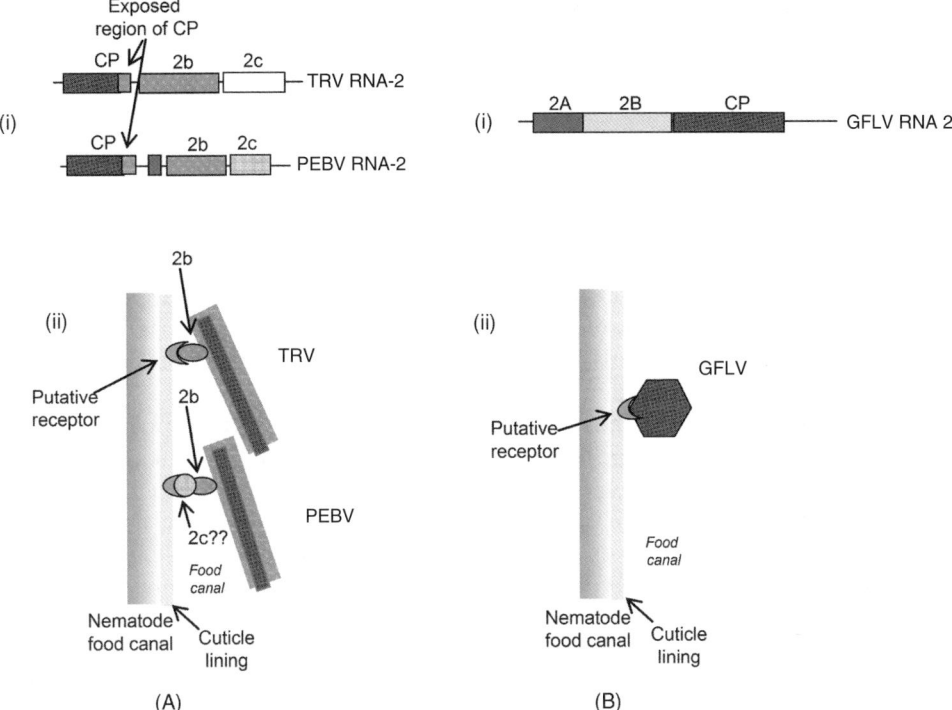

Fig. 23.5. Models for interaction between tobraviruses and GFLV with the nematode food canal. (A) Tentative model for interaction of tobravirus particles with the lining of nematode food canal. Part (i) shows the organization of proteins encoded by RNA-2 of TRV and PEBV. Similarly shaded regions designate sequence similarity. An exposed approximate 15 aa region at the C-terminus of the CP that binds to 2b *in vitro* is indicated by the arrows. Part (ii) shows hypothetical models for interaction of TRV and PEBV with the lining of the food canal. Proteins are shaded as in part (i). TRV may only require the 2b protein for attachment of particles to the lining of the food canal. PEBV requires both the 2b and 2c proteins for transmission; however, the role of 2c is not known and is only incorporated into the model to point out its putative role in attachment (see text for references and additional details). (B) Hypothetical model for interaction between GFLV and the lining of the nematode food canal. The genome structure of GFLV RNA-2 is shown in part (i) and the tentative model for transmission in part (ii).

above). A hypothetical model for tobravirus interaction with the nematode stylet is shown in Fig. 23.5A.

Transmission of nepoviruses

Specific Longidorid species have been identified as nematode vectors for different nepoviruses. Several nepoviruses are nematode-transmitted, with *Tobacco ringspot virus* (TRSV), *Tomato ringspot virus* (ToRSV), *Arabis mosaic virus* (ArMV) and *Grapevine fanleaf virus* (GFLV) being examples (Brown *et al.*, 1995; Taylor and Brown, 1997; Plumb, 2002). Early studies using serology and pseudorecombinants indicated that the CPs of nematode-transmitted viruses contain molecular determinants for transmission (Harrison and Murant, 1977).

Immunofluorescence studies of TRSV retention sites within the nematode *Xiphinema americanum* indicated that particles were only present within the lumen of the stylet extension and oesophagus. Particle acquisition increased during a 0–22 day period and this increase correlated with increased transmission. The greatest increase in fluorescence and transmission was at 0–5 days. Virus was found to be most prominent between the junction of the stylet and stylet extension, and the junction of the oesophagus and intestine. Fluorescence was not found in the stylet (Wang and Gergerich, 1998).

In-depth molecular studies of nematode-mediated nepovirus transmission have been hampered by difficulties in obtaining infectious clones of these viruses. However, infectious clones of *Grapevine fanleaf virus* (GFLV) have been produced, enabling molecular studies of its transmission. GFLV has a bipartite genome encapsidated in an icosahedral shell. RNA-2 encodes a polyprotein that is cleaved into three mature proteins (2A, 2B and 2C) (Fig. 23.5B). The 2A protein is a replication-associated factor, while 2B serves as a viral movement protein and 2C is the CP. Hybrids of infectious clones of GFLV (transmitted by *Xiphinema index*) were constructed in which segments of GFLV RNA-2 were replaced by the corresponding segments from the closely related nepovirus, ArMV (transmitted by *X. diversicaudatum*). In initial studies, it was found that the CP as well as the last nine residues of the 2B gene were required for successful transmission. Further mutational analyses indicated that the last nine residues of the 2B gene are dispensable. Thus, in the case of GFLV, the CP is sufficient for transmission (Andret-Link *et al.*, 2004).

Transmission by *Olpidium*

Two species of *Olpidium* are known to transmit plant viruses: *O. bornovanus* and *O. brassicae* (Table 23.1). *Olpidium* spp. (hereafter referred to as olpidium) is a member of the true fungi in the *Phylum Chytridiomycota*. The life cycle of *Olpidium* spp. involves the production of motile zoospores as a means of dispersal, along with the production of resting spores that enable long-term survival. The body of the zoospore is ellipsoidal and has a single flagellum (Campbell, 1996; Plumb, 2002; Rochon *et al.*, 2004). The zoospore body and flagellum are surrounded by a membrane sheath and an external matrix containing specific mannose- and fucose-containing oligosaccharides and glycoproteins (Kakani *et al.*, 2003; Rochon *et al.*, 2004). Infection of roots occurs when zoospores encyst upon a root cell. A hole in the cyst and a papillum produced by the root cell in response to infection provide a means for the cyst protoplast to enter the host cell and initiate infection. Thalli will develop into zoosporangia within the infected cells, followed by the formation of new zoospores. These zoospores are then released via an exit tube into the soil or water medium surrounding root cells. The thallus may also go on to develop into an environmentally stable resting spore, which can germinate under proper conditions to produce motile zoospores (Campbell, 1996; Rochon *et al.*, 2004).

Modes of *Olpidium* transmission

Two modes of acquisition, termed *in vitro* and *in vivo* acquisition, have been recognized for *Olpidium*-transmitted viruses (Campbell, 1996; Rochon *et al.*, 2004) (Table 23.1). Viruses transmitted following *in vitro* acquisition are various members of the *Tombusviridae* (Table 23.1) (Campbell, 1996; Rochon *et al.*, 2004). *In vitro* acquisition occurs when virus particles attach to the membrane of the motile zoospore in the soil or water medium and then become transmitted to roots when zoospores infect a root cell. It is believed that uptake of the virus into the root cell occurs when virus, attached to or possibly internalized by the zoospore plasmalemma (Fig. 23.6), enters root cells during infection. Virus acquired in this manner is not believed to become incorporated into zoospores during co-infection; however, they are found in a relatively stable association

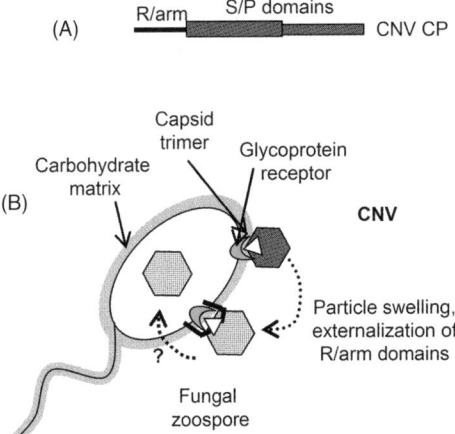

Fig. 23.6. Tentative model for interaction of CNV particles with zoospores of *O. bornovanus*. (A) Structure of the CNV CP showing the R and arm domains and the shell (S) and protruding (P) domains. (B) Hypothetical CNV–zoospore interactions. The trimer region of the capsid is shown as interacting with a putative glycoprotein receptor; however the direct role of the trimer has not been established (see text). CNV is then shown to undergo a conformational change wherein the R–arm regions are externalized, potentially stabilizing interaction with the zoospore membrane. It is not known if particles undergo a conformational change prior to or during the process of binding (see text). Similarly, it is not known if CNV enters the zoospore as part of the transmission process.

at or near the surface of the resting spore. Highly purified virus particles are sufficient for transmission, suggesting that a non-virion helper component is not required for efficient transmission.

The other mode of *Olpidium* transmission, that occurs following '*in vivo* acquisition' (Table 23.1), involves virus uptake by zoospores during virus infection within host root cells. Virus is found within the resting spore and likely within zoospores as well. The association of *Olpidium*-transmitted viruses within the highly stable resting spore enables the virus to remain in a viruliferous state in the soil for years to decades. Viruses transmitted in this manner include the filamentous ophioviruses and varicosaviruses (Campbell, 1996; Rochon *et al.*, 2004). Lettuce big vein disease which infects lettuce worldwide is perhaps the most well-known disease resulting from transmission by olpidium (Lot *et al.*, 2002).

CNV CP components involved in transmission

The CNV particle has T=3 icosahedral symmetry and consists of 180 identical CP subunits. Studies of several CNV CP mutants with decreased efficiency of transmission showed that specific amino acids located in or near a cavity on the particle pseudo-threefold axis (the trimer) decreases the efficiency of transmission as well as zoospore attachment, without significantly affecting particle accumulation in plants (Kakani *et al.*, 2001). However, it is not clear if the amino acids are directly involved in mediating zoospore attachment or if they induce a conformational change in the particle that affects transmission. A subsequent study showed that CNV particles undergo conformational changes upon zoospore attachment (Kakani *et al.*, 2004). Partial proteolytic digestion of zoospore-bound particles *in vitro* indicates that the conformational state of bound particles is similar to that of swollen CNV since they have similar digestion profiles. The swollen conformation is a structural state that can be induced in many spherical viruses and results from electrostatic repulsion of negatively charged residues in the three asymmetric subunits at the pseudo-threefold axis. The observation that zoospore-bound CNV is in a swollen-like conformation might suggest that the mutations that reside at the pseudo-threefold axis may decrease transmission through their effect on the ability of CNV to adopt the proper conformation required for transmission.

The swollen state of many spherical viruses as well as CNV is accompanied by externalization of the normally inward facing CP RNA binding domain (R) and arm (Kakani *et al.*, 2004). A proline (Pro) to glycine (Gly) mutation thought to mediate externalization of the arm domain has been demonstrated to nearly eliminate CNV transmission (Kakani *et al.*, 2004). Mutant particles were found to be resistant to mild proteolytic digestion, as were zoospore-bound Pro to Gly mutant particles. Thus, it appears that mutation of the Pro residue regulates externalization of the R–arm regions and thereby interferes with adoption of the proper conformational change required for transmission (Kakani *et al.*, 2004). In addition, the inability of particles to externalize the arms could affect stable binding of particles to the zoospore (Kakani *et al.*, 2004). It is known that during attachment of poliovirus particles (which are structurally and evolutionarily related to CNV particles) to host cellular membranes, particles undergo a conformational change involving externalization of CP regions structurally analogous to the CNV R and arm domains. Thus, it is possible that the initial interaction of CNV with zoospores may be functionally similar to that of poliovirus, and moreover, that the conserved structural features reflect, in part, conservation of function.

Zoospore components involved in transmission

Further studies have suggested the involvement of a putative receptor on the zoospore membrane in CNV attachment (Kakani *et al.*, 2003). Binding of CNV to zoospores was found to be saturable and occurred more efficiently with vector than with non-vector zoospores. In addition, binding of CNV to zoospores is competitively inhibited by two other viruses transmitted by *O. bornovanus* (D. Rochon, 2006). Treatment of zoospores with protease and periodate reduced binding of CNV (without having a noticeable effect on zoospore structure), suggesting the involvement of glycoprotein(s) in CNV attachment. Furthermore, binding of CNV to zoospores was competitively and specifically inhibited by mannose- and fucose-containing oligosaccharides (the two major sugars on the zoospore surface) *in vitro*, suggesting that CNV attachment to zoospores involves recognition of specific sugars. CNV was found to bind several zoospore proteins in overlay assays using electrophoretically separated zoospore proteins (Kakani *et al.*, 2003). The nature of these proteins and their putative involvement in CNV attachment and transmission remain to be determined. A hypothetical model for CNV transmission by zoospores is shown in Fig. 23.6.

Plasmodiophorid Transmission

Plasmodiophorids have been variably classified as fungi or protists; however, recent molecular evidence indicates that plasmodiophorids are protists and are not related to true fungi (Plumb, 2002; Rush, 2003; Rochon *et al.*, 2004). Three species of plasmodiophorids are known vectors of plant viruses. These are *Polymyxa graminis*, *P. betae* and *Spongospora subterranea*. They are responsible for transmission of a variety of plant viruses with distinctive genome organizations, including bymoviruses (*Family Potyviridae*), benyviruses, pecluviruses, furoviruses and pomoviruses (members of unassigned families) (Table 23.1) (Plumb, 2002; Rush, 2003; Rochon *et al.*, 2004). These viruses are associated with several important widespread diseases, including rhizomania of sugarbeet, *Potato mop top virus*-induced tuber necrosis and several cereal diseases.

The benyviruses, pecluviruses, furoviruses and pomoviruses are all rod-shaped viruses with varying lengths, and the bymovirus particle is a flexuous filament (Kanyuka et al., 2003; Rochon et al., 2004). The genome structures and proteins produced by these viruses are highly diverse (Rochon et al., 2004). As with the aphid-transmitted luteovirids, the terminator codons for CPs of members of the benyviruses, furoviruses and pomoviruses can be read through, producing a protein with a C-terminal extension (RTD). The CP RTD has been hypothesized to play a role in transmission based on the observation that loss of transmissibility of the respective viruses is associated with various sized deletions in this region (Adams et al., 2001; Plumb, 2002; Rochon et al., 2004). In the case of the benyviruses and pomoviruses, the RTD is present at one extremity of the particle and there is experimental evidence that it contains transmission determinants. In some bymoviruses, it has been shown that a non-structural protein (P1), which shows sequence similarity to the potyvirus HC-Pro, undergoes deletion, and this too is associated with a loss of transmissibility (Plumb, 2002; Rochon et al., 2004).

Although some amino acid sequence identity can be discerned among the CPs of the benyviruses, furoviruses, pecluviruses and pomoviruses, limited or no sequence identity exists among the CP RTD, even among members of the same genus. This has restricted attempts to delimit regions involved in transmission. A KTER motif in the CPs of two benyviruses has been implicated in transmission efficiency (Tamada et al., 1996). However, this motif has not been found to be conserved among the group as a whole. A comparison of the RTDs of the benyviruses, furoviruses and pomoviruses, as well as the P2 protein of the bymoviruses, has shown that they each contain two putative transmembrane regions, one of which becomes deleted in non-transmissible viruses (Adams et al., 2001). It was speculated that the transmembrane regions of these proteins may assist in movement of virus from the host cytoplasm and across the plasmodial membrane during infection (Adams et al., 2001).

The bymovirus CP sequence does not share sequence similarity with that of other plasmodiophorid transmitted viruses. Rather, it is most similar to that of other members of the *Potyviridae,* raising questions regarding its role in plasmodiophorid transmission. It is interesting that the N-terminal region of the P2 protein of bymoviruses contains amino acid sequence similarity with the CP of several rod-shaped viruses, including furoviruses, benyviruses, pecluviruses, pomoviruses, and to a lesser extent, with other rod-shaped viruses such as tobamoviruses, hordei and tobraviruses (D. Rochon, unpublished data) (Dessens et al., 1995). Thus, it might be speculated that this region of P2 reflects the evolution of bymovirus transmission as being via acquisition of the CP gene of a benyvirus or furovirus-like ancestor for transmission by *Polymyxa*. This hypothesis could be initially investigated by determining if some bymovirus particles consist of P2 protein (either exclusively or as part of the bymovirus particle) and, if so, if the absence of the P2 protein in particles is associated with the loss of transmissibility.

Conclusions

The evolution of plant viruses is associated with adoption of a variety of transmission mechanisms for successful dissemination (Table 23.1). Nevertheless, some commonalities exist. These include the high specificity of virus–vector interactions and the likelihood that one or more ligand–receptor interactions underlie this specificity. In addition, the virus particle is required for transmission and it is likely that, in each case, the major CP species of the particle plays an important role.

Interestingly, little correlation exists between the mode of transmission adopted by various viruses and either their overall particle structure (rod, sphere, etc.) or the virus family with which they are associated. Similarly, viruses within a family may utilize a variety of vectors. Finally, a similar mode of transmission may occur among different vectors (i.e. nematodes, aphids, etc.). Many of these anomalies can be seen to be a reflection of a more complex interplay that not only involves the virus and its vector, but also the host.

Several differences that exist among the various molecular aspects of transmission include: (i) transmission of some viruses occurs via direct interaction of the virion with its vector, wherein the virion is comprised of single CP subunit; (ii) transmission sometimes requires a minor CP readthrough product, wherein the RTD plays an essential role in transmission; (iii) transmission of some viruses occurs via direct interaction of the virion with the vector; however, unlike that involving a single virion CP subunit, the virions of these viruses are composed of one or more minor viral-encoded species that facilitate transmission; and (iv) one or more exogenous viral-encoded, non-virion helper components are required to facilitate transmission of some viruses by bridging an interaction between virus particles and the vector. As additional research is conducted, further basic similarities and differences among molecular aspects of transmission will certainly be found.

Speculations and Future Directions

Recent research on molecular aspects of virus–vector interaction has yielded a wealth of new information. However, as in all quality research endeavours, the answers lead to several new and important questions. The questions are vital to achieving further in-depth insights into the complex but orchestrated processes that underlay virus–vector interactions.

Many virus diseases are controlled by the application of pesticides to reduce vector populations. Several of these pesticides have been shown to be costly, toxic and ecologically unsound. Research on virus–vector interactions will undoubtedly assist in the development of new approaches for managing virus disease in a more sustainable manner. As vector-mediated virus dissemination plays a major role in initiation and maintenance of virus infection of crops, methods for interfering with this interaction should play a major role in management of disease control.

Recent research on molecular aspects of virus transmission has, expectedly, raised a number of questions for future investigations into virus–vector interactions. Such questions include the following:

1. Do viruses that appear to require a single CP species in the virion actually contain additional minor virion components that facilitate transmission? It appears in the case of CMV, GFLV and CNV that a single CP species is sufficient for transmission. However, is it possible that other viral-encoded (or host-encoded) (see below) protein components might increase the efficiency of transmission? Considering the increasing number and variety of viruses that are being found to contain minor virion and helper components, this area of research may be fruitful in providing a more complete picture of factors required for efficient transmission.

2. Do virions contain host-encoded components that facilitate vector attachment and transmission? In the case of luteovirids, symbionins produced by bacterial endosymbionts are weakly associated with particles during circulative transmission and are hypothesized to play a non-specific but important role in mediating virus stability, transport or conformational change for successful transmission (see section on luteovirid transmission). However, there is currently no information regarding the possible role of virion-associated host components in non-circulative transmission. Incorporation of host components into the virion can be envisioned as a potential means to facilitate transmission by adopting the normal cellular role of the protein for aspects of the transmission process. Identification of host proteins associated with virions may be a new avenue for providing additional insight into vector transmission.

3. Do viruses undergo conformational change as a stage in the transmission process? Conformational change appears to be fundamental to the function of many proteins and is also known to be important in the early stages of host-cell infection of many animal viruses (Kakani *et al.*, 2004). In the case of CNV, conformational change is associated with zoospore attachment and appears to be essential for transmission (Kakani *et al.*, 2004). Although conformational change in CNV might be spontaneous (as a result of the 'breathing process') (Rochon *et al.*, 2004), it may be induced by zoospore components during attachment or by an unidentified virion-associated viral- or host-encoded component. As described above for the luteovirids, symbionin has been postulated to play a role in conformational change. Could one or more of the multiple proteins associated with the tails of LIYV, BYV and CTV be involved in particle conformational changes associated with transmission? Similarly, could the helper component proteins of potyviruses, caulimoruses and tobraviruses provide this additional function?

4. Do vectors play the additional role of assisting in virus uncoating for the initial stages of host-cell infection? As stated above, CNV undergoes a conformational change upon zoospore attachment that is similar to the swelling process proposed to be involved in virus uncoating. Thus, is it possible that in this case, the zoospore vector not only assists in CNV dis-

semination, but can also be seen as assisting in particle uncoating required for root cell infection? If so, can other virus–vector interactions similarly contribute to particle disassembly? This finding would expand the role of the vector beyond its current primary role in providing virus dissemination and entry into plant cells.

5. How do viruses that are transmitted in a non-circulative mode dissociate from the lining of the stylet or foregut? This question has been repeatedly raised by many researchers and deserves attention since it is a critical aspect of successful inoculation following virion acquisition. Is association of virus with the food canal of its vector simply inherently unstable? As described previously, virus that is attached at the distal part of the stylet near the salivary canal is the most efficiently transmitted of the particles bound to the food canal. Can the physical properties of salivary secretions contribute to particle release (i.e. pH, ions, etc.) by destabilizing virus–receptor interactions or interactions between the site of attachment and the various helper components? Could proteolytic digestion or the presence of other proteins in saliva (or already in association with the particle) contribute to release through cleavage or via a normal role in destabilization of virus–receptor interaction?

Applications to Virus Disease Management

A variety of control measures exist for reducing diseases caused by plant viruses. In general, control involves the concurrent use of various methods aimed at reducing sources of infection. One such method is the use of pesticides (insecticides, nematicides, fungicides, etc.) to reduce vector populations. Other strategies involve the use of virus resistant cultivars, including transgenic plant lines that have been engineered for virus resistance (Hull, 2002). However, despite the existence of diverse control strategies, many virus diseases remain difficult to manage. Measures to reduce virus infection therefore require the ongoing development of new strategies.

Pesticides can be effective in controlling vector-mediated virus disease. However, their use is limited by a number of factors. These include: the development of pesticide resistance; re-emergence of vectors following loss of pesticide activity; the known or potential toxic effects of pesticides on humans and other animals and disruption of ecological balance. Indeed, many effective pesticides have been banned due to their toxicity and negative ecological effects (Hull, 2002). The importance of vectors in the initiation and dissemination of virus disease in a crop therefore requires the development of new and sustainable strategies for controlling vector-mediated virus disease.

Recent research into virus–vector interactions holds promise for new methods to control vector transmission. However, as this research is still in its infancy, further in-depth analyses are required for the development of sound and rational methods. Nevertheless, some potential strategies aimed to interfere with virus–vector interactions can be envisioned and are briefly outlined below.

As described in this chapter, viruses likely interact with their vectors in a highly specific ligand–receptor fashion at one or more stages of the transmission process. Broadly speaking, these interactions include virus particle–vector interactions, helper component–vector interactions and helper component–virus interactions, etc. (see Figs 23.1–23.6). The specificity of these interactions raises the possibility that synthetic inhibitors, designed to either mimic the receptor and prevent ligand binding, or mimic the ligand and prevent receptor binding, might be useful as agents to interfere with vector transmission. Ligand–receptor mimics have been, and continue to be, developed for interfering with animal virus infection at the level of attachment to their cellular receptors (Moscona, 2005).

Similarly, evidence that some viruses may require conformational change during the transmission process (see above) suggests that compounds that inhibit this conformational change might also assist in reducing vector-mediated virus spread. Well-known inhibitors of poliovirus infection have been found to act by preventing the particle from adopting the proper conformational change required for receptor-mediated entry of the virus into host cells (Goncalves et al., 2007). The design of inhibitors that act in a similar fashion may be useful in disrupting virus–vector interactions as well.

The various inhibitors that have been useful for interfering with virus–receptor interactions may be chemically based or could be peptide mimics. For peptides, transgenic plant approaches could be utilized to express peptides in leaves or roots. For chemically based peptides, it may be possible to apply the inhibitor to leaves or to soil depending on the virus–vector combination being targeted.

Most certainly, further detailed studies on molecular aspects of virus receptor–ligand interactions are required prior to assessing the potential applicability of an 'inhibitor-based' virus disease control strategy. Such research would include precise identification of both virus- and vector-encoded proteins that interact during the transmission process. Further detailed structural analyses of these components would likely be required to facilitate the design of the inhibitors. Peptide libraries could be screened for initial identification of potential peptide inhibitors. Once designed, the potential effectiveness of inhibitors could be tested using current laboratory methods that assess transmission efficiency. In the case of CNV, certain commercially available mannose-containing oligosaccharides were found to be effective at preventing virus–zoospore interactions in *in vitro* assays (Kakani et al., 2003). Structural analysis of the sugar-binding sites on the virion will be valuable in designing higher affinity sugars. In addition, the information gained from these studies will further contribute to assessing 'inhibitor-based' strategies for reducing vector-mediated spread of plant viruses in nature.

Acknowledgements

I am indebted to my present and previous collaborators, Ron Reade, Kishore Kakani, Elizabeth Hui and Marjorie Robbins, who through thoughtful discussion and research have contributed their insight and enthusiasm into studies of virus–vector interactions. I am grateful to Jane Theilmann for her editorial assistance. I am also indebted to the authors of the many excellent recent reviews that were used to consolidate this summary (see text). I thank Melissa Rochon-Weis for her assistance in preparing illustrations.

References

Adams, M.J., Antoniw, J.F. and Mullins, J.G. (2001) Plant virus transmission by plasmodiophorid fungi is associated with distinctive transmembrane regions of virus-encoded proteins. *Archives of Virology* 146, 1139–1153.

Agranovsky, A.A., Lesemann, D.E., Maiss, E., Hull, R. and Atabekov, J.G. (1995) 'Rattlesnake' structure of a filamentous plant RNA virus built of two capsid proteins. *Proceedings of the National Academy of Sciences of the United States of America* 92, 2470–2473.

Akula, S.M., Pramod, N.P., Wang, F.Z. and Chandran, B. (2002) Integrin alpha3beta1 (CD 49c/29) is a cellular receptor for Kaposi's sarcoma-associated herpesvirus (KSHV/HHV-8) entry into the target cells. *Cell* 108, 407–419.

Alzhanova, D.V., Prokhnevsky, A.I., Peremyslov, V.V. and Dolja, V.V. (2007) Virion tails of Beet yellows virus: coordinated assembly by three structural proteins. *Virology* 359, 220–226.

Andret-Link, P., Schmitt-Keichinger, C., Demangeat, G., Komar, V. and Fuchs, M. (2004) The specific transmission of *Grapevine fanleaf virus* by its nematode vector *Xiphinema index* is solely determined by the viral coat protein. *Virology* 320, 12–22.

Atreya, P., Lopez-Moya, J., Chu, M., Atreya, C.D. and Pirone, T.P. (1995) Mutational analysis of the coat protein N-terminal amino acids involved in potyvirus transmission by aphids. *Journal of General Virology* 76, 265–270.

Bandla, M., Campbell, L., Ullman, D. and Sherwood, J. (1998) Interaction of tomato spotted wilt tospovirus (TSWV) glycoproteins with a thrips midgut protein, a potential cellular receptor for TSWV. *Phytopathology* 88, 98–104.

Blanc, S., Lopez-Moya, J.J., Wang, R., Garcia-Lampasona, S., Thornbury, D.W. and Pirone, T.P. (1997) A specific interaction between coat protein and helper component correlates with aphid transmission of a potyvirus. *Virology* 231, 141–147.

Blanc, S., Ammar, E.D., Garcia-Lampasona, S., Dolja, V.V., Llave, C., Baker, J. and Pirone, T.P. (1998) Mutations in the potyvirus helper component protein: effects on interactions with virions and aphid stylets. *Journal of General Virology* 79, 3119–3122.

Blanch, E.W., Robinson, D.J., Hecht, L. and Barron, L.D. (2001) A comparison of the solution structures of tobacco rattle and tobacco mosaic viruses from Raman optical activity. *Journal of General Virology* 82, 1499–1502.

Bowman, V.D., Chase, E.S., Franz, A.W., Chipman, P.R., Zhang, X., Perry, K.L., Baker, T.S. and Smith, T.J. (2002) An antibody to the putative aphid recognition site on cucumber mosaic virus recognizes pentons but not hexons. *Journal of Virology* 76, 12250–12258.

Brault, V., Perigon, S., Reinbold, C., Erdinger, M., Scheidecker, D., Herrbach, E., Richards, K. and Ziegler-Graff, V. (2005) The polerovirus minor capsid protein determines vector specificity and intestinal tropism in the aphid. *Journal of Virology* 79, 9685–9693.

Brown, D.J.F., Robertson, W.M. and Trudgill, D.L. (1995) Transmission of viruses by plant nematodes. *Annual Review of Phytopathology* 33, 223–249.

Bruyere, A., Brault, V., Ziegler-Graff, V., Simonis, M.T., Van den Heuvel, J.F., Richards, K., Guilley, H., Jonard, G. and Herrbach, E. (1997) Effects of mutations in the beet western yellows virus readthrough protein on its expression and packaging and on virus accumulation, symptoms, and aphid transmission. *Virology* 230, 323–334.

Campbell, R. (1996) Fungal transmission of plant viruses. *Annual Review of Phytopathology* 34, 87–108.

Chay, C.A., Gunasinge, U.B., Dinesh-Kumar, S.P., Miller, W.A. and Gray, S.M. (1996) Aphid transmission and systemic plant infection determinants of barley yellow dwarf luteovirus-PAV are contained in the coat protein readthrough domain and 17-kDa protein, respectively. *Virology* 219, 57–65.

Dessens, J.T., Nguyen, M. and Meyer, M. (1995) Primary structure and sequence analysis of RNA2 of a mechanically transmitted barley mild mosaic virus isolate: an evolutionary relationship between bymo- and furoviruses. *Archives of Virology* 140, 325–333.

Drucker, M., Froissart, R., Hebrard, E., Uzest, M., Ravallec, M., Esperandieu, P., Mani, J.C., Pugniere, M., Roquet, F., Fereres, A. and Blanc, S. (2002) Intracellular distribution of viral gene products regulates a complex mechanism of cauliflower mosaic virus acquisition by its aphid vector. *Proceedings of the National Academy of Sciences of the United States of America* 99, 2422–2427.

Gildow, F. (1999) Luteovirus transmission and mechanisms regulating vector-specificity. In: Smith, H. and Barker, H. (eds) *The Luteoviridae*. CAB International, Wallingford, UK, pp. 88–112.

Gildow, F., Reavy, B., Mayo, M., Duncan, G., Woodford, J., Lamb, J. and Hay, R. (2000) Aphid acquisition and cellular transport of *Potato leafroll virus*-like particles lacking P5 readthrough protein. *Phytopathology* 90, 1153–1161.

Goncalves, R.B., Mendes, Y.S., Soared, M.R., Katpally, U., Smith, T.J., Silva, J.L. and Oliveira, A.C. (2007) VP4 protein from human rhinovirus 14 is released by pressure and locked in the capsid by the antiviral compound WIN. *Journal of Molecular Biology* 366, 295–306.

Gray, S. and Gildow, F.E. (2003) Luteovirus-aphid interactions. *Annual Review of Phytopathology* 41, 539–566.

Gray, S.M. and Banerjee, N. (1999) Mechanisms of arthropod transmission of plant and animal viruses. *Microbiology and Molecular Biology Reviews* 63, 128–148.

Harrison, B. and Murant, A. (1977) Nematode transmissibility of pseudo-recombinant isolates of tomato black ring virus. *Annals of Applied Biology* 86, 209–212.

He, X., Harper, K., Grantham, G., Yang, C.H. and Creamer, R. (1998) Serological characterization of the 3'-proximal encoded proteins of beet yellows closterovirus. *Archives of Virology* 143, 1349–1363.

Holeva, R.C. and MacFarlane, S.A. (2006) Yeast two-hybrid study of tobacco rattle virus coat protein and 2b protein interactions. *Archives of Virology* 151, 2123–2132.

Hull, R. (2002) *Matthews' Plant Virology*. Academic Press, London.

Kakani, K., Sgro, J.Y. and Rochon, D. (2001) Identification of specific cucumber necrosis virus coat protein amino acids affecting fungus transmission and zoospore attachment. *Journal of Virology* 75, 5576–5583.

Kakani, K., Robbins, M. and Rochon, D. (2003) Evidence that binding of cucumber necrosis virus to vector zoospores involves recognition of oligosaccharides. *Journal of Virology* 77, 3922–3928.

Kakani, K., Reade, R. and Rochon, D. (2004) Evidence that vector transmission of a plant virus requires conformational change in virus particles. *Journal of Molecular Biology* 338, 507–517.

Kanyuka, K., Ward, E. and Adams, M.J. (2003) *Polmyxa graminis* and the cereal viruses it transmits: a research challenge. *Molecular Plant Pathology* 4, 393–406.

Lee, L., Kaplan, I.B., Ripoll, D.R., Liang, D., Palukaitis, P. and Gray, S.M. (2005) A surface loop of the potato leafroll virus coat protein is involved in virion assembly, systemic movement, and aphid transmission. *Journal of Virology* 79, 1207–1214.

Leh, V., Jacquot, E., Geldreich, A., Haas, M., Blanc, S., Keller, M. and Yot, P. (2001) Interaction between the open reading frame III product and the coat protein is required for transmission of cauliflower mosaic virus by aphids. *Journal of Virology* 75, 100–106.

Li, C., Cox-Foster, D., Gray, S.M. and Gildow, F. (2001) Vector specificity of barley yellow dwarf virus (BYDV) transmission: identification of potential cellular receptors binding BYDV-MAV in the aphid, *Sitobion avenae*. *Virology* 286, 125–133.

Liu, S., He, X., Park, G., Josefsson, C. and Perry, K.L. (2002) A conserved capsid protein surface domain of cucumber mosaic virus is essential for efficient aphid vector transmission. *Journal of Virology* 76, 9756–9762.

Lot, H., Campbell, R., Souche, S., Milne, R. and Roggero, P. (2002) Transmission by *Olpidium brassicae* of Mirafiori lettuce virus and Lettuce big-vein virus, and their roles in Lettuce big-vein etiology. *Phytopathology* 92, 288–293.

MacFarlane, S.A. (2003) Molecular determinants of the transmission of plant viruses by nematodes. *Molecular Plant Pathology* 4, 211–215.

MacFarlane, S.A., Wallis, C.V. and Brown, D.J. (1996) Multiple virus genes involved in the nematode transmission of pea early browning virus. *Virology* 219, 417–422.

Medeiros, R.B., Ullman, D.E., Sherwood, J.L. and German, T.L. (2000) Immunoprecipitation of a 50-kDa protein: a candidate receptor component for tomato spotted wilt tospovirus (Bunyaviridae) in its main vector, *Frankliniella occidentalis*. *Virus Research* 67, 109–118.

Moreno, A., Hebrard, E., Uzest, M., Blanc, S. and Fereres, A. (2005) A single amino acid position in the helper component of cauliflower mosaic virus can change the spectrum of transmitting vector species. *Journal of Virology* 79, 13587–13593.

Moscona, A. (2005) Entry of parainfluenza virus into cells as a target for interrupting childhood respiratory disease. *Journal of Clinical Investigations* 115, 1688–1698.

Ng, J.C. and Falk, B.W. (2006a) *Bemisia tabaci* transmission of specific *Lettuce infectious yellows virus* genotypes derived from *in vitro* synthesized transcript-inoculated protoplasts. *Virology* 352, 209–215.

Ng, J.C. and Falk, B.W. (2006b) Virus–vector interactions mediating nonpersistent and semipersistent transmission of plant viruses. *Annual Review of Phytopathology* 44, 183–212.

Ng, J.C., Josefsson, C., Clark, A.J., Franz, A.W. and Perry, K.L. (2005) Virion stability and aphid vector transmissibility of Cucumber mosaic virus mutants. *Virology* 332, 397–405.

Ng, J.C.K. and Perry, K.L. (2004) Transmission of plant viruses by aphid vectors. *Molecular Plant Pathology* 5, 505–511.

Pekosz, A. and Gonzalez-Scarano, F. (1996) The extracellular domain of La Crosse virus G1 forms oligomers and undergoes pH-dependent conformational changes. *Virology* 225, 243–247.

Peng, Y.H., Kadoury, D., Gal-On, A., Huet, H., Wang, Y. and Raccah, B. (1998) Mutations in the HC-Pro gene of zucchini yellow mosaic potyvirus: effects on aphid transmission and binding to purified virions. *Journal of General Virology* 79, 897–904.

Perry, K.L., Zhang, L. and Palukaitis, P. (1998) Amino acid changes in the coat protein of cucumber mosaic virus differentially affect transmission by the aphids *Myzus persicae* and *Aphis gossypii*. *Virology* 242, 204–210.

Pirone, T.P. and Blanc, S. (1996) Helper dependent vector transmission of plant viruses. *Annual Review of Phytopathology* 34, 227–247.

Plisson, C., Drucker, M., Blanc, S., German-Retana, S., Le Gall, O., Thomas, D. and Bron, P. (2003) Structural characterization of HC-Pro, a plant virus multifunctional protein. *Journal of Biological Chemistry* 278, 23753–23761.

Plisson, C., Uzest, M., Drucker, M., Froissart, R., Dumas, C., Conway, J., Thomas, D., Blanc, S. and Bron, P. (2005) Structure of the mature P3-virus particle complex of cauliflower mosaic virus revealed by cryo-electron microscopy. *Journal of Molecular Biology* 346, 267–277.

Plumb, R. (2002) Plant virus–vector interactions. *Advances in Botanical Research*, Vol. 36. Academic Press, New York.

Reinbold, C., Gildow, F.E., Herrbach, E., Ziegler-Graff, V., Goncalves, M.C., van Den Heuvel, J.F. and Brault, V. (2001) Studies on the role of the minor capsid protein in transport of *Beet western yellows virus* through *Myzus persicae*. *Journal of General Virology* 82, 1995–2007.

Rochon, D., Kakani, K., Robbins, M. and Reade, R. (2004) Molecular aspects of plant virus transmission by *Olpidium* and plasmodiophorid vectors. *Annual Review of Phytopathology* 42, 211–241.

Ruiz-Ferrer, V., Boskovic, J., Alfonso, C., Rivas, G., Llorca, O., Lopez-Abella, D. and Lopez-Moya, J. (2005) Structural analysis of *Tobacco etch potyvirus* HC-pro oligomers involved in aphid transmission. *Journal of Virology* 79, 3758–3765.

Rush, C.M. (2003) Ecology and epidemiology of benyviruses and plasmodiophorid vectors. *Annual Review of Phytopathology* 41, 567–592.

Satyanarayana, T., Gowda, S., Ayllon, M.A. and Dawson, W.O. (2004) Closterovirus bipolar virion: evidence for initiation of assembly by minor coat protein and its restriction to the genomic RNA 5' region. *Proceedings of the National Academy of Sciences of the United States of America* 101, 799–804.

Seddas, P. and Boissinot, S. (2006) Glycosylation of beet western yellows virus proteins is implicated in the aphid transmission of the virus. *Archives of Virology* 151, 967–984.

Seddas, P., Boissinot, S., Strub, J.M., Van Dorsselaer, A., Van Regenmortel, M.H. and Pattus, F. (2004) Rack-1, GAPDH3, and actin: proteins of *Myzus persicae* potentially involved in the transcytosis of beet western yellows virus particles in the aphid. *Virology* 325, 399–412.

Shukla, D.D., Strike, P.M., Tracy, S.L., Gough, K.H. and Ward, C.W. (1988) N and C termini of the coat protein of potyviruses are surface-located: the N terminus contains the major virus-specific epitopes. *Journal of General Virology* 69, 1497–1508.

Smith, T.J., Chase, E., Schmidt, T. and Perry, K.L. (2000) The structure of cucumber mosaic virus and comparison to cowpea chlorotic mottle virus. *Journal of Virology* 74, 7578–7586.

Stenger, D.C., Hein, G.L., Gildow, F., Horken, K.M. and French, R. (2005) Plant virus HC-Pro is a determinant of eriophyid mit transmission. *Journal of Virology* 79, 9054–9061.

Tamada, T., Schmitt, C., Saito, M., Guilley, H., Richards, K. and Jonard, G. (1996) High resolution analysis of the readthrough domain of beet necrotic yellow vein virus readthrough protein: a KTER motif is important for efficient transmission of the virus by *Polymyxa betae*. *Journal of General Virology* 77, 1359–1367.

Taylor, C. and Brown, D. (1997) *Nematode Vectors of Plant Viruses*. CAB International, Wallingford, UK.

Terradot, L., Souchet, M., Tran, V. and Giblot Ducray-Bourdin, D. (2001) Analysis of a three-dimensional structure of *Potato leafroll virus* coat protein obtained by homology modeling. *Virology* 286, 72–82.

Tian, T., Rubio, L., Yeh, H.H., Crawford, B. and Falk, B.W. (1999) Lettuce infectious yellows virus: in vitro acquisition analysis using partially purified virions and the whitefly *Bemisia tabaci*. *Journal of General Virology* 80, 1111–1117.

Torrance, L., Andreev, I.A., Gabrenaite-Verhovskaya, R., Cowan, G., Makinen, K. and Taliansky, M.E. (2006) An unusual structure at one end of potato potyvirus particles. *Journal of Molecular Biology* 357, 1–8.

van den Heuvel, J.F., Bruyere, A., Hogenhout, S.A., Ziegler-Graff, V., Brault, V., Verbeek, M., van der Wilk, F. and Richards, K. (1997) The N-terminal region of the luteovirus readthrough domain determines virus binding to Buchnera GroEL and is essential for virus persistence in the aphid. *Journal of Virology* 71, 7258–7265.

van den Heuvel, J.F., Hogenhout, S.A. and van der Wilk, F. (1999) Recognition and receptors in virus transmission by arthropods. *Trends in Microbiology* 7, 71–76.

Van Regenmortel, M., Fauguet, C., Bishop, D., Carstens, E., Estes, M., Lemon, S., Maniloff, J., Mayo, M., McGeoch, D., Pringle, C. and Wickner, R. (2000) *Virus Taxonomy-Seventh Report of the International Committee on Taxonomy of Viruses*. Academic Press, New York.

Vassilakos, N., Vellios, E.K., Brown, E.C., Brown, D.J. and MacFarlane, S.A. (2001) Tobravirus 2b protein acts in trans to facilitate transmission by nematodes. *Virology* 279, 478–487.

Visser, P.B. and Bol, J.F. (1999) Nonstructural proteins of *Tobacco rattle virus* which have a role in nematode-transmission: expression pattern and interaction with viral coat protein. *Journal of General Virology* 80, 3273–3280.

Wang, S. and Gergerich, R.C. (1998) Immunofluorescent localization of *Tobacco ringspot* nepovirus in the vector nematode *Xiphinema americanum*. *Phytopathology* 88, 885–889.

Whitfield, A.E., Ullman, D.E. and German, T.L. (2004) Expression and characterization of a soluble form of tomato spotted wilt virus glycoprotein GN. *Journal of Virology* 78, 13197–13206.

Whitfield, A.E., Ullman, D.E. and German, T.L. (2005a) Tomato spotted wilt virus glycoprotein G(C) is cleaved at acidic pH. *Virus Research* 110, 183–186.

Whitfield, A.E., Ullman, D.E. and German, T.L. (2005b) Tospovirus–thrips interactions. *Annual Review of Phytopathology* 43, 459–489.

Index

Antiviral peptides 307–308, 363–365
Antiviral proteins 366–367
Aphids 536–539
Avirulence genes and proteins 2, 3–4
    in bacteria 29–30, 182
    in nematodes 74–75
    in viruses 401–402

Bacterial genome sequences 464–466, 489–490
Bacterial plant pathogens 16–17, 165–167
    adhesin production 482
    antifreeze protein 483
    avoidance 166–169
    biofilm formation 480–481
    control strategies 462–463, 468–470
    detection 172–179
    dissemination 167–171
    features of 467
    harpins 30
    insertion elements 184–185
    isolation from tissues 171
    lactone secretion 18–21
    molecular detection 165–194, 485–486
    oligopeptide signals 21–22
    pathogenicity-related genes 180–183, 470–471
    pigment production 480–483
    signaling molecules 23–24, 475–477
    toxin production 182, 479–480
    virulence mechanisms 17–18, 39
Bacterial type III secretory systems 17–18, 24–34
    effect of environment 31–32
    inhibition of 471–472, 477–478
    regulation of 32–39
    role of chaperones 30–31
Biological control
    mechanisms 518–531
    bacterial pathogens 48, 463, 484–485
    ecological factors 520–521
    fungal pathogens 518–523

Cationic antimicrobial peptides 301–320
    antimicrobial activity 303
    categories 305–306
        cysteine-rich peptides 305
        helical peptides 305
        sheet peptides 305
    mechanisms of action 306–308
        antibacterial 306–307, 471
        antifungal 307
        antiviral 307–308
    in plants 309
    production of 303

Cationic antimicrobial peptides (*continued*)
   synthetic 309
   in transgenic plants 310–315
Cyst nematodes 60–62
   identification of 209–211
   parasitism by 61–62, 72, 75
   syncytium formation 61–62, 71

Diagnostics 157–159, 186–187, 205–215, 227–249, 285–291
   see also Polymerase chain reaction
Disease resistance genes 2–4
   against viruses 367–368, 395–415
      characterization of 403–410
   in barley 6
   fungal disease control 8, 330, 333
   gene arrays 6
   gene-for-gene concept 2, 74–75, 401–402
   guard hypothesis 4, 401–402
   molecular breeding 326, 329–330
   receptor-like kinases 3, 403–404
   in tobacco 398–399, 406–408
   in tomato 3, 6, 9, 407–410
   in wheat 9, 322, 338, 348
Disease suppressive soils 498–501, 509–512
DNA arrays 151–153, 189, 215, 243, 507–508
DNA barcoding 156–157, 215–217

Epistasis 330
Expressed sequence tags 63, 67, 76, 325, 351

Flocculosin 525–527
Fluorescence *in situ* hybridization 509
Fluorescent pseudomonads 510–512
Fungal plant pathogens 146–164, 321–357
   disease epidemics 322–323

Gene clusters in bacteria 180–182, 470–471
Gene mapping 331–333, 351

Green fluorescent protein 520, 523–524

Hypersensitive response 402–403
   bacterial-induced 17, 29–30, 180
   kinase activity 5
   nematode-induced 75
   virus-induced 367–368, 403

Insect vectors 532–551
   see also Aphids
   see also Thrips

Loop-mediated isothermal amplification 237–238

Marker-assisted breeding
   methods 321, 323–353
   amplified fragment length polymorphism 325
   cleaved amplified polymorphic sequence 325
   random amplified polymorphic DNA 325
   restriction fragment length polymorphism 325
   sequence characterized amplified region 325
   simple sequence repeats 325
   single-nucleotide polymorphism 325–326
Mitochondrial DNA 204–207, 216
Mixed virus infections 379, 384–388, 429
Molecular breeding 321–357
   application of biotechnology 323–324
   breeding strategies 336–339
   gene pyramiding 337–339
   selection 343–349
   discovery of markers 326–327
   effect of environment 329–330
   marker development 327–329
   marker validation 332–334
Molecular diagnostics 146–164, 165–194, 195–226

bacterial plant pathogens
165–194, 485–487
fungal plant pathogens 146–164
nematode plant
pathogens 195–226
phytoplasmas 250–276
practical applications 157–159,
188–189, 217–218, 244,
290–291, 486
validation 156, 186–187,
240–242
viral plant pathogens 227–249
Molecular fingerprinting of soil
microbes 504–506, 510
Multidrug resistance efflux
pumps 472–473
inhibition of 472–473

Nematodes 58–56
application of biotechnology
58–59
cyst nematodes 60–62, 66,
69–70, 76
DNA sequencing 204–205
ectoparasites and
endoparasites 59–60
identification of 195–197
isoelectric focusing method
197–198
mass spectrometry
198–199
restriction fragment length
polymorphism 200
metabolic reprogramming 68
molecular diagnostics 195–226
oesophageal glands 60–63, 73
pathogenesis 59–62
preservation of 201
recovery from soil 202–203
root-knot nematodes 60–62, 66,
70, 76
stylet penetration 60, 66
virus vectors 534, 551–554
Nonhost disease resistance 399–400
role of reactive oxygen
species 400
Nucleic acid extraction methods
148–149, 172–173,
201–204, 230–231,
291–293, 503–504

Papaya ringspot virus 437–440
genome organization 437–438
molecular characterization 437
symptoms 437
Parasitism genes in nematodes 61–73
cell wall degrading enzymes
62–64, 66–67
protein degradation 71–72
signaling peptides 68–71
Pathogen-free planting materials
169–171, 289–291
Phytoplasmas 250–276
characteristics of 251
classification 256–257
control 261–262
economic significance 254–256
almond 255–256
china tree 255
fruit trees 256–261
grapevine 256, 260–261
palm trees 254–255
papaya 255
energy requirement 268–269
extrachromosomal DNA 266–267
genome characterization 263–265
genome sequencing
projects 262–265
geographic distribution 253–254
identification of 257–258
DNA-based techniques
257–258
immunological methods 257
ribosomal protein genes
265–266
molecular diversity 260–262
nucleic acids extraction 262–263
recognized species 260
symptoms of diseases 251–253
systematics 258–259
Plant defense mechanisms 1–15
Plant defensins 309
Polymerase chain reaction 147–148,
159, 173–179, 205–214
application to diagnostics
157–159, 186–187,
205–215, 285–291
for bacterial detection 172–179,
485–486
BIO-PCR 174–175
competitive PCR 176–177
conventional PCR 173

Polymerase chain reaction (*continued*)
    immunocapture-PCR 174
    multiplex PCR 175–176, 486
    nested PCR 175
    PCR-ELISA 174
    real-time PCR 177–179,
        506–507
    repetitive PCR 183–184
  for fungal detection 151–152
  for nematode identification 200,
    205–215
    multiplex PCR 210–211
    PCR-RFLP 205–210
    PCR-SSCP 211
    Real-time PCR 212–214
  primer design 179–185, 200–201,
    206–211, 231–232, 258,
    293–294
  for phytoplasma detection 258,
    265–266
  sensitivity 185, 214–215, 238–243
  for viroid detection 285–288
    multiplex RT-PCR 287
    real-time RT-PCR 287–288
  for virus detection 233–237
    conventional PCR 233
    cooperational PCR 234–235
    multiplex PCR 235–236
    nested PCR 233–234
    PCR-ELISA 234
    real-time PCR 236–237
Potato cyst nematode 70–71, 73,
    197–198, 212–213
Programmed cell death 44–45, 368
Proteomics 10–11, 351
Pseudozyma species 522–527

Quantitative trait loci 330–333
Quorum sensing 16–57
  description of systems 19–20
  inhibition of 49–50, 474
  modulation 39–45
  quenching 40, 46

Real-time PCR *see* Polymerase chain
    reaction
RNA interference (RNAi) 9–10, 59,
    77–78, 91, 103, 134–135, 521
RNA silencing 376–378

  mechanism 377–378, 399, 419
  in transgenic plants 428
  viral suppressors of 376, 379–383
Ribosomal RNA 153–155, 180, 200,
    204–211, 216, 258, 265, 486,
    505, 525–526
Ribosome-inactivating
    proteins 366–367
Root-knot nematodes 60–62
  changes in auxin levels 68
  giant-cell formation 61–62, 70
  identification of 197–198,
    206–209

Seed certification 168–169, 290
Seedborne bacterial pathogens
    168–170, 468
Signal transduction pathways 4–6,
    402
  in bacteria 17
  jasmonic acid 7
  mitogen-activated protein
    kinases 4
  salicylic acid-mediated 7, 68, 403
Soil microbial communities 147,
    499–501
  microbial diversity
    assessment 499–503
  nucleic acid-based analyses
    504–509, 513
Subtractive hybridization 183

Tobacco mosaic virus 396–399
  coat protein-mediated
    resistance 360–361, 375
  disease resistance genes 398–399
  genomic structure 396–398
  replication complex 397
  transmission 398
Thrips 549–550
Transgenic plants 360–369, 384–389,
    416–432
  bacterial disease resistance 8,
    44–47, 310–313
  environmental safety 417,
    419, 429–431, 444–446,
    448–451
  fungal disease resistance 8, 310,
    312–314

nematode resistance 77–78
papaya 436–461
   coat protein-mediated virus
      resistance 439
   cultivar development
      443–446
   deregulation of 454–457
   field evaluation 441–442
   greenhouse evaluation
      440–441
   impact on yield 446–448
   regulatory approval 444–446
virus resistance 95–96, 100–103,
   310, 314–315, 358–369,
   416–436, 437–461

Viroids 125–146, 277–300
   biological assays 280–281
   cell-to-cell movement 135
   characteristics 279
   classification 279–280
   discovery of 126–128
   intracellular transport 130–131
   management 138–139, 278,
      288–291
   mixed infection with viruses 287
   molecular detection 277–300
      imprint hybridization 285
      microtitre plate method 286
      molecular probes 283–284
      polyacrylamide gel
         electrophoresis
         281–282
      return-PAGE 282–283
      reverse transcription PCR
         285–287
      temperature gradient gel
         electrophoresis 283
   in ornamentals 289, 296
   pathogenesis 136–138
   phloem movement 135–136
   recognized species 129
   recovery from tissues 286–287
   replication cycle 131–133
   resistance to 140
   structural features 130
   symptomless infection 278,
      289, 295
   symptoms in plants 277–278,
      281, 289

Virulence genes in bacteria 180–183,
   470–471, 478
   regulation of 473–474
Viruses 87–108, 109–124, 227–249,
   358–373, 532–567
   antiviral strategies 101–103,
      358–373, 374–394
   control strategies 228, 359–360,
      418–419, 561–562
   evolution 109–124, 420
   genetic variability 110–112
      mutation 110–111, 113–115
      reassortment 112, 116–118
      recombination 111–112,
         116–117, 417, 420,
         423–425, 449–450
   host resistance 88–89, 93–94,
      395–415
   identification of 228–230
      nucleic acid hybridization
         229
      reverse transcription PCR
         229–230, 233–234
   movement proteins 92, 95–96, 97,
      362–363, 397–398
   replication complex 89–90,
      94, 397
   replication cycles 91–101
      DNA viruses 91–97
      host transmembrane proteins
         94–96
      plant cell cycle regulation
         98–99
      role of kinases 100–101
      RNA dependent RNA
         polymerase 91, 113
      RNA viruses 91–97
      translation factors 92–93
   replication in yeast 90–91
   transmission of 532–567
Virus disease resistance 358–373,
   374–394, 416–435
   antibody-mediated 365–366
   antiviral protein-mediated
      366–367
   cystatin-mediated 367
   endogenous resistance
      genes 403–406
   genetic engineering 359–360
   movement protein-mediated
      362–363

Virus disease resistance (*continued*)
   pathogen-derived coat-protein
      mediated 360–363,
      375–376, 418, 421–426
   mechanisms 360–362,
      375–376
      tobacco mosaic virus
         360–361, 375
      tomato spotted wilt virus
         361–362
   peptide-mediated 363–365
   R-gene-mediated 368, 403–410
   RNA-mediated 376–378, 383–384
Virus recombination 419–432
   effect of selection pressure
      426–427
   experimental evidence 422–424
   factors affecting 426–428
   field observations 424–426,
      429–430
   influence of viral sequence 427
   potential risks 420, 422, 429–431
   in transgenic plants 420–431,
      449–450

Virus transmission 532–567
   cucumovirus 541–542
   caulimovirus 543–545
   fungi 532–533, 554–557
   insects 532–536
   luteovirus 545–549
   modes of transmission 536–551
      nonpersistent 538–542
      persistent 545–551
      semipersistent 542–545
   nematodes 532, 551–554
      nepovirus 553–554
      tobravirus 552–553
   potyvirus 539–541
   protists 557–558
   tospovirus 549–551

Yeast two-hybrid screens 10, 70,
   89–90, 364, 401
Yeasts 522–527
   antibiotic production 525–527
   efficacy of 523
   use in biological control 522–523